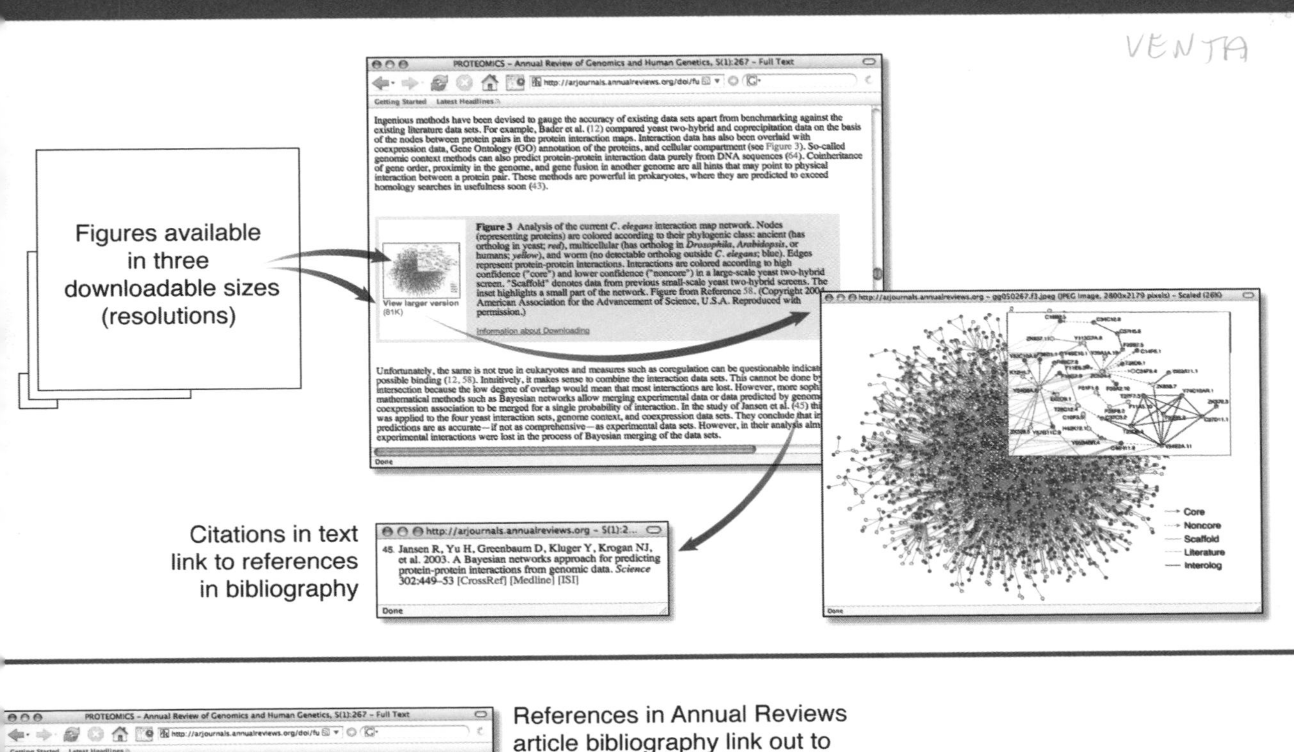

Figures available in three downloadable sizes (resolutions)
Citations in text link to references in bibliography
PROTEOMICS - Annual Review of Genomics and Human Genetics, 5(1):267 - Full Text
45. Jansen R, Yu H, Greenbaum D, Kluger Y, Krogan NJ, et al. 2003. A Bayesian networks approach for predicting protein-protein interactions from genomic data. Science 302:449–53 [CrossRef] [Medline] [ISI]
Core
Noncore
Scaffold
Literature
Interolog

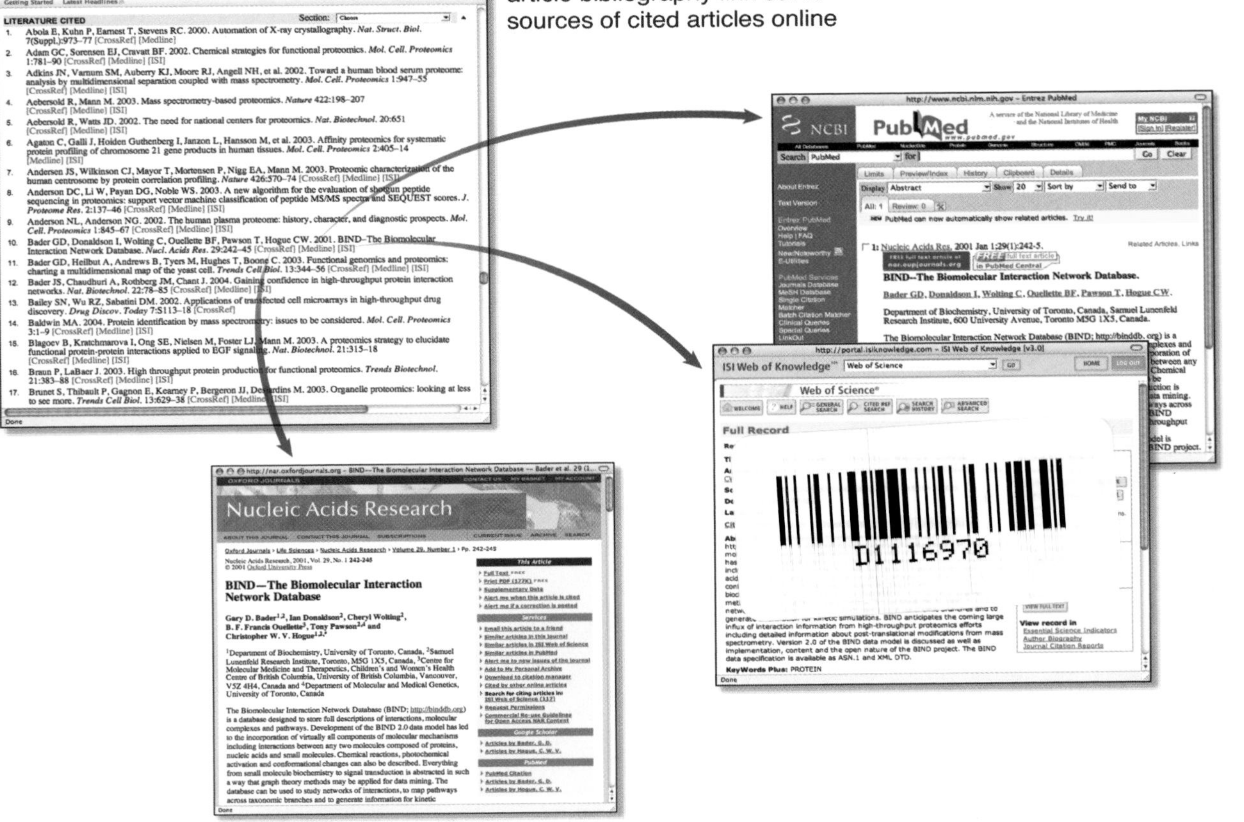

References in Annual Reviews article bibliography link out to sources of cited articles online
LITERATURE CITED
NCBI
PubMed
ISI Web of Knowledge
Web of Science
Nucleic Acids Research
BIND—The Biomolecular Interaction Network Database
D1116970

Annual Review of Genomics
and Human Genetics

Annual Review of Genomics and Human Genetics

Volume 9, 2008

Aravinda Chakravarti, *Co-Editor*
Johns Hopkins University School of Medicine

Eric Green, *Co-Editor*
National Human Genome Research Institute

www.annualreviews.org • science@annualreviews.org • 650-493-4400

Annual Reviews
4139 El Camino Way • P.O. Box 10139 • Palo Alto, California 94303-0139

Annual Reviews
Palo Alto, California, USA

International Standard Serial Number: 1527-8204
International Standard Book Number: 978-0-8243-3709-4

∞ The paper used in this publication meets the minimum requirements of American National Standards for Information Sciences—Permanence of Paper for Printed Library Materials, ANSI Z39.48-1992. The paper contains 20% postconsumer recycled content.

TYPESET BY APTARA
PRINTED AND BOUND BY SHERIDAN BOOKS, INC., CHELSEA, MICHIGAN

Annual Review of
Genomics and
Human Genetics

Volume 9, 2008

Contents

Indexes

Errata

An online log of corrections to *Annual Review of Genomics and Human Genetics* articles may be found at http://genom.annualreviews.org/

Related Articles

From the ***Annual Review of Biophysics and Biomolecular Structure***, Volume 37 (2008)

Riboswitches: Emerging Themes in RNA Structure and Function
Rebecca K. Montange and Robert T. Batey

Calorimetry and Thermodynamics in Drug Design
Jonathan B. Chaires

Protein Design by Directed Evolution
Christian Jäckel, Peter Kast, and Donald Hilvert

The Protein Folding Problem
Ken A. Dill, S. Banu Ozkan, M. Scott Shell, and Thomas R. Weikl

Molecular Biology of Single Molecules in Living Bacteria
X. Sunney Xie, Paul J. Choi, Gene-Wei Li, Nam Ki Lee, and Giuseppe Lia

Structural Principles from Large RNAs
Stephen R. Holbrook

From the ***Annual Review of Genetics***, Volume 41 (2007)

Epigenetic Control of Centromere Behavior
Karl Ekwall

Immunoglobulin Somatic Hypermutation
Grace Teng and F. Nina Papavasiliou

Chromosome Fragile Sites
Sandra G. Durkin and Thomas W. Glover

MHC, TSP, and the Origin of Species: From Immunogenetics to Evolutionary Genetics
Nikolas Nikolaidis, Akie Sato, and Jan Klein

DNA Transposons and the Evolution of Eukaryotic Genomes
Cedric Feschotte and Ellen J. Pritham

Mammalian Meiotic Recombination Hot Spots
Peter Calabrese, Norman Arnheim, and Irene Tiemann-Boege

Orthology and Functional Conservation in Eukaryotes
Kara Dolinski and David Botstein

Use of Y Chromosome and Mitochondrial DNA Population Structure in Tracing Human Migrations
Peter A. Underhill and Toomas Kivisild

From the ***Annual Review of Immunolog***, Volume 26 (2008)

Forward Genetic Dissection of Immunity to Infection in the Mouse
S.M. Vidal, D. Malo, J.-F. Marquis, and P. Gros

The APOBEC3 Cytidine Deaminases: An Innate Defensive Network Opposing Exogenous Retroviruses and Endogenous Retroelements
Ya-Lin Chiu and Warner C. Greene

From the ***Annual Review of Medicine***, Volume 59 (2008)

Therapy of Marfan Syndrome
Daniel P. Judge and Harry C. Dietz

Genetic Susceptibility to Type 2 Diabetes and Implications for Antidiabetic Therapy
Allan F. Moore and Jose C. Florez

Array-Based DNA Diagnostics: Let the Revolution Begin
Arthur L. Beaudet and John W. Belmont

Inherited Mitochondrial Diseases of DNA Replication
William C. Copeland

Expanded Newborn Screening: Implications for Genomic Medicine
Linda L. McCabe and Edward R.B. McCabe

Genetic Determinants of Aggressive Breast Cancer
Alejandra C. Ventura and Sofia D. Merajver

Cancer Epigenetics: Modifications, Screening, and Therapy
Einav Nili Gal-Yam, Yoshimasa Saito, Gerda Egger, and Peter A. Jones

Asthma Genetics: From Linear to Multifactorial Approaches
Stefano Guerra and Fernando D. Martinez

The Effect of Toll-Like Receptors and Toll-Like Receptor Genetics in Human Disease
Stavros Garantziotis, John W. Hollingsworth, Aimee K. Zaas, and David A. Schwartz

EGFR Tyrosine Kinase Inhibitors in Lung Cancer: An Evolving Story
Lecia V. Sequist and Thomas J. Lynch

From the ***Annual Review of Neuroscience***, Volume 31 (2008)

Annual Reviews is a nonprofit scientific publisher established to promote the advancement of the sciences. Beginning in 1932 with the *Annual Review of Biochemistry*, the Company has pursued as its principal function the publication of high-quality, reasonably priced *Annual Review* volumes. The volumes are organized by Editors and Editorial Committees who invite qualified authors to contribute critical articles reviewing significant developments within each major discipline. The Editor-in-Chief invites those interested in serving as future Editorial Committee members to communicate directly with him. Annual Reviews is administered by a Board of Directors, whose members serve without compensation.

Human Telomere Structure and Biology

Harold Riethman

The Wistar Institute, Philadelphia, Pennsylvania 19104; email: Riethman@wistar.org

Annu. Rev. Genomics Hum. Genet. 2008. 9:1–19

First published online as a Review in Advance on May 8, 2008

The *Annual Review of Genomics and Human Genetics* is online at genom.annualreviews.org

This article's doi:
10.1146/annurev.genom.8.021506.172017

1527-8204/08/0922-0001$20.00

Key Words

telomere length regulation, telomere epigenetics, segmental duplication, copy number variation, structural variation

Abstract

Human telomeric DNA is complex and highly variable. Subterminal sequences are associated with *cis*-acting determinants of allele-specific $(TTAGGG)_n$ tract length regulation and may modulate susceptibility of $(TTAGGG)_n$ tracts to rapid deletion events. More extensive subtelomeric DNA tracts are filled with segmental duplications and segments that vary in copy number, leading to highly variable subtelomeric allele structures in the human population. RNA transcripts encoded in telomere regions include multicopy protein-encoding gene families and a variety of noncoding RNAs. One recently described family of $(UUAGGG)_n$-containing subterminal RNAs appears to be critical for telomere integrity; these RNAs associate with telomeric chromatin and are regulated by RNA surveillance factors including human homologs of the yeast Est1p protein. An increasingly detailed and complete picture of telomeric DNA sequence organization and structural variation is essential for understanding and tracking allele-specific subterminal and subtelomeric features critical for human biology.

DDR: DNA damage response

The DNA at each human chromosome terminus is a simple repeat sequence tract $(TTAGGG)_n$, typically 5 kb to 15 kb in length in somatic cells (63). Loss of the tracts typically occurs gradually with each cell division, but sometimes tract loss is more rapidly mediated by exonuclease activities or by deletion. Lengthening of the tracts can be carried out either enzymatically by telomerase or by a recombination-based mechanism (15, 43, 67). In human somatic cells the average bulk telomere length is partly an inherited trait (78), but the replicative history of the cell lineage and environmental factors such as exposure to stress and oxidative damage are also key elements of length regulation (4, 30, 90). A single-stranded, G-rich overhang at the end of the $(TTAGGG)_n$ tracts (76) is involved in the formation of a t-loop secondary structure (40), believed to be required for proper telomere function. Disruption of t-loop or associated telomeric chromatin structures by interference with protein components required for their formation and maintenance (24, 48), by loss of the G-rich overhang itself, or by a critically short tract of $(TTAGGG)_n$ can each lead to telomere dysfunction and induction of the DNA damage response (DDR) pathway (23). Depending on the cellular context (especially the presence or absence of a functional p53 pathway), such telomere loss or uncapping can trigger cellular senescence, apoptosis, and/or genome instability.

Human telomerase is composed of the protein subunits hTERT and Dyskerin and a telomere template-containing RNA subunit, hTR (21); the holoenzyme binds to the ends of existing $(TTAGGG)_n$ tracts and adds additional $(TTAGGG)_n$ repeats in a template-independent fashion. Telomerase activity is highest in germ cells and in totipotent stem cells, is present in lower but detectable amounts in adult stem cell populations, and is absent or undetectable in most differentiated cell types (44). Ectopic expression of hTERT halts DDR signaling in senescent cells and permits (and in some cases apparently promotes) continued replication and expansion of the cell populations (17, 66, 77). Although $(TTAGGG)_n$ tracts are stabilized and often lengthened in hTERT-transfected cells (17, 81), hTERT overexpression without telomere lengthening can also induce stem cell proliferation in mice (33, 73), suggesting that hTERT-dependent cellular lifespan extension is mediated by more than simple $(TTAGGG)_n$ tract lengthening. For example, preferential recruitment of hTERT to short telomeres may somehow block DDR pathway activation without a requirement for telomere lengthening (49).

The replicative capacity of cells is limited by telomere length, implicating short telomere length in age-related diseases (42, 51). Supporting this notion, heterozygous germline mutations in the telomerase protein component (hTERT) or RNA component (hTR) are associated with abnormally short telomeres in humans and can lead to autosomal dominant forms of the diseases Dyskeratosis congenita (characterized by aplastic anemia and bone marrow failure, as well as predisposition to cancers of the skin, hematopoeitic system, and oral mucosa) and idiopathic pulmonary fibrosis (leading to respiratory failure) (3, 88). Because anticipation occurs in autosomal dominant dyskeratosis congenita (more severe and earlier occurring phenotypes are seen in successive generations), the disease must be caused by the short telomeres rather than by the telomerase mutation (2). Thus, short telomeres limit tissue renewal capacity in somatic tissues, even in the presence of telomerase (44), and can cause both congenital and age-related disease in humans.

TERMINAL $(TTAGGG)_n$ TRACTS AND IMMEDIATE SUBTERMINAL SEQUENCES

Measurements of average human telomere $(TTAGGG)_n$ tract lengths in normal cells have ranged from greater than 20 kb in germline tissues to approximately 2 kb in senescing cells (52). In a recent study, human germline terminal tract lengths displayed a wide individual-specific variation range (from less than 9 kb to greater than 17 kb), consistent with genetic differences in the factors that determine telomere length settings in individuals (9). In addition to interindividual variation of bulk average telomere lengths, the lengths of $(TTAGGG)_n$ tracts vary from telomere to telomere within individual cells (39, 50, 94) and between alleles at the same telomere (12, 27). These individual-specific patterns of relative telomere-specific $(TTAGGG)_n$ tract lengths are regulated in part by *cis*-acting factors (18, 37, 38), and these patterns appear to be defined in the zygote and maintained throughout life (39).

Although massive ectopic overexpression of telomerase may erase allele-specific $(TTAGGG)_n$ tract length differences (18), these patterns are maintained in vivo through the germ line (38) and in physiologically relevant telomerase-expressing settings including human lymphocytes and cancer cells (18). In senescing human primary cell lines, DNA-damage response factors accumulate at a small number of telomeres; the telomeres with DDR signals correspond to those with the shortest $(TTAGGG)_n$ tract lengths in a particular cell line (61). These observations are consistent with the idea that the shortest telomere or a small subset of the shortest telomeres in a cell will determine the onset of senescence or apoptosis (95), and that individual-specific patterns of allele-specific $(TTAGGG)_n$ tract lengths may thus be crucial for the biological functions of telomeres and the effects of telomere attrition and dysfunction.

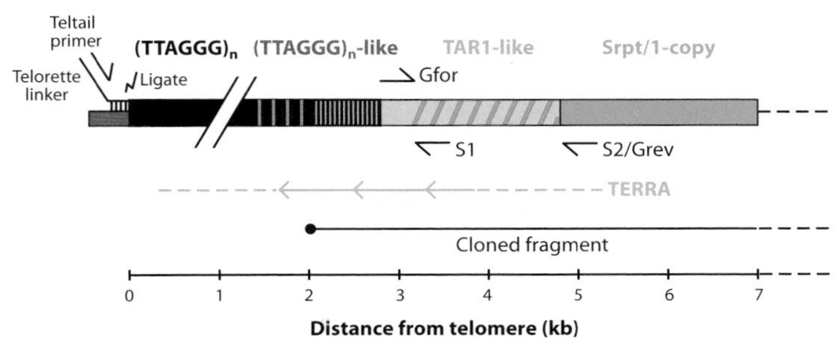

Figure 1

Model of terminal and subterminal human DNA. The single-stranded, G-rich 3′ overhang is shown in red; double-stranded $(TTAGGG)_n$ repeat DNA [5′-$(TTAGGG)_n$-3′ toward the telomere end, 5-$(CCCTAA)_n$-3′ strand toward the centromere] is shown in black. The lighter gray blocks interspersed with black represent the most centromeric part of the terminal repeat tract, where blocks of variant repeats related to $(TTAGGG)_n$ are usually present and interspersed with canonical $(TTAGGG)_n$ repeats in allele-specific patterns that are in linkage disequilibrium with the 2–4 kb of adjacent subterminal DNA (10, 11, 22). Noncanonical terminal repeats found in the centromeric part of the tract include $(TGAGGG)_n$, $(TCAGGG)_n$, and $(TTGGGG)_n$ (10, 11, 22) as well as many others [e.g., $(TTCGGG)_n$, $(TTAGGGG)_n$, $(TCGGGG)_n$] (92; H. Riethman, unpublished observations), and can extend telomerically for at least 1 kb into the tract (11). The pink region represents variably-sized (0–2 kb) sequence segments that bear similarity to the telomere-associated repeat 1 (TAR1) repeat family (19) and the blue region represents subterminal DNA that can be either 1-copy (six telomeres) or fall into one of the six families of subterminal repeat homology recently characterized (1). The orange arrow represents the position and orientation of telomeric repeat-containing RNA (TERRA) molecules, for which both the 5′ and 3′ end positions remain undefined (6). Also indicated are the positions of the linker (Telorette Linker) and primers (Teltail, S1, and S2 primers) for single telomere length analysis (STELA) assays (12) and primers for subterminal genotyping assays (Gfor and Grev).

Cis-acting regulators of allele-specific $(TTAGGG)_n$ tract length would be expected to depend upon subtelomeric sequences immediately adjacent to terminal $(TTAGGG)_n$ tracts. These sequences have been determined for the reference sequences of some chromosome arms (69), and a model of human terminal and subterminal DNA based upon these completed reference sequences is shown in **Figure 1**. All subterminal human DNA fragments cloned in bacteria (fosmids, cosmids) and yeast [half–yeast artificial chromosomes (YACs)] have lost the terminal portion of the $(TTAGGG)_n$ tract and retain 300 to 800 bp of the centromere-adjacent region of the simple repeat tract; the cloning-associated deletions appear to be specific for the distal $(TTAGGG)_n$ regions and do not involve rearrangement of subterminal sequences (H. Riethman, unpublished observation). Most of the known human $(TTAGGG)_n$-adjacent sequences are related in some fashion to the telomere-associated repeat (TAR) sequences originally described by Brown and coworkers (19, 70). Parts of the canonical TAR1 sequence are present within 2 kb of the beginning of nearly all sequenced $(TTAGGG)_n$-adjacent DNA, although the part and degree of similarity can vary substantially. TAR1 similarity can also be found adjacent to many of the internal $(TTAGGG)_n$-like sequences in subtelomeric repeat (Srpt) regions, and more distantly related copies of TAR1-like sequence are often found in

STELA: single telomere length analysis

pericentromere regions (1). Interestingly, one known case of a naturally occuring segment of $(TTAGGG)_n$-adjacent DNA lacking TAR1 similarity is an allele of 17p shown by single telomere ligation-associated PCR [single telomere length analysis (STELA)] to be shorter than most other telomeres (18). Several instances of terminally deleted chromosomes healed by $(TTAGGG)_n$ addition have been noted (89); these chromosomes can be transmitted stably through the human germline, but potential effects of non-native cis elements upon allele-specific $(TTAGGG)_n$ tract length regulation in the deleted chromosome have not been investigated.

Whereas the minimum length of the human terminal repeat tract was estimated at 2 kb using bulk telomere measurements in senescing cells (52), the higher-resolution STELA assay found a wide variability of telomere lengths at individual chromosome ends in senescing fibroblasts, with a minimum modal terminal tract length of 300 bp found on an allele of 17p in senescent cell populations (18). STELA (12) involves a ligation-mediated PCR between the end of the chromosome and priming sites in the subterminal region (S1 and S2 in **Figure 1**). A telorette linker with a stretch of homology to the $(TTAGGG)_n$ overhang aligns with and is ligated to the C-strand. A primer site within the linker is used to prime PCR between the linker and subterminal priming sites. The products are the result of single-molecule PCRs of the intervening DNA, and their size is dependent upon the terminal tract lengths of single telomeres. STELA has been used successfully with single-telomere specificity and with allele-specific primers to detect and measure single telomere lengths up to approximately 20 kb, including a detailed analysis of variation in human sperm telomeres (9, 12, 18). Working assays have been developed for some alleles of the XpYp, 2p, 11q, 12q, and 17p telomeres (18); as long as a primer with allele specificity can be found an appropriate distance from the chromosome end the assay could in theory be extended to any given subterminal allele. A major limitation in the use of STELA has been the lack of single-telomere and single-allele specificity for most telomeres, given our currently incomplete knowledge of subterminal sequences. STELA is unique in that it provides a molecular tool for investigating single-telomere lengths and length regulation without preparing metaphase chromosomes [as in metaphase quantitative fluorescence in situ hybridization (Q-FISH)] and can be used to measure telomere length variation and telomere stability on an allele-by-allele basis.

STELA-like end-ligation protocols have been used to identify the terminal nucleotides of both C and G strands on human telomeres (76), and to analyze in molecular detail the DNA sequences associated with both extremely short telomeres and telomere fusion structures (20, 92). STELA was used to identify the minimal critical telomere lengths in the context of analyzing telomere-telomere fusions; all fusion events lacked $(TTAGGG)_n$ sequences greater than 12.8 repeats, with greater than half lacking all canonical repeats at the fusion junction. In a second study, STELA was used to determine telomere lengths in a telomerase-positive cancer cell line with compromised checkpoint pathways (92). A distinct class of apparently capped, very short telomeres (t-stumps) coexisted in the cancer cells along with the longer telomeres; the t-stumps contained a minimal stretch of seven canonical $(TTAGGG)_n$ repeats. It was suggested that this stretch is sufficient for effective capping by TRF1/TRF2 and that telomerase may have a protective capping function at t-stumps, separable from its elongation function (92). As these examples illustrate, the new, high-resolution structural information on individual telomeres made possible by STELA will likely have a major impact on how researchers approach mechanistic studies of telomere length regulation and telomere chromatin studies. It is therefore quite important to understand detailed subterminal sequence organization to analyze the impact of subterminal structure and variation on telomere dynamics and, from a pragmatic experimental standpoint, to enable development of a much wider array of allele-specific STELA assays for studying individual telomeres in the context of small amounts of DNA from nondividing cells.

SUBTERMINAL SEQUENCE BLOCKS

Adjacent to most of the terminal $(TTAGGG)_n$ sequences in the human reference sequence and to many internal $(TTAGGG)_n$ sequences are segmental duplications termed subterminal duplicons (1). Six subterminal duplicon families (A-F) have been distinguished (**Table 1**). Together with six 1-copy DNA $(TTAGGG)_n$-adjacent regions (7q, 8q, 11q, 12q, 18q, and Xp/Yp), these duplicon families represent the global set of sequences occupying the DNA space immediately cis to terminal $(TTAGGG)_n$ tracts. As such, they are among the sequences most likely to directly impact terminal $(TTAGGG)_n$ tract regulation (18), and are the first non$(TTAGGG)_n$ sequences expected to be affected by telomere dysfunction, aberrant telomere replication, and telomere instability.

Table 1 shows the telomere and the defining subterminal segment size for these six duplicon families, as well as the copy number for each family. The copies are categorized according to those that occur in other subterminal regions [<25 kb from any known terminal $(TTAGGG)_n$ tract], those that occur in subtelomeric repeat regions but are not subterminal, and those that occur in nonsubtelomeric regions. The genomic locations of nonsubtelomeric copies of subterminal duplicons suggest sites of ancestral telomere-associated chromosome rearrangements, including a well-documented telomere fusion at 2q13-q14 (46) and ancestral inversion of a chromosome arm followed by duplication of pericentromeric sequences (1).

The relationship between subterminal duplicon copies within a family and between several related subterminal families is complex and broadly consistent with an earlier model of subtelomere structure featuring a subterminal compartment with more active recombinational features than the larger centromerically positioned subtelomere duplications (32). In particular, many of the subterminal intrafamily and cross-family homology regions are relatively short, their positions within the subterminal blocks vary, and they are located at different

Table 1 Subterminal duplicons

Subterminal block	Telomere	Size (kb)	Duplicated blocks	Location	Percent identity	Named transcripts
A	2p	7	6	Subterminal	91.74–92.46	RPL23AP7-related, FAM41C
			12	Subtelomeric	91.24–92.65	
			1	Nonsubtelomeric	91.8	
B	4p	17	10	Subterminal	90.67–98.39	RPL23AP7-related, FAM41C
			16	Subtelomeric	90.57–93.66	
			1	Nonsubtelomeric	91.9	
C	9p	10	6	Subterminal	98.29–99.00	MGC13005-related, DDX11-related, WASH-related
			1	Nonsubtelomeric	98.27	
D	10q	22	10	Subterminal	90.7–96.65	RPL23AP7-related, FAM41C
			15	Subtelomeric	91.68–96.09	
			2	Nonsubtelomeric	93.69–95.80	
E	17p	21	5	Subterminal	95.97–97.16	
F	18p	15	1	Subterminal	99.00	
			1	Subtelomeric	93.58	
			8	Nonsubtelomeric	91.19–94.27	

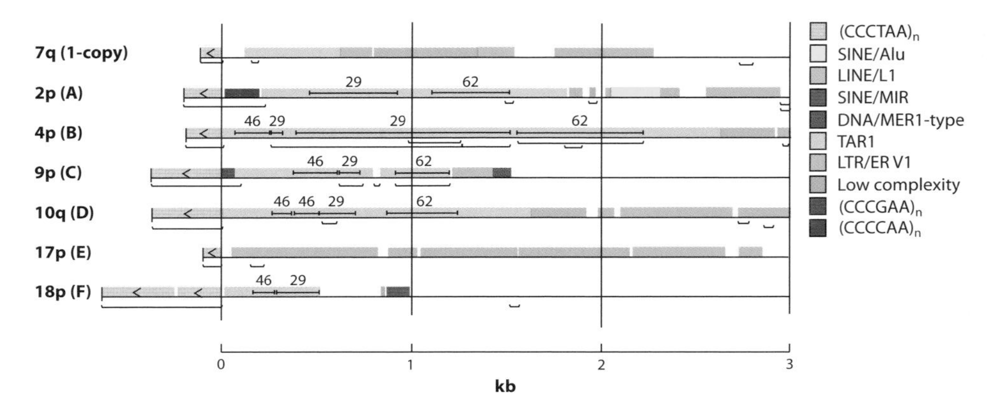

Figure 2

Telomere-subterminal duplicon boundaries for a sample 1-copy subtelomere and a representative of each of the six classes of subterminal duplicons. The beginning of the $(TTAGGG)_n$ tracts are at the left, and repeat classes recognized by RepeatMasker (**http://www.repeatmasker.org**) (79) are annotated. Clusters of variable number of tandem repeats (VNTRs) are designated by consensus repeat size (shown in bp); the 29mer VNTR is CpG rich, variable in length, hypomethylated in sperm, and hypermethylated in somatic cells. This VNTR is present in many but not all telomere-subterminal boundaries. SINE, short interspersed nuclear element; LINE, long interspersed nuclear element; MIR, mammalian-wide interspersed repeat family; MER, medium reiteration repeat family; TAR, telomere-associated repeat; LTR, long terminal repeat; ER, endogenous retrovirus.

distances from the terminal $(TTAGGG)_n$ tract. In addition, there are several alternative organizations of repetitive elements within these subterminal blocks. For example, there are very clear differences in the terminal-subterminal boundary regions for each of the 1-copy telomeres and for each class of subterminal duplicon families (**Figure 2**), as well as more subtle but distinct differences from telomeres within duplicon families (1).

Further refinement of the classification of these subterminal families appears feasible and will benefit from more extensive sampling of $(TTAGGG)_n$-adjacent sequences from additional alleles. Most subterminal duplicon sequences are clearly more divergent than the large duplicons that exist more centromerically (**Figure 3**), both in nucleotide sequence similarity and in sequence organization. This divergence might be exploited to develop subterminal allele-specific PCR assays (e.g., assays defined by primers Gfor and Grev in **Figure 1**) to track some of these sequences genetically in the context of total genomic DNA, in addition to generating additional allele-specific STELA assays. Fosmid libraries prepared from the DNA of unrelated individuals as part of the structural variation initiative (29) are proving to be a rich source of new subterminal sequences, both for closing subterminal gaps in the existing reference sequence and for isolating and characterizing new allelic variants of subterminal sequences.

WASP proteins: Wiscott-Aldrich Syndrome protein

WASH proteins: Wiscott-Aldrich Syndrome protein and Scar homolog

SUBTERMINAL TRANSCRIPTS

Several transcript families exist in subterminal duplicon blocks. Many family members appear to be noncoding, pseudogene copies of known genes; however, some transcripts contain open reading frames and could encode protein (1). Very recently, a particularly abundant and variable subterminal transcript family was shown to correspond to the 3′ ends of genes encoding a new subclass of the Wiscott-Aldrich Syndrome Protein (WASP) family, termed Wiscott-Aldrich Syndrome protein and Scar homolog (WASH) proteins (54).

The WASH protein family is ancient, conserved from amoeba through human lineages, and shown to be essential for survival in Drosphila (54). WASH genes are single copy in most lower organisms and early primates, but multicopy in gorilla, chimps, and humans, with the highest copy numbers in human. Like the WASP proteins, WASH proteins appear to be involved in cytoskeletal organization and signal transduction, and function as actin polymerization regulation factors. The human WASH gene family structures predict short forms and long forms of the protein similar to those found in Entamoeba; the 3′ exons encoding the short form are embedded in subterminal block C (**Table 1**), and the full length forms start near a CpG island approximately 20 kb from telomeres within a polymorphically distributed segmental duplication (58). The WASH family varies widely in both dosage and telomere distribution in individual genomes, and usually terminates less than 5 kb from the start of the terminal $(TTAGGG)_n$ tract; thus, individual telomeric transcription sites for this family might be differentially susceptible to position effects depending upon both local telomeric chromatin/heterochromatin status and on chromosome-specific telomere lengths. The subtelomeric/subterminal genomic location, as a hotspot of DNA break and repair (55, 71), has likely contributed to the production of many copies and combinations of the short and long forms of WASH gene family members in very recent primate and human evolution. It is intriguing to speculate on potential roles this gene family may have played in the evolution of primate and human-specific phenotypes, as well as potential roles in the somatic evolution of cancer genomes, where telomere loss or dysfunction is believed to play an initiating role (8, 59).

A recently discovered noncoding subterminal transcript that includes sequences from the terminal repeat may prove to be central to understanding telomere biology. The telomeric repeat-containing RNA (TERRA) molecules found in humans range in size from approximately 100 bp to at least 9 kb, are transcribed almost exclusively from the C-strand of subterminal DNA from multiple (perhaps all) telomeres, and associate with telomeric chromatin (6). The association of TERRA molecules with telomeric chromatin is regulated by protein effectors of the nonsense-mediated RNA decay pathway, which include the human homologs of the yeast Est1p protein; depletion of effectors increased the association of TERRA molecules with telomeric chromatin and caused loss of telomere DNA. A similar telomeric RNA molecule found in mouse was shown to be transcribed by DNA polymerase II, polyadenylated, and under developmental regulation (74). Evidence was presented suggesting that high levels of the mouse telomeric repeat-containing RNA could inhibit telomerase activity (74). Given the known role of noncoding RNAs in heterochromatin formation (83) and X-inactivation (91), the discovery of these telomeric repeat-containing RNAs seems likely to redefine our thinking with respect to the modulation of telomeric chromatin structure and telomeric heterochromatin formation.

TERRA: telomeric repeat-containing RNA

TELOMERE EPIGENETICS

Mouse studies have previously shown that mammalian telomere length and stability can be regulated by epigenetic factors (16); heterochromatic epigenetic marks such as trimethylated H3K9 and H4K20, heavily methylated CpGs, and HP1 localization have been identified at mouse telomeres. Loss of some of those marks occurs upon reduction of telomere $(TTAGGG)_n$ tract lengths to a critically short length in the absence of telomerase, and deregulation of telomere tract length and stability is evident in mouse knockouts affecting enzymes responsible for normally maintaining the heterochromatic marks (14, 36). Mouse telomeres appear to become much more recombinationally active in the absence of DNA methylation, implicating epigenetic changes in the activation of alternative lengthening of telomeres (ALT) pathways (36). In mouse, telomere heterochromatin appears to spread over significant distances, and may be capable of silencing subtelomeric genes via these position effects (65). Subtelomeric silencing is developmentally regulated in the mouse,

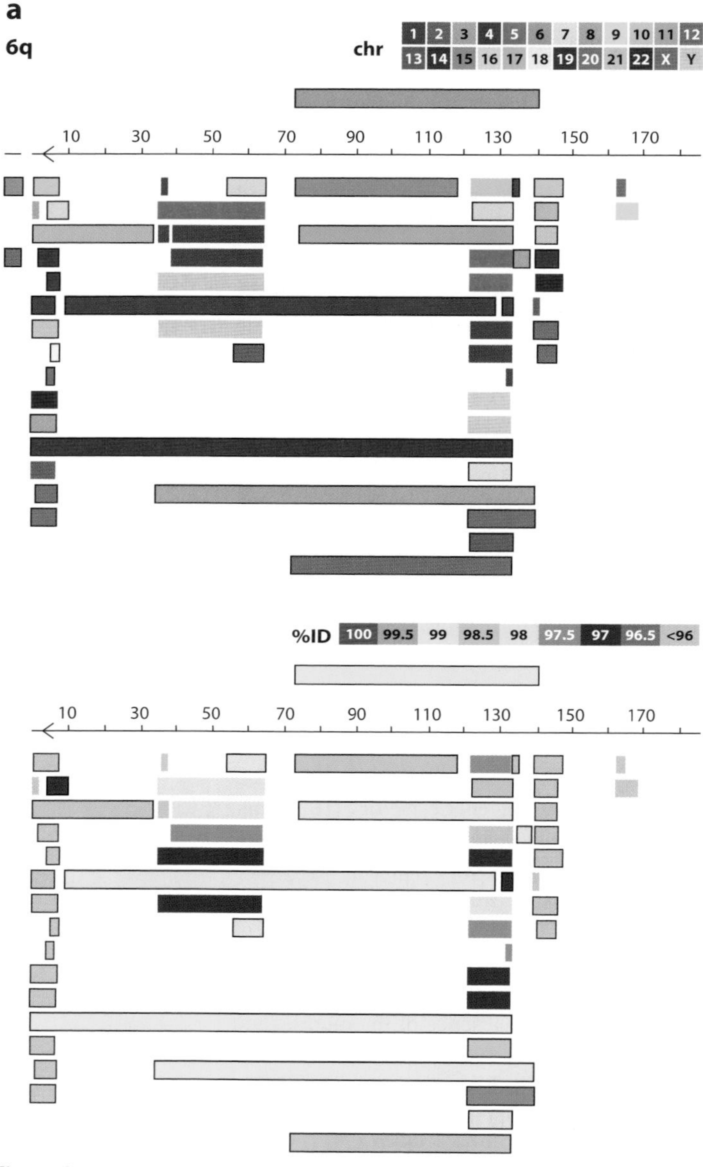

Figure 3

Duplicon analysis for two subtelomeres with large Srpt regions, 6q (*a*) and 16q (*b*, *next page*). The most distal regions of the 6q and 16q reference sequences are shown oriented with the telomeric end on the left. (*Top*) Duplicated genomic segments are identified by chromosome (*color*) and whether they are subtelomeric (*bounded rectangles*), nontelomeric (*unbounded rectangles*), or intrachromosomal (*located above the subtelomere coordinates*). Each rectangle represents a separate duplicon. (*Bottom*) Each of the pairwise alignments generated by the duplicon definitions was tabulated and analyzed according to percent nucleotide sequence identity. Our web site (**http://www.wistar.upenn.edu/Riethman**) contains analogous figures and underlying sequence alignments for the duplicon organization of each subtelomere region.

with low but detectable transcription in subtelomere regions of embryonic stem cells but a complete shutdown of subtelomeric transcription in embryonic fibroblasts and in adult tissues (34). Epigenetic modification of subterminal sequences may thus play a critical role in regulating telomere replication and telomerase-dependent elongation as well as recombination-associated phenomena such as ALT-based telomere elongation (67) and elevated levels of telomere-associated sister chromatid exchange (71).

However, it is important to note substantial differences between mouse and human telomeres. Telomeres of lab mice are very long relative to human telomeres, and there are no recognizable similarities between mouse and human subterminal DNA except for the simple repeat tracts themselves (H. Riethman, unpublished observations). Thus, $(TTAGGG)_n$ tract length regulation and its biological causes and consequences may vary considerably between the two species.

Little is currently known about the epigenetic status of human subterminal DNA other than it is hypomethylated in the germ line, methylated de novo in somatic tissues, and may contain noncanonical DNA modifications (25, 82). Hypomethylation of subterminal DNA in the human germline and suppression of meiotic recombination in the most distal 2–4 kb of telomere-adjacent DNA (11) are characteristics of satellite heterochromatin, but the active transcription of protein-encoding genes within a few kb of some human $(TTAGGG)_n$ tracts and the small telomere-adjacent linkage disequilibrium (LD) region suggest a very small telomeric heterochromatic region and perhaps a precisely regulated heterochromatic-euchromatic boundary region in human subterminal DNA. Human subterminal DNA sequences typically contain clusters of very CpG-rich sequences that are arranged in variably sized and organized stretches among separate telomeres (**Figure 2**); these sequences are among those differentially methylated in germline versus somatic cells, and their methylation patterns may reflect the telomeric heterochromatin-euchromatin transition for specific telomere alleles. Analysis of the dataset from a recent genome-wide location analysis of the insulator CCCTC-binding factor (CTCF) (13) showed an enrichment of this factor with subterminal DNA sequences immediately centromeric to this subterminal CpG island, possibly suggesting that this DNA is associated with a chromatin boundary region (H. Riethman, unpublished observations). Allele-specific $(TTAGGG)_n$ tract length regulation might be dependent upon differential positioning of the euchromatin/heterochromatin boundary within subterminal sequences, mediated in some fashion by the differential transcription and association of TERRA molecules.

EXTENDED HUMAN SUBTELOMERIC DNA REGIONS

Human subtelomeric segmental duplications (subtelomeric repeats) comprise approximately 25% of the most distal 500 kb and 80% of the most distal 100 kb in human DNA (47, 69). The overall size, sequence content, and organization of subtelomeric segmental duplications relative to the terminal $(TTAGGG)_n$ repeat tracts and to subtelomeric 1-copy DNA are different for each subtelomere (69). **Figure 4** illustrates a range of different types of subtelomeric sequence organization found in the human reference sequence. The bulk of Srpt sequences is confined to the most distal regions of the subtelomere, although there are several examples (2p, 2q, 3q, 5p, 7p, 8p, and 12p) where, in addition to a terminal block of Srpt, additional smaller segments are interspersed within the adjacent 1-copy DNA and segmentally duplicated DNA. Non-Srpt segmental duplication blocks are often found adjacent to Srpts, but display a highly variable pattern of content and distribution at each chromosome end (**Figure 4**). A significant number of gaps remain in the telomere-adjacent regions of the reference sequence (terminal gaps), and there is also an unusually high number of internal sequence gaps in subtelomere regions. Some of the gaps are due to incompletely sequenced half-YAC

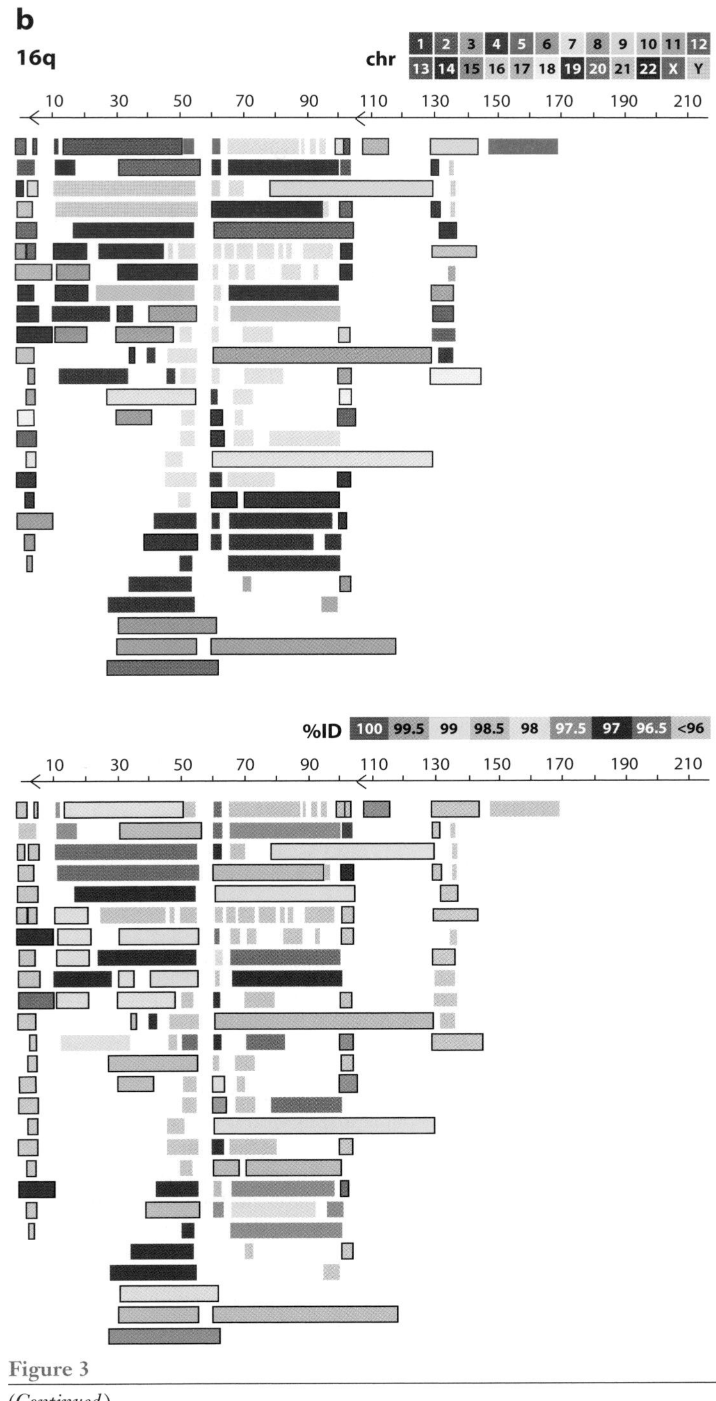

Figure 3

(Continued)

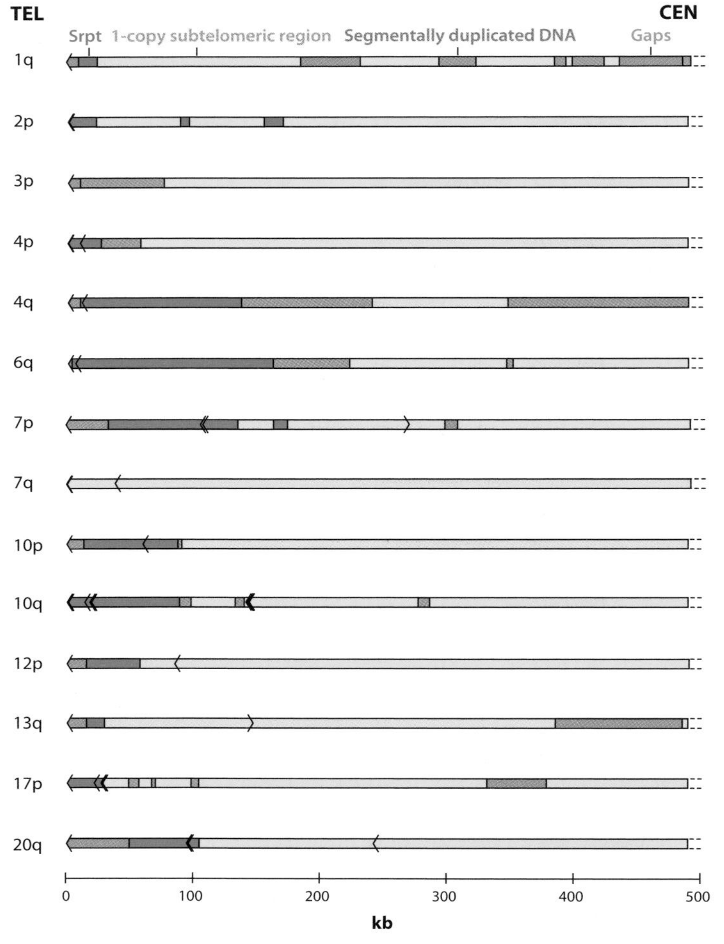

Figure 4

Examples of sequence organization in human subtelomeric DNA regions. The terminal $(TTAGGG)_n$ repeat tract typically consists of 5–15 kb of the simple repeat sequence $(TTAGGG)_n$, and is indicated by the black arrow at the 0 coordinate. The subtelomeric repeat (Srpt) region (*blue*) is composed of a mosaic patchwork of segmentally duplicated DNA tracts that occur in two or more subtelomere regions. The Srpt region can be less than 5 kb up to approximately 300 kb in size, depending on the specific telomere. Centromeric to the Srpt region is chromosome-specific genomic DNA, typically with a high GC content and high gene density. Stretches of segmentally duplicated DNA (sequence >90% identical, >1 kb in size) that are not present at multiple subtelomeric loci (*green*) are interspersed in a telomere-specific fashion with 1-copy subtelomeric regions (*yellow*). Short (50–250 bp) and often degenerate $(TTAGGG)_n$ tracts are interspersed within the Srpt region and occasionally within the adjacent 1-copy subtelomeric DNA (*internal black arrows*). Gaps in the reference sequence are shown in pink.

clones, whereas others are due to large-scale polymorphisms and allelic differences that preclude assembly of individual sequenced clones with adjacent clones derived from variant haplotypes.

Internal $(TTAGGG)_n$-Like Sequences

In yeast, interstitial telomere-like sequence islands in subtelomeric DNA can play important roles in subtelomeric recombination and transcription, telomere maintenance via the ALT pathway, and telomere healing (56, 80). Several studies have suggested similar roles for internal human $(TTAGGG)_n$-like sequence islands (5, 62, 72), including a hypothesized role as a boundary element for subtelomeric DNA compartments (32). $(TTAGGG)_n$-like islands are enriched approximately 25-fold in human subtelomeric regions (69). In addition, they tend to be both longer and more similar to perfect $(TTAGGG)_n$ tracts in subtelomeric DNA compared with elsewhere in the genome. From an evolutionary perspective, this might suggest that most subtelomeric interstitial $(TTAGGG)_n$ tracts arose more recently than those found elsewhere in the genome, originated via a separate mechanism than $(TTAGGG)_n$ islands found elsewhere (5), or are under some selective pressure to maintain similarity to $(TTAGGG)_n$ and a minimum tract length (32). Given the known functions of internal telomere sequences in yeast and the demonstrated ability of TRF1, TRF2, and hRAP1 to interact with very short TTAGGGTT motifs (26, 93), it is important to investigate further the abundance, allelic variation, and potential functions of internal $(TTAGGG)_n$ sequence islands in human subtelomere regions. There are currently no published data on the possible transcription or chromatin association of TERRA-like RNAs from internal $(TTAGGG)_n$-like islands.

SUBTELOMERIC DUPLICONS

Although subtelomeric regions of human chromosomes have long been known to contain mosaic patchworks of duplicons (28, 60, 69), recent genome-wide analyses of these regions have revealed new details. The patchworks of subtelomeric duplicons appear to arise from translocations involving the tips of chromosomes, followed by transmission of unbalanced chromosomal complements to offspring (55). A recent analysis of sister chromatid exchanges (SCEs) revealed highly elevated rates near telomeres [1600 times greater in the most distal 10 kb (including subterminal sequences and the telomere repeat itself) and 160 times greater in the adjacent 100-kb subtelomere regions (71)]. Together these studies indicate that subtelomeres are hotspots of DNA breakage and repair; these features are likely to be responsible for the generation of the complex interchromosomal duplication patterns and the rapid evolution of these genomic regions.

A detailed analysis of subtelomeric duplicons was completed recently (1). Duplicon blocks were defined by sequence similarity between segments of subtelomeric DNA from single telomeres and the assembled human genome. Each duplicon module was defined by a set of pairwise alignments with the query subtelomere sequence, and a percent nucleotide sequence identity for the nonmasked parts of each chained pairwise alignment was derived from the BLAST alignments. This analysis is summarized for two telomeres in **Figure 3**; the full set of modules for all telomeres, including the coordinates of their genomic alignments, is presented on our web site (**http://www.wistar.org/Riethman**).

Duplicons that occupy subtelomeric sequences are generally both larger and more abundant than those occurring elsewhere in the genome, consistent with the notion that subtelomeric location in humans is permissive for and/or somehow promotes large duplication events (1). Although smaller and fewer than subtelomeric copies, nonsubtelomeric copies of duplicons tend to cluster at relatively few pericentric and interstitial loci; these include 2q13-q14 (at the site where ancestral primate telomeres fused to form modern human chromosome 2) and a handful of additional sites also documented previously in genome-wide analyses of segmental duplications (7). These loci apparently represent sites susceptible to either donation or acceptance of these duplicated chromosome segments in recent evolutionary time.

ALT: alternative lengthening of telomeres

LD: linkage disequilibrium

SCE: sister chromatid exchange

The duplicon organization for the subtelomeres reveals several key features (1 and **http://www.wistar.org/Riethman**) (**Figure 3**). For any given segment of a subtelomere, the level of nucleotide sequence similarity with duplicated DNA depends entirely on the specific duplicon content and organization and does not necessarily correlate with its distance from the telomere terminus. Large duplicons with relatively high sequence similarity among family members cover a large proportion of the duplicated sequence space, but occupy only a subset of subtelomere regions and exist at variable distances from the terminal $(TTAGGG)_n$ tract. For subtelomeres that contain large duplicons, there is a consistent pattern of higher divergence in $(TTAGGG)_n$-adjacent subterminal sequence than in adjacent large duplicon regions. For subtelomeres that lack large duplicons, there is typically a much lower degree of sequence similarity throughout these subtelomeric duplication regions (often 90% to 96% nucleotide sequence identity).

A small group of subtelomeres fall outside of the aforementioned patterns. The 16p reference allele subtelomere and the Xq/Yq subtelomere have small, highly similar subterminal duplicons and more divergent adjacent subtelomeric duplicons, whereas the 2q, 12p, 17p, and 20q subtelomeres have moderately sized duplicons with <96% to 98.5% similarity throughout the duplicated regions. The 9p subtelomere has subterminal duplicons with high sequence similarity (98.5%–99%) and several large blocks of sequence that correspond to the 2qfus internal site and several internal loci on chromosome 9 (31 and **http://www.wistar.upenn.edu/riethman/telomere.cfm?tel=9p#Map**).

In each of the subtelomeres that contains duplicated sequences, the internal $(TTAGGG)_n$-like sequences are usually oriented toward the telomere and almost always colocalize to duplicon boundaries (1). The subtelomeric internal $(TTAGGG)_n$-like islands can reach up to 823 bp in size, but most are in the 150–200-bp range. Several of the $(TTAGGG)_n$-related motifs found in these islands are also commonly found in the proximal regions of telomere terminal tracts (e.g., TGAGGG, TCAGGG, TTGGGG) (H. Riethman & A. Ambrosini, unpublished data). A more detailed analysis of these interesting sequence islands and their comparison with a more comprehensive set of telomere-proximal sequences than is currently available might shed light on their origins and the relative timing of their internalization.

Subtelomeric Transcripts

Transcripts are generally distributed throughout human subtelomeric regions. In addition to the WASH gene family (54) described for subterminal DNA sequences and the TERRA RNAs that are transcribed from the regions spanning the subterminal DNA/$(TTAGGG)_n$ tract region, several other subtelomeric gene families have been identified that are embedded in duplicated subtelomere regions and/or sometimes contained in 1-copy subtelomere DNA. These include the immunoglobulin heavy chain genes (found at 14q), olfactory receptor genes [1-copy regions of 1q, 5q, 10q, and 15q as well as previously characterized subtelomeric repeat DNA (1p, 6p, 8p, 11p, 15q, 19p, and 3q) (84)], and zinc-finger genes (4p, 5q, 8p, 8q, 12q, 19q). Transcripts for multiple members of these gene families are found within many of the individual subtelomeric regions, along with transcripts for many additional gene families of unknown function. The abundance of gene families in subtelomeric regions is a common feature of many eukaryotes, and may reflect a common mechanism for generating duplications and utilizing subtelomeres for generating new genes.

Subtelomeric duplicons are known to harbor protein-encoding genes, predicted protein-encoding genes, pseudogenes, and many transcripts of unknown function. For several subtelomeric transcript families (IL9R, DUX4, FBXO25), functional evidence for protein expression from at least one transcript locus is available (41, 64, 87). However, for most transcript families the evidence for encoded protein function relies upon the existence of one or more actively transcribed loci with open reading frames predicted to encode evolutionarily conserved proteins (35, 53, 57, 86). Although these data strongly suggest that one or more members of each of these gene families encode functional protein, in most cases pseudogene copies of the respective gene family coexist among the duplicons. Delineating the evolution and function of individual members of subtelomeric gene families will be a major task for the near future. Because only a single reference sequence has been sampled and there is clearly abundant structural variation in subtelomeres, many additional members of these gene families certainly remain to be discovered in the human population.

SUBTELOMERIC STRUCTURAL VARIATION AND ITS FUNCTIONAL CONSEQUENCES

Large variant regions of subtelomeric DNA detected at many human telomeres were among the first examples of structural variation and copy number variation (CNV), which was subsequently found genome-wide (45, 75). The small subtelomere alleles sometimes differ from the large alleles by hundreds of kilobases of additional subtelomeric DNA, leading to huge allelic disparities in subtelomeric DNA sequence content and consequent separation of 1-copy genes from terminal $(TTAGGG)_n$ tracts. These large allelic disparities are postulated to have an impact not only on the dosage (and hence expression levels) of transcripts embedded within the duplicated sequences, but also on the transcription of genes in adjacent 1-copy regions. Although these polymorphisms are often due to mainly differential Srpt DNA content and organization on particular variant alleles, there are also cases of deletion/insertion polymorphisms of 1-copy DNA adjacent to the Srpt

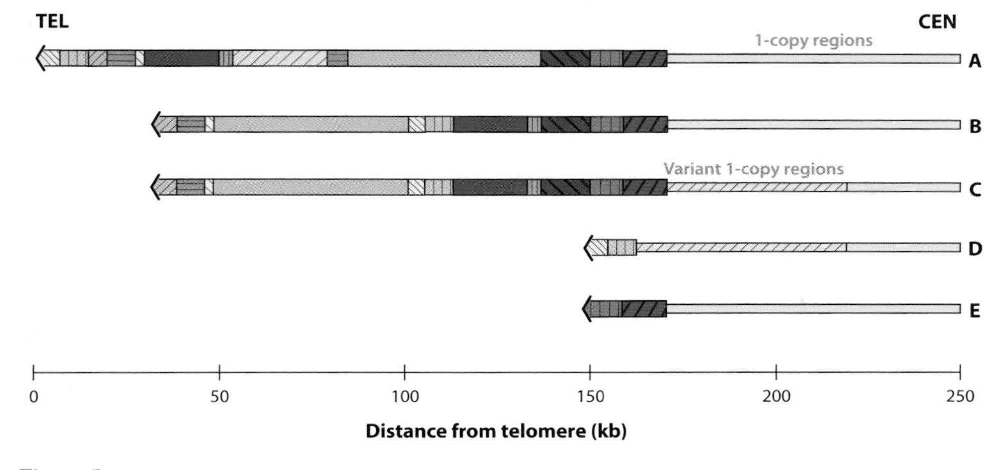

Figure 5

Examples of types of structural variation at a single human subtelomere. 1-copy regions are in yellow; modules of duplicon blocks are color-coded. Hashed yellow represents variant 1-copy regions. Alleles for a single subtelomeric locus can vary dramatically in size and sequence content, including large insertion-deletion polymorphisms within the 1-copy DNA adjacent to the segmental duplication regions.

regions. **Figure 5** illustrates several types of polymorphic subtelomere structures characterized to date. Any of these large allelic disparities might be expected to significantly affect subtelomeric transcript identity or abundance, either via copy number changes, by alteration of transcriptional regulation, or by rendering particular transcripts nonfunctional by insertion, deletion, or truncation.

The existence of large-scale subtelomere polymorphisms detected in these earlier studies was recently confirmed and extended in the much larger and comprehensive studies of Redon and coworkers (68). This group characterized the genomic DNA of 269 individuals [already genotyped for single nucleotide polymorphisms (SNPs) as part of the HapMap project] for CNV using two genome-scale array comparative genomic hybridization (CGH) platforms: the 500K Affymetrix SNP chip and a whole-genome tiling path (WGTP) array of bacterial artificial chrmosomes (BACs). CNV regions were detected at almost every subtelomere, and in most cases these CNV regions were adjacent to and/or were coextensive with the previously identified segmental duplication regions. However, the resolution of these platforms is generally >50 kb, and because of the shared duplicons near many telomeres, apparently locus-specific copy number changes will be influenced by overall subtelomeric duplicon number changes and are unlikely to be completely independent events; hence higher resolution methods are required to truly determine whether structural variations are occurring at particular subtelomeric loci, and whether they include 1-copy sequences or are purely due to copy number changes in segmental duplications. Fosmid end-sequence mapping strategies (85) applied to subtelomere regions will provide a much more powerful method for exploring and characterizing this variation, and will yield cloned fragments that correspond to the sites of variation to analyze at high resolution. For subtelomere alleles with large and complex duplications, existing and planned BAC libraries (29) will be needed to fully characterize these variants; however, the fosmid libraries can be used to define and characterize the boundaries of the large variant regions.

Structural variation is abundant in subterminal DNA, and is likely to have important consequences for expression of subterminal genes and TERRA RNAs as well as for regulation of $(TTAGGG)_n$ tract lengths, especially individual-specific patterns of allele-specific

$(TTAGGG)_n$ tract lengths that may be crucial for the biological functions of telomeres. Both CNV and alternative sequence organizations for DNA immediately adjacent to terminal $(TTAGGG)_n$ tracts are major components of this structural variation. Important near-term goals are to better understand the extent and nature of subterminal structural variation and to develop methods for tracking individual allelic variants in the context of total genomic DNA.

The construction and end-sequencing of fosmid libraries from the DNA of unrelated HapMap individuals as part of the Human Structural Variation initiative (29) have resulted in a remarkable resource for investigation of the highly variable subtelomeric regions of the human genome. Each fosmid end sequence (FES) library has roughly 12-fold clone coverage of the source genome. Importantly, because the libraries were prepared using sheared source DNA, terminal fosmids with one mate pair corresponding to part of the terminal $(TTAGGG)_n$ sequence tract can be identified. Preliminary studies in our lab have demonstrated the feasibility of isolating novel subterminal structural variants using this approach and have shown that although distal terminal $(TTAGGG)_n$ tracts are truncated in the clones, subterminal DNA remains intact (H. Riethman, unpublished observation). Many $(TTAGGG)_n$-adjacent sequences are still missing from the reference sequence, and a large number of structural variants carrying new common subterminal alleles remain to be discovered and characterized. Given the importance of subterminal sequences for telomere chromatin assembly and telomere length regulation, an effort to systematically identify and characterize subterminal sequences from common telomere alleles is warranted. Such an effort will fill a major gap in our knowledge of subtelomere structures in the human genome, and will permit the development of new tools for studying telomere biology.

SUMMARY POINTS

1. The DNA in human telomere $(TTAGGG)_n$ tracts is dynamically modulated by a balance of exonuclease- and replication-associated attrition and telomerase- and/or recombination-mediated lengthening. Depending on the cellular context, critically short tracts of $(TTAGGG)_n$ lead to cellular senescence, apoptosis, and/or genome instability. Aberrant regulation of $(TTAGGG)_n$ tract lengths is associated with cancer, aging, and age-related diseases.
2. Average $(TTAGGG)_n$ tract lengths vary substantially from individual to individual and have a large heritable component. Superimposed upon this bulk average $(TTAGGG)_n$ tract length variation are telomere-specific and allele-specific differences in tract length, which appear to be defined in the zygote and maintained throughout life, even in the presence of telomerase. *Cis*-acting factors regulate allele-specific $(TTAGGG)_n$ tract lengths.
3. Single telomere length analysis (STELA) (12) is providing high-resolution information on individual telomeres and novel approaches to study mechanisms of telomere length regulation and telomere loss. A detailed and complete understanding of subterminal sequence organization is needed to facilitate single-telomere resolution approaches to understand telomere dynamics and regulation, including studies of epigenetic changes that affect telomeres.
4. Subtelomeric DNA tracts are filled with segmental duplications and copy number variant DNA, which leads to highly variable subtelomeric allele structures in the human population. Subtelomeres are hotspots of DNA breakage and repair; these features are

likely to be responsible for the generation of complex interchromosomal duplication patterns and the rapid evolution of these genomic regions.

5. Transcription occurs throughout subtelomeric and subterminal DNA regions. Multicopy protein-encoding gene families and a variety of noncoding RNAs have been discovered. A distally located gene family encodes actin-interacting Wiscott-Aldrich Syndrome protein and Scar homolog (WASH) proteins (54), which are believed to function in cytoskeleton assembly regulation and signal transduction. A recently described family of $(UUAGGG)_n$-containing subterminal RNAs appears to be critical for telomere integrity; these non-coding RNAs associate with telomeric chromatin and are regulated by RNA surveillance factors including human homologs of the yeast Est1p protein (6).

FUTURE ISSUES

1. Subterminal DNA sequences immediately adjacent to terminal $(TTAGGG)_n$ tracts are characterized for only a handful of alleles. The roles of these sequences in the genetic and epigenetic regulation of telomere $(TTAGGG)_n$ tract length and stability render them critically important for understanding telomere biology. There is thus an urgent need to characterize common germline human subterminal alleles and subterminal structural variants with nucleotide sequence-level precision, to develop methods for analyzing them in human cells and in human populations, and to perhaps correlate particular variants and mutations in these sequences with human phenotypes and disease.
2. The large allelic disparities in subtelomeric DNA regions are postulated to have an impact both on the dosage (and hence expression levels) of transcripts embedded within the duplicated sequences and on telomere position-dependent transcriptional regulation. Discovering the biological functions of protein families encoded by these transcripts (e.g., the WASH proteins) is key to understanding their potential roles in human natural variation and primate evolution.
3. The recently discovered noncoding subterminal transcripts that include sequences from the terminal repeat may prove to be central to understanding telomere biology. Very little is currently known about the regulation and function of these molecules, and their study is likely to teach us much about telomeric chromatin structure and telomere length regulation in humans.

DISCLOSURE STATEMENT

The author is not aware of any biases that might be perceived as affecting the objectivity of this review.

ACKNOWLEDGMENTS

I thank Sheila Paul for helping with the preparation of the manuscript. Work from my lab was supported by NIH grants HG00567 and HG2933, and by the Commonwealth Universal Research Enhancement Program, PA Dept of Health.

LITERATURE CITED

1. Ambrosini A, Paul S, Hu S, Riethman H. 2007. Human subtelomeric duplicon structure and organization. *Genome Biol.* 8:R151
2. Armanios M, Chen JL, Chang YP, Brodsky RA, Hawkins A, et al. 2005. Haploinsufficiency of telomerase reverse transcriptase leads to anticipation in autosomal dominant dyskeratosis congenita. *Proc. Natl. Acad. Sci. USA* 102:15960–4
3. Armanios MY, Chen JJ, Cogan JD, Alder JK, Ingersoll RG, et al. 2007. Telomerase mutations in families with idiopathic pulmonary fibrosis. *N. Engl. J. Med.* 356:1317–26
4. Aviv A, Levy D, Mangel M. 2003. Growth, telomere dynamics and successful and unsuccessful human aging. *Mech. Ageing Dev.* 124:829–37
5. Azzalin CM, Nergadze SG, Giulotto E. 2001. Human intrachromosomal telomeric-like repeats: sequence organization and mechanisms of origin. *Chromosoma* 110:75–82
6. Azzalin CM, Reichenbach P, Khoriauli L, Giulotto E, Lingner J. 2007. Telomeric repeat containing RNA and RNA surveillance factors at mammalian chromosome ends. *Science* 318:798–801
7. Bailey JA, Gu Z, Clark RA, Reinert K, Samonte RV, et al. 2002. Recent segmental duplications in the human genome. *Science* 297:1003–7
8. Bailey SM, Murnane JP. 2006. Telomeres, chromosome instability and cancer. *Nucleic Acids Res.* 34:2408–17
9. Baird DM, Britt-Compton B, Rowson J, Amso NN, Gregory L, Kipling D. 2006. Telomere instability in the male germline. *Hum. Mol. Genet.* 15:45–51
10. Baird DM, Coleman J, Rosser ZH, Royle NJ. 2000. High levels of sequence polymorphism and linkage disequilibrium at the telomere of 12q: implications for telomere biology and human evolution. *Am. J. Hum. Genet.* 66:235–50
11. Baird DM, Jeffreys AJ, Royle NJ. 1995. Mechanisms underlying telomere repeat turnover, revealed by hypervariable variant repeat distribution patterns in the human Xp/Yp telomere. *EMBO J.* 14:5433–43
12. Baird DM, Rowson J, Wynford-Thomas D, Kipling D. 2003. Extensive allelic variation and ultrashort telomeres in senescent human cells. *Nat. Genet.* 33:203–7
13. Barski A, Cuddapah S, Cui K, Roh TY, Schones DE, et al. 2007. High-resolution profiling of histone methylations in the human genome. *Cell* 129:823–37
14. Benetti R, Garcia-Cao M, Blasco MA. 2007. Telomere length regulates the epigenetic status of mammalian telomeres and subtelomeres. *Nat. Genet.* 39:243–50
15. Blackburn EH. 2005. Telomeres and telomerase: their mechanisms of action and the effects of altering their functions. *FEBS Lett.* 579:859–62
16. Blasco MA. 2007. The epigenetic regulation of mammalian telomeres. *Nat. Rev. Genet.* 8:299–309
17. Bodnar AG, Ouellette M, Frolkis M, Holt SE, Chiu CP, et al. 1998. Extension of life-span by introduction of telomerase into normal human cells. *Science* 279:349–52
18. Britt-Compton B, Rowson J, Locke M, Mackenzie I, Kipling D, Baird DM. 2006. Structural stability and chromosome-specific telomere length is governed by cis-acting determinants in humans. *Hum. Mol. Genet.* 15:725–33
19. Brown WR, MacKinnon PJ, Villasante A, Spurr N, Buckle VJ, Dobson MJ. 1990. Structure and polymorphism of human telomere-associated DNA. *Cell* 63:119–32
20. Capper R, Britt-Compton B, Tankimanova M, Rowson J, Letsolo B, et al. 2007. The nature of telomere fusion and a definition of the critical telomere length in human cells. *Genes Dev.* 21:2495–508
21. Cohen SB, Graham ME, Lovrecz GO, Bache N, Robinson PJ, Reddel RR. 2007. Protein composition of catalytically active human telomerase from immortal cells. *Science* 315:1850–3
22. Coleman J, Baird DM, Royle NJ. 1999. The plasticity of human telomeres demonstrated by a hypervariable telomere repeat array that is located on some copies of 16p and 16q. *Hum. Mol. Genet.* 8:1637–46
23. d'Adda di Fagagna F, Reaper PM, Clay-Farrace L, Fiegler H, Carr P, et al. 2003. A DNA damage checkpoint response in telomere-initiated senescence. *Nature* 426:194–8
24. de Lange T. 2005. Shelterin: the protein complex that shapes and safeguards human telomeres. *Genes Dev.* 19:2100–10

25. de Lange T, Shiue L, Myers RM, Cox DR, Naylor SL, et al. 1990. Structure and variability of human chromosome ends. *Mol. Cell. Biol.* 10:518–27
26. Deng Z, Atanasiu C, Burg JS, Broccoli D, Lieberman PM. 2003. Telomere repeat binding factors TRF1, TRF2, and hRAP1 modulate replication of Epstein-Barr virus OriP. *J. Virol.* 77:11992–2001
27. der-Sarkissian H, Bacchetti S, Cazes L, Londono-Vallejo JA. 2004. The shortest telomeres drive karyotype evolution in transformed cells. *Oncogene* 23:1221–8
28. Der-Sarkissian H, Vergnaud G, Borde YM, Thomas G, Londono-Vallejo JA. 2002. Segmental polymorphisms in the proterminal regions of a subset of human chromosomes. *Genome Res.* 12:1673–8
29. Eichler EE, Nickerson DA, Altshuler D, Bowcock AM, Brooks LD, et al. 2007. Completing the map of human genetic variation. *Nature* 447:161–5
30. Epel ES, Blackburn EH, Lin J, Dhabhar FS, Adler NE, et al. 2004. Accelerated telomere shortening in response to life stress. *Proc. Natl. Acad. Sci. USA* 101:17312–5
31. Fan Y, Newman T, Linardopoulou E, Trask BJ. 2002. Gene content and function of the ancestral chromosome fusion site in human chromosome 2q13–2q14.1 and paralogous regions. *Genome Res.* 12:1663–72
32. Flint J, Bates GP, Clark K, Dorman A, Willingham D, et al. 1997. Sequence comparison of human and yeast telomeres identifies structurally distinct subtelomeric domains. *Hum. Mol. Genet.* 6:1305–13
33. Flores I, Cayuela ML, Blasco MA. 2005. Effects of telomerase and telomere length on epidermal stem cell behavior. *Science* 309:1253–6
34. Gao Q, Reynolds GE, Innes L, Pedram M, Jones E, et al. 2007. Telomeric transgenes are silenced in adult mouse tissues and embryo fibroblasts but are expressed in embryonic stem cells. *Stem Cells* 25:3085–92
35. Gianfrancesco F, Falco G, Esposito T, Rocchi M, D'Urso M. 2001. Characterization of the murine orthologue of a novel human subtelomeric multigene family. *Cytogenet. Cell Genet.* 94:98–100
36. Gonzalo S, Jaco I, Fraga MF, Chen T, Li E, et al. 2006. DNA methyltransferases control telomere length and telomere recombination in mammalian cells. *Nat. Cell Biol.* 8:416–24
37. Graakjaer J, Bischoff C, Korsholm L, Holstebroe S, Vach W, et al. 2003. The pattern of chromosome-specific variations in telomere length in humans is determined by inherited, telomere-near factors and is maintained throughout life. *Mech. Ageing Dev.* 124:629–40
38. Graakjaer J, Der-Sarkissian H, Schmitz A, Bayer J, Thomas G, et al. 2006. Allele-specific relative telomere lengths are inherited. *Hum. Genet.* 119:344–50
39. Graakjaer J, Pascoe L, Der-Sarkissian H, Thomas G, Kolvraa S, et al. 2004. The relative lengths of individual telomeres are defined in the zygote and strictly maintained during life. *Aging Cell* 3:97–102
40. Griffith JD, Comeau L, Rosenfield S, Stansel RM, Bianchi A, et al. 1999. Mammalian telomeres end in a large duplex loop. *Cell* 97:503–14
41. Hagens O, Minina E, Schweiger S, Ropers HH, Kalscheuer V. 2006. Characterization of FBX25, encoding a novel brain-expressed F-box protein. *Biochim. Biophys. Acta* 1760:110–8
42. Harley CB, Futcher AB, Greider CW. 1990. Telomeres shorten during ageing of human fibroblasts. *Nature* 345:458–60
43. Harley CB, Vaziri H, Counter CM, Allsopp RC. 1992. The telomere hypothesis of cellular aging. *Exp. Gerontol.* 27:375–82
44. Hiyama E, Hiyama K. 2007. Telomere and telomerase in stem cells. *Br. J. Cancer* 96:1020–4
45. Iafrate AJ, Feuk L, Rivera MN, Listewnik ML, Donahoe PK, et al. 2004. Detection of large-scale variation in the human genome. *Nat. Genet.* 36:949–51
46. Ijdo JW, Lindsay EA, Wells RA, Baldini A. 1992. Multiple variants in subtelomeric regions of normal karyotypes. *Genomics* 14:1019–25
47. International Human Genome Sequencing Consortium. 2004. Finishing the euchromatic sequence of the human genome. *Nature* 431:931–45
48. Karlseder J, Smogorzewska A, de Lange T. 2002. Senescence induced by altered telomere state, not telomere loss. *Science* 295:2446–9
49. Kim M, Xu L, Blackburn EH. 2003. Catalytically active human telomerase mutants with allele-specific biological properties. *Exp. Cell Res.* 288:277–87
50. Lansdorp PM, Verwoerd NP, van de Rijke FM, Dragowska V, Little MT, et al. 1996. Heterogeneity in telomere length of human chromosomes. *Hum. Mol. Genet.* 5:685–91

51. Lee HW, Blasco MA, Gottlieb GJ, Horner JW 2nd, Greider CW, DePinho RA. 1998. Essential role of mouse telomerase in highly proliferative organs. *Nature* 392:569–74
52. Levy MZ, Allsopp RC, Futcher AB, Greider CW, Harley CB. 1992. Telomere end-replication problem and cell aging. *J. Mol. Biol.* 225:951–60
53. Linardopoulou E, Mefford HC, Nguyen O, Friedman C, van den Engh G, et al. 2001. Transcriptional activity of multiple copies of a subtelomerically located olfactory receptor gene that is polymorphic in number and location. *Hum. Mol. Genet.* 10:2373–83
54. Linardopoulou EV, Parghi SS, Friedman C, Osborn GE, Parkhurst SM, Trask BJ. 2007. Human subtelomeric WASH genes encode a new subclass of the WASP family. *PLoS Genet.* 3:e237
55. Linardopoulou EV, Williams EM, Fan Y, Friedman C, Young JM, Trask BJ. 2005. Human subtelomeres are hot spots of interchromosomal recombination and segmental duplication. *Nature* 437:94–100
56. Lundblad V. 2002. Telomere maintenance without telomerase. *Oncogene* 21:522–31
57. Mah N, Stoehr H, Schulz HL, White K, Weber BH. 2001. Identification of a novel retina-specific gene located in a subtelomeric region with polymorphic distribution among multiple human chromosomes. *Biochim. Biophys. Acta* 1522:167–74
58. Martin-Gallardo A, Lamerdin J, Sopapan P, Friedman C, Fertitta AL, et al. 1995. Molecular analysis of a novel subtelomeric repeat with polymorphic chromosomal distribution. *Cytogenet. Cell Genet.* 71:289–95
59. Meeker AK, Hicks JL, Iacobuzio-Donahue CA, Montgomery EA, Westra WH, et al. 2004. Telomere length abnormalities occur early in the initiation of epithelial carcinogenesis. *Clin. Cancer Res.* 10:3317–26
60. Mefford HC, Trask BJ. 2002. The complex structure and dynamic evolution of human subtelomeres. *Nat. Rev. Genet.* 3:91–102
61. Meier A, Fiegler H, Munoz P, Ellis P, Rigler D, et al. 2007. Spreading of mammalian DNA-damage response factors studied by ChIP-chip at damaged telomeres. *EMBO J.* 26:2707–18
62. Mondello C, Pirzio L, Azzalin CM, Giulotto E. 2000. Instability of interstitial telomeric sequences in the human genome. *Genomics* 68:111–7
63. Moyzis RK, Buckingham JM, Cram LS, Dani M, Deaven LL, et al. 1988. A highly conserved repetitive DNA sequence, $(TTAGGG)_n$, present at the telomeres of human chromosomes. *Proc. Natl. Acad. Sci. USA* 85:6622–6
64. Ostlund C, Garcia-Carrasquillo RM, Belayew A, Worman HJ. 2005. Intracellular trafficking and dynamics of double homeodomain proteins. *Biochemistry (Mosc).* 44:2378–84
65. Pedram M, Sprung CN, Gao Q, Lo AW, Reynolds GE, Murnane JP. 2006. Telomere position effect and silencing of transgenes near telomeres in the mouse. *Mol. Cell. Biol.* 26:1865–78
66. Petersen T, Niklason L. 2007. Cellular lifespan and regenerative medicine. *Biomaterials* 28:3751–6
67. Reddel RR. 2003. Alternative lengthening of telomeres, telomerase, and cancer. *Cancer Lett.* 194:155–62
68. Redon R, Ishikawa S, Fitch KR, Feuk L, Perry GH, et al. 2006. Global variation in copy number in the human genome. *Nature* 444:444–54
69. Riethman H, Ambrosini A, Castaneda C, Finklestein J, Hu XL, et al. 2004. Mapping and initial analysis of human subtelomeric sequence assemblies. *Genome Res.* 14:18–28
70. Royle NJ, Hill MC, Jeffreys AJ. 1992. Isolation of telomere junction fragments by anchored polymerase chain reaction. *Proc. Biol. Sci.* 247:57–67
71. Rudd MK, Friedman C, Parghi SS, Linardopoulou EV, Hsu L, Trask BJ. 2007. Elevated rates of sister chromatid exchange at chromosome ends. *PLoS Genet.* 3:e32
72. Ruiz-Herrera A, Garcia F, Azzalin C, Giulotto E, Egozcue J, et al. 2002. Distribution of intrachromosomal telomeric sequences (ITS) on *Macaca fascicularis* (Primates) chromosomes and their implication for chromosome evolution. *Hum. Genet.* 110:578–86
73. Sarin KY, Cheung P, Gilison D, Lee E, Tennen RI, et al. 2005. Conditional telomerase induction causes proliferation of hair follicle stem cells. *Nature* 436:1048–52
74. Schoeftner S, Blasco MA. 2008. Developmentally regulated transcription of mammalian telomeres by DNA-dependent RNA polymerase II. *Nat. Cell Biol.* 10:228–36
75. Sebat J, Lakshmi B, Troge J, Alexander J, Young J, et al. 2004. Large-scale copy number polymorphism in the human genome. *Science* 305:525–8
76. Sfeir AJ, Chai W, Shay JW, Wright WE. 2005. Telomere-end processing the terminal nucleotides of human chromosomes. *Mol. Cell* 18:131–8

77. Shay JW, Wright WE. 2007. Hallmarks of telomeres in ageing research. *J. Pathol.* 211:114–23
78. Slagboom PE, Droog S, Boomsma DI. 1994. Genetic determination of telomere size in humans: a twin study of three age groups. *Am. J. Hum. Genet.* 55:876–82
79. Smit AFA, Green P. *RepeatMasker* [**http://www.repeatmasker.org**]
80. Stavenhagen JB, Zakian VA. 1994. Internal tracts of telomeric DNA act as silencers in *Saccharomyces cerevisiae*. *Genes Dev.* 8:1411–22
81. Steinert S, Shay JW, Wright WE. 2000. Transient expression of human telomerase extends the life span of normal human fibroblasts. *Biochem. Biophys. Res. Commun.* 273:1095–8
82. Steinert S, Shay JW, Wright WE. 2004. Modification of subtelomeric DNA. *Mol. Cell. Biol.* 24:4571–80
83. Sugiyama T, Cam HP, Sugiyama R, Noma K, Zofall M, et al. 2007. SHREC, an effector complex for heterochromatic transcriptional silencing. *Cell* 128:491–504
84. Trask BJ, Friedman C, Martin-Gallardo A, Rowen L, Akinbami C, et al. 1998. Members of the olfactory receptor gene family are contained in large blocks of DNA duplicated polymorphically near the ends of human chromosomes. *Hum. Mol. Genet.* 7:13–26
85. Tuzun E, Sharp AJ, Bailey JA, Kaul R, Morrison VA, et al. 2005. Fine-scale structural variation of the human genome. *Nat. Genet.* 37:727–32
86. van Geel M, Eichler EE, Beck AF, Shan Z, Haaf T, et al. 2002. A cascade of complex subtelomeric duplications during the evolution of the hominoid and Old World monkey genomes. *Am. J. Hum. Genet.* 70:269–78
87. Vermeesch JR, Petit P, Kermouni A, Renauld JC, Van Den Berghe H, Marynen P. 1997. The IL-9 receptor gene, located in the Xq/Yq pseudoautosomal region, has an autosomal origin, escapes X inactivation and is expressed from the Y. *Hum. Mol. Genet.* 6:1–8
88. Vulliamy T, Marrone A, Goldman F, Dearlove A, Bessler M, et al. 2001. The RNA component of telomerase is mutated in autosomal dominant dyskeratosis congenita. *Nature* 413:432–5
89. Wilkie AO, Lamb J, Harris PC, Finney RD, Higgs DR. 1990. A truncated human chromosome 16 associated with αthalassaemia is stabilized by addition of telomeric repeat $(TTAGGG)_n$. *Nature* 346:868–71
90. Wright WE, Shay JW. 2002. Historical claims and current interpretations of replicative aging. *Nat. Biotechnol.* 20:682–8
91. Wutz A, Gribnau J. 2007. X inactivation Xplained. *Curr. Opin. Genet. Dev.* 17:387–93
92. Xu L, Blackburn EH. 2007. Human cancer cells harbor T-stumps, a distinct class of extremely short telomeres. *Mol. Cell* 28:315–27
93. Zhou J, Chau CM, Deng Z, Shiekhattar R, Spindler MP, et al. 2005. Cell cycle regulation of chromatin at an origin of DNA replication. *EMBO J.* 24:1406–17
94. Zijlmans JM, Martens UM, Poon SS, Raap AK, Tanke HJ, et al. 1997. Telomeres in the mouse have large interchromosomal variations in the number of T_2AG_3 repeats. *Proc. Natl. Acad. Sci. USA* 94:7423–8
95. Zou Y, Sfeir A, Gryaznov SM, Shay JW, Wright WE. 2004. Does a sentinel or a subset of short telomeres determine replicative senescence? *Mol. Biol. Cell* 15:3709–18

Infectious Disease in the Genomic Era

Xiaonan Yang,[1,2] Hongliang Yang,[2,3,*] Gangqiao Zhou,[4] and Guo-Ping Zhao[1,2]

[1]Shanghai-MOST Key Laboratory of Health and Disease Genomics, Chinese National Human Genome Center at Shanghai and National Engineering Center for BioChip at Shanghai, Shanghai 201203, China; email: gpzhao@sibs.ac.cn

[2]Laboratory of Microbial Molecular Physiology, Institute of Plant Physiology and Ecology, Shanghai Institute for Biological Sciences, Chinese Academy of Sciences, Shanghai 200032, China

[3]Department of Microbiology and Parasitology, Shanghai Jiao Tong University School of Medicine, Shanghai 200025, China

[4]State Key Laboratory of Proteomics, Beijing Proteome Research Center, Beijing Institute of Radiation Medicine, Beijing 102206, China

Annu. Rev. Genomics Hum. Genet. 2008. 9:21–48

First published online as a Review in Advance on May 9, 2008

The *Annual Review of Genomics and Human Genetics* is online at genom.annualreviews.org

This article's doi:
10.1146/annurev.genom.9.081307.164428

1527-8204/08/0922-0021$20.00

*Current address: Mycobacteria Research Laboratories, Department of Microbiology, Immunology, and Pathology, Colorado State University, Ft. Collins, Colorado 80523

Key Words

pathogen, bacteria, virus

Abstract

After half a century of success in combating infectious diseases with vaccination and antibiotics, emerging and reemerging epidemics present a new threat to human health. Meanwhile, the rapid pace of viral and microbial genomics research, largely based on the success of genomics technologies, offers new data-generating platforms and a revolutionary knowledge base for better understanding the diseases and the associated pathogens. Systematic molecular biology studies using genomics information and technologies have helped to elucidate mechanisms of virulence and pathogenicity, whereas genomics-based medical genetic studies have been used to better understand pathogen susceptibility. This progress may lead to the development of effective and safe vaccines in the future. Here we highlight the ongoing historical transition in the field of infectious disease research and clinical practice in the new era of genomics.

INTRODUCTION TO GENOMICS AND EMERGING INFECTIOUS DISEASE

By the 1960s, sanitation improvement, childhood immunization, and an ever-increasing number of antibiotics had notably reduced the morbidity and mortality of infectious disease, at least in the developed countries and regions of the world (158). By the mid to late 1970s it was believed that infectious diseases soon would be a problem of the past; therefore, mechanistic studies of infectious disease and efforts to control infectious disease had, at the very least, lost the prominence they had enjoyed a generation or two earlier (100, 162).

However, infectious diseases did not disappear. Besides the recurrence of the previously well-controlled infectious diseases, novel etiological pathogens have been reported continuously since the 1970s (169). In 1994, these infectious diseases were defined as "reemerging or emerging infectious diseases" (158). Globalization of the economy, high-speed international travel, and ever-booming tourism have greatly intensified the situation since the past decade (54). The rapid spread of novel infectious diseases such as severe acute respiratory syndrome (SARS) shows the frailty of the global public health care system, which in turn further affects global tourism and trade (194). Besides worrying about the naturally occurring diseases, experts in security and public health also worry that publicly available scientific information and advanced genetic technologies could be misused to create weapons for bioterrorism, which may already be a real threat (176). All these situations support the idea that the battle against infectious diseases is not over and highlight the fact that we are not adequately prepared for dealing with such problems.

Fortunately, globalization not only increases the speed and scale of the spread of infectious disease far beyond our previous experiences, but also promotes the rapid development of science and technology. The globalized efforts in genomics research that were started in the late 1980s and quickly expanded in the 1990s have offered the world new concepts and regimens for high-speed and large-scale control of emerging and reemerging diseases.

The term genomics was coined in 1986 for a new scientific discipline (106), one that studies the primary genetic makeup of a living organism by focusing on both the sequence structure and the functional annotation of the entire genome (67, 106). Although the history of genomics research can be traced back to the late 1970s and the development of DNA sequencing technology (168) and continued in the early 1990s with the initiation of the Human Genome Project (HGP) (120), the beginning of the so-called genomic era should be credited to the completion of the first entire genomic sequence of a free-living organism, the pathogenic bacterium *Haemophilus influenzae* (57). This revolutionary whole-genome sequencing (WGS) technology greatly accelerated the progress of the HGP, which was officially completed on April 14, 2003. An accurate composite reference sequence representing the entire human genome is now available at **http://www.ncbi.nlm.nih.gov/genome/guide/human/**.

Two major technology outbreaks led to the success of WGS technology. The high-throughput (HTP) sequencing technology platform with a reasonable length of high-quality reads provided the basic data that matched well with the capability of high-speed computer-supported bioinformatics platforms for sequence assembly and gene annotation. As of December 21, 2007, more than 600 genomes of bacteria and several eukaryotic microorganisms were completely sequenced with these technology platforms (**http://www.ncbi.nlm.nih.gov/genomes/static/eub_g.html**). Significantly, most of the bacteria selected for sequencing are pathogens. Sequencing of the much smaller viral genomes has become routine for infectious disease studies such as epidemiology. Many mammalian genomes are now available, including mouse (*Mus musculus*) (133), dog (*Canis familiaris*) (118), and chimpanzee (*Pan troglodytes*) (105). These genome sequences are extremely valuable for disease-related genetic studies.

The rapid availability of complete and accurate pathogen genome sequences greatly improved the knowledge of pathogen variation at the real epidemiological scale. Accurate estimations of evolutionary parameters are central to resolving key problems of pathogen evolution, including the roles of natural selection versus genetic drift, the origins of the pandemic pathogen, and the ecology of the pathogen in its natural reservoir. In addition, comparative genomics facilitated the identification of virulence or pathogenicity determinants (220). Understanding the function of genes and other parts of the genome, known as functional genomics, certainly enhances the reverse genetics approach, which may lead to the characterization of potential drug targets, vaccine candidates, and diagnostic/prognostic markers (202).

Here we provide an overview of the current status and future trends of research and clinical practice against infectious diseases in the current genomic era. Bacterial and viral pathogens are intrinsically different in their genomic makeup (size and mechanism of inheritance) and mechanisms of pathogenicity. Because recent reviews have accounted for most aspects of bacterial infectious disease–related genomics, we mainly focus here on the genomics of viral infectious diseases.

EMERGING BACTERIAL INFECTIOUS DISEASES STUDIED WITH GENOMICS-BASED DIAGNOSTIC AND TYPING TECHNOLOGY

The identification of *Legionella pneumophila* in 1977 as the etiologic pathogen of the Philadelphia epidemic of 1976 was an important discovery (167). Since that time, besides the identification of the Shiga toxin–producing *Escherichia coli* strain O157:H7 (STEC O157) in 1982 (94, 159) and the bioterrorist attack with *Bacillus anthracis* (85), most of the bacterial-based endemics have been naturally reemerging cases, such as the 1992 cholera endemic caused by serogroup O139 in India and Bangladesh (155) and the 2005 summer *Streptococcus suis* outbreak in Sichuan, China, in which 204 people were infected, resulting in 38 deaths (216). Multiple sequence alignment analyses confirmed that the highly virulent strains of *S. suis* serotype 2 were the causative agents of the outbreak of streptococcal toxic shock syndrome (189).

The most significant reemerging bacterial infection is tuberculosis (TB), which is caused by *Mycobacterium tuberculosis* (124). *M. tuberculosis* infects one-third of the population worldwide and causes more than two million deaths each year (172). TB was once considered to be well controlled by BCG vaccination (199) and effective antibiotic treatment (4), but it is now reemerging owing to the extensive occurrence of multiple-drug-resistant *M. tuberculosis* strains (175), the poor performance of the conventional vaccine (at least for adults) (71), and the ever-increasing number of immunocompromised patients (200). The determination of the genomic sequence of *M. tuberculosis* has revolutionized TB research, contributed to major advances in understanding the evolution and pathogenesis of *M. tuberculosis*, and facilitated the development of new diagnostic typing techniques (4, 53, 124). Several recent studies adopted synonymous single nucleotide polymorphisms (SNPs) to study deep phylogenetic lineages for *M. tuberculosis* isolates globally (56, 73). These studies uniformly showed that distinct geographic regions of the world have variable distributions of lineages; certain regions have a major lineage that is a minor contributor elsewhere. Deciphering the genome sequence of *M. tuberculosis* also contributed to the determination of bacterial virulence, which could facilitate the discovery of a vaccine for *M. tuberculosis* (53, 71).

The Gram-negative bacillus *Yersinia pestis*, the etiological agent that causes bubonic and pneumonic plague, is considered to be one of the most dangerous and deadly pathogenic bacteria in the world (146). Three major plague pandemics are thought to have been caused by three biovars of *Y. pestis*, but these associations have yet to be confirmed with the evolutionary history of this largely monomorphic bacterium (46). In addition, plague is typical for natural focus-based diseases in that plague primarily affects more than 200 species of animals as hosts and more than 80 species of fleas as vectors in nature within specific geographic regions and is independent of human behavior. Because *Y. pestis* has evolved too recently to allow the accumulation of extensive sequence diversity, microevolution of the *Y. pestis* genomes was studied via high-resolution methods (i.e., using SNP) that have been applied to monomorphic species on samples collected from different geographic foci (5). These studies have led to the development of comprehensive molecular signatures to define genomvars (205, 224), instead of the classical biovars, for better understanding the intraspecies genomic variability and the adaptive microevolution of *Y. pestis*, as well as its speciation from the closely related *Yersinia pseudotuberculosis* (223). A recent comparative genomic study showed that some changes in gene families encoding kinases, proteases, and transporters are illustrative of the evolutionary jump from the free-living enteropathogen *Y. pseudotuberculosis* to the obligate host-borne blood pathogen *Y. pestis* (69).

Prompt diagnosis and proper typing of the endemic or epidemic strains are essential for effective control of infectious diseases (221). A large amount of bacterial genomic data greatly strengthens and simplifies pathogen diagnosis and typing as well as the study of the mechanisms of virulence and antibiotic resistance. These molecular typing technologies, in contrast to the classical phenotyping methods, allow the discrimination of variations among strains within a species, the elucidation of the route of contamination, the identification of the source of spoilage with fine molecular markers, as well as the analysis of epidemics relevant to the phylogenetics of the pathogen. Polymerase chain rection (PCR)-based genotyping such as random amplified polymorphic DNA (RAPD) (62) and amplified fragment length polymorphism (AFLP) (89) as well as restriction-based pulsed field electrophoresis (PFGE) (160) and restriction fragment length polymorphism (RFLP) (164) techniques have been broadly used in both research and clinical practice. Recently, several sequence-based approaches, such as those using simple sequence repeats (SSRs) (134), multilocus sequence typing (MLST) (165), SNPs (15), and combinations of these techniques (40) have greatly improved bacterial pathogenicity and epidemiology studies. Another great advantage of these sequence-based methods is that their outputs can be easily compiled into databases together with the WGS results and can be communicated among the laboratories involved (59).

HTP assay techniques such as DNA microarrays were quickly developed on the basis of large microbial genomic databases and have been widely used in bacterial typing and pathogen identification since their introduction (61). One successful example is the development of a DNA microarray that can identify *Staphylococcus aureus*, *E. coli*, and *Pseudomonas aeruginosa* strains and differentiate them from closely related Gram-positive and Gram-negative bacterial strains that cause bloodstream infections (37).

The impact of genomics upon bacterial infectious diseases (42, 59, 127), the aspects of evolution and ecology revealed by bacterial pathogenomics (142), and nucleotide-associated bacterial detection/identification approaches (59, 94) were recently reviewed. For topics related to bacterial vaccine development and drug resistance, readers are referred to previous reviews (41, 129, 174).

THE USE OF GENOMICS TO STUDY THE OUTBREAK OF VIRAL INFECTIOUS DISEASES

Owing to the close interactive relationship with their mammalian host cells, viral infectious diseases are more difficult to treat than diseases of bacterial origin. In addition, the extremely small genome sizes, short replication cycles, and lack of comprehensive replication proofreading mechanisms in RNA viruses make viral genomes highly variable. This variation in genomes provides the basis for survival and adaptation to environmental changes (particularly in the host). Therefore, in recent years most of the emerging infectious diseases have been viral rather than bacterial in origin (**Table 1**). These same characteristics of viruses also make research on viral pathogens in the genomic era much easier than research on bacteria. It is in this context that we will emphasize the impact of genomics upon pathogen identification and detection, pathogen evolution and epidemiology, natural reservoir identification, virulence determination, and host susceptibility (see **Table 2** for more information). We summarize these aspects by providing case studies of influenza A viruses.

Viral Pathogen Identification for Emerging Infectious Diseases

Viruses were the very first living organisms to have their genomes completely sequenced (57). Nowadays, a rapidly increasing number of viral genomic sequences together with the development of sequencing technology greatly strengthen pathogen identification, which is the first priority in controlling infectious disease. The quick identification of Nipah virus (NiV) and the novel coronavirus that caused the SARS epidemic clearly demonstrates the contributions of genomics to pathogen identification.

NiV is a recently emergent paramyxovirus that causes fatal encephalitis in humans. NiV was first discovered in 1998 in Malaysia and

Table 1 Recent emerging and reemerging infectious diseases and pathogens[a]

Year of major outbreak	Viral disease/pathogen	Bacterial disease/pathogen
2005		*Streptococcus suis*
2004	Avian influenza (human cases)	
2003	Severe acute respiratory syndrome (SARS)-CoV	
	Monkeypox virus	
2002		*Meningococcal meningitis*
2001	Dengue/dengue hemorrhagic fever (DHF)	
1999	Nipah virus	
	West Nile virus	
1997	H5N1 (avian influenza A virus)	
1996	New variant Creutzfelt-Jacob disease	
	Australian bat lyssavirus	
1995	Human herpesvirus 8 (Kaposi's sarcoma virus)	
1994	Sabia virus	
	Hendra virus	
1993	Hantavirus pulmonary syndrome (Sin Nombre virus)	
1992		*Vibrio cholerae* O139
1991	Guanarito virus	
1989	Hepatitis C	
1988	Hepatitis E	
	Human herpesvirus 6	
1983	Human immunodeficiency virus (HIV)	
1982		*Escherichia coli* O157:H7
		Lyme borreliosis (*Borrelia burgdorferi*)
	Human T-lymphotropic virus type-2	
1980	Human T-lymphotropic virus	

[a]Data derived from the World Health Organization (**http://www.who.int/research/en/**).

Table 2 Further reading regarding the impact of genomics on viral infectious diseases

Aspects of the study of infectious disease	Related articles
Identification of pathogen	74, 104, 125, 166
Molecular epidemiology of pathogen	86, 103, 136, 157, 221
Identification of viral reservoir	179, 209
Identification of virulence	104, 177, 193, 220
Host susceptibility	84, 97, 99

Singapore (18). More recently (2001 to 2005), outbreaks were detected in Bangladesh and South Asia (20, 74, 90). When the first case occurred in pig farmers near the city of Ipoh, Malaysia, the high fever and encephalitis (18) led to the assumption that Japanese encephalitis virus (JEV) infection was the cause, because it is transmitted by *Culex* mosquitoes and known to replicate in pigs (51). However, neither mosquito control nor JEV vaccination appeared to affect the course of the Malaysia outbreak (18). Later, the viruses isolated from patient samples were identified by their appearance to belong to the family of *Paramyxoviridae*, which does not include JEV (35, 51). NiV may cause cells to clump together in giant multinucleate cells or syncytia and reacts with antibodies against the Hendra virus (HeV), which also produces syncytia, causes encephalitis in humans, and was isolated from Australian pigs in 1994 (48). These characteristics indicated a similarity between the two viruses.

The genome of the new isolate was completely sequenced (18,246 nucleotides) and researchers confirmed it was a novel member of the *Paramyxoviridae* and designated it as NiV (34, 76). The genome of NiV is 12 nucleotides longer than that of HeV, and NiV and HeV have the largest genomes within the family. The 3′ and 5′ termini of the NiV genome are nearly identical to the genomic termini of HeV and share sequence homology with those of other members of the subfamily *Paramyxovirinae* (76). The coding region of the NiV virus shares 70% to 78% nucleotide identity with the coding region of HeV (34). Phylogenetic analysis of nucleoprotein gene sequences shows that NiV and HeV form a distinct cluster within the subfamily *Paramyxovirinae* and probably represent a new genus in this subfamily (34). These comparative genomics results match well with the former virological discoveries and confirmed the initial speculation that HeV and NiV are closely related but distinct viruses (34, 76). One should also note that the broad species tropisms and the ability to cause fatal disease in both animals and humans are the two major characteristics that distinguish HeV and NiV from all other known paramyxoviruses (74). It was thus finally confirmed that NiV but not JEV was the pathogen that caused the fatal encephalitis outbreak in Southeast Asia in 1998 (19).

SARS was the first emerging infectious disease of the 21st century, pandemic in almost half of the world. Starting from the November 2002 outbreak of "atypical pneumonia" in Guangdong, China, which was later designated SARS (221), the disease quickly spread to affect more than 8000 people in 25 countries and regions within less than half a year. By the end of July 2003, SARS had killed 774 people (184). SARS was a novel infectious disease with an unknown etiological agent that was rapidly spread by air travel, so the panic caused was enormous (121).

An international collaboration helped to identify the causative agent of SARS (11), which was first isolated from autopsy and nasopharyngeal aspiration samples from patients. A cytopathic effect was shown in Vero E6 cell inoculations and coronavirus was observed via electron microscopy. Effective seroconversion tests and sequencing of a short conserved fragment of the genome promptly identified this virus as a novel coronavirus responsible for SARS (SARS-CoV) (47, 144). The whole genome was completely sequenced at an unprecedented speed from the isolates of Canada, strain Tor2 (123) and Vietnam, strain Urbani (161). Phylogenetic analysis revealed that SARS-CoV was a novel coronavirus and was only moderately related to other two human CoVs, HCoV-OC43 and HCoV-229E, known to cause common colds and lower respiratory tract infections and diarrhea in humans (123). Meanwhile, the complete genomes of five other strains (BJ01–BJ04,

GZ01) were sequenced almost to completion by Chinese sequencing centers (9, 153). By April 29, 2003, ten isolates of SARS-CoV had been sequenced (52).

With the prompt availability of a vast amount of genomic sequence information about the etiologic agents, the diagnosis of viral infectious diseases and the detection/identification of viral pathogens via specific nucleotide sequence is becoming one of the most attractive choices for confirmation of etiologic agents (154); the so-called nucleic acid testing (NAT) methods to detect the viral genome have extended the diagnostic repertoire of virological laboratories considerably in recent years, and have proved to be superior to conventional techniques in many circumstances (45). NAT can be used to detect the viral genomes of the human hepatitis B virus (HBV), human hepatitis C virus (HCV), and the human immunodeficiency virus (HIV), which are difficult to culture in vitro (16). NAT is also the preferred method for quantification of the viral load by detection of genomic markers (2). NAT may also be used to characterize the viral genomes (genotyping) to detect virus variants (208). One attractive example of NAT is the GreeneChipPm, a panmicrobial microarray comprising 29,455 60mer oligonucleotide probes for vertebrate viruses, bacteria, fungi, and parasites (141). Researchers used this assay in the analysis of nasopharyngeal aspirates, blood, urine, and tissue from patients with various infectious diseases and confirmed the presence of viruses and bacteria identified by other methods; for example, researchers implicated *Plasmodium falciparum* in an unexplained fatal case of hemorrhagic fever–like disease during the Marburg hemorrhagic fever outbreak in Angola in 2004–2005 (135).

Meanwhile, the conventional immunological methods of diagnosis also benefit from genomic information for the quick availability of recombinant antigens; these methods are still widely in use owing to their high specificity and simple instrumentation (178). In most cases, both DNA and protein detection methods are used in combination for different clinical and research purposes (214); this was the case for SARS-CoV (47). Besides diagnosis based on S protein antibody detection enzyme-linked immunosorbent assay (ELISA) (79) and the conserved sequence of orf1b targeted RT-PCR (149), an antigen-capturing ELISA was developed to establish N protein concentrations in the samples with a specificity of 99.9% (25). This technique was used to analyze 420 serum samples taken from 317 SARS patients, and researchers found that the protein could be detected as early as day 1 after the onset of symptoms and until day 18. This technology identified an asymptomatic SARS patient in the second phase of the SARS outbreak, which further supported the scenario of direct transmission from animals to humans.

Molecular Epidemiology and the Evolution of Pathogens

Understanding the evolution of a pathogen is crucial for reconstructing its origin, deciphering its interaction with the host, and developing effective control strategies. Improvements in bioinformatics and epidemiological analysis, as well as a greatly expanded genomics sequence database of all kinds of pathogens, provided an unprecedented opportunity for investigating these long-standing questions.

West Nile Virus (WNV) was first isolated from the blood of a febrile woman in the West Nile province of Uganda in 1937 (14). WNV belongs to the family *Flaviviridae* and is a member of the JEV serocomplex (151). Mainly transmitted by *Culex* mosquitoes, WNV is not normally communicable among people (151). During the period from 1937 to 1999, epidemics of WNV infection occurred only occasionally, and the infection in humans, horses, and birds was generally either asymptomatic or mild; neurological disease and death were rare (101). Since its introduction to North America, WNV has gained great notoriety and become the most significant cause of epidemic encephalitis in the Western hemisphere (24). As of March 2007, 23,975 cases of human WNV infection (including 9843 cases of neuron-invasive disease and 962 cases that resulted in fatalities) occurred in the United States (103).

In 1999, researchers sequenced the whole genome of a flavivirus isolated from the brain of a dead Chilean flamingo (WN-NY99) in New York City (108). Phylogenetic analysis based on the envelope region of the flavivirus from other species, including mosquitoes and two fatal human cases, demonstrated that WNV circulated in natural transmission cycles and was responsible for the outbreak of human encephalitis in New York, the first reported introduction of WNV into the Western hemisphere (108). Moreover, researchers observed a high degree of similarity (99.8% amino acid identity) between the United States WNV (WN-NY99) and a virus isolated in Israel in 1998 (WN-Israel198) (108). The phylogenetic analysis was later extended to cover the entire genome of 15 WNV isolates from Africa, the Middle East, Eastern Europe, and the United States. This analysis indicated that the eight US strains displayed a high degree of sequence conservation (≥99.9% amino acid identity). The WN-Israel198 strain is very closely related to the WNVs isolated in the US (≥99.8% amino acid identity) (107). Therefore, the high degree of complete genomic sequence similarity between WNV isolates from the United States and those from Israel further supports the hypothesis that this epidemic is attributable to a WNV line that has been circulating in the Mediterranean region since 1998 (108).

Since then, the distribution of WNV has expanded to the 48 contiguous United States, as well as Mexico and Canada (78). On the basis of the well-documented location and year of the onset virus isolations (108), together with the genomic sequences, researchers made genetic and phenotypic comparisons among isolates from the onset of the North American WNV epizootic and isolates from all subsequent years and over a broad geographic distribution (43, 49). Consequently, genomic sequences of isolates from 1999 and 2000 provide a genetic baseline, allowing for the identification of novel mutations to the genomes of more recently isolated strains, which can then be used to infer phylogenetic relationships among isolates and to give a better understanding of how this virus has evolved since its introduction in 1999.

In 2004, Davis and coworkers reported a dominant genotype distinct from the 1999 genotype (44). This genotype was denoted as the North American or WN02 genotype, and is now the only WNV genotype recognized in the United States. Dominance in a genetic variant may possibly be due to the enhanced mosquito transmission efficiency of this variant in *Culex pipien*, one of the main vectors in the northeastern United States. In 2007, four nucleotide substitutions in the 3′ noncoding region (3′NCR) evolved rapidly in the ancestor of this single clade. These four nucleotide changes are predicted to have a major impact on the secondary structure of the 3′NCR, which in turn may influence virulence; for instance, these changes may have an effect on viral replication (92). Recently, analyses of the sequences of the 156 envelope protein coding regions further supported the disappearance of the introduced genotype and suggested this new dominant genotype has reached peak prevalence in North America (182).

Acquired immunodeficiency syndrome (AIDS) was first recognized in the United States in 1981 following an increase in the incidence of usually rare opportunistic infections that were caused by a general immune deficiency in homosexual men (156). HIV was first isolated in 1983 (7). The application of phylogenetic methods to epidemiologic studies on HIV and RNA viruses is possible thanks to their rapid evolution, which allows the accurate reconstruction of transmission events even within closely related clusters (110). There are two types of HIV, HIV-1 and HIV-2 (81). Around the world the predominant virus is HIV-1; generally, without specifying the type of virus, HIV and HIV-1 are used interchangeably. Researchers performing phylogenetic analyses of globally circulating viral strains have identified three distinct HIV-1 groups: the major group M, the outlier group O, and the new group N. More than

90% of HIV-1 infections belong to group M. During its spread among humans, the M group has developed an extraordinary degree of genetic diversity (188). To genetically classify global group M HIV-1 isolates, researchers initially developed a subtyping system based on *env* and *gag* sequences; recently, the sequencing of *pol* segments has become more common, mainly owing to its usefulness for detecting drug resistance–associated mutations (66).

More than 1000 full-length HIV-1 genomes are currently available in public databases (111). In 2000, a more comprehensive system based on complete genome analysis categorized the viruses into nine pure subtypes (A, B, C, D, F, G, H, J, and K) (195). HIV-1 genetic diversity in the global pandemic continues to evolve. Occasionally, two viruses of different subtypes are observed to encounter each other in the cell of an infected person and recombination of their genomes generates a new hybrid virus (a process called viral sex). Many of these new strains do not survive for long, but those that infect more than one person are known as circulating recombinant forms (CRFs). Identifying a new CRF requires the characterization of at least three epidemiologically unlinked viruses with identical mosaic structures and at least two of them should be characterized in near full-length genomes (>8 kb). The HIV database has reported 32 CRFs to date (**http://www.hiv.lanl.gov/content/sequence/HIV/CRFs/CRFs.html**).

The pervasive role of recombination as a major driving force in the generation of diversity in the HIV-1 pandemic is becoming evident and is particularly visible in areas in which different genetic forms meet, referred to as geographic recombination hotspots. The importance of superinfection and its impact on HIV-1 diversification and propagation has been reviewed in detail (195).

Although genomic sequences of a large number of epidemiological isolates are essential for pathogen evolution studies, the related biological information is equally critical. When SARS evolved as an endemic atypical pneumonia in 2003, the field workers of Guangdong Centers for Disease Control routinely collected patients' biological samples with their epidemiological information. These detailed epidemiology data became essential when whole-genome sequences of these early phase virus isolates were used for comparative genomic analysis. The molecular epidemiology study of SARS became the basis for understanding the evolution of SARS-CoV during the whole period of the pandemic.

After the first phase of sequencing of the SARS-CoV genome (52), Ruan and coworkers (163) were the first to analyze the epidemiology information of more than ten isolates related or unrelated to the isolate from Hong Kong onset patient A and compare this epidemiology data with the phylogenetic relationship of the viral genomic sequences. Further sequencing efforts added approximately 50 more genomes sampled from very early to the very late phases of the epidemic from different geographic locations as well as sequences of viral samples isolated from palm civets (see below). These sequences bridged not only the time and geographical gaps of the SARS epidemiology but also the phylogenetic gaps of SARS-CoV evolution (30). This comparative genomic study delineated the direction of SARS-CoV evolution and analyzed its intrinsic variation rates together with the variations in response to the selection pressure from the environment (e.g., host switch), which, in turn, supported further efforts to control the disease (221).

Obviously, because it is a respiratory tract infectious disease, the route of transmission is critical for the control of SARS. Dozens of critical amino acid variations in the Spike protein responsible for cross-host transmission were identified on the basis of phylogenetic analysis of the viral evolving process over two outbreaks, one year apart; two variations were characterized via the use of molecular biology studies that employed a pseudotyped model system. These studies implicated that some isolates in the palm civet had already evolved to acquire the ability to cross the entry barrier between palm civets and humans (221).

Viral Natural Reservoir Identification

The identification of the natural reservoir and possible intermediate hosts of pathogens is critical for understanding the transmission mode of the virus, designing a long-term disease control strategy, and preventing future reintroduction. This task has proven to be extremely challanging (13), but is quickly improving owing to the extensive use of genomics technology.

In the 1998 Malaysia outbreak, infected pigs were believed to be intermediate hosts of NiV (35). AbuBakar and coworkers (1) provided the molecular evidence showing that at least two major strains of NiV of pigs were circulating during the 1998 NiV outbreak in Malaysia (NiV-Seremban and NiV-Sungai Buloh). A high sequence similarity (>99%) between the NiV genomes from pigs and humans was noted (1). Phylogenetic analysis of the deduced amino acid sequences showed that the NiV-Seremban pig isolates were identical to the human NiV isolates CDC and UMMC2; the NiV-Sungai Buloh pig isolates were identical to the human NiV isolates UMMC1 and UM-0128. For both pig isolates, the NiV-Seremban differed from NiV-Sungai Buloh only by a Ser1645Phe change within the polymerase protein (L) (1).

Researchers also suspected that flying foxes (fruit bats) might be the natural reservoir of NiV, as they are for HeV (75). The presence of neutralizing antibodies against NiV in the island flying foxes (*Pteropus hypomelanus*) and the Malayan flying foxes (*Pteropus vampyrus*) (213) was in concordance with the isolation of NiV from the urine of, and the fruits partially eaten by, the island flying foxes (36). Besides the immunofluorescence identifications, the bat NiV isolates were confirmed by sequencing all the PCR amplicons of a 11,200-nt contiguous genomic fragment including the F and G genes that showed 99% sequence identity with those from human (Nipah-H); these genes differed only by six nucleotides (36). In addition, both NiV pig isolates differed from the flying fox NiV isolates at three amino acid positions: Thr30Ile, Pro206Leu, and Met348Thr in the coding regions of nucleoprotein (N), phosphoprotein (P), and fusion protein (F), respectively (1). These results confirmed that the infections in humans during the Malaysian southern outbreak originated from infected pigs, which are believed to be intermediate hosts, and fruit bats in the genus *Pteropus* (flying foxes), which are believed to be the natural reservoir of NiV (1, 36). Recently, bat-to-human and human-to-human transmission of NiV was reported in Bangladesh and India (20, 31, 72). In 2007, vertical transmission and fetal replication of NiV was reported in an experimentally infected cat (132).

The origin of HIV from African primates was proposed in 1990 when a retrovirus, the simian immunodeficiency virus (SIV), isolated from the common chimpanzee (*P. troglodytes*, SIVcpz) in 1989 (143), was found to be mostly similar in genomic sequence and organization to HIV-1 (91). Since then, five lines of evidence have accumulated to demonstrate the zoonotic transmission of primate lentiviruses to humans: (*a*) the similarities of their genomic organization, (*b*) phylogenetic relatedness, (*c*) prevalence in the natural host, (*d*) geographic coincidence, and (*e*) plausible routes of transmission. However, the route of transmission is controversial and dating transmission to humans is problematic, especially because of frequent recombination (156).

The oral polio vaccination (OPV) hypothesis (148) suggested that the ancestral strain of the M group emerged as a result of the vaccination of approximately one million people, living largely in the Democratic Republic of Congo from 1957 to 1960, with an oral vaccine against polio virus that had allegedly been cultured in chimpanzee kidneys (88, 157). This vaccination is claimed to have enabled the transfer of chimpanzee SIV to humans. Conversely, phylogenetic analysis of HIV-1 sequences indicated that group M originated before the vaccination campaign (102), supporting a model of natural transfer from chimpanzees to humans (157). Further phylogenetic analysis that employed genomic sequences of viruses obtained from the Congo as well as 223 strains representing the global diversity of HIV-1 (including all known subtypes) revealed that many Congo lineages fall basal to the origin of each subtype as currently defined by the phylogeny of global strains (122). The Congo and global phylogenies differ significantly in that the former show no more subtype structure than the complex phylogenetic trees for the global strains simulated under a model of exponential population growth (152). These studies thus indicated that the structure of HIV-1 phylogenies was the result of epidemiological processes acting within human populations alone, and was not due to multiple cross-species transmission initiated by oral polio vaccination.

In 1999, Gao and coworkers first documented the origin of HIV-1 in the chimpanzee *Pan troglodytes troglodytes* (60). Two chimpanzee subspecies exist in Africa, the central *P. t. troglodytes* and the eastern *P. t. schweinfurthii*, and both harbor SIVcpz. Their respective viruses form two highly divergent (but subspecies-specific) phylogenetic lineages. All HIV-1 strains known to infect humans, including groups M, N, and O, evolved from just one of these SIVcpz lineages, found in *P. t. troglodytes* (SIVcpz*Ptt*) (98), on at least three separate occasions. SIVcpz, the progenitor of HIV-1, arose as a recombinant of ancestors of SIV lineages presently infecting red-capped mangabeys and *Cercopithecus* monkeys in west-central Africa (6).

To identify host-specific adaptations in HIV-1, the inferred ancestral sequences of HIV-1 groups M, N, and O were compared with 12 full-length genome sequences of SIVcpz*Ptt* and four of the outlying but closely related SIVcpz*Pts* (from *P. t. schweinfurthii*). This analysis (206) revealed that a single locus (Gag-30) within the *gag*-encoded matrix protein (p17) that was completely conserved among SIVcpz*Ptt* strains (Met) underwent a conservative replacement by Leu in one lineage of SIVcpz*Pts* but changed radically to Arg in all three lineages leading to HIV-1. Site-directed mutagenesis studies were conducted to test the replication of these variants in both human and chimpanzee $CD4^+$ T lymphocytes. Remarkably, viruses encoding Met30 p17 replicated to higher titers than viruses encoding Lys30 p17 in chimpanzee T cells, but the opposite was found in human T cells. These observations provided compelling evidence for host-specific adaptation during the emergence of HIV-1 and identified the viral matrix protein as a modulator of viral fitness following transmission to the human host (206).

Further analysis using HIV-1 gene sequences of recovered archival samples from some of the earliest known Haitian AIDS patients showed that HIV-1 group M subtype B likely moved from Africa to Haiti around 1966. A non-Haitian subtype B clade emerged after a single migration of the virus out of Haiti around 1969 (65). The spread of the virus was likely driven by ecological rather than evolutionary factors. These results suggested that HIV-1 circulated cryptically in the United States for approximately 12 years before the official recognition of AIDS in 1981. Haiti appears to have the oldest HIV/AIDS epidemic outside sub-Saharan Africa and the most genetically diverse subtype B epidemic.

Ebola hemorrhagic fever (EHF) is a febrile hemorrhagic illness. Ebola virus (EBOV), a single, negative-stranded RNA virus of the family *Filoviridae* in the order *Mononegavirales* (112), gained public notoriety in the past decade largely as a consequence of the highly publicized isolation of a new EBOV species in a suburb of Washington, D.C. in 1989 (17), together with the dramatic clinical presentation of EHF and the high case-fatality rate in Africa (near 90% in some outbreaks) (68, 82).

Following the discovery of EBOV in 1976, and again after the 1994 case in the Ivory Coast and the 1995 outbreak in the Congo (formerly Zaire), intensive efforts were made to identify the natural reservoir for EBOV. However, neither potential hosts nor arthropod vectors were identified (reviewed in 68). Fruit bats (*Hypsignathus monstrosus*, *Epomops franqueti*, and *Myonycteris torquata*) were suggested to be the natural reservoir of the Zaire distinct subtype (EBOV-Z) in the 2005 study (113). EBOV-Z RNA was found in liver and spleen samples of the bats, whereas in other animals, EBOV-specific antibodies were detected in serum, perhaps indicating that the bats were recently infected and had not yet developed detectable immune responses. These data supported the earlier findings of replication and circulation of high titers of EBOV in experimentally infected fruit bats and insectivorous bats in the absence of illness (187). However, the high titers shown in the experimental bat model raised questions as to why virus isolation has not been achieved from any of the naturally infected bat species, particularly because virus isolation from clinical material is easily achieved for filoviruses, which usually cause the Marburg hemorrhagic fever in animals (135). The virus titers were either very low in these naturally infected animals (consistent with the need for nested RT-PCR to detect EBOV-Z-specific nucleic acid) or a specific physiological or environmental stimulus may be needed to stimulate virus infection (113). Therefore, determination of the natural reservoir for EBOV requires further laboratory and ecological investigations.

On the basis of the experience of studying the origin of NiV, an animal origin of SARS-CoV was proposed in the very early days of the epidemic, immediately after the new virus pathogen was identified (221). A coronavirus with greater than 99% nucleotide identity to the human SARS-CoV was identified from specimens collected from animals, particularly Himalayan palm civets (*Paguma larvata*) and raccoon dogs (*Nyctereutes procyonoides*) from a live wild-game animal market in Shenzhen City, China (70). Simultaneously, many workers who handled animals in these game markets were shown to have antibodies against SARS-CoV, even though they had no history of a SARS-like disease (70, 215). Phylogenetic analysis covering genomic sequences of more than 60 representative strains of SARS-CoV isolated from all phases of the 2002/2003 SARS epidemic indicated that the early phase viral genomes are much more similar to that of the virus from palm civets (221). One year later, the molecular epidemiology studies of the 2003/2004 community outbreak in Guangzhou City, China clearly inferred that the same SARS-CoV can infect both animals and humans, although the symptoms were mild and no human-to-human transmission was found (183). The phylogenetic analysis also suggested that the SARS-CoV genome was still evolving in palm civets or other related animals, which again supported the hypothesis that the palm civet is likely an important intermediate carrier but not the true natural reservoir; therefore, an unidentified natural reservoir for SARS-CoV should exist (96, 201).

Extensive surveillance studies for wild animals, mainly bats, led to the detection of a SARS-like-CoV present in at least three different species of the Chinese horseshoe bat of the family *Rhinolophidae*, a common insectivorous species found in China (116). The genome sequence of SARS-like-CoVs from horseshoe bats showed 88%–92% nucleotide identities to the genome of human/civet SARS-CoV. Phylogenetic analysis demonstrated that this SARS-like-CoV forms a cluster with SARS-CoV that is distantly related to the known group 2 coronaviruses and was thus designated as a group 2b coronavirus (116). Most of the differences between the bat SARS-like-CoV and SARS-CoV genomes were observed in the Spike genes, ORF3 and ORF8, which are the regions where the most variations were also observed among human and civet SARS-CoV genomes over the 2002–2004 period in Guangdong.

A detailed analysis of the amino acid sequences of S proteins among four SARS-like-CoVs and the SARS-CoVs in human and civet indicated that the putative S2 domains of all these viruses were highly similar to each other (92%–96%) and all the amino acid residues that are critical for SARS virus fusion to the cell membrane located in the S2 domain were conserved, suggesting that these viruses possessed the same fusion mechanism during infection of their host cells (198, 212). However, the putative S1 domain of SARS-like-CoV has little homology to that of SARS-CoVs, especially in the N-terminal region and the receptor-binding domain (RBD) (210). In addition, all the ORF8s of bat SARS-like-CoV are similar to that of SARS-CoV isolated from the palm civets (183) and some early-phase patients with a 29-nt molecular marker sequence, which was found to be deleted in most of the later phase human SARS-CoVs (192). This genomic variation feature further suggests that the SARS-like-CoV likely shares a common ancestor with SARS-CoV, but dramatic variations in S and some other related genes must have occurred for the virus to become infectious in human and palm civets.

Functional Genomics for the Analysis of Virulence Mechanisms

Genomics technology provides a systematic research platform of integrated molecular biology approaches for studying the virulence and pathogenesis of a pathogenic virus. In the case of SARS-CoV, the elucidation of the two critical amino acid residues in the RBD of the S protein (see above) was successfully achieved by two different approaches: the structure-to-function approach (198, 212) and the evolution-to-function approach (218). For the more deadly EBOV, a function-to-evolution approach was adopted.

The EBOV genome is almost 19 kb long and encodes seven viral proteins (112, 166). EBOV was experimentally adapted to model vertebrates for the initial efficacy studies under the rationale that the adaptation of the virus to another species may result in mutations in the virus genome, and may also change the pathogenesis of the disease compared with the condition in natural hosts (77). When mouse-adapted virus genomic sequence was compared with strain EBOV-Z from which it was derived (Zaire Ebola virus precursor), one nucleotide insertion in the intergenic region between VP30 and the matrix protein (VP24) as well as eight nucleotide changes, resulting in five amino acid changes, were identified in the mouse-adapted virus, i.e., Ser683Gly in the nucleoprotein (NP), Thr10493Ile in VP24, Ala3163Val in the phosphoprotein (VP35), and Phe14380Leu/Ile16174Val in the polymerase (12). Conversely, when the mouse-adapted virus genomic sequence was compared with that of another unrelated Zaire strain, EBOV-Z-76, only three nucleotide variations were found: a synonymous change in VP40 and two changes (nucleotide 6231 and 6774) in the glycoprotein (GP), both resulting in Pro-Ser changes. In addition, there is no amino acid change between EBOV-Z and EBOV-Z-76. The promising aspects of these findings were the lack of amino acid changes in GP relative to the precursor (EBOV-Z) (77) on the one hand, and the Ala3163Val change in VP35 on the other hand; these results, although not conclusive, suggest that VP35 may be an interferon (IFN) antagonist (8).

Mutations found in mouse-adapted EBOV-Z were introduced into the backbone of the wild-type viral cDNA and virulence was tested in the mouse-adapted reverse genetics system. It was thus found that mutations in VP24 and in NP were primarily responsible for the acquisition of high virulence of the adapted Mayinga strain in mice (50). Moreover, the role of these proteins in virulence correlates with their ability to evade type I interferon–stimulated antiviral responses (50). These findings suggested a critical role for overcoming the interferon-induced antiviral state in the pathogenicity of EBOV and offered new insights into the pathogenesis of EBOV infection.

Host Susceptibility: Genetic Association Studies

The pathogenesis of viral infectious disease is also affected to a certain extent by human genetic susceptibility. The human leukocyte antigen (HLA) complex often plays an important role in determining the susceptibility to infectious diseases (83). HLA class I gene products present antigenic peptides to T cells, initiating an immune response and the removal of foreign materials (171). Genes that encode HLA class I molecules are highly polymorphic, apparently as a result of natural selection processes that enable mammals to resist a wide variety of pathogens (93).

The development of HTP systems for SNP genotyping using information from the International Haplotype Mapping Project (191) and the use of comparative genomics to identify regions that are involved in the host-pathogen interaction have provided new opportunities in HIV susceptibility research. In addition to genotyping technology for selective large genomic regions or the whole genome, success in HIV-1 host susceptibility studies depends mainly on the identification of appropriate phenotypes and the availability of large human cohorts for genetics studies. The effect of SNPs and haplotypes within the *PPIA* gene (8 kb), encoding a ubiquitous cytoplasmic protein, peptidyl prolyl isomerase A, on HIV-1 infection and disease progression in five HIV-1 longitudinal history cohorts were examined. Two promoter SNPs (SNP3 and SNP4) in perfect linkage disequilibrium were associated with more rapid $CD4^+$ T-cell loss in African Americans and more rapid AIDS progression among European Americans. In addition, SNP5 (1650A/G), located in the 5′ UTR previously shown to be associated with higher ex vivo HIV-1 replication, was found to be more frequent in HIV-1-positive individuals than in highly exposed uninfected individuals (3). Another population-based genetic association study has led to the identification of various genetic factors that affect HIV-1/AIDS, and has provided unique insights into the interaction of HIV-1 with the host (140).

Susceptibility to SARS is considered to be a multifactorial and polygenic event with environmental, pathogenic, and host genetic components (84). People with HLA class I (Cw*0801) and HLA class I (B*4601) alleles were reported to be more susceptible to SARS-CoV infection in a population of Taiwanese health care workers (28, 138), whereas people with HLA class II alleles did not show any association. However, this finding was not confirmed in a Hong Kong population, where HLA class I (B*0703) and class II (DRB1*0301) alleles were associated with susceptibility to and resistance to SARS-CoV infection, respectively (28). In Hong Kong populations, homozygosity of *CLEC4M*, which encodes L-SIGN (liver/lymph node-specific ICAM-3 grabbing nonintegrin), a receptor for many viruses, including SARS, was recently shown to be associated with protection against SARS-CoV infection in both genetic and functional studies (22). However, the results in three case-control samples of northern Chinese individuals and another study of a Hong Kong population did not support the findings of significant association between the *CLEC4M* polymorphism and SARS risk (222).

Several other association studies also suggested that the *CD14* (217), *ICAM3* (21), *IFN-*γ (32), and *MxA* (80) genes are related to SARS susceptibility. However, it is important to bear in mind that association studies require replication in independent populations. Although the significant associations between these genes and SARS susceptibility derive from a biologically based a priori hypothesis, these initial findings should be independently verified in other subpopulations of ethnic Chinese origin or in populations of different ancestry, such as Caucasians.

Studies of two candidate SARS susceptibility genes were replicated in multiple populations. Genetic haplotypes associated with low serum mannose-binding lectin (MBL) were reported to be associated with SARS in a collection of case-control sample sets from a Hong Kong population; the findings were then confirmed in a northern Chinese population by another independent study, suggesting that MBL deficiency may be a susceptibility factor for the acquisition of SARS (219). Recently, Ng and coworkers (139) reported that a functional c28g nucleotide polymorphism in the promoter of the regulated upon activation, normal T cell-expressed and secreted (*RANTES*) gene showed significant association with the severity of SARS in both Beijing and Hong Kong populations.

Influenza A Virus: A Case Study

Background information. Avian influenza is a contagious disease of animals caused by avian influenza viruses (AIV) that normally infect only birds and, less commonly, horses, pigs, and chimpanzees. However, some of the subtypes can infect humans, causing significant mortality and morbidity throughout the world (130). AIVs are single-stranded, negative-sense RNA viruses of the family *Orthomyxoviridae*, and wild waterfowl serve as the natural reservoir (181).

Three phylogenetically and antigenically distinct viral types, A, B, and C, circulate globally in human populations. Both influenza A and B viruses contain eight RNA genomic segments (PB2, PB1, PA, HA, NP, NA, M1/2, NS1/2), whereas influenza C virus contains only seven RNA genomic segments. Among these viruses, type A viruses exhibit the greatest genomic genetic diversity (total length ~13 kb), infect the widest range of host species, and cause the vast majority of severe disease in humans, including the great pandemics.

The most widespread and researched subtypes of type A virus are those that infect human beings, including H1N1 (202), H2N2 (170), H3N2 (87, 137), H5N1 (150), H7N7 (58), H1N2 (64), H9N2 (145), H7N2 (109), H7N3 (119), and H10N7 (181). Recently, researchers found that viruses of the H5 and H7 subtypes are becoming highly pathogenic and transmissible to mammals, including humans, albeit at low frequencies (181). The highly pathogenic H5N1 viruses have spread and caused outbreaks in poultry since the outbreak in 1959 in Scotland (225). In 1996, researchers isolated the H5N1 virus for the first time from mildly diseased domestic geese in south China (A/Goose/Guangdong/1/96 virus); in 1997, the H5N1 virus was contracted by a 3-year-old boy in Hong Kong (H5N1/97), leading to his death, which was the first observed transmission from birds to humans (38, 186). High-virulence H5N1 proliferated virulently, mainly in poultry, concomitantly infecting humans sporadically and, as of April 10, 2007, 271 people worldwide had been infected and 59.1% of them died (150). There have been few reports of probable human-to-human transmission of avian influenza since the 1997 Hong Kong outbreak [e.g., one case of apparent child-to-mother transmission in Thailand (203)]. No community human-to-human transmission has been reported as yet. Intensified surveillance of patients by RT-PCR assay in northern Vietnam also led researchers to believe that the local virus strains may be adapting to humans (196). In addition, Kyoko and colleagues (180) found that the avian-derived H5N1 virus can be transmitted from birds to humans and replicated efficiently only in the lower region of the respiratory tract, where the avian virus receptor is prevalent. This restriction may contibute to the inefficient human-to-human transmission of the H5N1 virus.

Recent progress in genomics studies. Since 2005, large-scale genomic sequencing for influenza A virus has been carried out in conjunction with the H5N1 outbreaks in humans and poultry. To date, this global effort has generated approximately 2500 complete genomic sequences of influenza A virus (86). Currently, two major influenza virus genome databases are available for public access: the Influenza Virus Database Beijing (IVDB), **http://influenza.genomics.org.cn** (23), and the Influenza Virus Resource database (IVR), **http://www.ncbi.nlm.nih.gov/genomes/FLU/FLU.html** (87), which is the more comprehensive and well-updated between the two. From these resources, one may easily obtain the sequence information (at both nucleotide and amino acid levels) of the influenza viruses that have been isolated and identified all over the world.

Advances in genome sequencing aid in the tracking of viral biodiversity and in conducting molecular epidemiological surveys, but also create intellectual challenges. The difficulties are obviously due to the complicated mutation process of AIV over time, across global foci, and inside different hosts.

The segmented genome of the influenza virus facilitates reassortment between isolates that coinfect the same host cell. Severe influenza pandemics may occur following a sudden antigenic shift—when a reassortment event generates a novel combination of hemagglutinin (HA; 16 subtypes) and neuraminidase (NA; 9 subtypes) antigens to which the population is immunologically naive (136). Reassortment between different subtypes was a fundamental cause of the human pandemics of 1957 (H2N2) and 1968 (H3N2), during which the virus also acquired a new basic polymerase 1 (PB1) segment (64). By sequencing 209 influenza isolates (207/H3N2, 2/H1N2) (64), researchers discovered an epidemiologically significant reassortment that explained the appearance during the 2003–2004 season of the Fujian/411/2002-like strain, against which the existing vaccine had limited effectiveness. Reassortment can also occur among internal segments and among human strains of the same subtype (136, 207). A detailed phylogenetic analysis of 413 complete H3N2 genomes from New York in the United States, sampled over a 7-year period, revealed 14 reassortment events that were identified on the basis of incongruent phylogenetic trees of HA, NA, and concatenated internal proteins (137).

A potential risk of the H5N1 endemic is the reassortment between the H5N1 virus and currently circulating human influenza virus, which has been proposed to cause the next pandemic (181). H5N1 was not thought to infect humans until the 1997 index case in Hong Kong was reported. Genomic sequencing showed that except for the H5 gene, all other genes of this semianthropophilic virulent strain are almost identical to those of an ordinary colocated strain subclinically harbored by a feral migrating duck, A/teal/Hong-Kong/W312/97 (H6N1) (29). It has been speculated that the reassortment virus infected domestic ducks that could readily facilitate its proliferation because they belong to the same host genus as the teals. Further genetic changes within domestic duck populations gave rise to H5N1 strains with amplified pathogenicity toward ducks, chickens, and humans (26). Comparisons of all eight RNA segments from those virus classes revealed greater than 99% sequence identity among them (185). It has been stated that, "acting as a silent reservoir for the H5N1 virus, domestic ducks may have acquired an important new role in the transmission of this virus to other poultry and, possibly, to humans as well" (181).

Some reassortment events are undetectable by phylogenetic analysis because they do not lead to major differences in tree topology, or they involve parental isolates that have not been sampled or result in unfit progeny (207). Therefore, more sophisticated methods were developed to estimate both the rate of reassortment, particularly relative to the rate of mutation for each nucleotide, and the background phylogenetic history of the viral genome in the face of such frequent reassortment (see 136 for further information).

Because the human immune response to viral infection is not completely cross protective, natural selection favors amino acid variants of the major antigen proteins, HA and NA, that allow the virus to evade immunity, infect more hosts, and proliferate. This continuous change in antigenic structure through time is called antigenic drift (136). Although antigenic changes in HA are clearly important determinants of viral fitness, the antigenic evolution of HA1 seems to be more clustered than continuous (147). In addition, whole-genome phylogenies show the coexistence of multiple viral lineages, particularly on a limited spatial and temporal scale (136, 137). Therefore, the transition among antigenic types does not always proceed in a simple linear manner and reassortment among coexisting lineages is relatively frequent. For these reasons, predicting the path of influenza virus evolution from sequence data alone is inherently difficult (136). Consequently, to address the evolutionary and epidemiological processes that drive antigenic drift and the timescale on which this process occurs, a far larger sample of influenza virus genomes with greater resolution in both time and space is clearly required.

The variations in HA not only may affect its antigenicity but also are related to pathogenesis. A survey of the HA cleavage sites of H5 and H7 from many AIV isolates of high or low pathogenicity indicated that almost all the highly pathogenic H5 and H7 viruses bear HA cleavage sites significantly different from that of the consensus low pathogenic sequence (R-X-T-R/G), with a continuous peptide of 4–5 more basic amino acids adjacent to the N-terminal vicinity of the cleavage site (39).

Recently, plasmid-based reverse genetics allowed the generation of recombinant viruses that contain the HA and/or NA from the Spanish 1918 influenza pandemic virus rescued in the H1N1 or H3N2 influenza virus background (202). This study found that the cleavage of the HA of the 1918 virus, a step critical for virulent infection, occurs via a novel neuraminidase facilitation mechanism different from the mechanism employed by currently identified genetic variations of HA and NA responsible for the high-virulence phenotype (202). Simultaneously, phylogenetic analysis of the heterotrimeric polymerase complex (PB1, PB2, and PA) of the 1918 virus (190) indicated that the polymerase sequences from the 1918 human influenza virus differed from the consensus sequences of the avian virus at only a small number of amino acids, consistent with the hypothesis that the 1918 virus was derived from an avian source shortly before the pandemic. Furthermore, when compared with avian viral sequences, the nucleotide sequences of the 1918 virus polymerase gene have more synonymous differences than expected, suggesting evolutionary distance from known avian strains (190). These studies demonstrated how genomic analysis can provide crucial insights into critical questions about the virulence and etiology of a catastrophic disease event.

Accurate estimations of evolutionary rates at both the nucleotide and amino acid levels are central to resolving questions about the evolution of the influenza virus. Whole-genome comparative studies allowed the identification of synonymous and nonsynonymous mutations among the different viral isolates. A recent study indicated that despite the fact that unvaccinated poultry do not typically mount a strong immune response, higher rates of nonsynonymous substitution in the HA segment of H5N1 were observed in some domestic poultry species (27). These observations indicate that the selective pressures to adapt to a new host can lead to rapid evolution, even in the absence of immune selection.

In addition to HA and NA genes, other internal viral genes may also be responsible for pathogenicity and virulence. The nonstructural protein 1 (NS1), encoded by segment 8 of the viral genome, was shown to play an important role in antagonizing the host innate immune response (63). This regulatory protein has multiple ways to attenuate antiviral responses, such as inhibiting RNA processing through 2′-5′ oligo A synthetase (OAS) (128), interfering with transcription factors (i.e., IRF3, IRF7, and NF-κB) by interacting with retinoic acid-inducible gene I (RIG-I) (126), and inhibiting the Ser/Thr protein kinase (PKR), which may lead to the inhibition of host protein synthesis and other pathways (115). Alignment comparisons of deduced amino acid sequences of the NS1 gene helped to categorize the NS genes of H5N1 into two groups. Group A consists of the viruses isolated before 2000 and the primary structure contains a five-amino-acid residue stretch, A/T IAS V/S/R/L, at position 80–84 of the NS (211), which is missing in most isolates of Group B, the viruses collected in 2000–2004 (204). Although some H5N1 viruses isolated in 2000–2004 lack the five-amino-acid deletion, all new H5N1 strains collected from Thailand, Vietnam, and China in 2004 consistently contained this five-amino-acid deletion (204, 226). This result suggested the direct transmission of AIV to humans (204).

Moreover, other point mutations of NS1 were also implicated in virulence. A D92E point mutation of NS1 of the H5N1/97 influenza viruses increased viral resistance to cytokines such as IFN-α/β and TNF-α (173). Replacement of Pro to Ser at position 42 (P42S) dramatically enhanced the virulence of the H5N1 virus in mice (95). Li and coworkers (115) recently demonstrated that a recombinant virus expressing the GS/GD/1/96 NS1 protein with Ala149 is able to antagonize the induction of IFN in chicken embryo fibroblasts (CEFs), whereas a Val149 substitution has no such effect. Furthermore, Zhu and colleagues (227) found that the deletion of amino acids 191–195 of the NS1 protein is critical for the attenuation of the A/swine/Fujian/1/03 H5N1 virus in chickens and that this deletion affects the ability of the virus to antagonize IFN induction in host cells. The importance of sequence variation in NS1, in which most H5N1 viruses contain a C-terminal sequence motif that mediates binding to various cellular proteins, remains to be determined.

Vaccine development. In 2007, the U.S. Food and Drug Administration (FDA) announced approval of the first human vaccine against the H5N1 influenza virus (**http://www.fda.gov/bbs/topics/NEWS/2007/NEW01611.html**). The egg-grown inactivated vaccine was obtained from a human strain and is intended for immunization of people aged 18 through 64. The seed virus for the production of the vaccine was generated from the human isolate of the influenza A/Vietnam/1203/2004 (H5N1) virus (197). The gene segments encoding HA and NA were derived from the A/Vietnam/2004 virus, and all other genes were derived from the A/PR/8/34 virus, a laboratory strain commonly used as a platform for influenza vaccines. The HA gene was modified to replace a stretch of six basic amino acids at the cleavage site between HA1 and HA2 associated with high pathogenicity in birds with an avirulent avian sequence. The resulting influenza rgA/Vietnam/1203/2004 × A/PR/8/34 influenza (H5N1) virus is antigenically identical to the wild-type A/Vietnam/1203/2004 virus and reaches high titers in eggs. The approval of this vaccine is an important step forward in protecting humans against avian influenza.

With the support of the FDA, the U.S. National Institutes of Health, and other government agencies, manufacturers (e.g., Sanofi Pasteur Inc.) are working to develop a next-generation influenza vaccine to elicit enhanced immune responses at lower doses by using technologies intended to boost the immune response (55). Meanwhile, the approval and availability of this vaccine will enhance national readiness and the nation's ability to protect those at increased risk of exposure (197).

TRANSLATIONAL RESEARCH: FROM GENOMICS POTENTIAL TO CLINICAL REALITY

Genomics, its large-scale, HTP data generating/analysis technology, and its holistic/integrative concept for research, is changing the world of biology and medicine by creating a new scientific discipline termed systems medical biology (131). For the first time in history humans can quickly and accurately identify the pathogens that cause infectious diseases, even if the pathogen is novel, and also efficiently investigate the origins and evolution of a pathogen (33, 221). On the basis of this knowledge and these techniques, diagnosis and epidemiological studies as well as vaccine candidates and drug targets can be developed at an unprecedented speed, which will eventually provide human beings with new weapons to combat emerging or reemerging infectious diseases (41, 59, 86, 94, 129, 142, 193, 220).

However, genomics still needs further development and genomics alone cannot meet the needs of patients directly. Genomics data collecting technology must be further revolutionized to make the assays faster, cheaper, and more reliable. Genomics data analysis technology must be further developed to integrate different levels of data, from the genomics of pathogens and hosts to the biology, physiology, pathology, and epidemiology information that will make real contributions to the clinical or subclinical prediction, prevention, and treatment of disease. It is essential to integrate the efforts of field epidemiologists, clinical practitioners, and scientific researchers into the translational research scheme to realize the great potential of genomics in the fight against infectious disease.

DISCLOSURE STATEMENT

The authors are not aware of any biases that might be perceived as affecting the objectivity of this review.

ACKNOWLEDGMENTS

We thank Dr. Youjia Cao and Dr. Yuelong Shu for their important suggestions and critical reading of this manuscript. This work was supported by the State High Technology Development Program (863, Grant No. 2003AA208407), the State Key Program for Basic Research (973, Grant No. 2003CB514101), and the European Commission grant EPISARS (N°511063).

LITERATURE CITED

1. AbuBakar S, Chang LY, Ali AR, Sharifah SH, Yusoff K, Zamrod Z. 2004. Isolation and molecular identification of Nipah virus from pigs. *Emerg. Infect. Dis.* 10:2228–30
2. Allain JP. 2004. Occult hepatitis B virus infection: implications in transfusion. *Vox Sang.* 86:83–91
3. An P, Wang LH, Hutcheson-Dilks H, Nelson G, Donfield S, et al. 2007. Regulatory polymorphisms in the cyclophilin A gene, *PPIA*, accelerate progression to AIDS. *PLoS Pathog.* 3:e88
4. Arcus VL, Lott JS, Johnston JM, Baker EN. 2006. The potential impact of structural genomics on tuberculosis drug discovery. *Drug Discov. Today* 11:28–34
5. Auerbach RK, Tuanyok A, Probert WS, Kenefic L, Vogler AJ, et al. 2007. *Yersinia pestis* evolution on a small timescale: comparison of whole genome sequences from North America. *PLoS ONE* 2:e770
6. Bailes E, Gao F, Bibollet-Ruche F, Courgnaud V, Peeters M, et al. 2003. Hybrid origin of SIV in chimpanzees. *Science* 300:1713
7. Barre-Sinoussi F, Chermann JC, Rey F, Nugeyre MT, Chamaret S, et al. 1983. Isolation of a T-lymphotropic retrovirus from a patient at risk for acquired immune deficiency syndrome (AIDS). *Science* 220:868–71
8. Basler CF, Wang X, Muhlberger E, Volchkov V, Paragas J, et al. 2000. The Ebola virus VP35 protein functions as a type I IFN antagonist. *Proc. Natl. Acad. Sci. USA* 97:12289–94
9. Bi S, Qin E, Xu Z, Li W, Wang J, et al. 2003. Complete genome sequences of the SARS-CoV: the BJ Group (Isolates BJ01-BJ04). *Genomics Proteomics Bioinformatics* 1:180–92
10. Bleiber G, May M, Martinez R, Meylan P, Ott J, et al. 2005. Use of a combined ex vivo/in vivo population approach for screening of human genes involved in the human immunodeficiency virus type 1 life cycle for variants influencing disease progression. *J. Virol.* 79:12674–80
11. Bonn D. 2003. Closing in on the cause of SARS. *Lancet Infect. Dis.* 3:268
12. Bray M, Davis K, Geisbert T, Schmaljohn C, Huggins J. 1998. A mouse model for evaluation of prophylaxis and therapy of Ebola hemorrhagic fever. *J. Infect. Dis.* 178:651–61
13. Bricaire F, Bossi P. 2006. [Emerging viral diseases]. *Bull. Acad. Natl. Med.* 190:597–608; discussion 609, 625–27
14. Brinton MA. 2002. The molecular biology of West Nile Virus: a new invader of the western hemisphere. *Annu. Rev. Microbiol.* 56:371–402
15. Brookes AJ. 1999. The essence of SNPs. *Gene* 234:177–86
16. Busch MP, Glynn SA, Stramer SL, Strong DM, Caglioti S, et al. 2005. A new strategy for estimating risks of transfusion-transmitted viral infections based on rates of detection of recently infected donors. *Transfusion* 45:254–64
17. CDCP. 1989. Ebola virus infection in imported primates–Virginia, 1989. *MMWR* 38:831–32, 837–38
18. CDCP. 1999. Outbreak of Hendra-like virus–Malaysia and Singapore, 1998–1999. *MMWR* 48:265–69
19. CDCP. 1999. Update: outbreak of Nipah virus–Malaysia and Singapore, 1999. *MMWR* 48:335–37
20. Chadha MS, Comer JA, Lowe L, Rota PA, Rollin PE, et al. 2006. Nipah virus-associated encephalitis outbreak, Siliguri, India. *Emerg. Infect. Dis.* 12:235–40

21. Chan KY, Ching JC, Xu MS, Cheung AN, Yip SP, et al. 2007. Association of *ICAM3* genetic variant with severe acute respiratory syndrome. *J. Infect. Dis.* 196:271–80
22. Chan VS, Chan KY, Chen Y, Poon LL, Cheung AN, et al. 2006. Homozygous L-SIGN (*CLEC4M*) plays a protective role in SARS coronavirus infection. *Nat. Genet.* 38:38–46
23. Chang S, Zhang J, Liao X, Zhu X, Wang D, et al. 2007. Influenza Virus Database (IVDB): an integrated information resource and analysis platform for influenza virus research. *Nucleic Acids Res.* 35:D376–80
24. Charrel RN, de Lamballerie X. 2004. [West Nile virus, an emerging arbovirus]. *Presse Med.* 33:1521–26
25. Che XY, Hao W, Wang Y, Di B, Yin K, et al. 2004. Nucleocapsid protein as early diagnostic marker for SARS. *Emerg. Infect. Dis.* 10:1947–49
26. Chen H, Deng G, Li Z, Tian G, Li Y, et al. 2004. The evolution of H5N1 influenza viruses in ducks in southern China. *Proc. Natl. Acad. Sci. USA* 101:10452–57
27. Chen R, Holmes EC. 2006. Avian influenza virus exhibits rapid evolutionary dynamics. *Mol. Biol. Evol.* 23:2336–41
28. Chen YM, Liang SY, Shih YP, Chen CY, Lee YM, et al. 2006. Epidemiological and genetic correlates of severe acute respiratory syndrome coronavirus infection in the hospital with the highest nosocomial infection rate in Taiwan in 2003. *J. Clin. Microbiol.* 44:359–65
29. Chin PS, Hoffmann E, Webby R, Webster RG, Guan Y, et al. 2002. Molecular evolution of H6 influenza viruses from poultry in Southeastern China: prevalence of H6N1 influenza viruses possessing seven A/Hong Kong/156/97 (H5N1)-like genes in poultry. *J. Virol.* 76:507–16
30. Chinese SMEC. 2004. Molecular evolution of the SARS coronavirus during the course of the SARS epidemic in China. *Science* 303:1666–69
31. Choi C. 2004. Nipah's return. The lethal "flying fox" virus may spread between people. *Sci. Am.* 291:A21–22
32. Chong WP, Ip WK, Tso GH, Ng MW, Wong WH, et al. 2006. The interferon γ gene polymorphism +874 A/T is associated with severe acute respiratory syndrome. *BMC Infect. Dis.* 6:82
33. Chow KY, Hon CC, Hui RK, Wong RT, Yip CW, et al. 2003. Molecular advances in severe acute respiratory syndrome-associated coronavirus (SARS-CoV). *Genomics Proteomics Bioinformatics* 1:247–62
34. Chua KB, Bellini WJ, Rota PA, Harcourt BH, Tamin A, et al. 2000. Nipah virus: a recently emergent deadly paramyxovirus. *Science* 288:1432–35
35. Chua KB, Goh KJ, Wong KT, Kamarulzaman A, Tan PS, et al. 1999. Fatal encephalitis due to Nipah virus among pig-farmers in Malaysia. *Lancet* 354:1257–59
36. Chua KB, Lek Koh C, Hooi PS, Wee KF, Khong JH, et al. 2002. Isolation of Nipah virus from Malaysian Island flying-foxes. *Microbes Infect.* 4:145–51
37. Cleven BE, Palka-Santini M, Gielen J, Meembor S, Kronke M, Krut O. 2006. Identification and characterization of bacterial pathogens causing bloodstream infections by DNA microarray. *J. Clin. Microbiol.* 44:2389–97
38. Cohen J. 1997. Infectious disease: The flu pandemic that might have been. *Science* 277:1600–1
39. Cyranoski D. 2003. Vaccine sought as bird flu infects humans. *Nature* 422:6
40. Danin-Poleg Y, Somer L, Cohen LA, Diamant E, Palti Y, Kashi Y. 2006. Towards the definition of pathogenic microbe. *Int. J. Food Microbiol.* 112:236–43
41. Danzig L. 2006. Reverse vaccinology–in search of a genome-derived meningococcal vaccine. *Vaccine* 24(Suppl. 2):S2–11–2
42. Davidsen T, Koomey M, Tonjum T. 2007. Microbial genome dynamics in CNS pathogenesis. *Neuroscience* 145:1375–87
43. Davis CT, Beasley DWC, Guzman H, Siirin M, Parsons RE, et al. 2004. Emergence of attenuated West Nile virus variants in Texas, 2003. *Virology* 330:342–50
44. Davis CT, Ebel GD, Lanciotti RS, Brault AC, Guzman H, et al. 2005. Phylogenetic analysis of North American West Nile virus isolates, 2001–2004: Evidence for the emergence of a dominant genotype. *Virology* 342:252–65
45. Domiati-Saad R, Scheuermann RH. 2006. Nucleic acid testing for viral burden and viral genotyping. *Clin. Chim. Acta.* 363:197–205
46. Drancourt M, Roux V, Dang LV, Tran-Hung L, Castex D, et al. 2004. Genotyping, Orientalis-like *Yersinia pestis*, and plague pandemics. *Emerg. Infect. Dis.* 10:1585–92

47. Drosten C, Gunther S, Preiser W, van der Werf S, Brodt HR, et al. 2003. Identification of a novel coronavirus in patients with severe acute respiratory syndrome. *N. Engl. J. Med.* 348:1967–76
48. Eaton BT, Broder CC, Wang LF. 2005. Hendra and Nipah viruses: pathogenesis and therapeutics. *Curr. Mol. Med.* 5:805–16
49. Ebel GD, Carricaburu J, Young D, Bernard KA, Kramer LD. 2004. Genetic and phenotypic variation of West Nile virus in New York, 2000–2003. *Am. J. Trop. Med. Hyg.* 71:493–500
50. Ebihara H, Takada A, Kobasa D, Jones S, Neumann G, et al. 2006. Molecular determinants of Ebola virus virulence in mice. *PLoS Pathog.* 2:e73
51. Enserink M. 1999. New virus fingered in Malaysian epidemic. *Science* 284:407, 409–10
52. Enserink M, Vogel G. 2003. Infectious diseases. Hungry for details, scientists zoom in on SARS genomes. *Science* 300:715–17
53. Ernst JD, Trevejo-Nunez G, Banaiee N. 2007. Genomics and the evolution, pathogenesis, and diagnosis of tuberculosis. *J. Clin. Invest.* 117:1738–45
54. Fauci AS. 2001. Infectious diseases: considerations for the 21st century. *Clin. Infect. Dis.* 32:675–85
55. Fauci AS. 2006. Pandemic influenza threat and preparedness. *Emerg. Infect. Dis.* 12:73–77
56. Filliol I, Motiwala AS, Cavatore M, Qi W, Hazbon MH, et al. 2006. Global phylogeny of *Mycobacterium tuberculosis* based on single nucleotide polymorphism (SNP) analysis: insights into tuberculosis evolution, phylogenetic accuracy of other DNA fingerprinting systems, and recommendations for a minimal standard SNP set. *J. Bacteriol.* 188:759–72
57. Fleischmann RD, Adams MD, White O, Clayton RA, Kirkness EF, et al. 1995. Whole-genome random sequencing and assembly of *Haemophilus influenzae* Rd. *Science* 269:496–512
58. Fouchier RA, Schneeberger PM, Rozendaal FW, Broekman JM, Kemink SA, et al. 2004. Avian influenza A virus (H7N7) associated with human conjunctivitis and a fatal case of acute respiratory distress syndrome. *Proc. Natl. Acad. Sci. USA* 101:1356–61
59. Fournier PE, Drancourt M, Raoult D. 2007. Bacterial genome sequencing and its use in infectious diseases. *Lancet Infect. Dis.* 7:711–23
60. Gao F, Bailes E, Robertson DL, Chen Y, Rodenburg CM, et al. 1999. Origin of HIV-1 in the chimpanzee *Pan troglodytes troglodytes*. *Nature* 397:436–41
61. Garaizar J, Rementeria A, Porwollik S. 2006. DNA microarray technology: a new tool for the epidemiological typing of bacterial pathogens? *FEMS Immunol. Med. Microbiol.* 47:178–89
62. Gaviria Rivera AM, Priest FG. 2003. Molecular typing of *Bacillus thuringiensis* serovars by RAPD-PCR. *Syst. Appl. Microbiol.* 26:254–61
63. Geiss GK, Salvatore M, Tumpey TM, Carter VS, Wang X, et al. 2002. Cellular transcriptional profiling in influenza A virus-infected lung epithelial cells: the role of the nonstructural NS1 protein in the evasion of the host innate defense and its potential contribution to pandemic influenza. *Proc. Natl. Acad. Sci. USA* 99:10736–41
64. Ghedin E, Sengamalay NA, Shumway M, Zaborsky J, Feldblyum T, et al. 2005. Large-scale sequencing of human influenza reveals the dynamic nature of viral genome evolution. *Nature* 437:1162–66
65. Gilbert MT, Rambaut A, Wlasiuk G, Spira TJ, Pitchenik AE, Worobey M. 2007. The emergence of HIV/AIDS in the Americas and beyond. *Proc. Natl. Acad. Sci. USA* 104:18566–70
66. Gonzales MJ, Machekano RN, Shafer RW. 2001. Human immunodeficiency virus type 1 reverse-transcriptase and protease subtypes: classification, amino acid mutation patterns, and prevalence in a northern California clinic-based population. *J. Infect. Dis.* 184:998–1006
67. Groisman EA, Ehrlich SD. 2003. Genomics. A global view of gene gain, loss, regulation and function. *Curr. Opin. Microbiol.* 6:479–81
68. Groseth A, Feldmann H, Strong JE. 2007. The ecology of Ebola virus. *Trends Microbiol.* 15:408–16
69. Gu J, Neary JL, Sanchez M, Yu J, Lilburn TG, Wang Y. 2007. Genome evolution and functional divergence in *Yersinia*. *J. Exp. Zool. B* 308:37–49
70. Guan Y, Zheng BJ, He YQ, Liu XL, Zhuang ZX, et al. 2003. Isolation and characterization of viruses related to the SARS coronavirus from animals in southern China. *Science* 302:276–78
71. Gupta UD, Katoch VM, McMurray DN. 2007. Current status of TB vaccines. *Vaccine* 25:3742–51
72. Gurley ES, Montgomer JM, Jahangir Hossain M, Bell M, Abul Kalam Azad, et al. 2007. Person-to-person transmission of Nipah virus in a Bangladeshi community. *Emerg. Infect. Dis.* 13:7

73. Gutacker MM, Mathema B, Soini H, Shashkina E, Kreiswirth BN, et al. 2006. Single-nucleotide polymorphism-based population genetic analysis of *Mycobacterium tuberculosis* strains from 4 geographic sites. *J. Infect. Dis.* 193:121–28
74. Halpin K, Mungall BA. 2007. Recent progress in henipavirus research. *Comp. Immunol. Microbiol. Infect. Dis.* 30:287–307
75. Halpin K, Young PL, Field HE, Mackenzie JS. 2000. Isolation of Hendra virus from pteropid bats: a natural reservoir of Hendra virus. *J. Gen. Virol.* 81:1927–32
76. Harcourt BH, Tamin A, Halpin K, Ksiazek TG, Rollin PE, et al. 2001. Molecular characterization of the polymerase gene and genomic termini of Nipah virus. *Virology* 287:192–201
77. Hart MK. 2003. Vaccine research efforts for filoviruses. *Int. J. Parasitol.* 33:583–95
78. Hayes EB, Gubler DJ. 2006. West Nile virus: Epidemiology and clinical features of an emerging epidemic in the United States. *Annu. Rev. Med.* 57:181–94
79. Haynes LM, Miao C, Harcourt JL, Montgomery JM, Le MQ, et al. 2007. Recombinant protein-based assays for detection of antibodies to severe acute respiratory syndrome coronavirus spike and nucleocapsid proteins. *Clin. Vaccine Immunol.* 14:331–33
80. He J, Feng D, de Vlas SJ, Wang H, Fontanet A, et al. 2006. Association of SARS susceptibility with single nucleic acid polymorphisms of OAS1 and MxA genes: a case-control study. *BMC Infect. Dis.* 6:106
81. Heeney JL, Dalgleish AG, Weiss RA. 2006. Origins of HIV and the evolution of resistance to AIDS. *Science* 313:462–66
82. Hensley LE, Jones SM, Feldmann H, Jahrling PB, Geisbert TW. 2005. Ebola and Marburg viruses: pathogenesis and development of countermeasures. *Curr. Mol. Med.* 5:761–72
83. Hill AV. 1998. The immunogenetics of human infectious diseases. *Annu. Rev. Immunol.* 16:593–617
84. Hill AV. 2006. Aspects of genetic susceptibility to human infectious diseases. *Annu. Rev. Genet.* 40:469–86
85. Hoffmaster AR, Fitzgerald CC, Ribot E, Mayer LW, Popovic T. 2002. Molecular subtyping of *Bacillus anthracis* and the 2001 bioterrorism-associated anthrax outbreak, United States. *Emerg. Infect. Dis.* 8:1111–16
86. Holmes EC. 2007. Viral evolution in the genomic age. *PLoS Biol.* 5:e278
87. Holmes EC, Ghedin E, Miller N, Taylor J, Bao Y, et al. 2005. Whole-genome analysis of human influenza A virus reveals multiple persistent lineages and reassortment among recent H3N2 viruses. *PLoS Biol.* 3:e300
88. Hooper E. 1999. *The River: A Journal to the Source of HIV and AIDS*. London: Penguin
89. Horvath R, Dendis M, Schlegelova J, Ruzicka F, Benedik J. 2004. A combined AFLP-multiplex PCR assay for molecular typing of *Escherichia coli* strains using variable bacterial interspersed mosaic elements. *Epidemiol. Infect.* 132:61–65
90. Hsu VP, Hossain MJ, Parashar UD, Ali MM, Ksiazek TG, et al. 2004. Nipah virus encephalitis reemergence, Bangladesh. *Emerg. Infect. Dis.* 10:2082–87
91. Huet T, Cheynier R, Meyerhans A, Roelants G, Wain-Hobson S. 1990. Genetic organization of a chimpanzee lentivirus related to HIV-1. *Nature* 345:356–59
92. Hughes AL, Piontkivska H, Foppa I. 2007. Rapid fixation of a distinctive sequence motif in the 3′ noncoding region of the clade of West Nile virus invading North America. *Gene* 399:152–61
93. Hughes AL, Yeager M. 1998. Natural selection at major histocompatibility complex loci of vertebrates. *Annu. Rev. Genet.* 32:415–35
94. Hyytia-Trees EK, Cooper K, Ribot EM, Gerner-Smidt P. 2007. Recent developments and future prospects in subtyping of foodborne bacterial pathogens. *Future Microbiol.* 2:175–85
95. Jiao P, Tian G, Li Y, Deng G, Jiang Y, et al. 2008. A single amino acid substitution in the NS1 protein changes the pathogenicity of H5N1 avian influenza viruses in mice. *J. Virol.* 82:1146–54
96. Kan B, Wang M, Jing H, Xu H, Jiang X, et al. 2005. Molecular evolution analysis and geographic investigation of severe acute respiratory syndrome coronavirus-like virus in palm civets at an animal market and on farms. *J. Virol.* 79:11892–900
97. Karupiah G, Panchanathan V, Sakala IG, Chaudhri G. 2007. Genetic resistance to smallpox: lessons from mousepox. *Novartis Found. Symp.* 281:129–36; discussion 136–40, 208–9
98. Keele BF, Van Heuverswyn F, Li Y, Bailes E, Takehisa J, et al. 2006. Chimpanzee reservoirs of pandemic and nonpandemic HIV-1. *Science* 313:523–26

99. Kiryluk K, Martino J, Gharavi AG. 2007. Genetic susceptibility, HIV infection, and the kidney. *Clin. J. Am. Soc. Nephrol.* 2(Suppl. 1):S25–35
100. Kleinbaum DG, Kupper LL, Morgenstern H. 1982. *Epidemiologic Research: Principles and Quantitative Methods*. New York: Van Nostrand Reinhold
101. Komar N. 2003. West Nile virus: Epidemiology and ecology in North America. *Adv. Virus Res.* 61:185–234
102. Korber B, Muldoon M, Theiler J, Gao F, Gupta R, et al. 2000. Timing the ancestor of the HIV-1 pandemic strains. *Science* 288:1789–96
103. Kramer LD, Styer LM, Ebel GD. 2008. A global perspective on the epidemiology of West Nile virus. *Annu. Rev. Entomol.* 53:61–81
104. Kuiken T, Fouchier R, Rimmelzwaan G, Osterhaus A. 2003. Emerging viral infections in a rapidly changing world. *Curr. Opin. Biotechnol.* 14:641–46
105. Kuroki Y, Toyoda A, Noguchi H, Taylor TD, Itoh T, et al. 2006. Comparative analysis of chimpanzee and human Y chromosomes unveils complex evolutionary pathway. *Nat. Genet.* 38:158–67
106. Kuska B. 1998. Beer, Bethesda, and biology: how "genomics" came into being. *J. Natl. Cancer Inst.* 90:93
107. Lanciotti RS, Ebel GD, Deubel V, Kerst AJ, Murri M, et al. 2002. Complete genome sequences and phylogenetic analysis of West Nile Virus strains isolated from the United States, Europe, and the Middle East. *Virology* 298:96–105
108. Lanciotti RS, Roehrig JT, Deubel V, Smith J, Parker M, et al. 1999. Origin of the West Nile virus responsible for an outbreak of encephalitis in the northeastern United States. *Science* 286:2333–37
109. Lee CW, Lee YJ, Senne DA, Suarez DL. 2006. Pathogenic potential of North American H7N2 avian influenza virus: a mutagenesis study using reverse genetics. *Virology* 353:388–95
110. Leitner T, Escanilla D, Franzen C, Uhlen M, Albert J. 1996. Accurate reconstruction of a known HIV-1 transmission history by phylogenetic tree analysis. *Proc. Natl. Acad. Sci. USA* 93:10864–69
111. Leitner T, Foley B, Hahn B, Marx P, McCutchan F, eds. et al. HIV Sequence Compendium 2005. *Theoretical Biol. Biophys. Group, Los Alamos Natl. Lab., NM, LA-UR 06–0680*
112. Leroy EM, Baize S, Mavoungou E, Apetrei C. 2002. Sequence analysis of the GP, NP, VP40 and VP24 genes of Ebola virus isolated from deceased, surviving and asymptomatically infected individuals during the 1996 outbreak in Gabon: comparative studies and phylogenetic characterization. *J. Gen. Virol.* 83:67–73
113. Leroy EM, Kumulungui B, Pourrut X, Rouquet P, Hassanin A, et al. 2005. Fruit bats as reservoirs of Ebola virus. *Nature* 438:575–76
114. Li KS, Guan Y, Wang J, Smith GJ, Xu KM, et al. 2004. Genesis of a highly pathogenic and potentially pandemic H5N1 influenza virus in eastern Asia. *Nature* 430:209–13
115. Li S, Min JY, Krug RM, Sen GC. 2006. Binding of the influenza A virus NS1 protein to PKR mediates the inhibition of its activation by either PACT or double-stranded RNA. *Virology* 349:13–21
116. Li W, Shi Z, Yu M, Ren W, Smith C, et al. 2005. Bats are natural reservoirs of SARS-like coronaviruses. *Science* 310:676–79
117. Li Z, Jiang Y, Jiao P, Wang A, Zhao F, et al. 2006. The NS1 gene contributes to the virulence of H5N1 avian influenza viruses. *J. Virol.* 80:11115–23
118. Lindblad-Toh K, Wade CM, Mikkelsen TS, Karlsson EK, Jaffe DB, et al. 2005. Genome sequence, comparative analysis and haplotype structure of the domestic dog. *Nature* 438:803–19
119. Lister S, Knott C, Hammond P. 2006. Low pathogenic H7N3 avian influenza. *Vet. Rec.* 158:771
120. Little P. 1992. Human Genome Project: Mapping the way ahead. *Nature* 359:367–68
121. Liu C. 2003. The battle against SARS: a Chinese story. *Aust. Health Rev.* 26:3–13
122. Los Alamos Natl. Lab. 2001. *HIV Sequence Database*. **http://hiv-web.lanl.gov**
123. Marra MA, Jones SJ, Astell CR, Holt RA, Brooks-Wilson A, et al. 2003. The genome sequence of the SARS-associated coronavirus. *Science* 300:1399–404
124. Mathema B, Kurepina NE, Bifani PJ, Kreiswirth BN. 2006. Molecular epidemiology of tuberculosis: current insights. *Clin. Microbiol. Rev.* 19:658–85
125. McCormack JG. 2005. Hendra and Nipah viruses: new zoonotically-acquired human pathogens. *Respir. Care Clin. N. Am.* 11:59–66

126. Mibayashi M, Martinez-Sobrido L, Loo YM, Cardenas WB, Gale M Jr., Garcia-Sastre A. 2007. Inhibition of retinoic acid-inducible gene I-mediated induction of beta interferon by the NS1 protein of influenza A virus. *J. Virol.* 81:514–24
127. Millar BC, Xu J, Moore JE. 2007. Molecular diagnostics of medically important bacterial infections. *Curr. Issues Mol. Biol.* 9:21–39
128. Min JY, Krug RM. 2006. The primary function of RNA binding by the influenza A virus NS1 protein in infected cells: Inhibiting the 2′-5′ oligo (A) synthetase/RNase L pathway. *Proc. Natl. Acad. Sci. USA* 103:7100–5
129. Monaghan RL, Barrett JF. 2006. Antibacterial drug discovery–then, now and the genomics future. *Biochem. Pharmacol.* 71:901–9
130. Monto AS. 2005. The threat of an avian influenza pandemic. *N. Engl. J. Med.* 352:323–25
131. Morel NM, Holland JM, van der Greef J, Marple EW, Clish C, et al. 2004. Primer on medical genomics. Part XIV: Introduction to systems biology–a new approach to understanding disease and treatment. *Mayo Clin. Proc.* 79:651–58
132. Mungall BA, Middleton D, Crameri G, Halpin K, Bingham J, et al. 2007. Vertical transmission and fetal replication of Nipah virus in an experimentally infected cat. *J. Infect. Dis.* 196:812–16
133. Mural RJ, Adams MD, Myers EW, Smith HO, Miklos GL, et al. 2002. A comparison of whole-genome shotgun-derived mouse chromosome 16 and the human genome. *Science* 296:1661–71
134. Nakamura Y, Leppert M, O'Connell P, Wolff R, Holm T, et al. 1987. Variable number of tandem repeat (VNTR) markers for human gene mapping. *Science* 235:1616–22
135. Ndayimirije N, Kindhauser MK. 2005. Marburg hemorrhagic fever in Angola–fighting fear and a lethal pathogen. *N. Engl. J. Med.* 352:2155–57
136. Nelson MI, Holmes EC. 2007. The evolution of epidemic influenza. *Nat. Rev. Genet.* 8:196–205
137. Nelson MI, Simonsen L, Viboud C, Miller MA, Taylor J, et al. 2006. Stochastic processes are key determinants of short-term evolution in influenza a virus. *PLoS Pathog.* 2:e125
138. Ng MH, Lau KM, Li L, Cheng SH, Chan WY, et al. 2004. Association of human-leukocyte-antigen class I (B*0703) and class II (DRB1*0301) genotypes with susceptibility and resistance to the development of severe acute respiratory syndrome. *J. Infect. Dis.* 190:515–18
139. Ng MW, Zhou G, Chong WP, Lee LW, Law HK, et al. 2007. The association of *RANTES* polymorphism with severe acute respiratory syndrome in Hong Kong and Beijing Chinese. *BMC Infect. Dis.* 7:50
140. O'Brien SJ, Nelson GW. 2004. Human genes that limit AIDS. *Nat. Genet.* 36:565–74
141. Palacios G, Quan PL, Jabado OJ, Conlan S, Hirschberg DL, et al. 2007. Panmicrobial oligonucleotide array for diagnosis of infectious diseases. *Emerg. Infect. Dis.* 13:73–81
142. Pallen MJ, Wren BW. 2007. Bacterial pathogenomics. *Nature* 449:835–42
143. Peeters M, Honore C, Huet T, Bedjabaga L, Ossari S, et al. 1989. Isolation and partial characterization of an HIV-related virus occurring naturally in chimpanzees in Gabon. *AIDS* 3:625–30
144. Peiris JS, Lai ST, Poon LL, Guan Y, Yam LY, et al. 2003. Coronavirus as a possible cause of severe acute respiratory syndrome. *Lancet* 361:1319–25
145. Peiris M, Yuen KY, Leung CW, Chan KH, Ip PL, et al. 1999. Human infection with influenza H9N2. *Lancet* 354:916–17
146. Perry RD, Fetherston JD. 1997. *Yersinia pestis*–etiologic agent of plague. *Clin. Microbiol. Rev.* 10:35–66
147. Plotkin JB, Dushoff J, Levin SA. 2002. Hemagglutinin sequence clusters and the antigenic evolution of influenza A virus. *Proc. Natl. Acad. Sci. USA* 99:6263–68
148. Poinar H, Kuch M, Paabo S. 2001. Molecular analyses of oral polio vaccine samples. *Science* 292:743–44
149. Poon LL, Chan KH, Wong OK, Cheung TK, Ng I, et al. 2004. Detection of SARS coronavirus in patients with severe acute respiratory syndrome by conventional and real-time quantitative reverse transcription-PCR assays. *Clin. Chem.* 50:67–72
150. Proenca-Modena JL, Macedo IS, Arruda E. 2007. H5N1 avian influenza virus: an overview. *Braz. J. Infect. Dis.* 11:125–33
151. Prowse CV. 2003. An ABC for West Nile virus. *Transfus. Med.* 13:1–8
152. Pybus OG, Rambaut A, Harvey PH. 2000. An integrated framework for the inference of viral population history from reconstructed genealogies. *Genetics* 155:1429–37

153. Qin L, Xiong B, Luo C, Guo ZM, Hao P, et al. 2003. Identification of probable genomic packaging signal sequence from SARS-CoV genome by bioinformatics analysis. *Acta Pharmacol. Sin.* 24:489–96
154. Quan PL, Palacios G, Jabado OJ, Conlan S, Hirschberg DL, et al. 2007. Detection of respiratory viruses and subtype identification of influenza A viruses by GreeneChipResp oligonucleotide microarray. *J. Clin. Microbiol.* 45:2359–64
155. Rabbani GH, Mahalanabis D. 1993. New strains of *Vibrio cholerae* O139 in India and Bangladesh: lessons from the recent epidemics. *J. Diarrhoeal Dis. Res.* 11:63–66
156. Rambaut A, Posada D, Crandall KA, Holmes EC. 2004. The causes and consequences of HIV evolution. *Nat. Rev. Genet.* 5:52–61
157. Rambaut A, Robertson DL, Pybus OG, Peeters M, Holmes EC. 2001. Human immunodeficiency virus. Phylogeny and the origin of HIV-1. *Nature* 410:1047–48
158. Reingold AL. 2000. Infectious disease epidemiology in the 21st century: will it be eradicated or will it reemerge? *Epidemiol. Rev.* 22:57–63
159. Riley LW, Remis RS, Helgerson SD, McGee HB, Wells JG, et al. 1983. Hemorrhagic colitis associated with a rare *Escherichia coli* serotype. *N. Engl. J. Med.* 308:681–85
160. Romling U, Tummler B. 1991. The impact of two-dimensional pulsed-field gel electrophoresis techniques for the consistent and complete mapping of bacterial genomes: refined physical map of *Pseudomonas aeruginosa* PAO. *Nucleic Acids Res.* 19:3199–206
161. Rota PA, Oberste MS, Monroe SS, Nix WA, Campagnoli R, et al. 2003. Characterization of a novel coronavirus associated with severe acute respiratory syndrome. *Science* 300:1394–99
162. Rothman KJ. 1986. *Modern Epidemiology*. Boston: Little, Brown & Co.
163. Ruan YJ, Wei CL, Ee AL, Vega VB, Thoreau H, et al. 2003. Comparative full-length genome sequence analysis of 14 SARS coronavirus isolates and common mutations associated with putative origins of infection. *Lancet* 361:1779–85
164. Rudner R, Studamire B, Jarvis ED. 1994. Determinations of restriction fragment length polymorphism in bacteria using ribosomal RNA genes. *Methods Enzymol.* 235:184–96
165. Sails AD, Swaminathan B, Fields PI. 2003. Utility of multilocus sequence typing as an epidemiological tool for investigation of outbreaks of gastroenteritis caused by *Campylobacter jejuni*. *J. Clin. Microbiol.* 41:4733–39
166. Sanchez A, Rollin PE. 2005. Complete genome sequence of an Ebola virus (Sudan species) responsible for a 2000 outbreak of human disease in Uganda. *Virus Res.* 113:16–25
167. Sanford JP. 1979. Legionnaires' disease: one person's perspective. *Ann. Intern. Med.* 90:699–703
168. Sanger F, Nicklen S, Coulson AR. 1977. DNA sequencing with chain-terminating inhibitors. *Proc. Natl. Acad. Sci. USA* 74:5463–67
169. Satcher D. 1995. Emerging infections: getting ahead of the curve. *Emerg. Infect. Dis.* 1:1–6
170. Schafer JR, Kawaoka Y, Bean WJ, Suss J, Senne D, Webster RG. 1993. Origin of the pandemic 1957 H2 influenza A virus and the persistence of its possible progenitors in the avian reservoir. *Virology* 194:781–88
171. Segal S, Hill AVS. 2003. Genetic susceptibility to infectious disease. *Trends Microbiol.* 11:445–48
172. Sen J. 2001. Taking toll of TB. *Trends Immunol.* 22:297–98
173. Seo SH, Hoffmann E, Webster RG. 2002. Lethal H5N1 influenza viruses escape host antiviral cytokine responses. *Nat. Med.* 8:950–54
174. Seshadri R, Samuel J. 2005. Genome analysis of *Coxiella burnetii* species: insights into pathogenesis and evolution and implications for biodefense. *Ann. NY Acad. Sci.* 1063:442–50
175. Shah NS, Wright A, Bai GH, Barrera L, Boulahbal F, et al. 2007. Worldwide emergence of extensively drug-resistant tuberculosis. *Emerg. Infect. Dis.* 13:380–87
176. Sharp PA. 2005. 1918 flu and responsible science. *Science* 310:17
177. Shi PY. 2003. Genetic systems of West Nile virus and their potential applications. *Curr. Opin. Investig. Drugs* 4:959–65
178. Shi Y, Yi Y, Li P, Kuang T, Li L, et al. 2003. Diagnosis of severe acute respiratory syndrome (SARS) by detection of SARS coronavirus nucleocapsid antibodies in an antigen-capturing enzyme-linked immunosorbent assay. *J. Clin. Microbiol.* 41:5781–82
179. Shi Z, Hu Z. 2008. A review of studies on animal reservoirs of the SARS coronavirus. *Virus Res.* 133:74–87

180. Shinya K, Ebina M, Yamada S, Ono M, Kasai N, Kawaoka Y. 2006. Avian flu: influenza virus receptors in the human airway. *Nature* 440:435–36
181. Shoham D. 2006. Review: molecular evolution and the feasibility of an avian influenza virus becoming a pandemic strain–a conceptual shift. *Virus Genes* 33:127–32
182. Snapinn KW, Holmes EC, Young DS, Bernard KA, Kramer LD, Ebel GD. 2007. Declining growth rate of West Nile virus in North America. *J. Virol.* 81:2531–34
183. Song H-D, Tu C-C, Zhang G-W, Wang S-Y, Zheng K, et al. 2005. Cross-host evolution of severe acute respiratory syndrome coronavirus in palm civet and human. *Proc. Natl. Acad. Sci. USA* 102:2430–35
184. Stadler K, Masignani V, Eickmann M, Becker S, Abrignani S, et al. 2003. SARS–beginning to understand a new virus. *Nat. Rev. Microbiol.* 1:209–18
185. Suarez DL, Perdue ML, Cox N, Rowe T, Bender C, et al. 1998. Comparisons of highly virulent H5N1 influenza A viruses isolated from humans and chickens from Hong Kong. *J. Virol.* 72:6678–88
186. Subbarao K, Klimov A, Katz J, Regnery H, Lim W, et al. 1998. Characterization of an avian influenza A (H5N1) virus isolated from a child with a fatal respiratory illness. *Science* 279:393–96
187. Swanepoel R, Leman PA, Burt FJ, Zachariades NA, Braack LE, et al. 1996. Experimental inoculation of plants and animals with Ebola virus. *Emerg. Infect. Dis.* 2:321–25
188. Takeb EY, Kusagawa S, Motomura K. 2004. Molecular epidemiology of HIV: tracking AIDS pandemic. *Pediatr. Int.* 46:236–44
189. Tang J, Wang C, Feng Y, Yang W, Song H, et al. 2006. Streptococcal toxic shock syndrome caused by *Streptococcus suis* serotype 2. *PLoS Med.* 3:e151
190. Taubenberger JK, Reid AH, Lourens RM, Wang R, Jin G, Fanning TG. 2005. Characterization of the 1918 influenza virus polymerase genes. *Nature* 437:889–93
191. Telenti A, Goldstein DB. 2006. Genomics meets HIV-1. *Nat. Rev. Microbiol.* 4:865–73
192. The Chinese SMEC. 2004. Molecular evolution of the SARS coronavirus during the course of the SARS epidemic in China. *Science* 303:1666–69
193. Theriault S, Groseth A, Artsob H, Feldmann H. 2005. The role of reverse genetics systems in determining filovirus pathogenicity. *Arch. Virol. Suppl.* 2005:157–77
194. Thiermann A. 2004. Emerging diseases and implications for global trade. *Rev. Sci. Tech. Off. Int. Epiz.* 23:701–8
195. Thomson MM, Najera R. 2005. Molecular epidemiology of HIV-1 variants in the global AIDS pandemic: an update. *AIDS Rev.* 7:210–24
196. Tran TH, Nguyen TL, Nguyen TD, Luong TS, Pham PM, et al. 2004. Avian influenza A (H5N1) in 10 patients in Vietnam. *N. Engl. J. Med.* 350:1179–88
197. Treanor JJ, Campbell JD, Zangwill KM, Rowe T, Wolff M. 2006. Safety and immunogenicity of an inactivated subvirion influenza A (H5N1) vaccine. *N. Engl. J. Med.* 354:1343–51
198. Tripet B, Howard MW, Jobling M, Holmes RK, Holmes KV, Hodges RS. 2004. Structural characterization of the SARS-coronavirus spike S fusion protein core. *J. Biol. Chem.* 279:20836–49
199. Trunz BB, Fine P, Dye C. 2006. Effect of BCG vaccination on childhood tuberculous meningitis and miliary tuberculosis worldwide: a meta-analysis and assessment of cost-effectiveness. *Lancet* 367:1173–80
200. Tsenova L, Harbacheuski R, Sung N, Ellison E, Fallows D, Kaplan G. 2007. BCG vaccination confers poor protection against *M. tuberculosis* HN878-induced central nervous system disease. *Vaccine* 25:5126–32
201. Tu C, Crameri G, Kong X, Chen J, Sun Y, et al. 2004. Antibodies to SARS coronavirus in civets. *Emerg. Infect. Dis.* 10:2244–48
202. Tumpey TM, Basler CF, Aguilar PV, Zeng H, Solorzano A, et al. 2005. Characterization of the reconstructed 1918 Spanish influenza pandemic virus. *Science* 310:77–80
203. Ungchusak K, Auewarakul P, Dowell SF, Kitphati R, Auwanit W, et al. 2005. Probable person-to-person transmission of avian influenza A (H5N1). *N. Engl. J. Med.* 352:333–40
204. Viseshakul N, Thanawongnuwech R, Amonsin A, Suradhat S, Payungporn S, et al. 2004. The genome sequence analysis of H5N1 avian influenza A virus isolated from the outbreak among poultry populations in Thailand. *Virology* 328:169–76
205. Vogler AJ, Keys CE, Allender C, Bailey I, Girard J, et al. 2007. Mutations, mutation rates, and evolution at the hypervariable VNTR loci of *Yersinia pestis*. *Mutat. Res.* 616:145–58

206. Wain LV, Bailes E, Bibollet-Ruche F, Decker JM, Keele BF, et al. 2007. Adaptation of HIV-1 to its human host. *Mol. Biol. Evol.* 24:1853–60
207. Wan XF, Chen G, Luo F, Emch M, Donis R. 2007. A quantitative genotype algorithm reflecting H5N1 Avian influenza niches. *Bioinformatics* 23:2368–75
208. Weber B. 2005. Genetic variability of the S gene of hepatitis B virus: clinical and diagnostic impact. *J. Clin. Virol.* 32:102–12
209. Wong S, Lau S, Woo P, Yuen KY. 2007. Bats as a continuing source of emerging infections in humans. *Rev. Med. Virol.* 17:67–91
210. Wong SK, Li W, Moore MJ, Choe H, Farzan M. 2004. A 193-amino acid fragment of the SARS coronavirus S protein efficiently binds angiotensin-converting enzyme 2. *J. Biol. Chem.* 279:3197–201
211. Xu X, Subbarao, Cox NJ, Guo Y. 1999. Genetic characterization of the pathogenic influenza A/Goose/Guangdong/1/96 (H5N1) virus: similarity of its hemagglutinin gene to those of H5N1 viruses from the 1997 outbreaks in Hong Kong. *Virology* 261:15–19
212. Xu Y, Lou Z, Liu Y, Pang H, Tien P, et al. 2004. Crystal structure of severe acute respiratory syndrome coronavirus spike protein fusion core. *J. Biol. Chem.* 279:49414–19
213. Yob JM, Field H, Rashdi AM, Morrissy C, van der Heide B, et al. 2001. Nipah virus infection in bats (order Chiroptera) in peninsular Malaysia. *Emerg. Infect. Dis.* 7:439–41
214. Yoshikawa A, Gotanda Y, Itabashi M, Minegishi K, Kanemitsu K, Nishioka K. 2005. HBV NAT positive [corrected] blood donors in the early and late stages of HBV infection: analyses of the window period and kinetics of HBV DNA. *Vox Sang.* 88:77–86
215. Yu D, Li H, Xu RH, He JF, Al E. 2003. Prevalence of IgG antibody to SARS-associated coronavirus in animal traders-Guangdong Province, China. *MMWR* 52:986–87
216. Yu H, Jing H, Chen Z, Zheng H, Zhu X, et al. 2006. Human *Streptococcus suis* outbreak, Sichuan, China. *Emerg. Infect. Dis.* 12:914–20
217. Yuan FF, Boehm I, Chan PK, Marks K, Tang JW, et al. 2007. High prevalence of the CD14–159CC genotype in patients infected with severe acute respiratory syndrome-associated coronavirus. *Clin. Vaccine Immunol.* 14:1644–45
218. Zhang CY, Wei JF, He SH. 2006. Adaptive evolution of the spike gene of SARS coronavirus: changes in positively selected sites in different epidemic groups. *BMC Microbiol.* 6:88
219. Zhang H, Zhou G, Zhi L, Yang H, Zhai Y, et al. 2005. Association between mannose-binding lectin gene polymorphisms and susceptibility to severe acute respiratory syndrome coronavirus infection. *J. Infect. Dis.* 192:1355–61
220. Zhang R, Zhang CT. 2006. The impact of comparative genomics on infectious disease research. *Microbes Infect.* 8:1613–22
221. Zhao GP. 2007. SARS molecular epidemiology: a Chinese fairy tale of controlling an emerging zoonotic disease in the genomics era. *Philos. Trans. R. Soc. London Ser. B* 362:1063–81
222. Zhi L, Zhou G, Zhang H, Zhai Y, Yang H, et al. 2007. Lack of support for an association between *CLEC4M* homozygosity and protection against SARS coronavirus infection. *Nat. Genet.* 39:692–94
223. Zhou D, Han Y, Song Y, Huang P, Yang R. 2004. Comparative and evolutionary genomics of *Yersinia pestis*. *Microbes. Infect.* 6:1226–34
224. Zhou D, Han Y, Song Y, Tong Z, Wang J, et al. 2004. DNA microarray analysis of genome dynamics in *Yersinia pestis*: insights into bacterial genome microevolution and niche adaptation. *J. Bacteriol.* 186:5138–46
225. Zhou H, Jin M, Chen H, Huag Q, Yu Z. 2006. Genome-sequence analysis of the pathogenic H5N1 avian influenza A virus isolated in China in 2004. *Virus Genes* 32:85–95
226. Zhou JY, Shen HG, Chen HX, Tong GZ, Liao M, et al. 2006. Characterization of a highly pathogenic H5N1 influenza virus derived from bar-headed geese in China. *J. Gen. Virol.* 87:1823–33
227. Zhu Q, Yang H, Chen W, Cao W, Zhong G, et al. 2008. A naturally occurring deletion in its NS gene contributes to the attenuation of an H5N1 swine influenza virus in chickens. *J. Virol.* 82:220–28

NOTE ADDED IN PROOF

While this manuscript was in preparation, a recent reverse genetic study indicated that recombinant AIV strains containing NS1 C-terminal sequences of X-S/T-X-V, such as the 1918 H1N1 and the high-virulence H5N1, were highly pathogenic in the infected mice compared to the virulence of the wild-type A/WSN/33 virus (background).

Jackson, D., M. J. Hossain, et al. 2008. A new influenza virus virulence determinant: the NS1 protein four C-terminal residues modulate pathogenicity. *Proc. Natl. Acad. Sci. USA* 105: 4381–86.

ENU Mutagenesis, a Way Forward to Understand Gene Function

Abraham Acevedo-Arozena,[1] Sara Wells,[2] Paul Potter,[2] Michelle Kelly,[1] Roger D. Cox,[1] and Steve D.M. Brown[1]

[1]MRC Mammalian Genetics Unit, Harwell, Oxfordshire, OX11 0RD, United Kingdom; email: a.acevedo@har.mrc.ac.uk, m.kelly@har.mrc.ac.uk, r.cox@har.mrc.ac.uk, s.brown@har.mrc.ac.uk

[2]MRC Mary Lyon Center, Harwell, Oxfordshire, OX11 0RD, United Kingdom; email: s.wells@har.mrc.ac.uk, p.potter@har.mrc.ac.uk

Annu. Rev. Genomics Hum. Genet. 2008. 9:49–69

First published online as a Review in Advance on May 12, 2008

The *Annual Review of Genomics and Human Genetics* is online at genom.annualreviews.org

This article's doi:
10.1146/annurev.genom.9.081307.164224

1527-8204/08/0922-0049$20.00

Key Words

phenotype-driven screen, phenotyping, gene-driven screen

Abstract

Arguably, the main challenge for contemporary genetics is to understand the function of every gene in a mammalian genome. The mouse has emerged as a model for this task because its genome can be manipulated in a number of ways to study gene function or mimic disease states. Two complementary genetic approaches can be used to generate mouse models. A reverse genetics or gene-driven approach (gene to phenotype) starts from a known gene and manipulates the genome to create genetically modified mice, such as knockouts. Alternatively, a forward genetics or phenotype-driven approach (phenotype to gene) involves screening mice for mutant phenotypes without previous knowledge of the genetic basis of the mutation. N-ethyl-N-nitrosourea (ENU) mutagenesis has been widely used for both approaches to generate mouse mutants. Here we review progress in ENU mutagenesis screening, with an emphasis on creating mouse models for human disorders.

ES cells: embryonic stem cells

KO: knockout

N-ethyl-N-nitrosourea (ENU): a chemical used for randomly mutating the genome

INTRODUCTION

The analysis of organisms that carry genetic mutations has been fundamental to research into human disease states. The mouse has emerged as a key organism in this endeavor. The mouse is easy to breed with a relatively short generation time and well-studied genetics, including a complete genome sequence and a variety of embryonic stem (ES) cell–based gene manipulation technologies. Moreover, many of the pathological consequences of genetic mutation appear to parallel the disease states seen in humans. Indeed, many aspects of murine biology, from basic physiology to behavior, can be mapped directly to humans. In many cases it has been possible to recreate the disease-causing mutation in mice. Different genetic approaches have been used to generate mouse models of human disease. Creating a null or altered mutation in mice involves the manipulation of mouse embryonic stem cells, resulting in so-called knockout (KO) mice in which a specific gene has been inactivated or knockin mice in which a more subtle gene alteration has been introduced. However, these gene-driven strategies to some degree require a priori assumptions about the likely function of a gene and its role in disease. By contrast, phenotype-driven approaches, including chemical mutagenesis, offer a complementary way of producing mouse models of human diseases. In this approach, point mutations are generated at random throughout the genome and mutant mice are analyzed for disease phenotypes of interest. No assumptions are made about the underlying genetic causes of the disease, and therefore we can expect the phenotype-driven approach to be an efficient method for revealing novel genes or genetic pathways involved in pathobiological mechanisms. As well as the generation of null alleles, partial loss-of-function and gain-of-function alleles may be recovered, thereby potentially mirroring the genetic alterations seen in human patients (4, 20, 51, 59, 66).

In this review we discuss the application of mouse mutagenesis strategies for screening for models of human disease. Because the majority of mutants are identified by virtue of their phenotype, we also describe the wide use of phenotype-driven screens and compare and contrast the phenotypes seen in mutagenesis programs to those obtained by gene-targeting experiments.

ENU MUTAGENESIS IN THE MOUSE

Actions of ENU

In the late 1970s, Bill Russell of the Oak Ridge National Laboratories discovered that the chemical N-ethyl-N-nitrosourea, commonly known by the acronym ENU, is the most potent agent for causing mutations in mice. He stated: "Ethylnitrosourea is clearly the mutagen of choice for the production of any kind of desired new gene mutation in mice" (86).

ENU acts primarily through its alkylating moiety, transferring its ethyl group to any of a number of identified nucleophilic nitrogen or oxygen sites on each of the four deoxyribonucleotides (51). The transferred ethyl group constitutes a DNA adduct that during cell proliferation and DNA replication results in heritable mutations (6, 70). Of course, this assault on DNA structure does not go unchallenged by the cellular DNA repair mechanisms, including O^6-alkylguanine-DNA alkyltransferase and nucleotide excision repair, which play an important role in the final spectrum and extent of the mutations induced. ENU treatment is largely ineffectual below a threshold dose because of the DNA repair mechanisms of the cell (27, 91). Once these systems reach saturation, the mutation rate increases in a dose-dependent manner to a level where it is detrimental to the fertility (spermatogonial mutations) or the health (somatic mutations) of the treated animal and therefore decreases the ability of the animal to pass on the mutations to its offspring. Systems that are deficient for repair pathways (such as O^6-alkylguanine-DNA alkyltransferase, nucleotide excision, and mismatch repair mechanisms) display increased mutagenicity with ENU, implicating their roles in modulating the cellular affects of this chemical in intact organisms (9, 15).

The action of ENU on DNA results, in the vast majority of cases, in a nucleotide substitution. However, in a minority of cases the cause of the mutation has been identified as a small deletion (51). Point mutations induced by ENU exhibit a strong bias toward specific nucleotide substitutions. By far the most common site for the occurrence of mutations is at A-T base pairs. Between 70% and 85% of all ENU-induced nucleotide substitutions are estimated to be either A-T to T-A transversions or A-T to G-C transitions (69, 70, 94). Conversely, the G-C to C-G transversion event is rarely seen (1). Following translation into proteins, these substitutions result in approximately 70% nonsynonymous changes of which approximately 65% are missense changes and the remainder are nonsense or splice mutations (51, 94).

The strong bias of ENU toward specific base substitutions may eventually be a limiting factor because it obviously causes a particular class of amino acid change more effectively than others, resulting in some genes or domains being mutated more frequently than others depending upon their nucleotide composition. Therefore, because of the low frequency of G-C to C-G transversions, some amino acid changes will be underrepresented. There may also be a bias involving the strand location or context of the altered residue; a thymine on the nontranscribed strand (92, 94) or sites flanked by a G or C residue within genes with a higher G + C content (5) may be favored sites of alkylation. However, it is evident from the widespread successful use of ENU over the past three decades that these biases have yet to limit the successful identification of new mutant models (4).

ENU Treatment and Mutation Rates

ENU is usually administered by a sequence of intraperitoneal injections to adult male mice. The doses and injection regimes of ENU necessary for the most efficient induction of heritable mutations in different strains of mice have largely been elucidated (42, 61, 84, 105). A typical treatment schedule might comprise two or three injections a week apart at a dose of approximately 80–100 mg/kg for an adult male mouse. However, the optimum dose and regime vary according to the strain of mouse being used (50). As a mutagenic agent, ENU is toxic and carcinogenic, so an increase in dose may result in a higher mutation load but it may also cause a concomitant reduction in the viability of the mouse or its reproductive success (50, 105).

The mutation rate for any given gene varies significantly according to the gene size; larger genes will have a higher chance of being mutated by virtue of their larger size. Sequence-based approaches estimate that an optimal dose protocol results in one mutation every 1–1.5 Mb (18, 19, 63, 74, 87).

A specific-locus test in which ENU-treated males are mated to females of a test strain carrying homozygous alleles for easily detectable loss-of-function mutations (such as eye and coat color) has been used to estimate the phenotypic hit rate of ENU mutagenesis. On the basis of these tests optimum dosing protocols are estimated to result in one loss-of-function mutation in a specific gene for every 700 gametes (42). Extrapolating these data to estimate projected rates for observing a particular disease state is difficult because there are many confounding issues that affect the detection of mutant phenotypes. These issues include the effectiveness of the screening assay and the number of genes within a system that when mutated might cause a disease endpoint. Predicting mutation rates on the basis of phenotypic assessment is also complicated by the fact that ENU treatment can result in many different types of alleles besides point mutations, including hypomorphs, hypermorphs, neomorphs, and splice variants (69).

Missense mutation: a point mutation that results in the alteration of an amino acid in the protein product of the gene

Hypomorph: a mutation that leads to a partial loss but not complete abolition of gene function

Hypermorph: a mutation that leads to an increase in normal gene function

Neomorph: a mutation that leads to a dominant gain of function

BREEDING STRATEGIES

One of the primary sites for the mutagenic action of ENU in the mouse is the premeiotic spermatogonial stem cell (77, 85). Following injection, male mice undergo a period of sterility owing to the depletion of differentiated

G1: first generation of mutagenized mice

G3: third generation of mutagenized mice

spermatogonia caused by the ENU treatment (85). Indeed, the period of sterility is widely used as an indicator of effective treatment. A return of fertility prior to 10 weeks after injection indicates that the ENU treatment may not have been fully effective.

The breeding schemes used for propagation of mutations vary according to the allelic characteristics required (whether screening for dominant or recessive mutations) and the strains required for subsequent gene mapping. The ENU-treated males (G0) are crossed with wild-type females to produce G1 individuals that can be assayed for dominant mutations (**Figure 1**). To screen for recessive mutations, pedigrees are bred by intercrossing the offspring of a G1 individual (G2) or crossing them back to the original G1, thus making homozygous (homozygosing) mutations in a proportion of the resultant G3 offspring (**Figure 2**).

Because ENU acts randomly at multiple sites throughout the genome, a single G1 individual is estimated to contain approximately 30–50 potential functional mutations throughout its genome (55). Therefore, in subsequent generations of mice, bred by virtue of their phenotype, the mutant line will also retain some residual mutations that can potentially confound physiological analysis. However, this is rarely a problem in practical terms because the mutation load decreases by approximately one half every generation. Moreover, once the mutation has been mapped to a region of a chromosome approximately 5 Mb in size, the likelihood of a second mutation in this area is also very low (55). Elimination of residual mutations in a mutant line can be further enhanced by techniques such as marker-assisted selection, which allows the selection of breeding stock that contains the smallest amount of DNA from the strain originally mutagenized (101) and therefore the lowest number of additional mutations.

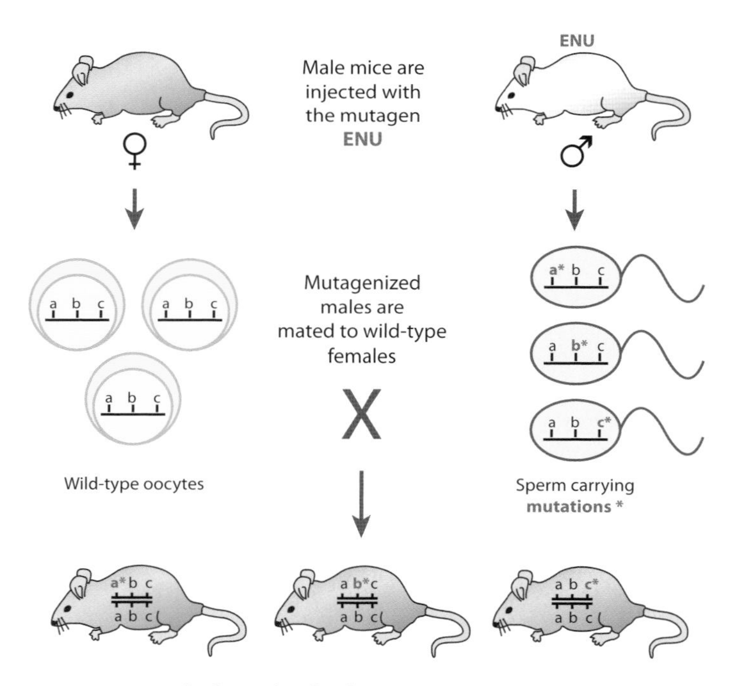

Figure 1

Breeding mice to screen for dominant mutations. Male mice are treated with N-ethyl-N-nitrosourea (ENU) and after a period of sterility are mated to wild-type females. The point mutations induced in the DNA carried by spermatogonial cells are therefore transmitted to the G1 generation, which is subsequently screened for aberrant phenotypes. Coat colors are shown as different to stress that the mutagenized strain and the wild-type females should be from different inbred strains so the next backcross can be used for mapping purposes.

SCREENING FOR MUTANTS: PHENOTYPE DRIVEN APPROACH

Phenotype-Driven Approaches for Screening for Mutant Models

Many researchers have used forward genetics—the phenotype-driven approach—to identify mouse models of disease states (43, 46, 67, 71). The fundamental principle behind phenotype-driven mutagenesis screens is that no assumptions are made about the genetic basis or the etiopathology of a particular disorder; models are identified purely on the basis of whether they express particular phenotypes associated with human conditions. Thus, within the constraints of phenotyping capabilities, it is theoretically possible to generate murine models of almost any phenotype or disease. Such screens

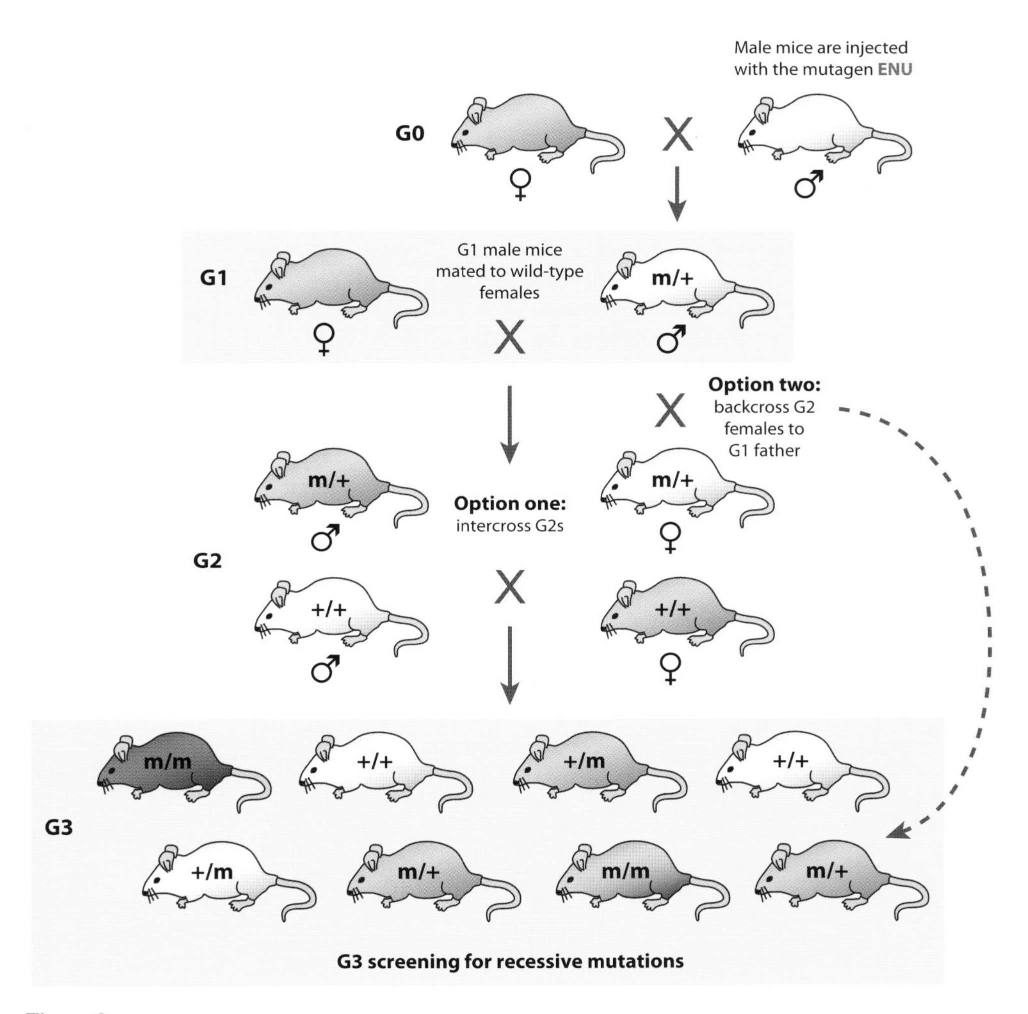

Figure 2

Breeding recessive pedigrees. Male mice are treated with N-ethyl-N-nitrosourea (ENU) and after a period of sterility are mated to wild-type females. G1 male mice, heterozygous for N-ethyl-N-nitrosourea (ENU)-induced mutations, are mated to wild-type females. Their offspring (G2) are then either intercrossed or mated back to the original G1 to homozygose mutations. Recessive and dominant mutations can then be detected in the resultant G3 progeny. Coat colors are shown as different to stress that the mutagenized strain and the wild-type females should be from different inbred strains so G3 mice can be used for mapping purposes.

can also identify novel pathways, genes, or functions for genes involved in disease pathogenesis. The development of more standardized and validated protocols for mouse phenotyping has and will continue to advance the development of mouse screening (10, 11).

Early screens employed visible phenotypes or lethality to identify mutations (86), but increasingly complex screening systems are being applied to mutagenesis programs. A variety of strains, ENU dosing regimes, breeding strategies, and phenotypic screens have been employed successfully by a number of researchers, resulting in a wide variety of mouse models. Different strains present different ENU sensitivities but the genetic background might also sensitize different strains to develop certain phenotypes. Outlined below are examples of the various strategies used in selected screens past and present.

Phenotyping Pipelines

As a typical example, the initial pipeline of mice generated at the Medical Research Council (MRC) in Harwell, United Kingdom involved mutagenizing BALB/c males and crossing them to C3H mice, the offspring of which were screened at 12 weeks for dominant mutations, including those identified in dysmorphology, SHIRPA (Smith Kline Beecham, Harwell, Imperial College, Royal London Hospital phenotype assessment), behavioral screens, and clinical chemistry screens (67). The SHIRPA screening protocol is a phenotyping platform designed to detect outliers in motor neuron, muscle, CNS, and sensory function (80). In total more than 28,600 progeny were screened, 193 mutations were confirmed as being inherited, and 86 mutations were mapped. Since 2004 this pipeline has concentrated on using C57BL/6 as the mutagenized strain. The screening strategies employed at MRC Harwell emphasize the wide range of phenotyping tests that can be used to maximize the identification of mutants from a single pipeline of mice; each mouse goes through a battery of primarily noninvasive screens (**Figure 3**). Other large-scale screens have taken place at the Jackson Laboratory, United States (17), HMGU (HelmholtzZentrum Munchen), Germany (46), and Riken, Japan (87).

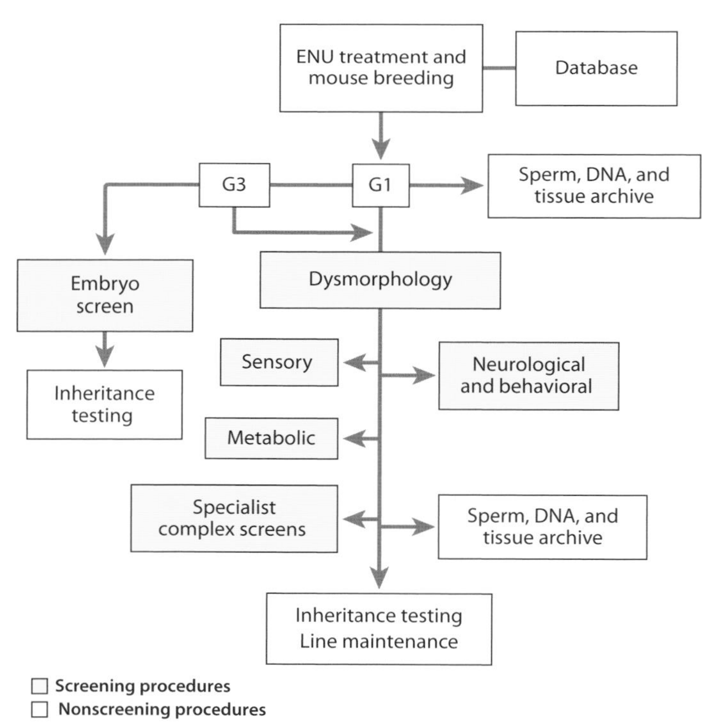

Figure 3

Mutagenesis pipelines. To maximize the use of potential mutants, mice bred from N-ethyl-N-nitrosourea (ENU)-treated individuals can be screened using tests and assays from a number of disciplines. Sperm, DNA, and tissue can be archived from G1 males, whereas pedigrees screened for recessive embryonic phenotypes can also be examined for postnatal abnormalities.

In addition to the standard mutagenesis pipelines that employ hybrid mice, some programs involve mutagenizing and breeding using a single pure inbred background. This strategy is sometimes advantageous in the identification of mutants, because on an inbred background less intrinsic phenotypic variation would be expected. However, an outcross is required to map the mutation, and therefore the mutant phenotype must be robust enough to be identifiable on a hybrid background.

The value of genetically altered mice as models for disease relies heavily on downstream research capabilities to fully characterize the animals generated. To maximize the potential of mouse resources a number of centers have developed pipelines of tests in which cohorts of mice are comprehensively phenotyped through a battery of largely noninvasive complementary tests. The concept of a mouse clinic has emerged, pioneered by the German Mouse Clinic (31), where a range of phenotyping platforms and expertise is available. There are significant advantages to mouse clinics that employ robust and standardized screens such as the European Mouse Phenotyping Resource of Standardized Screens (EMPReSS; **http://empress.har.mrc.ac.uk**). This screen is being used by the European Mouse Disease Clinic (EUMODIC; **http://www.eumodic.org/**) consortium to enable high-throughput, primary phenotyping of large numbers of mutant mouse lines, and the data are available on a public access database (**http://www.europhenome.org**).

Subsequent to the primary screens, mouse lines might undergo specialist secondary phenotyping analysis to further characterize the mutant phenotype. Examples include auditory brainstem response (ABR) for deafness mutants (89), X-ray analysis for skeletal deficits (45), screens for vision deficits (12, 39, 95), and comprehensive metabolic phenotyping (44, 47). However, more sophisticated tests can be employed as a primary screen, so the division between primary and secondary phenotyping is increasingly blurred.

The development of new and more advanced mouse phenotyping tests relevant to human disease states is pivotal. There is currently much progress on adapting technologies used for human and veterinary diagnosis to the mouse. This progress is exemplified by the current application of imaging techniques such as magnetic resonance imaging to define mouse anatomical lesions, both in the adult and embryo (73, 88) and dual energy X-ray analysis (DEXA) to measure mouse body composition (76).

Hierarchical phenotype screens that employ primary screens and more sophisticated secondary and tertiary screens have been used to identify mutants that are affected in virtually all body systems. Sometimes simple screens have revealed unexpected phenotypes that represent novel models of human disease. For example, high-throughput screening that employs a high-frequency tone burst from a clickbox and assesses the Preyer reflex has uncovered a new class of mutations that result in the development of otitis media (middle ear inflammatory disease) (37, 72). Similarly simple dysmorphology analyses have revealed many useful neurological, sensory, and motor mutants (30).

Phenotype screen: a test for a specific biological characteristic

As an example, for neurological mutations large-scale phenotype screens have traditionally focused on overt phenotypes such as tremors, gait abnormalities (ataxias), locomotor activity, or sensorimotor gating. These ongoing screens have been successful in creating new mouse models of complex neurological disorders (**Table 1**) and implicated new genes in different processes of brain biology from development to neurodegeneration.

Table 1 ENU-induced mutants with adult neurological and behavioral phenotypes

Gene	Mutation type	Phenotype	Reference
Af4	Missense	Cerebellar degeneration/ataxia	(49)
Cacna1a	Missense	Ataxia	(108)
Chd7	Multiple	CHARGE syndrome	(8)
Clock	Splicing	Circadian homeostasis	(60)
Disc1	Missense (2)	Depression and schizophrenia	(16)
Dync1h1	Missense (2)	Motor neuron degeneration	(36)
Fblx3	Missense (2)	Circadian homeostasis	(32)
Gars	Missense	Grip-strength deficit, neuronal loss	E.M.C. Fisher, personal communication
Glra1	Missense	Startle disease	(98)
Kcnq2	Missense	Seizures	(54)
Lyst	Multiple	Cerebellar degeneration	(83)
Pcdh15	Splicing	Cochlear-ganglion degeneration	(104)
Pde6b	Multiple	Retinal degeneration	(39)
Pmp22	Multiple	Peripheral neuropathy	(48)
Quaking	Multiple	Dysmyelination	(68)
Rab3a	Missense	Circadian homeostasis	(52)
Scn8a	Missense (3)	Hind limb paralysis	(12)
Tuba1	Missense	Hyperactivity	(56)

One example of a more complex primary screen in which mutations have been identified is modification of the environmental conditions by manipulation of light/dark cycles to identify genes involved in circadian rhythm homeostasis (3). Mutations that demonstrate insulin resistance were identified by challenging mice with a bolus of glucose and determining the clearance dynamics (97).

Recessive pedigree screening. A screen for immunological phenotypes was undertaken at the John Curtain School of Medical Research in Canberra, Australia (65). This recessive screen primarily employed fluorescence-activated cell sorting (FACS) and immune challenges to probe recessive pedigrees for abnormal immune cell development and defective or altered adaptive immune responses (100). Initially, this screen employed a breeding strategy in which each recessive pedigree was derived from a single ENU injected male but, to increase the number of potential mutations per pedigree and to screen for mutations on the X chromosome, recessive pedigrees were generated by crossing unrelated male and female G1 mice and intercrossing their progeny (reviewed in 65 and 100). The phenotypic screen consisted of a basic FACS screen of immune cell populations, with a particular emphasis on screening for abnormal development of T cell subsets and immune challenges. The immunizations employed were designed to examine the nature of the humoral immune response and determine whether one, both, or neither of the Th1 and Th2 responses were affected, as well as determine the amplitude, specificity, and memory induction of the immunoglobulin response. From 185 pedigrees a series of 11 mutants were identified in genes involved in T cell development (71), one of which affected a novel gene. *San Roque* mutants harboring a mutation on the E3 ubiquitin ligase *Roquin* led to the identification of this protein as a key regulator of the mechanism that inhibits autoreactive T cell responses (99). Further screens for autoimmune biomarkers, with a particular interest in markers of systemic lupus erythematosus (SLE) such as antinuclear antibodies (ANA) and renal pathology, were also employed (100).

Researchers at the Scripps Research Institute used both in vitro and in vivo phenotyping to characterize a number of mutants that influence the innate immune response. A modification of the recessive pedigree strategy was employed in a screen in which micropedigrees of six offspring from each G1 male were screened. Such pedigrees provide a theoretical 50% chance of identifying a recessive mutation. The reduced likelihood of identifying mutations is balanced by lower costs and the ability to carry out more intensive and time-consuming screening protocols, including susceptibility to infection and tumor challenge with in vitro analysis of Toll-like receptor function (43).

Screening for embryonic mutants. Together with the screening of adult mice, screens for recessive mutations that affect mouse embryonic development have been widely used. As mammals, mice, and humans share the majority of their key developmental programs, it is possible to model human developmental defects such as heart or lung malformations in the mouse. Different laboratories have focused on different embryo stages or systems, creating a vast array of new mouse developmental mutants. Researchers at Brigham and Women's Hospital used a screen for abnormalities in organogenesis at embryonic day (E) 18.5 to identify several new mutant lines with defects ranging from cleft palate to hypoplastic lungs and diaphragm (41). Researchers at the Memorial Sloan-Kettering Cancer Center used a screen for genes that control early mammalian embryogenesis at E9.5 to identify new genes essential for neural tube closure or Sonic Hedgehog signaling (26).

Together with genome-wide mutagenesis strategies, region-specific screens for embryonic defects have been carried out. Researchers at the Baylor College of Medicine pioneered the use of balancer chromosomes (40) in phenotypic screens. This approach enables the rapid localization of mutations in a manner similar to that achieved with chromosomal deletions (78).

The balancer chromosome contains a large inversion that prevents recombination across the region and is combined with visible markers (e.g., ear color and coat texture). Use of the balancer chromosome allows a breeding regime in which G3 progeny are either all homozygous for the mutant allele or carry the mutant allele on one chromosome *in trans* to the balancer chromosome, thus enabling the easy identification of homozygous lethal mutations and their subsequent maintenance. The initial screen identified mutations leading to defects in fertility, craniofacial and cardiovascular development, and hematopoeisis (58). All 88 mutations in the balancer chromosome region were mapped, tripling the number of models with mutations in this region, whereas only 12 of the 142 mutations occurring outside the balancer region were mapped within the same time frame (40).

Researchers at MRC Harwell have used another region-specific strategy. Monosomy of the human chromosome 6p terminal region results in a variety of congenital malformations including brain, craniofacial, and organogenesis abnormalities. A mouse line with a deletion of a large block of proximal mouse chromosome 13, syntenic to human 6p (Del(13)Svea36H), was used to identify recessive mouse mutations relevant to the human conditions. One of the genes located in the region, a Forkhead transcription factor (Foxq1), acts as a deletion marker because homozygous null mutants present a glossy coat. This enables the use of a two-generation genetic screen to identify viable and lethal recessive mutations within the deletion. Using such a strategy, 13 lines were recovered with phenotypes relevant to the human 6p conditions, including two loci that produce holoprosencephaly and another two that produce anophthalmia (7).

PHENOTYPE-DRIVEN SCREENS: REVEALING LOCI INVOLVED IN COMPLEX DISEASE

For many human diseases the genetic basis is complex or multifactorial and likely results from alterations in a number of genes. Nevertheless, null alleles or ENU mutants can provide valuable single-gene models of apparently complex or multifactorial disorders. For example, the *Jeff* mutant, carrying a mutation in *Fbxo11* (a specificity component of the SCF E3 ubiquitin ligase), provides an excellent example of how ENU mutagenesis can create a single-gene model of a complex, multifactorial disease such as otitis media (38). Moreover, Fbxo11 has recently been demonstrated to show nominal evidence of association with the development of chronic otitis media in an American population (90). ENU mutations recovered in Disrupted in Schizophrenia 1 (*Disc1*), which has been previously implicated in schizophrenia, appear to provide models of both depression and schizophrenia (16).

It is also possible to investigate the complex multifactorial nature of disease and elaborate the components of the genetic pathways involved by the use of sensitized/modifier or challenge screens. The number of disease models available continues to increase exponentially owing to the ongoing large-scale KO projects as well as large-scale ENU programs. We can take advantage of the increasing plethora of disease mouse models available to try to identify new genes/pathways involved in the pathology of disease via the use of ENU sensitized/modifier or challenge screens (**Figure 4**). ENU mutagenesis offers a hypothesis-generating tool to try to uncover new disease mechanisms that might be useful in finding new therapeutic avenues, because quite often upstream and downstream genes make better drug targets than the primary genetic lesion. This type of approach has been successfully used in invertebrate models and is now starting to be pursued in the mouse (13, 82); researchers are exploiting genetic, pharmacological, or environmental modifications.

Depending on the nature of the mouse models available and the type of screen applied, a number of different strategies can be applied. For dominant modifier screens, inbred males (for example C57BL/6) can be mutagenized and then outcrossed to hemi- or heterozygous mutant females on a different genetic

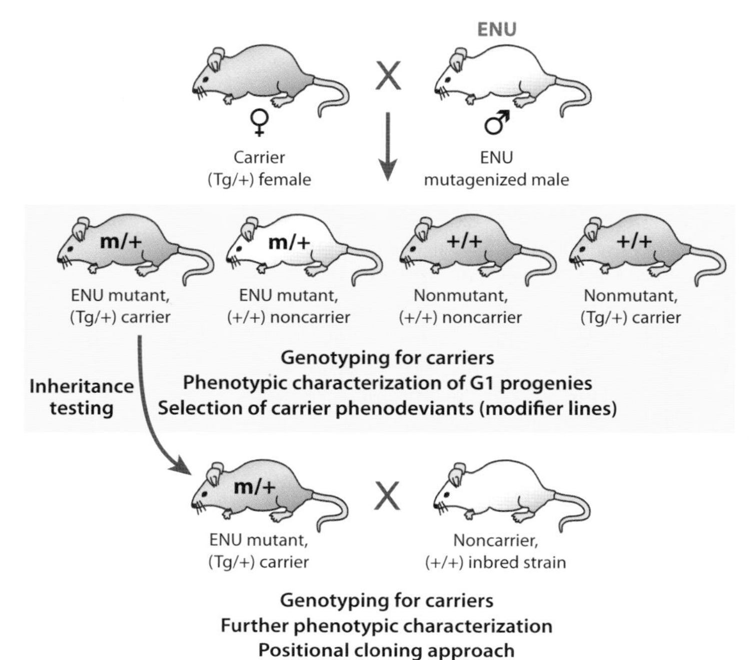

Figure 4

Dominant modifier screens. Mutagenized male mice are crossed to carrier females. In this example, females are hemizygous for a transgene that makes them a disease model. G1 progenies are genotyped for the transgene and screened with a phenotype-driven approach for a modification of the phenotype seen normally on the disease model. Phenodeviants from the G1 should be transgenic carriers and are backcrossed to the nonmutagenized strain for inheritance testing. Coat colors are shown as different to stress that the mutagenized strain and the carrier females should be from a different inbred strain to use the next backcross for mapping purposes.

background. If homozygosity is required to observe the disease phenotype, homozygous mutant males can be mutagenized and crossed to homozygous mutant females. However, the mutant males and females must be on different genetic backgrounds to be able to pursue a positional cloning approach in the second generation. Recessive modifier screens are possible, but they are far more laborious because three generations of breeding are required.

In addition to genetic modifications, the screen can involve a challenge to the mice such as a manipulation of environmental conditions or treatment of the mice. For example, mutations may become apparent only when infectious challenges are employed, thus providing genetic models of host susceptibility to pathogens. Another important parameter is diet. Many human conditions, such as diabetes and atherosclerosis, are exacerbated by diet and thus the identification of models of genetic susceptibility requires screens of mice reared on high-fat diets.

Phenodeviant: an individual or mutant line that shows a deviation from the expected phenotype

We consider here a number of issues involved with modifier and challenge screens, and focus in particular on screens for neurological disorders and infection susceptibility.

Challenge Screens: Infection Susceptibility

Infection susceptibility is controlled by the host immune system. To assess infection susceptibility, a challenge with the pathogen is required. Mouse models offer a good opportunity to dissect the genetics of the host immune response to an infection. Hence, a challenge screen of ENU-mutagenized mice offers the opportunity to dissect molecular pathways involved in the host response to the infection. This kind of approach has been employed on a model of viral infection using mouse cytomegalovirus (MCMV) as the infectious agent (21). To specify mutations affecting the innate (early) immune response an early time point after inoculation was employed as the cutoff point for resistance. To detect phenodeviants that did not succumb within five days to the viral challenge the viral load was determined by measuring plaque-forming units within the spleen, which led to the identification of mice with a reduced innate response. With such a challenge there is a significant risk that mortality will prevent breeding from the susceptible mutants; here the use of pedigrees enables further breeding from the G1 and G2 parents, and also generates enough offspring to map the mutation. The screen identified eight mutations in all; five result in mortality within 2–4 days postinfection, and a further three result in high viral loads by day 5. Some of the mutants are also susceptible to vesicular stomatitus virus, indicating a fundamental defect in the immune response. One of the mutants (*Domino*) was the result of a point mutation in a critical signaling molecule (STAT1) in a domain not previously identified as important in influencing phosphorylation of STAT 1, again emphasizing the advantage of a phenotype-driven screen with no preconceptions about the genetic basis. A second mutant from this screen, *Jinx*, has resulted in a novel model of a human disease, type 3 familial hemophagocytic lymphohistocytosis; mutations in MUNC13–4 are known to be associated with disease, but evidence for an environmental trigger was lacking (22). *Jinx* mice have a mutation in the murine ortholog of MUNC13–4, Unc 13d, but only develop symptoms comparable to the human condition when challenged with lymphocytic choriomeningitis virus (LCMV), thus providing strong evidence of an infectious trigger in the human disease.

Modifier Screens: Neurodegenerative Disorders

The availability of a number of mouse models of neurodegenerative disorders, together with the fact that the pathology of these diseases is not well understood, makes these diseases an ideal target for modifier screens. Even in the case of a paradigmatic dominant disorder such as Huntington's disease (HD), where the disease is caused by a mutation in a single gene, we still do not fully understand how the primary genetic lesion leads to such devastating effects (35). A modifier screen would enable the identification of new genes that suppress or enhance the disease phenotype. ENU can create a range of mutations (from loss- to gain-of-function) and it is possible that different mutations in the same gene might act as either suppressors or enhancers. Both types of modifiers are quite valuable: Suppressors make obvious new drug targets, but enhancers are also worthy of further exploration; together with suppressors they might highlight new disease genes or pathways and lead to new therapeutic targets. As a proof of principle, an ENU mutation (*Loa*) in the cytoplasmic dynein heavy chain (*Dync1h1*, 36) acts as a modifier of two neurodegenerative disorders. *Loa* acts as an enhancer of HD, enhancing the overall pathological and behavioral Huntington's phenotype as well as shortening survival. The *Loa* mutation appears to enhance the HD phenotype by impairing autophagic clearance of mutant Huntingtin (75). In contrast, *Loa* acts as a suppressor of the SOD1 G93A model of amyotrophic lateral sclerosis (ALS), a form of motor neuron disease. *Loa* delays disease progression, significantly increases lifespan, and leads to a complete recovery of the axonal transport deficits in motorneurons of SOD1 mice (57).

Similar to the human diseases, the available neurodegeneration mouse models present progressive disorders; using an ENU strategy we can identify genes that would affect either age at onset and/or the disease progression rate. This distinction is important because, in some diseases like ALS, emerging evidence suggests that different sets of genes and indeed different cell types might be involved in the onset and the progression of the disease (24).

As for any mutagenesis strategy, the phenotype analysis is critical for this modifier screen approach to succeed. Mouse models of neurological disorders usually present complex phenotypes, including a variety of pathological and behavioral symptoms that range from motor to cognitive deficits (81). Because disease phenotypes are complex, investigators often focus on individual traits or endophenotypes (53). Examples of endophenotypes include prepulse inhibition deficits in schizophrenia, tremor onset in HD, and grip-strength deficits in ALS. The ability to dissect a complex neurological phenotype into simpler, less variant endophenotypes makes the analysis of the mutagenized cohorts a simpler task. Different disease-associated genes would likely be involved in the modification of different endophenotypes, allowing us to disentangle the genotype-phenotype correlations. Still, by its very nature, behavioral analysis is complex and variable. The ability to find single phenodeviants using an ENU modifier screen may be compromised by the intrinsic variation of the phenotype in the disease mouse model. Importantly, in contrast to other mutagenesis strategies, all mice, including non-ENU mutants, will

Penetrance: the proportion of individuals carrying a mutation that displays the related phenotype

present a disease phenotype of their own. It is therefore critical to extensively characterize the disease-associated phenotype to choose the most appropriate phenotypes for the modifier screen. This should be done on the same genetic background as the G1 population to minimize and account for phenotypic variation. The baseline data obtained provide the boundaries of the normal phenotypic variation of the nonmutagenized population of the mouse model. Phenodeviants would represent single mice that present an extreme phenotype that reflects the modulation of the complex disease-associated phenotype over time.

MAPPING AND CLONING MUTATIONS

Once an ENU mutant is selected and inheritance is established, a positional cloning approach is used to identify the gene responsible for the new phenotype. Although each G1 animal carries on average approximately 30–50 functional mutations, in the vast majority of cases the mutant phenotype has been proven to be the result of a mutation in the coding region of a single gene (5). For inheritance testing in dominant screens, a backcross to the nonmutagenized strain is used. Depending on the screening strategy, progenies from this cross might be useful for mapping purposes; if on the G0 and inheritance testing crosses the nonmutagenized strain is different from the mutagenized strain, then progenies are from a second generation backcross and can be used for mapping. If on the contrary the same inbred strain is used for the G0 mating, then inheritance can be tested on the G1 inbred background, but two extra generations of mating to a different inbred strain would be needed to generate usable mutant mice for mapping. Regardless of the strategy, penetrance can be inferred through inheritance testing; it is not unusual for ENU mutations to present with incomplete penetrance. Assuming that a new dominant phenotype is caused by a single fully penetrant mutation, 50% of the progenies of an affected animal would inherit the mutation if it lies on an autosome. X-linked mutations can be identified only on female founders because on G1 males the X chromosome would arise from the nonmutagenized female.

Once inheritance and penetrance is established, a genome-wide scan can be used for the initial mapping of the causative mutation. Usually, a small number of mutant animals are selected. The analysis is simple because the mutation must be linked with a region of the genome arising from the inbred strain used for the ENU injections. Currently, single nucleotide polymorphisms (SNPs) are the marker of choice owing to the availability of large databases with SNP information from many mouse inbred strains (107). The strategy is basically the same for dominant and recessive screenings. The only difference is that on recessive screens G3 mice would be used and both dominant and recessive mutations might be identified.

Initial linkage is usually obtained to a large region of approximately 20 Mb. For fine mapping, the generation, genotyping, and phenotyping of a large number of backcross or intercross progeny would be needed to narrow down the critical region to less than one Mb. A candidate-gene approach might be followed at any stage of the process to ultimately clone the causative mutation.

SCREENING FOR MUTANTS: GENE-DRIVEN APPROACH

A recent plethora of successful genome-wide association studies in the human population has led to the identification of a number of genes as possible candidates for genetic disease (25, 28, 79, 93, 96, 106), and raises the question of how to assess the involvement of these candidates in model systems. The success of traditional phenotype-driven ENU screens raised the possibility of carrying out a gene-driven approach using ENU and identifying mutations within genes of specific interest in DNA archives established either from ENU-mutagenized mice or embryonic stem cell archives treated with ENU (**Figure 5**). Mutation detection in these archives can be

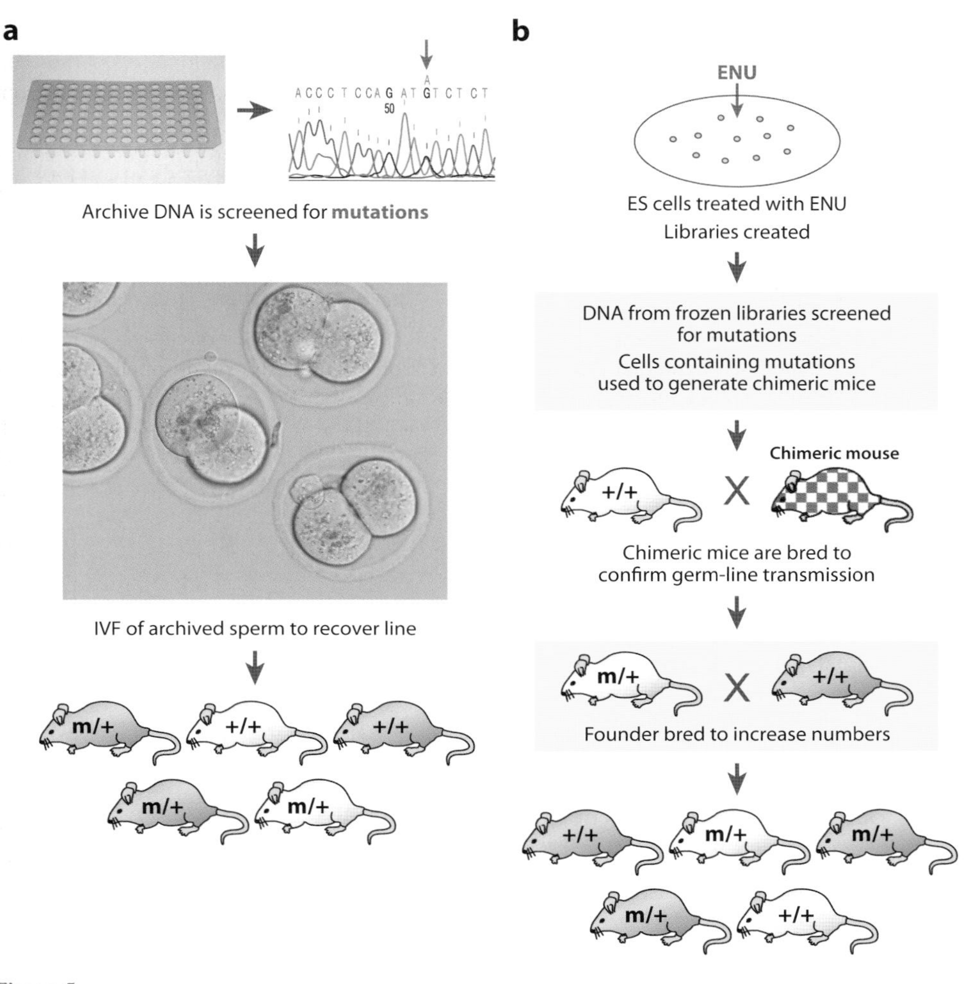
a
ACCCTCCAGATGTCTCT
50
Archive DNA is screened for mutations
IVF of archived sperm to recover line
m/+
+/+
+/+
m/+
m/+
b
ENU
ES cells treated with ENU
Libraries created
DNA from frozen libraries screened for mutations
Cells containing mutations used to generate chimeric mice
Chimeric mouse
+/+
X
Chimeric mice are bred to confirm germ-line transmission
m/+
X
+/+
Founder bred to increase numbers
+/+
m/+
m/+
m/+
+/+

Figure 5

Gene-driven screens for N-ethyl-N-nitrosourea (ENU) mutations. Methods for the screening of ENU mutants on genes of interest are shown. (*a*) A DNA archive of ENU-treated G1 male mice is screened and mutations are recovered by in vitro fertilization (IVF) from the G1 frozen sperm. (*b*) A library of ENU-treated ES cells is screened for ENU nutations. Once a mutated clone is identified, ES cells are microinjected into a mouse blastocyst to create mutant mice that carry the ENU-induced mutation.

carried out relatively rapidly using a number of techniques, including temperature gradient capillary electrophoresis (TGCE), denaturing high-performance liquid chromatography (DHPLC), and direct sequencing (74). The following section discusses two methods available for gene-driven ENU screens: parallel sperm and DNA archives and the ES cell archive.

Sperm and DNA Archives

The first method involves the establishment of parallel archives of DNA and frozen sperm from the G1 progenies of mutagenized males. The DNA archive is screened for mutations in a gene of interest and subsequently the frozen sperm is used to rederive any mice that carry mutations. In 2002 an archive of more than 2000 DNA samples was screened for four genes.

IVF: in vitro fertilization

The screen identified a number of silent, missense, and stop mutations (18). Two missense and one stop mutation were identified in the Connexin 26 gene. The stop mutation was recovered by in vitro fertilization (IVF) and shown to fail to complement a knockout mutation at the same locus. In 2005 an ENU archive of 4000 DNA samples was screened for mutations in 11 different genes; 38 mutations were identified and nine animals were recovered (63). Similarly, 15 mutations, three of which were recovered using IVF, were identified following an archive screen of 1700 DNA samples to identify mutations in three genes belonging to the G protein–coupled receptor superfamily; in total three genes were screened (34)

Many research centers continue to expand their archives, aiming to increase the potential recovery of mutant alleles at any gene of interest. For example, the Harwell DNA archive (74) has grown to include 7500 DNA samples. To date, a total of 49 genes have been screened, 66 mutations have been identified, and 41 mice have been rederived by IVF. Two of these mutations were identified in the nicotinamide nucleotide transhydrogenase (NNT) gene, which has been linked to type 2 diabetes. Both of these mutations were found to result in mice that showed impaired glucose tolerance and reduced insulin secretion (29).

The DNA archive held at Harwell [frozen embryo and sperm archive (FESA)] is one of a number of archives that contain ENU-mutagenized DNA. Other research centers include the RIKEN Bioresource Center, The Australian Phenomics facility, and the German ENU-mouse mutagenesis screening project.

Embryonic Stem Cell Archives

An alternative ENU gene-driven approach is to carry out mutagenesis in ES cells. Mutagenized ES cells can be screened for mutations in a gene of interest and a parallel frozen cell archive can then be used to generate mice by blastocyst microinjection. There are several advantages to this method: Doses of mutagen can be optimized and adjusted without considering viability and animal welfare. Large-scale screens can be carried out using mutagenized ES cells without having to breed and house mice. Another advantage of this method is the possibility of undertaking phenotyping directly on ES cells, either prior to or following differentiation. In 2000 a number of studies (14, 64) showed that ES cells can be efficiently mutagenized in culture and then used to generate chimeras that are capable of germ line transmission. In 2002 ENU-treated ES cells were screened for mutations in *Smad2* and *Smad4* (103). The researchers screened a total of 2060 mutagenized ES cells and identified 29 clones with mutations in the genes of interest. All the mutations were single nucleotide substitutions; 62% were identified as missense alterations in the protein, 7% were found to affect splicing, and the remaining 31% were found to be silent. In 2005 a screen was carried out to identify splice mutants from a library of 40,000 ENU-treated ES cells. A splice mutation in exon 18 of the *Kit* gene was identified and showed germline transmision (33). At least one publicly available resource of ENU-mutagenized ES cells exists: The library held at the University of North Carolina currently banks 3072 ES cell lines.

Advances in high-throughput sequencing raise the possibility that in the not too distant future it will be economically as well as technically feasible to sequence the entire genome of every mouse and ES cell line contained within the mutant archives, generating a database that can be searched to identify an allelic series of mutants for any gene of interest.

ENU AS A COMPLEMENT TO THE LARGE-SCALE KNOCKOUT PROJECTS FOR THE FUNCTIONAL CHARACTERIZATION OF THE MOUSE GENOME

With the ultimate aim of characterizing the function of every gene, three complementary large-scale international projects are currently underway: In Europe, these include the European Conditional Mouse Mutagenesis Program

(EUCOMM) (**http://www.eucomm.org**) (2), the North American Conditional Mouse Mutagenesis Project (NORCOMM) (**http://norcomm.phenogenomics.ca/index.htm**), and the Knockout Mouse Project (KOMP) (**http://www.nih.gov/science/models/mouse/knockout**). The goal of KOMP is to create conditional knockouts for every gene in the mouse genome. The phenotypic characterization of the thousands of lines obtained would present the community with a formidable task; to this end, in Europe an international consortium (EUMODIC, see above) will be characterizing a number of lines created through the EUCOMM project.

ENU mutants represent a powerful complement to the large-scale KO projects. Two main advantages make ENU a perfect companion to the KO projects. The first is its ability to create all kinds of mutations: from loss-of-function (97), gain-of-function (109), and hypomorphs and hypermorphs (102) to dominant negative mutations (62). The second is the availability of sperm archives of ENU-mutagenized males (18) and mutagenized ES cells, allowing the generation of allelic series of point mutations for every mouse gene. Creating an allelic series of mutants for any given gene has the potential to reveal not only novel functionality, but also might point toward critical functional domains within the gene structure.

There are many examples in the literature of ENU mutants (**Table 2**) that show distinct phenotypes from their loss-of-function (KO) counterparts, highlighting ENU's powerful ability to unravel gene function. Examples include ENU mutants that show more robust or different phenotypes in the same system when compared with the corresponding KO. Alternatively, ENU may reveal completely novel phenotypes that affect different tissue systems. A paradigmatic case of a more robust phenotype is the *Clock* gene. This gene was first highlighted as a master regulator of circadian homeostasis via an ENU mutant (102). Years later, the KO mouse surprisingly showed only a mild circadian phenotype (23). An example of gene function revealed in a novel system is the *Af4* gene, a gene previously implicated from studies of the KO in lymphocyte development. However, an ENU mutant showed cerebellar degeneration, revealing a role in Purkinje cells (49).

CONCLUDING REMARKS

In the past few years, the benefits of ENU have ranged widely, from the generation of novel models of human disorders to a better understanding of gene function and complex biological systems. One of the main advantages of ENU mutagenesis is its versatility: ENU mutagenesis generates point mutations that have the potential to reveal a wide array of mutant effects. Moreover, ENU mutagenesis can be used in forward and reverse genetics approaches. Ultimately, ENU mutagenesis can lead to the

Table 2 Differences in ENU versus knockout phenotypes

Gene	ENU phenotype	Knockout phenotype
Af4	Small size, ataxia, and cataracts Mild T cell maturation deficits	Impaired B/T cell development Some small size
Clock	Profound circadian deficits	Mild circadian impairment
Col4a1	Abnormal skin pigmentation and retinal morphology	Embryonic lethal
Disc1	Major depression and schizophrenia-like disorder	Schizophrenia-like disorder
Dynchc1	Adult motor neuron degeneration	Embryonic lethal
Egfr	Dark skin pigmentation	Embryonic or perinatal lethality
Evi1	Otitis media	Embryonic lethal/oncogene
G6pdx	Decreased angiogenesis and abnormal redox activity	Embryonic lethal, X-linked
Plcg2	Severe spontaneous inflammation and autoimmunity	Decrease in mature B cells
Rxra	Alopecia, dermal cysts, and impaired Th2 response	Embryonic lethal

production of an allelic series of mutants in any gene. Given this versatility, ENU will remain an important part of the mouse genetic toolkit for the foreseeable future. Large-scale phenotype-driven approaches will continue to generate novel mouse models of human disorders. As a complement to large-scale KO projects, ENU will be vital in elaborating the full range of functionality associated with every gene in the mammalian genome. As more and more candidate disease genes are revealed from genome-wide association studies, these loci will need to be functionally validated and the genetic network in which they act elaborated. In the mouse, the availability of ENU mutants as well as null alleles will make this critical task achievable.

SUMMARY POINTS

1. A major goal of the genomic era is to functionally characterize every single gene in the genome.
2. N-ethyl-N-nitrosourea (ENU) offers a powerful, unbiased, hypothesis-generating tool to identify new gene functions and create new mouse models of human disorders.
3. Because ENU creates point mutations, it might model some human disease states better than null alleles. Mutants can be identified either with previous gene knowledge (gene-driven screens) or with no prior assumptions (phenotype-driven approach).

FUTURE ISSUES

1. Large-scale phenotype-driven approaches will likely continue to produce invaluable new mouse models of human disorders.
2. Advances in phenotyping screening techniques together with the availability of systematic phenotypic pipelines will likely increase even further the possibilities of modeling human diseases in the mouse.
3. Small-scale modifier, sensitized, or challenge screens will start to provide much needed new gene targets and systems for drug discovery in a wide variety of diseases.
4. As a complement to the large-scale knockout projects, ENU models will likely continue to play a critical role in dissecting and understanding gene functions.

DISCLOSURE STATEMENT

The authors are not aware of any biases that might be perceived as affecting the objectivity of this review.

ACKNOWLEDGMENTS

We would like to thank Dr. Pat Nolan and Dr. Francesca Mackenzie for critical reading of the manuscript.

LITERATURE CITED

1. Augustin M, Sedlmeier R, Peters T, Huffstadt U, Kochmann E, et al. 2005. Efficient and fast targeted production of murine models based on ENU mutagenesis. *Mamm. Genome* 16(6):405–13

2. Auwerx J, Avner P, Baldock R, Ballabio A, Balling R, et al. 2004. The European dimension for the mouse genome mutagenesis program. *Nat. Genet.* 36:925–27
3. Bacon Y, Ooi A, Kerr S, Shaw-Andrews L, Winchester L, et al. 2004. Screening for novel ENU-induced rhythm, entrainment and activity mutants. *Genes Brain Behav.* 3(4):196–205
4. Balling R. 2001. ENU mutagenesis: analyzing gene function in mice. *Annu. Rev. Genomics Hum. Genet.* 2:463–92
5. Barbaric I, Wells S, Russ A, Dear TN. 2007. Spectrum of ENU-induced mutations in phenotype-driven and gene-driven screens in the mouse. *Environ. Mol. Mutagen.* 48(2):124–42
6. Bielas JH, Heddle JA. 2000. Proliferation is necessary for both repair and mutation in transgenic mouse cells. *Proc. Natl. Acad. Sci. USA* 97(21):11391–96
7. Bogani D, Willoughby C, Davies J, Kaur K, Mirza G, et al. 2005. Dissecting the genetic complexity of human 6p deletion syndromes by using a region-specific, phenotype-driven mouse screen. *Proc. Natl. Acad. Sci. USA* 102(35):12477–82
8. Bosman EA, Penn AC, Ambrose JC, Kettleborough R, Stemple DL, Steel KP. 2005. Multiple mutations in mouse *Chd7* provide models for CHARGE syndrome. *Hum. Mol. Genet.* 14(22):3463–76
9. Bronstein SM, Cochrane JE, Craft TR, Swenberg JA, Skopek TR. 1991. Toxicity, mutagenicity, and mutational spectra of N-ethyl-N-nitrosourea in human cell lines with different DNA repair phenotypes. *Cancer Res.* 51(19):5188–97
10. Brown SD, Chambon P, de Angelis MH, Eumorphia Consortium. 2005. EMPReSS: standardized phenotype screens for functional annotation of the mouse genome. *Nat. Genet.* 37(11):1155
11. Brown SD, Hancock JM, Gates H. 2006. Understanding mammalian genetic systems: the challenge of phenotyping in the mouse. *PLoS Genet.* 2(8):e118
12. Buchner DA, Seburn KL, Frankel WN, Meisler MH. 2004. Three ENU-induced neurological mutations in the pore loop of sodium channel Scn8a (Na(v)1.6) and a genetically linked retinal mutation, rd13. *Mamm. Genome* 15(5):344–51
13. **Carpinelli MR, Hilton DJ, Metcalf D, Antonchuk JL, Hyland CD, et al. 2004. Suppressor screen in *Mpl*$^{-/-}$ mice: *c-Myb* mutation causes supraphysiological production of platelets in the absence of thrombopoietin signaling. *Proc. Natl. Acad. Sci. USA* 101(17):6553–58**

13. The first cloned modifier from an ENU modifier screen.

14. Chen Y, Yee D, Dains K, Chatterjee A, Cavalcoli J, et al. 2000. Genotype-based screen for ENU-induced mutations in mouse embryonic stem cells. *Nat. Genet.* 24(3):314–17
15. Claij N, van der Wal A, Dekker M, Jansen L, te Riele H. 2003. DNA mismatch repair deficiency stimulates N-ethyl-N-nitrosourea-induced mutagenesis and lymphomagenesis. *Cancer Res.* 63(9):2062–66
16. Clapcote SJ, Lipina TV, Millar JK, Mackie S, Christie S, et al. 2007. Behavioral phenotypes of *Disc1* missense mutations in mice. *Neuron* 54(3):387–402
17. Clark AT, Goldowitz D, Takahashi JS, Vitaterna MH, Siepka SM, et al. 2004. Implementing large-scale ENU mutagenesis screens in North America. *Genetica* 122(1):51–64
18. **Coghill EL, Hugill A, Parkinson N, Davison C, Glenister P, et al. 2002. A gene-driven approach to the identification of ENU mutants in the mouse. *Nat. Genet.* 30(3):255–56**

18. Gene-driven screens made possible with the availability of sperm archives of ENU-mutagenized males.

19. Concepcion D, Seburn KL, Wen G, Frankel WN, Hamilton BA. 2004. Mutation rate and predicted phenotypic target sizes in ethylnitrosourea-treated mice. *Genetics* 168(2):953–59
20. Cox R, Brown SD. 2003. Rodent models of genetic disease. *Curr. Opin. Genet. Dev.* 13(3):278–83
21. Crozat K, Georgel P, Rutschmann S, Mann N, Du X, et al. 2006. Analysis of the MCMV resistome by ENU mutagenesis. *Mamm. Genome* 17(5):398–406
22. Crozat K, Hoebe K, Ugolini S, Hong NA, Janssen E, et al. 2007. *Jinx*, an MCMV susceptibility phenotype caused by disruption of *Unc13d*: a mouse model of type 3 familial hemophagocytic lymphohistiocytosis. *J. Exp. Med.* 204(4):853–63
23. Debruyne JP, Noton E, Lambert CM, Maywood ES, Weaver DR, Reppert SM. 2006. A clock shock: mouse CLOCK is not required for circadian oscillator function. *Neuron.* 50(3):465–77
24. Di Giorgio FP, Carrasco MA, Siao MC, Maniatis T, Eggan K. 2007. Non-cell autonomous effect of glia on motor neurons in an embryonic stem cell-based ALS model. *Nat. Neurosci.* 10(5):608–14
25. Duerr RH, Taylor KD, Brant SR, Rioux JD, Silverberg MS, et al. 2006. A genome-wide association study identifies *IL23R* as an inflammatory bowel disease gene. *Science* 314(5804):1461–63

26. Eggenschwiler JT, Espinoza E, Anderson KV. 2001. Rab23 is an essential negative regulator of the mouse Sonic hedgehog signalling pathway. *Nature* 412(6843):194–98
27. Favor J. 1998. The mutagenic activity of ethylnitrosourea at low doses in spermatogonia of the mouse as assessed by the specific-locus test. *Mutat. Res.* 405(2):221–26
28. Frayling TM, Timpson NJ, Weedon MN, Zeggini E, Freathy RM, et al. 2007. A common variant in the *FTO* gene is associated with body mass index and predisposes to childhood and adult obesity. *Science* 316(5826):889–94
29. Freeman H, Shimomura K, Horner E, Cox RD, Ashcroft FM. 2006. Nicotinamide nucleotide transhydrogenase: a key role in insulin secretion. *Cell Metab.* 3(1):35–45
30. Fuchs H, Schughart K, Wolf E, Balling R, Hrabé de Angelis M. 2000. Screening for dysmorphological abnormalities—a powerful tool to isolate new mouse mutants. *Mamm. Genome* 11(7):528–30
31. Gailus-Durner V, Fuchs H, Becker L, Bolle I, Brielmeier M, et al. 2005. Introducing the German Mouse Clinic: open access platform for standardized phenotyping. *Nat. Methods* 2(6):403–4
32. Godinho SI, Maywood ES, Shaw L, Tucci V, Barnard AR, et al. 2007. The after-hours mutant reveals a role for Fbxl3 in determining mammalian circadian period. *Science* 316(5826):897–900
33. Greber B, Lehrach H, Himmelbauer H. 2005. Mouse splice mutant generation from ENU-treated ES cells–a gene-driven approach. *Genomics* 85(5):557–62
34. Grosse J, Tarnow P, Römpler H, Schneider B, Sedlmeier R, et al. 2006. N-ethyl-N-nitrosourea-based generation of mouse models for mutant G protein-coupled receptors. *Physiol. Genomics* 26(3):209–17
35. Gusella JF, Macdonald ME. 2006. Huntington's disease: seeing the pathogenic process through a genetic lens. *Trends Biochem. Sci.* 31(9):533–40
36. Hafezparast M, Klocke R, Ruhrberg C, Marquardt A, Ahmad-Annuar A, et al. 2003. Mutations in dynein link motor neuron degeneration to defects in retrograde transport. *Science* 300(5620):808–12
37. Hardisty RE, Erven A, Logan K, Morse S, Guionaud S, et al. 2003. The deaf mouse mutant *Jeff* (*Jf*) is a single gene model of otitis media. *J. Assoc. Res. Otolaryngol.* 4(2):130–38
38. Hardisty-Hughes RE, Tateossian H, Morse SA, Romero MR, Middleton A, et al. 2006. A mutation in the F-box gene, *Fbxo11*, causes otitis media in the *Jeff* mouse. *Hum. Mol. Genet.* 15(22):3273–79
39. Hart AW, McKie L, Morgan JE, Gautier P, West K, et al. 2005. Genotype-phenotype correlation of mouse *Pde6b* mutations. *Invest. Ophthalmol. Vis. Sci.* 46(9):3443–50
40. Hentges KE, Justice MJ. 2004. Checks and balancers: balancer chromosomes to facilitate genome annotation. *Trends Genet.* 20(6):252–59
41. **Herron BJ, Lu W, Rao C, Liu S, Peters H, et al. 2002. Efficient generation and mapping of recessive developmental mutations using ENU mutagenesis. *Nat. Genet.* 30(2):185–89**

41. An excellent example of G3 screening for developmental mutations.

42. Hitotsumachi S, Carpenter DA, Russell WL. 1985. Dose-repetition increases the mutagenic effectiveness of N-ethyl-N-nitrosourea in mouse spermatogonia. *Proc. Natl. Acad. Sci. USA* 82(19):6619–21
43. Hoebe K, Beutler B. 2005. Unraveling innate immunity using large scale N-ethyl-N-nitrosourea mutagenesis. *Tissue Antigens* 65(5):395–401
44. Hough TA, Nolan PM, Tsipouri V, Toye AA, Gray IC, et al. 2002. Novel phenotypes identified by plasma biochemical screening in the mouse. *Mamm. Genome* 13(10):595–602
45. Hough TA, Polewski M, Johnson K, Cheeseman M, Nolan PM, et al. 2007. A novel mouse model of autosomal semidominant adult hypophosphatasia has a splice site mutation in the tissue nonspecific alkaline phosphatase gene *Akp2*. *J. Bone Miner. Res.* 22:1397–407
46. **Hrabé de Angelis MH, Flaswinkel H, Fuchs H, Rathkolb B, Soewarto D, et al. 2000. Genome-wide, large-scale production of mutant mice by ENU mutagenesis. *Nat. Genet.* 25(4):444–47**

46. The first of two examples of large-scale mutagenesis programs in the mouse.

47. Inoue M, Sakuraba Y, Motegi H, Kubota N, Toki H, et al. 2004. A series of maturity onset diabetes of the young, type 2 (MODY2) mouse models generated by a large-scale ENU mutagenesis program. *Hum. Mol. Genet.* 13(11):1147–57
48. Isaacs AM, Davies KE, Hunter AJ, Nolan PM, Vizor L, et al. 2000. Identification of two new *Pmp22* mouse mutants using large-scale mutagenesis and a novel rapid mapping strategy. *Hum. Mol. Genet.* 9(12):1865–71
49. Isaacs AM, Oliver PL, Jones EL, Jeans A, Potter A, et al. 2003. A mutation in *Af4* is predicted to cause cerebellar ataxia and cataracts in the robotic mouse. *J. Neurosci.* 23(5):1631–37

50. **Justice MJ, Carpenter DA, Favor J, Neuhauser-Klaus A, Hrabé de Angelis M, et al. 2000. Effects of ENU dosage on mouse strains. *Mamm. Genome* 11(7):484–88**

50. Describes the different doses and regimes required for different strains of mice when mutagenizing with ENU.

51. Justice MJ, Noveroske JK, Weber JS, Zheng B, Bradley A. 1999. Mouse ENU mutagenesis. *Hum. Mol. Genet.* 8(10):1955–63
52. Kapfhamer D, Valladares O, Sun Y, Nolan PM, Rux JJ, et al. 2002. Mutations in *Rab3a* alter circadian period and homeostatic response to sleep loss in the mouse. *Nat. Genet.* 32(2):290–95
53. Kas MJ, Van Ree JM. 2004. Dissecting complex behaviours in the postgenomic era. *Trends Neurosci.* 27(7):366–69
54. Kearney JA, Yang Y, Beyer B, Bergren SK, Claes L, et al. 2006. Severe epilepsy resulting from genetic interaction between *Scn2a* and *Kcnq2*. *Hum. Mol. Genet.* 15(6):1043–48
55. Keays DA, Clark TG, Flint J. 2006. Estimating the number of coding mutations in genotypic- and phenotypic-driven N-ethyl-N-nitrosourea (ENU) screens. *Mamm. Genome* 17(3):230–38
56. Keays DA, Tian G, Poirier K, Huang GJ, Siebold C, et al. 2007. Mutations in α-tubulin cause abnormal neuronal migration in mice and lissencephaly in humans. *Cell* 128(1):45–57
57. Kieran D, Hafezparast M, Bohnert S, Dick JR, Martin J, et al. 2005. A mutation in dynein rescues axonal transport defects and extends the life span of ALS mice. *J. Cell Biol.* 169(4):561–67
58. **Kile BT, Hentges KE, Clark AT, Nakamura H, Salinger AP, et al. 2003. Functional genetic analysis of mouse chromosome 11. *Nature* 425(6953):81–86**

58. The first example of the use of a balancer chromosome in the mouse for region-specific mutagenesis.

59. Kile BT, Hilton DJ. 2005. The art and design of genetic screens: mouse. *Nat. Rev. Genet.* 6(7):557–67
60. King DP, Zhao Y, Sangoram AM, Wilsbacher LD, Tanaka M, et al. 1997. Positional cloning of the mouse circadian clock gene. *Cell* 89(4):641–53
61. Lyon MF, Morris T. 1966. Mutation rates at a new set of specific loci in the mouse. *Genet. Res.* 7(1):12–17
62. Masuya H, Nishida K, Furuichi T, Toki H, Nishimura G, et al. 2007. A novel dominant-negative mutation in *Gdf5* generated by ENU mutagenesis impairs joint formation and causes osteoarthritis in mice. *Hum. Mol. Genet.* 16(19):2366–75
63. Michaud EJ, Culiat CT, Klebig ML, Barker PE, Cain KT, et al. 2005. Efficient gene-driven germ-line point mutagenesis of C57BL/6J mice. *BMC Genomics* 6:164
64. Munroe RJ, Bergstrom RA, Zheng QY, Libby B, Smith R, et al. 2000. Mouse mutants from chemically mutagenized embryonic stem cells. *Nat. Genet.* 24(3):318–21
65. Nelms KA, Goodnow CC. 2001. Genome-wide ENU mutagenesis to reveal immune regulators. *Immunity* 15(3):409–18
66. Nolan PM, Hugill A, Cox RD. 2002. ENU mutagenesis in the mouse: application to human genetic disease. *Brief Funct. Genomics Proteomics* 1(3):278–89
67. **Nolan PM, Peters J, Strivens M, Rogers D, Hagan J, et al. 2000. A systematic, genome-wide, phenotype-driven mutagenesis programme for gene function studies in the mouse. *Nat. Genet.* 25(4):440–43**

67. The second of the first two examples of large-scale mutagenesis in the mouse.

68. Noveroske JK, Hardy R, Dapper JD, Vogel H, Justice MJ. 2005. A new ENU-induced allele of mouse quaking causes severe CNS dysmyelination. *Mamm. Genome* 16(9):672–82
69. Noveroske JK, Weber JS, Justice MJ. 2000. The mutagenic action of N-ethyl-N-nitrosourea in the mouse. *Mamm. Genome* 11(7):478–83
70. O'Neill JP. 2000. DNA damage, DNA repair, cell proliferation, and DNA replication: how do gene mutations result? *Proc. Natl. Acad. Sci. USA* 97(21):11137–39
71. Papathanasiou P, Goodnow C. 2005. Connecting mammalian genome with phenome by ENU mouse mutagenesis: gene combinations specifying the immune system. *Annu. Rev. Genet.* 5(39):241–62
72. Parkinson N, Hardisty-Hughes RE, Tateossian H, Tsai HT, Brooker D, et al. 2006. Mutation at the *Evi1* locus in *Junbo* mice causes susceptibility to otitis media. *PLoS Genet.* 2(10):e149
73. Pieles G, Geyer SH, Szumska D, Schneider J, Neubauer J, et al. 2007. MicroMRI-HREM pipeline for high-throughput, high-resolution phenotyping of murine embryos. *J. Anat.* 211:132–37
74. Quwailid MM, Hugill A, Dear N, Vizor L, Wells S, et al. 2004. A gene-driven ENU-based approach to generating an allelic series in any gene. *Mamm. Genome* 15(8):585–91
75. Ravikumar B, Acevedo-Arozena A, Imarisio S, Berger Z, Vacher C, et al. 2005. Dynein mutations impair autophagic clearance of aggregate-prone proteins. *Nat. Genet.* 37(7):771–76

76. Reed DR, Bachmanov AA, Tordoff MG. 2007. Forty mouse strain survey of body composition. *Physiol. Behav.* 91(5):593–600
77. Rinchik EM. 1991. Chemical mutagenesis and fine-structure functional analysis of the mouse genome. *Trends Genet.* 7(1):15–21
78. Rinchik EM, Carpenter DA, Selby PB. 1990. A strategy for fine-structure functional analysis of a 6- to 11-centimorgan region of mouse chromosome 7 by high-efficiency mutagenesis. *Proc. Natl. Acad. Sci. USA* 87(3):896–900
79. Rioux JD, Xavier RJ, Taylor KD, Silverberg MS, Goyette P, et al. 2007. Genome-wide association study identifies new susceptibility loci for Crohn disease and implicates autophagy in disease pathogenesis. *Nat. Genet.* 39(5):596–604
80. Rogers DC, Fisher EM, Brown SD, Peters J, Hunter AJ, Martin JE. 1997. Behavioral and functional analysis of mouse phenotype: SHIRPA, a proposed protocol for comprehensive phenotype assessment. *Mamm. Genome* 8(10):711–13
81. Rubinsztein DC. 2002. Lessons from animal models of Huntington's disease. *Trends Genet.* 18(4):202–9
82. Rubio-Aliaga I, Soewarto D, Wagner S, Klaften M, Fuchs H, et al. 2007. A genetic screen for modifiers of the delta1-dependent notch signaling function in the mouse. *Genetics* 175(3):1451–63
83. Rudelius M, Osanger A, Kohlmann S, Augustin M, Piontek G, et al. 2006. A missense mutation in the WD40 domain of murine Lyst is linked to severe progressive Purkinje cell degeneration. *Acta Neuropathol.* 112(3):267–76
84. Russell LB, Hunsicker PR, Russell WL. 2007. Comparison of the genetic effects of equimolar doses of ENU and MNU: while the chemicals differ dramatically in their mutagenicity in stem-cell spermatogonia, both elicit very high mutation rates in differentiating spermatogonia. *Mutat. Res.* 616(1–2):181–95
85. Russell WL, Hunsicker PR, Carpenter DA, Cornett CV, Guinn GM. 1982. Effect of dose fractionation on the ethylnitrosourea induction of specific-locus mutations in mouse spermatogonia. *Proc. Natl. Acad. Sci. USA* 79(11):3592–93
86. **Russell WL, Kelly EM, Hunsicker PR, Bangham JW, Maddux SC, Phipps EL. 1979. Specific-locus test shows ethylnitrosourea to be the most potent mutagen in the mouse. *Proc. Natl. Acad. Sci. USA* 76(11):5818–19**

86. This paper marked a milestone in mouse mutagenesis when it announced that ENU is the most effective mutagen for creating new gene mutations in the mouse.

87. Sakuraba Y, Sezutsu H, Takahasi KR, Tsuchihashi K, Ichikawa R, et al. 2005. Molecular characterization of ENU mouse mutagenesis and archives. *Biochem. Biophys. Res. Commun.* 336(2):609–16
88. Schneider JE, Böse J, Bamforth SD, Gruber AD, Broadbent C, et al. 2004. Identification of cardiac malformations in mice lacking *Ptdsr* using a novel high-throughput magnetic resonance imaging technique. *BMC Dev. Biol.* 4:16
89. Schwander M, Sczaniecka A, Grillet N, Bailey JS, Avenarius M, et al. 2007. A forward genetics screen in mice identifies recessive deafness traits and reveals that pejvakin is essential for outer hair cell function. *J. Neurosci.* 27:2163–75
90. Segade F, Daly KA, Allred D, Hicks PJ, Cox M, et al. 2006. Association of the FBXO11 gene with chronic otitis media with effusion and recurrent otitis media: the Minnesota COME/ROM Family Study. *Arch. Otolaryngol. Head Neck Surg.* 132(7):729–33
91. Shibuya T, Murota T, Horiya N, Matsuda H, Hara T. 1993. The induction of recessive mutations in mouse primordial germ cells with N-ethyl-N-nitrosourea. *Mutat. Res.* 290(2):273–80
92. Skopek TR, Walker VE, Cochrane JE, Craft TR, Cariello NF. 1992. Mutational spectrum at the *Hprt* locus in splenic T cells of B6C3F1 mice exposed to *N*-ethyl-*N*-nitrosourea. *Proc. Natl. Acad. Sci. USA* 89(17):7866–70
93. Sladek R, Rocheleau G, Rung J, Dina C, Shen L, et al. 2007. A genome-wide association study identifies novel risk loci for type 2 diabetes. *Nature* 445(7130):881–85
94. Takahasi KR, Sakuraba Y, Gondo Y. 2007. Mutational pattern and frequency of induced nucleotide changes in mouse ENU mutagenesis. *BMC Mol. Biol.* 8:52
95. Thaung C, West K, Clark BJ, McKie L, Morgan JE, et al. 2002. Novel ENU-induced eye mutations in the mouse: models for human eye disease. *Hum. Mol. Genet.* 11(7):755–67
96. Todd JA, Walker NM, Cooper JD, Smyth DJ, Downes K, et al. Genetics of Type 1 Diabetes in Finland, Simmonds MJ, Heward JM, Gough SC; Wellcome Trust Case Control Consortium, Dunger DB,

Wicker LS, Clayton DG. 2007. Robust associations of four new chromosome regions from genome-wide analyses of type 1 diabetes. *Nat. Genet.* 39(7):857–64

97. Toye AA, Moir L, Hugill A, Bentley L, Quarterman J, et al. 2004. A new mouse model of type 2 diabetes, produced by N-ethyl-nitrosourea mutagenesis, is the result of a missense mutation in the glucokinase gene. *Diabetes* 53(6):1577–83
98. Traka M, Seburn KL, Popko B. 2006. *Nmf11* is a novel ENU-induced mutation in the mouse glycine receptor alpha 1 subunit. *Mamm. Genome* 17(9):950–55
99. Vinuesa CG, Cook MC, Angelucci C, Athanasopoulos V, Rui L, et al. 2005. A RING-type ubiquitin ligase family member required to repress follicular helper T cells and autoimmunity. *Nature* 435(7041):452–58
100. Vinuesa CG, Goodnow CC. 2004. Illuminating autoimmune regulators through controlled variation of the mouse genome sequence. *Immunity* 20(6):669–79
101. Visscher PM, Haley CS, Thompson R. 1996. Marker-assisted introgression in backcross breeding programs. *Genetics* 144(4):1923–32
102. Vitaterna MH, King DP, Chang AM, Kornhauser JM, Lowrey PL, et al. 1994. Mutagenesis and mapping of a mouse gene, Clock, essential for circadian behavior. *Science* 264(5159):719–25
103. Vivian JL, Chen Y, Yee D, Schneider E, Magnuson T. 2002. An allelic series of mutations in *Smad2* and *Smad4* identified in a genotype-based screen of *N*-ethyl-*N*- nitrosourea-mutagenized mouse embryonic stem cells. *Proc. Natl. Acad. Sci. USA* 99(24):15542–47
104. Washington JL 3rd, Pitts D, Wright CG, Erway LC, Davis RR, Alagramam K. 2005. Characterization of a new allele of Ames waltzer generated by ENU mutagenesis. *Hear. Res.* 202(1–2):161–69
105. Weber JS, Salinger A, Justice MJ. 2000. Optimal N-ethyl-N-nitrosourea (ENU) doses for inbred mouse strains. *Genesis* 26(4):230–33
106. Wellcome Trust Case Control Consortium, et al. 2007. Association scan of 14,500 nonsynonymous SNPs in four diseases identifies autoimmunity variants. *Nat Genet.* 39(11):1329–37
107. Wiltshire T, Pletcher MT, Batalov S, Barnes SW, Tarantino LM, et al. 2003. Genome-wide single-nucleotide polymorphism analysis defines haplotype patterns in mouse. *Proc. Natl. Acad. Sci. USA* 100(6):3380–85
108. Xie G, Clapcote SJ, Nieman BJ, Tallerico T, Huang Y, et al. 2007. Forward genetic screen of mouse reveals dominant missense mutation in the P/Q-type voltage-dependent calcium channel, CACNA1A. *Genes Brain Behav.* 6(8):717–27
109. Yu P, Constien R, Dear N, Katan M, Hanke P, et al. 2005. Autoimmunity and inflammation due to a gain-of-function mutation in phospholipase Cγ2 that specifically increases external Ca^{2+} entry. *Immunity* 22(4):451–65

Clinical Utility of Contemporary Molecular Cytogenetics

Bassem A. Bejjani and Lisa G. Shaffer

Signature Genomic Laboratories, LLC, Spokane, Washington 99202;
email: bejjani@signaturegenomics.com

Annu. Rev. Genomics Hum. Genet. 2008. 9:71–86

First published online as a Review in Advance on May 12, 2008

The *Annual Review of Genomics and Human Genetics* is online at genom.annualreviews.org

This article's doi:
10.1146/annurev.genom.9.081307.164207

1527-8204/08/0922-0071$20.00

Key Words

array CGH, chromosome abnormality, CNV, genomic disorder, microarray

Abstract

The development of microarray-based comparative genomic hybridization (array CGH) methods represents a critical new advance in molecular cytogenetics. This new technology has driven a technical convergence between molecular diagnostics and clinical cytogenetics, questioned our naïve understanding of the complexity of the human genome, revolutionized the practice of medical genetics, challenged conventional wisdom related to the genetic bases of multifactorial and sporadic conditions, and is poised to impact all areas of medicine. The use of contemporary molecular cytogenetic techniques in research and diagnostics has resulted in the identification of many new syndromes, expanded our knowledge about the phenotypic spectrum of recognizable syndromes, elucidated the genomic bases of well-established clinical conditions, and refined our view about the molecular mechanisms of some chromosomal aberrations. Newer methodologies are being developed, which will likely lead to a new understanding of the genome and its relationship to health and disease.

INTRODUCTION

Historical Overview

FISH: fluorescence in situ hybridization

Microarray: libraries of clones placed on a solid surface such as a glass slide to allow clone identification on the basis of chromosomal location

sSMC: small supernumerary marker chromosome

The era of modern clinical cytogenetics is generally thought to have begun in 1956 with Tjio & Levan's (95) observation that normal human cells contain 46 chromosomes. This discovery was made possible in 1952 by Hsu's preparation of nuclear pellets in a hypotonic solution that caused the nuclei to swell, allowing the chromosomes to spread out (37). It is inspiring to think how far we have progressed since these modest beginnings 50 years ago.

Within a few years of these breakthroughs, discoveries were made quickly in identifying numerical anomalies associated with specific abnormal phenotypes: In 1959, Lejeune and coworkers (49) discovered that Down syndrome was caused by trisomy 21, and in 1960, Patau (65) and Edwards and colleagues (25) recognized that what became their eponymous syndromes are caused by trisomy 13 and 18, respectively. In 1960 Nowell (64) described a "minute chromosome" (the Philadelphia chromosome) that was found consistently in "chronic granulocytic leukemia," which Rowley later showed is caused by a translocation between chromosomes 9 and 22 (74). In 1970, Caspersson (15) reported that Quinacrine mustard caused chromosomes to exhibit light and dark bands along their length, thus ushering in the era of human chromosomal banding and making it possible to determine breakpoints associated with translocations and other structural abnormalities in human genetics. Within a few years, Yunis (105) described a method to study human chromosomes with high resolution, allowing increasingly subtle abnormalities of the human genome to be discovered and linked to specific abnormal phenotypes.

Since then, progressive improvements in cytogenetic techniques and the introduction of molecular methods in the form of fluorescence in situ hybridization (FISH) and microarray-based comparative genomic hybridization (array CGH) have revolutionized this field and extended its applications to many areas of medicine other than medical genetics, although it is here that microarray technology has had its greatest impact. In only a few years, many new syndromes have been discovered, the phenotypes of existing ones have been expanded, and a new appreciation of the diversity of the human genome is now emerging. Contemporary molecular cytogenetics is also transforming other areas of medicine: Recent applications of high-density arrays have led to the identification of specific genetic alterations in sporadic and supposedly multifactorial conditions such as cancer, Alzheimer, autism, and even infectious disease.

The Evolution of Contemporary Molecular Cytogenetics

Since its inception, clinical cytogenetics has been an integral part of the investigation of suspected genetic conditions. The type of tissue studied may differ depending on the reason for investigation, but the characteristic banding patterns indicative of specific chromosomes are largely consistent. Although many aberrations are revealed through cytogenetic studies, conventional cytogenetic analysis cannot reliably detect rearrangements of genomic segments smaller than 3–5 million base pairs (Mb). Furthermore, microscopic examination of the chromosomes may not reveal the chromosomal origin of small supernumerary marker chromosomes (sSMCs) because their limited size is insufficient to provide a recognizable characteristic banding pattern, may not identify subtle rearrangements of the subtelomeric regions (29, 43), and may not elucidate the complexity of some rearrangements, especially in solid tumors and leukemias.

The introduction of FISH circumvented some of the limitations of traditional cytogenetics and became an integral part of a comprehensive clinical cytogenetic evaluation (96). FISH permits the determination of the number and location of specific DNA sequences, both in metaphase chromosomes and in interphase nuclei, significantly simplifying the preparation and evaluation of samples. The applications of FISH in the clinical setting include screening for aneuploidy in prenatal specimens, searching for microdeletions in contiguous gene syndromes, evaluating rearrangements of the subtelomeric regions in nonspecific mental retardation, and defining gene rearrangements (fusion genes) in leukemias and lymphomas (96). Typically, these rearrangements are difficult or impossible to visualize with conventional banding techniques owing to their small size or undetectable changes in banding patterns within the altered segments.

A variety of probe types may be used in FISH to investigate specific genomic segments. For example, unique sequence probes (single-copy probes) are used to identify deletions or duplications associated with contiguous gene syndromes or other syndromes caused by microrearrangements of unique loci (79), repetitive sequence probes unique to each centromere are used to identify the copy number of specific chromosomes or to identify the chromosomal origin of sSMCs, and whole chromosome painting probes are useful for characterizing translocations or other complex rearrangements, excluding inversions. However, the efficient use of FISH dictates that the patient either exhibits features consistent with a clinically recognizable syndrome with a known chromosomal etiology or demonstrates an abnormal karyogram that requires further molecular characterization (e.g., marker chromosome). This is because single FISH probes reveal gains, losses, or rearrangement of only the targeted segments and do not provide information on the rest of the genome. Thus, FISH analyses do not detect abnormalities distinct from the genomic segments for which probes have been designed and used. Furthermore, the clinical cytogenetics laboratory must rely heavily on clinical direction from the physician evaluating the patient. In most cases, the clinician's judgment influences or directly determines the choice of probes for establishing a diagnosis.

Comparative genomic hybridization (CGH) represents a variation on FISH technology with the clear advantage of revealing imbalances across the whole genome. In CGH, a DNA sample is extracted from an individual with a known, typically normal karyotype (control) and compared to the DNA sample obtained from an individual with an unknown karyotype or a known abnormal karyotype (subject). These two DNA specimens are differentially labeled with different fluorochromes and allowed to cohybridize onto metaphase chromosomes prepared from an individual with a normal karyotype. Differences between the respective fluorescent intensities along the length of any given chromosome represent gains or losses of segments of DNA from the subject relative to the control (51). This technology has many of the same limitations found in conventional cytogenetics, because the substrates for analysis are tightly condensed metaphase chromosomes. Thus, the resolution of CGH is limited to that of metaphase chromosomes—approximately 5–10 Mb for most clinical applications (41, 42, 53).

To achieve the simultaneous examination of multiple loci at a higher resolution, CGH technology has been modified and applied to an array of DNA targets fixed to a solid support (66). Arrays have been constructed with a variety of DNA targets ranging from oligonucleotides (25–80 bp) (62, 78) to bacterial artificial chromosomes (BACs) (80–200 kb) (8). The main advantage of array CGH is the ability to simultaneously detect aneuploidy, deletions, duplications, or amplifications of any locus represented on the array. In contrast, a multiple-probe FISH analysis on metaphase chromosomes may detect aneuploidy and microdeletions, but its ability to detect duplications is limited. Indeed, duplications that involve segments smaller than 1.5 Mb may be routinely missed even by FISH of interphase nuclei (82). Microarray analysis can identify any segmental imbalance (aneuploidy, deletion, duplication) of the loci represented on the microarray; the resolution is limited only by the size of the insert used and the distance between clones. In addition, dense single nucleotide polymorphism (SNP)-based oligonucleotide arrays can identify single nucleotide changes, small DNA copy changes (on the order of a few kb), and large DNA copy changes (up to the level of

Comparative genomic hybridization (CGH): competitive hybridization of differentiating, fluorescently labeled DNA to detect relative excesses or deficiencies of chromosome regions

Bacterial artificial chromosome (BAC): human DNA segment inserted into a bacterial vector; can be copied in bacterial cells

SNP: single nucleotide polymorphism

whole chromosome aneuploidy); this technology therefore offers, on a single platform, the functionality that was afforded only by sequencing (to detect single base changes or a few hundred nucleotide changes), Southern blotting (to detect DNA changes on the order of 10–40 kb), pulse field gel electrophoresis (PFGE) (to detect DNA changes on the order of a few hundred kb to a few Mb), FISH (to detect DNA changes on the order of 35 kb or larger), and traditional G-banded chromosome analysis (to detect DNA changes larger than 5 Mb).

Mosaicism: a mixture of two or more cell lines in the same specimen or individual; i.e., a chromosome abnormality seen in a subset of cells

NAHR: nonallelic homologous recombination

LCR: low-copy repeat

The impact of array CGH on the practice of medical genetics has been transformative. This technology has already been used to expand the phenotypes of existing conditions, identify the reciprocal products of known abnormalities, determine the genomic lesions in known conditions, discover new syndromes, appreciate the prevalence of mosaicism in individuals with developmental delay, and ascertain the unexpected frequency and effect of copy number variants across the genome.

CLINICAL UTILITY OF CONTEMPORARY MOLECULAR CYTOGENETICS

Genetics

Expanding the phenotypes of existing conditions. "Recognizable syndromes" are recognizable because they exhibit, to the trained clinician, a constellation of signs and symptoms that arouse sufficient suspicion to cause the clinician to order a test that will confirm the clinical diagnosis. This age-old pattern of medical practice creates a loop that includes the patient, the clinician, and the laboratory and, in doing so, reinforces these recognizable features, cements them to the syndrome, and makes the clinician more confident in his/her diagnostic skills. This approach to diagnostics, however, ignores that identical molecular lesions may result in clinical features that are only partially present in, or sometimes completely absent from, the narrow spectrum of recognizable features that make that syndrome identifiable (5). With the application of array CGH to individuals with nonspecific developmental delay (DD) and/or mental retardation (MR), with or without dysmorphic features (DF), it is now clear that many recognizable microdeletions and microduplication syndromes have a much wider spectrum of clinical presentation than was previously appreciated (46, 80). Such variation in the phenotype of individuals with identical molecular lesions at the locus of interest has long been acknowledged in some conditions such as the DiGeorge/velo-cardio-facial syndrome (VCFS) spectrum (30), but a new appreciation of this clinical variation is emerging from other genomic disorders (20, 32). A more complete understanding of the full clinical spectrum of these disorders will be achieved as the use of array CGH in the clinic becomes more prevalent and as correlations of these clinical findings with the genomic lesions are made. Existing website resources such as DECIPHER (DatabasE of Chromosomal Imbalance and Phenotype in Humans using Ensembl Resources; **http://www.sanger.ac.uk/PostGenomics/decipher/**) (2) may facilitate widespread appreciation of such phenotypic variability.

Identifying the reciprocal products of known conditions. Many recognizable microdeletion syndromes are caused by nonallelic homologous recombination (NAHR) mediated by flanking low-copy repeat (LCR) sequences (83). This mechanism predicts that the prevalence of the reciprocal duplication product in the population should be equal to the frequency of deletion (57). However, duplications have not been observed until fairly recently (26, 33, 69, 93, 104), likely because, in general, individuals with duplications tend to have a milder phenotype than those with the complementary deletions (9, 11, 12, 26, 68, 93, 104); this milder phenotype may not lead to clinical investigation. The recent characterization of these duplication conditions and the prerequisite for FISH investigation of a clinical suspicion (94) make the ascertainment of individuals with these conditions less likely by traditional approaches. The introduction of array CGH in clinical practice has virtually eliminated all the technical impediments of traditional cytogenetics and FISH and allowed the detection of such conditions with relative—but not complete—independence from the clinician's diagnostic judgment. Therefore, recent reviews of cohorts of patients ascertained with array CGH showed that the frequency of these duplications is much higher than heretofore appreciated (24, 55, 80, 81). As array CGH becomes the primary method of testing individuals with even mild DD/MR, the frequency of microduplications at the common microdeletion syndrome loci will likely increase.

Determining the genomic lesions in known conditions. Array CGH is proving to be a powerful instrument for the elucidation of the genomic etiology of known conditions (47). The discovery of a candidate gene for CHARGE syndrome (stands for Coloboma of the eye, Heart defects, Atresia of the choanae, Retardation of growth and/or development, Genital and/or urinary abnormalities, and Ear abnormalities and deafness) provides an illustration of the power of array CGH to study genetic disorders that have resisted years of investigation with traditional methods. Following several reports in which researchers unsuccessfully scanned the genomes of CHARGE subjects with CGH and microsatellite analysis, Vissers and colleagues (99) hybridized cell lines from two subjects with CHARGE syndrome onto a 1-Mb genome-wide array. After confirming a ~5-Mb deletion on chromosome 8q12 in one subject, with a BAC tiling resolution array, the authors hybridized DNA from a subject with CHARGE syndrome and an apparently balanced translocation. Array CGH detected a complex rearrangement at the 8q12 breakpoint, comprising two deletions that overlapped that of the first deletion subject. After determining the 2.3-Mb smallest region of overlap (SRO), the authors screened 17 other individuals with CHARGE, none of whom had copy-number changes. Sequencing of the nine genes in the SRO revealed 10 heterozygous mutations in *CHD7* in the 17 individuals without microdeletions, suggesting that deletion or mutation of this gene was causative for CHARGE (99). Array CGH is particularly useful in the diagnosis of rare syndromes (91, 106) because isolated reports are unlikely to elicit a pattern of malformation that will warrant clinical suspicion.

Discovering new syndromes. The discovery of new syndromes is perhaps the most dramatic impact of contemporary molecular cytogenetics on current medical practice. The use of array CGH has allowed the discovery of many new syndromes in the last few years and has revitalized clinical cytogenetics (6, 45, 61, 70, 71, 85, 86, 88, 89, 98). These syndromes were identified via the use of arrays that were built based on specific architectural features of the human genome (86), are targeted to specific areas of the genome such as the pericentromeric regions (6), have an arbitrary coverage of the genome (71), have a high-density tiling path (45), or are targeted to areas of suspected clinical significance (85). Regardless of how these arrays were built or used, many new syndromes will be discovered as the use of array CGH in the clinic continues.

Appreciating the prevalence of mosaicism in individuals with developmental delay. Even though the effect of mosaicism on embryonic development and pregnancy outcome is not entirely clear, mosaic chromosomal imbalances have been shown to affect the development of in vitro–generated preimplantation embryos (10). However, the detection of mosaicism in only 5% of aneuploid spontaneous miscarriages between 6–20 weeks gestation (34) and in only 1–2% of viable pregnancies screened by chorionic villus sampling (CVS) (48, 101) indicates that the incidence of mosaicism decreases through the first and second trimesters of pregnancy and is even rarer in live births. This dramatic reduction in mosaicism from the early stages of embryonic development through the late stages of clinically established pregnancies suggests that there is significant selection against mosaicism. Nevertheless, detecting low-level

Copy-number variant (CNV): Submicroscopic alteration resulting in relative excess or deficiency of a chromosome region in comparison with control populations

mosaicism for clinically significant chromosome abnormalities remained a pressing diagnostic challenge for conventional cytogenetic testing until the advent of array CGH. The first systematic study of mosaicism in a large cohort identified mosaicism in as little as 3% of the cells on the basis of metaphase counts (7). Array CGH has revealed a surprising prevalence of mosaicism in unselected populations of individuals who are referred for evaluation owing to DD, MR, and DF; the frequency of mosaicism among such individuals may be as high as 8–10% (7, 80). Perhaps most significant is that array CGH does not require the stimulation of cell cultures by phytohemagglutinin (PHA), as classically done in clinical cytogenetics, which may distort the percentage of mosaic cells and inhibit the detection of some mosaic abnormalities by chromosome analysis (7, 18).

Ascertaining the unexpected frequency of copy number variants across the genome. The use of array CGH has established that genomic copy number variations (CNV) are very common and involve a surprisingly large proportion of the genome (19, 36, 38, 54, 60, 72, 77, 87, 97, 102). Although some of these CNVs likely contribute to susceptibility to common diseases such as Alzheimer, autism, cancer, and glomerulonephritis (3, 56, 73, 76), the significance of most remains unclear (1, 52). The use of array CGH in research and clinical studies will further characterize and catalog genomic CNVs in healthy individuals as a foundation for assessing their putative implications for disease-associated phenotypes relevant to genetic conditions and to common complex disorders. The CNVs reported to date are documented in the TCAG (The Centre for Applied Genomics) Database of Genomic Variants (1), which contains almost 4000 copy number variant loci. However, information on CNVs and their clinical implications remains incomplete (90). This is likely due to a number of factors. First, the size distributions of CNVs detected is dependent on the technology used (early CNVs were detected by BAC arrays and tend to be of the order of approximately 100 kb or larger). Second, given the probable differences in frequency of particular CNVs in different populations, the variety of different CNVs observable in a given study may be significantly limited by the number and ethnic origin of individuals examined. Third, the majority of CNVs might be rare variants, occurring with a minor allele frequency <5%. Finally, the number of false-positive and false-negative CNVs in different studies may be variable, depending on the methodology, the sample quality, and the sample source (22).

Sporadic Disease

By blurring the line between the definition of a "molecular" test and a "cytogenetic" test, contemporary cytogenetics has caused convergence between these two specialties and made possible the investigation and diagnosis of conditions that were previously in the province of one or the other, but rarely in both. It has been known for more than a decade that what was once thought to be a mendelian condition, and usually assumed to be caused by point mutations or small rearrangements on the order of a few nucleotides of a single gene, may be caused by large genomic rearrangements, sometimes as large as a few megabases, that affect the copy number of one or more contiguous genes both on the affected DNA segment (16, 40, 44, 59, 75) and sometimes, through position effect, at loci quite distant from the affected segment (28). In addition, what is traditionally thought of as "sporadic" may be due to a chromosomal rearrangement. The frequencies of de novo chromosomal events are orders of magnitude larger than spontaneous point mutations and are likely to represent the overwhelming majority of "sporadic" disease due to genetic defects (84). Furthermore, this awareness of the frequency of such rearrangements should make clinicians sensitive to the possibility that many sporadic diseases may be caused by chromosomal alterations and perhaps multifactorial and complex traits may be caused or facilitated by de novo or inherited CNVs that are now detectable with contemporary cytogenetic methods.

The application of high-resolution genome analysis in research and clinical laboratories will uncover the genomic basis of many such disorders and will allow for better correlation of the many known CNVs with specific phenotypes. Although this promises to be a very challenging exercise (58), much work has already been initiated in cancer, neurological and neuropsychological conditions, infectious diseases, and others, suggesting that the clinical utility and applicability of such investigations cannot be too distant.

Cancer

Although somatic mutations, acquired chromosomal rearrangements of various sizes, and genomic instability are well known to be associated with oncogenesis and tumor progression (14, 50, 103), this review covers constitutional anomalies that cause or predispose to the development of specific malignancies in individuals or families. In our study of 8789 individuals with MR/DD/DF with array CGH (80), we identified 12 individuals with constitutional deletions (**Figure 1**) and four individuals with constitutional duplications at specific tumor suppressor genes but with no outward

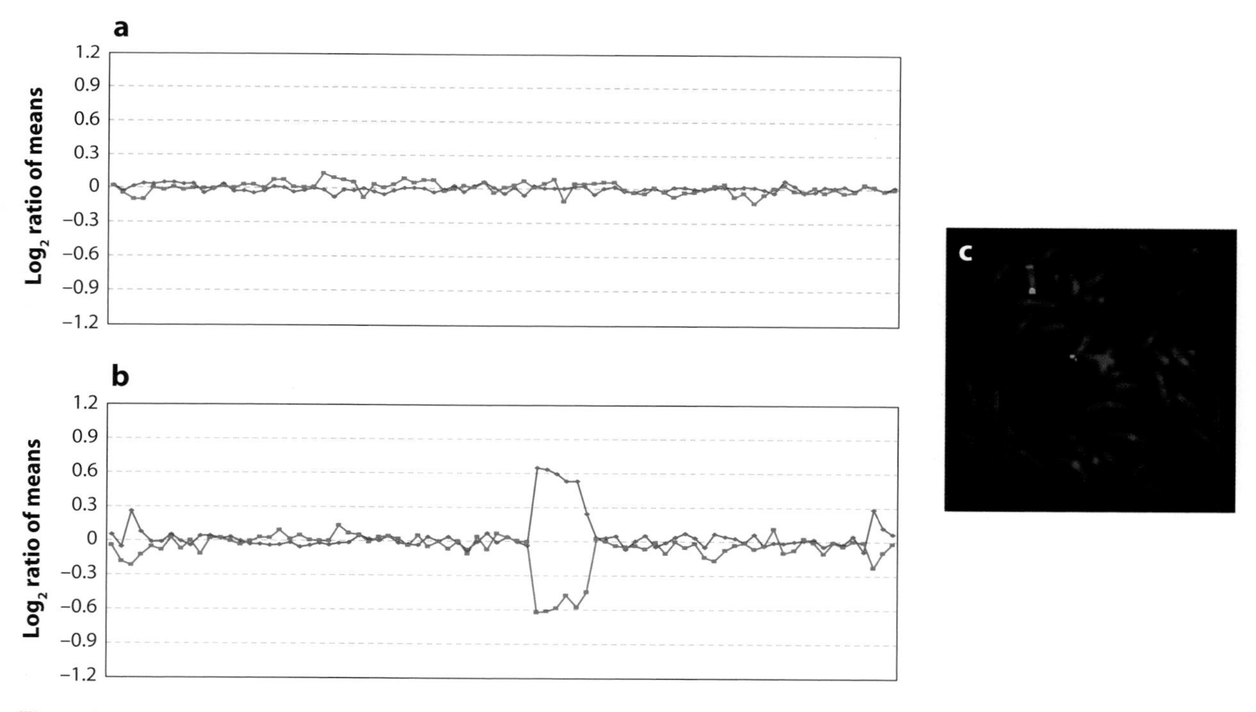

Figure 1

Microarray profile for deletion of 5q22.1–5q22.2 identified by microarray-based comparative genomic hybridization (array CGH). Each clone represented on the array is arranged along the x-axis according to its location on the chromosome with the most distal/telomeric p-arm clones on the left and the most distal/telomeric q-arm clones on the right. The blue lines represent the $\log_2$ ratios from the first experiment (control Cy5/patient Cy3), whereas the pink plots represent the $\log_2$ ratios obtained from the second experiment in which the dyes have been reversed (patient Cy5/control Cy3). (*a*) Plot for a normal chromosome 5 showing a ratio of 0 on a $\log_2$ scale for all clones. (*b*) Plot showing a single-copy loss of six bacterial artificial chromosome (BAC) clones at 5q22.1–5q22.2, approximately 1.8 Mb in size. This deleted region includes the adenomatosis polyposis coli (APC) locus as well as a region proximal to the APC locus. Single-copy losses of individual clones on both the distal p-arm and distal q-arm are normal copy-number variants. (*c*) Fluorescence in situ hybridization (FISH) analysis using a BAC clone from the APC region (RP11–107C15, *red*), confirming a deletion at 5q22.2. The subtelomeric probe to the short arm of chromosome 5 (*green*) was used as a control and showed a normal hybridization pattern.

clinical signs of cancer. Further investigation with colonoscopy of one individual with deletion of the APC locus on 5q22.1–5q22.2 identified thousands of colonic polyps and established the diagnosis of adenomatous polyposis of the colon, consistent with a clinical diagnosis of Gardner syndrome. A search for extracolonic tumors revealed thyroid carcinoma. Management consisted of successful total colectomy and thyroidectomy and close surveillance, resulting in a favorable outcome (35). Other studies with array CGH of constitutional CNVs associated with syndromic and nonsyndromic cancers have been performed and led to the identification of loci that cause or predispose to the emergence of cancers either as isolated anomalies or as part of recognizable cancer syndromes (4, 13, 23). As array CGH is used more broadly, our understanding of the many CNVs will evolve, allowing a better appreciation of the effect of these CNVs on the predisposition to, or causation of, cancer. This improved understanding will translate into clinical applications.

Neurological and Neuropsychological Conditions

Although many genomic disorders have been associated with neurological deficits/neurodevelopmental delays (**Table 1**), researchers have recently reported CNV-related findings in nonsyndromic neurodegenerative conditions. Novel findings in Alzheimer disease, Parkinson disease (PD), and autism are particularly significant. Rovelet-Lecrux and colleagues (73) reported five families with autosomal dominant Alzheimer disease in whom affected individuals have duplications at the amyloid precursor protein (APP) locus on chromosome 21 ranging from 0.58 to 6.37 Mb in size and containing 5 to 12 genes centromeric to the Down syndrome critical region. They estimated that as many as 8% of autosomal dominant early-onset Alzheimer disease (ADEOAD) patients carry this duplication. The frequency of this duplication in ADEOAD is as much as half that of missense mutations in APP (73). Despite an earlier report of three French individuals with sporadic Alzheimer disease and a duplication at the APP locus (21), the frequency of this duplication in sporadic Alzheimer disease is still unknown. The use of high-density arrays in the investigation of large cohorts of individuals with sporadic Alzheimer disease with dense coverage of the APP locus will help confirm this two-decades–old observation, will establish the frequency of such duplication in

Table 1 Common genomic disorders associated with neurological deficits/neurodevelopmental delays

Syndrome	OMIM #	Chromosome location
Williams-Beuren syndrome	194050	7q11.23
Williams-Beuren region duplication syndrome	609757	7q11.23
Angelman syndrome	105830	15q11-q13
Prader-Willi syndrome	176270	15q12
15q11-q13 duplication	None	15q11-q13
Miller-Dieker syndrome	247200	17p13.3
Smith-Magenis syndrome	182290	17p11.2
Potocki-Lupski syndrome	610883	17p11.2
Neurofibromatosis type 1	162200	17q11.2
DiGeorge syndrome	188400	22q11.2
Microduplication 22q11.2	608363	22q11.2
Rett syndrome	312750	Xq28
MECP2 duplication	None	Xq28
Pelizaeus-Merzbacher disease	312080	Xq22

this condition, and may identify other CNVs of clinical significance to Alzheimer disease.

PD is the second most common neurodegenerative condition after Alzheimer disease. The genetic basis of the majority of cases of sporadic PD is unknown, but some families with autosomal dominant PD have mutations in the α-synuclein (*SNCA*) locus on chromosome 4q21 (67), whereas other cases are caused by DNA copy gains (duplication and triplication) at the *SNCA* locus (17, 27, 39, 63, 92). Families with duplication of *SNCA*, in contrast to those with triplication, exhibit symptoms that resemble idiopathic Parkinson disease, which has a late age of onset and progresses slowly and in which neither cognitive decline nor dementia are prominent (17, 39, 63). These findings show that the severity of the familial Parkinson phenotype is dependent upon SNCA gene dosage, suggesting that the regulation of α-synuclein levels has potential significance in the pathogenesis and treatment of sporadic PD. The use of contemporary cytogenetic techniques in the case of familial PD may not only establish a diagnosis of an adult-onset neurodegenerative disease, but also allow for accurate prognostic counseling.

Efforts to map autism genes by linkage studies, association studies, and chromosomal abnormalities associated with the autism spectrum phenotype identified dozens of loci that span all chromosomes (100). Despite this confusing picture, a few loci are more commonly seen in individuals with autism. Chromosome regions of interest (CROIs) that would merit further investigation are not equally distributed across the human genome but tend to cluster on specific regions. These regions include chromosome 2 band q37, chromosome 5 band p14–p15, several locations on the long arm of chromosome 7, chromosome 15 band q11–q13, chromosome 11 band q25, chromosome 16 band q22.3, chromosome 17 band p11.2, chromosome 18 bands q21.1 and q23, chromosome 22 bands q11.2 and q13.3, and chromosome X band p22 (100). More recently, the first high-resolution array CGH analysis on a large cohort of families with autism, including 118 families with a single affected individual, 47 with multiple affected individuals, and 99 unaffected control families, found that de novo CNVs were significantly associated with autism (76). Such CNVs were found in 10% of individuals with sporadic autism, in 3% of patients with an affected first-degree relative, and in 1% of controls, establishing that de novo germline mutations are a more significant risk factor for autism than previously recognized (76).

CONCLUSIONS

The development of array CGH represents the latest advancement in molecular cytogenetics, which has revolutionized the field of clinical cytogenetics and expanded the repertoire of conditions that can be tested and that were, until recently, the concern of other disciplines. In addition to some malignancies and neurological conditions (reviewed above), there is a clear indication that constitutional chromosomal alterations that are detectable by these contemporary cytogenetic techniques may affect many organ systems and that CNVs may play important roles in multifactorial conditions. For example, it has been known that segmental duplications in the human genome are enriched for genes involved in immunity and that phenotypic consequences could result from these CNVs. In the last few years, it became clear that predisposition to HIV1/AIDS is influenced by segmental duplications containing the CC chemokine ligand 3-like 1 gene (*CCL3L1*) on 17q (31). A more striking example is the finding that a CNV involving the human *FCGR3B* (Fc fragment of IgG, low affinity IIIb, receptor), an ortholog of rat *Fcgr3*, is associated with glomerulonephritis in systemic lupus erythematosus and predisposes to immunologically mediated renal disease in both mammalian species (3). This work points to the importance of genome plasticity in the evolution of genetically complex phenotypes, including susceptibility to common human disease, and also highlights a potential role for the high-density molecular cytogenetic tools of the present and future in the diagnosis of such conditions.

With these new powerful diagnostic instruments comes great responsibility. Although our understanding of the CNVs and their roles in health and disease is improving, it is still rudimentary. Even with targeted arrays to genes of well-characterized function, it is common to identify alterations of unclear clinical significance that cause considerable interpretive difficulties for the laboratory and create a significant counseling dilemma for the clinician. As our understanding of genome architecture improves, diagnostic arrays for genome inspection will continue to require regular updating to stay abreast of all established novel regions responsible for abnormal human phenotypes. However, diagnostic laboratories should remain at least one step behind the cutting edge of research because they are concerned with the welfare of patients, not study subjects. With the dizzying pace of this revolution, it is the duty of the clinical diagnostician to protect patients from the inevitable mistakes that occur in the explorative world of research while providing the latest advancements in diagnostics. Attaining this delicate balance is a challenge.

Contemporary cytogenetics has quickly integrated the new array technologies into clinical use. Well-designed arrays for clinical use allow for straightforward interpretation and are likely to provide diagnoses in a substantial number of currently undiagnosed cases of human genetic disorders and—in the near future—in other multifactorial conditions. Array CGH has already transformed the practice of both medical genetics and clinical cytogenetics and is about to do the same for all areas of medicine, ushering in the new era of genomic medicine. New chromosomal syndromes have been identified, the chromosomal etiologies of known syndromes have been uncovered, and clinical phenotypes of chromosome imbalances have been expanded. The contemporary cytogenetics laboratory already has fewer microscopes than only a few years ago, and cytogeneticists are learning how to manipulate genomic DNA in addition to metaphase chromosomes. The scrupulous band-by-band analysis of metaphase chromosomes is quickly becoming obsolete and is being replaced with a cursory inspection of the metaphase spread to exclude apparently balanced translocations or inversions of large chromosomal segments. All these dramatic changes require the design of better training programs to prepare the trainees in cytogenetics to face the future challenges of this exciting discipline. These changes are also forcing medical schools and residency programs to redesign their curricula to incorporate courses in genetics and genomics, making genomic medicine a reality.

SUMMARY POINTS

1. The development of microarray-based comparative genomic hybridization (array CGH) represents the latest advancement in molecular cytogenetics.
2. This new technology highlights the complexity of the human genome.
3. The identification of increasingly more subtle DNA copy gain or loss is also redefining our presumed understanding of some multifactorial conditions.
4. The success of these contemporary molecular cytogenetic techniques has resulted in an exponential increase in genetic information that has far exceeded our ability to understand and use this flood of information in the clinical setting.
5. This technology is an unprecedented reversal of the usual order of the practice and progress of medicine, where clinical suspicion and medical acumen often direct laboratory investigation and suggest specific genetic lesions and mechanisms.
6. Clinicians are now able to scrutinize the genome for guidance in their clinical practice.

FUTURE ISSUES

1. Our understanding of copy number variations (CNVs) and their roles in health and disease will improve.
2. As our understanding of genome architecture improves, diagnostic arrays for genome inspection will establish novel regions responsible for abnormal human phenotypes.
3. Well-designed arrays for clinical use will diagnose many novel human genetic disorders and—in the near future—in other multifactorial conditions.
4. Array CGH is about to transform all areas of medicine, and will usher in a new era of genomic medicine.
5. All these dramatic changes require the design of better training programs to prepare the trainees in cytogenetics to face the future challenges of this exciting discipline.
6. These changes are also forcing medical schools and residency programs to redesign their curricula to incorporate courses in genetics and genomics, making genomic medicine a reality.

DISCLOSURE STATEMENT

The authors have ownership, receive consulting fees, and sit on the Members' board of Signature Genomic Laboratories.

ACKNOWLEDGMENTS

We thank Blake C. Ballif and Aaron P. Theisen (Signature Genomic Laboratories, LLC, Spokane, WA) for editing our manuscript.

LITERATURE CITED

1. The Centre for Applied Genomics. (TCAG) 2007. *Database of Genomic Variants—a curated catalogue of structural variation in the human genome*. **http://projects.tcag.ca/variation/**
2. Wellcome Trust Sanger Inst. 2007. *DECIPHER—DatabasE of Chromosomal Imbalance and Phenotype in Humans using Ensembl Resources*. **http://www.sanger.ac.uk/PostGenomics/decipher/**
3. Aitman TJ, Dong R, Vyse TJ, Norsworthy PJ, Johnson MD, et al. 2006. Copy number polymorphism in *Fcgr3* predisposes to glomerulonephritis in rats and humans. *Nature* 439:851–55
4. Ammerlaan AC, de Bustos C, Ararou A, Buckley PG, Mantripragada KK, et al. 2005. Localization of a putative low-penetrance ependymoma susceptibility locus to 22q11 using a chromosome 22 tiling-path genomic microarray. *Genes Chromosomes Cancer* 43:329–38
5. Anguiano A, Oates RD, Amos JA, Dean M, Gerrard B, et al. 1992. Congenital bilateral absence of the vas deferens. A primarily genital form of cystic fibrosis. *JAMA* 267:1794–97
6. Ballif BC, Hornor SA, Jenkins E, Madan-Khetarpal S, Surti U, et al. 2007. Discovery of a previously unrecognized microdeletion syndrome of 16p11.2-p12.2. *Nat. Genet.* 39:1071–73
7. Ballif BC, Rorem EA, Sundin K, Lincicum M, Gaskin S, et al. 2006. Detection of low-level mosaicism by array CGH in routine diagnostic specimens. *Am. J. Med. Genet. A* 140:2757–67
8. Bejjani BA, Saleki R, Ballif BC, Rorem EA, Sundin K, et al. 2005. Use of targeted array-based CGH for the clinical diagnosis of chromosomal imbalance: Is less more? *Am. J. Med. Genet. A* 134:259–67

9. Berg JS, Brunetti-Pierri N, Peters SU, Kang SH, Fong CT, et al. 2007. Speech delay and autism spectrum behaviors are frequently associated with duplication of the 7q11.23 Williams-Beuren syndrome region. *Genet. Med.* 9:427–41
10. Bielanska M, Tan SL, Ao A. 2002. Chromosomal mosaicism throughout human preimplantation development in vitro: incidence, type, and relevance to embryo outcome. *Hum. Reprod.* 17:413–19
11. Brewer C, Holloway S, Zawalnyski P, Schinzel A, FitzPatrick D. 1998. A chromosomal deletion map of human malformations. *Am. J. Hum. Genet.* 63:1153–59
12. Brewer C, Holloway S, Zawalnyski P, Schinzel A, FitzPatrick D. 1999. A chromosomal duplication map of malformations: regions of suspected haplo- and triplolethality–and tolerance of segmental aneuploidy–in humans. *Am. J. Hum. Genet.* 64:1702–8
13. Bruder CE, Hirvela C, Tapia-Paez I, Fransson I, Segraves R, et al. 2001. High resolution deletion analysis of constitutional DNA from neurofibromatosis type 2 (NF2) patients using microarray-CGH. *Hum. Mol. Genet.* 10:271–82
14. Cahill DP, Kinzler KW, Vogelstein B, Lengauer C. 1999. Genetic instability and darwinian selection in tumours. *Trends Cell Biol.* 9:M57–60
15. Caspersson T, Zech L, Johansson C, Modest EJ. 1970. Identification of human chromosomes by DNA-binding fluorescent agents. *Chromosoma* 30:215–27
16. Chance PF, Alderson MK, Leppig KA, Lensch MW, Matsunami N, et al. 1993. DNA deletion associated with hereditary neuropathy with liability to pressure palsies. *Cell* 72:143–51
17. Chartier-Harlin MC, Kachergus J, Roumier C, Mouroux V, Douay X, et al. 2004. α-synuclein locus duplication as a cause of familial Parkinson's disease. *Lancet* 364:1167–69
18. Cheung SW, Shaw CA, Scott DA, Patel A, Sahoo T, et al. 2007. Microarray-based CGH detects chromosomal mosaicism not revealed by conventional cytogenetics. *Am. J. Med. Genet. A* 143:1679–86
19. Conrad DF, Andrews TD, Carter NP, Hurles ME, Pritchard JK. 2006. A high-resolution survey of deletion polymorphism in the human genome. *Nat. Genet.* 38:75–81
20. Cytrynbaum CS, Smith AC, Rubin T, Weksberg R. 2005. Advances in overgrowth syndromes: clinical classification to molecular delineation in Sotos syndrome and Beckwith-Wiedemann syndrome. *Curr. Opin. Pediatr.* 17:740–46
21. Delabar JM, Goldgaber D, Lamour Y, Nicole A, Huret JL, et al. 1987. β amyloid gene duplication in Alzheimer's disease and karyotypically normal Down syndrome. *Science* 235:1390–92
22. de Smith AJ, Tsalenko A, Sampas N, Scheffer A, Yamada NA, et al. 2007. Array CGH analysis of copy number variation identifies 1284 new genes variant in healthy white males: implications for association studies of complex diseases. *Hum. Mol. Genet.* 16:2783–94
23. de Stahl TD, Hartmann C, de Bustos C, Piotrowski A, Benetkiewicz M, et al. 2005. Chromosome 22 tiling-path array-CGH analysis identifies germ-line- and tumor-specific aberrations in patients with glioblastoma multiforme. *Genes Chromosomes Cancer* 44:161–69
24. de Vries BB, Pfundt R, Leisink M, Koolen DA, Vissers LE, et al. 2005. Diagnostic genome profiling in mental retardation. *Am. J. Hum. Genet.* 77:606–16
25. Edwards JH, Harnden DG, Cameron AH, Crosse VM, Wolff OH. 1960. A new trisomic syndrome. *Lancet* 1:787–90
26. Ensenauer RE, Adeyinka A, Flynn HC, Michels VV, Lindor NM, et al. 2003. Microduplication 22q11.2, an emerging syndrome: clinical, cytogenetic, and molecular analysis of thirteen patients. *Am. J. Hum. Genet.* 73:1027–40
27. Farrer M, Kachergus J, Forno L, Lincoln S, Wang DS, et al. 2004. Comparison of kindreds with parkinsonism and alpha-synuclein genomic multiplications. *Ann. Neurol.* 55:174–79
28. Feuk L, Marshall CR, Wintle RF, Scherer SW. 2006. Structural variants: changing the landscape of chromosomes and design of disease studies. *Hum. Mol. Genet.* 15(Spec No 1):R57–66
29. Flint J, Knight S. 2003. The use of telomere probes to investigate submicroscopic rearrangements associated with mental retardation. *Curr. Opin. Genet. Dev.* 13:310–16
30. Goldberg R, Motzkin B, Marion R, Scambler PJ, Shprintzen RJ. 1993. Velo-cardio-facial syndrome: a review of 120 patients. *Am. J. Med. Genet.* 45:313–19
31. Gonzalez E, Kulkarni H, Bolivar H, Mangano A, Sanchez R, et al. 2005. The influence of *CCL3L1* gene-containing segmental duplications on HIV-1/AIDS susceptibility. *Science* 307:1434–40

32. Gropman AL, Elsea S, Duncan WC Jr, Smith AC. 2007. New developments in Smith-Magenis syndrome (del 17p11.2). *Curr. Opin. Neurol.* 20:125–34
33. Hassed SJ, Hopcus-Niccum D, Zhang L, Li S, Mulvihill JJ. 2004. A new genomic duplication syndrome complementary to the velocardiofacial (22q11 deletion) syndrome. *Clin. Genet.* 65:400–4
34. Hassold T. 1982. Mosaic trisomies in human spontaneous abortions. *Hum. Genet.* 61:31–35
35. Heald B, Moran R, Milas M, Eng C. 2007. Familial adenomatous polyposis in a patient with unexplained mental retardation. *Nat. Clin. Pract. Neurol.* 3:694–700
36. Hinds DA, Stuve LL, Nilsen GB, Halperin E, Eskin E, et al. 2005. Whole-genome patterns of common DNA variation in three human populations. *Science* 307:1072–79
37. Hsu TC. 1952. Mammalian chromosomes in vitro: 1. The karyotype of man. *J. Hered.* 43:167–72
38. Iafrate AJ, Feuk L, Rivera MN, Listewnik ML, Donahoe PK, et al. 2004. Detection of large-scale variation in the human genome. *Nat. Genet.* 36:949–51
39. Ibanez P, Bonnet AM, Debarges B, Lohmann E, Tison F, et al. 2004. Causal relation between α-synuclein gene duplication and familial Parkinson's disease. *Lancet* 364:1169–71
40. Inoue K, Osaka H, Sugiyama N, Kawanishi C, Onishi H, et al. 1996. A duplicated PLP gene causing Pelizaeus-Merzbacher disease detected by comparative multiplex PCR. *Am. J. Hum. Genet.* 59:32–39
41. Kirchhoff M, Gerdes T, Rose H, Maahr J, Ottesen AM, Lundsteen C. 1998. Detection of chromosomal gains and losses in comparative genomic hybridization analysis based on standard reference intervals. *Cytometry* 31:163–73
42. Kirchhoff M, Rose H, Lundsteen C. 2001. High resolution comparative genomic hybridisation in clinical cytogenetics. *J. Med. Genet.* 38:740–44
43. Knight SJ, Flint J. 2004. The use of subtelomeric probes to study mental retardation. *Methods Cell Biol.* 75:799–831
44. Konrad M, Saunier S, Heidet L, Silbermann F, Benessy F, et al. 1996. Large homozygous deletions of the 2q13 region are a major cause of juvenile nephronophthisis. *Hum. Mol. Genet.* 5:367–71
45. Koolen DA, Vissers LE, Pfundt R, de Leeuw N, Knight SJ, et al. 2006. A new chromosome 17q21.31 microdeletion syndrome associated with a common inversion polymorphism. *Nat. Genet.* 38:999–1001
46. Krepischi-Santos AC, Vianna-Morgante AM, Jehee FS, Passos-Bueno MR, Knijnenburg J, et al. 2006. Whole-genome array-CGH screening in undiagnosed syndromic patients: old syndromes revisited and new alterations. *Cytogenet. Genome Res.* 115:254–61
47. Lalani SR, Stockton DW, Bacino C, Molinari LM, Glass NL, et al. 2003. Toward a genetic etiology of CHARGE syndrome: I. A systematic scan for submicroscopic deletions. *Am. J. Med. Genet. A* 118:260–66
48. Ledbetter DH, Zachary JM, Simpson JL, Golbus MS, Pergament E, et al. 1992. Cytogenetic results from the U.S. Collaborative Study on CVS. *Prenat. Diagn.* 12:317–45
49. Lejeune J, Gautier M, Turpin R. 1959. Study of somatic chromosomes from 9 mongoloid children. *C. R. Acad Sci.* 248:1721–22
50. Lengauer C, Kinzler KW, Vogelstein B. 1998. Genetic instabilities in human cancers. *Nature* 396:643–49
51. Levy B, Dunn TM, Kaffe S, Kardon N, Hirschhorn K. 1998. Clinical applications of comparative genomic hybridization. *Genet. Med.* 1:4–12
52. Levy S, Sutton G, Ng PC, Feuk L, Halpern AL, et al. 2007. The diploid genome sequence of an individual human. *PLoS Biol.* 5:e254
53. Lichter P, Joos S, Bentz M, Lampel S. 2000. Comparative genomic hybridization: uses and limitations. *Semin. Hematol.* 37:348–57
54. Locke DP, Sharp AJ, McCarroll SA, McGrath SD, Newman TL, et al. 2006. Linkage disequilibrium and heritability of copy-number polymorphisms within duplicated regions of the human genome. *Am. J. Hum. Genet.* 79:275–90
55. Lu X, Shaw CA, Patel A, Li J, Cooper ML, et al. 2007. Clinical implementation of chromosomal microarray analysis: summary of 2513 postnatal cases. *PLoS ONE* 2:e327
56. Lucito R, Suresh S, Walter K, Pandey A, Lakshmi B, et al. 2007. Copy-number variants in patients with a strong family history of pancreatic cancer. *Cancer Biol. Ther.* (In press)
57. Lupski JR. 1998. Genomic disorders: structural features of the genome can lead to DNA rearrangements and human disease traits. *Trends Genet.* 14:417–22

58. Lupski JR. 2007. Genomic rearrangements and sporadic disease. *Nat. Genet.* 39:S43–47
59. Lupski JR, de Oca-Luna RM, Slaugenhaupt S, Pentao L, Guzzetta V, et al. 1991. DNA duplication associated with Charcot-Marie-Tooth disease type 1A. *Cell* 66:219–32
60. McCarroll SA, Hadnott TN, Perry GH, Sabeti PC, Zody MC, et al. 2006. Common deletion polymorphisms in the human genome. *Nat. Genet.* 38:86–92
61. Mikhail FM, Descartes M, Piotrowski A, Andersson R, de Stahl TD, et al. 2007. A previously unrecognized microdeletion syndrome on chromosome 22 band q11.2 encompassing the BCR gene. *Am. J. Med. Genet. A* 143:2178–84
62. Ming JE, Geiger E, James AC, Ciprero KL, Nimmakayalu M, et al. 2006. Rapid detection of submicroscopic chromosomal rearrangements in children with multiple congenital anomalies using high density oligonucleotide arrays. *Hum. Mutat.* 27:467–73
63. Nishioka K, Hayashi S, Farrer MJ, Singleton AB, Yoshino H, et al. 2006. Clinical heterogeneity of α-synuclein gene duplication in Parkinson's disease. *Ann. Neurol.* 59:298–309
64. Nowell PC, Hungerford DA. 1960. A minute chromosome in human chronic granulocytic leukemia. *Science* 132:1488–501
65. Patau K, Smith DW, Therman E, Inhorn SL, Wagner HP. 1960. Multiple congenital anomaly caused by an extra autosome. *Lancet* 1:790–3
66. Pinkel D, Segraves R, Sudar D, Clark S, Poole I, et al. 1998. High resolution analysis of DNA copy number variation using comparative genomic hybridization to microarrays. *Nat. Genet.* 20:207–11
67. Polymeropoulos MH, Lavedan C, Leroy E, Ide SE, Dehejia A, et al. 1997. Mutation in the alpha-synuclein gene identified in families with Parkinson's disease. *Science* 276:2045–47
68. Potocki L, Bi W, Treadwell-Deering D, Carvalho CM, Eifert A, et al. 2007. Characterization of Potocki-Lupski syndrome (dup(17)(p11.2p11.2)) and delineation of a dosage-sensitive critical interval that can convey an autism phenotype. *Am. J. Hum. Genet.* 80:633–49
69. Potocki L, Chen KS, Park SS, Osterholm DE, Withers MA, et al. 2000. Molecular mechanism for duplication 17p11.2- the homologous recombination reciprocal of the Smith-Magenis microdeletion. *Nat. Genet.* 24:84–87
70. Rajcan-Separovic E, Harvard C, Liu X, McGillivray B, Hall JG, et al. 2007. Clinical and molecular cytogenetic characterisation of a newly recognised microdeletion syndrome involving 2p15–16.1. *J. Med. Genet.* 44:269–76
71. Redon R, Baujat G, Sanlaville D, Le Merrer M, Vekemans M, et al. 2006. Interstitial 9q22.3 microdeletion: clinical and molecular characterisation of a newly recognised overgrowth syndrome. *Eur. J. Hum. Genet.* 14:759–67
72. Redon R, Ishikawa S, Fitch KR, Feuk L, Perry GH, et al. 2006. Global variation in copy number in the human genome. *Nature* 444:444–54
73. Rovelet-Lecrux A, Hannequin D, Raux G, Le Meur N, Laquerriere A, et al. 2006. APP locus duplication causes autosomal dominant early-onset Alzheimer disease with cerebral amyloid angiopathy. *Nat. Genet.* 38:24–26
74. Rowley JD. 1973. Letter: A new consistent chromosomal abnormality in chronic myelogenous leukaemia identified by quinacrine fluorescence and Giemsa staining. *Nature* 243:290–93
75. Saunier S, Calado J, Heilig R, Silbermann F, Benessy F, et al. 1997. A novel gene that encodes a protein with a putative src homology 3 domain is a candidate gene for familial juvenile nephronophthisis. *Hum. Mol. Genet.* 6:2317–23
76. Sebat J, Lakshmi B, Malhotra D, Troge J, Lese-Martin C, et al. 2007. Strong association of de novo copy number mutations with autism. *Science* 316:445–49
77. Sebat J, Lakshmi B, Troge J, Alexander J, Young J, et al. 2004. Large-scale copy number polymorphism in the human genome. *Science* 305:525–28
78. Selzer RR, Richmond TA, Pofahl NJ, Green RD, Eis PS, et al. 2005. Analysis of chromosome breakpoints in neuroblastoma at subkilobase resolution using fine-tiling oligonucleotide array CGH. *Genes Chromosomes Cancer* 44:305–19
79. Shaffer LG. 1997. Diagnosis of microdeletion syndromes by fluorescence in situ hybridization (FISH). In *Current Protocols in Human Genetics*, ed. NC Dracopoli, JL Haines, BR Korf, DT Moir, CC Morton, et al., pp. 1–14. New York: Wiley

80. Shaffer LG, Bejjani BA, Torchia B, Kirkpatrick S, Coppinger J, Ballif BC. 2007. The identification of microdeletion syndromes and other chromosome abnormalities: Cytogenetic methods of the past, new technologies for the future. *Am. J. Med. Genet. C* 145:335–45
81. Shaffer LG, Kashork CD, Saleki R, Rorem E, Sundin K, et al. 2006. Targeted genomic microarray analysis for identification of chromosome abnormalities in 1500 consecutive clinical cases. *J. Pediatr.* 149:98–102
82. Shaffer LG, Kennedy GM, Spikes AS, Lupski JR. 1997. Diagnosis of CMT1A duplications and HNPP deletions by interphase FISH: implications for testing in the cytogenetics laboratory. *Am. J. Med. Genet.* 69:325–31
83. Shaffer LG, Ledbetter DH, Lupski JR. 2001. Molecular cytogenetics of contiguous gene syndromes: mechanisms and consequences of gene dosage imbalance. In *Metabolic and Molecular Basis of Inherited Disease*, ed. CR Scriver, AL Beaudet, WS Sly, D Valle, B Childs, et al., pp. 1291–324. New York: McGraw Hill
84. Shaffer LG, Lupski JR. 2000. Molecular mechanisms for constitutional chromosomal rearrangements in humans. *Annu. Rev. Genet.* 34:297–329
85. Shaffer LG, Theisen A, Bejjani BA, Ballif BC, Aylsworth AS, et al. 2007. The discovery of microdeletion syndromes in the postgenomic era: review of the methodology and characterization of a new 1q41q42 microdeletion syndrome. *Genet. Med.* 9:607–16
86. Sharp AJ, Hansen S, Selzer RR, Cheng Z, Regan R, et al. 2006. Discovery of previously unidentified genomic disorders from the duplication architecture of the human genome. *Nat. Genet.* 38:1038–42
87. Sharp AJ, Locke DP, McGrath SD, Cheng Z, Bailey JA, et al. 2005. Segmental duplications and copy-number variation in the human genome. *Am. J. Hum. Genet.* 77:78–88
88. Sharp AJ, Selzer RR, Veltman JA, Gimelli S, Gimelli G, et al. 2007. Characterization of a recurrent 15q24 microdeletion syndrome. *Hum. Mol. Genet.* 16:567–72
89. Shaw-Smith C, Pittman AM, Willatt L, Martin H, Rickman L, et al. 2006. Microdeletion encompassing *MAPT* at chromosome 17q21.3 is associated with developmental delay and learning disability. *Nat. Genet.* 38:1032–37
90. Shianna KV, Willard HF. 2006. Human genomics: in search of normality. *Nature* 444:428–29
91. Shieh JT, Aradhya S, Novelli A, Manning MA, Cherry AM, et al. 2006. Nablus mask-like facial syndrome is caused by a microdeletion of 8q detected by array-based comparative genomic hybridization. *Am. J. Med. Genet. A* 140:1267–73
92. Singleton AB, Farrer M, Johnson J, Singleton A, Hague S, et al. 2003. α-Synuclein locus triplication causes Parkinson's disease. *Science* 302:841
93. Somerville MJ, Mervis CB, Young EJ, Seo EJ, del Campo M, et al. 2005. Severe expressive-language delay related to duplication of the Williams-Beuren locus. *N. Engl. J. Med.* 353:1694–701
94. Stankiewicz P, Beaudet AL. 2007. Use of array CGH in the evaluation of dysmorphology, malformations, developmental delay, and idiopathic mental retardation. *Curr. Opin. Genet. Dev.* 17:182–92
95. Tijo HJ, Levan A. 1956. The chromosome numbers of man. *Hereditas* 42:1–6
96. Trask BJ. 1991. Fluorescence in situ hybridization: applications in cytogenetics and gene mapping. *Trends Genet.* 7:149–54
97. Tuzun E, Sharp AJ, Bailey JA, Kaul R, Morrison VA, et al. 2005. Fine-scale structural variation of the human genome. *Nat. Genet.* 37:727–32
98. Varela MC, Krepischi-Santos AC, Paz JA, Knijnenburg J, Szuhai K, et al. 2006. A 17q21.31 microdeletion encompassing the *MAPT* gene in a mentally impaired patient. *Cytogenet. Genome Res.* 114:89–92
99. Vissers LE, van Ravenswaaij CM, Admiraal R, Hurst JA, de Vries BB, et al. 2004. Mutations in a new member of the chromodomain gene family cause CHARGE syndrome. *Nat. Genet.* 36:955–57
100. Vorstman JA, Staal WG, van Daalen E, van Engeland H, Hochstenbach PF, Franke L. 2006. Identification of novel autism candidate regions through analysis of reported cytogenetic abnormalities associated with autism. *Mol. Psychiatry* 11:18–28
101. Wang BB, Rubin CH, Williams J 3rd. 1993. Mosaicism in chorionic villus sampling: an analysis of incidence and chromosomes involved in 2612 consecutive cases. *Prenat. Diagn.* 13:179–90
102. Wong KK, deLeeuw RJ, Dosanjh NS, Kimm LR, Cheng Z, et al. 2007. A comprehensive analysis of common copy-number variations in the human genome. *Am. J. Hum. Genet.* 80:91–104

103. Wood LD, Parsons DW, Jones S, Lin J, Sjöblom T, et al. 2007. The genomic landscapes of human breast and colorectal cancers. *Science* 318:1108–13
104. Yobb TM, Somerville MJ, Willatt L, Firth HV, Harrison K, et al. 2005. Microduplication and triplication of 22q11.2: a highly variable syndrome. *Am. J. Hum. Genet.* 76:865–76
105. Yunis JJ. 1976. High resolution of human chromosomes. *Science* 191:1268–70
106. Zweier C, Peippo MM, Hoyer J, Sousa S, Bottani A, et al. 2007. Haploinsufficiency of TCF4 causes syndromal mental retardation with intermittent hyperventilation (Pitt-Hopkins Syndrome). *Am. J. Hum. Genet.* 80:994–1001

The Role of Aminoacyl-tRNA Synthetases in Genetic Diseases*

Anthony Antonellis and Eric D. Green

Genome Technology Branch, National Human Genome Research Institute, National Institutes of Health, Bethesda, Maryland 20892; email: egreen@nhgri.nih.gov

Annu. Rev. Genomics Hum. Genet. 2008. 9:87–107

First published online as a Review in Advance on May 9, 2008

The *Annual Review of Genomics and Human Genetics* is online at genom.annualreviews.org

This article's doi:
10.1146/annurev.genom.9.081307.164204

1527-8204/08/0922-0087$20.00

Key Words

Charcot-Marie-Tooth disease, neurodegeneration, peripheral neuropathy, protein synthesis, translation

Abstract

Aminoacyl-tRNA synthetases (ARSs) are ubiquitously expressed, essential enzymes responsible for performing the first step of protein synthesis. Specifically, ARSs attach amino acids to their cognate tRNA molecules in the cytoplasm and mitochondria. Recent studies have demonstrated that mutations in genes encoding ARSs can result in neurodegeneration, raising many questions about the role of these enzymes (and protein synthesis in general) in neuronal function. In this review, we summarize the current knowledge of genetic diseases that are associated with mutations in ARS-encoding genes, discuss the potential pathogenic mechanisms underlying these disorders, and point to likely areas of future research that will advance our understanding about the role of ARSs in genetic diseases.

tRNA: transfer ribonucleic acid

Aminoacyl-tRNA synthetases (ARSs): the family of enzymes responsible for performing aminoacylation reactions in the cytoplasm and mitochondria

BACKGROUND ON AMINOACYL-tRNA SYNTHETASES

The transfer of biological information from DNA to RNA to protein is critical for the survival and propagation of cells, tissues, and organisms. One key component of this central dogma is protein translation, which involves using the genetic code to translate genetic information (in the form of messenger RNA) to produce protein. The first essential step of protein translation involves covalently attaching an amino acid to its cognate transfer RNA (tRNA). This process (often referred to as tRNA charging) is performed by a highly specialized group of enzymes, the aminoacyl-tRNA synthetases (ARSs) (11) (**Figure 1**). There is at least one ARS enzyme designated for each amino acid. Reflecting their fundamental importance for cellular life, ARSs are ubiquitously expressed enzymes that are present in species ranging from bacteria to humans.

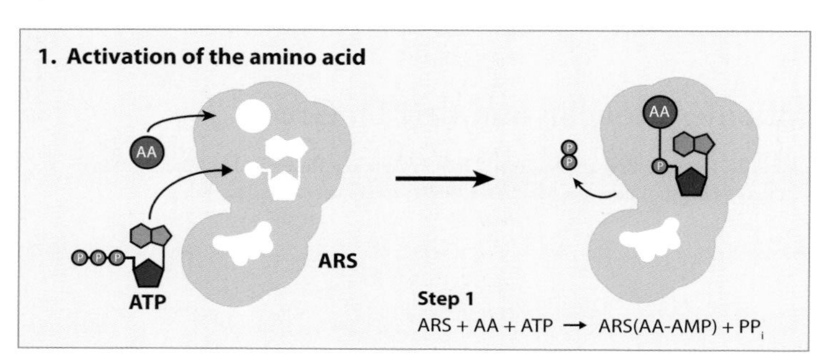

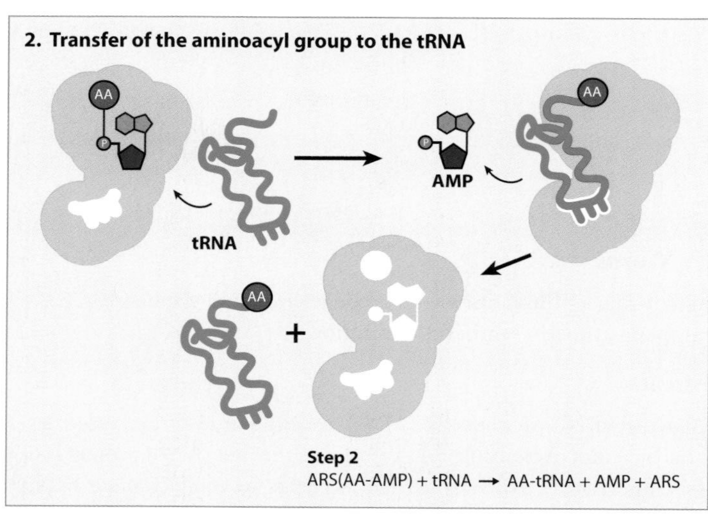

Figure 1

The two-step aminoacylation reaction. Each aminoacyl-tRNA synthetase (ARS) charges a specific tRNA molecule with its cognate amino acid (AA) via a two-step enzymatic reaction. In the first step, the ARS binds the AA and an ATP molecule to form the aminoacyl adenylate (AA-AMP) intermediate, and a pyrophosphate molecule (PP_i) is released. In the second step, a tRNA molecule binds the ARS via the anticodon binding domain (*white space on ARS with three extensions*), and the AA is transferred to the tRNA. An AMP molecule is then released, followed by the charged tRNA. The ARS is then free to charge another tRNA molecule. The chemical equation for each step is provided in the lower right in each case.

In a typical human cell, tRNAs are charged (or aminoacylated) in three locations. First, this process takes place in the nucleus to ensure that nuclear-encoded tRNAs become charged (30); the resulting tRNA:amino acid complexes are then exported to the cytoplasm. Second, tRNAs are charged by ARSs in the cytoplasm, and the tRNA:amino acid complexes are then transported to the ribosome for nascent polypeptide elongation (29). The cytoplasmic ARSs are then free to charge additional tRNA molecules, which allows the cycle to continue. Finally, because protein translation also occurs in the mitochondria, ARS activity is required in these organelles (7). To facilitate the latter process, nuclear-encoded ARSs are imported into the mitochondria; this step involves the use of specific localization signals within the ARS proteins. Thus, each ARS can be categorized as cytoplasm-specific, mitochondria-specific, or bifunctional (i.e., involved in charging tRNA molecules in both locations) (**Table 1**). ARSs can also be categorized into two groups on the basis of their protein structural characteristics. Group I ARSs contain a Rossman (parallel β-sheet nucleotide-binding) fold in their catalytic domain, whereas Group II ARSs share other homologous amino acid sequence motifs, termed motif 1, 2, and 3 (3).

Thirty-six ARSs perform all the required aminoacylation of tRNAs in humans: 16 act exclusively in the cytoplasm, 17 act exclusively in the mitochondria, and 3 are bifunctional (**Table 1**). The reason for the discrepancy between the number of available cytoplasmic ARSs (i.e., 19) and the total number of amino

Table 1: Human aminoacyl-tRNA synthetases (ARSs)

Cytoplasmic

Gene symbol	Name	RefSeq number	Location
AARS	alanyl-tRNA synthetase	NM_001605	16q22
CARS	cysteinyl-tRNA synthetase	NM_001751	11p15.5
DARS	aspartyl-tRNA synthetase	NM_001349	2q21.3
EPRS	glutamyl-prolyl-tRNA synthetase	NM_004446	1q41-q42
FARSA	phenylalanyl-tRNA synthetase alpha subunit	NM_004461	19p13.2
FARSB	phenylalanyl-tRNA synthetase beta subunit	NM_005687	2q36.1
HARS	histidyl-tRNA synthetase	NM_002109	5q31.3
IARS	isoleucyl-tRNA synthetase	NM_002161	9q21
LARS	leucyl-tRNA synthetase	NM_020117	5q32
MARS	methionyl-tRNA synthetase	NM_004990	12q13.2
NARS	asparaginyl-tRNA synthetase	NM_004539	18q21.2-q21.3
RARS	arginyl-tRNA synthetase	NM_002887	5q35.1
SARS	seryl-tRNA synthetase	NM_006513	1p13.1-p13.3
TARS	threonyl-tRNA synthetase	NM_152295	5p13.2
VARS	valyl-tRNA synthetase	NM_006295	6p21.3
WARS	tryptophanyl-tRNA synthetase	NM_004184	14q32.31
YARS	tyrosyl-tRNA synthetase	NM_003680	1p35.1

Mitochondrial

Gene symbol	Name	RefSeq number	Location
AARS2	alanyl-tRNA synthetase 2	NM_020745	6p21.1
CARS2	cysteinyl-tRNA synthetase 2	NM_024537	13q34
DARS2	aspartyl-tRNA synthetase 2	NM_018122	1q25.1
EARS2	glutamyl-tRNA synthetase 2	NM_001083614	16p12.1
FARS2	phenylalanyl-tRNA synthetase 2	NM_006567	6p25.1
HARS2	histidyl-tRNA synthetase 2	NM_012208	5q31.3
IARS2	isoleucyl-tRNA synthetase 2	NM_018060	1q41
LARS2	leucyl-tRNA synthetase 2	NM_015340	3p21.3
MARS2	methionyl-tRNA synthetase 2	NM_138395	2q33.1
NARS2	asparaginyl-tRNA synthetase 2	NM_024678	11q14.1
PARS2	prolyl-tRNA synthetase 2	NM_152268	1p32.2
RARS2	arginyl-tRNA synthetase 2	NM_020320	6q16.1
SARS2	seryl-tRNA synthetase 2	NM_017827	19q13.2
TARS2	threonyl-tRNA synthetase 2	NM_025150	1q21.2
VARS2	valyl-tRNA synthetase 2	NM_020442	6p21.3
WARS2	tryptophanyl-tRNA synthetase 2	NM_015836	1p13.1-p13.3
YARS2	tyrosyl-tRNA synthetase 2	NM_001040436	12p11.21

Bifunctional

Gene symbol	Name	RefSeq number	Location
GARS	glycyl-tRNA synthetase	NM_002047	7p15
KARS	lysyl-tRNA synthetase	NM_005548	16q23-q24
QARS	glutaminyl-tRNA synthetase	NM_005051	3p21.1-p21.3

Aminoacylation: the biochemical reaction involving the covalent attachment of an amino acid to its cognate tRNA molecule

GARS: human glycyl-tRNA synthetase

YARS: human tyrosyl-tRNA synthetase

MSC: multiple synthetase (ARS) complex

Noncanonical function: secondary function unrelated to the primary or canonical function of a protein in a cell, tissue, or organ

acids (i.e., 20) relates to the fact that one ARS (glutamyl-prolyl-tRNA synthetase) is responsible for charging $tRNA^{Glu}$ and $tRNA^{Pro}$ molecules in the cytoplasm with glutamic acid and proline, respectively (6). Thirty-seven known or predicted ARS genes are encoded in the human nuclear genome (**Table 1**). The nomenclature for human ARS genes, in most cases, involves the use of the single-letter amino acid code followed by '*ARS*' (e.g., the gene symbol for glycyl-tRNA synthetase is *GARS*). In the case of mitochondria-specific ARSs, a '2' is added at the end (e.g., the gene symbol for the mitochondria-specific tyrosyl-tRNA synthetase is *YARS2*).

An interesting feature of ARS biology is that some of these proteins have a secondary, noncanonical function (26, 33). In humans, these functions include a wide array of activities that range from transcriptional control to angiogenic signaling (**Table 2**). Secondary ARS activities have also been demonstrated in other organisms, including a putative role for GlyRS (the yeast ortholog of human GARS) in mRNA formation (21). Such secondary functions are particularly relevant to consider when studying the role of ARSs in human disease (see below). Furthermore, cell-imaging and biochemical studies have revealed that a subset of ARSs reside in multiple synthetase complexes (MSCs) in the cytoplasm of human cells (26, 33). To date, nine of the human cytoplasmic ARSs have been localized to MSCs; however, the presence/absence of the other ten cytoplasmic ARSs in such complexes has not been completely established. Although the precise role of these MSCs remains unclear, two possibilities have been proposed (33). First, MSCs may increase the efficiency of tRNA charging by providing subcellular sites for translation. Second, MSCs may play a role in regulating the secondary functions of ARSs, or may, as a whole, have a discrete secondary (noncanonical) function. A more complete determination of the ARS content of MSCs in various cell types may provide additional insight about these possibilities.

Mutations in specific ARS genes are responsible for tissue-specific disease phenotypes in humans and mice. This is particularly surprising in light of the ubiquitous nature of ARS expression and function. Here, we present a summary of the current knowledge of the role of ARSs in human diseases and discuss the

Table 2: ARS secondary functions

Species	ARS	Activity
Homo sapiens	EPRS	Silences translation
Homo sapiens	KARS	Regulates transcription Involved in HIV packaging Promotes inflammatory response
Homo sapiens	MARS	Promotes rRNA biogenesis
Homo sapiens	QARS	Inhibits apoptosis
Homo sapiens	WARS	Inhibits angiogenic signaling
Homo sapiens	YARS	Promotes angiogenic signaling Stimulates immune cells
Saccharomyces cerevisiae	GlyRS	Involved in mRNA 3'-end formation
Saccharomyces cerevisiae	LeuRS	Involved in group I intron splicing
Neurospora crassa	TyrRS (mitochondrial)	Involved in group I intron splicing
Escherichia coli	alaS	Represses its own transcription
Escherichia coli	tyrS	Represses its own translation

(Table adapted from References 26 and 33)

potential pathogenic mechanisms that underlie these disorders.

DISEASES CAUSED BY MUTATIONS IN AMINOACYL-tRNA SYNTHETASE GENES

The first report of a disease-causing mutation in an ARS gene was published in 2003 (1). Since then, disease-causing mutations in three additional ARS genes have been reported (15, 22, 37) (**Figure 2** and **Table 3**). Surprisingly, these mutations are associated mostly with diseases of the nervous system, including encephalopathy, cerebellar ataxia, and peripheral neuropathy. These data point to the importance of ARSs in the development and function of neurons, and suggest that a broader set of (or perhaps

Peripheral neuropathy: a heterogeneous group of peripheral nerve disorders generally characterized by motor and/or sensory impairment in the distal extremities

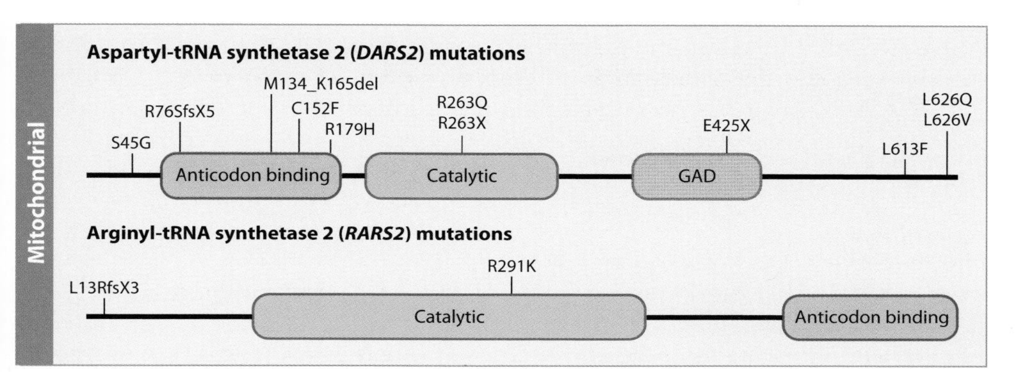

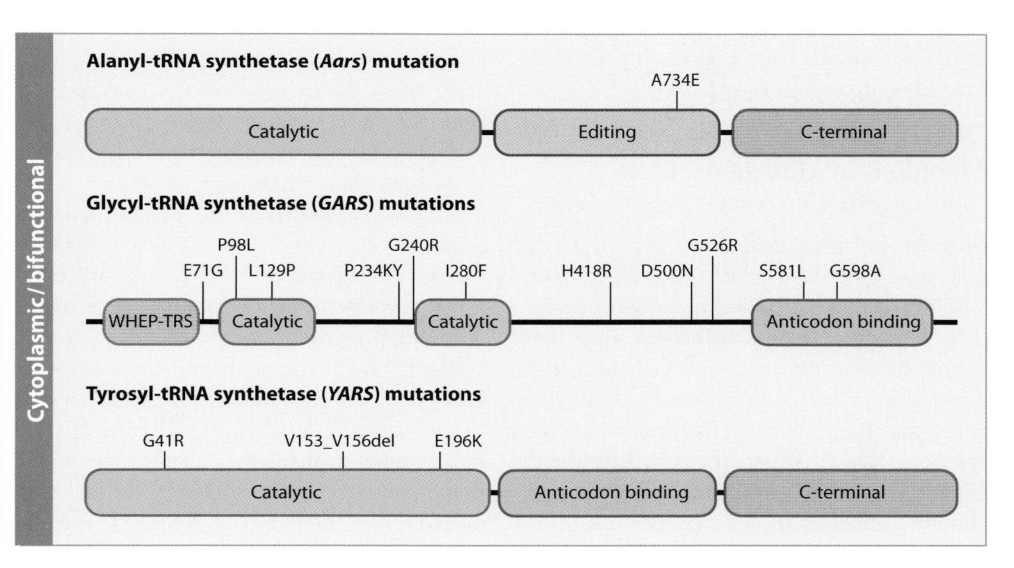

Figure 2

Location of known aminoacyl-tRNA synthetase (ARS) mutations relative to the corresponding protein sequences. The known functional domains of aspartyl-tRNA synthetase 2 (DARS2), arginyl-tRNA synthetase 2 (RARS2), alanyl-tRNA synthetase (Aars), glycyl-tRNA synthetase (GARS), and tyrosyl-tRNA synthetase (YARS) include the core catalytic domain (*red*), tRNA anticodon-binding domain (*blue*), C-terminal domain (*green*), and the editing domain (*purple*). Additional domains of unknown function (*tan*) are also shown. The relative position of each mutation in the mitochondrial (*top*) and cytoplasmic/bifunctional (*bottom*) ARS proteins is indicated. Note that P98L and P234KY GARS were identified in fruit fly and mouse, respectively; here, the position is shown relative to the human protein.

Table 3: Disease phenotypes associated with ARS mutations

Species	Gene	Phenotype	Mode of inheritance	No. of mutations	Mutation class	References
Human	*GARS*	Charcot-Marie-Tooth disease type 2D (CMT2D) and distal spinal muscular atrophy type V	Autosomal dominant	9	Missense	1, 11, 20, 40
Human	*YARS*	Dominant intermediate Charcot-Marie-Tooth disease (DI-CMT)	Autosomal dominant	3	Missense, in-frame deletion	22
Human	*DARS2*	Leukoencephalopathy with brain stem and spinal cord involvement and lactate elevation	Autosomal recessive	11	Missense, nonsense, in-frame deletion, splice site	37
Human	*RARS2*	Infantile encephalopathy, strongly affecting the cerebellum	Autosomal recessive	2 inherited together	Missense, splice site	15
Mouse	*Aars*	Cerebellar ataxia and hair follicle dystrophy	Autosomal recessive	1	Missense	25
Mouse	*Gars*	Motor and sensory neuropathy accompanied by abnormal neuromuscular function	Autosomal dominant	1	Missense	39
Fruit fly	*gars*	Severe defect in terminal arborization of axons and dendrites	Autosomal recessive (see text for details)	2	Missense, P-element insertion	9

all) ARS genes may play important roles in inherited neurological disease.

Mutations in Mitochondrial Aminoacyl-tRNA Synthetase Genes

Mitochondria are the electric companies of the cell, producing ATP via the respiratory chain. The importance of these organelles is underscored by the semi-independence they have from the rest of the eukaryotic cell; mitochondria contain their own genome and synthesize or import all the elements required for protein synthesis, which is carried out independently from that in the cytoplasm. Inherited mitochondrial diseases are a heterogeneous group of disorders caused by mutations in genes that encode proteins or tRNAs involved in mitochondrial function (42). The tissues typically affected in these diseases are highly metabolic (e.g., muscle cells and neurons), reflecting the central requirement of ATP to support proper cellular functioning.

DARS2: human mitochondrial aspartyl-tRNA synthetase

Leukoencephalopathy with brain stem and spinal cord involvement and lactate elevation (LBSL; OMIM No. 611105) (**Table 3**) is an early onset, autosomal recessive disease characterized by cerebellar ataxia, spasticity, and variable cognitive impairment (45). Lactate elevation in the central nervous system white matter is a hallmark of this disease. Because lactate elevation can be associated with mitochondrial dysfunction, these features suggested a role for the mitochondria in disease onset. Through the genetic analyses of sib pairs concordant for LBSL, 11 disease-causing mutations were identified in the aspartyl-tRNA synthetase 2 gene (*DARS2*) (37) (**Figure 2**). The most common of these is a splice-site mutation in intron 2 that results in the deletion of exon 3, a frameshift (R76SfsX5), and the premature termination of the protein. DARS2 charges $tRNA^{Asp}$ molecules in the mitochondria, and homozygous deleterious mutations would presumably be lethal. Two findings help to clarify this situation. First, no patients have been found that are homozygous for the same *DARS2* mutation (37). Because the DARS2 holoenzyme is a homodimer, dimers carrying two different mutations may have more activity than those carrying the same mutation (and thus might not yield a lethal phenotype). Second, functional analyses of a subset of *DARS2* missense mutations revealed that the tRNA-charging activity is severely reduced, but not ablated (37). Interestingly, overall mitochondrial function in patient fibroblasts and lymphoblasts (i.e., non-neuronal cells) is seemingly normal. Thus, whereas residual DARS2 activity remains in patient cells, disease onset might be a consequence of DARS2 activity falling below some key threshold that is required in neuronal cells.

Pontocerebellar hypoplasia (PCH; OMIM No. 607596) (**Table 3**) comprises a heterogeneous group of disorders characterized by a severe reduction in cerebellum and brainstem size (4). Three children from a consanguineous family with autosomal recessive PCH and multiple defects in the mitochondrial respiratory chain were reported, as detected by testing muscle tissue and skin biopsies (15). Homozygosity mapping and DNA sequencing revealed that each patient was homozygous for two variants in the arginyl-tRNA synthetase 2 gene (*RARS2*) (**Figure 2**): a missense change (R291K) and a substitution in an intron 2 splice site (IVS2+5a→g). Functional analyses revealed that most (but not all) of the *RARS2* transcripts in these patients lack exon 2, which would be predicted to cause a frameshift (L13RfsX15) and a loss of enzyme function. In support of this prediction, patient fibroblasts were found to have severely reduced levels of $tRNA^{Arg}$ molecules; presumably, the residual enzyme activity is due to a subset of transcripts that get properly spliced. Thus, similar to *DARS2* mutations, *RARS2* mutations severely reduce, but do not completely ablate, mitochondrial $tRNA^{Arg}$ charging (15). However, in the case of *RARS2* mutations, mitochondrial dysfunction is evident and appears to be systemic. It is striking that both *DARS2* mutations and *RARS2* mutations give rise to phenotypes whose features are restricted to the central nervous system.

An Alanyl-tRNA Synthetase Mutation in a Mouse Model of Ataxia

Cerebellar ataxia is a general term for primary movement disorders caused by dysfunction of the cerebellum. The precise phenotype often depends on the population(s) of cerebellar cells affected and, in the case of the inherited forms, the nature of the genetic lesion (41). Purkinje cells in the cerebellum are mainly responsible for the output of signals that are important for movement. Not surprisingly, impaired function of these cells causes defects in coordination, and has been associated with ataxia in humans. The mouse mutant sticky (*sti*) is characterized by the autosomal recessive inheritance of rough, unkempt fur and tremors that progress to severe ataxia (25) (**Table 3**). Histological and biochemical studies revealed that the ataxia appears to be caused by apoptosis of cerebellar Purkinje cells, which begins at three weeks of age and is complete by one year. Genetic analyses mapped the *sti* trait to the region of the mouse genome containing the alanyl-tRNA synthetase gene (*Aars*); subsequently, a homozygous *Aars* missense mutation (A734E) was identified in *sti* mice (25) (**Figure 2**). The role of defective *Aars* in the *sti* phenotype was confirmed by genetic complementation studies using a transgene that expresses wild-type *Aars*; normal Purkinje cells were found in *sti* mice carrying this transgene, indicating that wild-type *Aars* is required for survival of these cells. The latter notion is further supported by the finding that heterozygous mice (*sti*/+) have normal Purkinje cells.

Aars is a homodimer that charges $tRNA^{Ala}$ molecules in the cytoplasm and bears three functional domains: an N-terminal catalytic domain, an internal editing domain, and a C-terminal domain of unknown function (**Figure 2**). The A734E mutation resides within the Aars editing domain, which is responsible for hydrolyzing incorrectly incorporated noncognate amino acids from $tRNA^{Ala}$ molecules. It was thus postulated that the A734E mutation interferes with the editing capacity of Aars, and this hypothesis was

RARS2: human mitochondrial arginyl-tRNA synthetase

Aars: mouse alanyl-tRNA synthetase

Charcot-Marie-Tooth disease (CMT): a group of inherited motor and sensory peripheral neuropathies

supported by the results of a series of experiments conducted by Lee and coworkers (25). First, wild-type and A734E Aars were shown to both be able to charge $tRNA^{Ala}$ molecules with alanine in vitro, suggesting that the disease phenotype is unlikely to be caused by decreased $Ala\text{-}tRNA^{Ala}$ levels. Second, wild-type and mutant (*sti/sti*) mouse embryonic fibroblasts (MEFs) were tested for sensitivity to increased levels of noncognate amino acids in the culture media. The rationale for this experiment was that increased environmental levels of noncognate amino acids should lead to cell death if there is an editing-defective ARS (owing to the incorporation of incorrect amino acids into nascent peptide chains), a result previously observed for the *Escherichia coli* Aars ortholog (5). This experiment revealed a drastic increase in mutant (but not wild-type) MEF cell death with increased levels of serine (but not four other noncognate amino acids: alanine, glycine, histidine, and methionine) in the culture media (25). Furthermore, mutant (but not wild-type) MEFs stained positive for ubiquitin, suggesting that impaired Aars editing leads to the incorporation of noncognate amino acids and downstream protein misfolding. To confirm this, wild-type and A734E Aars were tested for their abilities to produce misacylated $Ser\text{-}tRNA^{Ala}$ and to deacylate (or edit) $Ser\text{-}tRNA^{Ala}$. Both human and mouse wild-type Aars produce much less $Ser\text{-}tRNA^{Ala}$ than A734E Aars. In the case of the human enzyme, wild-type AARS deacylates $Ser\text{-}tRNA^{Ala}$, whereas A734E AARS has a severe reduction in this capacity.

Impaired Aars editing would presumably lead to ubiquitous misincorporation of amino acids (most likely serine) in all alanine-bearing proteins. The downstream effect of this may be impaired protein folding and inappropriate protein aggregation, which offers an attractive explanation for the neuronal phenotype in *sti* mice that is consistent with other neurodegenerative diseases. To test this hypothesis, Lee and colleagues (25) performed histological studies on wild-type and *sti* cerebella, which revealed a number of findings consistent with the presence of impaired folding in *sti* mice. Specifically, the cerebella of *sti* mice have the following: (*a*) globular structures in the cytoplasm consistent with protein accumulation; (*b*) ubiquitin-positive inclusions in the cytoplasm, axons, dendrites, and nucleolus, suggesting an accumulation of proteins targeted for degradation; (*c*) induced levels of HSP70 chaperone family members, indicating a cellular response to impaired protein folding; and (*d*) expression of the endoplasmic reticulum stress–related chaperone BiP and the transcription factor CCAAT/enhancer-binding protein (C/EBP) homologous protein (CHOP). This last set of findings is particularly relevant because endoplasmic reticulum stress is a known cellular response to misfolded proteins (48) and may explain the observed apoptosis of Purkinje cells in *sti* mice. Thus, a loss of function in Aars editing appears to lead to a downstream toxic gain of function by proteins produced in Purkinje cells.

Glycyl-tRNA Synthetase Mutations in Patients with Charcot-Marie-Tooth Disease Type 2D and Distal Spinal Muscular Atrophy Type V

Inherited peripheral neuropathies are a heterogeneous group of diseases characterized by impaired motor function and sensory loss in the extremities (31). The pathology of these disorders centers around the peripheral nerve, which is composed of myriad cell types that are responsible for interpreting stimuli from and responding to the external environment. A critical interaction for proper peripheral nerve function involves the nerve cell axon and the myelinating Schwann cell. Peripheral neuropathies are thus divided into two major subtypes on the basis of these two components (14). Axonal peripheral neuropathies have an impairment of the peripheral nerve axon, whereas demyelinating peripheral neuropathies have a defect in the Schwann cell and associated myelin sheath. Charcot-Marie-Tooth disease type 2D (CMT2D; OMIM No. 601472) (**Table 3**) and distal spinal muscular atrophy type V (dSMA-V; OMIM No. 600794) (**Table 3**) are two related, autosomal dominant axonal neuropathies characterized by a phenotype that is generally more severe in the upper extremities (10, 16, 19, 34, 36).

Consistent with the phenotypic overlap between CMT2D and dSMA-V, both disorders are caused by mutations in the glycyl-tRNA synthetase gene (*GARS*) (1). To date, nine disease-causing *GARS* mutations have been identified (**Figure 2**), all of which lead to missense changes at highly conserved amino acid residues (1, 12, 20, 40). GARS is a homodimer that ligates glycine to $tRNA^{Gly}$ molecules in the cytoplasm and mitochondria; this enzyme is thus essential for translation in both cellular locations. Complete deletion of the GARS-encoding gene is lethal in yeast, plants, flies, and mice (39, 43, 44).

Various studies examined how defects in GARS lead to CMT2D and dSMA-V (2). For example, analysis of one of the disease-causing *GARS* mutations (G240R) demonstrated that the mutant allele is expressed and yields a normal amount of the GARS protein. Further functional analyses of five *GARS* mutations revealed two loss-of-function characteristics. First, yeast complementation assays were performed that modeled human mutations in the yeast *GARS* ortholog (*GRS1*). Three out of five mutations (L129P, H418R, and G526R) severely affect cell viability in the absence of wild-type *GRS1*; cells expressing both mutant and wild-type *GRS1* show normal viability (2). Second, transfection of cultured mammalian neuronal and non-neuronal cells with constructs expressing wild-type and mutant *GARS* tagged with enhanced green fluorescent protein (EGFP) revealed that the wild-type GARS protein associates with cytoplasmic granules. In contrast, three out of five mutant forms of GARS (L129P, G240R, and H418R) are not associated with such granules, and instead display a more diffuse cellular-localization pattern (2) (**Figure 3**). The presence of GARS-associated granules in vivo was confirmed by immunohistochemical analysis of normal peripheral nerve; GARS-staining granules were detected in axons of the anterior horn and root, dorsal horn and root, and sural nerve, as well as in non-neuronal cells (2).

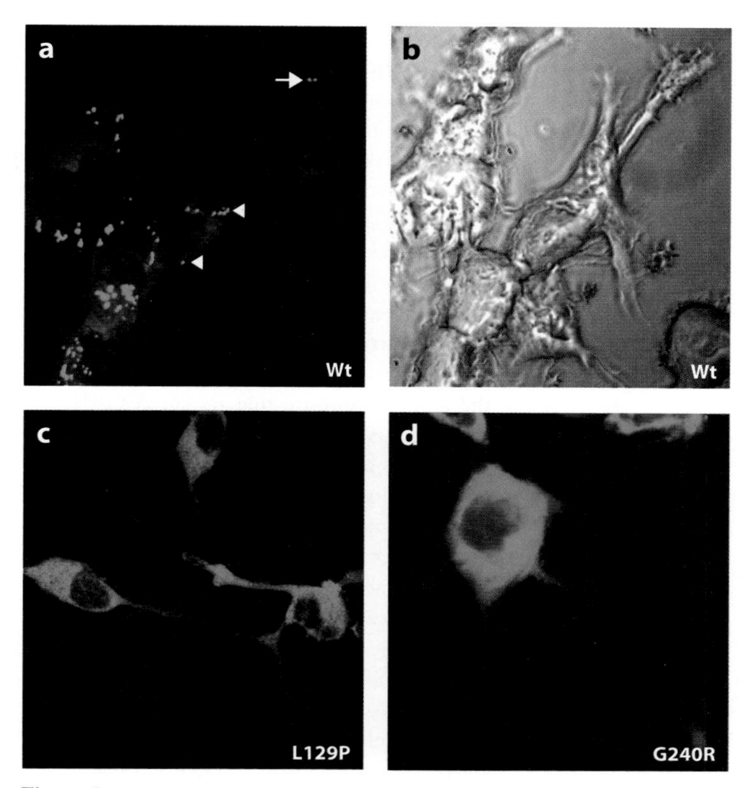

Figure 3

Mislocalization of mutant forms of glycyl-tRNA synthetase (GARS). Differentiated neurons expressing wild-type (Wt) GARS-enhanced green fluorescent protein (EGFP) (*a* and *b*), L129P GARS-EGFP (*c*), or G240R GARS-EGFP (*d*) were examined by fluorescence microscopy. Wild-type GARS-EGFP–associated granules within the cell body and neurite projections are indicated with arrowheads and the arrow (*a*), respectively. Merged images from fluorescence and differential interference contrast microscopy (*b*) reveal cell morphology and neurite-projection paths of cells expressing wild-type GARS-EGFP. Note the more diffuse signal and lack of distinct granules associated with L129P and G240R GARS-EGFP (*c* and *d*, respectively). Copyright 2006 by the Society for Neuroscience.

Together, these findings are consistent with a loss-of-function mechanism for *GARS* mutations; however, one of the studied mutations (E71G) did not show a loss of function in either of the above two analyses. This raises the possibility that *GARS* mutations are pathogenic by another mechanism or that E71G results in a loss of function by a mechanism that has not been assayed.

In another study of the cellular localization of GARS variants (32), mouse neuroblastoma cells (N2a) were transfected with constructs

EGFP: enhanced green fluorescent protein

expressing wild-type or mutant human GARS tagged with the V5 epitope. Immunostaining showed that wild-type GARS is diffusely localized in these neuronal cells, including in the neurite projections of differentiated N2a cells. In contrast, seven mutant forms of human GARS (L129P, P234KY, G240R, H418R, D500N, G526R, and S581L) vary with respect to their localization in N2a neurite projections; in all cases, they are present at lower levels in these structures compared with wild-type GARS. On the basis of these in vitro studies, the authors propose that impaired GARS localization is detrimental to neurons and that GARS has a noncanonical role in axons (32); however, non-neuronal cells were not tested in this study, making it unclear whether these findings are neuron specific. Further, the results of this study are quite different from those of the cellular-localization analyses discussed above. Specifically, one analysis (2) revealed that wild-type GARS is associated with granules in vitro and in vivo (observed in both non-neuronal and neuronal cells, including axons), and that only a subset of mutant forms of GARS is not associated with granules. In contrast, another analysis (32) revealed that wild-type GARS has a more diffuse localization pattern in vitro, and that all tested mutant forms of GARS have an altered localization pattern in cultured neurons compared with wild-type GARS. The discrepancies between these two studies need to be resolved before any definitive conclusions about the effect of *GARS* mutations on the cellular localization of the encoded protein can be made.

Seburn and colleagues (39) identified an ENU-induced mouse mutant that carries a mutation in the mouse ortholog of the glycyl-tRNA synthetase gene (*Gars*). The heterozygous mutation (P278KY in the mouse Gars sequence, which corresponds to P234KY in human GARS; we refer to this mutation as P234KY) (**Figure 2**) causes severe neuromuscular dysfunction by three weeks of age and a shortened life span; most mice do not survive past eight weeks. Histological and electrophysiological analyses of these $Gars^{P234KY/+}$ mice revealed a length-dependent, degenerative loss of motor neurons in the neuromuscular junctions that is consistent with a severe axonopathy; the axon loss is also observed in the sensory nerves (39). The phenotype of these mice is consistent with axonal CMT (CMT2), and thus the $Gars^{P234KY/+}$ mouse can be regarded as a model for CMT2D. Interestingly, the $Gars^{P234KY/P234KY}$ homozygote has an embryonic lethal phenotype; however, the embryonic day of lethality remains undetermined.

The mouse P234KY mutation alters a highly conserved amino acid residue that is located just six amino acids away from one of the known human *GARS* mutations (G240R) (**Figure 2**). The embryonic lethality seen with $Gars^{P234KY/P234KY}$ mice suggests a loss-of-function mechanism. To test this, aminoacylation assays were performed using recombinant wild-type and P234KY Gars, as well as brain homogenates from wild-type and $Gars^{P234KY/+}$ mice (39). These in vitro analyses revealed that P234KY Gars has a similar ability to charge tRNA as the wild-type protein. Further studies were performed with a mouse *Gars* null allele (generated by a gene-trap strategy). In $Gars^{+/null}$ mice, tissue *Gars* mRNA levels are roughly twofold lower than those in $Gars^{+/+}$ mice, and Gars enzyme activity (in brain homogenates) is roughly threefold lower than that in wild-type mice. Thus, the $Gars^{null}$ allele is associated with a loss of function in terms of mRNA expression and Gars activity. Interestingly, histological and electrophysiological examination of $Gars^{+/null}$ mice failed to reveal any of the abnormal phenotypic features seen with $Gars^{P234KY/+}$ mice (39). These results indicate that haploinsufficiency is not the mechanism accounting for the pathologic features seen in $Gars^{P234KY/+}$ mice. Finally, crosses of $Gars^{P234KY/+}$ and $Gars^{+/null}$ mice failed to produce any $Gars^{P234KY/null}$ offspring, indicating that this genotype yields an embryonic lethal phenotype. The embryonic lethality seen with both the $Gars^{P234KY/P234KY}$ and $Gars^{P234KY/null}$ genotypes suggests that either P234KY is associated with some loss-of-function mechanism not yet assayed (see discussion below) or a gain of function that is more severe in the absence of the wild-type allele (39).

Two groups examined the crystal structure of GARS, which provided information about the spatial positioning of amino acid residues in the catalytic core, C-terminal anticodon binding domain, and dimer interface (8, 47). Localizing known disease-causing mutations on the GARS crystal structure revealed that eight mutations (E71G, L129P, P234KY, G240R, I280F, H418R, D500N, and G526R) are positioned in the catalytic core and two mutations (S581L and G598A) reside within the anticodon-binding domain (8, 32); however, E71G and I280F may also sit at positions that interact with $tRNA^{Gly}$ (8). Ten of the 11 mutations are positioned at regions of interface between monomers (8, 32); although the position of D500N cannot be fully established relative to the crystal structure, this mutation is thought to reside within an insertion domain that may interact with $tRNA^{Gly}$ (8). These protein structural studies suggest that disease-causing *GARS* mutations may impair GARS enzyme activity and/or dimer formation. To examine this further, in vitro aminoacylation assays were used to test eight GARS mutations (8, 32, 47). Five mutations (L129P, G240R, H418R, S581L, and G526R) are associated with severely depleted or absent tRNA-charging activity, whereas the other three mutations (E71G, P234KY, and D500N) appear to be associated with full activity. Thus, impaired in vitro tRNA-charging activity is not a hallmark of mutant forms of GARS.

To study dimer formation, V5-tagged wild-type and mutant forms of GARS were each expressed in mouse N2a cells (32). Relative to wild-type GARS, two mutations (L129P and G240R) diminished dimer formation and three mutations (D500N, G526R, and S581L) enhanced dimer formation; finally, two mutations (P234KY and H418R) did not appear to affect dimer formation. Thus, although the majority of GARS mutations alter amino acids that reside at the dimer interface, only a subset appears to affect dimer formation (32).

A mutation in the *Drosophila melanogaster* ortholog of the glycyl-tRNA synthetase gene (*gars*) was identified in a mosaic forward genetic screen for defects in the morphology of olfactory projection neurons (9). The homozygous mutation P98L (**Figure 2** and **Table 3**) causes severe defects in the ability to initiate and maintain stable dendrite terminals and axon arbors. These data, in conjunction with the fact that human *GARS* mutations are associated with a neuronal-specific phenotype, suggest that the P98L mutation (*a*) results in impaired $tRNA^{Gly}$ charging, with axons showing sensitivity to the resulting protein-synthesis defect; (*b*) is associated with a loss of some neuron-specific, secondary function of gars; or (*c*) is severely toxic to neurons, similar to the situation proposed for the mouse $Gars^{P234KY}$ mutation (39).

Several lines of evidence suggest that the *Drosophila* P98L mutation acts via a loss-of-function mechanism (9). First, P98L is lethal when flies are ubiquitously homozygous for the mutation (i.e., carry two copies of $gars^{P98L}$ in all cells). Second, homozygous deletion of the *gars* gene in projection neurons results in an identical phenotype as that seen in flies homozygous for P98L. Finally, a wild-type *gars*-expressing transgene fully restores a normal projection-neuron phenotype in flies carrying the P98L mutation. To establish whether the neuronal phenotype is specific to *gars* mutations, two additional genes encoding cytoplasmic ARSs (*wars* and *qars*) were deleted in a mosaic fashion, and mutant cell clones were visualized in projection neurons; virtually identical results were obtained as with *gars* mutations (9). These results have important implications for understanding the role of ARS genes in neurodegenerative disease. First, the finding that independently deleting three *Drosophila* ARS genes yields the same neuronal phenotype argues against a specific, noncanonical function of gars in neurons. In addition, these *Drosophila* studies illustrate the importance of cytoplasmic ARSs in the development and function of axons, and this parallels the association between *GARS* mutations and length-dependent axonal neuropathies in humans.

The *Drosophila* model was further used to examine two human *GARS* mutations (9). Constructs expressing wild-type or mutant (E71G or L129P) human *GARS* were introduced into wild-type projection neurons; neither mutant nor wild-type *GARS* caused any morphological changes, indicating that *GARS* mutations are not associated with a toxic gain of function in *Drosophila*. Next, similar experiments were performed with projection neurons that were homozygous for deleted *gars*. Wild-type *GARS* fully restored a normal projection-neuron phenotype. In contrast, L129P *GARS* completely failed to restore and E71G *GARS* only partially restored a normal phenotype. These data suggest that both the E71G and L129P mutations cause a loss of function at some level (note that this represents the first data supporting a loss-of-function mechanism for the E71G mutation).

Tyrosyl-tRNA Synthetase Mutations in Patients with Dominant Intermediate Charcot-Marie-Tooth Disease

The identification of *GARS* mutations as the genetic cause of CMT2D and dSMA-V immediately raised two distinct possibilities. First, peripheral neurons may be particularly sensitive to *GARS* mutations owing to some specific requirement for GARS. Alternatively, peripheral neurons may be particularly sensitive to ARS defects in general, perhaps reflecting a heightened requirement for tRNA charging and protein synthesis. Support for the latter possibility came from the discovery that mutations in the tyrosyl-tRNA synthetase gene (*YARS*) are responsible for dominant intermediate CMT (DI-CMT; OMIM No. 608323) (**Table 3**), an autosomal dominant subtype of CMT that has both axonal and demyelinating features (22, 23). YARS is a homodimer responsible for charging $tRNA^{Tyr}$ molecules in the cytoplasm; encoded by a separate gene, YARS2 charges $tRNA^{Tyr}$ molecules in the mitochondria (**Table 1**). Two missense *YARS* mutations (G41R and E196K) and one deletion that removes four amino acids (V153_V156del) in YARS (**Figure 2**) were found in patients with DI-CMT (22).

Biochemical analyses revealed that both the G41R and E196K YARS mutations lead to a decreased ability to carry out the first step of aminoacylation (**Figure 1**) compared to wild-type YARS (22). Yeast complementation studies were also performed by expressing different forms of human *YARS* and its yeast ortholog (*TYS1*). In yeast cells expressing one copy of *TYS1*, G41R was lethal, whereas E196K impaired cell growth. When mutant *TYS1* was overexpressed in the presence of wild-type *TYS1*, both G41R and E196K impaired cell growth, suggesting that the mutant monomer has a dominant-negative effect on the holoenzyme.

Cell-imaging studies revealed that endogenous YARS is associated with granular structures in the neurite projections of cultured neurons as well as in primary embryonic motor neurons (22). This localization pattern was not observed in non-neuronal cells, suggesting that these granular structures might have a neuron-specific function. Compared to wild-type YARS tagged with EGFP, the G41R and E196K forms of YARS (also tagged with EGFP) show a marked reduction in granule localization and a generally more diffuse cellular staining pattern.

In short, analyses of *YARS* and *GARS* mutations have yielded analogous results, and a loss-of-function mechanism is suggested in both cases. However, one notable difference is that GARS-associated granules are seemingly cell-type ubiquitous, whereas YARS-associated granules appear to be neuron specific.

Autoantibodies to Aminoacyl-tRNA Synthetases

The idiopathic inflammatory myopathies are a group of rare diseases characterized by the inflammation of muscle tissue in conjunction with one or more additional pathologic features (e.g., interstitial lung disease). These disorders have been associated with autoantibodies to components of the translation machinery, including six cytoplasmic ARSs (HARS, AARS, GARS, IARS, TARS, and

NARS) (27). Although details of the pathogenic mechanism of idiopathic inflammatory myopathy associated with anti-ARS antibodies have yet to be elucidated, several ideas have been proposed. One prominent hypothesis revolves around a common N-terminal domain found in ARSs—an α-helical, coiled-coil structure with putative roles in stabilizing ARS:tRNA interactions and in anchoring ARSs within MSCs (17). One possibility is that an insult to muscle cells disturbs the translational apparatus, thereby exposing the ARS N-terminal domain to the intra- and extracellular environment and leading to a primary immune response (13, 35).

It remains to be determined if there is any pathogenic overlap between diseases associated with anti-ARS antibodies and those associated with mutations in ARS genes (see above and **Table 3**). Because many ARS gene mutations are associated with neuromuscular dysfunction brought on by an axonopathy, it is tempting to search for a link between these two groups of disorders at the neuromuscular junction. However, there has been no report of an inflammatory myopathy in a patient(s) harboring an ARS gene mutation, even in the case of mutations that yield a mislocalized ARS protein (which might elicit an immune response to an exposed N-terminal region). In this regard, it would be interesting to see whether any disease-associated amino acid changes reside in ARS regions that are the target of autoantibodies.

PATHOGENIC MECHANISMS OF AMINOACYL-tRNA SYNTHETASE MUTATIONS IN PERIPHERAL NEUROPATHIES

Proper functioning of the protein translation machinery is critical for virtually all aspects of normal physiology. Not surprisingly, mutations in the genes encoding proteins responsible for translation are associated with myriad phenotypes (38). Mutations in ARS-encoding genes have the potential to interfere with the first step of protein synthesis, and thus pose the risk of interfering with the accurate read out of the genetic code. A subset of these mutations is associated with fairly straightforward molecular pathologies. For example, two genes encoding mitochondrial ARSs (*DARS2* and *RARS2*) have been implicated in encephalopathies, and these diseases are thought to involve loss-of-enzyme activity on the basis of several lines of evidence (15, 37). Impaired tRNA charging in mitochondria would presumably lead to respiratory chain defects, which is directly supported by functional studies of *RARS2* mutations. Other mitochondrial diseases are known to affect highly metabolic tissues (such as neurons) (42), and the encephalopathy seen in patients with *DARS2* and *RARS2* mutations is consistent with this observation.

In another example, a mutation in the mouse *Aars* gene causes a loss of function in the ability of Aars to edit incorrectly charged $tRNA^{Ala}$ molecules (25). This defect leads to a downstream toxic gain of function; specifically, proteins with incorrectly incorporated amino acids do not fold properly and subsequently aggregate, leading to apoptosis of Purkinje cells.

In contrast to these two examples, the pathogenic mechanism that leads to peripheral neuropathy in the case of *GARS* and *YARS* mutations remains unclear. Several general mechanistic classes remain as possibilities: (*a*) loss of primary enzyme function; (*b*) toxic gain of function; and (*c*) impaired secondary, noncanonical function. Each of these possibilities is discussed further below.

Loss of Primary Enzyme Function: Do Axons Have a Special Requirement for Protein Synthesis?

One explanation for the axonal neuropathy associated with *GARS* and *YARS* mutations is that these mutations impair tRNA charging in a fashion that affects protein translation throughout the nerve cell (**Figure 4**, Mechanism 1a). Neurons (particularly those with very long axons, such as those in peripheral nerves) may be more susceptible to protein-translation defects than non-neuronal cells. Two general mechanisms can be envisioned for how

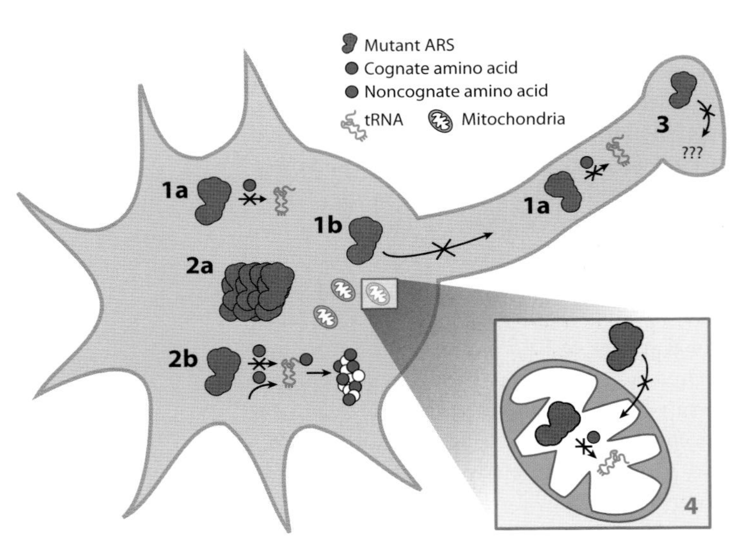

Figure 4

Possible mechanisms by which aminoacyl-tRNA synthetase (ARS) mutations cause neurodegeneration. A neuron is schematically depicted with the cell body on the left and an axon protruding toward the right. ARS mutations might lead to a number of cellular defects, including the following: impaired tRNA charging in the cell body and axon (Mechanism 1a); impaired localization of the ARS, resulting in its deficiency in the axon (Mechanism 1b); a toxic gain of function due to aggregation of mutant ARS (Mechanism 2a); a toxic gain of function due to misincorporation of amino acids into nascent peptides, causing global protein aggregation (Mechanism 2b); a loss of some axon-specific, secondary function (Mechanism 3); and mitochondrial dysfunction due to impaired tRNA charging within the mitochondria or impaired import of the ARS into the mitochondria, leading to decreased ATP concentration in the cell body and axon (Mechanism 4).

mutations interfere with the enzyme activity of an ARS. First, a mutation could impair the ability of the enzyme to complete the aminoacylation reaction (**Figure 4**, Mechanism 1a); indeed, seven of the ten tested *GARS* and *YARS* mutations appear to cause impaired tRNA charging by the respective ARS (see above). Second, a mutation could alter the cellular localization of the ARS (**Figure 4**, Mechanism 1b), which might represent a loss of function (albeit in a cell compartment–specific manner). Such a mechanism has been reported for other genetic diseases (e.g., the classic F508del mutation and cystic fibrosis) (24). Some *YARS* and *GARS* mutations lead to altered localization of the encoded protein (2, 22). Importantly, both wild-type ARSs were found within axons on the basis of in vitro and in vivo studies, suggesting that ARSs have a critical function in these structures.

Neurodegeneration: a general term for the progressive impairment or death of neurons, including axonopathy, neuronopathy, and cell death

In the likely event that *GARS* and *YARS* mutations act via the same pathogenic mechanism, there are three potential arguments against a loss-of-function etiology underlying their respective diseases. First, a subset of mutant forms of GARS appears to charge tRNA in a normal fashion, and a subset of *GARS* mutations modeled in yeast appears to be fully functional (2, 32, 39). However, these mutations may be associated with other loss-of-function characteristics, such as impaired localization within a neuron (see above). Second, one GARS mutation (E71G) charges tRNA normally, is fully functional when modeled in yeast, and has a normal cellular localization pattern (2, 32). However, studies in fruit flies show that E71G GARS is less effective than wild-type GARS in rescuing the effects of deleted *gars* (9); thus, the E71G mutation likely acts via a loss-of-function property. For example, E71G GARS may associate with granules in neurons, but these granules may not be efficiently delivered to the axon. Finally, haploinsufficiency is an uncommon mechanism for an autosomal dominant neurodegenerative disease. Indeed, this mechanism would appear to be contrary to the findings from *Gars*$^{+/-}$ mice, which are phenotypically normal (39); at the same time, the fact that mice have peripheral axons that span a much shorter distance than those in humans cannot be ignored. In the case of *YARS* mutations, this issue is at least partially clarified by data that indicate a dominant-negative mechanism (22). Such a mechanism involves the incorporation of a mutant YARS monomer into a homodimer along with the wild-type monomer, with a concomitant loss of function. In the case of *GARS* mutations, the story is less clear. Although the GARS protein is also a homodimer, analyses fail to reveal a dominant-negative effect when *GARS* mutations are modeled in yeast (2). To summarize, all but two *GARS* and *YARS* mutations have been shown in some way to be associated with a loss of function; I280F and G598A GARS have yet to be fully characterized.

The most straightforward explanation for these observations is that GARS and YARS are required in axons for local tRNA charging, and the absence or dysfunction of these proteins is specifically detrimental to axons. Importantly, this explanation can account for variants (e.g., GARS E71G, P234KY, and D500N) that are fully able to charge tRNA—these mutant forms may be functional but may not properly localize to axons. For example, P234KY Gars in brain homogenates is able to charge tRNA (39). Although some have argued that these data rule out a loss-of-function mechanism for P234KY Gars, the utilized assay did not specifically test for tRNA charging in axons, and two independent assays show altered cellular localization of this mutant form of Gars (32; A. Antonellis and E.D. Green, unpublished data). To clarify this situation further, assays that directly examine tRNA charging in axons are needed, preferably those that study neurons in a model vertebrate system that harbors *GARS* or *YARS* mutations.

It is relevant to consider these various issues within the broader context of protein translation in neurons. A central axiom of neuroscience has been that protein translation takes place in both the neuronal cell body and the dendrites, but that the requirement for axonal proteins is handled via fast axonal transport of membrane-bound and excreted proteins and slow axonal transport of cytoplasmic proteins (28). Historically, a major argument against the latter idea has been that slow axonal transport is not predicted to be sufficient to meet the demand for axonal proteins, particularly in the long axons of peripheral nerves (18). It has, therefore, been postulated that axonal proteins are synthesized by local translation. Recent evidence that certain mRNA populations important for axonal function are specifically transported to axons, that components of the translational machinery are present in axons, and that deletion of (or mutations in) genes important for protein translation are specifically detrimental to axogenesis supports the notion that local translation occurs in the axon (9, 18). Given this evidence, it is reasonable to postulate that *GARS* and *YARS* mutations act via a loss-of-function mechanism that results in altered levels of charged tRNA molecules, which in turn would negatively impact local protein translation in the axon. In the case of a typical non-neuronal cell, this impairment may not be sufficiently detrimental to lead to an abnormal phenotype. However, a highly metabolic peripheral neuron (with a particularly long axon) may simply be more vulnerable to the disruption of protein translation, leading to abnormal axon-specific sequelae.

It is important to point out that a loss of function associated with *GARS* mutations would also affect mitochondrial function. That is, if GARS enzyme activity is impaired within the mitochondria, or if mutant GARS is unable to be imported into the mitochondria (**Figure 4** inset, Mechanism 4), then the concentration of ATP in both the cell body and axon would be affected. Indeed, this notion is supported by the association between mutations in a gene encoding a protein that is important for mitochondrial function (MFN2) and another axonal form of CMT (CMT2A) (49). Although mitochondrial dysfunction is unlikely to reflect the situation with *YARS* mutations (YARS2 charges $tRNA^{Tyr}$ molecules in the mitochondria), the presence of mitochondrial dysfunction brought on by *GARS* mutations cannot be ruled out. It would thus be appropriate to examine more fully mitochondrial function in CMT2D/dSMA-V patients or the $Gars^{P234KY/+}$ mouse.

Toxic Gain of Function: Do Aminoacyl-tRNA Synthetase Gene Mutations Fit a More Common Model of Neurodegenerative Disease?

A common model for the molecular pathology underlying neurodegenerative diseases involves a toxic gain of function via the inappropriate aggregation of a protein (46) (**Figure 4**, Mechanism 2a). In that regard, it is possible that mutant forms of GARS and YARS do not fold properly and then aggregate in a fashion that becomes toxic to neurons. There are several pieces of evidence that could support such a model: (*a*) The majority of *GARS* and *YARS* mutations reflect missense amino acid changes; (*b*) the mutant forms of the protein appear to be expressed and are stable; and (*c*) in the majority of cases, the mutant proteins mislocalize in neurons (2, 22, 32). In terms of the latter finding, it is possible that mislocalized proteins would then be available to aggregate and/or interfere with cellular processes critical for neuron function.

Interestingly, the mouse *Aars* mutation is predicted to cause incorporation of incorrect amino acids into newly synthesized proteins, impaired folding and aggregation of these proteins, and consequent apoptosis of the neuron (25) (**Figure 4**, Mechanism 2b). In this case, an impairment in the primary function of the enzyme leads to a downstream toxic gain of function. It is possible that *GARS* and *YARS* mutations have a similar effect; in the environment of decreased GARS or YARS charging capacity, nascent polypeptides may become truncated or synthesized with incorrect amino acids. The study of mouse strains that overexpress mutant *GARS* or *YARS* could be used to examine the above hypotheses. Specifically, spinal cord sections from such mice could be analyzed as was performed with brain tissue from mutant *Aars* mice, looking for GARS or YARS aggregation (supporting Mechanism 2a in **Figure 4**) or more general protein aggregation (supporting Mechanism 2b in **Figure 4**).

Impaired Secondary, Noncanonical Function: Do ARSs Have Undiscovered Functions in Neurons?

The discovery that ARSs have secondary functions unrelated to tRNA charging (**Table 2**) opens up other possibilities for the consequences of *GARS* and *YARS* mutations in neurons. In fact, it is possible that these proteins have a secondary function in these cells (**Figure 4**, Mechanism 3) and that disease-causing mutations interfere with this function. For example, GARS and YARS might act as signaling molecules in neurons, analogous to the role of YARS and WARS in angiogenesis (**Table 2**). The existence of neuron-specific granules associated with YARS in vitro indirectly supports such a hypothesis (22); these structures may have a neuron-specific function. In contrast, the localization of GARS in granules appears to be ubiquitous, including in axons (2); these findings might suggest that the presence of GARS-associated granules in axons relates to its ubiquitous tRNA-charging activity. More detailed studies of these GARS- and YARS-containing granules are needed to establish the functions of these structures in axons. For example, the identification of proteins that physically associate with GARS and YARS in a neuron-specific and translation-independent manner would strongly suggest the presence of an important noncanonical function of these proteins. Alternatively, the association of GARS and YARS with other ARSs and components of the translational machinery would strongly argue against such a function. Of relevance, the neuron-specific phenotypes associated with *DARS2* and *RARS2* mutations (15, 37) suggest that neurons are particularly susceptible to impaired tRNA charging, at least in the mitochondria.

AMINOACYL-tRNA SYNTHETASE GENES: NEW CANDIDATES FOR NEURODEGENERATIVE DISEASES

The discovered link between mutations in ARS genes and neuron-specific phenotypes has opened up an exciting new area of neurogenetics. Specifically, multiple studies now suggest important roles for ARSs in maintaining neuronal health, which is leading to new research avenues. To begin with, all ARS genes can be considered candidates for harboring mutations responsible for uncharacterized neurological diseases. Global studies that examine all ARS genes (**Table 1**) in patients with various neurodegenerative diseases might provide substantially more insights about the role of ARSs in human genetic disease. Further, known

Table 4: Non-synonymous SNPs in human ARS genes

Gene	Amino acid change	Average heterozygosity	Accession number
AARS	D275G	0.19	rs11537667
EPRS	P296A	0.09	rs35999099
	E308D	0.20	rs2230301
	H893P	0.15	rs5030751
	I1043V	0.07	rs5030752
	N1399T	0.09	rs34559775
FARS (alpha)	Q341R	0.04	rs35087277
FARS (beta)	I585V	0.26	rs7185
GARS	T214I	0.05	rs34886142
	Q334R	0.01	rs17159287
IARS	K1182E	0.46	rs556155
KARS	T595S	0.26	rs6834
LARS	K1088R	0.30	rs10988
RARS	I3V	0.41	rs244903
	Y397F	0.29	rs2305734
TARS	M1V	0.04	rs16891129
	D21G	0.10	rs34334786
VARS	P51R	0.45	rs2607015
	H139R	0.04	rs17207531
	R181C	0.08	rs35196751
	P626S	0.07	rs11531
	P1008L	0.03	rs1076827
AARS2	I339V	0.22	rs324136
	M850V	0.05	rs35783144
CARS2	K440E	0.03	rs965189
	Q555P	0.15	rs1043886
DARS2	K196R	0.10	rs35515638
FARS2	A11S	0.49	rs2224391
	S57C	0.05	rs34382405
	N280S	0.32	rs11243011
IARS2	I522V	0.04	rs11800305
LARS2	K727N	0.10	rs36054230
	E831D	0.04	rs9827689
NARS2	N87T	0.45	rs10501429
PARS2	R28S	0.10	rs11577368
	N235S	0.26	rs2270004
RARS2	K291R	0.10	rs17850652
	I331V	0.08	rs3757370
SARS2	H8R	0.05	rs33988199
	T35A	0.22	rs34264048
	S83L	0.05	rs34050897
VARS2	Q337R	0.02	rs3218820
	R449W	0.48	rs2249464
	Q917R	0.48	rs9394021
	T965A	0.10	rs2252863
	Q1049R	0.21	rs4678
WARS2	P267A	0.25	rs3790549
YARS2	G191V	0.18	rs11539445

missense amino acid changes in ARSs, such as those already cataloged in polymorphism databases (**Table 4**), should be studied for a possible role in neurodegenerative disease, both as the direct cause and as modifiers of phenotypes caused by mutations in other genes. Finally, a more complete view of the range of functions of ARSs in axons is needed, so as to provide an intellectual framework to develop therapeutics for certain neurodegenerative diseases. Overall, these research areas are at a very nascent stage, and future discoveries will undoubtedly reveal a more complex and exciting role for ARSs in human genetic diseases.

SUMMARY POINTS

1. Aminoacyl-tRNA synthetases (ARSs) are a family of essential enzymes that are ubiquitously expressed and charge tRNA molecules in both the cytoplasm and mitochondria.
2. Despite their critical and global role in protein synthesis, mutations in genes encoding mitochondrial and cytoplasmic ARSs result in somewhat restricted neurological phenotypes, such as peripheral neuropathies, encephalopathy, and ataxia.
3. Mutations in mitochondrial ARSs that are associated with encephalopathy appear to cause a loss of enzyme function, leading to severely reduced tRNA-charging ability.
4. A mutation in one cytoplasmic ARS (Aars) that is associated with ataxia appears to cause a loss of function in the editing capacity of the enzyme, leading to a downstream toxic gain of function.
5. The pathological mechanism(s) underlying mutations in ARS-encoding genes that are associated with peripheral neuropathies remains unclear.
6. A subset of ARSs has been localized to axons in vitro and in vivo, suggesting that these proteins play an important role in axon function.

FUTURE ISSUES

1. Additional research is needed to establish the pathogenic mechanism underlying all disease-causing ARS mutations.
2. It is important to establish the role(s) of ARSs in axons and the effect of ARS mutations on this role(s).
3. All ARS-encoding genes represent viable candidates for harboring mutations that underlie neurodegenerative phenotypes.

DISCLOSURE STATEMENT

The authors are not aware of any biases that might be perceived as affecting the objectivity of this review.

ACKNOWLEDGMENTS

The authors thank William Motley for helpful suggestions on the manuscript.

LITERATURE CITED

1. Antonellis A, Ellsworth RE, Sambuughin N, Puls I, Abel A, et al. 2003. Glycyl tRNA synthetase mutations in Charcot-Marie-Tooth disease type 2D and distal spinal muscular atrophy type V. *Am. J. Hum. Genet.* 72:1293–99
2. Antonellis A, Lee-Lin SQ, Wasterlain A, Leo P, Quezado M, et al. 2006. Functional analyses of glycyl-tRNA synthetase mutations suggest a key role for tRNA-charging enzymes in peripheral axons. *J. Neurosci.* 26:10397–406
3. Arnez JG, Moras D. 1997. Structural and functional considerations of the aminoacylation reaction. *Trends Biochem. Sci.* 22:211–16
4. Barth PG. 1993. Pontocerebellar hypoplasias. An overview of a group of inherited neurodegenerative disorders with fetal onset. *Brain Dev.* 15:411–22
5. Beebe K, Ribas De Pouplana L, Schimmel P. 2003. Elucidation of tRNA-dependent editing by a class II tRNA synthetase and significance for cell viability. *EMBO J.* 22:668–75
6. Berthonneau E, Mirande M. 2000. A gene fusion event in the evolution of aminoacyl-tRNA synthetases. *FEBS Lett.* 470:300–4
7. Bonnefond L, Fender A, Rudinger-Thirion J, Giege R, Florentz C, Sissler M. 2005. Toward the full set of human mitochondrial aminoacyl-tRNA synthetases: characterization of AspRS and TyrRS. *Biochemistry* 44:4805–16
8. Cader MZ, Ren J, James PA, Bird LE, Talbot K, Stammers DK. 2007. Crystal structure of human wildtype and S581L-mutant glycyl-tRNA synthetase, an enzyme underlying distal spinal muscular atrophy. *FEBS Lett.* 581:2959–64
9. Chihara T, Luginbuhl D, Luo L. 2007. Cytoplasmic and mitochondrial protein translation in axonal and dendritic terminal arborization. *Nat. Neurosci.* 10:828–37
10. Christodoulou K, Kyriakides T, Hristova AH, Georgiou DM, Kalaydjieva L, et al. 1995. Mapping of a distal form of spinal muscular atrophy with upper limb predominance to chromosome 7p. *Hum. Mol. Genet.* 4:1629–32
11. Delarue M. 1995. Aminoacyl-tRNA synthetases. *Curr. Opin. Struct. Biol.* 5:48–55
12. Del Bo R, Locatelli F, Corti S, Scarlato M, Ghezzi S, et al. 2006. Coexistence of CMT-2D and distal SMA-V phenotypes in an Italian family with a *GARS* gene mutation. *Neurology* 66:752–54
13. Dohlman JG, Lupas A, Carson M. 1993. Long charge-rich α-helices in systemic autoantigens. *Biochem. Biophys. Res. Commun.* 195:686–96
14. Dyck PJ, Lambert EH. 1968. Lower motor and primary sensory neuron diseases with peroneal muscular atrophy. II. Neurologic, genetic, and electrophysiologic findings in various neuronal degenerations. *Arch. Neurol.* 18:619–25
15. Edvardson S, Shaag A, Kolesnikova O, Gomori JM, Tarassov I, et al. 2007. Deleterious mutation in the mitochondrial arginyl-transfer RNA synthetase gene is associated with pontocerebellar hypoplasia. *Am. J. Hum. Genet.* 81:857–62
16. Ellsworth RE, Ionasescu V, Searby C, Sheffield VC, Braden VV, et al. 1999. The *CMT2D* locus: refined genetic position and construction of a bacterial clone-based physical map. *Genome Res.* 9:568–74
17. Francin M, Kaminska M, Kerjan P, Mirande M. 2002. The N-terminal domain of mammalian lysyl-tRNA synthetase is a functional tRNA-binding domain. *J. Biol. Chem.* 277:1762–69
18. Giuditta A, Kaplan BB, van Minnen J, Alvarez J, Koenig E. 2002. Axonal and presynaptic protein synthesis: new insights into the biology of the neuron. *Trends Neurosci.* 25:400–4
19. Ionasescu V, Searby C, Sheffield VC, Roklina T, Nishimura D, Ionasescu R. 1996. Autosomal dominant Charcot-Marie-Tooth axonal neuropathy mapped on chromosome 7p (CMT2D). *Hum. Mol. Genet.* 5:1373–75
20. James PA, Cader MZ, Muntoni F, Childs AM, Crow YJ, Talbot K. 2006. Severe childhood SMA and axonal CMT due to anticodon binding domain mutations in the *GARS* gene. *Neurology* 67:1710–12
21. Johanson K, Hoang T, Sheth M, Hyman LE. 2003. GRS1, a yeast tRNA synthetase with a role in mRNA 3′ end formation. *J. Biol. Chem.* 278:35923–30

22. Jordanova A, Irobi J, Thomas FP, Van Dijck P, Meerschaert K, et al. 2006. Disrupted function and axonal distribution of mutant tyrosyl-tRNA synthetase in dominant intermediate Charcot-Marie-Tooth neuropathy. *Nat. Genet.* 38:197–202
23. Jordanova A, Thomas FP, Guergueltcheva V, Tournev I, Gondim FA, et al. 2003. Dominant intermediate Charcot-Marie-Tooth type C maps to chromosome 1p34-p35. *Am. J. Hum. Genet.* 73:1423–30
24. Kartner N, Augustinas O, Jensen TJ, Naismith AL, Riordan JR. 1992. Mislocalization of Δ F508 CFTR in cystic fibrosis sweat gland. *Nat. Genet.* 1:321–27
25. Lee JW, Beebe K, Nangle LA, Jang J, Longo-Guess CM, et al. 2006. Editing-defective tRNA synthetase causes protein misfolding and neurodegeneration. *Nature* 443:50–55
26. Lee SW, Cho BH, Park SG, Kim S. 2004. Aminoacyl-tRNA synthetase complexes: beyond translation. *J. Cell Sci.* 117:3725–34
27. Levine SM, Rosen A, Casciola-Rosen LA. 2003. Anti-aminoacyl tRNA synthetase immune responses: insights into the pathogenesis of the idiopathic inflammatory myopathies. *Curr. Opin. Rheumatol.* 15:708–13
28. Levitan IB, Kaczmarek LK. 2002. *The Neuron: Cell and Molecular Biology*. New York: Oxford Univ. Press
29. Lodish H, Berk A, Zipursky SL, Matsudaira P, Baltimore D, Darnell JE. 2000. *Molecular Cell Biology*. New York: Freeman
30. Lund E, Dahlberg JE. 1998. Proofreading and aminoacylation of tRNAs before export from the nucleus. *Science* 282:2082–85
31. Murakami T, Garcia CA, Reiter LT, Lupski JR. 1996. Charcot-Marie-Tooth disease and related inherited neuropathies. *Medicine* 75:233–50
32. Nangle LA, Zhang W, Xie W, Yang XL, Schimmel P. 2007. Charcot-Marie-Tooth disease-associated mutant tRNA synthetases linked to altered dimer interface and neurite distribution defect. *Proc. Natl. Acad. Sci. USA* 104:11239–44
33. Park SG, Ewalt KL, Kim S. 2005. Functional expansion of aminoacyl-tRNA synthetases and their interacting factors: new perspectives on housekeepers. *Trends Biochem. Sci.* 30:569–74
34. Pericak-Vance MA, Speer MC, Lennon F, West SG, Menold MM, et al. 1997. Confirmation of a second locus for CMT2 and evidence for additional genetic heterogeneity. *Neurogenetics* 1:89–93
35. Raben N, Nichols R, Dohlman J, McPhie P, Sridhar V, et al. 1994. A motif in human histidyl-tRNA synthetase which is shared among several aminoacyl-tRNA synthetases is a coiled-coil that is essential for enzymatic activity and contains the major autoantigenic epitope. *J. Biol. Chem.* 269:24277–83
36. Sambuughin N, Sivakumar K, Selenge B, Lee HS, Friedlich D, et al. 1998. Autosomal dominant distal spinal muscular atrophy type V (dSMA-V) and Charcot-Marie-Tooth disease type 2D (CMT2D) segregate within a single large kindred and map to a refined region on chromosome 7p15. *J. Neurol. Sci.* 161:23–28
37. Scheper GC, van der Klok T, van Andel RJ, van Berkel CG, Sissler M, et al. 2007. Mitochondrial aspartyl-tRNA synthetase deficiency causes leukoencephalopathy with brain stem and spinal cord involvement and lactate elevation. *Nat. Genet.* 39:534–39
38. Scheper GC, van der Knaap MS, Proud CG. 2007. Translation matters: protein synthesis defects in inherited disease. *Nat. Rev. Genet.* 8:711–23
39. Seburn KL, Nangle LA, Cox GA, Schimmel P, Burgess RW. 2006. An active dominant mutation of glycyl-tRNA synthetase causes neuropathy in a Charcot-Marie-Tooth 2D mouse model. *Neuron* 51:715–26
40. Sivakumar K, Kyriakides T, Puls I, Nicholson GA, Funalot B, et al. 2005. Phenotypic spectrum of disorders associated with glycyl-tRNA synthetase mutations. *Brain* 128:2304–14
41. Taroni F, DiDonato S. 2004. Pathways to motor incoordination: the inherited ataxias. *Nat. Rev. Neurosci.* 5:641–55
42. Taylor RW, Turnbull DM. 2005. Mitochondrial DNA mutations in human disease. *Nat. Rev. Genet.* 6:389–402
43. Turner RJ, Lovato M, Schimmel P. 2000. One of two genes encoding glycyl-tRNA synthetase in *Saccharomyces cerevisiae* provides mitochondrial and cytoplasmic functions. *J. Biol. Chem.* 275:27681–88
44. Uwer U, Willmitzer L, Altmann T. 1998. Inactivation of a glycyl-tRNA synthetase leads to an arrest in plant embryo development. *Plant Cell* 10:1277–94

45. van der Knaap MS, van der Voorn P, Barkhof F, Van Coster R, Krageloh-Mann I, et al. 2003. A new leukoencephalopathy with brainstem and spinal cord involvement and high lactate. *Ann. Neurol.* 53:252–58
46. Wood JD, Beaujeux TP, Shaw PJ. 2003. Protein aggregation in motor neurone disorders. *Neuropathol. Appl. Neurobiol.* 29:529–45
47. Xie W, Nangle LA, Zhang W, Schimmel P, Yang XL. 2007. Long-range structural effects of a Charcot-Marie-Tooth disease-causing mutation in human glycyl-tRNA synthetase. *Proc. Natl. Acad. Sci. USA* 104:9976–81
48. Xu C, Bailly-Maitre B, Reed JC. 2005. Endoplasmic reticulum stress: cell life and death decisions. *J. Clin. Invest.* 115:2656–64
49. Zuchner S, Mersiyanova IV, Muglia M, Bissar-Tadmouri N, Rochelle J, et al. 2004. Mutations in the mitochondrial GTPase mitofusin 2 cause Charcot-Marie-Tooth neuropathy type 2A. *Nat. Genet.* 36:449–51

A Bird's-Eye View of Sex Chromosome Dosage Compensation

Arthur P. Arnold, Yuichiro Itoh, and Esther Melamed

Department of Physiological Science and Laboratory of Neuroendocrinology of the Brain Research Institute, University of California, Los Angeles, California 90095;
email: arnold@ucla.edu; yitoh@physci.ucla.edu; emelamed@ucla.edu

Annu. Rev. Genomics Hum. Genet. 2008. 9:109–27

First published online as a Review in Advance on May 19, 2008

The *Annual Review of Genomics and Human Genetics* is online at genom.annualreviews.org

This article's doi:
10.1146/annurev.genom.9.081307.164220

1527-8204/08/0922-0109$20.00

Key Words

Z chromosome, gene expression, sex difference, evolution, male hypermethylated (MHM), chicken

Abstract

Intensive study of a few genetically tractable species with XX/XY sex chromosomes has produced generalizations about the process of sex chromosome dosage compensation that do not fare well when applied to ZZ/ZW sex chromosome systems, such as those in birds. The inherent sexual imbalance in dose of sex chromosome genes has led to the evolution of sex-chromosome-wide mechanisms for balancing gene dosage between the sexes and relative to autosomal genes. Recent advances in our knowledge of avian genomes have led to a reexamination of sex-specific dosage compensation (SSDC) in birds, which is less effective than in known XX/XY systems. Insights about the mechanisms of SSDC in birds also suggest similarities to and differences from those in XX/XY species. Birds are thus offering new opportunities for studying dosage compensation in a ZZ/ZW system, which should shed light on the evolution of SSDC more broadly.

INTRODUCTION

Sex chromosomes have attracted a great deal of attention for more than a century (70, 92) because, unlike autosomes, they are not a matched pair. One sex chromosome is often much larger and more gene rich than the other, which is often atrophied and gene poor. These differences have enormous ramifications for the organization and evolution of the two chromosomes, and indeed for the rest of the genome (28, 41, 71, 98). From the standpoint of gene dose, the sex chromosomes present two main "difficulties." One issue is that the sex chromosomes, unlike autosomes, are differentially represented in the two sexes, which leads to sexual bias in selection pressures that operate on the sex chromosomes but not on the autosomes. These pressures result in alterations in gene content relative to the autosomes and differences in sex-biased gene expression. The second problem is that a twofold difference in genomic dose of one sex chromosome means that if the gene dosage difference is not mitigated, gene networks in the two sexes cannot function equally, and hence not optimally, in at least one sex. The two main vertebrate heteromorphic sex chromosome systems, XX/XY and ZZ/ZW, are considered here. In XX/XY systems, the male is heterogametic XY and the female is homogametic XX, whereas in ZZ/ZW systems the female is ZW heterogametic. The vast majority of molecular genetic research has been carried out on three genetically tractable XX/XY systems: mammals (mouse and human), *Drosophila*, and *Caenorhabditis elegans*. Accordingly, the theories concerning sex chromosome evolution and dosage compensation have disproportionately arisen from work on XX/XY systems, and have been tested almost exclusively in those systems.

The recent emergence of genome resources for birds, including the sequencing of two avian genomes (19, 102), now allows for the first time a global approach to the study of sex chromosome dosage compensation and evolution in ZZ/ZW systems. The early returns from global studies of dosage compensation challenge existing ideas that arose from the study of XX/XY systems. The novel avian perspective is likely to continue to offer fresh insights that might not previously have been possible if we were limited to the study of XX/XY systems.

THEORIES OF SEX CHROMOSOME EVOLUTION

Why Are Sex Chromosomes Heteromorphic?

XX/XY and ZZ/ZW systems have dramatic similarities. Both often have one large (homogametic) sex chromosome (X or Z, in mammals and birds with ~1000 genes) and one small (heterogametic) sex chromosome (Y or W, typically containing only a few dozen genes). The repeated, independent, and convergent evolution of the mismatched sex chromosome pair in totally unrelated animal taxa likely reflects a common series of evolutionary events. The initiating event is thought to have been the emergence of a dominant sex-determining allele on an autosome in a species lacking well-differentiated sex chromosomes (16, 17). For example, in the therian ancestors of marsupial and eutherian (placental) mammals, the *Sox3* gene probably sustained a series of mutations that transformed it into the *Sry* gene, the dominant testis-determining gene on the Y chromosome of marsupial and eutherian mammals (41, 78). Accumulation of other male-advantage genes near the male-determining mutation would have differentiated one region of the proto-Y from the proto-X, and led to a loss of X-Y recombination of that segment. The subsequent inexorable march to degeneration of the Y would have been the result of accumulation of deleterious mutations, caused by Muller's ratchet, genetic hitchhiking, population size effect, and background selection (98, 106). As genes were lost from the Y, corresponding regions of the X would have been stranded as single-copy segments in the heterogametic sex, whereas the homogametic sex would have had two copies of these segments. Moreover, the dosage of the stranded X genes relative to their interacting partners on the autosomes [X to autosomal (A) ratio] would have been different in the two sexes, so that reciprocal regulatory effects of X and A genes would have been disrupted in one or both sexes. The loss of Y genes would therefore have set up a male-specific selection pressure to increase expression of X genes to a level comparable to that of females (or at least to a level at which it no longer had deleterious functional consequences), and/or a female-specific selection pressure to reduce X gene expression until deleterious functional consequences were reduced.

These ideas are attractive because they represent a unified theory to explain the evolution of diverse dosage compensation systems among XX/XY species. In mammals, a female-specific mechanism evolved to reduce transcription of one entire X chromosome (X-inactivation) in each female cell, leaving one active X chromosome. However, evidence suggests that the expression of genes across the single active X chromosome of both sexes in mammals is also increased by an unknown mechanism so that X gene expression occurs at a level comparable to autosomal expression (9, 46, 68). Thus, balance is achieved between the sexes in X gene expression, and between X and A expression within each sex. In *Drosophila*, both types of balance are achieved because of the evolution of a mechanism in males to increase X gene expression to the level of females. In *C. elegans*, the XX hermaphrodite reduces expression of X genes on both X chromosomes to the level of expression of the XO male, but both sexes upregulate the expression of X genes to a level comparable to the expression of A genes. The strongest argument for the critical importance of sex chromosome dosage compensation is that these three unrelated groups have each evolved a different but elaborate set of genetic tricks to solve the same problem.

Although most authors seem to suggest that the degeneration of the proto-Y or proto-W chromosome is an almost unavoidable consequence of a dominant sex-determining mutation on that chromosome, we note that some sex chromosomes are not well differentiated in species with genetic sex determination. Among birds, the ancient ratites (flightless birds such as the ostrich, emu, and rhea) have poorly differentiated sex chromosomes (87, 96), even though these chromosomes have had as much chance to differentiate as those of other birds. Although the Z and W chromosomes in these species are different, the wholesale degeneration of the W has not occurred and remains unexplained.

Gene Networks and Dosage Compensation

Genes operate in complex networks in which tens of thousands of genes have positive and negative regulatory influences on one another. Gene networks show scale-free properties (5), meaning that a minority of genes (hub genes) have widespread connections or reciprocal regulatory interactions with many other genes, but progressively larger sets of genes have progressively smaller numbers of connections. Most genes influence the expression of only a small number of other genes. Therefore, removing a random gene abruptly from a stable gene network can have an enormous deleterious influence if the gene is a hub gene on which many other genes rely for proper regulation, but typically will have relatively little effect when one of the more common non-hub genes is removed (48, 104). Many individual genes play roles that are redundant with others, or are functionally important in restricted environments, so there is no obvious phenotypic disadvantage if their copy number is reduced. Accordingly, variability in the copy number of individual genes or small segments of the genome is maintained with significant frequency in wild and laboratory populations (22, 32) and is often compatible with survival.

In contrast, monosomy or trisomy for a whole large chromosome usually presents a major problem; in humans, for example, either is usually lethal. This is probably because differences in dose over a large genomic segment are more likely to involve differences in dose of critical hub genes. The more hub genes involved, the larger the problem. Thus, one can imagine that as the Y chromosome proceeded along its inexorable degenerative path, the reduction in copy number of many or most X genes in males had relatively small phenotypic effects with minor effects on fitness. This view suggests that the reduction in dose of a minority of X genes drove the selection of complex molecular mechanisms to adjust X dosage in one sex or the other.

Specialization of Gene Content on the Sex Chromosomes

Other forces have also contributed to altering the gene content of the sex chromosomes because they are differentially represented in the two sexes. For example, genes involved in late stages of spermatogenesis are absent from the X chromosome of *C. elegans* and mammals because the X chromosome is silenced at later stages of spermatogenesis (41, 77).

More significant in the context of avian dosage compensation is the effect of sexually antagonistic mutations, which increase the fitness of one sex more than the other (28, 34, 79, 98). Sexually antagonistic alleles on autosomes have a net fitness that is the average fitness between males and females. Because the alleles are passed equally to males and females, if the loss of fitness to one sex substantially decreases the gain to the other sex, the allele will not become fixed. However, this balance is upset in the sex chromosomes because they are unequally represented in the two sexes. Recessive X mutations that benefit males produce a phenotype in the hemizygous male that is selected for even if the mutations are deleterious to females, because the phenotype would not be expressed in females until the allele's frequency in the population reaches the point at which homozygous females become more abundant (79). The allele will therefore likely be fixed as a polymorphism at frequencies that promote its survival in males but are not yet sufficient to produce significant loss of fitness in females (34). The disadvantage to females may set up selection pressure to downregulate the gene specifically in females, for example in vertebrate species if ovarian hormones or other female-specific factors evolve the ability to reduce the gene's expression. Thus, sex differences in gene expression can be the result of sex-specific selection pressures resulting from sexual antagonism and the differential representation of X genes in the two sexes. The selection of sexually antagonistic alleles can lead to some masculinization of the X chromosome, defined as an increase in frequency of male-benefit genes on the X relative to autosomes, which may be observed as an increase in male-biased genes (defined as genes with higher expression in males). Perhaps more significantly, the X chromosome is also subject to opposing pressures to be feminized (accumulation of female-benefit alleles) or demasculinized (loss of male-benefit alleles). The X chromosome spends twice as much of its evolutionary history in XX females than in XY males, so X genes (especially those controlling dominant traits) are subject to greater selection pressure to benefit females even if they lower fitness for males. Again, the accumulation of female-benefit male-detriment alleles on the X chromosome would set up selection pressure to reduce expression of the genes specifically in males (for example, if the gene evolves sensitivity to male-specific factors that lower expression). In ZZ/ZW systems, similar selection pressures are predicted to masculinize the Z chromosome, enriching it in male-biased genes, or lead to polymorphisms involving recessive Z alleles of female-biased genes.

In addition, the X chromosome seems to be subject to unusually high rates of retrotransposition in *Drosophila* and mammals (7, 29, 52), in which copies of genes are moved both from the X chromosome to autosomes and in the reverse direction. The retroposition of genes might in part mitigate the effects of sex-specific selection pressures. For example, as the Y chromosome diverged from the X chromosome, the sex-specific selection pressures on newly hemizygous X genes would be reduced if an autosomal copy of the gene were present at a functional dose. Sexual antagonism on the sex chromosomes can also set up selection pressures on autosomal genes that are redundant with or influence the function of sex chromosome genes

(91, 107) For example, if a male-benefit male-biased allele arises on the Z chromosome of birds, changes in regulation of a redundant autosomal gene could compensate for the deleterious effect of the male-benefit Z allele in females.

Evidence suggests that these processes have had a significant influence on the X chromosome gene content and bias in gene expression. For example, the X chromosome of *Drosophila* has an overrepresentation of female-biased genes in gonads and fewer genes with male-biased expression in all tissues (41, 72, 91, 98). In *C. elegans*, the X chromosome contains significantly fewer germline-intrinsic and sperm-enriched genes (98). The mammalian X chromosome is enriched for genes involved in reproductive, brain, and muscle function and those involved in the early stages of spermatogenesis, all of which have been explained as the result of selection for male-benefit alleles driven by natural or sexual selection (41, 51).

Types of Dosage Compensation

Here we define dosage compensation as any mechanism that reduces the disparity in the expression of genes that differ in copy number between groups of animals. Thus, if the male to female (M:F) ratio of expression of an X-linked gene is significantly higher than 0.5 in a mammal, we conclude that some process has acted to bring the ratio above the simple 1:2 ratio expected on the basis of genomic dose. One type of dosage compensation is sex-specific dosage compensation (SSDC), which we define as an evolved molecular mechanism, present in only one sex, that reduces the sexual disparity in the expression of sex chromosome genes. Examples of SSDC include female-specific X-inactivation in mammals, male-specific upregulation of X genes in *Drosophila*, and hermaphrodite-specific downregulation of X chromosome expression in *C. elegans*. In each of these cases, SSDC is chromosome wide, although it appears to have evolved gradually, perhaps on a gene-by-gene basis as the X chromosome became differentiated from the degenerating Y (30, 38, 90). Importantly, SSDC is triggered by a mechanism originating on the sex chromosomes that measures the number of X (or Z) chromosomes directly or indirectly, so that a sex-specific regulation of sex chromosome genes can be initiated.

Another mechanism of dosage compensation is sometimes called autosomal dosage compensation (10, 24, 25), which we call network dosage compensation (NDC) here because it operates on all chromosomes. Because genes are embedded in networks that involve complex regulatory interactions, their expression is influenced by processes such as negative feedback, autoregulation, and competition for limited regulatory factors (8–11, 106). These forces, which are not chromosome specific, usually buffer the effect of a change in genomic dose (10, 49, 99) and therefore reduce the expression ratio to a level below the ratio of genomic dose. In mouse models of Down syndrome, for example, in which the mice have three copies of genes homologous to segments of human chromosome 21, the average ratio of expression of genes (relative to control mice) is often less than 3:2 as expected from a simplistic model in which expression is proportional to genomic dose (33, 97). Similarly, *Drosophila* strains differing in copy number of a specific autosomal segment show highly variable ratios of expression among genes in the segment, which does not track copy number. For example, flies with three copies of a chromosomal segment have approximately 1.5 times higher expression of genes in that segment than flies with one copy of the segment (42, 106). These results strongly support a quantitatively significant NDC mechanism that applies throughout the genome. How big is the NDC effect? How much NDC can one expect in the complete absence of an evolved SSDC mechanism? Unfortunately, relatively few studies have been conducted that bear on this question, and the results are highly variable. We define the percentage NDC effect as 100 (1 - Pe/Pd), where Pe is the percentage difference in expression and Pd is the percentage difference in gene copy number. For example, in the case of trisomy versus disomy, genes compensated 60%

would be expressed at a 20% higher level in the trisomic state relative to disomic state, instead of the expected 50% higher level based on a 3:2 gene dose. In a variety of papers measuring gene expression in mice with known differences in copy number, the percent NDC ranged from 0% to 80%, a rather wide range (4, 31, 33, 58, 74, 83, 97). In *Drosophila* the percent NDC was roughly 50%–60% (42, 106), indicating that if species exist that completely lack SSDC, then their sex ratio of X or Z gene expression is expected to be considerably less than 2:1. However, the NDC effect appears not to be large enough (100%) on average to completely compensate for the 2:1 sex difference in genomic dose of all genes on a large chromosome such as the eutherian X chromosome; otherwise there would have been insufficient evolutionary pressure for SSDC mechanisms to evolve.

Although SSDC mechanisms in eutherian mammals, *Drosophila*, and *C. elegans* all effectively balance X and autosomal expression in the two sexes and balance X gene expression in males and females, the molecular mechanisms are different in each case. When multiple X chromosomes are present in mammals, a counting mechanism (57) triggers the expression of the noncoding Xist RNA, which accumulates on the X chromosome that is to be inactivated. Xist expression triggers a cascade of epigenetic mechanisms that silences the majority of genes in one X chromosome in each female cell (14, 21, 45, 75, 82). Some (15%–25%) genes escape inactivation, at least in humans (15), suggesting that there is no strong selection pressure to apply the SSDC mechanism perfectly across the entire chromosome. In *Drosophila*, dosage compensation results from mechanisms that measure the expression ratio of X to A genes and trigger the expression of *sex lethal* (*Sxl*) in females (23, 35, 63, 86). If the X:A ratio is 0.5, as in males, the *Sxl* transcript, normally expressed in females, is not expressed. In the absence of SXL protein, a dosage compensation complex (DCC) of proteins accumulates on the X chromosome. Among other epigenetic modifications, the DCC causes acetylation of histone 4 at lysine 16, resulting in increased transcription of X genes to the level of females (13). Two noncoding RNAs (ncRNAs), roX1 and roX2, are associated with the DCC and are essential for dosage compensation. In *C. elegans* (XX hermaphrodite versus XO male), X genes are expressed from both X chromosomes but at a reduced level (30). Downregulation of genes on both X chromosomes in XX cells is initiated by SDC-2 (sex determination and dosage compensation defect-2), which assembles the DCC. The DCC is recruited to multiple cis-acting regions of the X and spreads out along the chromosome from its initial binding sites (30). This results in suppression of transcription on the X. Although the SSDC mechanisms of mammals and *C. elegans* both involve reduction of X expression in the homogametic sex in addition to bisexual increase in X expression so that it reaches the level of autosomal expression, the mechanisms for balancing X and autosomal expression are not known (9, 42, 68).

DOSAGE COMPENSATION IN BIRDS

Dosage Compensation is Less Effective in Birds Than in XX/XY Systems

By 1967, only a few years after the identification of the avian W chromosome, and prior to the identification of any specific protein product encoded by the avian Z chromosome, Ohno (70) had already concluded that dosage compensation does not occur in birds. The evidence was that both Z chromosomes of males appear euchromatic along their entire length, a Barr body is absent, late replication of the Z chromosome does not occur (84), and some Z-linked mutant phenotypes are expressed more in ZZ males than ZW females. For example, the Z-linked white bar mutation in chickens produces bars that are twice as wide in ZZ males than ZW females, and ZZ males heterozygous for the mutation have the same phenotype as hemizygous ZW females (20, 70). The first study of sex differences in Z gene expression (6) showed that the aconitase protein was expressed at twice the level in males than in females, supporting ıe conclusion that no dosage compensation ccurred. That view was left unchallenged for vo decades, when studies of mRNA expression f approximately a dozen Z genes found that lthough some Z genes escape compensation nd are expressed at M:F ratios near 2, others ave ratios near 1 and hence show evidence of ome compensation (55, 64). Several other find-ıgs fueled the idea that dosage compensation weak. For example, studies of sexual dimor-hism in gene expression found a dispropor-onate number of Z genes among male-biased enes (1, 2, 18, 85, 100), and genes with male-pecific expression were disproportionately Z-nked rather than autosomal in gene expres-ion databases (50, 89). Moreover, studies using NA FISH (fluorescent in situ hybridization) ıat measured the site of transcription for six genes in ZZ male chickens showed biallelic xpression, leading to the conclusion that one ntire Z chromosome is not inactivated in the ame manner as occurs in mammals (54, 55).

The first global assessments of Z chromo-ome dosage compensation in birds (27, 47) ere enabled by the sequencing of the chicken enome (102), which provided unprecedented ıformation on chromosomal linkage of genes nd led to the availability of high-quality gene xpression microarrays. We measured mRNA evels from three somatic tissues of male and emale chick embryos at 14 days of incubation, omparing expression of approximately 1000 Z enes with approximately 17,000 A genes (47). he M:F ratios of expression of A genes varied round a mean of 1. In contrast, M:F ratios for genes were strongly shifted toward higher ex-ression in males, with means between 1.24 in ver and 1.4 in brain (**Figure 1**). Similar results ere found in measurements of smaller num-ers of genes in the zebra finch.

The methods used to estimate M:F ratios ffect the ratios that are observed. Microar-ay measurements have some nonlinearities and ackground hybridization (44) and less dynamic ange than more labor-intensive methods such s quantitative RT-PCR. These nonlinearities kely reduce M:F ratios. For example, microar-ay measurements of Y gene expression in mam-

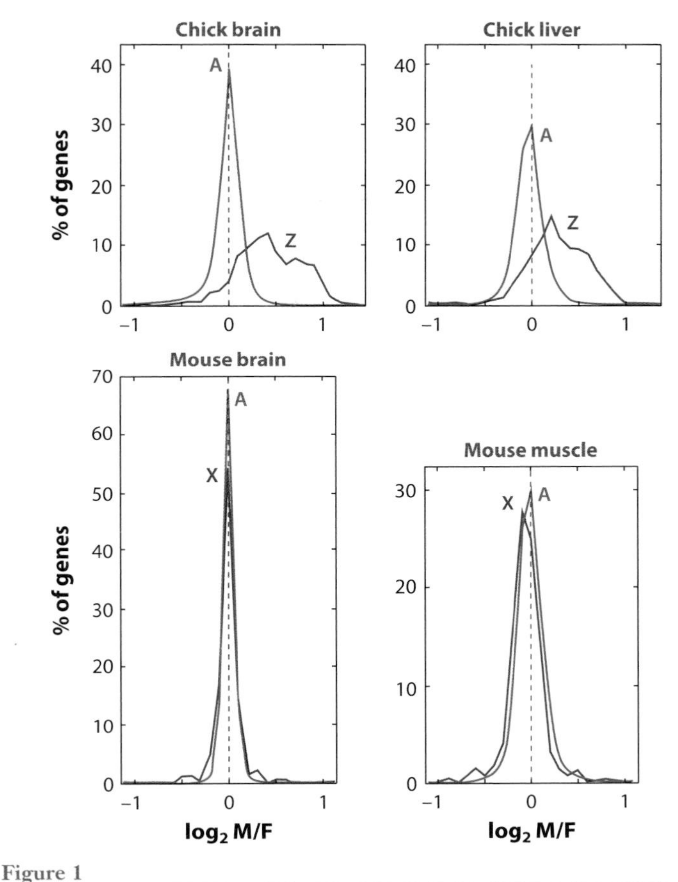

Figure 1

Distributions of male:female (M:F) ratios of mRNA expression (measured by microarrays) are shown for two chick and two mouse tissues, comparing autosomal (A) and Z genes for chick, and A and X genes for mouse. In mouse, both X and A genes show M:F ratios centered around 1 ($\log_2$ ratio 0), whereas in the chick Z genes show much higher ratios. From Reference 47 with permission.

mals show M:F ratios far below the true ratio of infinity, probably because of background hy-bridization in females (26, 80, 105). In general, M:F ratios in chicken measured by quantitative RT-PCR are approximately 10%–20% higher than those obtained from microarrays (27, 47, 85). Thus, the mean M:F ratios for chicken Z genes in somatic tissues could be as much as 50%–60% higher than for autosomes. Quanti-tative estimation of the amount of dosage com-pensation is best done in tissues in which males and females have a similar distribution of cell types in which the genes are presumably func-tioning in the same gene networks; an example is somatic but not gonadal tissues. In this manner differences in gene expression influenced by cell type do not confound comparison of the two sexes. However, because evolutionary pressures leading to dosage compensation and sexual bias in expression can operate most importantly in individual tissues (often in gonads) or under specific developmental or environmental conditions, measurement of sexual bias in gene expression needs to be performed in many tissues, environments, and developmental stages. Global analyses of dosage compensation so far have been performed in only a few tissues of chick embyros (27, 47).

The contrast between global assessments of dosage compensation for birds and mammals is striking (47). Comparable global measurements of M:F ratios of gene expression in adult mouse (**Figure 1**) and adult human somatic tissues (47) show only small discrepancies in the distribution of M:F ratios for X and A genes. In some cases the X genes are expressed at a slightly lower ratio than A genes. In brain the X genes show a greater amount of sexual dimorphism in both directions than A genes, confirming a previous report (105) indicating that the X chromosome of mammals contains more sexually dimorphic brain genes than the average autosome. The X versus A difference in amount of sexual dimorphism can be explained by the processes, reviewed above, that favor sex bias in expression. However, the percentage of X genes showing unusual amounts of sex bias is small. Clearly, mammalian dosage compensation mechanisms (SSDC plus NDC plus perhaps other unknown mechanisms) are globally effective in bringing M:F ratios of X gene expression in line with autosomal expression. It is interesting that the M:F ratios show different variability across mouse somatic tissues (adipose and liver are more sexually dimorphic in gene expression than whole brain), but that the variance in M:F ratios among X genes closely matches that for A genes in each tissue (**Figure 1**). The explanation probably is not that X and A genes evolved virtually identical percentages of genes that respond in a precisely similar graded fashion to sex-specific regulatory signals (for example, equal variance in regulation by gonadal hormones, which are the predominant signals controlling sexually dimorphic gene expression in mammals). Rather, the variance of X and A M:F ratios more likely match each other because of network interactions, i.e., the reciprocal regulation of X and A genes that underlies NDC in each tissue. The same NDC mechanisms can be invoked to explain the close matching of X to A gene dose in various XX/XY tissues (9), rather than evolution of many precise matching mechanisms for each transcript (90).

In chicks the mean Z:A ratios of mRNA expression are near 1 for ZZ males (range across tissues is 0.92 to 1.03) and near 0.8 for ZW females (range is 0.77 to 0.88) (47). These ratios are comparable to those in mammals, in which male and female X:A ratios are near 1 (range approximately 0.8 to 1.2) across many tissues, developmental stages, species, and sexes (68). *Drosophila* and *C. elegans* also achieve X:A ratios near 1 (42). If the need to balance Z:A (or X:A) gene dosage is the primary force that drives the evolution of dosage compensation mechanisms, as has been argued (46), then it would appear that some mechanism achieves compensation in birds, almost as well as in mammals. Alternatively, the striking bird-mammal difference in M:F ratio curves (**Figure 1**) and the unusual sex difference in Z:A ratios suggest that birds should be considered to be qualitatively different from the three well-known XX/XY systems because the dosage compensation mechanisms are relatively ineffective.

Is the amount of dosage compensation found in birds within the range that can be explained entirely by NDC? As reviewed above, there are relatively few papers that estimate the magnitude of compensation produced by NDC. Moreover, the NDC effect has been estimated most often in models of trisomy rather than monosomy, where the percentage compensation caused by NDC may be greater. Nevertheless, the amount of dosage compensation found for Z genes (roughly 50% compensation) is within the range of the size of the NDC effect previously measured. Also, it is important o ask whether the amount of compensation of roteins encoded by Z genes mirrors that found or Z transcripts. To our knowledge, only two Z roteins have been measured, and each shows igher expression in males (6, 18).

The contrast of the widespread concentraion of sex-biased genes on the avian Z chronosome but not on the X chromosomes of nammals and other XX/XY species may lead o two conclusions. Either SSDC is poorly eveloped or the pattern reflects gene speialization caused by masculinization of the Z hromosome due to selection of male-benefit nale-biased genes. Some evidence suggests that he latter idea, based on sexual antagonism, robably does not provide the entire explanaion. First, the degree of concentration of sexiased genes on the Z chromosome is much reater than that observed to date in any XX/XY ystem. Those systems show statistically signifiantly different frequencies of sex-biased genes n the X chromosome versus autosomes, but he differences are on the order of a few perent of genes and are not as dramatic as on he avian Z (72). To our knowledge no theoies predict that birds should have a dramatially greater effect of sexual antagonism on the Z chromosome, resulting in greater sex bias on he Z than on the X chromosome. Second, the nale-biased expression of Z genes extends even o housekeeping genes that are needed in all ells and are thus unlikely to show sex-specific unctional specialization or sexual antagonism. Thus, mitochondrial and ribosomal genes have listinctly higher M:F ratios on the Z than on utosomes (47) (**Figure 2**). Nevertheless, sexially biased expression as a response to sexual ntagonism cannot be ruled out and probably ontributes to high Z M:F ratios along with the ack of SSDC. No matter which of these facors is responsible, the large difference between M:F ratios on Z versus A is unexpected on the asis of the study of XX/XY systems. SSDC is educed or absent rather than critical and ubiqitous as previously thought (36, 42, 68), and/or here is an unexpected degree of selection for nale-benefit alleles on the Z chromosome. Eiher conclusion is remarkable.

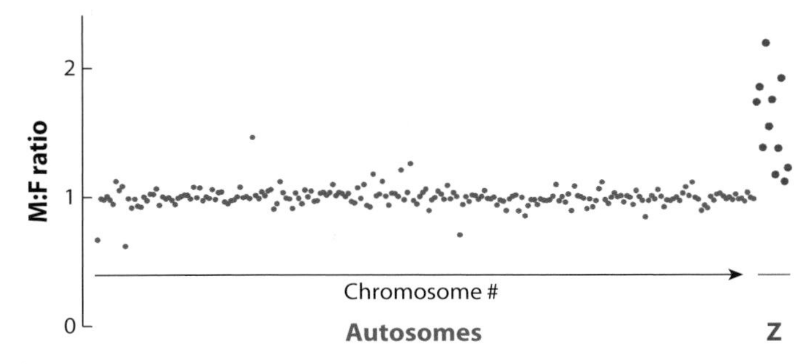

Figure 2

Male:female (M:F) ratios of expression of housekeeping (mitochondrial and ribosomal) genes in chick embryo, mapped by position in the genome. The autosomal positions are mapped by chromosome number (chromosome 1 to the left), and contrasted with ratios on the Z chromosome. Housekeeping genes may be unlikely to include male-benefit genes, but these genes have higher M:F ratios if they reside on the Z chromosome, suggesting that lack of dosage compensation rather than clustering of male-benefit genes better explains the high M:F ratio of Z genes. Data from Reference 47.

Regional Differences in Dosage Compensation on the Z Chromosome

When M:F ratios are mapped by gene position along the Z chromosome, intriguing patterns emerge (65) (**Figure 3**). Genes with high and low M:F ratios are found all along the chromosome, but Zq shows significantly higher M:F ratios than Zp. When the map is smoothed by averaging M:F ratios in a sliding window of genes, a region of low average M:F ratios (valley) is found 25–35 Mb from the telomere on Zp. In addition, one or two broad peaks of high M:F ratios are found toward the distal part of the Zq map. These variations are far from random and are unexpected by chance (65). The valley is not the result of a small number of genes with especially low M:F ratios, but rather reflects the absence of genes with high M:F ratios. For example, the valley includes a statistically unusual, uninterrupted string of genes with lower M:F ratios: 27 genes below 1.5 in brain, 25 genes below 1.2 in liver, and 23 genes below 1.4 in heart (data from Reference 65). The valleys in the curves for the three tissues occur at the same location on the Z chromosome, suggesting that regulation of M:F ratios of Z genes may be controlled in part by factors that are not tissue specific and may influence clusters of genes.

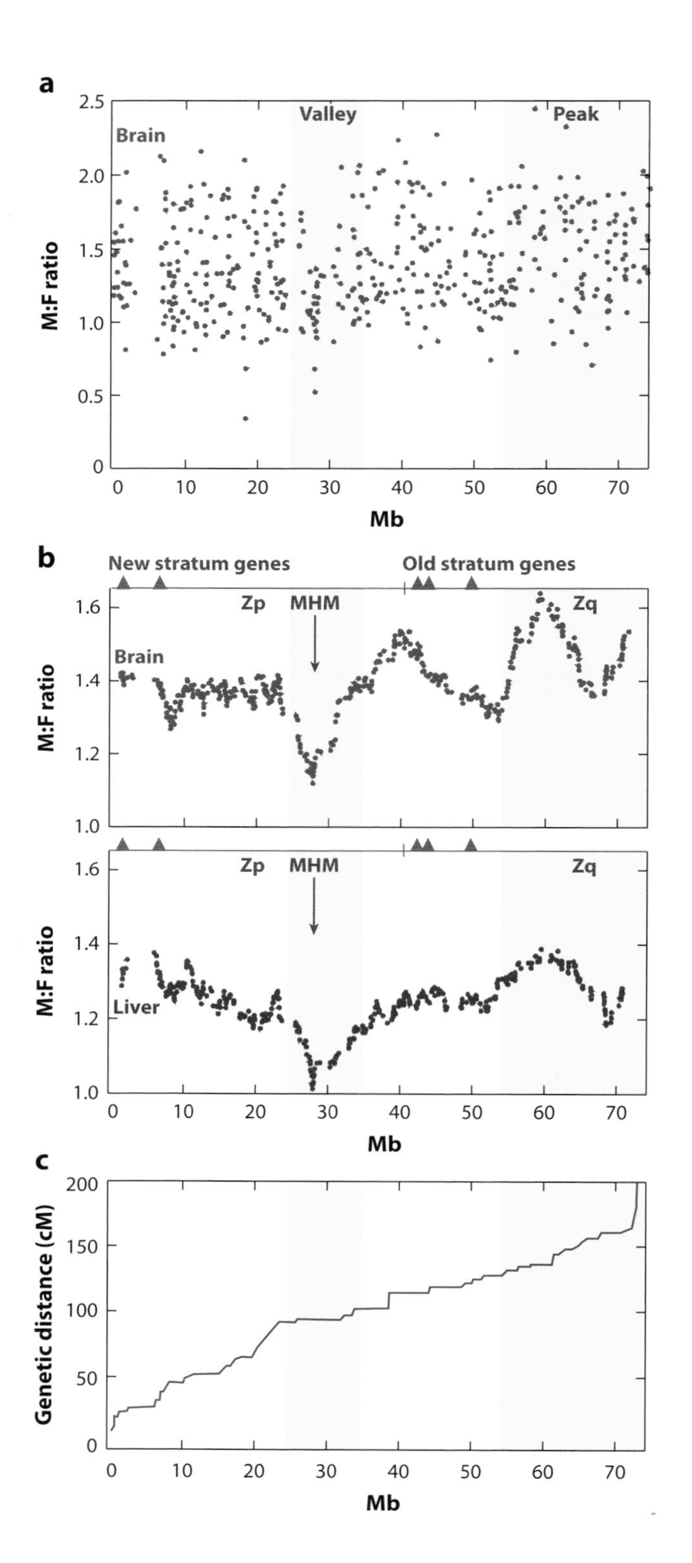

What is special about valley genes, and wh do they have unusually low average M:F ra tios? Importantly, the valley occurs precisel at a region of female-specific modification c Z chromatin and female-specific expression c a Z ncRNA (12, 93, 94). The ncRNA is ex pressed only in females from the MHM (mal hypermethylated) locus, probably because th DNA at the MHM locus is hypermethylate and transcriptionally silenced in ZZ males. Th MHM RNA is approximately 9.5 kb in lengt and is composed of a 2.2-kb repeat. Treatmen of males with the DNA methyltransferase in hibitor 5-azacytidine reduces the methylatio at the MHM locus, and the MHM ncRNA i then expressed in male cells (94). MHM ncRN accumulates along the female Z chromosom near its site of transcription. At the same re gion of the female but not male Z chromo some, histone 4 is acetylated at lysine 16 (12 The expression of the MHM ncRNA appear to be from the sense but not antisense stran and occurs in ZZW autosomal triploid and ZV diploid chickens, but not in ZZZ triploids an ZZ diploids. Thus, Teranishi and coworker (94) suggested that a W chromosome factor i necessary to prevent DNA methylation of th MHM locus, and by extension to trigger th

Figure 3

Regional specialization of the chicken Z chromosome. (*a*) Male:female (M:F) ratios of expression are plotted according to Z gene position for all brain genes. (*b*) The running average of 30 M: ratios (plotted at the position of the fifteenth gene) i mapped across the Z chromosome for brain and live The curve dips at the MHM (male hypermethylated locus (*valley*) and is higher at the end of Zq (*peak*). Valleys and peaks are similar in diverse somatic tissues such as brain and liver. Data taken from Reference 65. Strata on the Z chromosome have been estimated on the basis of degree of divergence of five Z-W genes, the positions of which are shown as triangles. The old and new stratum genes are mapped. Data taken from Reference 43. (*c*) The graph of genetic (in centimorgans, cM) versus physical distance (in megabases, Mb) based on linkage studies shows that a recombinatio cold spot, or flat portion of the curve, occurs at the MHM valley. Reprinted from Reference 101 with the permission of S. Karger AG, Basel.

acetylation of H4K16 in this region (12; see also 40). However, in those experiments the presence of the W chromosome is confounded by Z:A ratio, so it is equally feasible that a Z:A ratio below 1 in ZW diploid or ZZW triploid females leads to expression of the MHM RNA, and/or a Z:A ratio of 1 causes methylation of the MHM locus in ZZZ triploid or ZZ diploid males.

Because acetylation of H4K16 is associated with transcriptional activation in a variety of systems, including upregulation of the male's single X chromosome to compensate dosage in *Drosophila* (13), Bisoni and colleagues (12) postulated that the female-specific histone modifications in the MHM region control SSDC. This idea is supported by the clustering of relatively compensated genes in this specific region of the Z chromosome (65) (**Figure 3**). Moreover, the involvement of a female-specific ncRNA at the same locus is reminiscent of the critical involvement of other sex-specific ncRNAs in SSDC, such as *Xist* in mammals and *roX* genes in *Drosophila*. These findings together lead to the theory that MHM ncRNA attracts proteins that modify the Z chromatin in the MHM valley, causing acetylation of H4K16 and probably other modifications in ZW females. Alternatively, the MHM ncRNA may directly modify the chromatin structure in the MHM valley (81). If either mechanism is confirmed, then birds may have evolved an SSDC mechanism that is flylike in that it involves upregulation of the X/Z chromosome in the heterogametic sex. However, numerous other questions remain. What is the molecular signal that initiates the male-specific hypermethylation of the MHM locus? Is the DNA methylation of the MHM locus primary in controlling its expression (suggesting the existence of a male-specific molecular SSDC mechanism as in mammals or *C. elegans*), or is hypermethylation of MHM a secondary consequence of the absence of H4K16 acetylation of the MHM valley in males? Is SSDC triggered by a W chromosome factor, a Z chromosome counting mechanism (Z:A ratio), or both? Who is doing the compensating? Are Z genes upregulated in females, downregulated in males, or both?

The results reviewed so far imply that a mechanism of SSDC has evolved in birds, but the mechanism is not chromosome wide as in mammals, *Drosophila*, and *C. elegans*. Although dosage-compensated genes are concentrated in the region of the MHM locus, they also occur at other regions of the Z chromosome. If the MHM ncRNA is central to SSDC in birds, are compensated genes outside of the MHM valley also regulated at a distance by factors encoded within the MHM valley?

A priori one expects that dosage differences are critical for only a minority of genes, those for which a sexual imbalance in expression is strongly disadvantageous. These would presumably be the hub genes, for which sex differences in expression would have widespread secondary effects on expression in diverse molecular pathways. Evidence in favor of this idea is that Z genes that are dosage compensated show different gene ontology categories from those that are not dosage compensated (27, 65). Compensated genes are enriched for genes involved in development and physiological processes, whereas uncompensated genes are enriched in other categories, including catalytic activity.

Intriguingly, the MHM valley rather closely corresponds to a Z chromosome recombination cold spot, a region of infrequent crossing over during meiosis (101) (**Figure 3**). Similarly, the Xist locus in humans is at the end of a recombination cold spot (67). Recombination cold spots are thought to occur at regions containing specialized DNA sequences, methylated DNA, transcriptional inactivity, and/or specific chromatin status (69, 73). DNA methylation, a feature of the male MHM locus in ZZ male chickens and the Xist locus in female humans, decreases recombination (59). The methylation of repetitive DNA in the MHM region of male chickens may function to inhibit inappropriate heterologous recombination of dispersed DNA repeats in and near the MHM locus to protect the stability of the genome in this region.

Many genes in the MHM valley are found in a syntenic region and in the same gene order in platypus, human, and cow (**Figure 4**). Most

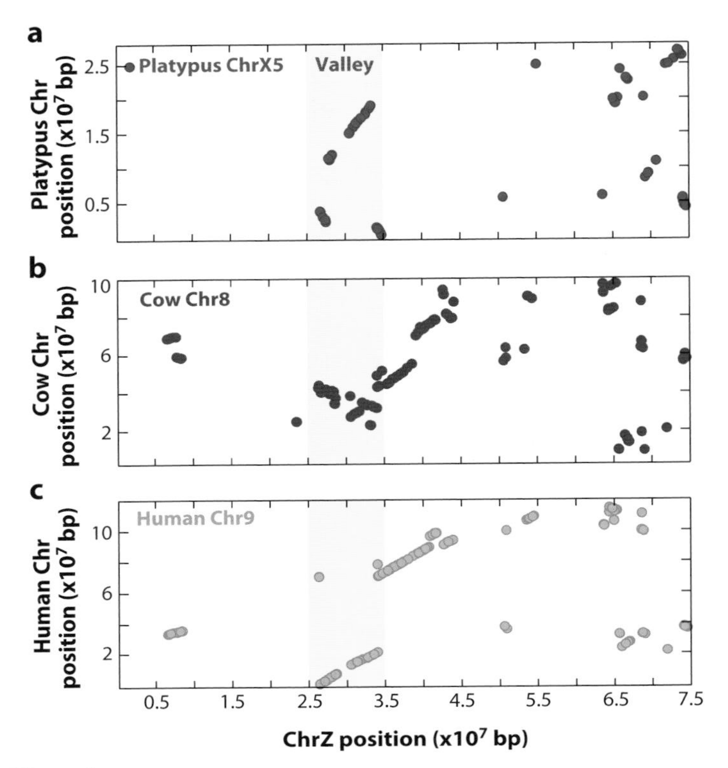

Figure 4

Conserved synteny between the chicken MHM (male hypermethylated) valley and mammalian chromosomes. The position of chicken Z chromosome genes are plotted against the positions of orthologous genes on the chromosome with maximal homology to the chicken MHM valley. In general, MHM valley genes are syntenic with similar gene order on one chromosome in these species. Valley genes are primarily located on sex chromosome X5 of platypus, a monotreme mammal that diverged from therian mammals 210 million years ago, and thus may represent a segment of chromosome from an ancestral precursor to avian ZZ/ZW and monotreme XX/XY sex chromosomes. Data from Reference 65.

interesting is that 65% of MHM valley genes are found on one of the five X chromosomes of the platypus (65, 78). Is it possible that the MHM valley genes represent a remnant of an ancestral sex chromosome that gave rise to both avian and monotreme sex chromosomes? If so, could monotremes share some of the dosage compensation mechanisms that are seen in birds? Approximately 80% of MHM valley genes are also still syntenic in humans and cows, although not as much in other mammalian species, giving rise to the speculation that some factors might have inhibited the breakup of these genes in the 300 million years since the split between human and avian lineages.

Strata on the Sex Chromosomes

The mammalian X and Y are stratified, reflecting an evolutionary history in which segments were added to the sex chromosomes in successive waves, followed by attrition of those genes on the Y (39, 41). Thus, the X and Y chromosome ceased recombination in discrete steps (56). The oldest segment of the human X chromosome contains the primary genes responsible for sex determination and SSDC, possibly because the SSDC mechanism had the longest time to evolve on that segment. The oldest stratum probably arose after the mutation of *Sox 3*, which produced the male-determining gene *Sry* on the proto-Y chromosome. Also in the oldest stratum is *Xist*, responsible for SSDC among eutherians. Newer strata contain genes added to the eutherian X before or after the split from marsupials. The different degrees of divergence of X-Y paralogous genes in each stratum indicate the time since that layer ceased to recombine in X and Y (56). Moreover, the newest X strata in humans contain genes that most frequently escape X inactivation (15), whereas the oldest stratum is more completely dosage compensated. The gradient in dosage compensation suggests that applying the Xist-mediated SSDC mechanism to all new X genes has taken tens of millions of years in each stratum (38). Therefore, the selection pressure to compensate every last X gene seems not to be strong.

The chicken Z chromosome also shows signs of stratification, although much less information is available. On the basis of the degree of divergence of five Z-W paralogous genes, Handley and coworkers (43) found that three genes on Zq near the centromere seem to represent an older stratum that ceased recombination ~102–170 Mya (million years ago), whereas two genes within 7 Mb of the Zp telomere represent a newer stratum with a divergence time of ~58–85 Mya (**Figure 3**). Unfortunately the relative time since loss of recombination of the MHM valley region has not yet been estimated. The MHM valley includes the putative Z sex-determining gene DMRT1, a testis-determining gene that is present in two copies in males but in one copy in females (88, 94). Moreover, because SSDC evolves slowly, older segments of the Z chromosome might harbor the best-compensated genes. It would make sense if the MHM valley represents the oldest stratum containing the primary SSDC mechanism, clusters of compensated genes, and the origin of the sex-determining mutation, just as occurs in the mammalian X chromosome (65). An alternative view suggests the opposite relation, that older strata contain more poorly compensated genes (27). In animals with poor SSDC, the higher expression of Z genes in males is the default condition, which would promote fixation of male-benefit alleles on the Z chromosome. That process would have had more time to occur on the older strata. Thus, the higher M:F ratios on Zq, relative to Zp, might be related to the presumed older age of Zq. Resolution of this issue requires more information, including a more fine-grained mapping of strata, clarification of the role of the MHM ncRNA in SSDC, and discovery of the sex-determining mechanism of birds.

Our current hypothesis is that SSDC has evolved in birds, but is predominantly restricted to the MHM region and has not spread across the Z chromosome as has happened in mammalian, *Drosophila*, and *C. elegans* XX/XY systems. Thus, the MHM valley can be seen as a possible stratum, differentiated from other Z regions on the basis of the amount of dosage compensation. Why is there no effective SSDC mechanism beyond the MHM valley? One answer may be that major portions of the avian Z-W sex chromosomes have ceased recombination only relatively recently (58–85 Mya) (43) compared with segments of the eutherian X chromosome (41). However, although the eutherian X-Y divergence may have begun prior to the avian Z-W divergence, the most recently diverged segments of the human X-Y chromosome appear to have separated 40 and 100 Mya (103). Thus, there may not have been a large bird-human difference in divergence time of the most recently added segments. Nevertheless, the highly dimorphic Z and W found among most birds diverged approximately 80 Mya from those of the ratites, which have poorly differentiated sex chromosomes, so that most of the genes on the W probably were lost only in that period. Thus, it is possible that differences in the history of avian and eutherian sex chromosomes can account for some of the bird-mammal differences. Further work on bird dosage compensation mechanisms promises interesting insights, especially because of the availability for study of poorly differentiated Z and W chromosomes of the ratites (61, 96).

Who Is Doing the Compensating?

If there is a regional or even chromosome-wide but selective mechanism of SSDC in birds, is the heterogametic female upregulating Z gene expression to match that of males (a fly-like mechanism) or do males downregulate genes (a mammalian or wormlike mechanism)? So far there are only hints at an answer. The existence of female-specific acetylation of H4K16 in the MHM valley strongly suggests that there is transcriptional activation in this region in females but not in males. Conversely, Z genes that are unbiased are expressed at levels in males and females that are below mean autosomal expression, suggesting that the male may be compensating (27). Z gene expression is positively correlated with M:F ratio in males but uncorrelated in females (lower in compensated or unbiased genes than in uncompensated or male-biased genes), a pattern that is compatible with a male-specific mechanism of regulation (65). However, resolution of this question requires better understanding of the molecular mechanisms of compensation.

What Is Special About ZZ/ZW Systems?

Limited evidence suggests that the relative lack of SSDC may also occur in other ZZ/ZW systems. Most Lepidoptera (butterflies and moths), like birds, are ZZ/ZW (37, 95). Quantitative RT-PCR was used to assess the M:F ratios of expression of 14 contiguous genes from a 340-kb segment of the Z chromosome of the silkworm moth *Bombyx mori* (53). All but one gene showed higher expression in males. However, the median M:F ratio was 2.6 (range 0.5–9.7), which is too high to reflect simply a failure of dosage compensation. Rather, the genes measured may be exceptionally male biased, perhaps clustered in a specialized region of the Z. Still, there is little evidence so far suggesting that dosage compensation occurs in Lepidoptera. More information is needed.

Although numerous theories point to differences between XX/XY and ZZ/ZW systems in the degree and nature of sexual selection or sexual conflict (3, 60, 62, 66, 76), it is far from clear that the nature of sexual antagonism is so fundamentally different in female heterogametic systems that it should favor a Z chromosome that is loaded with male-benefit genes, in contrast to more modest loading in XX/XY systems. On the basis of present information, we suggest that SSDC in birds has evolved in the MHM region, but for unknown reasons did not spread across the entire Z chromosome, as happened in the three major XX/XY systems that have been studied. In the absence of a chromosome-wide SSDC mechanism, the deleterious effects of dosage imbalance are therefore likely to have been mitigated in non-MHM regions by several factors, including the following: (*a*) Some Z genes have individually evolved sensitivity to sex-specific regulatory factors that cause dosage compensation of particularly dosage-critical Z genes such as hub genes. The regulatory factors may include the Z-specific SSDC mechanism (e.g., MHM-mediated), gonadal hormones present in one sex, or uncompensated Z genes that are constitutively higher in male cells. These factors have evolved control over Z genes to downregulate them in males and/or upregulate them in females. The effects of this sex-specific regulation are seen in the low M:F ratios of a minority of Z genes dispersed along the Z chromosome (**Figure 3**). These local, gene-specific regulatory interactions evolved gene by gene as the W chromosome degenerated, leaving Z genes stranded in the hemizygous state in females. (*b*) Adjustments in autosomal genes also occurred to offset the sexual imbalance in the functional effect of the Z genes expressed at different levels in the two sexes (91). For example, mutations would have been favored to replace a dosage-critical Z gene by a related autosomal gene in a functional pathway. In this context it is interesting that female-biased genes have evolved more rapidly during avian evolution, which could reflect in part alterations in female genetic networks that offset the male bias of some Z genes (62). (*c*) The higher expression of some Z genes in males probably reflects some masculinization of the Z chromosome (increase in content of male-benefit genes), because the same pressure accounts for the femininization or demasculinization of the X chromosome in XX/XY systems. However, in XX/XY systems the specialization effect is quantitative, a statistical shift, not a qualitative change in the gene content of the X chromosome (72). Therefore, it seems unlikely that sexual antagonism alone would have produced a Z chromosome with such striking male bias as is observed. Conversely, in species lacking chromosome-wide SSDC, the selection for male-benefit Z alleles might be more efficient. The deleterious effects for females of male-benefit Z alleles would be naturally reduced because of their hemizygous state in females (27), so some Z genes may have rapidly evolved functions that favor males.

SUMMARY AND CONCLUSIONS

Z genes in birds show average M:F ratios that are roughly 50% higher than those for autosomal genes. These ratios are less than the 2:1 M:F ratio of Z gene copy number, suggesting that some sort of dosage compensation occurs. One likely source of dosage compensation is the process of NDC, which is probably quantitatively sufficient to explain 50% compensation. However, the regional concentration of compensated genes near the MHM locus, together with the sex-specific DNA methylation and histone modifications in that region, strongly suggest that some SSDC also occurs. A minority of Z genes located all along the Z chromosomes shows dosage compensation (near sexual equality), suggesting that the deleterious effects of the genomic dosage difference for hub genes were eliminated by sex-specific adjustments in gene expression [either controlled by MHM, an undiscovered SSDC mechanism, or by gene-specific (and probably tissue-specific) compensatory adjustments]. For other non-sexually antagonistic genes that retain a male bias in expression, the deleterious effects of the sex difference in gene dose may have been reduced by the evolution of autosomal adjustments that reduce the functional sex difference, for example by female-specific upregulation of autosomal genes with functions redundant with the male-biased Z gene (107), or changes in the gene network to make it insensitive to the sex difference. Solving the dosage problem for the most dosage-critical genes might have been enough for birds to forego the evolution of a more effective chromosome-wide mechanism of dosage compensation. Selection for male-benefit alleles may also have contributed to the frequency of male-biased genes on the Z.

Although one can find ways to explain away the greater sex bias of sex chromosome expression in ZZ/ZW birds than in known XX/XY systems, the essential problem is that no theoretical framework exists that explains when SSDC evolves and when it does not. What differences between birds and mammals, or between ZZ/ZW and XX/XY systems, account for the different path of evolution of SSDC mechanisms?

It is remarkable that forty years ago Ohno (70) was more right than wrong in his conclusion that birds do not show dosage compensation. He summed up as follows: "One may conclude that the dosage compensation mechanism for sex-linked genes is a luxury which is desirable to have, but it is definitely not a *sine qua non* of successful speciation. Conversely, one may take the stand that the failure to evolve an effective dosage compensation mechanism for numerous Z-linked genes is one major reason why birds have not escaped the status of feathered reptiles" (70, p. 147). The status of these magnificent "feathered reptiles" is exalted in our view, because birds (both sexes) are amazingly adaptable across extremes of environments, and have highly evolved functions (e.g., flight, spatial orientation and memory, vocal learning) that are truly impressive and often superior to our own. Thus, we are more attracted to Ohno's former rather than latter option, that dosage compensation, at least at the transcriptional level, is not a prerequisite for success for all species with heteromorphic sex chromosomes. The study of dosage compensation and evolution of the sex chromosomes in ZZ/ZW systems has barely begun, and there are more questions than answers. Thankfully, the genomic resources available to address these questions in birds have increased significantly with the sequencing of the chicken and zebra finch genomes, which have catalyzed further advances in other molecular and cytological resources. Thus, the study of birds will likely provide some of the answers. It's going to be exciting.

DISCLOSURE STATEMENT

The authors are not aware of any biases that might be perceived as affecting the objectivity of this review.

ACKNOWLEDGMENTS

Thanks to Jennifer Marshall Graves, Brian Oliver, Jake Lusis, and Eric Schadt for discussions and Kathy Kampf for assistance.

LITERATURE CITED

1. Agate RJ, Choe M, Arnold AP. 2004. Sex differences in structure and expression of the sex chromosome genes *CHD1Z* and *CHD1W* in zebra finches. *Mol. Biol. Evol.* 21:384–96
2. Agate RJ, Grisham W, Wade J, Mann S, Wingfield J, et al. 2003. Neural not gonadal origin of brain sex differences in a gynandromorphic finch. *Proc. Natl. Acad. Sci. USA* 100:4873–78
3. Albert AYK, Otto SP. 2005. Sexual selection can resolve sex-linked sexual antagonism. *Science* 310:119–21
4. Amano K, Sago H, Uchikawa C, Suzuki T, Kotliarova SE, et al. 2004. Dosage-dependent over-expression of genes in the trisomic region of Ts1Cje mouse model for Down syndrome. *Hum. Mol. Genet.* 13:1333–40
5. Barabasi AL, Oltvai ZN. 2004. Network biology: understanding the cell's functional organization. *Nat. Rev. Genet.* 5:101–13
6. Baverstock PR, Adams M, Polkinghorne RW, Gelder M. 1982. A sex-linked enzyme in birds–Z-chromosome conservation but no dosage compensation. *Nature* 296:763–66
7. Betran E, Emerson JJ, Kaessmann H, Long M. 2004. Sex chromosomes and male functions: where do new genes go? *Cell Cycle* 3:873–75
8. Birchler JA, Bhadra U, Bhadra MP, Auger DL. 2001. Dosage-dependent gene regulation in multicellular eukaryotes: implications for dosage compensation, aneuploid syndromes, and quantitative traits. *Dev. Biol.* 234:275–88
9. Birchler JA, Fernandez HR, Kavi HH. 2006. Commonalities in compensation. *BioEssays* 28:565–68
10. Birchler JA, Riddle NC, Auger DL, Veitia RA. 2005. Dosage balance in gene regulation: biological implications. *Trends Genet.* 21:219–26
11. Birchler JA, Veitia RA. 2007. The gene balance hypothesis: from classical genetics to modern genomics. *Plant Cell* 19:395–402
12. Bisoni L, Batlle-Morera L, Bird AP, Suzuki M, McQueen HA. 2005. Female-specific hyperacetylation of histone H4 in the chicken Z chromosome. *Chromosome Res.* 13:205–14
13. Bone JR, Lavender J, Richman R, Palmer MJ, Turner BM, Kuroda MI. 1994. Acetylated histone H4 on the male X chromosome is associated with dosage compensation in *Drosophila*. *Genes Dev.* 8:96–104
14. Boumil RM, Lee JT. 2001. Forty years of decoding the silence in X-chromosome inactivation. *Hum. Mol. Genet.* 10:2225–32
15. Carrel L, Willard HF. 2005. X-inactivation profile reveals extensive variability in X-linked gene expression in females. *Nature* 434:400–4
16. Charlesworth B. 1991. The evolution of sex chromosomes. *Science* 251:1030–33
17. Charlesworth D, Charlesworth B, Marais G. 2005. Steps in the evolution of heteromorphic sex chromosomes. *Heredity* 95:118–28
18. Chen X, Agate RJ, Itoh Y, Arnold AP. 2005. Sexually dimorphic expression of *trkB*, a Z-linked gene, in early posthatch zebra finch brain. *Proc. Natl. Acad. Sci. USA* 102:7730–35
19. Clayton DF, Arnold AP. 2008. Studies of songbirds in the age of genetics: What to expect from genomic approaches in the next 20 years. In *Biology of Birdsong*, ed. P Zeigler, P Marler. New York: Ann. NY Acad. Sci., pp. 369–378
20. Cock AG. 1964. Dosage compensation and sex-chromatin in non-mammals. *Genet. Res.* 5:354
21. Cohen HR, Royce-Tolland ME, Worringer KA, Panning B. 2005. Chromatin modifications on the inactive X chromosome. *Prog. Mol. Subcell. Biol.* 38:91–122
22. Cutler G, Marshall LA, Chin N, Baribault H, Kassner PD. 2007. Significant gene content variation characterizes the genomes of inbred mouse strains. *Genome Res.* 17:1743–54
23. Deng X, Meller VH. 2006. Non-coding RNA in fly dosage compensation. *Trends Biochem. Sci.* 31:526–32
24. Devlin RH, Holm DG, Grigliatti TA. 1982. Autosomal dosage compensation in *Drosophila melanogaster* strains trisomic for the left arm of chromosome 2. *Proc. Natl. Acad. Sci. USA* 79:1200–4
25. Devlin RH, Holm DG, Grigliatti TA. 1988. The influence of whole-arm trisomy on gene expression in *Drosophila*. *Genetics* 118:87–101
26. Dewing P, Shi T, Horvath S, Vilain E. 2003. Sexually dimorphic gene expression in mouse brain precedes gonadal differentiation. *Brain Res. Mol. Brain Res.* 118:82–90
27. Ellegren H, Hultin-Rosenberg L, Brunström B, Dencker L, Kultima K, Scholz B. 2007. Faced with inequality: Chicken does not have a general dosage compensation of sex-linked genes. *BMC Biology* 5:40

28. Ellegren H, Parsch J. 2007. The evolution of sex-biased genes and sex-biased gene expression. *Nat. Rev. Genet.* 8:689–98
29. Emerson JJ, Kaessmann H, Betran E, Long M. 2004. Extensive gene traffic on the mammalian X chromosome. *Science* 303:537–40
30. Ercan S, Giresi PG, Whittle CM, Zhang X, Green RD, Lieb JD. 2007. X chromosome repression by localization of the *C. elegans* dosage compensation machinery to sites of transcription initiation. *Nat. Genet.* 39:403–8
31. Esposito G, Imitola J, Lu J, De Filippis D, Scuderi C, et al. 2008. Genomic and functional profiling of human Down syndrome neural progenitors implicates S100B and aquaporin 4 in cell injury. *Hum. Mol. Genet.* 17:440–57
32. Estivill X, Armengol L. 2007. Copy number variants and common disorders: filling the gaps and exploring complexity in genome-wide association studies. *PLoS Genet.* 3:1787–99
33. Fitzpatrick DR. 2005. Transcriptional consequences of autosomal trisomy: primary gene dosage with complex downstream effects. *Trends Genet.* 21:249–53
34. Gibson JR, Chippindale AK, Rice WR. 2002. The X chromosome is a hot spot for sexually antagonistic fitness variation. *Proc. R. Soc. London Ser. B* 269:499–505
35. Gilfillan GD, Dahlsveen IK, Becker PB. 2004. Lifting a chromosome: dosage compensation in *Drosophila melanogaster*. *FEBS Lett.* 567:8–14
36. Gilfillan GD, Straub T, de Wit E, Greil F, Lamm R, et al. 2006. Chromosome-wide gene-specific targeting of the *Drosophila* dosage compensation complex. *Genes Dev.* 20:858–70
37. Goldsmith MR, Shimada T, Abe H. 2005. The genetics and genomics of the silkworm, *Bombyx mori*. *Annu. Rev. Entomol.* 50:71–100
38. Graves JAM, Disteche CM. 2007. Does gene dosage really matter? *J. Biol.* 6:1
39. Graves JAM. 1995. The origin and function of the mammalian Y chromosome and Y-borne genes—an evolving understanding. *BioEssays* 17:311–19
40. Graves JAM. 2003. Sex and death in birds: a model of dosage compensation that predicts lethality of sex chromosome aneuploids. *Cytogenet. Genome Res.* 101:278–82
41. Graves JAM. 2006. Sex chromosome specialization and degeneration in mammals. *Cell* 124:901–14
42. Gupta V, Parisi M, Sturgill D, Nuttall R, Doctolero M, et al. 2006. Global analysis of X-chromosome dosage compensation. *J. Biol.* 5:3
43. Handley LJ, Ceplitis H, Ellegren H. 2004. Evolutionary strata on the chicken Z chromosome: Implications for sex chromosome evolution. *Genetics* 167:367–76
44. He YD, Dai H, Schadt EE, Cavet G, Edwards SW, et al. 2003. Microarray standard data set and figures of merit for comparing data processing methods and experiment designs. *Bioinformatics* 19:956–65
45. Heard E. 2004. Recent advances in X-chromosome inactivation. *Curr. Opin. Cell Biol.* 16:247–55
46. Heard E, Disteche CM. 2006. Dosage compensation in mammals: fine-tuning the expression of the X chromosome. *Genes Dev.* 20:1848–67
47. Itoh Y, Melamed E, Yang X, Kampf K, Wang S, et al. 2007. Dosage compensation is less effective in birds than in mammals. *J. Biol.* 6:2
48. Jeong H, Mason SP, Barabasi AL, Oltvai ZN. 2001. Lethality and centrality in protein networks. *Nature* 411:41–42
49. Kacser H, Burns JA. 1981. The molecular basis of dominance. *Genetics* 97:639–66
50. Kaiser VB, Ellegren H. 2006. Nonrandom distribution of genes with sex-biased expression in the chicken genome. *Evolution* 60:1945–51
51. Khil PP, Camerini-Otero RD. 2005. Molecular features and functional constraints in the evolution of the mammalian X chromosome. *Crit. Rev. Biochem. Mol. Biol.* 40:313–30
52. Khil PP, Oliver B, Camerini-Otero RD. 2005. X for intersection: retrotransposition both on and off the X chromosome is more frequent. *Trends Genet.* 21:3–7
53. Koike Y, Mita K, Suzuki MG, Maeda S, Abe H, et al. 2003. Genomic sequence of a 320-kb segment of the Z chromosome of *Bombyx mori* containing a *kettin* ortholog. *Mol. Genet. Genomics* 269:137–49
54. Kuroda Y, Arai N, Arita M, Teranishi M, Hori T, et al. 2001. Absence of Z-chromosome inactivation for five genes in male chickens. *Chromosome Res.* 9:457–68

55. Kuroiwa A, Yokomine T, Sasaki H, Tsudzuki M, Tanaka K, et al. 2002. Biallelic expression of Z-linked genes in male chickens. *Cytogenet. Genome Res.* 99:310–14
56. Lahn BT, Page DC. 1999. Four evolutionary strata on the human X chromosome. *Science* 286:964–67
57. Lee JT. 2005. Regulation of X-chromosome counting by *Tsix* and *Xite* sequences. *Science* 309:768–71
58. Lyle R, Gehrig C, Neergaard-Henrichsen C, Deutsch S, Antonarakis SE. 2004. Gene expression from the aneuploid chromosome in a trisomy mouse model of Down syndrome. *Genome Res.* 14:1268–74
59. Maloisel L, Rossignol JL. 1998. Suppression of crossing-over by DNA methylation in *Ascobolus*. *Genes Dev.* 12:1381–89
60. Mank JE, Axelsson E, Ellegren H. 2007. Fast-X on the Z: rapid evolution of sex-linked genes in birds. *Genome Res.* 17:618–24
61. Mank JE, Ellegren H. 2007. Parallel divergence and degradation of the avian W sex chromosome. *Trends Ecol. Evol.* 22:389–91
62. Mank JE, Hultin-Rosenberg L, Axelsson E, Ellegren H. 2007. Rapid evolution of female-biased, but not male-biased, genes expressed in the avian brain. *Mol. Biol. Evol.* 24:2698–706.
63. Manolakou P, Lavranos G, Angelopoulou R. 2006. Molecular patterns of sex determination in the animal kingdom: a comparative study of the biology of reproduction. *Reprod. Biol. Endocrinol.* 4:59
64. McQueen HA, McBride D, Miele G, Bird AP, Clinton M. 2001. Dosage compensation in birds. *Curr. Biol.* 11:253–57
65. Melamed E, Arnold AP. 2007. Regional differences in dosage compensation on the chicken Z chromosome. *Genome Biol.* 8:R202
66. Miller PM, Gavrilets S, Rice WR. 2006. Sexual conflict via maternal-effect genes in ZW species. *Science* 312:73
67. Nagaraja R, MacMillan S, Kere J, Jones C, Griffin S, et al. 1997. X chromosome map at 75-kb STS resolution, revealing extremes of recombination and GC content. *Genome Res.* 7:210–22
68. Nguyen DK, Disteche CM. 2006. Dosage compensation of the active X chromosome in mammals. *Nat. Genet.* 38:47–53
69. Nishant KT, Rao MR. 2006. Molecular features of meiotic recombination hot spots. *BioEssays* 28:45–56
70. Ohno S. 1967. *Sex Chromosomes and Sex-Linked Genes*. New York: Springer-Verlag
71. Oliver B, Parisi M. 2004. Battle of the Xs. *BioEssays* 26:543–48
72. Parisi M, Nuttall R, Naiman D, Bouffard G, Malley J, et al. 2003. Paucity of genes on the *Drosophila* X chromosome showing male-biased expression. *Science* 299:697–700
73. Petes TD. 2001. Meiotic recombination hot spots and cold spots. *Nat. Rev. Genet.* 2:360–69
74. Phillips JL, Hayward SW, Wang Y, Vasselli J, Pavlovich C, et al. 2001. The consequences of chromosomal aneuploidy on gene expression profiles in a cell line model for prostate carcinogenesis. *Cancer Res.* 61:8143–49
75. Plath K, Fang J, Mlynarczyk-Evans SK, Cao R, Worringer KA, et al. 2003. Role of histone H3 lysine 27 methylation in X inactivation. *Science* 300:131–35
76. Reeve HK, Pfennig DW. 2003. Genetic biases for showy males: are some genetic systems especially conducive to sexual selection? *Proc. Natl. Acad. Sci. USA* 100:1089–94
77. Reinke V. 2004. Sex and the genome. *Nat. Genet.* 36:548–49
78. Rens W, O'Brien PCM, Grutzner F, Clarke O, Graphodatskaya D, et al. 2007. The multiple sex chromosomes of platypus and echidna are not completely identical and several share homology with the avian Z. *Genome Biol.* 8:R243
79. Rice WR. 1984. Sex chromosomes and the evolution of sexual dimorphism. *Evolution* 38:735–42
80. Rinn JL, Rozowsky JS, Laurenzi IJ, Petersen PH, Zou KY, et al. 2004. Major molecular differences between mammalian sexes are involved in drug metabolism and renal function. *Dev. Cell* 6:791–800
81. Rodriguez-Campos A, Azorin F. 2007. RNA is an integral component of chromatin that contributes to its structural organization. *PLoS ONE* 2:e1182
82. Salstrom JL. 2007. X-inactivation and the dynamic maintenance of gene silencing. *Mol. Genet. Metab.* 92:56–62
83. Saran NG, Pletcher MT, Natale JE, Cheng Y, Reeves RH. 2003. Global disruption of the cerebellar transcriptome in a Down syndrome mouse model. *Hum. Mol. Genet.* 12:2013–19

84. Schmid M, Enderle E, Schindler D, Schempp W. 1989. Chromosome banding and DNA replication patterns in bird karyotypes. *Cytogenet. Cell Genet.* 52:139–46
85. Scholz B, Kultima K, Mattsson A, Axelsson J, Brunstrom B, et al. 2006. Sex-dependent gene expression in early brain development of chicken embryos. *BMC Neuroscience* 7:12
86. Schubeler D. 2006. Dosage compensation in high resolution: global up-regulation through local recruitment. *Genes Dev.* 20:749–53
87. Shetty S, Griffin DK, Graves JAM. 1999. Comparative painting reveals strong chromosome homology over 80 million years of bird evolution. *Chromosome Res.* 7:289–95
88. Smith CA, Roeszler KN, Hudson QJ, Sinclair AH. 2007. Avian sex determination: what, when and where? *Cytogenet. Genome Res.* 117:165–73
89. Storchova R, Divina P. 2006. Nonrandom representation of sex-biased genes on chicken Z chromosome. *J. Mol. Evol.* 63:676–81
90. Straub T, Becker PB. 2007. Dosage compensation: the beginning and end of generalization. *Nat. Rev. Genet.* 8:47–57
91. Sturgill D, Zhang Y, Parisi M, Oliver B. 2007. Demasculinization of X chromosomes in the *Drosophila* genus. *Nature* 450:238–41
92. Sturtevant AH. 1965. *A History of Genetics*. New York: Harper & Row
93. Teranishi M, Mizuno S. 2004. The male hypermethylation (MHM) region on the chicken Z chromosome: female-specific transcription and its biological implication. In *Chromosomes Today*, ed. M Schmid, I Nanda, 14:45–53. Dordrecht: Kluwer Acad.
94. Teranishi M, Shimada Y, Hori T, Nakabayashi O, Kikuchi T, et al. 2001. Transcripts of the MHM region on the chicken Z chromosome accumulate as non-coding RNA in the nucleus of female cells adjacent to the *DMRT1* locus. *Chromosome Res.* 9:147–65
95. Traut W, Marec F. 1996. Sex chromatin in lepidoptera. *Q. Rev. Biol.* 71:239–56
96. Tsuda Y, Nishida-Umehara C, Ishijima J, Yamada K, Matsuda Y. 2007. Comparison of the Z and W sex chromosomal architectures in elegant crested tinamou (*Eudromia elegans*) and ostrich (*Struthio camelus*) and the process of sex chromosome differentiation in palaeognathous birds. *Chromosoma* 116:159–73
97. Vacik T, Ort M, Gregorova S, Strnad P, Blatny R, et al. 2005. Segmental trisomy of chromosome 17: A mouse model of human aneuploidy syndromes. *Proc. Natl. Acad. Sci. USA* 102:4500–5
98. Vallender EJ, Lahn BT. 2004. How mammalian sex chromosomes acquired their peculiar gene content. *BioEssays* 26:159–69
99. Veitia RA. 2003. Nonlinear effects in macromolecular assembly and dosage sensitivity. *J. Theor. Biol.* 220:19–25
100. Wade J, Tang YP, Peabody C, Tempelman RJ. 2005. Enhanced gene expression in the forebrain of hatchling and juvenile male zebra finches. *J. Neurobiol.* 64:224–38
101. Wahlberg P, Stromstedt L, Tordoir X, Foglio M, Heath S, et al. 2007. A high-resolution linkage map for the Z chromosome in chicken reveals hot spots for recombination. *Cytogenet. Genome Res.* 117:22–29
102. Wallis JW, Aerts J, Groenen MA, Crooijmans RP, Layman D, et al. 2004. A physical map of the chicken genome. *Nature* 432:761–64
103. Waters PD, Delbridge ML, Deakin JE, El-Mogharbel N, Kirby PJ, et al. 2005. Autosomal location of genes from the conserved mammalian X in the platypus (*Ornithorhynchus anatinus*): implications for mammalian sex chromosome evolution. *Chromosome Res.* 13:401–10
104. Winzeler EA, Shoemaker DD, Astromoff A, Liang H, Anderson K, et al. 1999. Functional characterization of the *S. cerevisiae* genome by gene deletion and parallel analysis. *Science* 285:901–6
105. Yang X, Schadt EE, Wang S, Wang H, Arnold AP, et al. 2006. Tissue-specific expression and regulation of sexually dimorphic genes in mice. *Genome Res.* 16:995–1004
106. Zhang Y, Oliver B. 2007. Dosage compensation goes global. *Curr. Opin. Genet. Dev.* 17:113–20
107. Zhang Y, Sturgill D, Parisi M, Kumar S, Oliver B. 2007. Constraint and turnover in sex-biased gene expression in the genus *Drosophila*. *Nature* 450:233–37

Linkage Disequilibrium and Association Mapping

B.S. Weir

Department of Biostatistics, University of Washington Seattle, WA 98195-7232, USA

Annu. Rev. Genomics Hum. Genet. 2008. 9:129–42

First published online as a Review in Advance on May 27, 2008

The *Annual Review of Genomics and Human Genetics* is online at genom.annualreviews.org

This article's doi:
10.1146/annurev.genom.9.081307.164347

1527-8204/08/0922-0129$20.00

Key Words

Case-control, gene-gene correlation, Hardy-Weinberg, three-locus disequilibrium, trend test

Abstract

Linkage disequilibrium refers to the association between alleles at different loci. The standard definition applies to two alleles in the same gamete, and it can be regarded as the covariance of indicator variables for the states of those two alleles. The corresponding correlation coefficient ρ is the parameter that arises naturally in discussions of tests of association between markers and genetic diseases. A general treatment of association tests makes use of the additive and nonadditive components of variance for the disease gene. In almost all expressions that describe the behavior of association tests, additive variance components are modified by the squared correlation coefficient ρ^2 and the nonadditive variance components by ρ^4, suggesting that nonadditive components have less influence than additive components on association tests.

INTRODUCTION

The current surge of interest in locating genes that affect human health on the basis of associations with marker genes has brought a new need to characterize the joint frequencies of sets of alleles at different loci. The underlying parameter is that of linkage disequilibrium even though that quantity is traditionally restricted to alleles at two loci located on the same gamete. This review phrases the joint behavior of allele frequencies in terms of correlation coefficients and shows how these coefficients determine the behavior of association tests. Under appropriate assumptions this approach leads to a treatment of many loci, especially for situations with only two alleles per locus.

HARDY-WEINBERG EQUILIBRIUM

It is convenient to introduce notation and concepts by reviewing the means to describe the association of alleles at a single locus. For locus **A** it is helpful to define an indicator variable x that takes the value 1 for allele A and 0 for any other allele. If alleles within individuals are indexed by j, $j = 1, 2$ then the expected values of interest involve the allele frequency p_A and the homozygote frequency P_{AA}:

$$\begin{aligned}\varepsilon(x_j) &= p_A, \quad j = 1, 2\\ \varepsilon\left(x_j^2\right) &= p_A, \quad j = 1, 2\\ \varepsilon(x_1 x_2) &= P_{AA}.\end{aligned}$$

The variance of x_j is $p_A(1 - p_A)$ for $j = 1, 2$ and the covariance and correlation coefficients for x_1 and x_2 are $\text{Cov}(x_1, x_2) = P_{AA} - p_A^2$ and $\text{Corr}(x_1, x_2) = (P_{AA} - p_A^2)/p_A(1 - p_A)$. If expectation refers to the average over samples from the current population [i.e., "statistical sampling" as opposed to the "genetic sampling" that includes variation among populations (27)] this last quantity is usually referred to as the within-population inbreeding coefficient f_A or F_{IS} and it allows the frequencies for genotypes AA, Aa, aa in the two-allele case to be parameterized as

$$\begin{aligned}P_{AA} &= p_A^2 + f_A p_A p_a\\ P_{Aa} &= 2p_A p_a(1 - f_A)\\ P_{aa} &= p_a^2 + p_A p_a f_A.\end{aligned}$$

This leads to bounds on f_A: $\max(-p_A/p_a, -p_a/p_A) \leq f_A \leq 1$. An alternative parameterization introduces $D_A = f_A p_A(1 - p_A)$.

The maximum-likelihood estimate of f_A is $\hat{f}_A = 1 - \tilde{P}_{Aa}/(2\tilde{p}_A\tilde{p}_a)$ where tildes indicate sample values of allele or genotype frequencies. If this quantity is estimated from a sample of n genotypes it has a large-sample variance (3) of

$$\begin{aligned}\text{Var}(\hat{f}_A) = \frac{1}{2np_A p_a}&(1 - f_A)[2p_A p_a(1 - f_A)\\ &\times (1 - 2f_A) + f_A(2 - f_A)].\end{aligned}$$

The usual goodness-of-fit chi-square test statistic for testing the hypothesis $H_0 : f_A = 0$, i.e., Hardy-Weinberg equilibrium, is $X_A^2 = n\hat{f}_A^2$. This is distributed as $\chi^2_{(1)}$ when the hypothesis is true, and it is also obtained by assuming that $\hat{f}_A$ is normally distributed. Although Fisher's exact test (7) is generally preferred for small samples, the chi-square test has the advantage of simplifying power calculations. When the Hardy-Weinberg hypothesis is not true, the test statistic has a noncentral chi-square distribution with one degree of freedom (df) and noncentrality parameter $\lambda = nf_A^2$. To reach 90% power with a 5% significance level, for example, it is necessary that $\lambda \geq 10.5$ (27). In this one-df case, the noncentrality value follows from percentiles of the standard normal distribution. If z_x is the xth percentile of the standard normal, than for significance level α and power $1 - \beta$, $\lambda = (z_{\alpha/2} + z_\beta)^2$ as can be checked for $\alpha = 0.05, z_{\alpha/2} = -1.96$ and $1 - \beta = 0.90, z_\beta = -1.28$.

Discussions of Hardy-Weinberg equilibrium have been used in the context of testing for selective neutrality (12, 25) and to challenge the statistical analysis of forensic DNA profiles (2, 26). More recently, Hardy-Weinberg testing has been used as a quality-control measure when large numbers of single nucleotide polymorphisms (SNPs) are typed (6, 34). This activity requires a clear understanding of the distribution of the test statistic, especially for sample sizes only in the hundreds (32; R. Rohlfs and B.S. Weir, manuscript submitted) and it should also recognize the correlation between test statistics for different loci. I show below that Hardy-Weinberg testing can also be used to detect marker-disease association if the testing is confined to people affected with the disease.

GAMETIC DATA

Gametic Linkage Disequilibrium

There is a natural extension of the one-locus discussion to pairs of loci **A** and **B**. Indicator variables x and y take the value 1 for alleles A and B respectively at those loci and are zero for other alleles. If gametes (the material received from each parent) within individuals are indexed by j, $j = 1, 2$ then the expected values of interest involve allele frequencies p_A, p_B and gamete frequencies P_{AB}

$$\varepsilon(x_j) = p_A, \quad \varepsilon(y_j) = p_B, \quad j = 1, 2$$
$$\varepsilon\left(x_j^2\right) = p_A, \quad \varepsilon\left(y_j^2\right) = p_B, \quad j = 1, 2$$
$$\varepsilon(x_j y_j) = P_{AB}, \quad j = 1, 2.$$

The covariance of x_j and y_j introduces the gametic linkage disequilibrium parameter D_{AB}: $\mathrm{Cov}(x_j, y_j) = P_{AB} - p_A p_B = D_{AB}$. Note the analogy to the one-locus deviations $D_A = P_{AA} - p_A^2$, $D_B = P_{BB} - p_B^2$. In the two-allele case, the bounds on D_{AB} are $\max(-p_A p_B, -p_a p_b) \leq D_{AB} \leq \min(p_A p_b, p_a p_B)$. Because the variances of x_j and y_j are $p_A(1 - p_A)$ and $p_B(1 - p_B)$ respectively, the squared correlation of the indicator variables is $\rho_{AB}^2 = D_{AB}^2/[p_A(1-p_A)p_B(1-p_B)]$.

If observed gamete frequencies are available, then the maximum-likelihood estimate of D_{AB} is $\hat{D}_{AB} = \tilde{P}_{AB} - \tilde{p}_A\tilde{p}_B$. The large-sample variance of estimates made from random samples of $2n$ gametes is

$$\mathrm{Var}(\hat{D}_{AB}) = \frac{1}{2n}\left[p_A(1 - p_A)p_B(1 - p_B) + (1 - 2p_A)(1 - 2p_B)D_{AB} - D_{AB}^2\right].$$

The sample value of the squared correlation coefficient is written as r_{AB}^2 and the usual 2×2 contingency-table test statistic for testing the hypothesis H_0: $\rho_{AB} = 0$, i.e., no linkage disequilibrium, is $X_{AB}^2 = nr_{AB}^2$. The same statistic follows by assuming $\hat{D}_{AB}$ is normally distributed.

Estimates of the inbreeding coefficients at two loci are correlated because of linkage disequilibrium between them. The covariance of estimates $\hat{f}_A$, $\hat{f}_B$ at loci **A, B** in the situation where the parameters f_A, f_B are zero is $D_{AB}^2/[np_A(1 - p_A)p_B(1-p_B)]$ so their correlation is ρ_{AB}^2. When f_A, f_B are not zero, the covariance involves associations between three or four of the two alleles at each of the two loci (31).

Marker-Trait Associations

The current interest in whole-genome association studies is based on the premise that a significant degree of marker-disease association implies significant linkage disequilibrium, which may indicate that the loci involved are physically close in the genome. Whether the design is case-control or case-cohort, there is a common underlying theory to support this premise. Suppose locus **T** with alleles T, t affects susceptibility to a disease and that marker locus **M** with alleles M, m can be scored. Under the assumption of random union of gametes, leading to Hardy-Weinberg equilibrium, the frequencies of the nine genotypic classes are shown in **Table 1**.

Table 1 Frequencies of the nine genotypic classes for trait and marker loci

	TT	***Tt***	***tt***
MM	$P_{MMTT} = P_{MT}^2$	$P_{MMTt} = 2P_{MT}P_{Mt}$	$P_{MMtt} = P_{Mt}^2$
Mm	$P_{MmTT} = 2P_{MT}P_{mT}$	$P_{MmTt} = 2P_{MT}P_{mt} + 2P_{mT}P_{Mt}$	$P_{Mmtt} = 2P_{Mt}P_{mt}$
mm	$P_{mmTT} = P_{mT}^2$	$P_{mmTt} = 2P_{mT}P_{mt}$	$P_{mmtt} = P_{mt}^2$

$P_{MT} = p_M p_T + D_{MT}$, $P_{mT} = p_m p_T - D_{MT}$, $P_{Mt} = p_M p_t - D_{MT}$, $P_{mt} = p_m p_t + D_{MT}$.

It is helpful to introduce variables X and G for loci **M** and **T**. The values of X will be assigned for the marker whereas the values of G represent the genetic contributions to measured trait variables or to disease status. For a measured trait, the Gs can be regarded as the expected values of the trait for each trait genotype. For disease status, the Gs are the probabilities of being affected for each trait genotype. In either case, the Hardy-Weinberg assumption provides the following expressions for the means and variances:

$$\begin{aligned}
\varepsilon(X) = \mu_X &= p_M^2 X_{MM} + 2p_M p_m X_{Mm} + p_m^2 X_{mm} \\
\varepsilon(G) = \mu_G &= p_T^2 G_{TT} + 2p_T p_t G_{Tt} + p_t^2 G_{tt} \\
\mathrm{Var}(X) &= \sigma_{A_X}^2 + \sigma_{D_X}^2 \\
\mathrm{Var}(G) &= \sigma_{A_T}^2 + \sigma_{D_T}^2.
\end{aligned}$$

Quantitative geneticists refer to σ_A^2 and σ_D^2 as additive and dominance components of variance and they reflect the separate and joint effects on a trait of the two alleles an individual has at a locus. They can be expressed as

$$\begin{aligned}
\sigma_{A_X}^2 &= 2p_M p_m [p_M(X_{MM} - X_{Mm}) \\
&\quad + p_m(X_{Mm} - X_{mm})]^2 \\
\sigma_{A_T}^2 &= 2p_T p_t [p_T(G_{TT} - G_{Tt}) + p_t(G_{Tt} - G_{tt})]^2 \\
\sigma_{D_X}^2 &= p_M^2 p_m^2 (X_{MM} - 2X_{Mm} + X_{mm})^2 \\
\sigma_{D_T}^2 &= p_T^2 p_t^2 (G_{TT} - 2G_{Tt} + G_{tt})^2
\end{aligned}$$

and these lead to the following expression for the covariance of X and G:

$$\mathrm{Cov}(G, X) = \rho_{MT}\sigma_{A_T}\sigma_{A_X} + \rho_{MT}^2\sigma_{D_T}\sigma_{D_X}.$$

If either variable X or G is wholly additive ($\sigma_{D_X}^2 = 0$ or $\sigma_{D_T}^2 = 0$) then the correlation between them is $\rho_{XG} = \rho_{MT}$ and if either is wholly nonadditive ($\sigma_{A_X}^2 = 0$ or $\sigma_{A_T}^2 = 0$) then $\rho_{XG} = \rho_{MT}^2$.

Measured Traits

Suppose Y is a measured risk variable, such as body mass index, such that $Y = G + E$ where G is the genetic effect of locus **T** and E encompasses all other effects. These other effects are supposed to have mean zero and to be independent of both G and the marker variable X. Then

$$\begin{aligned}
\varepsilon(Y) &= \varepsilon(G) \\
\mathrm{Cov}(X, Y) &= \mathrm{Cov}(X, G) \\
\mathrm{Var}(Y) &= \sigma_{A_T}^2 + \sigma_{D_T}^2 + \sigma_E^2.
\end{aligned}$$

If a cohort of individuals is being studied, then trait values Y may be regressed on marker variables X. The regression coefficient is

$$\beta_{YX} = \frac{\mathrm{Cov}(X, Y)}{\mathrm{Var}(X)} = \frac{\rho_{MT}\sigma_{A_T}\sigma_{A_X} + \rho_{MT}^2\sigma_{D_T}\sigma_{D_X}}{\sigma_{A_X}^2 + \sigma_{D_X}^2}.$$

Variable X may be chosen to be additive, e.g., the number of M alleles, and then $\beta_{YX} = \rho_{MT}\sigma_{A_T}/(2p_M p_m)$. Alternatively, the marker variable can be made to have a zero additive variance, e.g., $X_{MM} = p_m$, $X_{Mm} = 0$, $X_{mm} = p_M$, and then $\beta_{YX} = \rho_{MT}^2\sigma_{D_T}/(p_M p_m)^2$. A significant regression coefficient implies a significant linkage disequilibrium measure ρ_{MT} between marker and disease loci. The signal is expected to be stronger with an additive marker variable because $\rho_{MT} \geq \rho_{MT}^2$ and it is usual that $\sigma_{A_T}^2 \geq \sigma_{D_T}^2$. Even so, there may be merit in performing both tests because they provide separate indications of additive and nonadditive effects of disease alleles.

It may be more convenient to work with the correlation of X and Y. For an additive marker variable

$$\mathrm{Corr}(X, Y) = \rho_{XY} = \rho_{MT} h_{Y_T}$$

where $h_{Y_T}^2 = \sigma_{A_T}^2/(\sigma_{A_T}^2 + \sigma_{D_T}^2 + \sigma_E^2)$ is the heritability of trait Y due to locus **T**. Sample values r_{XY} for the correlation ρ_{XY} can be transformed to normal variables with Fisher's transformation,

$$z = \frac{1}{2}\ln\left(\frac{1 + r_{XY}}{1 - r_{XY}}\right).$$

For samples of n individuals, z has a normal distribution with mean $\ln[(1+\rho_{XY})/(1-\rho_{XY})]/2$ and variance $1/(n-3)$. For an α% significance level, the hypothesis H_0: $\rho_{XY} = 0$ (i.e., $\rho_{MT} = 0$) is rejected when $z \leq z_{\alpha/2}$ or $z \geq z_{1-\alpha/2}$. When H_0 is true the probability of rejection is α. A 5% significance level, $\alpha = 0.05$, has $z_{\alpha/2} = -1.96$. When multiple SNPs are used, a much lower significance level is appropriate; for $\alpha = 10^{-7}$ the critical value is $z_{\alpha/2} = -5.32$.

Standard theory for correlation coefficients provides that, for $(1-\beta)\%$ power, the necessary sample size n is approximately

$$n = \left[\frac{2(z_{\alpha/2} + z_\beta)}{\ln\left(\frac{1+\rho_{XY}}{1-\rho_{XY}}\right)} \right]^2 + 3.$$

For 80% power ($z_\beta = -0.85$) and significance level 10^{-7}, a SNP with $\rho^2_{MT} = 0.8$ to the disease gene and a trait with heritability $h^2_{Y_T} = 0.2$, this size is approximately 218. It can be shown (B.S. Weir and Q. Zhao, manuscript in preparation) that this approach is at least as powerful as those based on the noncentral F or t-distributions (13, 14, 22).

An alternative to regression is analysis of variance as a process for comparing trait means among the three genotype classes. The expected trait values are

$$\begin{aligned}
\varepsilon(Y \mid MM) &= \frac{1}{P_{MM}}[P_{MMTT}G_{TT} + P_{MMTt}G_{Tt} \\
&\quad + P_{MMtt}G_{tt}] \\
&= \frac{1}{p_M^2}\left[P^2_{MT}G_{TT} + 2P_{MT}P_{Mt}G_{Tt}\right. \\
&\quad \left. + P^2_{Mt}G_{tt}\right] \\
&= \mu_G + 2\rho_{MT}\sigma_{A_T}\sqrt{\frac{2p_m}{p_M}} \\
&\quad + \rho^2_{MT}\sigma_{D_T}\frac{p_m}{p_M} \\
\varepsilon(Y \mid Mm) &= \mu_G + \frac{2(p_m - p_M)\rho_{MT}\sigma_{A_T}}{\sqrt{p_M p_m}} \\
&\quad - \rho^2_{MT}\sigma_{D_T} \\
\varepsilon(Y \mid mm) &= \mu_G - 2\rho_{MT}\sigma_{A_T}\sqrt{\frac{2p_M}{p_m}} \\
&\quad + \rho^2_{MT}\sigma_{D_T}\frac{p_M}{p_m}.
\end{aligned}$$

Therefore an analysis of variance will also test that $\rho_{MT} = 0$ and the test will be affected by both additive and dominance effects at the trait locus.

Dichotomous Traits: Case Only

The case-control approach starts with independent samples of people who are either affected or not affected with a disease and compares marker frequencies between the two groups. The framework of the previous section can be used provided the trait genetic values G are interpreted as the probabilities of an individual being affected. The mean value μ_G of G is the probability that a random person in the population has the disease. With the Hardy-Weinberg assumption still in effect, the MM marker frequency among cases is

$$\begin{aligned}
\Pr(MM \mid \text{Case}) &= \frac{1}{\mu_G}(P_{MMTT}G_{TT} \\
&\quad + P_{MMTt}G_{Tt} + P_{MMtt}G_{tt}) \\
&= p_M^2 + \frac{1}{\mu_G}\{2p_M D_{MT}[p_T(G_{TT} \\
&\quad - G_{Tt}) + p_t(G_{Tt} - G_{tt})] \\
&\quad + D^2_{MT}(G_{TT} - 2G_{Tt} + G_{tt})\}.
\end{aligned}$$

The expressions for all three marker genotypes can be written as

$$\begin{aligned}
\Pr(MM \mid \text{Case}) &= p_M^2 + \frac{1}{\mu_G}\left[p_M\rho_{MT}\sigma_{A_T}\sqrt{2p_Mp_m}\right. \\
&\quad \left. + \rho^2_{MT}\sigma_{D_T}p_Mp_m\right] \\
\Pr(Mm \mid \text{Case}) &= 2p_Mp_m + \frac{1}{\mu_G}\left[(p_m\right. \\
&\quad - p_M)\rho_{MT}\sigma_{A_T}\sqrt{2p_Mp_m} \\
&\quad \left. - 2\rho^2_{MT}\sigma_{D_T}p_Mp_m\right] \\
\Pr(mm \mid \text{Case}) &= p_m^2 + \frac{1}{\mu_G}\left[-p_m\rho_{MT}\sigma_{A_T}\sqrt{2p_Mp_m}\right. \\
&\quad \left. + \rho^2_{MT}\sigma_{D_T}p_Mp_m\right].
\end{aligned}$$

Combining the genotypic frequencies gives the following allele frequencies:

$$\begin{aligned}
\Pr(M \mid \text{Case}) &= p_M + \frac{\rho_{MT}\sigma_{A_T}}{2\mu_G}\sqrt{2p_Mp_m} \\
\Pr(m \mid \text{Case}) &= p_m - \frac{\rho_{MT}\sigma_{A_T}}{2\mu_G}\sqrt{2p_Mp_m}.
\end{aligned}$$

These results suggest the use of case-only testing for Hardy-Weinberg equilibrium at marker loci as a way to detect linkage disequilibrium between marker and disease loci (16). The inbreeding coefficient at the marker locus in the case population is

$$f = \frac{2p_Mp_m\rho^2_{MT}\left(2\mu_G\sigma_{D_T} - \sigma^2_{A_T}\right)}{(2\mu_Gp_M + \rho_{MT}\sigma_{A_T}\sqrt{2p_Mp_m})(2\mu_Gp_m - \rho_{MT}\sigma_{A_T}\sqrt{2p_Mp_m})}.$$

The power of this test depends on nf^2, which is proportional to ρ_{MT}^4 so the power will decrease quickly as ρ_{MT} decreases. It is common for investigators to assume a multiplicative disease model (i.e., additive on a log scale): $G_{TT} = G, G_{Tt} = G\gamma, G_{tt} = G\gamma^2$ but that assumption leads to Hardy-Weinberg equilibrium at marker loci among cases because then $2\mu_G\sigma_{D_T} = \sigma_{A_T}^2$. The quantity γ is referred to as the relative risk.

Dichotomous Traits: Case-Control

An argument similar to that in the previous section provides the marker genotype frequencies among controls as follows:

$$\begin{aligned}
\Pr(MM\,|\,\text{Control}) &= p_M^2 - \frac{1}{1-\mu_G} \\
&\quad \times \left[p_M\rho_{MT}\sigma_{A_T}\sqrt{2p_Mp_m} + \rho_{MT}^2\sigma_{D_T}p_Mp_m\right] \\
\Pr(Mm\,|\,\text{Control}) &= 2p_Mp_m - \frac{1}{1-\mu_G}\left[(p_m - p_M)\rho_{MT}\sigma_{A_T}\sqrt{2p_Mp_m} - 2\rho_{MT}^2\sigma_{D_T}p_Mp_m\right] \\
\Pr(mm\,|\,\text{Control}) &= p_m^2 - \frac{1}{1-\mu_G} \\
&\quad \times \left[-p_m\rho_{MT}\sigma_{A_T}\sqrt{2p_Mp_m} + \rho_{MT}^2\sigma_{D_T}p_Mp_m\right].
\end{aligned}$$

Combining the genotypic frequencies to give allele frequencies

$$\begin{aligned}
\Pr(M\,|\,\text{Control}) &= p_M - \frac{\rho_{MT}\sigma_{A_T}}{2(1-\mu_G)}\sqrt{2p_Mp_m} \\
\Pr(m\,|\,\text{Control}) &= p_m + \frac{\rho_{MT}\sigma_{A_T}}{2(1-\mu_G)}\sqrt{2p_Mp_m}.
\end{aligned}$$

There is also a departure from Hardy-Weinberg equilibrium in the control population but for rare diseases (small μ_G) the departure may be too small to detect.

The simplest case-control test compares marker allele frequencies between the two samples and it is clearly equivalent to testing that $\rho_{MT} = 0$ because δ_M, the difference in marker allele M frequencies between cases and controls, is

$$\begin{aligned}
\delta_M &= \Pr(M\,|\,\text{Case}) - \Pr(M\,|\,\text{Control}) \\
&= \frac{\rho_{MT}\sigma_{A_T}\sqrt{2p_Mp_m}}{2\mu_G(1-\mu_G)}.
\end{aligned}$$

The test is not affected by nonadditivity at the disease locus. If the allelic counts for M, m in cases and controls are laid out in a 2×2 table (**Table 2**), the contingency-table chi-square test statistic has one df. In the notation of that table, the test statistic is

$$X^2 = \frac{(a+b+c+d)(ad-bc)^2}{(a+b)(a+c)(b+d)(c+d)}.$$

Inserting parametric values, e.g., $a = n_{\text{Case}}p_{M|\text{Case}}$, into the case-control test statistic gives the noncentrality parameter λ. If the denominator in this expression is approximated by assuming marker allele frequencies are the same in cases and controls, then

$$\lambda = \frac{n_{\text{Case}}n_{\text{Control}}}{2(n_{\text{Case}}+n_{\text{Control}})}\frac{\delta_M^2}{p_Mp_m}.$$

This parameter, and hence the power of the test, is maximized for a fixed total sample size when $n_{\text{Case}} = n_{\text{Control}} = n$. In that case, if $p_M = p_m = 0.5$, $\lambda = n\delta_M^2$. Using the previous relation $\lambda = (z_{\alpha/2} + z_{1-\beta})^2$ for one-df tests, 80% power is achieved with 10^{-7} significance level when the numbers of cases and controls are at least $38.1/\delta_M^2$. This result does not require explicit specification of the disease model but for the multiplicative model described above with relative risk γ the difference in marker allele frequencies between cases and controls when $p_M = p_m = p_T = p_t = 0.5$ is, approximately, $\delta_M = \rho_{MT}(\gamma-1)/(\gamma+1)$. In the case of $\rho_{MT}^2 = 0.8$, the necessary sample sizes are $47.6(\gamma+1)^2/(\gamma-1)^2$ and these are $n = 5762$ for a relative risk of 1.2.

Table 2 Allele counts in cases and controls

	Case	Control	Total
M	$a = n_{M\|\text{Case}}$	$b = n_{M\|\text{Control}}$	$a+b = n_M$
m	$c = n_{m\|\text{Case}}$	$d = n_{m\|\text{Control}}$	$c+d = n_m$
Total	$a+c = 2n_{\text{Case}}$	$b+d = 2n_{\text{Control}}$	$a+b+c+d = 2(n_{\text{Case}}+n_{\text{Control}})$

GENOTYPIC DATA

Contingency Table Test

Tests based on a 2×2 table of marker allele counts in cases and controls can detect only additive effects at trait loci. Additive and nonadditive effects can be detected jointly with tests based on a 3×2 table of marker genotype counts in cases and controls (**Table 3**) but the extra df can reduce power and there may be problems with small expected counts in some cells of the table. The effects of genotyping errors are especially acute for cells with small expected counts. It is usual instead to use the single-df Armitage trend test based on a score statistic U (20).

Trend Test

In terms of the notation on **Table 3** the score statistic U for trend tests is defined as

$$U = \frac{n_{\text{Case}} n_{\text{Control}}}{n_{\text{Case}} + n_{\text{Control}}} [X_{MM}(\tilde{P}_{MM\mid\text{Case}} - \tilde{P}_{MM\mid\text{Control}}) + X_{Mm}(\tilde{P}_{Mm\mid\text{Case}} - \tilde{P}_{Mm\mid\text{Control}}) + X_{mm}(\tilde{P}_{mm\mid\text{Case}} - \tilde{P}_{mm\mid\text{Control}})]$$

where $\tilde{P}_{MM\mid\text{Case}} = n_{MM\mid\text{Case}}/n_{\text{Case}}$ and $\tilde{P}_{MM\mid\text{Control}} = n_{MM\mid\text{Control}}/n_{\text{Control}}$, etc. With random sampling, the case and control counts are multinomially distributed and the expected value of U is

$$\varepsilon(U) = W\big[X_{MM}(P_{MM\mid\text{Case}} - P_{MM\mid\text{Control}}) + X_{Mm}(P_{Mm\mid\text{Case}} - P_{Mm\mid\text{Control}}) + X_{mm}(P_{mm\mid\text{Case}} - P_{mm\mid\text{Control}})\big]$$

where $W = n_{\text{Case}} n_{\text{Control}}/(n_{\text{Case}} + n_{\text{Control}})$. This can be written as

$$\varepsilon(U) = W \frac{\rho_{MT}\sigma_{A_T}\sigma_{A_X} + \rho_{MT}^2 \sigma_{D_T}\sigma_{D_X}}{\mu_G(1-\mu_G)}.$$

The variance of U is

$$\text{Var}(U) = \frac{W^2}{n_{\text{Case}}} \big[(X_{MM}^2 P_{MM\mid\text{Case}} + X_{Mm}^2 P_{Mm\mid\text{Case}} + X_{mm}^2 P_{mm\mid\text{Case}}) - (X_{MM}P_{MM\mid\text{Case}} + X_{Mm}P_{Mm\mid\text{Case}} + X_{mm}P_{mm\mid\text{Case}})^2\big]$$
$$+ \frac{W^2}{n_{\text{Control}}} \big[(X_{MM}^2 P_{MM\mid\text{Control}} + X_{Mm}^2 P_{Mm\mid\text{Control}} + X_{mm}^2 P_{mm\mid\text{Control}}) - (X_{MM}P_{MM\mid\text{Control}} + X_{Mm}P_{Mm\mid\text{Control}} + X_{mm}P_{mm\mid\text{Control}})^2\big].$$

Table 3 Genotype counts in cases and controls

Genotype	MM	Mm	mm	Total
Variable	X_{MM}	X_{Mm}	X_{mm}	
Case counts	$n_{MM\mid\text{Case}}$	$n_{Mm\mid\text{Case}}$	$n_{mm\mid\text{Case}}$	n_{Case}
Control counts	$n_{MM\mid\text{Control}}$	$n_{Mm\mid\text{Control}}$	$n_{mm\mid\text{Control}}$	n_{Control}
Total counts	n_{MM}	n_{Mm}	n_{mm}	$n_{\text{Case}} + n_{\text{Control}}$

Under the hypothesis of no association the marker genotype frequencies do not depend on disease status so $\varepsilon(U) = 0$ and

$$\text{Var}(U) = W\big[(X_{MM}^2 P_{MM} + X_{Mm}^2 P_{Mm} + X_{mm}^2 P_{mm}) - (X_{MM}P_{MM} + X_{Mm}P_{Mm} + X_{mm}P_{mm})^2\big] = W(\sigma_{A_X}^2 + \sigma_{D_X}^2).$$

Assuming normality for U, the test statistic $X^2 = U^2/\widehat{\text{Var}}(U)$ is distributed as $\chi_{(1)}^2$ under the hypothesis $H_0 : \rho_{MT} = 0$. In this expression the variance is estimated using $\tilde{P}_{MM} = (n_{MM\mid\text{Case}} + n_{MM\mid\text{Control}})/(n_{\text{Case}} + n_{\text{Control}})$, etc. It is usual to consider a linear trend test, e.g., $X_{MM} = 2$, $X_{Mm} = 1$, $X_{mm} = 0$, so that $\sigma_{D_X}^2 = 0$ and the test is for additive effects at the disease locus. Setting $X_{MM} = p_m$, $X_{Mm} = 0$, $X_{mm} = p_M$ gives $\sigma_{A_X}^2 = 0$ and a test for nonadditive effects.

When the Hardy-Weinberg equilibrium relation holds in the combined case plus control data, it can be shown (20) that the linear trend test statistic is identical to the allelic case-control test statistic described above. The two tests will have the same power.

Composite Linkage Disequilibrium

Linkage disequilibrium refers to the association of alleles at two or more loci carried on the same gamete but data are generally collected at the genotypic level. Provided each locus has genotypic frequencies in Hardy-Weinberg equilibrium this method is not a problem as gametic frequencies can be estimated from genotypic frequencies (9, 29). The missing data aspect of unknown phase of multiple heterozygotes (not knowing if *AaBb* genotypes result from the union of *AB* and *ab* gametes or the union of *Ab* and *aB* gametes, for example) is overcome by adding the Hardy-Weinberg assumption that genotype frequencies are products of gamete frequencies as illustrated in **Table 1**.

It is possible, however, to avoid the Hardy-Weinberg assumption by modifying the definition of linkage disequilibrium for alleles A, B at loci **A, B** to become the difference $(P_{AB} + P_{A,B} - 2p_Ap_B)$. Whereas P_{AB} is the frequency of AB gametes, $P_{A,B}$ is the probability of an individual carrying alleles A and B on different gametes. The nongametic linkage disequilibrium is defined as $D_{A,B} = P_{A,B} - p_Ap_B$ and the meaning of this quantity is best explained with indicator variables. For gamete $j, j = 1, 2$ within an individual, x_j, y_j are one if the **A, B** alleles are A, B respectively and zero otherwise. The expected values of x_1y_2 or x_2y_1 are $P_{A,B}$ as compared to the expected values of x_1y_1 or x_2y_2 being P_{AB}. The utility of the nongametic frequency comes from the sum $P_{AB} + P_{A,B}$, which is the expected value of $(x_1+x_2)(y_1+y_2)/2$: half the number of A alleles multiplied by the number of B alleles carried by an individual. The covariance of these two numbers is termed the composite linkage disequilibrium (28) and it is written algebraically as

$$\Delta_{AB} = \frac{1}{2}\mathrm{Cov}(x_1 + x_2, y_1 + y_2)$$
$$= P_{AB} + P_{A,B} - 2p_Ap_B.$$

Now the variance of $(x_1 + x_2)$ is $\mathrm{Var}(x_1) + 2\mathrm{Cov}(x_1, x_2) + \mathrm{Var}(x_2)$ or $2p_A(1 - p_A)(1 + f_A)$ and there is a similar expression for the variance of $(y_1 + y_2)$. Therefore, the correlation coefficient $\rho^c_{A,B}$ for $(x_1 + x_2)$ and $(y_1 + y_2)$, the numbers of A and B alleles, is

$$\rho^c_{AB} = \frac{\Delta_{AB}}{\sqrt{p_A(1 - p_A)(1 + f_A)p_B(1 - p_B)(1 + f_B)}}.$$

An immediate benefit of this approach is that Δ_{AB} and ρ^c_{AB} can be estimated directly from genotypic data without any Hardy-Weinberg assumptions or numerical iteration. Because

$$\tilde{P}_{AB} = \tilde{P}_{AABB} + \frac{1}{2}\tilde{P}_{AABb} + \frac{1}{2}\tilde{P}_{AaBB} + \frac{1}{2}\tilde{P}_{AB/ab}$$
$$\tilde{P}_{A,B} = \tilde{P}_{AABB} + \frac{1}{2}\tilde{P}_{AABb} + \frac{1}{2}\tilde{P}_{AaBB} + \frac{1}{2}\tilde{P}_{Ab/aB}$$

the maximum likelihood estimate of Δ_{AB} is

$$\hat{\Delta}_{AB} = 2\tilde{P}_{AABB} + \tilde{P}_{AABb} + \tilde{P}_{AaBB} + \frac{1}{2}\tilde{P}_{AaBb} - 2\tilde{p}_A\tilde{p}_B.$$

The variance of this estimate depends on the degree of association among all four alleles at two loci carried by an individual. If the three- and four-allele disequilibria are ignored the variance for an estimate obtained from a sample of n genotypes, when $\Delta_{AB} = 0$, reduces to

$$\mathrm{Var}(\hat{\Delta}_{AB}) = \frac{1}{n}[p_A(1 - p_A)(1 + f_A)p_B(1 - p_B)(1 + f_B)].$$

It is generally assumed that each locus has equal departures from Hardy-Weinberg so $f_A = f_B$. Asymptotic normality of the maximum-likelihood estimate suggests that $n(r^c_{AB})^2$ has a $\chi^2_{(1)}$ distribution, as does $2nr^2_{AB}$. If there is Hardy-Weinberg equilibrium and the parametric values of D_{AB} and Δ_{AB}, or of ρ_{AB} and ρ^c_{AB}, are equal then there is an advantage in working with the gametic measure D_{AB}. If the sample size n is such that the gametic test has 90% power to detect the value ρ^2_{AB} with significance level α, then the same value of n gives only 63% power to detect the same value with the composite test.

A complete discussion of linkage disequilibria for genotypic data does require one to take account of association among sets of three or four alleles. The one-locus parameterization for AA homozygotes, $P_{AA} = p^2_A + f_Ap_A(1 - p_A) = p^2_A + D_A$, is expanded for two-locus homozygotes to

$$P_{AABB} = P_{AA}P_{BB} + 2p_AD_{ABB} + 2p_BD_{AAB} + 2p_Ap_B\Delta_{AB} + \Delta_{AABB}$$

where

$$D_{AAB} = P_{AAB} - p_A\Delta_{AB} - p_BD_A - p^2_Ap_B$$
$$D_{ABB} = P_{ABB} - p_B\Delta_{AB} - p_AD_B - p_Ap^2_B$$

Table 4 **Frequencies of two-locus genotypes allowing for Hardy-Weinberg disequilibrium***

Genotype	**1**	$2D_{ABB}$	$2D_{AAB}$	$2\Delta_{AB}$	Δ^2_{AB}	D_B	D_A	$D_A D_B$	Δ_{AABB}
AABB	$p_A^2 p_B^2$	p_A	p_B	$p_A p_B$	1	p_A^2	p_B^2	1	1
AABb	$2p_A^2 \pi_B$	$-2p_A$	τ_B	$p_A \tau_B$	-2	$-2p_A^2$	$2\pi_B$	-2	-2
AAbb	$p_A^2 p_b^2$	p_A	$-p_b$	$-p_A p_b$	1	p_A^2	p_b^2	1	1
AaBB	$2\pi_A p_B^2$	τ_A	$-2p_B$	$\tau_A p_B$	-2	$2\pi_A$	$-2p_B^2$	-2	-2
AaBb	$4\pi_A \pi_B$	$-2\tau_A$	$-2\tau_B$	$\tau_A \tau_B$	4	$-4\pi_A$	$-4\pi_B$	4	4
Aabb	$2\pi_A p_b^2$	τ_A	$2p_b$	$-\tau_A p_b$	-2	$2\pi_A$	$-2p_b^2$	-2	-2
aaBB	$p_a^2 p_B^2$	$-p_a$	p_B	$-p_a p_B$	1	p_a^2	p_B^2	1	1
aaBb	$2p_a^2 \pi_B$	$2p_a$	τ_B	$-p_a \tau_B$	-2	$-2p_a^2$	$2\pi_B$	-2	-2
aabb	$p_a^2 p_b^2$	$-p_a$	$-p_b$	$p_a p_b$	1	p_a^2	p_b^2	1	1

*$\pi_A = p_A p_a, \pi_B = p_B p_b \tau_A = (p_a - p_A), \tau_B = (p_b - p_B)$.

and Δ_{AABB} is defined to complete the genotype expression. The complete decomposition for all nine two-locus genotypes shown in **Table 4** (30) is a generalization of the values shown in **Table 1**. The nine genotypic frequencies have been parameterized with eight parameters: two allele frequencies, two departures from Hardy-Weinberg equilibrium, two three-allele disequilibria, and composite disequilibria for two alleles and for four alleles. The association mapping results discussed so far have assumed Hardy-Weinberg equilibrium when it is reasonable to suppose that the three- and four-allele disequilibria are zero.

The absence of Hardy-Weinberg equilibrium will affect tests for marker-trait association even if the tests are confined to marker alleles. Noting that now $\Pr(\text{Case}) = [(p_T^2 + D_T)G_{TT} + (2p_T p_t - 2D_T)G_{Tt} + (p_t^2 + D_T)G_{tt}]$, the marker genotype frequencies among cases and controls become

$$\Pr(MM \mid \text{Case}) = p_M^2 + D_M + \frac{2}{\mu_G}(p_M \Delta_{MT} + D_{MMT})\frac{\sigma_{A_T}}{\sqrt{2p_T p_t}} + \frac{1}{\mu_G}\left(\Delta^2_{MT} + 2p_M D_{MTT} + \Delta_{MMTT}\right)\frac{\sigma_{D_T}}{p_T p_t}$$

$$\Pr(Mm \mid \text{Case}) = 2p_M p_m - 2D_M + \frac{2}{\mu_G}(p_m - p_M)\Delta_{MT} - 2D_{MMT}\frac{\sigma_{A_T}}{\sqrt{2p_T p_t}} - \frac{2}{\mu_G}\left(\Delta^2_{MT} - (p_m - p_M)D_{MTT} + \Delta_{MMTT}\right)\frac{\sigma_{D_T}}{p_T p_t}$$

$$\Pr(mm \mid \text{Case}) = p_m^2 + D_M + \frac{2}{\mu_G}(-p_m \Delta_{MT} + D_{MMT})\frac{\sigma_{A_T}}{\sqrt{2p_T p_t}} + \frac{1}{\mu_G}\left(\Delta^2_{MT} - 2p_m D_{MTT} + \Delta_{MMTT}\right)\frac{\sigma_{D_T}}{p_T p_t}.$$

The marker allele frequencies are

$$\Pr(M \mid \text{Case}) = p_M + \frac{\Delta_{MT}}{\mu_G}\frac{\sigma_{A_T}}{\sqrt{2p_T p_t}} + \frac{D_{MTT}}{\mu_G}\frac{\sigma_{D_T}}{p_T p_t}$$

$$\Pr(M \mid \text{Control}) = p_M - \frac{\Delta_{MT}}{1 - \mu_G}\frac{\sigma_{A_T}}{\sqrt{2p_T p_t}} - \frac{D_{MTT}}{1 - \mu_G}\frac{\sigma_{D_T}}{p_T p_t}.$$

The analogy to the Hardy-Weinberg situation is clear, particularly when the trait alleles act additively ($\sigma^2_{D_T} = 0$): The allelic case-control test now addresses the hypothesis $\Delta_{MT} = 0$ rather than $D_{MT} = 0$. The test is valid even when there is Hardy-Weinberg disequilibrium in the population although then it addresses the association of marker and trait alleles whether they are on the same or different gametes. The hypothesis being tested for a nonadditive disease model is more complex. Hardy-Weinberg tests among only the cases will be affected by disequilibrium in the population as well as by marker-trait association.

MULTIPLE LOCI

It would seem unlikely that the complexity of the human genome could be captured by second-order statistics such as linkage disequilibrium or the correlation of allele frequencies at pairs of loci, but empirical evidence suggests that these statistics are indeed very useful. The very successful International HapMap Project (23) was based, in large part, on the finding that the genome consists of haplotype blocks that can be defined operationally with statistics such as r^2_{AB} for pairs of SNPs **A, B** (4). There has not been a systematic search for associations among alleles at more than two loci and part of the reason may be because of the difficulty of testing for such associations.

It is useful to extend the previous discussion to three loci, in part because it is the next step toward characterizing the genetic structure of the genome and in part because this extension provides the basis for interval mapping of disease genes. Returning to Hardy-Weinberg populations and the availability of gametic data, the classical definition of D_{ABC}, the linkage disequilibrium among alleles A, B, C at loci **A, B, C** respectively, is inherent in the expression

$$P_{ABC} = p_A p_B p_C + p_A D_{BC} + p_B D_{AC} + p_C D_{AB} + D_{ABC}.$$

Unlike the two-locus case where the range of values for disequilibrium coefficients includes zero, the range of values for D_{ABC} may not include zero (24) and then the hypothesis $H_0 : D_{ABC} = 0$ cannot be true.

Replacing population frequencies by their observed values and two-locus disequilibria by their maximum likelihood estimates in this last expression provides the maximum likelihood estimate of D_{ABC}. The large-sample variance of this estimate from $2n$ gametes is (27)

$$\begin{aligned} \mathrm{Var}(\hat{D}_{ABC}) = \frac{1}{2n} \{ & p_A(1-p_A)p_B(1 \\ & - p_B)p_C(1-p_C) + 6D_{AB}D_{BC}D_{AC} \\ & + p_A(1-p_A)\left[(1-2p_B)(1-2p_C)D_{BC} - D^2_{BC}\right] \\ & + p_B(1-p_B)\left[(1-2p_A)(1-2p_C)D_{AC} - D^2_{AC}\right] \\ & + p_C(1-p_C)\left[(1-2p_A)(1-2p_B)D_{AB} - D^2_{AB}\right] \\ & + D_{ABC}[(1-2p_A)(1-2p_B)(1-2p_C) \\ & - 2(1-2p_A)D_{BC} \\ & - 2(1-2p_B)D_{AC} - (1-2p_C)D_{AB} - D_{ABC}] \}. \end{aligned}$$

Although this result is correct and it gives the value of $p_A(1-p_A)p_B(1-p_B)p_C(1-p_C)/2n$ when all the disequilibria are zero, it can provide negative values if the disequilibria are assigned values outside the range of their validity. In particular, the variance can be negative if D_{ABC} is set to zero when the allele frequencies and two-locus disequilibria constrain the three-locus coefficient away from zero.

Another way of considering the significance of three-locus disequilibria is with goodness-of-fit tests on the eight gamete counts for two-allele loci. Hill (10) showed that the goodness-of-fit four-df chi-square test statistic for the hypothesis of complete independence ($H_0 : P_{ABC} = p_A p_B p_C$, etc.) when the sample has $2n$ chromosomes is:

$$\begin{aligned} X^2 = n \Bigg(& \frac{(\tilde{P}_{ABC} - \tilde{p}_A\tilde{p}_B\tilde{p}_C)^2}{\tilde{p}_A\tilde{p}_B\tilde{p}_C} \\ & + \frac{(\tilde{P}_{ABc} - \tilde{p}_A\tilde{p}_B\tilde{p}_c)^2}{\tilde{p}_A\tilde{p}_B\tilde{p}_c} + \frac{(\tilde{P}_{AbC} - \tilde{p}_A\tilde{p}_b\tilde{p}_C)^2}{\tilde{p}_A\tilde{p}_b\tilde{p}_C} \\ & + \frac{(\tilde{P}_{Abc} - \tilde{p}_A\tilde{p}_b\tilde{p}_c)^2}{\tilde{p}_A\tilde{p}_b\tilde{p}_c} + \frac{(\tilde{P}_{aBC} - \tilde{p}_a\tilde{p}_B\tilde{p}_C)^2}{\tilde{p}_a\tilde{p}_B\tilde{p}_C} \\ & + \frac{(\tilde{P}_{aBc} - \tilde{p}_a\tilde{p}_B\tilde{p}_c)^2}{\tilde{p}_a\tilde{p}_B\tilde{p}_c} \\ & + \frac{(\tilde{P}_{abC} - \tilde{p}_a\tilde{p}_b\tilde{p}_C)^2}{\tilde{p}_a\tilde{p}_b\tilde{p}_C} + \frac{(\tilde{P}_{abc} - \tilde{p}_a\tilde{p}_b\tilde{p}_c)^2}{\tilde{p}_a\tilde{p}_b\tilde{p}_c} \Bigg) \\ = {} & 2n\left(r^2_{AB} + r^2_{AC} + r^2_{BC} + r^2_{ABC}\right) \end{aligned}$$

where

$$r^2_{AB} = \frac{\tilde{D}^2_{AB}}{\tilde{p}_A\tilde{p}_a\tilde{p}_B\tilde{p}_b}, \quad r^2_{AC} = \frac{\tilde{D}^2_{AC}}{\tilde{p}_A\tilde{p}_a\tilde{p}_C\tilde{p}_c},$$
$$r^2_{BC} = \frac{\tilde{D}^2_{AB}}{\tilde{p}_B\tilde{p}_b\tilde{p}_C\tilde{p}_c}, \quad r^2_{ABC} = \frac{\tilde{D}^2_{ABC}}{\tilde{p}_A\tilde{p}_a\tilde{p}_B\tilde{p}_b\tilde{p}_C\tilde{p}_c}.$$

The four-df statistic X^2 has therefore been partitioned into four terms, the first three of which are the one-df statistics for testing two-locus disequilibria. This suggests that the fourth term is a one-df chi-square statistic for testing $H_0 : D_{ABC} = 0$. However, Hill (10) pointed out that the three two-locus disequilibria are not independent so the partitioning does not give four one-df test statistics.

More recently, alternative ways of characterizing multi-locus disequilibria have been considered in which the ordering of loci is taken into account (5, 8, 11). For loci **A**, **B**, **C** in that order it may be appropriate to define the absence of three-locus disequilibria in terms of first-order (i.e., not second-order) Markov chains. Allele states follow a Markov chain if the probability of the allele at locus **C** on any gamete depends on the allele at locus **B** but not on the allele at locus **A** on the same gamete: $\Pr(C \mid AB) = \Pr(C \mid B)$. Using this requirement in an expression for P_{AC}:

$$\begin{aligned} P_{AC} &= P_{ABC} + P_{AbC} = P_{AB}P_{BC}/p_B + P_{Ab}P_{bC}/p_b \\ &= p_A p_C + D_{AB}D_{BC}/(p_B p_b) \\ D_{AC} &= D_{AB}D_{BC}/(p_B p_b) \\ \rho_{AC} &= \rho_{AB}\rho_{BC}. \end{aligned}$$

This result is important for interval mapping. Also note that this definition of three-locus equilibrium implies that $D_{ABC} = (1 - 2p_B)D_{AB}D_{BC}/(p_B p_b)$ rather than $D_{ABC} = 0$.

Interval Mapping: Continuous Traits

The association tests described above are designed to identify the marker loci most highly associated with disease loci, presumably because they are closest to those loci. However, the tests do not indicate whether the markers are proximal or distal to the disease loci and an alternative is to search for pairs of markers that define an interval containing the disease gene. Schaid (21) points out that marker haplotypes can offer more power than single-marker tests when they capture ancestral genetic structure. Conditions under which either haplotype or single-marker tests may be more powerful were also discussed by Nielsen and coworkers (17). For disease locus **T** and marker loci **M**, **N** consider the partial regression coefficients for the trait on each marker while holding the other marker constant. If X and Y are variables defined for the **M** and **N** genotypes respectively,

Table 5 **Partial regression coefficient terms for trait *Y* (locus T) on two marker variables *X*, *Y* (loci M,N)**

		Order *TMN*	Order *MTN*	Order *MNT*
$\beta_{YX.Z}$	Additive	$\frac{\rho_{MT}\sigma_{A_T}}{\sqrt{2p_M p_m}}$	$\frac{1-\rho_{NT}^2}{1-\rho_{MN}^2}\frac{\rho_{MT}\sigma_{A_T}}{\sqrt{2p_M p_m}}$	0
	Nonadditive	$\frac{\rho_{MT}^2\sigma_{D_T}}{p_M p_m}$	$\frac{1-\rho_{NT}^4}{1-\rho_{MN}^4}\frac{\rho_{MT}^2\sigma_{D_T}}{p_M p_m}$	0
$\beta_{YZ.X}$	Additive	0	$\frac{1-\rho_{MT}^2}{1-\rho_{MN}^2}\frac{\rho_{NT}\sigma_{A_T}}{\sqrt{2p_N p_n}}$	$\frac{\rho_{NT}\sigma_{A_T}}{\sqrt{2p_N p_n}}$
	Nonadditive	0	$\frac{1-\rho_{MT}^4}{1-\rho_{MN}^4}\frac{\rho_{NT}^2\sigma_{D_T}}{p_N p_n}$	$\frac{\rho_{NT}^2\sigma_{D_T}}{p_N p_n}$

and if appeal is made to normal-distribution theory, the partial regression coefficients are

$$\beta_{YX.Z} = \frac{\sigma_{YX}\sigma_Z^2 - \sigma_{YZ}\sigma_{XZ}}{\sigma_X^2\sigma_Z^2 - \sigma_{XZ}^2}.$$

For additive marker variables, e.g., $X_{MM} = -1, X_{Mm} = 0, X_{mm} = 1, \sigma_{A_X}^2 = 2p_M p_m$ and $\sigma_{D_X}^2 = \sigma_{D_Z}^2 = 0$,

$$\beta_{YX.Z} = \frac{(\rho_{MT} - \rho_{NT}\rho_{MN})}{1 - \rho_{MN}^2}\frac{\sigma_{A_T}}{\sqrt{2p_M p_m}};$$

whereas for nonadditive marker variables, e.g., $X_{MM} = p_m, X_{Mm} = 0, X_{mm} = p_M, \sigma_{A_X}^2 = \sigma_{A_Z}^2 = 0$ and $\sigma_{D_X}^2 = p_M^2 p_m^2$,

$$\beta_{YX.Z} = \frac{(\rho_{MT}^2 - \rho_{NT}^2\rho_{MN}^2)}{1 - \rho_{MN}^4}\frac{\sigma_{D_T}}{p_M p_m}.$$

The terms $(\rho_{MT} - \rho_{NT}\rho_{MN})$ and $(\rho_{MT}^2 - \rho_{NT}^2\rho_{MN}^2)$ depend on the order of loci **M**, **N**, **T** as shown in **Table 5** under the Markov definition of no three-locus disequilibrium. Only if both partial regression coefficients $\beta_{YX.Z}$, $\beta_{YZ.X}$ are nonzero is there evidence that **T** lies in the interval [**M**, **N**]. The marker variables can be chosen to be sensitive to either additive or nonadditive gene action at **T**.

Interval Mapping: Case-Control

For independent samples of cases and controls for a disease, when gametic data are available, the usual allelic test can be extended to a test on gamete counts. Results equivalent to those of Nielsen & Weir (19) show that gametic frequencies in the two groups are

$$P_{MN\,|\,\text{Case}} = P_{MN} + \frac{1}{\mu_G}(p_N D_{MT} + p_M D_{NT} + D_{MNT})\frac{\sigma_{A_T}}{\sqrt{2p_T p_t}}$$

$$P_{MN\,|\,\text{Control}} = P_{MN} - \frac{1}{1-\mu_G}(p_N D_{MT} + p_M D_{NT} + D_{MNT})\frac{\sigma_{A_T}}{\sqrt{2p_T p_t}}.$$

A comparison of two-marker gametic frequencies between cases and controls provides a joint test of association of each marker separately ($D_{MT} = D_{NT} = 0$) plus the effects of the markers acting together ($D_{MNT} = 0$). By analogy to the case-only Hardy-Weinberg disequilibrium, the gametic linkage disequilibrium among cases is (17)

$$D_{MN\,|\,\text{Case}} = D_{MN} - \frac{D_{MT}D_{NT}}{\mu_G^2}\sqrt{\frac{\sigma_{A_T}^2}{2p_T p_t}} + \frac{D_{MNT}}{\mu_G}\frac{\sigma_{A_T}^2}{2p_T p_t}.$$

These equations suggest that marker gamete frequencies can depend on case/control status even if neither marker is separately associated with the disease. However, for loci in the order **MTN** under the Markov assumption, case/control status affects the linkage disequilibrium between pairs of markers only when both markers are associated with the disease.

DISCUSSION

This review has phrased the association among alleles at different loci in terms of correlation coefficients of indicator variables for those alleles. Attention has been given mainly to pairs of loci for which the squared correlation coefficient is the parameter that determines the behavior of most of the standard tests for marker-disease association. In the multiple regression framework, two-locus correlations also lead to partial regression or partial correlation coefficients. Regression methods were reviewed earlier by Schaid (21). The treatment of disequilibrium at multiple loci is simplified considerably under a (first-order) Markov assumption whereby the allelic state at a locus depends on the allelic state at only one adjacent locus. There is little empirical evidence about the range of applicability of this assumption, although the increasing details of fine-scale recombination patterns (15) suggest that it may be limited.

I restricted this discussion to the situation of two alleles per locus. More general treatments are possible: Nielsen and colleagues (17–19) worked with allelic association parameters δ_i for the ith marker allele when the disease locus had alleles T_j:

$$\delta_i = \sum_{j,j'} p_j D_{ij} G_{jj'}.$$

The generality of this approach is not as necessary when the markers are biallelic SNPs. The work of Nielsen and colleagues (17–19) invoked a separate linkage disequilibrium coefficient for every pair of alleles between two loci. Testing that this set of coefficients is zero may be conducted with the statistic $X^2 = \sum_{i,j} \hat{D}_{ij}^2/(\tilde{p}_i\tilde{p}_j)$. The df for this statistic is the product of the number of alleles minus one at each locus. Recently Zaykin and coworkers (33) have considered the more difficult problem of testing for the set of composite linkage disequilibrium coefficients in the multiple allele case.

I also restricted this discussion to random samples from populations or from people who either have or do not have a disease. The rich methodology of approaches that use family data (1) has not been mentioned. It can be noted, however, that those methods often invoke the recombination fraction c_{AB} between loci **A**, **B** in a way analogous to the use of the linkage disequilibrium parameter ρ_{AB}. In particular, the Markov assumption for allele frequencies at ordered loci **A**, **B**, **C**: $\rho_{AC} = \rho_{AB}\rho_{BC}$ plays the same role for random samples as does the assumption of no interference $(1-2c_{AC}) = (1-2c_{AB})(1-2c_{BC})$ in family data.

I have not restricted the present discussion in the choice of disease model. Although it is customary to work with the special cases of additive, multiplicative, dominant, or recessive diseases, greater insight may follow from allowing a general disease model. For disease genes with two alleles this leads to expressions that include the additive and nonadditive components of variance. Provided linkage disequilibrium is not complete so that ρ_{MT}, the correlation coefficient between marker and trait loci, is less than one and $\rho_{MT}^2 < \rho_{MT}$, it may be appropriate to ignore the nonadditive aspect of disease-allele action.

DISCLOSURE STATEMENT

The author is not aware of any biases that might be perceived as affecting the objectivity of this review.

ACKNOWLEDGMENTS

This work was supported in part by NIH grant GM 075091. Helpful comments on a draft of this paper were made by Drs. W.G. Hill and J.C. Wakefield.

LITERATURE CITED

1. Abecasis GR, Cookson WOC, Cardon LR. 2001. The power to detect linkage disequilibrium with quantitative traits in selected samples. *Am. J. Hum. Genet.* 68:1463–1474
2. Cohen JE, Lynch M, Taylor CE. 1991. Forensic DNA tests and Hardy-Weinberg equilibrium. *Science* 253:1037–1038
3. Curie-Cohen M. 1982. Estimates of inbreeding in a natural population: a comparison of sampling properties. *Genetics* 100:339–358
4. Daly MJ, Rioux JD, Schaffner SF, Hudson TJ, Lander ES. 2002. High resolution haplotype structure in the human genome. *Nat. Genet.* 29:229–232
5. Feng S. 2004. *Statistical Studies of Genomics Data.* PhD thesis. Raleigh, NC: North Carolina State University
6. Gomes I, Collins A, Lonjou C, Thomas NS, Wilkinson J, Watson M, Morton N. 1999. Hardy-Weinberg quality control. *Ann. Hum. Genet.* 63:535–538
7. Guo SW, Thompson EA. 1992. Performing the exact test of Hardy-Weinberg proportion for multiple alleles. *Biometrics* 48:361–372
8. Hernandez-Sanchez J, Haley CS, Woolliams JA. 2004. On the prediction of simultaneous inbreeding coefficients at multiple loci. *Genet. Res.* 83:113–120
9. Hill WG. 1974. Estimation of linkage disequilibrium in randomly mating populations. *Heredity* 33:229–239
10. Hill WG. 1976. Non-random association of neutral linked genes in finite populations. In *Population Genetics and Ecology*, ed. S Karlin, E Nevo. New York: Academic Press, pp. 339–376
11. Kim Y, Feng S, Zeng Z-B. 2008. Measuring and partitioning the high order linkage disequilibrium by multiple order Markov chains. *Genet. Epi.* 32:301–12
12. Lewontin RC, Cockerham CC. 1959. The goodness-of-fit test for detecting natural selection in random mating populations. *Evolution* 13:561–564
13. Luo ZW. 1998. Detecting linkage disequilibrium between a polymorphic marker locus and a trait locus in natural populations. *Heredity* 80:198–208
14. Luo ZW, Tao SH, Zeng Z-B. 2000. Inferring linkage disequilibrium between a polymorphic maker locus and a trait locus in natural populations. *Genetics* 156:457–467
15. McVean GA, Myers SR, Hunt S, Deloukas P, Bently DR, Donnelly P. 2004. The fine scale structure of recombination rate variation in the human genome. *Science* 304:581–584
16. Nielsen DM, Ehm MG, Weir BS. 1998. Detecting marker-disease association by testing for Hardy-Weinberg disequilibrium at a marker locus. *Am. J. Hum. Genet.* 63:1531–1540
17. Nielsen DM, Ehm MG, Zaykin DV, Weir BS. 2004. Effect of two- and three-locus linkage disequilibrium on the power to detect marker/phenotype associations. *Genetics* 168:1029–1040

18. Nielsen DM, Weir BS. 1999. A classical setting for associations between markers and loci affecting quantitative traits. *Genet. Res* 74:271–277
19. Nielsen DM, Weir BS. 2001. Association studies under general disease models. *Theor. Popn. Biol.* 60:253–263
20. Sasieni PD. 1997. From genotype to genes: Doubling the sample size. *Biometrics* 53:1253–1261
21. Schaid DJ. 2004. Evaluating associations of haplotypes with traits. *Genet. Epi.* 27:348–364
22. Tenesa A, Visscher PM, Carothers AD, Knott SA. 2005. Mapping quantitative trait loci using linkage disequilibrium: marker-versus trait-based methods. *Behav. Genet.* 35:219–228
23. The International HapMap Consortium. 2003. The International HapMap Project. *Nature* 426:789–796
24. Thomson G, Baur M. 1984. Third order linkage disequilibrium. *Tissue Antigens* 13:91–102
25. Wallace B. 1958. The comparison of observed and calculated zygotic distributions. *Evolution* 12:113–115
26. Weir BS. 1992. Population genetics in the forensic DNA debate. *Proc. Natl. Acad. Sci. USA* 89:11654–11659
27. Weir BS. 1996. *Genetic Data Analysis II.* Sunderland, MA: Sinauer. 376 pp.
28. Weir BS. 1979. Inferences about linkage disequilibrium. *Biometrics* 35:235–254
29. Weir BS, Cockerham CC. 1979. Estimation of linkage disequilibrium in randomly mating populations. *Heredity* 42:105–111
30. Weir BS, Cockerham CC. 1989. Complete characterization of disequilibrium at two loci. In *Mathematical Evolutionary Theory*, Ed. M.W. Feldman, pp. 86–110. Princeton Univ. Press
31. Weir BS, Hill WG, Cardon LR. 2004. Allelic association patterns for a dense SNP map. *Genet. Epi.* 24:442–450
32. Wigginton JE, Cutler DJ, Abecasis GR. 2005. A note on exact tests of Hardy-Weinberg equilibrium. *Am. J. Hum. Genet.* 76:887–893
33. Zaykin DV, Pudovkin A, Weir BS. 2008. Correlation based inference for linkage disequilibrium. (submitted)
34. Zou GY, Donner A. 2006. The merits of testing Hardy-Weinberg equilibrium in the analysis of unmatched case-control data: a cautionary note. *Ann. Hum. Genet.* 70:921–933

Positive Selection in the Human Genome: From Genome Scans to Biological Significance

Joanna L. Kelley[1] and Willie J. Swanson[2]

[1]Department of Human Genetics, University of Chicago, Chicago, IL 60637; email: jkelley@uchicago.edu

[2]Department of Genome Sciences, University of Washington, Seattle, Washington 98195; email: wswanson@gs.washington.edu

Annu. Rev. Genomics Hum. Genet. 2008. 9:143–60

First published online as a Review in Advance on May 27, 2008

The *Annual Review of Genomics and Human Genetics* is online at genom.annualreviews.org

This article's doi:
10.1146/annurev.genom.9.081307.164411

1527-8204/08/0922-0143$20.00

Key Words

adaptive evolution, population differentiation, amylase, FOXP2, lactase, Neanderthal

Abstract

Here we review the evidence for positive selection in the human genome and its role in human evolution and population differentiation. In recent years, there has been a dramatic increase in the use of genome-wide scans to identify adaptively evolving loci in the human genome. Attention is now turning to understanding the biological relevance and adaptive significance of the regions identified as being subject to recent positive selection. Examples of adaptively evolving loci are discussed, specifically *LCT* and *FOXP2*. Comprehensive studies of these loci also provide information about the functional relevance of the selected alleles. We discuss current studies examining the role of positive selection in shaping copy number variation and noncoding genomic regions and highlight challenges presented by the study of positive selection in the human genome.

INTRODUCTION

Haplotype: linked alleles that are inherited as a unit

A fundamental question in evolutionary biology is what distinguishes humans from our closest living relatives, chimpanzees? One way to uncover biologically relevant differences in our genetic makeup is to examine regions of the genome that have been subject to positive selection. In this review, we define positive selection as the process that increases the frequency of mutations that confer a fitness advantage to individuals carrying those mutations. The identification of adaptively evolving genes and regulatory regions may help to elucidate specific regions of the genome that have evolved during the past six million years to allow humans to develop language, use tools, and ultimately populate nearly every continent.

Historically, the study of positive selection in human evolution was motivated by a priori assumptions about selective pressure. For example, the Duffy Fy*O blood group mutation was shown to be under positive selection (28, 29), a discovery motivated by the finding that African individuals with the Fy*O blood group were resistant to malaria infection by *Plasmodium vivax* (52). Recently, the availability of large-scale polymorphism and divergence data has made it possible to scan the human genome for evidence of positive selection without making assumptions about the potential selective advantage at particular loci. Genome-wide scans for adaptations have identified hundreds of putatively selected genes in the human genome (1, 7, 9, 11, 37, 38, 40, 43, 54, 67, 79, 86, 89, 90; also see 5, 55, 68). These scans apply a range of statistical methods to divergence and variation data to investigate features of the genome such as haplotype lengths and allele frequency spectrum differences. As genotyping and large-scale DNA sequencing efforts produce additional data, these scans will become more common and will ultimately provide a catalog of regions that have adaptively evolved in the human lineage and in specific populations. However, this information will not indicate which loci represent true examples of positive selection, the biological relevance of the region under positive selection, or the time of onset of the selective pressures.

METHODS TO IDENTIFY POSITIVE SELECTION

Positive selection may be identified in two fundamental ways: Divergence data are used to identify positive selection between species, whereas polymorphism data are used to identify positive selection within a species. Each method detects selection on a different timescale. Divergence data are used to identify older selective events, whereas polymorphism data are used to identify recent selective events. The use of both divergence and polymorphism data may provide additional support for positive selection acting in a region; however, a signature of selection with one test is sufficient evidence for selection (54).

Divergence Data

For protein-coding regions, a clear signal of positive selection is an excess in the number of nonsynonymous substitutions (i.e., amino acid altering) per nonsynonymous sites (d_N) compared with the number of synonymous substitutions (i.e., silent changes) per synonymous sites (d_S) (92). This comparison is typically quantified by the ratio of d_N to d_S (d_N/d_S). In the absence of selection, the d_N/d_S ratio is expected to equal one because d_N and d_S are normalized to the number of sites. An average d_N/d_S ratio that is less than one is a signature of purifying selection—selection against nonsynonymous substitutions. A high d_N/d_S ratio can result from either positive selection or a lack of functional constraint. A stringent criterion for positive selection requires an average d_N/d_S ratio greater than one for the entire gene (92). Analysis of six human class I major histocompatibility complex (MHC) molecules showed them to have a d_N/d_S ratio averaged across all sites and lineages of 0.5, which is high but less than one (76). However, the subset of sites in the three-dimensional structure that comprise the antigen-recognition site had a d_N/d_S ratio significantly greater than one (34). This local variation in the d_N/d_S ratio can be examined using codon models that allow for variability in the d_N/d_S ratio between sites; the best-fitting model is identified using maximum likelihood ratio tests (57, 93, 94). These methods can identify subsets of sites that have been subjected to positive selection even when the d_N/d_S ratio averaged across all sites is less than one. One additional benefit of these analyses is the ability to identify specific amino acid sites that have been the target of selection (95), which can lead to explicit functional predictions such as identification of binding sites. A limit of these tests is that they cannot detect selection acting on regulatory changes, which have been hypothesized to be important in human evolution (44).

Population Data

The detection of recent positive selection requires the identification of regions displaying evidence of a selective sweep. A selective sweep is the increase in frequency of a beneficial allele that confers a fitness advantage to its carrier. Neutral alleles closely linked to the beneficial allele also rise in frequency, which is called the hitchhiking effect (**Figure 1*b***). A characteristic molecular signature of a selective sweep is the elimination of nucleotide variation in the region of the genome close to the beneficial allele (**Figure 1*c***). As new mutations subsequently arise in the swept region, the excess of rare alleles skews the site frequency spectrum. An incomplete or partial sweep, where the favored allele does not reach fixation, does not lead to as dramatic a deviation in the site frequency spectrum (71). However, the region surrounding an incomplete selective sweep will have a mix of long haplotypes with the selected allele and ancestral haplotypes of varying lengths (**Figure 1*b***). Thus, a consideration of different aspects of the data allows researchers to identify selective events that vary in their extent and timing (68).

Site frequency spectrum: frequency distribution of allele in a population

Signatures of adaptive events that occurred since the last ice age or during human migration out of Africa should still be identifiable from samples of extant human populations. A selective sweep takes approximately $2ln(2N)/s$ generations, where N is the effective population size and s is the selection coefficient (72). Using the estimated human effective population size of 10,000 (67) and 25 years per generation, sweeps with strong selection coefficients will be complete in ~10,000 years (s = 5%). In addition, the signature of selection will be identifiable for an additional amount of time, depending on the mutation and recombination rate in the region (71). Alleles with a lower

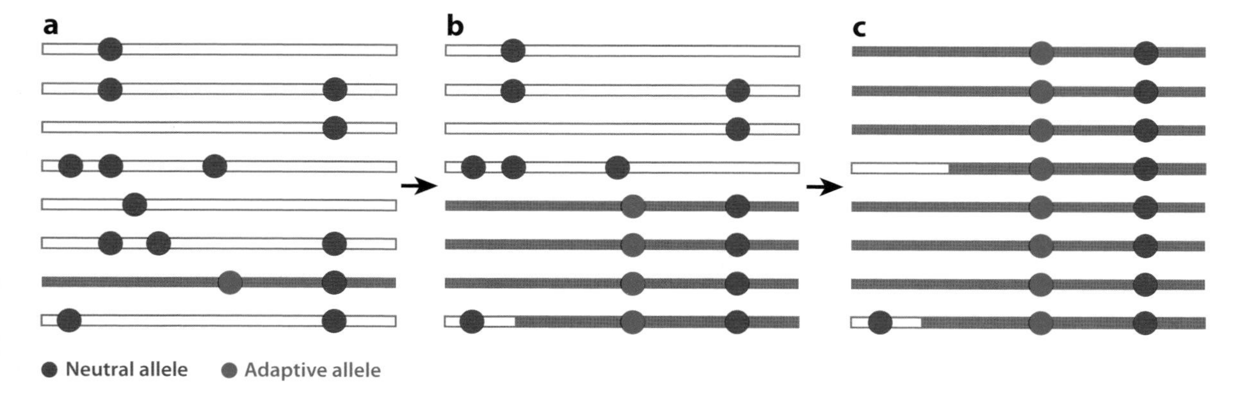

Figure 1

Selective sweep with recombination. The figure shows eight chromosomal regions with neutrally segregating alleles (*red*); the gray chromosome is the chromosome with the adaptive allele (*blue*). (*b*) A snapshot of the region during the sweep. (*c*) The result after a complete sweep in the region.

selection coefficient will take longer to reach fixation in the population. For example, an allele with s = 1% will take ~50,000 years to fix in the population. The time to completion of a selective sweep is largely dependent on the strength of selection. Therefore, sweeps that are presently incomplete are either recent sweeps with a strong selection coefficient or old sweeps with a weaker coefficient.

Derived allele: mutation from a common ancestor, usually determined by comparison to an outgroup

iHS: integrated haplotype score

SNP: single-nucleotide polymorphism

Site Frequency Spectrum

As depicted in **Figure 1*c***, following a complete selective sweep the site frequency spectrum is shifted relative to the neutral expectation (71). The swept region has very little variation, if any. The amount of variation remaining in the region is highly dependent on the recombination distance from the selected site. As new mutations accumulate after a complete selective sweep, there will be an excess of rare alleles in the swept region as compared with neutral regions that are unlinked to selected sites. Tajima's D is the most sensitive test for identifying regions with an excess of common alleles or an excess of rare alleles (78). However, Tajima's D is also sensitive to population demographics (64, 77). A test used to identify skews in the site frequency spectrum that is less affected by population demography is the composite likelihood test (42, 56, 90). The composite likelihood test uses the background pattern of variation to compare the likelihood of a neutral model versus a selective sweep model for a given genomic region. During a selective sweep, the hitchhiking effect drags variants to high or low frequency. Therefore, high-frequency derived alleles are a hallmark signature of a selective sweep. An excess of derived alleles can be measured using Fay and Wu's H (20).

Population Differentiation

Population differentiation, or allele frequency variation between populations, is largely determined by genetic drift in a population. When a locus is subjected to positive selection in a geographically restricted population, the allele frequencies around the selected locus change rapidly, leading to a high degree of population differentiation in the region (49). Therefore, a high degree of population differentiation (measured by F_{ST}, the fixation index) can be an indication of positive selection (2). However, F_{ST} alone does not indicate that a region has been adaptively evolving (22).

Haplotype Length

Hitchhiking due to positive selection results in an increase in the frequency of the haplotype on which the selected allele occurs. The selected haplotype is longer than expected because insufficient time has elapsed to allow recombination to reduce its length. Long high-frequency haplotypes are therefore indicative of the action of positive selection, whereas nonselected haplotypes vary in length. The observation of unusually frequent, long haplotypes within a population is measured by determining the relative haplotype homozygosity (67) or integrated haplotype score (iHS) (86). These statistics identify incomplete selective sweeps by requiring that the selected and nonselected haplotypes be present within the test population. More recently, test statistics have been developed to compare haplotype lengths between populations in an effort to identify complete selective sweeps (43, 69, 79).

GENOME-WIDE SCANS

Polymorphism-based scans aimed at detecting recent positive selection apply various combinations of the test statistics outlined above to publicly available genotype data (9, 37, 38, 40, 43, 69, 79, 86, 89). The two main genotype resources are Perlegen Sciences and the International HapMap Project, which have typed more than 1.5 and 3.1 million single-nucleotide polymorphisms (SNPs), respectively, in individuals with European, Asian, and African ancestry (33, 37, 38). Although the datasets are extensive, SNP genotype data are not free of ascertainment and data-collection biases. The ascertainment schemes, SNP density, and sample sizes differ between the two datasets. For example, for the Perlegen resource (33), SNPs for genotyping were ascertained by sequencing from an ethnically diverse panel of 24 individuals from the DNA Polymorphism Discovery Resource (13). By discovering polymorphisms in a diverse panel, common polymorphisms are more likely to be discovered; however, rare polymorphisms are less likely to be discovered and genotyped (12). In contrast, for the HapMap dataset, SNPs were specifically chosen to represent the most common polymorphisms in the human genome (37). Many of the test statistics were developed for application to sequence data, not genotype data. Therefore, it is important to take into account ascertainment biases when using these datasets to identify adaptively evolving regions because biases can skew population genetic measures (12). For example, if rare alleles are missed owing to either the SNP discovery method or the resequencing technology, statistics that rely on the site frequency spectrum will be skewed.

Many of the genome-wide scans for selection rely on using an empirical distribution of the test statistic to determine significance. Regions that fall in the tail of the distribution are considered candidate selection loci. Prior to the availability of genome-wide data, coalescent simulations were the most sophisticated method for determining whether or not a locus had significant evidence of positive selection [see Rosenberg & Nordborg (66) for a review of coalescent theory]. Coalescent simulations rely on assumptions about the underlying demographic histories, mutation rates, and recombination rates. An empirical distribution has the advantage in that it eliminates the need to assume the underlying demographic history, but the disadvantage is that it requires the assumption that positive selection is prevalent in human evolution. Choosing outliers using an empirical distribution assumes that an identifiable proportion of loci have been recently subject to positive selection (40) and will miss loci that are under positive selection but do not fall in the tail of the distribution (80). Moreover, although the set of candidate selection loci chosen from the tail of an empirical distribution may be enriched for true positives, it will also contain false positives.

The identification of regions by genome-wide scans depends on the test statistic used. The overlap between lists of identified regions is minimal, ranging from 8% to 27% (5; also see 55 for a discussion of the correlation between scans). Moreover, even when the test statistic identifies similar selective events (e.g., scans that identify unusually long haplotypes), the correspondence between candidate loci has so far been limited. A question that frequently arises is how prevalent is positive selection in the human genome? Evidence from divergence data (7) and site frequency spectrum data (90) suggests that approximately 10% of the genome is under positive selection. This estimate implies that selective sweeps are recurrent and frequent; therefore, parsing out recent and past selective events will be difficult. New theoretical models will be required to understand how the site frequency spectrum and haplotype lengths will be affected by multiple, overlapping sweeps in a region.

To summarize the functional representation of candidate loci, we have compiled population-specific lists of putatively selected genes. We further grouped the lists into three categories on the basis of the type of selective event: complete selective sweeps, intermediate sweeps, and fixed differences. We analyzed the PANTHER (Protein ANalysis THrough Evolutionary Relationships) classifications (81) to identify the categories of biological processes that are represented by candidate selection loci (**Figure 2**). The classes of genes identified are remarkably similar between the different sets of loci, suggesting that similar processes are recurrently selected. However, evaluation of the PANTHER classifications for over- and underrepresentation (82) of specific processes in the complete sweep versus partial sweep candidate loci lists revealed a few differences. The differences include, but are not limited to, an overrepresentation of loci involved in mRNA transcription in Asian complete sweeps, chemosensory perception and olfaction in

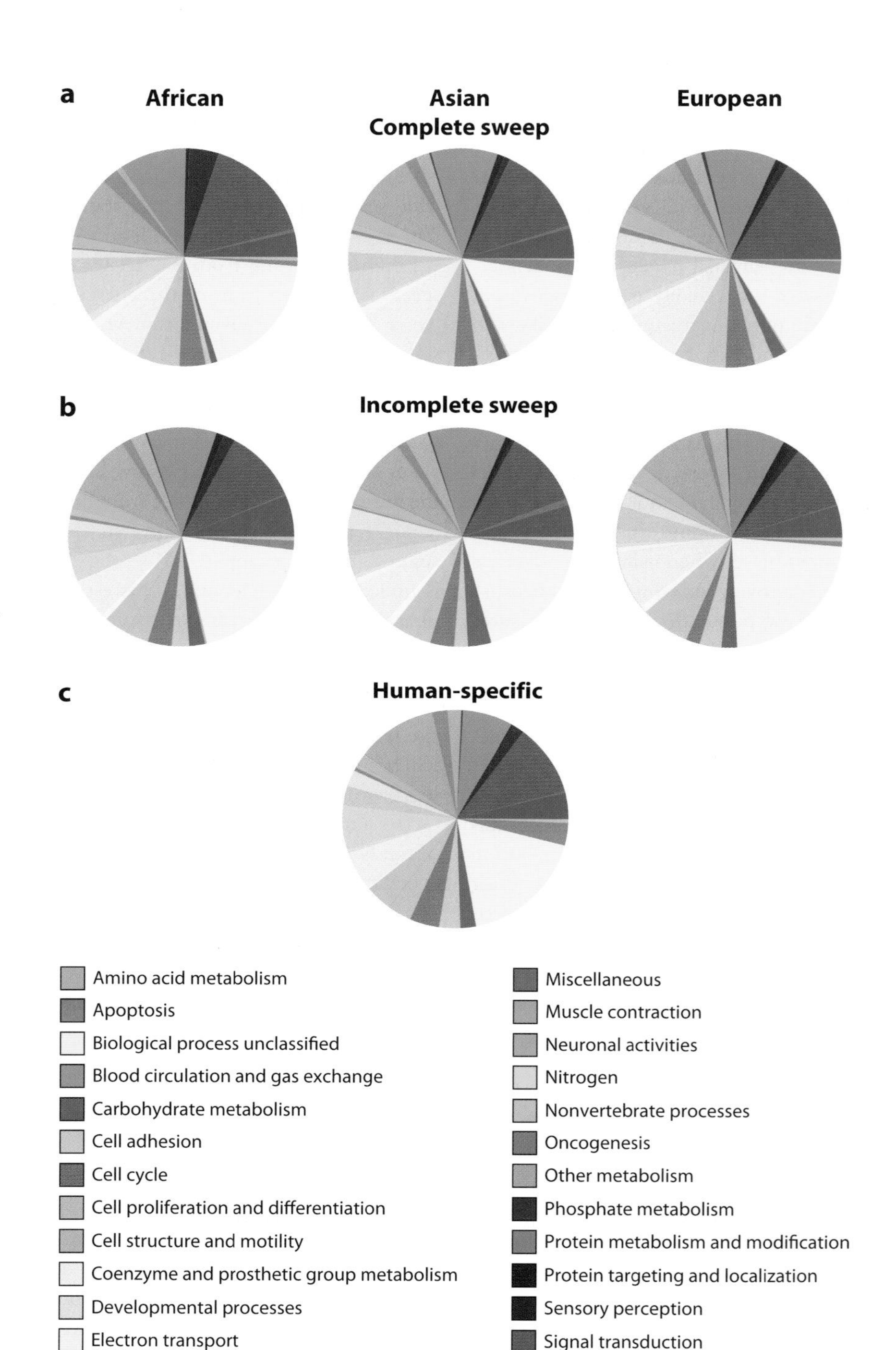
a
African
Asian
Complete sweep
European
b
Incomplete sweep
c
Human-specific
Amino acid metabolism
Apoptosis
Biological process unclassified
Blood circulation and gas exchange
Carbohydrate metabolism
Cell adhesion
Cell cycle
Cell proliferation and differentiation
Cell structure and motility
Coenzyme and prosthetic group metabolism
Developmental processes
Electron transport
Homeostasis
Immunity and defense
Intracellular protein traffic
Lipid, fatty acid, and steroid metabolism
Miscellaneous
Muscle contraction
Neuronal activities
Nitrogen
Nonvertebrate processes
Oncogenesis
Other metabolism
Phosphate metabolism
Protein metabolism and modification
Protein targeting and localization
Sensory perception
Signal transduction
Sulfur metabolism
Transport
Nucleoside, nucleotide, and nucleic acid metabolism

African complete sweeps, and cell cycle and signal transduction in European complete sweeps.

The rest of this review focuses on examples of existing techniques and highlights areas of recent interest where new techniques are needed. First, we outline two examples of loci that are under positive selection that have proposed biological relevance: lactase persistence and the lactase gene (*LCT*) (population-specific selection) and speech and the forkhead box P2 gene (*FOXP2*) (human-specific changes). We discuss areas that have only recently been considered, and where methods for detecting positive selection are being developed, specifically in relation to copy number variation and noncoding genomic regions.

Population-Specific Selection: Lactase Persistence and *LCT*

Lactase persistence is an example of positive selection acting in different populations, on several variants, over the past 7000 years. Lactase persistence is postulated to be an adaptive feature because it is a human-specific trait, differs between populations, and is correlated with the domestication of cattle (74). The ability to digest lactose is speculated to be adaptive in populations that have domesticated cattle because the selective advantage enables individuals to obtain nutrition from the milk of domesticated cows.

In Northern Europeans, a common haplotype that has been described contains *cis*-regulatory polymorphisms that affect *LCT* transcript abundance and thus lactase persistence (19, 46, 59, 84). The polymorphisms found to correlate with lactase persistence occur on a haplotype with high linkage disequilibrium extending over 1 Mb (62). Poulter and colleagues (62) speculate the C/T-13,910 allele to be causal; however, some individuals with lactase persistence lack the causal allele, suggesting that additional uncharacterized *cis*- or *trans*-acting polymorphisms exist.

Linkage disequilibrium: nonrandom association of alleles

Although the derived SNP was found to be important in lactase persistence, the evolutionary history of the allele had yet to be fully characterized. The *LCT* region appears to have undergone a selective sweep 2000–20,000 years ago (4), coinciding with the domestication of cattle. Bersaglieri and colleagues (4) used population differentiation (F_{ST}), the long-range haplotype test (67), and a novel p_{excess} statistic to look at the empirical distribution of population differentiation as evidence of positive selection acting in the region. The haplotype on which the derived C/T-13,910 SNP occurs is much longer than expected, given its frequency. The derived C/T-13,910 SNP allele frequency differs greatly between Europeans (77%), African Americans (~14%), and Chinese (0%), which is consistent with lactase persistence in each population. The absence of the European allele in non-European populations suggests the allele arose after the colonization of Europe. The high selection coefficient (between 0.014 and 0.15) distinguishes *LCT* as one of the most strongly selected loci in the human genome (4, 86). The length of the haplotype and the linkage disequilibrium in the region preclude the definitive identification of the -13910T allele as the causal variant.

Lactase persistence is prevalent in pastoral populations in Africa; however, the European variant is absent or at low frequency. This suggests that there is either a different causal variant in Europe or independent origins of lactase persistence in Africa. Tishkoff and colleagues (83) undertook a genotype-phenotype

Figure 2

Biological process representation of complied candidate selection loci. Distribution of candidate selection in PANTHER (Protein ANalysis THrough Evolutionary Relationships) biological process categories for loci identified for each population in scans for complete (*a*) and incomplete (*b*) selective sweeps, as well as loci identified under positive selection in the human lineage, using divergence data (*c*). Loci for analysis are pooled for complete sweeps (9, 40, 43, 69, 79, 90), intermediate sweeps (38, 86), and fixed differences (7, 54).

association study in 470 Africans from several different populations (43 ethnic groups) to identify the causal alleles for lactase persistence. Using a lactose tolerance test, they classified individuals as having lactase persistence, intermediate persistence, or nonpersistence. Lactase persistence was at the highest frequency in pastoral populations and lowest in a hunter-gatherer population, supporting the hypothesis that lactase persistence is correlated with cattle domestication. Three novel variants (G/C-14010, T/G-13915, and C/G-13907) were found to have a significant association with lactase persistence in various pastoral populations. The three newly discovered alleles occur on different haplotype backgrounds, which are distinct from the European-selected haplotype. The three SNPs account for ~20% of phenotypic variation, suggesting that other genetic and/or environmental factors affect lactase persistence in the populations.

To identify the effects of each SNP on promoter activity, the *LCT* core promoter was fused to 2 kb of each of the ancestral and derived haplotypes. In the in vitro transcription assay, all fusion constructs show nominal enhancer activity over the promoter's baseline activity. Promoters driven by the derived alleles showed an 18%–30% increase in expression over the fusions containing ancestral alleles. There was no expression difference between the derived fusions, as was the case for ancestral fusions. The linkage disequilibrium in the region extends for more than 1 Mb for haplotypes with any of the derived alleles. The iHS score for the C-14010 allele was highly significant compared with the iHS empirical distribution for HapMap Yoruba populations. The variants appear to have arisen in the past 1200–23,200 years, with selection coefficients in the range of 0.01 to 0.15. From association study, expression analysis, and presence of an extended haplotype, the G/C-14010 allele appears to be the causal SNP for lactase persistence in several African populations.

The Saudi Arabian population also has a high proportion of lactose tolerant individuals (15). The G-13915 derived allele was found to correlate with disaccharidase activities and the lactase:sucrase ratio, which are a proxy for lactase persistence in an urban Saudi Arabian population (36). Although Imtiaz and colleagues (36) did not specifically search for evidence of positive selection in the population, the presence of lactase persistence may shed light on the origin of some of the African haplotypes.

Knowing where and how transcription factors bind to the region may reveal how SNP differences affect promoter binding and drive differences in expression. Several of the alleles implicated in lactase persistence fall in the Oct-1 transcription factor-binding site (48) (**Figure 3**). The studies discussed above represent a detailed analysis of the evolution of lactase persistence and derived alleles that are associated with lactase persistence. Lactase persistence has independently evolved at least twice in geographically distinct populations. The study of *LCT* is a good example of linking positive selection and biological significance. The causal allele was predicted and functional

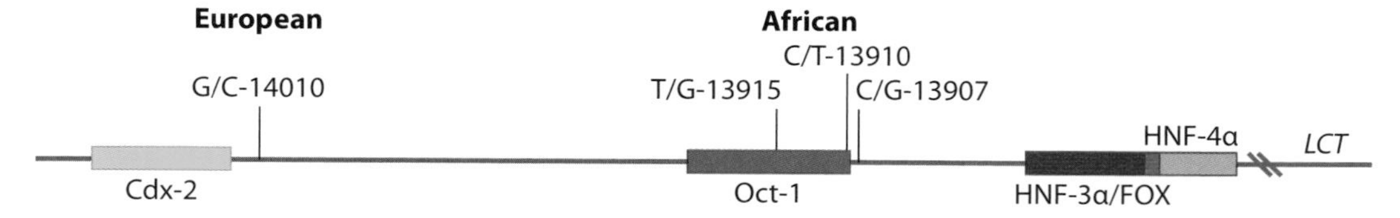

Figure 3

Locations of transcription factor-binding sites and predicted adaptive alleles upstream of *LCT*, the lactase gene. Three alleles were identified as potentially causal alleles in the African pastoral populations, whereas C/T-13910 was predicted to be the causal allele in Northern Europeans. Additionally, the T/G-13915 allele is correlated with lactase persistence in the Saudi Arabian population. The transcription factors and the sequence they bind in a supershift assay (48) are: HNF-4α (−13854 to −13830), HNF-3α and FOX (−13872 to −13848), Oct-1 and GAGA (−13933 to −13909), and Cdx-2 (−14040 to −14016).

follow-up demonstrated the significance of the variants. These types of detailed statistical and functional analysis are needed for characterizing loci identified by genome-wide scans.

HUMAN-SPECIFIC CHANGES: FOXP2

Cognition and the ability to communicate through language are attributes that set humans apart from other species. Because of its complexity, the evolution of cognition is undoubtedly influenced by many interacting factors: genetic, epigenetic, and environmental. The identification of a family with the inability to speak provided the first insight into the presence of a gene (*FOXP2*) that influences human speech (47, 50). It is important to note that the phenotype of individuals affected by *FOXP2* mutations is more complex than the presence or absence of speech. FOXP2 is a transcription factor that is highly conserved and is in the 5% tail of most conserved sequences in comparisons between human and mouse coding regions (18). Sequence comparison between extant primates and several other mammals revealed very few changes, notably two on the human lineage, one of which is predicted to have functional consequence because it creates a phosphorylation site (18). *FOXP2* appears to be adaptively evolving along the human lineage (96) and has evidence of a recent selective sweep on the basis of intronic variation revealed by Tajima's D comparison with 313 other genes (73) and comparison with standard, neutral coalescent simulations (18). Using a likelihood approach with the limited amount of available variation data, the selective sweep was dated to between 0 and 120,000 years ago, around the emergence of anatomically modern humans (39).

Krause and colleagues (45) sequenced the two human-specific derived alleles in Neanderthal DNA. The sequence results revealed that the Neanderthal sequence contains the two human-specific variants. Several possible explanations for the observations exist. To date, it has proven quite difficult to extract Neanderthal DNA from bone fragments. Contamination is a pervasive feature of Neanderthal sequence products (88). If the presence of the variants is not an artifact of contamination or DNA damage, the sites may have been fixed or segregating in the ancestral population prior to the human-Neanderthal split. Perhaps because the phenotype is so complex, other epistatic mutations had to occur, leading to the selective sweep in humans. The presence of the derived alleles in both humans and Neanderthals may reflect admixture between *H. neanderthalensis* and *H. sapiens*. Alternatively, the dating method used to estimate the age of the selective sweep in humans may be flawed.

The *FOXP2* case is an interesting example of positive selection because it incorporates both polymorphism and divergence data. Using a disease model provides some insight into the putative functional relevance. Finally, Neanderthal data demonstrate the need for more reliable methods to date the onset of selective pressures.

Positive Selection on Copy Number

The small number of amino acid differences between humans and chimpanzees has led to the suggestion that the observed phenotypic diversity may be due to regulatory changes (44). Expression differences exist between populations (35) and can confer different fitness advantages and thus be positively selected. Both copy number variation and noncoding substitutions can affect transcript levels. Therefore, positive selection can potentially act on copy number and on noncoding regions.

Copy number is a highly variable feature of the human genome and copy number polymorphisms exist within and between populations (35, 65, 70, 85). Recently, Perry and colleagues (60) correlated salivary amylase gene (*AMY1*) copy number with dietary starch prevalence. Amylase is an enzyme involved in the metabolism of starch. *AMY1* copy number varies considerably (27, 35), and copy number is positively correlated with salivary amylase protein expression (60). Seven populations were analyzed; four populations historically consumed a low-starch diet and the other three consumed a high-starch diet. Mean *AMY1* copy number was higher in the high-starch populations (60). Geographic population distribution and therefore common ancestry were not the best predictors of *AMY1* copy number. Dietary habits were found to more accurately predict copy number, suggesting that positive selection may be acting on *AMY1* copy number. To assess how copy number variation at the *AMY1* locus compares with other copy number variable loci in the human genome, Perry and colleagues (60) compared all copy number variable loci in the Yakut population to the same loci in the Japanese HapMap sample. This empirical distribution of copy number variation revealed that *AMY1* is more differentiated between the two populations than other copy number variable loci. The observed degree of population differentiation further supports the hypothesis of positive selection acting on *AMY1* copy number in high-starch diet populations. The same level of analysis was not conducted for the other high- and low-starch diet populations. The speculated adaptive phenotype is the production of additional AMY1 protein through increased copy number. The hypothesized fitness advantage of having higher salivary amylase expression is the increased digestion of starchy foods. The higher production of amylase may translate to higher amylase levels in the stomach and intestine (21) for continual enzymatic activity throughout digestion. In addition to the human population data, the fact that chimpanzees do not have multiple *AMY1* copies suggests that *AMY1* copy number has adaptively evolved in human populations. These lines of reasoning suggest that *AMY1* copy number has been under positive selection in recent human evolutionary history (60).

Additional information is necessary to provide evidence of positive selection acting on *AMY1* copy number. For example, detailed sequence analysis of each variant would facilitate the identification of a promoter for each *AMY1* copy and help determine whether each copy generates a transcript and functional protein product. Copy variants may have amino acid differences that affect function; for example, although bonobos appear to have additional *AMY1* copies compared with chimpanzees, the additional copies contain disrupted coding sequences (60). Characterizing copy number variation in additional populations, which may not be categorized into starch or nonstarch dietary preference, will provide additional information about selective pressure and perhaps support the hypothesis of positive selection acting on copy number.

There are several examples of copy number variation associated with disease susceptibility. Copy number variants may confer some selective advantage; however, this hypothesis has not been tested. For example, individuals with a low copy number, relative to the population average, of the CC chemokine ligand 3–like 1 gene (*CCL3L1*), have an increased susceptibility to HIV/AIDS (25). Low CCL3L1 copy number, along with specific CCR5 mutations, act together to increase susceptibility to HIV infection. Length polymorphisms in the mucin 6 gene (*MUC6*) are associated with *Helicobacter pylori* infection (53). Individuals with shorter MUC6 alleles are more susceptible to infection by the bacterium. The previous examples are those in which infection susceptibility depends on copy number. Further analysis may reveal evidence for recent positive selection.

Amylase copy number is an interesting potential example of positive selection acting on increased copy number, which is associated with enhanced starch digestion. Statistical methods to correlate copy number with positive selection have not yet been developed. Accurate and high-throughput detection and typing of copy number variants in multiple populations coupled with the development of new statistics may provide insight into copy number variation in human adaptation.

Positive Selection on Noncoding Genomic Regions

Noncoding polymorphisms that have conferred selective advantage in recent human history are identifiable by polymorphism-based genome-wide scans. For example, several scans have identified gene deserts with evidence for positive selection, which may reflect selection acting on uncharacterized regulatory regions (8, 86, 90). d_N/d_S methods are not applicable to noncoding regions of the DNA because it is not possible to partition each region into synonymous and nonsynonymous sites. Sophisticated methods have been developed to identify positive selection in noncoding regions on longer timescales (32, 41, 63). By comparing *cis*-regulatory containing regions with intronic sequence, Haygood and colleagues (32) found that neural and nutritional genes, which are essential to human differentiation from primates, have been subject to positive selection in the human lineage. By comparing substitution rates in conserved noncoding sequences (CNSs), Prabhakar and colleagues (63) found an excess of human-specific substitutions in CNSs near neuronal genes. CNSs are enriched for regulatory regions (31), which suggests that positive selection may be acting on neuronal regulatory regions (63).

Currently, promoter and enhancer annotation is limited; therefore, it is hard to parse out which regions of the noncoding genome are functionally relevant and which are evolving at neutral rates. Several approaches have been utilized to find human-specific or general regions of noncoding DNA that are under positive selection or have an elevated accumulation of substitutions along a branch (32, 41, 61, 63). These methods are particularly challenging to develop because of context-dependent mutation rate differences (6).

FUTURE DIRECTIONS AND CHALLENGES

Genomic scans of polymorphism and divergence data have led to the identification of hundreds of loci predicted to have been subject to positive selection in recent human history. Many of the regions identified by polymorphism-based genome-wide scans show population-specific signatures of selection. Using these results, researchers can begin to evaluate how selective signatures detected in the human genome relate to postagricultural selective pressures and to what extent they reflect older events that occurred along the human lineage after the human-chimpanzee split.

A major advantage of scanning the genome for positive selection is the ability to create an empirical distribution that is not sensitive to assumptions about the populations' demographic history. However, regardless of the type of distribution that is used, simulated or empirical, many selected loci will be inherently missed (80). Missed loci include those that have a lower selection coefficient or were selected on a different timescale and have recovered from the sweep to a point that is not in the tail of an empirical distribution. Conversely, false positives may be prevalent in the outliers of a distribution. Discriminating between true and false positives will require additional sequence analysis and, ultimately, finding the functional significance of the adaptive change.

Perhaps inflating the false-positive and false-negative rates, demography and ascertainment biases are confounders to studying positive selection with genome-wide scans. Demographic events perturb the site frequency spectrum in a very similar way to positive selection, making it difficult to distinguish between selection and changes in the underlying population structure (64). Ascertainment biases in the genotype data may also confound analyses by skewing the site frequency spectrum (12). The correspondence between candidate-selected loci is limited, even for methods that identify selective events on similar timescales. As it stands, each method is designed to deal with ascertainment biases and correct for potentially confounding demographic histories in a slightly different way. Each method is often evaluated with a highly specific set of simulated data, which is inconsistent between studies. Perhaps a consistent set of simulated data should be used to compare the multitude of scanning methods currently available.

Once a candidate region has been identified to be under positive selection the identification

Paleosequencing: DNA sequencing from extinct organisms

of the specific adaptive alleles in selected loci can be difficult, especially in regions whose signature of selection may extend over large genomic regions. However, identifying the causative SNP is imperative for dating the onset of selection in the region and understanding the biological relevance of the change. A simple strategy has been developed to identify the selected allele (69) on the basis of the following assumptions: i) the allele is newly arisen, ii) the allele is frequent only in the population with evidence for a selective sweep, and iii) the allele is annotated as nonsynonymous or in a conserved noncoding region. This strategy provides a starting point for identifying the causative alleles, yet its assumptions are conservative. For example, these strategies will miss cases where selection is acting on standing variation, migration has occurred between populations, or regulatory changes do not occur in conserved noncoding regions.

Once the specific adaptive allele has been identified, it becomes possible to determine the age of the selected allele as a proxy for the onset of selection. Precisely dating the onset of selective pressure has important implications for our understanding of human evolution. Modern humans are estimated to have arisen in Africa approximately 100,000 years ago and subsequently colonized the globe (39). Therefore, an accurate estimate of the age of selected alleles frames the anthropological context of selection and facilitates the identification of population-specific selective events. In this respect, the human-specific substitutions thought to be the target of selection in *FOXP2* are relevant. On the basis of variation among extant humans, the amino acid substitutions thought to be the target of selection were originally dated as having arisen in the past 200,000 years (18). This date was contradicted by evidence that the modern human allele is also found in Neanderthals, suggesting the allele predates the human-Neanderthal split ~400,000 years ago (45). This disparity highlights the need to assess the reliability of current methods in estimating the age of selected alleles and to develop new methods that integrate extant population data and Neanderthal data.

Paleosequencing the genome of *H. neanderthalensis* will provide insight into the selective events specific to the evolution of modern humans. The Neanderthals' most recent common ancestor to modern humans dates to ~400,000 years ago (58). The Neanderthal genome sequence could validate hypotheses about polymorphisms that have been predicted to be recent (<100,000 years ago) and uniquely human; it will also address the controversy surrounding the possibility of admixture between modern humans and Neanderthals (16, 87). However, these paleosequencing efforts are far from perfect. Human and bacterial DNA contamination, degradation due to double-strand breaks, chemical modification of the bases, and the limited specimen availability make analyses more difficult (26, 58, 88).

The sequencing of additional genomes provides more data for comparative genomics. For example, the macaque genome enables high-confidence ancestral state reconstruction for primates and provides a close outgroup to humans and chimpanzees (24). These data help to identify bursts of adaptive evolution along the human lineage. However, as compared with human, mouse, and chimpanzee sequence alignments for identifying human-specific positive selection (11), the replacement of the mouse sequence with the macaque sequence does not appear to provide much additional statistical power (24). For polymorphism studies, the Human Genome Diversity Project (HGDP) (10) provides a more extensive panel of individuals from several additional populations to understand population demography and selective events (14).

In addition to the vast amount of sequence data being generated for comparative genomics, high-throughput genomics is a cutting-edge technology to survey many individuals across the genome (see sidebar, Genome Scans and Balancing Selection). With 454 (51) and Solexa/Illumina (3) sequencing technologies, the costs of large population-based studies of human evolution are greatly diminished. Although the new technology reduces the ascertainment bias introduced from genotyping, the methods are not free of problems. Currently, the error rates are higher than traditional Sanger sequencing methods, including systematic errors in 454 sequencing that are not resolved by increasing read depth (97; J. Shendure, personal communication). Using high-throughput genomic techniques to sequence multiple individuals requires additional statistical consideration, including how read lengths and error types affect population genetic inferences.

One of the most important challenges presented by genome-wide scans is the determination of the biological relevance of a locus predicted to be under positive selection. Several methods can be used to determine the function of a locus. One method is to identify disease-causing mutations, as in the case of *FOXP2*. However, although disease mutations provide insight into the most extreme function of the gene, they are not necessarily indicative of how subtle allele changes will affect normal gene function. Using mouse models is another way to uncover gene function; however, for the loci that are most relevant to human population-specific adaptation, an appropriate mouse model or mouse phenotype may not exist.

In our view, the two most pressing issues in understanding positive selection in the human genome are accurately identifying the causative allele and understanding the functional relevance of the allele that has been subjected to positive selection. What are the anthropological and biological implications of the candidate selected locus? For some cases, such as lactase persistence and disease resistance, this is evident; in other cases, especially when the phenotype is complex, it is much more difficult. Identifying instances of human population-specific adaptation enables population geneticists to elucidate selective pressures in recent human history, characterize disease genes, and understand how human populations have adapted to new environmental challenges. However, before we can gain a clear understanding of these features, the challenges posed by genome-wide scans must be overcome.

GENOME SCANS AND BALANCING SELECTION

Genome-wide scans for positive selection have focused on identifying genes that have undergone selective sweeps, but miss other signatures of selection such as balancing selection, which acts to maintain polymorphisms. The focus on selective sweeps is likely due to single nucleotide polymorphism (SNP) ascertainment bias, which skews the site frequency spectrum toward higher frequency SNPs, confounding the ability to identify balancing selection. Reproductive genes are often subject to positive selection in comparisons between species (17, 75, 91). However, polymorphism-based genome-wide scans fail to identify reproductive proteins previously demonstrated to be adaptively evolving. Recent polymorphism surveys using complete resequencing data indicate that two reproductive proteins show the unusual feature of rapid divergence between species as well as a high number of amino acid polymorphisms within species, consistent with balancing selection (23, 30). The regions under selection between species are the same regions that demonstrate high levels of amino acid polymorphisms within humans, suggesting potential adaptive functionality related to fertilization. The high levels of polymorphisms make it unlikely that these loci would be identified by genome-wide scans that use genotyping data. However, with high-throughput sequencing technology and the prospect of complete resequencing of multiple individuals we can look forward to the identification of loci under different selective pressures, such as balancing selection.

SUMMARY POINTS

1. Genome-wide scans for evidence of positive selection in the human genome have identified hundreds of candidate selection genes.

2. Lactase persistence has independently evolved at least twice in geographically distinct populations. Alleles upstream of *LCT* (lactase gene) that are responsible for lactase persistence have recently been subjected to positive selection. Studies of *LCT* mutations, including the prediction of the causal allele and functional follow-up, provide a comprehensive example of how loci identified by polymorphism-based genome-wide scans should be verified.
3. *FOXP2* (forkhead box P2 gene) adaptively evolved along the human lineage and highlights the challenges presented by dating the selected alleles and using Neanderthal sequence data.
4. High *AMY1* (amylase gene) copy number correlates with high starch dietary preference. *AMY1* is the first locus studied suggesting that positive selection may act on copy number variation.
5. Researchers are currently developing methods that utilize divergence data to identify noncoding regions that have been subject to positive selection along the human lineage.
6. The next major step in studying adaptive evolution in the human genome will be to take candidate loci from genome scans and understand the biological significance of the selected allele.

FUTURE ISSUES

1. Next-generation DNA sequencing technologies will improve our ability to detect positive selection, particularly balancing selection. However, statistical methods must be developed to analyze the vast amount of data.
2. Increasing numbers of test statistics are available for application to genome-wide data. With the limited correspondence between genes identified by scans, a method to compare the sensitivity and specificity of scanning methods is becoming increasingly necessary.
3. Determining the biological relevance of a selected allele is challenging, especially when the phenotype is complex and the selective pressure was exerted thousands of years ago.
4. What is the appropriate strategy for identifying the causal allele?
5. Theoretical models and statistical methods are required to study selection acting on standing variation.
6. Accurately dating the selected allele requires more sophisticated methods than those that currently exist, including data from paleosequencing of Neanderthal DNA.

DISCLOSURE STATEMENT

The authors are not aware of any biases that might be perceived as affecting the objectivity of this review.

ACKNOWLEDGMENTS

We would like to acknowledge Cindy Desmarais, Kiran Dhillon, Carole Kelley, Graham McVicker, Molly Przeworski, and Matt Sandel for thoughtful contributions to this review. J.L.K. was supported by NSF DIG 0709660 and W.J.S. by NSF grant DEB-0716761 and NIH grants HD042563 and HD054631. In this review we tried to focus on a few specific examples; we apologize for not highlighting all the exciting examples of human adaptive evolution.

LITERATURE CITED

1. Akey JM, Eberle MA, Rieder MJ, Carlson CS, Shriver MD, et al. 2004. Population history and natural selection shape patterns of genetic variation in 132 genes. *PLoS Biol.* 2:e286
2. Akey JM, Zhang G, Zhang K, Jin L, Shriver MD. 2002. Interrogating a high-density SNP map for signatures of natural selection. *Genome Res.* 12:1805–14
3. Bentley DR. 2006. Whole-genome resequencing. *Curr. Opin. Genet. Dev.* 16:545–52
4. Bersaglieri T, Sabeti PC, Patterson N, Vanderploeg T, Schaffner SF, et al. 2004. Genetic signatures of strong recent positive selection at the lactase gene. *Am. J. Hum. Genet.* 74:1111–20
5. Biswas S, Akey JM. 2006. Genomic insights into positive selection. *Trends Genet.* 22:437–46
6. Blake RD, Hess ST, Nicholson-Tuell J. 1992. The influence of nearest neighbors on the rate and pattern of spontaneous point mutations. *J. Mol. Evol.* 34:189–200
7. Bustamante CD, Fledel-Alon A, Williamson S, Nielsen R, Hubisz MT, et al. 2005. Natural selection on protein-coding genes in the human genome. *Nature* 437:1153–7
8. Carlson AE, Westenbroek RE, Quill T, Ren D, Clapham DE, et al. 2003. CatSper1 required for evoked Ca^{2+} entry and control of flagellar function in sperm. *Proc. Natl. Acad. Sci. USA* 100:14864–8
9. Carlson CS, Thomas DJ, Eberle MA, Swanson JE, Livingston RJ, et al. 2005. Genomic regions exhibiting positive selection identified from dense genotype data. *Genome Res.* 15:1553–65
10. Cavalli-Sforza LL. 2005. The Human Genome Diversity Project: past, present and future. *Nat. Rev. Genet.* 6:333–40
11. Clark AG, Glanowski S, Nielsen R, Thomas P, Kejariwal A, et al. 2003. Positive selection in the human genome inferred from human-chimp-mouse orthologous gene alignments. *Cold Spring Harb. Symp. Quant. Biol.* 68:471–7
12. Clark AG, Hubisz MJ, Bustamante CD, Williamson SH, Nielsen R. 2005. Ascertainment bias in studies of human genome-wide polymorphism. *Genome Res.* 15:1496–502
13. Collins FS, Brooks LD, Chakravarti A. 1998. A DNA polymorphism discovery resource for research on human genetic variation. *Genome Res.* 8:1229–31
14. Conrad DF, Jakobsson M, Coop G, Wen X, Wall JD, et al. 2006. A worldwide survey of haplotype variation and linkage disequilibrium in the human genome. *Nat. Genet.* 38:1251–60
15. Cook GC, al-Torki MT. 1975. High intestinal lactase concentrations in adult Arabs in Saudi Arabia. *Br. Med. J.* 3:135–6
16. Currat M, Excoffier L. 2004. Modern humans did not admix with Neanderthals during their range expansion into Europe. *PLoS Biol.* 2:e421
17. Dorus S, Evans PD, Wyckoff GJ, Choi SS, Lahn BT. 2004. Rate of molecular evolution of the seminal protein gene *SEMG2* correlates with levels of female promiscuity. *Nat. Genet.* 36:1326–9
18. **Enard W, Przeworski M, Fisher SE, Lai CS, Wiebe V, et al. 2002. Molecular evolution of *FOXP2*, a gene involved in speech and language. *Nature* 418:869–72**

18. Initial study of FOXP2 selective sweep.

19. Enattah NS, Sahi T, Savilahti E, Terwilliger JD, Peltonen L, Jarvela I. 2002. Identification of a variant associated with adult-type hypolactasia. *Nat. Genet.* 30:233–7
20. Fay JC, Wu CI. 2000. Hitchhiking under positive Darwinian selection. *Genetics* 155:1405–13
21. Fried M, Abramson S, Meyer JH. 1987. Passage of salivary amylase through the stomach in humans. *Dig. Dis. Sci.* 32:1097–103
22. Gardner M, Williamson S, Casals F, Bosch E, Navarro A, et al. 2007. Extreme individual marker F_{ST} values do not imply population-specific selection in humans: the *NRG1* example. *Hum. Genet.* 121:759–62
23. Gasper J, Swanson WJ. 2006. Molecular population genetics of the gene encoding the human fertilization protein zonadhesin reveals rapid adaptive evolution. *Am. J. Hum. Genet.* 79:820–30
24. Gibbs RA, Rogers J, Katze MG, Bumgarner R, Weinstock GM, et al. 2007. Evolutionary and biomedical insights from the rhesus macaque genome. *Science* 316:222–34

25. Gonzalez E, Kulkarni H, Bolivar H, Mangano A, Sanchez R, et al. 2005. The influence of *CCL3L1* gene-containing segmental duplications on HIV-1/AIDS susceptibility. *Science* 307:1434–40
26. Green RE, Krause J, Ptak SE, Briggs AW, Ronan MT, et al. 2006. Analysis of one million base pairs of Neanderthal DNA. *Nature* 444:330–6
27. Groot PC, Bleeker MJ, Pronk JC, Arwert F, Mager WH, et al. 1989. The human α-amylase multigene family consists of haplotypes with variable numbers of genes. *Genomics* 5:29–42
28. Hamblin MT, Di Rienzo A. 2000. Detection of the signature of natural selection in humans: evidence from the Duffy blood group locus. *Am. J. Hum. Genet.* 66:1669–79
29. Hamblin MT, Thompson EE, Di Rienzo A. 2002. Complex signatures of natural selection at the Duffy blood group locus. *Am. J. Hum. Genet.* 70:369–83
30. Hamm D, Mautz BS, Wolfner MF, Aquadro CF, Swanson WJ. 2007. Evidence of amino acid diversity-enhancing selection within humans and among primates at the candidate sperm-receptor gene *PKDREJ*. *Am. J. Hum. Genet.* 81:44–52
31. Hardison RC. 2000. Conserved noncoding sequences are reliable guides to regulatory elements. *Trends Genet.* 16:369–72
32. Haygood R, Fedrigo O, Hanson B, Yokoyama KD, Wray GA. 2007. Promoter regions of many neural- and nutrition-related genes have experienced positive selection during human evolution. *Nat. Genet.* 39:1140–4
33. **Hinds DA, Stuve LL, Nilsen GB, Halperin E, Eskin E, et al. 2005. Whole-genome patterns of common DNA variation in three human populations. *Science* 307:1072–9**

33. Perlegen dataset for genome-wide scans.

34. Hughes AL, Nei M. 1988. Pattern of nucleotide substitution at major histocompatibility complex class I loci reveals overdominant selection. *Nature* 335:167–70
35. Iafrate AJ, Feuk L, Rivera MN, Listewnik ML, Donahoe PK, et al. 2004. Detection of large-scale variation in the human genome. *Nat. Genet.* 36:949–51
36. Imtiaz F, Savilahti E, Sarnesto A, Trabzuni D, Al-Kahtani K, et al. 2007. The T/G 13915 variant upstream of the lactase gene (*LCT*) is the founder allele of lactase persistence in an urban Saudi population. *J. Med. Genet.* 44:e89
37. International HapMap Consortium. 2005. A haplotype map of the human genome. *Nature* 437:1299–320
38. **International HapMap Consortium. 2007. A second generation human haplotype map of over 3.1 million SNPs. *Nature* 449:851–61**

38. HapMap dataset for genome-wide scans.

39. **Jobling MA, Hurles M, Tyler-Smith C. 2004. *Human evolutionary genetics: origins, peoples & disease*. New York: Garland Science. 523 pp.**

39. Comprehensive text about human evolution.

40. Kelley JL, Madeoy J, Calhoun JC, Swanson W, Akey JM. 2006. Genomic signatures of positive selection in humans and the limits of outlier approaches. *Genome. Res.* 16:980–9
41. Kim SY, Pritchard JK. 2007. Adaptive evolution of conserved noncoding elements in mammals. *PLoS Genet.* 3:1572–86
42. Kim Y, Stephan W. 2002. Detecting a local signature of genetic hitchhiking along a recombining chromosome. *Genetics* 160:765–77
43. Kimura R, Fujimoto A, Tokunaga K, Ohashi J. 2007. A practical genome scan for population-specific strong selective sweeps that have reached fixation. *PLoS ONE* 2:e286
44. King MC, Wilson AC. 1975. Evolution at two levels in humans and chimpanzees. *Science* 188:107–16
45. Krause J, Lalueza-Fox C, Orlando L, Enard W, Green RE, et al. 2007. The derived *FOXP2* variant of modern humans was shared with Neandertals. *Curr. Biol.* 17:1908–12
46. Kuokkanen M, Enattah NS, Oksanen A, Savilahti E, Orpana A, Jarvela I. 2003. Transcriptional regulation of the lactase-phlorizin hydrolase gene by polymorphisms associated with adult-type hypolactasia. *Gut* 52:647–52
47. Lai CS, Fisher SE, Hurst JA, Vargha-Khadem F, Monaco AP. 2001. A forkhead-domain gene is mutated in a severe speech and language disorder. *Nature* 413:519–23
48. Lewinsky RH, Jensen TG, Moller J, Stensballe A, Olsen J, Troelsen JT. 2005. T_{-13910} DNA variant associated with lactase persistence interacts with Oct-1 and stimulates lactase promoter activity in vitro. *Hum. Mol. Genet.* 14:3945–53
49. Lewontin RC, Krakauer J. 1973. Distribution of gene frequency as a test of the theory of the selective neutrality of polymorphisms. *Genetics* 74:175–95

50. MacDermot KD, Bonora E, Sykes N, Coupe AM, Lai CS, et al. 2005. Identification of FOXP2 truncation as a novel cause of developmental speech and language deficits. *Am. J. Hum. Genet.* 76:1074–80
51. Margulies M, Egholm M, Altman WE, Attiya S, Bader JS, et al. 2005. Genome sequencing in microfabricated high-density picolitre reactors. *Nature* 437:376–80
52. Miller LH, Mason SJ, Clyde DF, McGinniss MH. 1976. The resistance factor to *Plasmodium vivax* in blacks. The Duffy-blood-group genotype, FyFy. *N. Engl. J. Med.* 295:302–4
53. Nguyen TV, Janssen M Jr, Gritters P, te Morsche RH, Drenth JP, et al. 2006. Short mucin 6 alleles are associated with *H. pylori* infection. *World J. Gastroenterol.* 12:6021–5
54. Nielsen R, Bustamante C, Clark AG, Glanowski S, Sackton TB, et al. 2005. A scan for positively selected genes in the genomes of humans and chimpanzees. *PLoS Biol.* 3:e170
55. **Nielsen R, Hellmann I, Hubisz M, Bustamante C, Clark AG. 2007. Recent and ongoing selection in the human genome. *Nat. Rev. Genet.* 8:857–68**
56. Nielsen R, Williamson S, Kim Y, Hubisz MJ, Clark AG, Bustamante C. 2005. Genomic scans for selective sweeps using SNP data. *Genome Res.* 15:1566–75
57. Nielsen R, Yang Z. 1998. Likelihood models for detecting positively selected amino acid sites and applications to the HIV-1 envelope gene. *Genetics* 148:929–36
58. **Noonan JP, Coop G, Kudaravalli S, Smith D, Krause J, et al. 2006. Sequencing and analysis of Neanderthal genomic DNA. *Science* 314:1113–8**
59. Olds LC, Sibley E. 2003. Lactase persistence DNA variant enhances lactase promoter activity in vitro: functional role as a cis regulatory element. *Hum. Mol. Genet.* 12:2333–40
60. Perry GH, Dominy NJ, Claw KG, Lee AS, Fiegler H, et al. 2007. Diet and the evolution of human amylase gene copy number variation. *Nat. Genet.* 39:1256–60
61. Pollard KS, Salama SR, King B, Kern AD, Dreszer T, et al. 2006. Forces shaping the fastest evolving regions in the human genome. *PLoS Genet.* 2:e168
62. Poulter M, Hollox E, Harvey CB, Mulcare C, Peuhkuri K, et al. 2003. The causal element for the lactase persistence/nonpersistence polymorphism is located in a 1 Mb region of linkage disequilibrium in Europeans. *Ann. Hum. Genet.* 67:298–311
63. Prabhakar S, Noonan JP, Paabo S, Rubin EM. 2006. Accelerated evolution of conserved noncoding sequences in humans. *Science* 314:786
64. Przeworski M, Hudson RR, Di Rienzo A. 2000. Adjusting the focus on human variation. *Trends Genet.* 16:296–302
65. Redon R, Ishikawa S, Fitch KR, Feuk L, Perry GH, et al. 2006. Global variation in copy number in the human genome. *Nature* 444:444–54
66. Rosenberg NA, Nordborg M. 2002. Genealogical trees, coalescent theory and the analysis of genetic polymorphisms. *Nat. Rev. Genet.* 3:380–90
67. Sabeti PC, Reich DE, Higgins JM, Levine HZ, Richter DJ, et al. 2002. Detecting recent positive selection in the human genome from haplotype structure. *Nature* 419:832–7
68. Sabeti PC, Schaffner SF, Fry B, Lohmueller J, Varilly P, et al. 2006. Positive natural selection in the human lineage. *Science* 312:1614–20
69. Sabeti PC, Varilly P, Fry B, Lohmueller J, Hostetter E, et al. 2007. Genome-wide detection and characterization of positive selection in human populations. *Nature* 449:913–8
70. Sebat J, Lakshmi B, Troge J, Alexander J, Young J, et al. 2004. Large-scale copy number polymorphism in the human genome. *Science* 305:525–8
71. Simonsen KL, Churchill GA, Aquadro CF. 1995. Properties of statistical tests of neutrality for DNA polymorphism data. *Genetics* 141:413–29
72. Stephan W, Wiehe T, Lenz MW. 1992. The effect of strongly selected substitutions on neutral polymorphism: analytical results based on diffusion theory. *Theor. Popul. Biol.* 41:237–54
73. Stephens JC, Schneider JA, Tanguay DA, Choi J, Acharya T, et al. 2001. Haplotype variation and linkage disequilibrium in 313 human genes. *Science* 293:489–93
74. Swallow DM. 2003. Genetics of lactase persistence and lactose intolerance. *Annu. Rev. Genet.* 37:197–219
75. Swanson WJ, Vacquier VD. 2002. Rapid evolution of reproductive proteins. *Nat. Rev. Genet.* 3:137–44
76. Swanson WJ, Yang Z, Wolfner MF, Aquadro CF. 2001. Positive Darwinian selection drives the evolution of several female reproductive proteins in mammals. *Proc. Nat. Acad. Sci. USA* 98:2509–14

55. Review of genome-wide scans and statistical considerations.

58. Describes the Neanderthal genome.

77. Tajima F. 1989. The effect of change in population size on DNA polymorphism. *Genetics* 123:597–601
78. Tajima F. 1989. Statistical method for testing the neutral mutation hypothesis by DNA polymorphism. *Genetics* 123:585–95
79. Tang K, Thornton KR, Stoneking M. 2007. A new approach for using genome scans to detect recent positive selection in the human genome. *PLoS Biol.* 5:e171
80. Teshima KM, Coop G, Przeworski M. 2006. How reliable are empirical genomic scans for selective sweeps? *Genome Res.* 16:702–12
81. Thomas PD, Campbell MJ, Kejariwal A, Mi H, Karlak B, et al. 2003. PANTHER: a library of protein families and subfamilies indexed by function. *Genome Res.* 13:2129–41
82. Thomas PD, Kejariwal A, Guo N, Mi H, Campbell MJ, et al. 2006. Applications for protein sequence-function evolution data: mRNA/protein expression analysis and coding SNP scoring tools. *Nucleic Acids Res.* 34:W645–50
83. **Tishkoff SA, Reed FA, Ranciaro A, Voight BF, Babbitt CC, et al. 2007. Convergent adaptation of human lactase persistence in Africa and Europe. *Nat. Genet.* 39:31–40**

83. Finding of independent evolution of lactase persistence in Africa.

84. Troelsen JT, Olsen J, Moller J, Sjostrom H. 2003. An upstream polymorphism associated with lactase persistence has increased enhancer activity. *Gastroenterology* 125:1686–94
85. Tuzun E, Sharp AJ, Bailey JA, Kaul R, Morrison VA, et al. 2005. Fine-scale structural variation of the human genome. *Nat. Genet.* 37:727–32
86. Voight BF, Kudaravalli S, Wen X, Pritchard JK. 2006. A map of recent positive selection in the human genome. *PLoS Biol.* 4:e72
87. Wall JD, Hammer MF. 2006. Archaic admixture in the human genome. *Curr. Opin. Genet. Dev.* 16:606–10
88. Wall JD, Kim SK. 2007. Inconsistencies in Neanderthal genomic DNA sequences. *PLoS Genet.* 3:1862–6
89. Wang ET, Kodama G, Baldi P, Moyzis RK. 2006. Global landscape of recent inferred Darwinian selection for *Homo sapiens*. *Proc. Natl. Acad. Sci. USA* 103:135–40
90. Williamson SH, Hubisz MJ, Clark AG, Payseur BA, Bustamante CD, Nielsen R. 2007. Localizing recent adaptive evolution in the human genome. *PLoS Genet.* 3:e90
91. Wyckoff GJ, Wang W, Wu CI. 2000. Rapid evolution of male reproductive genes in the descent of man. *Nature* 403:304–09
92. Yang Z, Bielawski JP. 2000. Statistical methods for detecting molecular adaptation. *Trends Ecol. Evol.* 15:496–503
93. Yang Z, Nielsen R, Goldman N, Pedersen AM. 2000. Codon-substitution models for heterogeneous selection pressure at amino acid sites. *Genetics* 155:431–49
94. Yang Z, Swanson WJ, Vacquier VD. 2000. Maximum-likelihood analysis of molecular adaptation in abalone sperm lysin reveals variable selective pressures among lineages and sites. *Mol. Biol. Evol.* 17:1446–55
95. Yang Z, Wong WS, Nielsen R. 2005. Bayes empirical bayes inference of amino acid sites under positive selection. *Mol. Biol. Evol.* 22:1107–18
96. Zhang J, Webb DM, Podlaha O. 2002. Accelerated protein evolution and origins of human-specific features: Foxp2 as an example. *Genetics* 162:1825–35
97. Zwick ME. 2005. A genome sequencing center in every lab. *Eur. J. Hum. Genet.* 13:1167–8

RELATED RESOURCES

Haplotter: **http://hg-wen.uchicago.edu/selection/haplotter.htm**
Perlegen Sciences: **http:///www.perlegen.com/**
International HapMap Consortium: **http://www.hapmap.org/**
Human Genome Diversity Project: **http://www.stanford.edu/group/morrinst/hgdp.html**

The Current Landscape for Direct-to-Consumer Genetic Testing: Legal, Ethical, and Policy Issues

Stuart Hogarth,[1] Gail Javitt,[2] and David Melzer[3]

[1]Department of Social Sciences, Loughborough University, Loughborough LE11 3TU, United Kingdom; email: s.hogarth@lboro.ac.uk

[2]Genetics and Public Policy Center, Johns Hopkins University, Washington, DC 20036; email: gjavitt1@jhu.edu

[3]Epidemiology and Public Health Group, Peninsula Medical School, Exeter EX2 5DW, United Kingdom; email: david.melzer@pms.ac.uk

Annu. Rev. Genomics Hum. Genet. 2008. 9:161–82

The *Annual Review of Genomics and Human Genetics* is online at genom.annualreviews.org

This article's doi:
10.1146/annurev.genom.9.081307.164319

1527-8204/08/0922-0161$20.00

Key Words

personalized genomics, government regulation, DNA profiling

Abstract

This review surveys the developing market for direct-to-consumer (DTC) genetic tests and examines the range of companies and tests available, the regulatory landscape, the concerns raised about DTC testing, and the calls for enhanced oversight. We provide a comparative overview of the situation, particularly in the United States and Europe, by exploring the regulatory frameworks for medical devices and clinical laboratories. We also discuss a variety of other mechanisms such as general controls on advertising and consumer law mechanisms.

INTRODUCTION

DTC: direct to consumer

Direct-to-consumer (DTC) genetic testing is a growing phenomenon in the United States and (to a lesser extent) internationally. Harnessing the power of the Internet and the promise of the Human Genome Project, and fueled by the potential for profit and consumer interest in self-mediated healthcare, an increasing number of companies are starting to offer health-related genetic testing services directly to the public.

The advent of DTC genetic testing has sparked considerable alarm among geneticists, public health and consumer advocates, and governmental bodies (4, 36, 37, 46, 59, 77, 86, 90). Critics of DTC genetic testing have raised a number of concerns: the quality of the tests, the accuracy and adequacy of the information provided by companies, and the risk that consumers may be misled by false or misleading claims and may make harmful healthcare decisions on the basis of test results (36, 37, 48, 89). Some have asserted that genetic testing should take place only through a healthcare provider and with adequate counseling (4). Conversely, advocates of DTC testing—primarily representing purveyors of DTC tests—contend that a DTC approach enables greater consumer awareness of and access to tests. These tests can help them improve their health and make beneficial treatment and lifestyle decisions (10). These groups also claim that DTC testing provides a privacy advantage over testing through a healthcare provider (90). Little empirical evidence exists regarding the impact of DTC testing on the public.

Government oversight of DTC genetic testing—as with genetic testing generally—is quite limited. Most genetic tests are not subject to any type of government review before they are made available to the public. Federal requirements for genetic testing laboratories are general in nature and do not set specific standards for genetic tests. Thus, DTC companies face few barriers to market entry, and few governmental mechanisms exist to ensure that laboratories reliably obtain the correct result or that the tests they perform accurately predict phenotype.

The dynamic nature of the DTC marketplace makes it a somewhat difficult topic for a review article. Some DTC company websites that existed five years ago have disappeared, and those that are around today may not survive until this review's publication date. Despite the inherent fluidity of the marketplace, the commercial allure of DTC testing, coupled with the lack of regulatory barriers to market entry, has led to a steady stream of new entrants; currently more than two dozen DTC companies exist worldwide. Although some individual players may change, the phenomenon can be expected to grow in the absence of regulatory changes. Moreover, DTC offerings are expanding to include not only single-gene tests but also large-scale single-nucleotide polymorphism (SNP) profiling. Eventually, whole-genome sequencing may be offered affordably in a DTC fashion.

Given the expansive potential of DTC genetic testing, it is important to understand the regulatory framework in which DTC genetic tests are offered and the regulatory approaches that different countries have adopted. This review defines genetic testing, describes the types of genetic tests that are available, and explains the purposes for which they may be used. We then define DTC genetic testing, discuss the concerns that have been raised about specific tests (both tests currently offered and those expected to be offered in the future) discuss concerns about DTC marketing in general, and describe what is known empirically about consumer and provider awareness of DTC tests. Next, we summarize the regulation of DTC genetic testing in the United States and internationally, with particular focus on the European Union. We identify gaps in current regulations and their consequences, and discuss regulatory efforts that have been undertaken to address these gaps. Finally, we describe policy approaches that could be taken with respect to DTC testing and analyze the merits and drawbacks of such approaches.

DEFINING GENETIC TESTING

DTC testing has emerged amid a period of rapid growth in the number of available genetic tests. Today, genetic tests for more than 1200 diseases are available in a clinical setting and several hundred more are available in a research setting (25).

There is no internationally agreed upon definition of the term 'genetic test.' The term has been defined in various ways in United States state laws, by United States and international advisory committees examining genetic testing oversight (11, 44, 49, 50, 53, 77, 78), and in recently enacted federal legislation to prohibit genetic discrimination (28). For the purposes of this review, genetic test refers to an analysis of human DNA, RNA, protein(s), or metabolite(s) to diagnose or predict a heritable human disease; to guide treatment decisions, such as drug prescribing or dosing on the basis of an individual's genetic makeup; or to predict disease recurrence on the basis of data about multiple genes or their encoded products (e.g., RNA or proteins).

Over the past decade, genetic testing has become integral to diagnosing, predicting, and preventing disease. Depending on the condition under consideration, genetic testing may be recommended throughout the life cycle. Preimplantation genetic diagnosis (PGD) following in vitro fertilization can identify embryos with specific disease-causing mutations, such as Fanconi anemia, or desired genetic characteristics, such as HLA type, prior to transfer into a woman's uterus (45). Prenatal testing is performed to detect genetic abnormalities, such as Down syndrome, in a developing fetus (26). Newborn screening is performed to diagnose certain metabolic disorders, such as phenylketonuria (PKU), for which early intervention can prevent adverse consequences (26). Genetic tests can be used to confirm the diagnosis of monogenic diseases, such as cystic fibrosis (26), or to determine the risk of developing a particular disease or condition, such as hereditary breast, ovarian, or colon cancer (26). More recently, interest has increased in using genetic tests to predict response to medication (76), such as Her2/neu testing prior to prescribing the breast cancer drug Herceptin.

FDA: Food and Drug Administration

With the exception of genetic tests performed on samples obtained through invasive medical procedures (e.g., amniocentesis), any genetic test could, in theory, be offered directly to consumers. Although a small fraction of tests, which are available for more than 1500 diseases, are offered in this manner today, there is no technological barrier to offering a wide range of DTC genetic tests, whether diagnostic, predictive, or preventive. Tests offered over the Internet include some that are conducted as part of routine clinical practice, such as those for mutations that cause cystic fibrosis, hemochromatosis, and Fragile X syndrome, as well as many tests that have not yet been accepted into routine clinical practice. This is particularly the case for tests that purport to predict susceptibility to common complex conditions, such as cancer and heart disease.

DEFINING DIRECT-TO-CONSUMER

The term 'direct-to-consumer' has been used variously to refer to both advertising and sale of genetic tests. In the first instance, the availability of a test is advertised to the public, but the test must be ordered by, and the results delivered to, a healthcare provider. This situation is similar to that seen for prescription drug advertisements in the United States, although unlike prescription drugs, the genetic tests being advertised are not generally subject to premarket review or approval by the Food and Drug Administration (FDA), as discussed below. Although much policy discussion has concentrated on advertising, many DTC companies appear to be focusing their marketing budgets on efforts to gain favorable media coverage.

In the second instance, genetic tests are not only advertised directly to consumers, but the purchase of genetic testing services also is initiated at the consumer's request, and the results are delivered directly to the consumer, without the involvement of the consumer's healthcare provider. In some cases, the test may be, as a formal matter, "ordered" by a healthcare provider employed by the company to comply with legal requirements, but the company-employed provider does not establish a doctor-patient relationship with the consumer. These two distinct DTC models have raised different concerns.

Direct-to-Consumer Genetic Test Advertising

Few genetic tests are advertised to consumers but made available only through healthcare providers. The best-known and most controversial example in the United States involves Myriad's advertising campaigns for its BRACAnalysis test, which predicts predisposition to hereditary breast and ovarian cancer. Myriad launched a pilot advertising campaign in two cities in 2002 (63, 74) using television, print, and radio to "alert women with a family history of cancer to recent advances in cancer prevention and early disease detection" (70). The advertising campaign encouraged consumers to consult their physician about the genetic test.

Several studies have evaluated the impact of Myriad's ad campaign on awareness and uptake of the test, as well as public reaction to the advertising. The resulting data indicate that the campaign led to increased awareness of testing among providers and patients, an increase in the referral rate for genetic counseling services among low-risk women (69), and an increase in the number of tests ordered (8). The data do not indicate any negative psychological impact on patients or primary care providers as a result of the ads (69).

In September 2007, Myriad again stirred controversy when it launched a larger-scale television, radio, and print advertising campaign for its BRACAnalysis test in New York, Connecticut, Rhode Island, and Massachusetts. The ads urged women to discuss their family history with a doctor or to call a toll-free number to find out if they are good candidates for the test. Some physicians supported the campaign as a means to educate women and primary care physicians (87). Critics of the campaign raised concerns that the ads may create unnecessary anxiety and lead to overuse of the test (88).

Direct-to-Consumer Genetic Tests: Models of Provision

The majority of tests advertised DTC also are sold directly to consumers. For example, although Myriad does not sell its BRACAnalysis test directly to consumers, the test can be ordered by consumers directly from a third party company, DNA Direct, which employs its own healthcare providers, who order the test from Myriad on behalf of consumers. DNA Direct then communicates the results directly to consumers. This marketing scenario demonstrates just how blurred the lines between DTC advertising and DTC sale can become. DTC testing has also brought about the rise of the third-party intermediary, a company interposed between the patient and the laboratory that makes claims about the test but does not perform the testing.

Similar arrangements have been seen in the United Kingdom; for example, the company MediChecks (65) offers a wide range of DTC tests via the Internet in collaboration with the private pathology laboratory TDL. Scienta Health Center (81), a Toronto-based company, offers a range of tests developed and performed by the Austrian company Genosense (31). Genosense has partners across the globe; for instance, their tests were recently made available in the United Kingdom by a company called Genetic Health (27).

Both the number of companies and the variety of tests offered have grown since DTC genetic testing first began. In 2003 Gollust and coworkers (36) identified seven websites offering health-related DTC genetic testing for seven conditions.[1] Of these seven, four are no longer in business or are no longer offering DTC testing. Today more than two dozen websites (including three of the original seven) offer more than 50 health-related tests to consumers.[2]

Recent entrants to the sector may augur a fundamental change in the size and nature of the DTC genetic testing market (**Table 1**). Fueled by the diminishing cost of performing DNA microarray analysis and the rapid pace of scientific discovery in genome-wide association studies, companies in the United States (1, 71) and Europe (12) are offering tests based on data from very broad panels of SNPs that provide information about a variety of common diseases. These companies, many of which have significant venture capital financing and collaborations with the leading manufacturers of the microarrays, seek to establish an ongoing relationship with customers by providing periodic updates on health risk based on new scientific findings. Some of these companies reportedly also are trying to develop large-scale biobanks for research purposes, and some people have questioned whether these companies are making their intentions explicit or appropriately obtaining informed consent from their customers for research use of their DNA (35).

Market for Direct-to-Consumer Testing

Little is known about consumer awareness and uptake of DTC genetic tests. Although DTC testing companies undoubtedly maintain information regarding the number of consumers using their services, such information is proprietary. Understanding the level of consumer awareness and interest would be useful in considering the need for and appropriate tailoring of policy responses.

A 2007 study by Goddard and coworkers (34) assessed consumer and physician awareness of nutrigenomic tests and consumer use of such tests via two national surveys. They found that 14% of consumers were aware of nutrigenomic tests and 0.6% had used these tests. Consumers who were aware of the tests tended to be young and educated with a high income. A greater percentage of physicians (44%) were aware of nutrigenomic tests compared with the average consumer, although 41% of these physicians had never had a patient ask about such tests, and a majority (74%) had never discussed the results of a nutrigenomic test with a patient.

WHAT IS THE HARM?

The debate about DTC genetic testing occurs within a broader social and historical context. Clinical genetics arose in the shadow of the eugenics movement during the first half of the twentieth century, a dark legacy that underpins many fears concerning the use and misuse of genetic information (72). In response, the clinical practice of genetics has come to place an immense value on informed consent, confidentiality, and nondirectional counseling. Additionally, clinical geneticists place special importance on the act of diagnosis, because many of the diseases they diagnose are not treatable and therefore require special sensitivity in communicating test results to patients.

As a consequence of this unique cultural and clinical context, clinical geneticists generally argue that the most appropriate means of accessing genetic tests is through a medical consultation in which patients receive appropriate counseling and advice about the suitability of the test and its potential implications, expert interpretation of the test results, and guidance about actions to take as a consequence (5, 49). DTC testing challenges this longstanding tenet

[1] Gollust and coworkers included an additional seven sites that either permitted consumers to order tests but required results to be received by physicians, or required consumers to inquire about ordering information. For the purposes of this review, neither of these scenarios is considered DTC testing. Of the seven sites that Gollust and coworkers identified in this category, only two remain active today.

[2] Tests for some conditions, such as cystic fibrosis, are offered on more than one website. However, because it is not possible to determine which mutations or SNPs an individual company is testing for, each test offered by each company was counted separately to arrive at the estimate of 50 tests total.

Table 1 **Direct-to-consumer (DTC) testing companies**

Company	Tests offered	Delivery model
23andMe	Susceptibility testing for common diseases as well as ancestry testing	DTC via Internet
Acu-Gen Biolab, Inc.	Fetal DNA gender test	DTC via Internet
Consumer Genetics	Fetal gender; caffeine metabolism; alcohol metabolism; asthma drug response	DTC via Internet
Cygene Direct	Osteoporosis; athletic performance; glaucoma and macular degeneration; thrombosis	DTC via Internet
deCODE (Iceland)	Susceptibility testing for cancers, diabetes, heart disease, osteoporosis, and Parkinson's disease and others; also ancestry testing	DTC via Internet
Dermagenetics	Skin DNA profile; custom skin cream	DTC through spas and similar retailers
DNADirect (15)	α-1 antitrypsin deficiency; Ashkenazi Jewish carrier screening; blood clotting disorders; breast and ovarian cancer; colon cancer screening; cystic fibrosis; diabetes risk; drug response panel; hemochromatosis; infertility; recurrent pregnancy loss; tamoxifen	DTC via Internet; genetic counselors available by phone
G-Nostics (UK) (24)	Predisposition to nicotine addiction and response to nicotine replacement products	DTC via Internet and through pharmacies
Genelex	Pharmacogenetics testing; celiac disease; hemochromatosis; gum disease; nutritional genetic testing; DNA Diet™ consultation; weight loss system	DTC via Internet
Genetic Health (UK) (Tests are performed by Austrian test developer and laboratory Genosense)	For males: genetic predisposition to prostate cancer, thrombosis, osteoporosis, metabolic imbalances of detoxification, and chronic inflammation For females: genetic predisposition to breast cancer, bone metabolism (osteoporosis), thrombosis, cancer, and long-term exposure to estrogens Nutrigenetic test: test for range of genes that influence nutritional processes such as lipid and glucose metabolism Pharmacogenetic test: test for CYP450 genes, which influence how the liver metabolizes a large number of commonly prescribed drugs Premium Male Gene/Premium Female Gene: combine all the other tests except the nutrigenetic one	DTC via Internet; most services include a medical consultation
Geneticom (Netherlands)	Common disease risk	Not clear
Genosense (Austria)	Susceptibility tests	Do not offer DTC tests themselves but some of the institutions they partner with to order tests for consumers offer DTC testing (e.g., Genetic Health in United Kingdom)

(*Continued*)

Table 1 (*Continued*)

Company	Tests offered	Delivery model
Graceful Earth	Alzheimer (ApoE)	DTC via Internet
Health Tests Direct	More than 400 blood tests, including a few genetic tests (cystic fibrosis carrier screen, Factor V Leiden); others may also be available by calling	DTC via Internet
Health Check USA	A wide range of laboratory tests including the following genetic tests: celiac disease; factor V R2; factor V Leiden; hereditary hemochromatosis	DTC via Internet; as additional service, patient can request interpretation by board-certified physician; free genetic counseling offered by Kimball Genetics for physicians, patients, and families
Holistic Health	Nutrigenomic test: comprehensive methylation panel with methylation pathway analysis; company also sells a variety of nutritional supplements	Not described
Kimball	Wide range of well-established genetic tests	DTC via Internet but detailed telephone consultation with certified genetic counselor is mandatory; report is sent to physician and customer
MediChecks (UK)	Wide range of well-established genetic tests, from Factor V thrombosis risk to BRCA testing for breast cancer risk (most tests are performed by the private pathology laboratory TDL)	DTC via Internet but company recommends physician referral for high-impact tests such as BRCA
Medigenomix (Germany)	Thrombophilia and osteoporosis risk tests	DTC via Internet
Mygenome.com	Alzheimer's disease (genetic testing for common risk factors); drug sensitivities (genetic tests for genes that affect the safety and activity of many common prescription and over-the-counter drugs); cardiovascular disease (genetic tests differentiate treatable risk factors for heart disease and stroke); thrombosis (genetic tests identify risk factors for blood clots); pregnancy risk (genetic tests identify risk factors for complications of pregnancy); osteoporosis (genetic tests identify risk factors for osteoporosis and fractures)	Not clear
Navigenics	Risk analysis for more than 20 common diseases, such as prostate cancer and diabetes	DTC via Internet
Quixtar	Heart health; nutrigenic tests and supplements; also sells dietary supplement	DTC via Internet
Salugen	Nutrigenic tests and supplements	DTC sold through spas
Sciona	Heart health; bone health; insulin resistance; antioxidant/detoxification; inflammation	DTC via Internet
Smart Genetics	Prediction of HIV progression to AIDS	DTC via Internet; free counseling available
Suracell	DNA profile test that identifies inherited genetic aging profile, and a biomarker assessment test that measures DNA damage, oxidative stress, and free radical levels; personal genetic supplements for DNA repair and nutrition	DTC via Internet

Clinical validity: the accuracy with which a test predicts a particular clinical outcome

of clinical genetics. Unsurprisingly, the growth of DTC genetic testing has thus provoked considerable concern about the quality of service that is provided to patients.

Critics of DTC testing, steeped in the clinical genetics tradition, argue that without medical context and qualified counseling, consumers are vulnerable to being misled and to making inappropriate healthcare decisions. For example, although mutations in the *BRCA1* and *BRCA2* genes are highly predictive of breast cancer in women with a strong family history of the disease, they do not signal increased risk in women with no family history of disease. Adequate counseling is needed to explain this context, yet the DTC mode of test delivery may make it difficult to communicate this nuance, particularly because companies have a commercial interest in selling tests to the broadest population possible. Inappropriate *BRCA*-gene testing may lead to needless anxiety or, more seriously, to women seeking unnecessary medical interventions.

However, the change in delivery model from clinical encounter to consumer transaction is not the only cause of concern. Some of the newer genetic tests being offered move beyond testing for traditional Mendelian disorders—where the presence of certain gene variants is highly correlated with the development of the condition—and into the arena of more common complex diseases, where the relationship between specific genetic variants and disease is less clear. An early example was Alzheimer disease, where certain variants in the *APOE* gene are associated with only a moderately heightened risk of contracting the disease. More recent examples include *CYP450* testing to guide selection of antidepressant (selective serotonin reuptake inhibitor) medication, notwithstanding expert reports that found a lack of evidence supporting the clinical validity or utility of such testing (17, 60). Other examples include genetic tests that purport to predict the risk of diabetes, obesity, and osteoporosis and then make diet and lifestyle recommendations on the basis of the results. Many dismiss these claims as premature and not useful to the public (6, 39, 54, 73). Thus, DTC testing raises concerns about the quality of the tests and the associated testing services.

The rise of DTC genetic testing raises concerns about both consumer harms and social costs. Potential consumer harms include discrimination and stigmatization if the privacy of results is not adequately maintained, increased anxiety and needless medical interventions based on erroneous or misinterpreted test results that indicate increased risk of disease, and failure to take preventive measures based on false reassurance that one is at low risk of disease. DTC testing could pose an additional social cost in wasted scarce health resources if it leads to unnecessary visits to healthcare providers and genetic counselors and unnecessary medical tests and procedures.

Proponents of DTC testing argue that it is a means to increase consumer access to genetic tests and to empower consumers to make independent medical decisions, as well as an opportunity to educate consumers about their own health risks and the steps they can take to mitigate those risks. There is also an argument that some genetic tests pose fewer risks than others and that it may be appropriate to offer some tests DTC but others not. Arguments against DTC must address the charge of genetic exceptionalism—i.e., the concern that genetic information should not be subject to special regulation that is more stringent than regulation that is required of other types of healthcare or health-related information (80). DTC proponents also argue that DTC tests are meeting unmet demand, that there is a public appetite for information about the fruits of the Human Genome Project, and that lack of clinical uptake of new tests can be addressed by DTC advertising and test provision. Furthermore, these proponents argue that it would be paternalistic to prevent individuals from accessing information about their genomes (49).

Although these are certainly potential benefits of the DTC model, their realization requires that DTC testing is offered to the appropriate population, performed accurately, interpreted correctly, accompanied by adequate counseling, and governed by appropriate safeguards to protect the privacy of the information. However, as discussed in the next section, the current lack of regulation at all stages of the genetic testing process precludes the ability to fully assure that DTC genetic testing will benefit, rather than harm, the public.

REGULATION OF DIRECT-TO-CONSUMER GENETIC TESTING

Regulation of What?

For genetic testing to benefit an individual patient or consumer, the laboratory performing the test must be able to get the right answer as to whether a specific genetic variant is present or absent—so-called analytic validity. The genetic variant being analyzed also must correlate with a specific disease or condition in the patient (i.e., a phenotype) or with heightened risk of disease. This is called clinical validity. Finally, the test must provide information that is helpful to the individual being tested (e.g., in diagnosing, treating, or preventing the disease or condition). The latter is termed clinical utility, and is a somewhat controversial subject among geneticists because determining whether information may be useful to an individual patient can include subjective considerations, such as whether knowing the genetic basis for a disease that cannot be treated or prevented may nevertheless provide peace of mind to the patient. Determinations of clinical utility tend to be made by payers, both public and private, in considering whether or not to reimburse for testing, rather than before a test is offered to the public. In the DTC context, the consumer ultimately makes the decision about whether taking the test will be useful, but such a determination may be flawed if the information provided by the DTC company is false or misleading or is not explained adequately to the consumer.

In the United States and other countries, different entities regulate—or foreseeably could regulate—the analytical and clinical validity of genetic tests as well as the claims made about those tests. To date, there is only one example of a regulatory regime specifically designed to deal with DTC genetic tests. The Advisory Committee on Genetic Testing (ACGT), a governmental body in the United Kingdom, was established in 1996 for the purpose of considering public health and consumer protection issues around genetic testing in both the public and private sectors (43). The ACGT developed a Code of Practice for genetic testing services supplied directly to the public that established requirements regarding informed consent, genetic counseling, and provision of information on the validity and utility of tests to patients in an easily understandable format (3). The Code also established a system of compliance and monitoring under which suppliers planning to offer a DTC genetic testing service (or proposing an amendment to an existing service) first would present their proposal to the ACGT. Although compliance with the Code was voluntary, companies that did not comply faced the threat of a statutory alternative.

The ACGT was disbanded in 1999, and its responsibilities were passed to the newly formed Human Genetics Commission (HGC), the United Kingdom government's strategic advisory body on developments in human genetics. However, after enforcing the Code once in the case of the nutrigenetics company Sciona, the HGC concluded its position as a regulator was incompatible with its primary mission to offer independent strategic policy guidance to the government; thus the Code is no longer enforced. In the absence of a system designed specifically to regulate DTC testing, regulation of these tests falls under the existing regulatory mechanisms that cover clinical laboratories, medical devices, and fair trade and advertising practices.

Analytic validity: the accuracy with which a given laboratory test identifies a particular genetic variant

Clinical utility: the likelihood that using the test result(s) will lead to a beneficial outcome

ACGT: Advisory Committee on Genetic Testing

HGC: Human Genetics Commission

Regulation of Clinical Genetic Testing Laboratories

Laboratory oversight seeks to ensure the quality of the testing process by, for example, setting requirements for personnel training, specifying how and for what records must be maintained,

CMS: Centers for Medicare and Medicaid Services

CLIA: Clinical Laboratory Improvements Amendments

and assessing the analytical accuracy of tests performed by the laboratory. Laboratory oversight typically includes periodic inspection of the facility and its records and also may include other periodic assessments of quality, such as proficiency testing, which is a means to assess analytical accuracy. With few exceptions, the entities charged with regulating laboratories do not evaluate the clinical validity of the tests offered by the laboratories that they inspect.

United States Regulation

Federal regulation. The United States federal government exercises only limited oversight of laboratories that conduct genetic testing. The Centers for Medicare and Medicaid Services (CMS) implement and enforce the Clinical Laboratory Improvement Amendments of 1988 (CLIA) (9). CLIA applies to all clinical laboratories that operate or provide testing services in the United States. The statute defines a clinical laboratory as a "facility for the... examination of materials derived from the human body for the purpose of providing information for the diagnosis, prevention, or treatment of any disease or impairment of, or the assessment of the health of, human beings [9, §263a(a)]. The statute prohibits the solicitation or acceptance of "materials derived from the human body for laboratory examination or other procedure" unless CMS or a CMS-authorized entity issues the laboratory a certificate [9, §263a(b)].

CMS issued final regulations that implemented CLIA in 1992 (64). These regulations created "specialty areas" for laboratories that perform high-complexity tests, which specified personnel, quality assurance, and proficiency testing requirements for tests such as toxicology and immunology. Genetic testing, which was in its infancy at the time, was not included in these specialty areas. As a result, proficiency testing was never mandated for genetic testing laboratories.

As awareness grew about the potential role of genetic testing in healthcare, several expert panels considered what regulatory changes would ensure the smooth transition of genetic testing from research to practice (53). Key among the recommendations of these groups was that CMS create standards that focused specifically on genetic tests (44, 78). Although CMS indicated for several years that it was developing new regulations for genetic testing laboratories (18), the agency abruptly changed its mind in September 2006 (19). Thereafter, several advocacy groups petitioned the agency to issue regulations (47), but CMS denied the petition, citing cost concerns (79).

Inadequate oversight of clinical laboratories under CLIA has harmful implications for DTC testing (57, 58). Analytical validity is essential for accurate test results. Yet the absence of a specialty area means that consumers cannot have full confidence in DTC tests. Although this is true for all genetic tests, not just those ordered DTC, the fact that the consumer's healthcare provider is not involved in test interpretation increases the likelihood that analytical errors in testing will not be detected.

Inadequate transparency by CMS also leaves consumers in the dark about whether the laboratory providing results is CLIA certified, has conducted proficiency testing, and/or has received any negative inspection reports; the statute states that proficiency testing results must be made publicly available, but CMS does not have a system for sharing such information. Although CMS does make public information about laboratories that have had their certificates suspended or revoked, the most current information is at least two years out of date.

However, even if CLIA was appropriately implemented, it would provide only limited assurance about the quality of genetic tests. Under CLIA, CMS certifies laboratories but does not evaluate the clinical validity of the tests those laboratories offer, instead leaving it up to the laboratory director's determination. Thus, although CLIA could, if appropriately implemented and enforced, ensure analytic validity of genetic tests (including those offered DTC), CLIA is insufficient to ensure their clinical validity, at least as currently interpreted by CMS. There also is some doubt about whether CLIA covers tests that fall somewhere on the boundary between clinical tests and so-called lifestyle tests.

State regulation. States may choose to follow the requirements of CLIA or to implement a system of laboratory oversight that is equally or more stringent. Only the states of New York and Washington have opted out of the CLIA program in favor of a state-supervised alternative. Therefore, laboratories that test samples from patients in New York must be certified and inspected by New York in addition to their inspection under CLIA. Laboratories also may choose to be accredited by a private accrediting body with higher standards than CLIA, such as the College of American Pathologists.

State law also dictates whether healthcare provider authorization is required to obtain a laboratory test, including a genetic test. Some states explicitly authorize laboratories to accept samples from and deliver test results for specific tests (such as cholesterol or pregnancy tests) directly to patients, without authorization from a healthcare provider. Even when states prohibit DTC testing they may face difficulties in prohibiting the sale of DTC tests to consumers in their state, particularly when such sales are mediated through the Internet. Other states, such as New York, categorically prohibit all DTC testing. Still other states are silent on the issue, which leaves it up to individual laboratories to decide whether to offer DTC testing. Currently, 25 states and the District of Columbia permit DTC laboratory testing without restriction, whereas 13 states categorically prohibit it (29). DTC testing for certain specified categories of tests is permitted in 12 states; these laws would likely not permit DTC genetic tests. Even when a healthcare provider's order is required, the provider may have a conflict of interest if he or she is employed by the laboratory that offers the testing.

European Union regulation. The regulation of clinical laboratories in Europe is similarly patchy. Although, as discussed below, there have been efforts to harmonize regulations governing in vitro diagnostic (IVD) devices, as yet no common European requirements for laboratory quality assurance exist. In part, this may be because laboratory testing often is carried out within national healthcare services, a sphere of activity that is seen as the responsibility of member states. A recent Europe-wide survey revealed that very few laboratories have formal accreditation; up to 50% of laboratories surveyed do not undergo any official inspection (52). A survey conducted in 1997 suggested that in European Union countries where clinical genetics is well established, often a legal framework governed the service (40). A more recent survey of European countries found that seven had legislation: Austria, Belgium, France, Norway, Sweden, Switzerland, and the Netherlands (33) (**Table 2**). This legislation often covers the licensing of laboratories but sometimes relates to the clinics within which genetics is practiced.

IVD: in vitro diagnostic

In the European Union, there have been considerable efforts to harmonize oversight of laboratory quality assurance systems through a number of national, regional, and international schemes, culminating in the European Molecular Genetics Quality Network. Participants in the Network include 34 European countries and laboratories from Australia and the United States. These quality assurance initiatives have led to a broader project, EuroGentest (16), an ambitious attempt to move beyond the previous focus on laboratory quality assurance to develop a series of discrete but linked programs that deal with all aspects of quality in genetic testing services, from evaluation of the clinical validity and utility of tests to genetic counseling.

In the United Kingdom (and the European Union generally), laboratory regulation is not dealt with by statute but through a voluntary system of accreditation. In the United Kingdom, this system is run by Clinical Pathology Accreditation UK Ltd. (CPA), a body set up by the four main United Kingdom organizations of laboratory professionals and the United Kingdom Accreditation Service (UKAS). Laboratories that participate in the CPA scheme are inspected every five years. Although the scheme

Table 2 European legislation governing genetic testing

Country	Regulations
Austria	Gene Technology Act 1994 requires that labs that conduct predisposition and carrier status testing must be accredited; tests for diagnosis are exempt.
Belgium	1987 legislation stipulates that genetic diagnostic testing can be carried out only in the country's well-established genetics centers and with the provision of genetic counseling; state funding is provisional on the supply of detailed annual activity reports.
France	Legislation passed in 2000 states that testing can be done only by accredited personnel and labs and sets down guidelines that cover the reporting of results and confidentiality of records.
Norway	1994 act covers accreditation of institutions; it sets no restrictions on diagnostic testing but requires counseling for predictive, presymptomatic, and carrier-status testing and outlaws such testing on individuals 16 and under.
Netherlands	Legislation limits genetic testing to those institutions with a government license and places restrictions on commercial genetic testing.
Sweden	1988 act states that genetic testing can be carried out only with the permission of the National Board of Health and Welfare.

is voluntary, the majority of United Kingdom clinical laboratories participate. Within the National Health Service (NHS), all local pathology services are expected to be accredited by CPA or the equivalent, a requirement that includes NHS genetic laboratories (13). Membership in the recently established United Kingdom Genetic Testing Network also requires compliance with new European guidelines, evidence of internal quality control, and participation in external quality assessment.

With the exception of human immunodeficiency virus (HIV) testing, the United Kingdom has no restrictions on patients ordering DTC medical tests. The HIV Testing Kits and Services Regulations of 1992 made it an offense to sell, supply, or advertise for sale an HIV testing kit or a component part to a member of the public. In 2002, the Dutch government introduced regulations to restrict the availability of some tests so that they could be accessed only through doctors or pharmacists.

Regulation of Genetic Tests as Medical Devices

IVD genetic tests fall, at least in theory, under the broader statutory regimes for the regulation of medical devices. These regimes are far less onerous than those for pharmaceutical products, but they nevertheless share the goal of ensuring safety and effectiveness through a variety of regulatory mechanisms, including premarket review and regulation of labeling and promotional materials. Ensuring truth in labeling and truthful promotion—an honest account of the strengths and weakness of a test's performance—can be thought of as the fundamental function of premarket review. For high-risk tests the process may be more difficult. Regulators set out in some detail the types of clinical studies required to gain approval. Once a device is on the market, it is subject to postmarketing surveillance and to removal from the marketplace if it is found to be unsafe. However, as discussed below, although genetic tests are as a formal matter considered to be medical devices, for the most part they have been subject to far less regulation than other devices.

United States regulation. Currently, the FDA regulates test kits sold to clinical laboratories to perform testing pursuant to its authority to regulate IVD devices (55). Both genetic and nongenetic laboratory tests are considered to be IVDs if the components of the tests are bundled together, labeled for a particular use, and sold to a laboratory as a unit. Such kits must undergo successful premarket review before they may be commercially distributed. The amount and type of evidence that the FDA requires depend on the specific claims made by the test manufacturer. The FDA has to date reviewed approximately eight test kits that detect variants in human DNA or DNA expression products (55).

However, most genetic tests are developed in-house by clinical laboratories and do not use a test kit. These laboratory-developed tests (LDTs) use purchased individual components and/or components the laboratories make themselves. The FDA has historically exercised "enforcement discretion" with respect to LDTs. As a result, the FDA does not review the vast majority of genetic tests, and in particular does not assess clinical validity (55).

The FDA recently indicated its intention to require premarket review for a limited subset of LDTs known as in vitro diagnostic multivariate index assays (IVDMIAs) (22). IVDMIAs are LDTs that analyze laboratory data using an algorithm (an analytical tool) to generate a result for the purpose of diagnosing, treating, or preventing disease. The agency expressed heightened concern about these LDTs because they use proprietary methods to calculate a patient-specific result that cannot be independently derived, confirmed, or interpreted by a healthcare provider. The FDA cleared its first application for an IVDMIA in February 2007. The test analyzes gene expression products (mRNA) to determine the likelihood of breast cancer returning within five to ten years of a woman's initial cancer diagnosis (23).

Most genetic tests sold directly to consumers are LDTs and few, if any, would be considered IVDMIAs. Thus, most DTC genetic tests sold are not subject to any independent oversight to assure their clinical validity. Consumers offered these tests therefore have no means to distinguish those tests that have been shown to be useful in diagnosis or prediction of disease from those that lack adequate scientific support for the claims being made. Additionally, the FDA's ability to regulate claims about medical products is predicated on its regulation of the products themselves. When the FDA approves a test kit, it can also constrain the claims that can be made about a test's benefits and mandate the disclosure of information about a test's limitations or risks. In the absence of FDA oversight of LDTs, the agency has no mechanism to address false or misleading claims made by companies that sell DTC tests. As discussed below, a bill was introduced in the United States Congress in 2007 that would give the FDA explicit authority to regulate LDTs; however, as of the time of publication of this review the bill has not yet passed.

Laboratory-developed test (LDT): test developed for in-house use by a clinical laboratory

IVDMIA: in vitro diagnostic multivariate index assay

International Regulation

Other countries have taken varying approaches to LDT oversight. In the European Union, Sweden, and Australia, LDTs are included in device regulations (although there are exemptions in the European system for what are termed health institutions). In Canada, device regulators have sought legal opinion on whether they can regulate LDTs and have received a succession of conflicting opinions, the most recent of which suggests that such tests cannot be regulated as medical devices under current law.

Although Europe treats commercial LDTs as devices subject to the IVD Directive, it is not clear that this applies to LDTs performed by labs outside of Europe. For instance, the United States companies InterGenetics and Myriad have both made their tests available through third parties in the United Kingdom. These United Kingdom third parties collect the samples and return the results, but the test is performed in the United States. Both the United Kingdom regulator and the European Commission appear to consider such companies to be exempt from the IVD Directive. This is the opposite of the approach taken in the United States; a reference laboratory based in Europe would need FDA approval to market its test in the United States.

Although the European system treats LDTs as medical devices, most genetic tests nevertheless are not subject to independent premarket review in the European Union (42). This is because they are classified as low-risk and therefore exempt from review by an independent third party. In contrast, the United States,

Table 3 Risk Classification

Country/region	Risk categories*	Genetic tests	Premarket review
United States	I–III	II or III	Yes
Canada	I–IV	III	Yes
Australia	I–IV	II or III	Yes (class III)
Europe	I–III	I	No

*I = low risk, II = moderate-low risk, III = moderate-high risk, IV = high risk.

Canada, and Australia classify genetic tests that fall within the medical device regulations as moderate- to high-risk, and therefore generally require premarket review (**Table 3**). Thus, although Europe does not suffer the confusion over the regulatory status of LDTs that prevails in the United States, it still does not review genetic tests, including those sold directly to consumers, before they are marketed.

Regulation of Claims

In addition to laws that regulate medical devices, laws that prohibit false or misleading advertising claims could ensure that consumers receive accurate information. In the United States, Europe, and elsewhere, general laws govern the misleading promotion of goods and services and can be enforced either through private complaints to the courts by individuals or through the actions of statutory bodies, such as the Federal Trade Commission in the United States and the Office of Fair Trading in the United Kingdom. However, these mechanisms by and large have not been used to prohibit false and misleading claims about DTC genetic tests.

United States

Federal Trade Commission. The Federal Trade Commission (FTC) Act declares unlawful "unfair or deceptive acts or practices in or affecting commerce" and directs the Commission to prevent such activities (21). The statute also specifically prohibits the dissemination of false advertising to induce the purchase of drugs, devices, food, or cosmetics and defines the phrase "false advertisement" as "misleading in a material respect." The statute directs the agency to take into account not only representations made for the product, but also omissions of facts that are material given such representations.

To the extent that companies offering DTC genetic tests make claims of clinical validity without adequate scientific evidence, the FTC has the legal authority to bring an enforcement action to prohibit such claims from being made. However, the FTC has not pursued enforcement action against companies that make false or misleading claims about genetic tests even when it has received complaints about a specific test. The agency did, however, issue a consumer alert warning the public to be wary of the claims made by some companies, and advising that some results have meaning only in the context of a full medical evaluation (20).

Congress. The United States Congress also has the power to conduct investigations and hearings to uncover unfair and deceptive trade practices. In 2006, the Senate Special Committee on Aging held a hearing (85) concerning a report by the Government Accountability Office (GAO) regarding companies that offer nutrigenetic tests over the Internet (86). The GAO investigated four such companies by submitting DNA samples along with fictitious consumer profiles and analyzing the reports provided by the companies. The GAO found that although all four companies stated their tests were not intended to diagnose disease or predisposition to disease, all sent back results warning that the fictitious customers were at risk for a range of medical conditions, including type 2 diabetes, osteoporosis, cancer, heart disease, and brain aging. These predictions appeared to be independent of which DNA sample was sent in and which gene variants were present in the

FTC: Federal Trade Commission

GAO: Government Accountability Office

sample. The GAO also reported that several of the companies sent recommendations for "personalized" dietary supplements, which in reality contained ingredients similar to multivitamins that could be purchased in the drugstore, but that cost much more. Some of the claims for the supplements also were unproven, such as the claim that they could "promote DNA repair."

Following the Senate hearing, the FDA sent letters to several of the investigated companies expressing concern that their activities might be subject to FDA regulation, and requesting a meeting (38). It is unclear what subsequently transpired; however, these companies continue to offer their DTC tests.

States. State laws in the United States also prohibit unfair or deceptive trade practices within their borders, and these laws could be brought to bear on false or misleading claims about genetic tests. To date, these laws have not been used against any DTC test manufacturer. However, in the wake of Myriad's most recent campaign, the Connecticut attorney general issued a subpoena for information from the company on the basis of concern about the accuracy of some of the company's advertising claims (75).

Europe. The regulation of test claims in other countries varies, but instruments include voluntary advertising codes, statutes, and regulations. In the United Kingdom, the Medicines (Advertising) Regulations 1994 (SI 1932) appears to prohibit the advertising of DTC testing services (66, Regulation 9). However, the Medicines (Advertising) Amendment Regulations 2004 (SI 1480) amended the previous law and removed the prohibition on advertising over-the-counter medicinal products for the diagnosis of genetic disorders to the public (68). Still, the Advertising Standards Authority (ASA) (2) enforces a code of practice that requires that advertisements be "legal, decent, honest and truthful" and "capable of objective substantiation." Special rules apply to health products.

The ASA's powers were tested in 2003 when the HGC's Genetic Services subgroup complained about a product being sold by the Growth Hair Clinic (Genetic Hair). The ASA upheld the HGC's complaint that the advertisement implied that the product used genetic technology for hair restoration/grafting when in fact no such technology exists (43).

ASA: Advertising Standards Authority

Although the ASA enforces a voluntary code of practice, promotional claims also are governed by statute, in particular the Trade Descriptions Act (84). Under the Act, it is an offense for a person to apply a false (to a material degree) or misleading trade description to goods or services. The Act was tested in 2004 when the nongovernmental organization GeneWatch UK complained about claims made by g-Nostics regarding the NicoTest. GeneWatch was told that the Act did apply, and the complaint was passed to the Trading Standards office in Oxford (local to g-Nostics). Shortly after GeneWatch filed its complaint, g-Nostics modified its website claims regarding NicoTest. Thus, GeneWatch did not pursue its complaint (Personal communication with Helen Wallace, Director, GeneWatch UK).

Private Law Mechanisms

The judicial system also serves as a nonregulatory means to deter false or misleading claims that cause physical or financial harm to consumers. Whether in practice the threat of liability would deter false or misleading claims for DTC genetic tests is unclear. The consumer would first need to be aware that he or she was subject to false or misleading claims, and would have to demonstrate harm resulting from those claims. For example, the consumer would need to show that the test result led to some harmful action, and that the action was a foreseeable result of the misleading information. Emotional harm, such as added anxiety from being told one was at greater likelihood of developing a disease, would likely be an insufficient basis for receiving damages in the absence of more concrete injury. However, financial harm as a result of misstatements could be sufficient, and in fact is at the heart of the only lawsuit filed against a DTC test company so far. In February 2006, a class action lawsuit was filed against Massachusetts company Acu-Gen Biolabs for its Baby Gender Mentor test (56). The lawsuit, which was filed on behalf of women who had used the test and were dissatisfied with the incorrect results received, focused on the company's failure to honor its "200% money back guarantee" and its claim of "99.9% accuracy." In addition, the lawsuit claimed that the company provided incorrect medical advice to women about the health of their fetuses, which caused them emotional distress and led them to undergo unnecessary testing (5).

Private law mechanisms could—at least in theory—function in other countries as well. However, it is widely acknowledged that such consumer law mechanisms have significant limitations. In general, only a very small proportion of consumers with grievances pursue complaints against companies. The reasons for this include consumers' lack of appreciation of their legal rights, consumers' lack of resources (time, money, and the skills or experience to pursue legal claims), the limited number of legal firms practicing consumer law, and lack of access to experts able to provide evidence to support complaints (82).

PENDING LEGISLATION AND RECOMMENDATIONS

The year 2007 was marked both by a significant increase in the range of tests and test providers in the DTC testing market and by a renewal of the policy debate about the oversight of genetic tests in general and DTC genetic tests in particular. In the United States, two bills were introduced in Congress to strengthen government oversight over genetic tests, including those sold DTC. The Laboratory Test Improvement Act, introduced by Senator Edward Kennedy and Senator Gordon Smith, would grant explicit authority to the FDA to regulate LDTs as medical devices (62). DTC tests would have to undergo FDA review before being marketed. The Genomics and Personalized Medicine Act of 2007, introduced by Senator Barack Obama and Senator Richard Burr, would direct the Department of Health and Human Services (DHHS) Secretary to improve the safety and effectiveness of genetic tests (30). Under this bill, the Secretary would be required to commission a study from the Institute of Medicine that would make recommendations regarding the development of a "decision matrix" for use in determining which tests to regulate and how they should be regulated [30, §7(b)(3)]. The bill would direct the Centers for Disease Control to study the issue of DTC testing and its impact on consumers. The prospect for passage of either of these bills is uncertain.

In addition to these legislative efforts, the Secretary's Advisory Committee on Genetics, Health, and Society (SACGHS), which advises the DHHS Secretary on genetics policy issues, was asked by Secretary Michael Leavitt to produce a report on the oversight of genetic tests as part of his Personalized Healthcare Initiative (14). A draft report was issued for public comment in November 2007, and the final report was issued in April 2008 (77) (**Table 1**).

In the United Kingdom, the HGC published in 2007 *More Genes Direct*, a report that revisits the recommendations the Commission made in 2003 about the regulation of DTC genetic tests (50). Whether or not the government will respond remains to be seen, but the HGC now intends to facilitate the development of a code of practice, and other actions may be taken at a European level. A revision of the IVD Directive is said to be imminent, and there is much expectation that it will become more prescriptive and plug some of the current regulatory gaps, in part to address concerns about genetic tests (7, 61). As part of the Global Harmonization Task Force (GHTF), European regulators have been involved in the development of a new model for risk classification and conformity assessment of IVDs, which, if adopted by Europe, would see many new genetic tests subject to independent premarket review (32).

Another European body, the Council of Europe, also is driving policy in this area. In

2007, it published a draft Protocol on Genetic Testing that would require most genetic testing to be performed only "under individualized medical supervision." The impact of the Protocol will depend on how many states choose to formally ratify it and thereby accept it as a legally binding protocol.

Finally, in Australia the Therapeutic Goods Administration (TGA), the body responsible for licensing IVD devices, is revising its regulations, in part as a response to concerns expressed by successive Australian governments about the regulation of genetic tests. The TGA intends to implement regulatory mechanisms that will prohibit access to home use tests (self-testing) for serious disease markers, including genetic tests. As part of this process, the TGA has issued a guidance document about the regulation of nutrigenetic tests (83) (**See Supplemental Table 1**. Follow the **Supplemental Material link** from the Annual Reviews home page at **http://www.annualreviews.org**).

POLICY APPROACHES

Since the early 1990s, numerous expert bodies have convened to consider the oversight of genetic testing generally and, in recent years, DTC genetic testing specifically. These groups, typically established by the government, have issued myriad reports and recommendations (**Table 1**). Notwithstanding these prodigious efforts, little has actually changed in the oversight of genetic testing, while the number and range of tests has risen dramatically.

There is a wide range of possible policy approaches to DTC genetic testing—from prohibiting the sale of all such tests to permitting them without limitation. Policymakers must decide if the risks posed by DTC genetic tests require a targeted approach, or whether existing regulatory frameworks are adequate and adequately enforced. Policymakers also must consider whether greater control over the DTC testing market would be best achieved through enhanced regulation of the quality of genetic tests more broadly, or whether there are additional risks raised by DTC genetic tests that require distinct regulatory approaches.

Between the two extremes of a total ban and an unfettered market lies the intermediate option of permitting certain tests to be sold DTC by certain entities under certain conditions. For example, in its *Genes Direct* report, the Human Genetics Commission laid out an approach that would limit the sale of DTC genetic tests to a subset of tests considered appropriate to be offered without medical referral. Other restrictions that could be put in place include demanding that entities offering DTC tests meet licensure and quality control requirements, and insisting on heightened scrutiny (e.g., premarket review) of tests marketed directly to consumers.

Limitations also could be placed on DTC advertising, such as prior review of advertisements to ensure accuracy, limitation of advertising to certain types of tests and certain types of media (e.g., print and Internet versus television and radio), and requiring the disclosure of certain information in advertisements. Any restrictions on advertising would need to be consistent with a particular country's legal protections of commercial speech.

Some of these policy options may be pursued by enhancing the enforcement or scope of existing regulatory instruments; others might require new mechanisms, such as the ACGT's Code of Practice. Such codes could be developed by a group of companies or a trade association and voluntarily adopted, or could be developed and enforced by a government agency. As in the case of the Code of Practice, the government also can threaten to impose mandatory requirements if companies do not adhere to a voluntary code (39, 40, 82).

CONCLUSION

DTC genetic tests continue to proliferate while policies to ensure their quality have lagged behind. Establishing a coherent oversight system is challenging because of the different entities involved in oversight, the lack of existing regulations tailored to the DTC testing context, and the lack of agreement about the need for and type of oversight appropriate for DTC tests (41). The heterogeneity of tests offered and the range of delivery and promotional models further complicate the development of oversight mechanisms. The challenge for policymakers is to create standards that adequately protect consumers from harms associated with unsafe tests, while ensuring access to tests that are analytically and clinically valid in a manner that provides appropriate context and counseling. Regulatory requirements must be proportionate to the risks posed by the tests, and must recognize that some tests carry greater risks than others.

It is unlikely that we will see the emergence of a common, harmonized approach to DTC testing. However, efforts to strengthen oversight of genetic testing generally, which are underway in many countries, will, it is hoped, improve the quality of DTC genetic tests.

SUMMARY POINTS

1. The range of companies and tests in the direct-to-consumer (DTC) market is growing and seems likely to continue to increase.
2. Clinicians, scientists, consumers, and patient groups have raised concerns about various aspects of the DTC testing market.
3. DTC genetic testing is subject to a complex overlapping series of regulations, but there are significant gaps in the existing regulatory systems in both Europe and the United States.
4. A succession of policy reports have called for enhanced oversight of genetic tests in general and DTC genetic tests in particular, but thus far there has been limited policy action.

FUTURE ISSUES

1. The recent high-profile launch of a number of new consumer genetics companies has heightened concerns, as have renewed efforts to advertise DTC tests.
2. These developments have been matched by a renewed interest in oversight among policymakers, but it remains to be seen what action will result.

DISCLOSURE STATEMENT

The authors are not aware of any biases that might be perceived as affecting the objectivity of this review.

ACKNOWLEDGMENTS

The Genetics and Public Policy Center is supported at The Johns Hopkins University by The Pew Charitable Trusts and with research funding from the National Human Genome Research Institute.

Some of the research informing this paper was conducted by Stuart Hogarth and David Melzer with funding from the Wellcome Trust. The Trust played no part in the writing of this paper or the decision to submit it for publication.

LITERATURE CITED

1. 23andMe. 2007. **http://www.23andme.com**
2. Advert. Stand. Auth. (ASA). *Codes of Practice.* **http://www.asa.org.uk/asa/codes/**
3. Advis. Comm. Genet. Test. 1999. *Third Annual Report and Compendium of Guidance*. London: Dep. Health
4. Am. Coll. Med. Genet. 2008. *ACMG statement on direct-to-consumer genetic testing.* **http://www.acmg.net/AM/Template.cfm?Section=Policy_Statements&Template=/CM/ContentDisplay.cfm&ContentID=2975**
5. Babygenderinvestigation.com. 2007. *Allegations.* **http://www.babygenderinvestigation.com/ALLEGATIONS.html**
6. Baird P. 2002. Identification of genetic susceptibility to common diseases: the case for regulation. *Perspect. Biol. Med.* 45:516–28
7. Barton D. 2007. *EuroGentest workshop on genetic testing and the in vitro diagnostic devices directive. Meet. Rep.* **http://en.eurogentest.org//documents2/1193125494407/Final_Report_EuroGentest_IVDD_Workshop.pdf**
8. Cent. Dis. Control Prev. 2004. Genetic testing for breast and ovarian cancer susceptibility: evaluating direct-to-consumer marketing—Atlanta, Denver, Raleigh-Durham, and Seattle. *MMWR* 53:603–6
9. Clin. Lab. Improv. Amend. (CLIA) 1988. Public Law 100–578, codified at *U.S. Code* 42, §263a
10. Colliver V. 2007. Home DNA tests create medical, ethical quandaries. *San Francisco Chronicle*, Aug. 21: C1
11. Counc. Eur. Steer. Comm. Bioeth. 2007. *Draft Additional Protocol to the Convention on Human Rights and Biomedicine concerning genetic testing for health purposes and Draft explanatory report to the Additional Protocol to the Convention on Human Rights and Biomedicine concerning genetic testing for health purposes.* Strasbourg: Counc. Eur.
12. deCODEme. 2007. **http://www.deCodeme.com**
13. Dep. Health. 2003. *Our Inheritance, Our Future: Realising the Potential of Genetics in the NHS.* London: HMSO
14. Dep. Health Hum. Serv. 2007. *Personalized Health Care.* **http://www.dhhs.gov/myhealthcare/**
15. DNADirect. 2007. **http://www.dnadirect.com**
16. EuroGentest. 2007. **http://www.eurogentest.org/cocoon/egtorg/web/index.xhtml**
17. Eval. Genomic Appl. Pract. Prev. (EGAPP) Work. Group. 2007. Recommendations from the EGAPP Working Group: testing for cytochrome P450 polymorphisms in adults with nonpsychotic depression treated with selective serotonin reuptake inhibitors. *Genet. Med.* 9:819–25
18. *Fed. Regist.* (May) 2000. 65:25928
19. *Fed. Regist.* (April) 2006. 71:22595
20. Fed. Trade Comm. 2007. *Consumer Alert.* **http://www.ftc.gov/bcp/edu/pubs/consumer/health/hea02.shtm**
21. Fed. Trade Comm. Act 1914. Codified at United States Code, Title 15, Sections 45 et seq.
22. Food Drug Adm. (FDA). 2007. *FDA Draft guidance for industry, clinical laboratories, and FDA staff—In Vitro Diagnostic Multivariate Index Assays* **http://www.fda.gov/cdrh/oivd/guidance/1610.pdf**
23. Food Drug Adm. (FDA). 2007. *510(k) Substantial equivalence determination decision summary for mammaPrint (June 4).* **http://www.fda.gov/cdrh/reviews/K062694.pdf**
24. G-Nostics. 2007. NicoTest. **http://www.nicotest.com/**
25. GeneTests. 2007. **http://www.genetests.org**
26. GeneTests. 2007. *Uses of genetic testing.* **http://www.genetests.org/servlet/access?id=8888891&key=AceqTkMblNH-3&fcn=y&fw=Sv-&filename=/concepts/primer/primerusesof.html**
27. Genet. Health. 2007. **http://www.genetic-health.co.uk/**
28. Genet. Inf. Nondiscrimination Act of 2007. State. Adm. Policy: H.R. 493
29. Genet. Public Policy Cent. 2007. *Survey of direct-to-consumer testing statutes and regulations.* **http://www.dnapolicy.org/resources/DTCStateLawChart.pdf**
30. Genomics Pers. Med. Act of 2007, S.976, 110th Congr.
31. Genosense. 2007. **http://www.genosense.com/EN/Index.html**

32. Global Harmonization Task Force. 2007. *Study Group 1. Principles of In Vitro Diagnostic (IVD) Medical Devices Classification and Principles of Conformity Assessment for In Vitro Diagnostic (IVD) Medical Devices.* **http://www.ghtf.org/sg1/sg1-proposed.html**
33. Godard B, Kääriäinen H, Kristoffersson U, Tranebjaerg L, Coviello D, Aymé S. 2003. Provision of genetic services in Europe: current practices and issues. *Eur. J. Hum. Genet.* 11(Suppl. 2):13–48
34. Goddard KA, Moore C, Ottman D, Szegda KL, Bradley L, Khoury MJ. 2007. Awareness and use of direct-to-consumer nutrigenomic tests, United States, 2006. *Genet. Med.* 9:510–17
35. Goetz T. 2007. 23andMe will decode your DNA for $1,000: welcome to the age of genomics. *Wired Mag.* Issue 15(Nov. 17):12
36. Gollust SE, Hull SC, Wilfond BS. 2002. Limitations of direct-to-consumer advertising for clinical genetic testing. *JAMA* 288:1762
37. Gollust SE, Wilfond BS, Hull SC. 2003. Direct-to-consumer sales of genetic services on the Internet. *Genet. Med.* 5:332
38. Gutman SI. 2006. Pers. Comms. to Peter Vitulli, President, Sciona, Inc., Howard C. Coleman, Chairman, Genelex Corp., Timothy L. Ramsey, CEO, SureGene, LLC (letters on file with FDA)
39. Haga S, Khoury MJ, Burke W. 2003. Genomic profiling to promote a healthy lifestyle: not ready for prime-time. *Nat. Genet.* 34:347–50
40. Harris R, Reid M. 1997. 'Medical genetic services in 31 countries: an overview'. *Eur. J. Hum. Genet.* 5(Suppl. 2):3–21
41. Hogarth S, Liddell K, Ling T, Sanderson S, Zimmern R, Melzer D. 2007. Enhancing the regulation of genetic tests using responsive regulation. *Food Drug Law J.* 62:831–48
42. Hogarth S, Melzer D. 2007. *The IVD Directive and genetic testing: problems and proposals. A briefing presented to the 20th meeting of Competent Authorities.* Cambridge, UK: Cambridge Univ.
43. Hogarth S, Melzer D, Zimmern R. 2005. *The Regulation of Commercial Genetic testing in the UK, a Briefing for the Human Genetics Commission.* Cambridge, UK: Cambridge Univ.
44. Holtzman N, Watson M, eds. 1997. *Promoting Safe and Effective Genetic Testing in the United States: Final Report of the Task Force on Genetic Testing.* Baltimore: Johns Hopkins Univ. Press
45. Hudson K, Baruch S, Javitt G. 2005. Genetic testing of human embryos: ethical challenges and policy choices. In *Expanding Horizons in Bioethics*, ed. AW Galston, CZ Peppard, pp. 103–22. Dordrecht: Springer
46. Hudson K, Javitt G, Burke W, Byers P. 2007. ASHG statement on direct-to-consumer genetic testing in the United States. *Am. J. Hum. Genet.* 81:635–37
47. Hudson K, Terry S, Lurie P. 2006. *Petition for rulemaking.* **http://www.dnapolicy.org/resources/Petition_For_Rulemaking_September_2006.pdf**
48. Hull SC, Prasad K. 2001. Reading between the lines: direct-to-consumer advertising of genetic testing. *Hastings Cent. Rep.* 31:33–35
49. Hum. Genet. Comm. 2003. *Genes direct - Ensuring the effective oversight of genetic tests supplied directly to the public.* London: Dep. Health
50. Hum. Genet. Comm. 2007. *More genes direct: a report on developments in the availability, marketing and regulation of genetic tests supplied direct to the public.* London: Dep. Health
51. Hum. Genet. Comm. 2007. *Minutes of the Genes Direct follow-up meeting.* **http://www.hgc.gov.uk/Client/document.asp?DocId=125&CAtegoryId=8**
52. Ibaretta D, Bock A, Klein C, Rodriguez-Cerezo E. 2003. *Towards quality assurance and harmonisation of genetic testing services in the EU.* IPTS, Brussels
53. Institute of Medicine. 1994. *Assessing Genetic Risks: Implications for Health and Social Policy*, ed. LB Andrews, JE Fullarton, NA Holtzman, AG Motulsky. Washington, DC: Natl. Acad. Press
54. Janssens CAJW, Carolina Pardo M, Steyerberg EW, van Duijn CM. 2004. Revisiting the clinical validity of multiplex genetic testing in complex diseases. *Am. J. Hum. Genet.* 74:585–588
55. Javitt G. 2007. In search of a coherent framework: options for FDA oversight of genetic tests. *Food Drug Law J.* 62:617–52
56. Javitt G. 2006. Pink & Blue? The need for genetic test regulation is black and white. *Fertil. Steril.* 86:13–15
57. Javitt G, Hudson K. 2006. *Public Health at Risk: Failures in Oversight of Genetic Testing Laboratories.* Washington, DC: Genet. Public Policy Cent.

58. Javitt G, Hudson K. 2006. Federal neglect: regulation of genetic testing. *Issues Sci. Technol.* 22:59–66
59. Javitt G, Stanley E, Hudson K. 2004. Direct-to-consumer genetic tests, government oversight, and the first amendment: what the government can (and can't) do to protect the public's health. *Okla. Law Rev.* 57:251–302
60. Katsanis SH, Javitt G, Hudson K. 2008. A case study of personalized medicine. *Science* 320:53–54
61. Kenny M. 2007. The ambitious work programme of the European Commission. *Regul. Aff. J. Devices* 15:73–74
62. Lab. Test Improv. Act of 2007, S.736, 110th Congr.
63. Marcus A. 2003. *First ad campaign touts genetic screening for cancer. HealthScout News* Sept. 23. **http://www.lifeclinic.com/healthnews/article_view.asp?story=509235**
64. Medicare, Medicaid CLIA Programs. 1992. *Regulations Implementing the Clinical Laboratory Improvement Amendments of 1988 (CLIA)*, 57 *Fed. Regist.* 7002 (codified in 42 CFR Pt. 405, 410, 416, 417, 418, 440, 482, 483, 484, 485, 488, 491, 493 & 494)
65. MediChecks. 2007. *Health Screening Blood Tests*. **http://www.medichecks.com**
66. Medicines (Advert.) Amend. Regul. 2004. (SI 1480)
67. Medicines (Advert.) Regul. 1994. (SI 1932)
68. Medicines Healthcare Products Regulatory Agency. 2003. *Advertising Restrictions on Non-Prescription Medicines to be Swept Away*. London: MHRA
69. Mouchawar J, Hensley-Alford S, Laurion S, Ellis J, Kulchak-Rahm A, et al. 2005. Impact of direct-to-consumer advertising for hereditary breast cancer testing on genetic services at a managed care organization: a naturally-occurring experiment. *Genet. Med.* 7:191–97
70. Myriad Genetics Inc. 2002. *Myriad Genetics Launches Direct to Consumer Advertising Campaign for Breast Cancer Test*. **http://www.corporate-ir.net/ireye/ir_site.zhtml?ticker=mygn&script=413&layout=9&item_id=333030**
71. Navigenics. 2007. **http://www.navigenics.com**
72. Nuffield Counc. Bioeth. 1993. *Genetic Screening—Ethical Issues*. London: Nuffield Trust
73. Offit K. 2008. Genomic profiles for disease risk: predictive or premature? *J. Am. Med. Assoc.* 299:1353–55
74. Pearson H. 2003. Genetic test adverts under scrutiny. *Nature News*, 17 Mar. **http://www.nature.com/news/2003/030319/full/news030317-3.html**
75. Pollack A. 2007. A genetic test that very few need, marketed to the masses. *New York Times*, Sept. 11
76. Roses AD. 2004. Pharmacogenetics and drug development: the path to safer and more effective drugs. *Nat. Rev. Genet.* 5:645–56
77. Secretary's Advis. Comm. Genet., Health Soc. (SACGHS) 2008. *U.S. System of Oversight of Genetic Testing: A Response to the Charge of the Secretary of Health and Human Services*. Bethesda: Natl. Inst. Health
78. Secretary's Advis. Comm. Genet. Test. 2000. *Enhancing the Oversight of Genetic Tests: Recommendations of the SACGT*. Bethesda: Natl. Inst. Health
79. Smith DG. 2007. Letter to Kathy Hudson (On file Genet. and Public Policy Cent.) **http://www.dnapolicy.org/resources/CMSresponse8.15.07.pdf**
80. Suter SM. 2001. The allure and peril of genetics exceptionalism: do we need special genetics legislation? *Wash. Univ. Law Q.* 79:669–748
81. Scienta Health Center. **http://www.scientahealth.com**
82. Scott C, Black J. 2000. *Cranston's Consumers and the Law*. London: Butterworth's
83. Ther. Goods Adm. 2007. *The Regulation of Nutrigenetic Tests in Australia: Guidance Document*. **http://www.tga.gov.au/devices/ivd-nutrigenetic.htm**
84. Trade Descr. Act 1968
85. U.S. Congr. 2006. *At Home DNA tests: marketing scam or medical breakthrough: hearing before the Special Committee on Aging*, 109th Congr., 2n Sess. Senate Hear. 1090–707. Washington, DC
86. U.S. Gov. Account. Off. (GAO) 2006. *Nutrigenetic Testing: Tests Purchased from Four Web Sites Mislead Consumers: GAO-06-977, testimony before the Senate Special Committee on Aging (Statement of Gregory Kutz)*, pp. 2–27
87. Williams C. 2007. Cancer docs debate gene test ad campaign. *ABC News*, Sept. 12. **http://abcnews.go.com/Health/CancerPreventionAndTreatment/story?id=3588056&page=1**

88. Williams S. 2007. Myriad Genetics launches BRCA testing ad campaign in Northeast, *GPPC eNews*, Sept. 23. **http://www.dnapolicy.org/news.enews.article.nocategory.php?action=detail&newsletter_id=26&article_id=111**
89. Williams-Jones B. 2003. Where there's a web, there's a way: commercial genetic testing and the internet. *Community Genet.* 6:46–57
90. Wolfberg AJ. 2006. Genes on the web-direct-to-consumer marketing of genetic testing. *N. Engl. J. Med.* 355:543–45

Transcriptional Control of Skeletogenesis

Gerard Karsenty

Department of Genetics and Development, College of Physicians and Surgeons, Columbia University, New York, New York 10032; email: gk2172@columbia.edu

Annu. Rev. Genomics Hum. Genet. 2008. 9:183–96

The *Annual Review of Genomics and Human Genetics* is online at genom.annualreviews.org

This article's doi:
10.1146/annurev.genom.9.081307.164437

1527-8204/08/0922-0183$20.00

Key Words

skeletal dysplasia, osteoblast-specific transcription, chondrocyte-specific transcription

Abstract

The skeleton contains three specific cell types: chondrocytes in cartilage and osteoblasts and osteoclasts in bone. Our understanding of the transcriptional mechanisms that lead to cell differentiation along these three lineages has increased considerably in the past ten years. In the case of chondrocytes and osteoblasts advances have been made possible largely through the molecular elucidation of human skeletal dysplasias. This review discusses the key transcription factors that regulate skeletogenesis and highlights their function, mode of action, and regulation by other factors, with a special emphasis on how human genetics has contributed to this knowledge.

Chondrocyte: cell of mesenchymal origin present only in cartilage

Osteoblast: cell of mesenchymal origin present only in bone; responsible for bone formation and controls osteoclast differentiation

INTRODUCTION

The skeleton contains three specific cell types: chondrocytes of various size and shape in cartilage and osteoblasts and osteoclasts in bones. Whereas chondrocytes and osteoblasts are of mesenchymal origin, osteoclasts belong to the monocyte-macrophage cell lineage. In the past two decades extraordinary progress has been made in our understanding of cell differentiation in the skeleton and especially in the identification of transcription factors involved in these events. It is customary to highlight the contribution of animal models, and in particular of mouse genetic studies, in the progress that has been made in the past fifteen years toward the molecular understanding of these processes. This view is justified by the facts, yet it should not obscure the critical contributions of human genetics to the gain of knowledge in this field. This review revisits the main aspects of the transcriptional control of cell differentiation during skeletogenesis by showing how powerful each approach has been and discussing the important combination of these two approaches. Because human genetics has influenced our understanding of the transcriptional control of chondrocyte and osteoblast differentiation significantly, more than it has influenced our understanding of osteoclast differentiation, we focus here on chondrogenesis and osteogenesis.

HUMAN AND MOUSE GENETIC STUDIES OF CHONDROGENESIS

Chondrocytes are the first skeleton-specific cells to appear during embryonic development. Once undifferentiated mesenchymal cells aggregate to form mesenchymal condensations at the location of each future skeletal element, they acquire genetic characteristics of nonhypertrophic chondrocytes (22). The two main features of these resting and proliferating chondrocytes are that they express *Aggrecan* and *α_1(II) collagen*. As skeletogenesis proceeds, proliferating chondrocytes progressively exit the cell cycle, hypertrophy, and become bona fide hypertrophic chondrocytes (25). This latter subset of chondrocytes does not express *α_1(II) collagen* anymore, but instead expresses *α_1(x) collagen*. To date we know much more about the transcriptional control of the early part of chondrogenesis (differentiation of resting and proliferating chondrocytes) than about chondrocyte hypertrophy.

Although many molecular and mouse genetic studies followed, in all fairness this field took off when a human genetic disease marked by severe cartilage abnormalities called campomelic dysplasia was shown to be caused by an inactivating mutation in the gene encoding sex determining region Y (SRY)-box 9 (Sox9) (18, 71) (**Table 1**). This observation put an end to several years of molecular trial and error. Sox9 is a transcription factor that contains a high mobility group (HMG) box, a DNA binding domain that exhibits a high degree of homology with the DNA binding domain of the mammalian testis-determining factor SRY. Following this landmark discovery Sox9 was shown to regulate the expression of *Aggrecan* and *α_1(II) collagen* as well as the expression of *α_1(XI) collagen* and *cartilage-derived retinoic acid-sensitive protein* (*CD-RAP*), two other markers of nonhypertrophic chondrocytes (1, 5, 43, 50, 61, 76). Two experiments further established in vivo the critical importance of Sox9 during chondrogenesis. First, ectopic expression of *Sox9* in vivo is able to transactivate the *α_1(II) collagen* gene in cells not destined to become chondrocytes. Second, *$Sox9^{-/-}$* embryonic stem (ES) cells are always excluded from the chondrogenic condensations and do not express any of the molecular markers of nonhypertrophic chondrocytes (1, 2). Thus, the conjunction of human genetics, mouse genetics, and molecular studies helped to identify Sox9 as the master gene of chondrogenesis by showing that it controls proliferation and differentiation of nonhypertrophic chondrocytes (**Figure 1**). In addition, Sox9 seems to act as a negative regulator of chondrocyte hypertrophy (3, 27).

Other transcription factors are required or involved in the differentiation of nonhypertrophic chondrocytes; however, until now none have been linked to a particular skeletal

Table 1 Association between human diseases, mouse models, and genes regulating cell differentiation in the skeleton

Gene	Type of factor	Human disease(s)	Mouse model(s)	Process regulated
SOX9	Transcription factor	Campomelic dysplasia	Loss of function Ectopic expression	Chondrocyte differentiation
RUNX2	Transcription factor	Cleidocranial dysplasia	Loss of function Ectopic expression Overexpression	Chondrocyte and osteoblast differentiation
TWIST1	Transcription factor	Sathre-Chotzen syndrome	Loss of function Overexpression	Chondrocyte and osteoblast differentiation
FGFR3	Transmembrane receptor	Achondroplasia Thanatophoric dysplasia	Loss of function Gain of function	Chondrocyte differentiation
MSX2	Transcription factor	Boston-type craniosynostosis Enlarged parietalforamina	Loss of function	Osteoblast differentiation
SATB2	Nuclear matrix protein	Cleft palate	Loss of function	Osteoblast differentiation
RSK2	Protein kinase	Coffin-Lowry syndrome	Loss of function	Osteoblast differentiation and function
NF1	Ras-GTPase activating factor	Neurofibromatosis type I	Loss of function	Osteoblast differentiation and function

dysplasia. Two of these molecules, Sox5 and Sox6, belong to the same family of proteins as Sox9. Both of them can bind to Sox9 and increase Sox9 transactivation function in vitro, although neither Sox5 nor Sox6 has a transactivation domain (44). The importance of these two proteins was verified in vivo: Embryos lacking both Sox5 and Sox6 die at embryonic day 16.5 (E16.5) and display a failure of chondrocyte progenitor cells to differentiate into hypertrophic chondrocytes (44). Besides HMG box–containing transcription factors, hypoxia inducible factor-1 (Hif-1α), a basic helix-loop-helix (bHLH) domain-containing protein, favors chondrocyte survival by regulating the expression of *Vegf*, which encodes a secreted molecule required for vascular invasion of the forming bones (60).

Researchers have also begun to elucidate the transcriptional control of the transition of proliferating chondrocytes into hypertrophic chondrocytes. Here again progress was made largely from the study of genetically engineered mouse models for genes that were initially identified through the molecular elucidation of human skeletal dysplasia. The master gene of osteoblast differentiation, *Runt-related 2* (*Runx2*), whose identification and functional characterization are presented in greater detail below, has a broader role during skeletogenesis, as evidenced by the analysis of the *Runx2*$^{-/-}$ mice. Besides their osteoblast phenotype these mutant mice lack hypertrophic chondrocytes in some but not all skeletal elements (36).

Skeletal dysplasia: genetic disease of the skeleton that affects cell differentiation in all skeletal elements

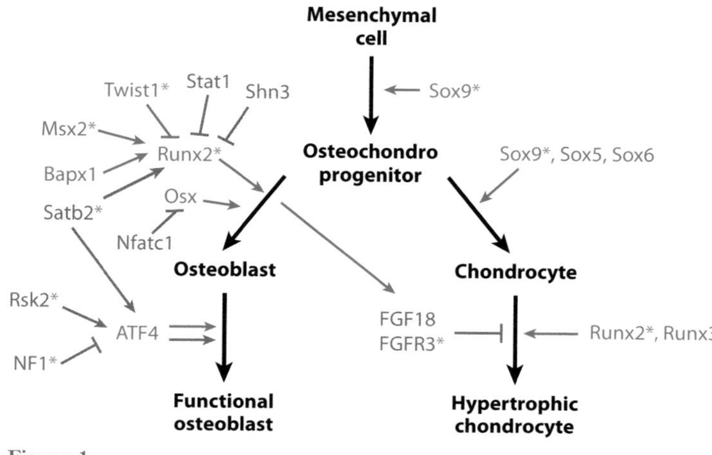

Figure 1

Schematic representation of the transcriptional control of cell differentiation along the chondrocyte and osteoblast lineages. Lines with arrowheads indicate a positive action and lines with bars indicate an inhibition. Regulation at the transcriptional level is shown in blue; regulation at the posttranscriptional level is shown in green. Red asterisks indicate genes whose mutation has been identified as disease-causing in humans.

That skeletal elements containing hypertrophic chondrocytes (such as ulnae) develop later than elements without hypertrophic chondrocytes (like humeri) indicates that this phenotype is not due to a mere delay in cell differentiation. Further analyses showed that *Runx2* is transiently expressed in prehypertrophic chondrocytes, the subpopulation of proliferating chondrocytes that gives rise to the hypertrophic chondrocytes (64). Moreover, constitutive expression of *Runx2* in these cells leads to ectopic chondrocyte hypertrophy in transgenic mice and allows chondrocyte hypertrophy to occur in *Runx2*$^{-/-}$ mice (64). What are the transcription factors that affect chondrocyte hypertrophy in skeletal elements where *Runx2* deletion does not affect chondrocyte hypertrophy? Genetic studies have shown that another member of the Runx2 family of transcription factors, Runx3, is important for chondrocyte hypertrophy. Indeed, mice lacking both Runx2 and Runx3 do not have recognizable hypertrophic chondrocytes or *$\alpha_1(X)$ collagen*–expressing cells anywhere in the skeleton (79). Therefore, two members of the Runx family of transcription factors, Runx2 and Runx3, are positive regulators of chondrocyte hypertrophy (**Figure 1**).

The study of another transcription factor, Twist-1, added some complexity to this picture. The function of Twist-1during skeletogenesis was also revealed via the molecular elucidation of a human skeletal dysplasia, Saethre-Chotzen syndrome (16, 26) (**Table 1**). Twist-1 is a nuclear protein containing at least two known functional structures: a bHLH domain and the Twist-box, a domain located at the C terminus that mediates the physical interaction of Twist-1 with Runx2 and inhibits the ability of this transcription factor to bind to DNA (4). Both Twist-1 and Runx2 are expressed in undifferentiated cells that form the perichondrium, a multilayer sheath of cells that surrounds the growth plate cartilage and inhibits chondrocyte hypertrophy. Analyses of loss- and gain-of-function mouse models showed that Twist-1 favors chondrocyte hypertrophy through its perichondrial expression and that this function requires Runx2 (4, 24). Further studies showed that Runx2 regulates positively the expression of *fibroblast growth factor 18* (*Fgf18*) in the perichondrium (24). FGF18, a secreted molecule, then activates FGFR3 signaling in chondrocytes, thereby inhibiting chondrocyte hypertrophy (46, 52) (**Figure 1**). Human geneticists identified the function of FGFR3 in chondrocytes in the 1990s because activating mutations in this gene cause achondroplasia and thanatophoric dysplasia (49, 55, 62, 66) (**Table 1**). Therefore, Runx2 has two opposite functions during chondrocyte hypertrophy. Initially, through its transient expression in prehypertrophic chondrocytes, Runx2 induces chondrocyte hypertrophy and sets up the scene for the next events of skeletogenesis, i.e., vascular invasion and osteoblast differentiation. Subsequently, through its constitutive expression in the cells of the perichondrium, Runx2 inhibits chondrocyte proliferation and hypertrophy, possibly to avoid premature bone formation.

In summary, it is remarkable that the two main transcriptional architects of chondrogenesis, Sox9 and Runx2, were identified through the joint efforts of human geneticists, molecular biologists, and mouse geneticists. As presented below, the same is true for the transcriptional control of osteoblast differentiation.

TRANSCRIPTIONAL CONTROL OF OSTEOBLAST DIFFERENTIATION

Multiple nuclear proteins contribute to the regulation of osteoblast differentiation and function. Some of them act throughout the skeleton, others act only in a subset of skeletal elements, and members of a third category modulate the activity of classical transcription factors. Remarkably, many molecular determinants of osteoblast differentiation and function were identified through human genetic studies as well as through mouse genetics and molecular studies.

Runx2, the Master Gene of Bone Formation

Runx2, a member of the Runt domain family of transcription factors, was shown to be the earliest and most powerful molecular determinant of osteoblast differentiation simultaneously by molecular biologists, mouse geneticists, and human geneticists. A classical cell-specific promoter-based search for osteoblast-specific transcription factors identified Runx2 as the factor binding to an osteoblast-specific *cis*-acting element in the promoter of the genes coding for osteocalcin, an osteoblast-specific hormone that regulates energy metabolism (11, 12, 42). Runx2 has all the molecular hallmarks of an osteoblast differentiation factor. In particular, Runx2 is expressed in cells prefiguring the skeleton as early as E10.5 (9). At that stage, these cells still have the capacity to differentiate into osteoblasts or chondrocytes and therefore are termed osteochondro progenitors (**Figure 1**). Subsequently, while its expression in differentiating chondrocytes decreases and eventually vanishes (at E16.5) Runx2 remains expressed at high levels in cells of the osteoblast lineage and of the perichondrium (4, 12, 24). Runx2 regulates many but not all genes that determine the osteoblast phenotype. Remarkably, forced expression of Runx2 in nonosteoblast cells is sufficient to induce the expression of many osteoblast-specific genes such as *Osteocalcin* (12). Consistent with its pattern of expression and its function in vitro, inactivation of both *Runx2* alleles in mice results in a mutant mouse deprived of osteoblasts throughout the skeleton (40, 53). Moreover, haploinsufficiency at the *Runx2* locus results in mice with hypoplastic clavicles and delayed closure of the fontanelles (47, 53). These latter abnormalities are similar to what is seen in a human skeletal dysplasia called cleidocranial dysplasia (CCD) (**Table 1**). Indeed, *Runx2* maps in the middle of the CCD locus and molecular studies demonstrated that most forms of CCD are due to inactivating mutations in *Runx2* (41, 47, 80). Therefore, this overwhelming molecular and genetic evidence accounts for the widely accepted view that Runx2 is the master gene of osteoblast differentiation. However, as presented above, the functions of Runx2 extend to other aspects of skeletogenesis.

Regulators of Runx2: A Tale of Human and Mouse Genetics

Given the critical functions that Runx2 exerts during skeletogenesis, it is not surprising that its activity, if not its expression, is tightly regulated. Many factors have been shown to affect the ability of Runx2 to bind to DNA and/or to regulate its transactivation function; the order in which they are presented here does not reflect a ranking of their biological importance but rather serves to illustrate how human genetics has been an engine in deciphering the transcriptional control of cell differentiation during skeletogenesis.

Molecularly Runx2 was identified on the basis of its ability to regulate the expression of *Osteocalcin*, an osteoblast-derived hormone expressed only in fully differentiated osteoblasts (12, 42). However, during mouse development *Osteocalcin* expression does not appear before E15.5, i.e., four to five days after Runx2 expression can be detected. One possible explanation for this paradox could be that during these four to five days Runx2 function is transiently inhibited by another nuclear protein. One way to identify such a protein is to use human genetics observations to single out candidate genes. Haploinsuffiency at the *Runx2* locus leads to delayed ossification of the skull bones (47, 53). Conceivably, inactivation of a gene whose function is to inhibit Runx2 function should thus lead to an increase in bone formation in the skull, a condition called craniosynostosis (72). Among the genes whose inactivation causes such a phenotype only one encodes for a nuclear protein, Twist-1. Indeed, haploinsufficiency at the *Twist-1* locus causes Saethre-Chotzen syndrome, a form of craniosynostosis (16, 26) (**Table 1**). Human genetics therefore identified Twist-1 as a candidate gene and opened the way for mouse genetics and molecular studies. These analyses verified that Twist-1 indeed delays osteogenesis via an inhibition of Runx2 activity (4). The researchers showed that (*a*) during early development *Twist-1* is transiently coexpressed with *Runx2* in cells destined to become osteoblasts and its expression disappears in these cells precisely when osteoblast differentiation is initiated; (*b*) the Twist box binds to the Runx2 runt domain (DNA binding domain) and inhibits Runx2 binding to DNA; (*c*) removing one allele of *Twist-1* from Runx2$^{+/-}$ mice is sufficient to correct the skull phenotype of each single heterozygous mutant mouse; and (*d*) an N-ethyl-N-nitrosourea (ENU) mutant in which the Twist box is disrupted displays premature osteoblast differentiation and acceleration of chondrocyte hypertrophy. That Twist-1 is an inhibitor of Runx2 would likely have been shown sooner or later but human genetics helped to quickly identify the best candidate that could inhibit Runx2 functions early during development (**Figure 1**).

Other regulators of Runx2 function have been identified. The role of the homeobox-containing protein muscle segment homeobox (msh) homolog (MSX2) during skeletal development was demonstrated when gain- and loss-of-function mutations were identified in human patients with either Boston-type craniosynostosis or enlarged parietal foramina, respectively (28, 73) (**Table 1**). Accordingly, *Msx2*-deficient mice display defective ossification of the skull and bones that develop by endochondral ossification (58). Because the expression of *Osteocalcin* and *Runx2* is strongly reduced in *Msx2*-deficient mice, Msx2 was proposed to act upstream of Runx2 in a transcriptional cascade that regulates osteoblast differentiation (**Figure 1**). A similar observation was made for mice lacking the homeodomain-containing transcription factor bagpipe homeobox gene 1 homolog (Bapx1) (68). These mice die at birth owing to severe dysplasia of the axial skeleton, whereas the appendicular skeleton is virtually unaffected. *Runx2* expression in *Bapx1*-deficient mice is strongly reduced in osteochondrogenic precursor cells of the prospective vertebral column, indicating that Bapx1 is required for *Runx2* expression specifically in these skeletal elements (**Figure 1**). Other nuclear proteins inhibit Runx2 function during osteoblast differentiation by interacting physically with the Runx2 DNA-binding domain. One of these proteins is signal transducer and activator of transcription 1 (Stat1), a transcription factor regulated by extracellular signaling molecules such as interferons. *Stat1*-deficient mice are viable but develop a high-bone-mass phenotype explained by enhanced bone formation (37). The increase of osteoblast differentiation and function in these mice is molecularly explained by the lack of a Stat1-mediated inhibition of the transcriptional activity of Runx2 (**Figure 1**). Interestingly, the physical interaction of both proteins is independent of Stat1 activation by phosphorylation. Stat1 has been proposed to act by inhibiting the translocation of Runx2 into the nucleus because overexpression of Stat1 in osteoblasts leads to cytosolic retention of Runx2 whereas nuclear translocation of Runx2 is much more prominent in *Stat1*-deficient osteoblasts (37). Schnurri 3 (Shn3) is another protein that interacts with Runx2 and acts by decreasing the availability of Runx2 in the nucleus (**Figure 1**). Shn3 is a zinc finger adapter protein originally thought to be involved in the VDJ recombination of immunoglobulin genes (74). A *Shn3*-deficient mouse model unexpectedly revealed a major function of this protein in bone formation. These mice display a severe adult-onset osteosclerotic phenotype owing to a cell-autonomous increase of bone matrix deposition (31). Interestingly, whereas several Runx2 target genes are expressed at higher rates in *Shn3*-deficient osteoblasts, *Runx2* expression is not affected by the absence of Shn3. However, Runx2 protein levels increase strikingly in *Shn3*-deficient osteoblasts. This latter finding is molecularly explained by the function of Shn3 as an adapter molecule linking Runx2 to the E3 ubiquitin ligase WW domain–containing protein 1 (WWP1) (31). The Shn3-mediated recruitment of WWP1 in turn leads to an enhanced proteasomal degradation of Runx2. This mechanism is best underscored by the finding that RNAi-mediated downregulation of WWP1 in osteoblasts leads to increased Runx2 protein levels and enhanced extracellular matrix mineralization, thereby virtually mimicking the defects observed in the absence of Shn3 (31). Altogether, these data identify Shn3 as a key regulator of Runx2 actions in vivo (**Figure 1**). Moreover, given the postnatal onset of the bone phenotype of the *Shn3*-deficient mice, compounds blocking the interaction of Runx2, Shn3, and WWP1 may serve as specific therapeutic agents for the treatment of bone loss diseases such as osteoporosis.

In addition to negative regulators of *Runx2* function, interacting factors that enhance Runx2 activity also exist. One of them is the nuclear matrix protein special AT-rich sequence-binding protein 2 (SATB2). The importance of this protein in skeletogenesis was first discovered in human patients with cleft palate who carry a heterozygous chromosomal translocation that inactivates the *SATB2* gene (17) (**Table 1**). The generation of a *Satb2*-deficient mouse model confirmed the importance of this gene in craniofacial development, skeletal patterning, and osteoblast differentiation (8). The latter function was in part attributed to an increased expression of *homeo box A2* (*Hoxa2*), a negative regulator of prechondrogenesis and bone formation (32), whose expression is repressed by the binding of Satb2 to an enhancer element present in the *Hoxa2* promoter (8). In addition to this type of action, there is also a Hoxa2-independent influence of Satb2 on the transcription of *Bone Sialoprotein* and *Osteocalcin*. Whereas in the case of *Bone Sialoprotein* Satb2 directly binds to an osteoblast-specific promoter element, the activation of *Osteocalcin* expression by Satb2 requires a physical interaction with Runx2 (8) (**Figure 1**). This requirement was demonstrated by cotransfection assays using an osteoblast-specific *Osteocalcin* promoter fragment and by coimmunoprecipitation experiments. Moreover, the synergistic action of Satb2 and Runx2 in osteoblasts was genetically confirmed by the generation of compound heterozygous mice lacking one allele of each gene (8). These results identified Satb2 as an important regulator of osteoblast differentiation in both mice and humans. Moreover, the finding that Satb2 also interacts with activating transcription factor-4 (ATF4), another transcription factor involved in the regulation of osteoblast differentiation and function that is discussed below (8), illustrates that the transcriptional network regulating bone formation is much more complex than previously anticipated.

Osterix, a Runx2-Dependent Osteoblast-Specific Transcription Factor Required for Bone Formation

Besides Runx2, there is at least one more transcription factor, termed Osterix (Osx), whose activity is absolutely required for osteoblast differentiation in mice (**Figure 1**). Osx is a zinc finger–containing transcription factor that is specifically expressed in osteoblasts of all skeletal elements (48). Inactivation of Osx in mice results in perinatal lethality owing to a complete absence of bone formation (48). Unlike *Runx2*-deficient mice whose skeleton is entirely nonmineralized, the *Osx*-deficient mice lack a mineralized matrix only in bones formed by intramembranous ossification. The *Osx*-deficient bones formed by endochondral ossification contain some mineralized matrix, although it resembles calcified cartilage, not mineralized bone matrix (48). This finding shows that Osx, unlike Runx2, is not required for chondrocyte hypertrophy, thereby demonstrating that Osx specifically induces osteoblast differentiation and bone formation in vivo (**Figure 1**). Comparative expression analyses by in situ hybridization further revealed that *Osx* is not expressed in *Runx2*-deficient embryos, whereas *Runx2* is normally expressed in *Osx*-deficient embryos (48). These results demonstrated that Osx acts downstream of Runx2 in the transcriptional cascade of osteoblast differentiation, and that Osx expression could be directly regulated by the binding of Runx2 to a responsive element in the promoter of the *Osx* gene (51). Unlike for *Runx2*, no mutations of the human *Osx* gene have been identified that would be associated with decreased bone formation. Moreover, in contrast to the steadily increasing knowledge about the function of Runx2 and its regulation by other molecules, the molecular mechanisms underlying the action of Osx in osteoblasts are less well understood. Nevertheless, one recent publication provides evidence for a contribution of Osx to the negative effects of nuclear factor of activated T cells (NFAT) inhibitors on bone mass (39). NFAT inhibitors, such as FK506 or cyclosporin A, are commonly used as immunosuppressants, for example after organ transplantation (45). However, this treatment is often accompanied by the development of osteopenia in the receptive patients (54). Likewise, treatment of mice with FK506 leads to decreased bone mass owing to impaired bone formation, and the same phenotype is observed in mice lacking the transcription factor Nfatc1 (39). The deduced role of Nfatc1 as a physiological activator of osteoblast differentiation and function can be molecularly explained by an interaction with Osx. In fact, both proteins synergistically stimulate the activity of an osteoblast-specific *α1(I)-Collagen* promoter fragment via the formation of an Nfatc1/Osx DNA-binding complex (39) (**Figure 1**). The complexity of the transcriptional control of osteoblast differentiation will likely increase further when more Osx-interacting molecules are identified.

AP1 Regulation of Osteoblast Differentiation and Function

Activator protein 1 (AP1) proteins are heterodimeric transcription factors composed of members of the Jun and Fos family of basic leucine zipper proteins (33). These proteins include the Jun proteins c-Jun, JunB, and JunD, as well as the Fos proteins c-Fos, Fra1, Fra2, and Fosb. Although AP1 transcription factors fulfill various functions in different cell types, it is striking that some family members play specific roles in bone remodeling, as demonstrated by several loss- or gain-of-function studies in mice (30, 70). For instance, the deletion of *c-Fos* from the mouse genome results in severe osteopetrosis owing to an arrest of osteoclast differentiation, whereas overexpression of *c-Fos* in transgenic mice results in osteosarcoma development (20, 21). Moreover, transgenic mice overexpressing either *fos-related antigen 1* (*Fra1*) or *ΔfosB*, a splice variant of *FosB*, display a severe osteosclerotic phenotype caused by increased osteoblast differentiation and function (29, 57). Likewise, mice lacking Fra1 in extraplacental tissues display an osteopenia associated with reduced bone formation, indicating a physiological role of Fra1 in osteoblasts (13). When the same approach was used to inactivate JunB in extraplacental tissues, thereby circumventing the embryonic lethality caused by a complete genomic deletion of *JunB*, the resulting mice developed a state of low bone turnover due not only to cell-autonomous defects of osteoblasts but also to abnormal osteoclast differentiation (35). Taken together, these data provide evidence for a role of AP1 transcription factors in the regulation of bone formation, although their connection to the other transcriptional regulators described above or below still needs further investigation. For instance, it is known from other cell types that Jun proteins can also interact with ATF family members, thus raising the possibility that heterodimerization with ATF4 may be one mechanism by which these proteins can regulate osteoblast-specific gene expression (6). Interestingly, David and coworkers (7) recently demonstrated that the osteosarcoma development of *c-Fos* transgenic mice is dramatically decreased in a *p90 ribosomal S6 protein kinase 2* (*Rsk2*)-deficient genetic background. This observation is molecularly explained by the lack of c-Fos phosphorylation by Rsk2, leading to increased proteosomal degradation. Thus, Rsk2 is apparently not only involved in the physiological regulation of bone formation via phosphorylation of ATF4, but also may have an influence on the development of osteosarcomas via phosphorylation of c-Fos.

Another mechanism by which AP1 family members might be involved in the regulation of bone formation was identified by the analysis of mouse models with impaired circadian regulation; this mechanism links AP1 family members to the regulation of bone mass by leptin (10, 19, 34). Mutant mice that lack components of the molecular clock, namely the *period* (*Per*) or *cryptochrome* (*Cry*) genes, display a high-bone-mass phenotype caused by increased bone formation and respond to intracerebroventricular infusion of leptin with an increase instead of a decrease in bone mass (19). These results suggest that the components of the molecular clock are involved in the regulation of bone formation via the sympathetic nervous system (SNS), a known mediator of leptin's actions on bone mass (34, 65). Interestingly, virtually all genes encoding members of the AP1 transcription factor family are expressed at higher levels in osteoblasts derived from mice lacking either the *Per* genes or the β2-adrenergic receptor Adrβ2, the receptor targeted by the SNS in osteoblasts, compared with wild type (19, 65). This increase is especially pronounced in the case of the *c-Fos* gene, whose expression can also be induced by the addition of isoproterenol, a sympathomimetic, in wild-type osteoblasts (19). In turn, c-Fos leads to a direct activation of *c-Myc* transcription, thereby indirectly increasing the intracellular levels of cyclin D1 and promoting osteoblast proliferation (19). These data demonstrate that the expression of AP1 components is activated via sympathetic signaling, and the activity of clock gene products counteracts this induction.

ATF4, a Regulator of Osteoblast Function Implicated in Two Human Diseases

This review ends with a discussion of the transcription factor ATF4, which seems to play the most important role in assuring that osteoblasts fulfill their function. In this case the biological importance of this factor was revealed by human genetics as much as by any other type of approach. Studies that led to its identification started with the following question: How can inactivation of a kinase decrease bone mass? *RSK2*, which encodes a kinase, is the gene mutated in Coffin-Lowry syndrome, an X-linked mental retardation condition associated with skeletal abnormalities (69) (**Table 1**). Likewise, *Rsk2*-deficient mice display decreased bone mass owing to impaired bone formation (78). In vitro kinase assays demonstrated that ATF4 is more strongly phosphorylated by Rsk2 than any other proposed substrate, and that this phosphorylation is undetectable in osteoblasts derived from *Rsk2*-deficient mice. The subsequent analysis of an *ATF4*-deficient mouse model revealed that this transcription factor plays several crucial roles in osteoblast differentiation and function (**Figure 1**); *ATF4*-deficient mice display a delayed skeletal development and thereafter develop a severe low-bone-mass phenotype caused by decreased bone formation (78).

Molecularly, ATF4 was identified as the factor binding to an osteoblast-specific element in the *Osteocalcin* promoter, thereby directly activating the transcription of the *Osteocalcin* gene (11, 59, 78). Moreover, ATF4 is required for proper synthesis of type I collagen (which seems to be the main mechanism whereby ATF4 regulates bone formation), although this function is not mediated by a transcriptional regulation of *type I collagen* expression (78). In fact, because type I collagen synthesis is specifically reduced in primary osteoblast cultures lacking ATF4, but this defect can be rescued by adding nonessential amino acids to the culture, ATF4 appears to be required for efficient amino acid import into osteoblasts, as described for other cell types (23). Reduced type I collagen synthesis was subsequently observed in mice lacking Rsk2, providing evidence that the diminished ATF4 phosphorylation in the absence of Rsk2 may contribute to the skeletal defects associated with Coffin-Lowry syndrome (78).

In addition to the role of ATF4 in bone formation, ATF4 regulates osteoclast differentiation and ultimately bone resorption through its expression in osteoblasts (14). This function is molecularly explained by the binding of ATF4 to the promoter of the *receptor activator of nuclear factor-KB ligand* (*Rankl*) gene, which encodes a factor secreted by osteoblasts that promotes osteoclast differentiation (67). Accordingly, *ATF4*-deficient mice have decreased osteoclast numbers owing to reduced *Rankl* expression. Most importantly, this function of ATF4 is involved in the control of bone resorption by the SNS (14). In fact, treatment of normal osteoblasts with isoprotenerol, a surrogate of sympathetic signaling, enhances osteoclastogenesis of cocultured bone marrow macrophages through an induction of osteoblastic *Rankl* expression (14). As expected, this effect is blunted when the osteoblasts are derived from mice lacking the β2-adrenergic receptor Adrβ2. However, the effect of isoprotenerol is also blunted by an inhibitor of protein kinase A, or by using osteoblasts derived from *ATF4*-deficient mice (14). Taken together, these results demonstrate that ATF4 is an important mediator of extracellular signals, such as β-adrenergic stimulation, in osteoblasts.

Thus, it is not surprising that the function of ATF4 is regulated mostly posttranslationally. For example, as mentioned above, ATF4 also interacts with other proteins, such as the nuclear matrix protein Satb2 (8). As described above, the proximal *Osteocalcin* promoter contains two osteoblast-specific elements that serve as binding sites for ATF4 and Runx2, respectively (11, 12, 59, 78). Owing to the proximity of both elements a physical interaction of the two proteins occurs that is stabilized by Satb2, which acts as a scaffold to enhance the synergistic activity of Runx2 and ATF4 in the regulation of *Osteocalcin* expression (8, 75).

Other aspects of ATF4 biology are also regulated posttranslationally. In fact, even the osteoblast-specificity of ATF4 function is not determined by osteoblast-specific *ATF4* expression, but by a selective accumulation of the ATF4 protein in osteoblasts, which is explained by the lack of proteosomal degradation (77). This concept is best demonstrated by the finding that the treatment of nonosteoblastic cell types with the proteasome inhibitor MG115 leads to an accumulation of the ATF4 protein, thereby resulting in ectopic *Osteocalcin* expression (77). These data provide the first evidence that the cell-specific function of a transcriptional activator can be achieved by a posttranslational mechanism. This finding is of general importance for our understanding of the transcriptional networks that control cellular differentiation and function because it may relate to other factors that regulate these processes.

As mentioned above and illustrated in the case of Coffin-Lowry syndrome, ATF4 biology illustrates how the molecular understanding of a disease-causing gene can translate to therapeutic interventions. This concept was further established by the finding of a link between an increased Rsk2-dependent phosphorylation of ATF4 and the development of the skeletal abnormalities in human patients suffering from neurofibromatosis (56, 63) (**Table 1**). This disease, primarily known for tumor development in the nervous system, is caused by inactivating mutations of the *neurofibromatosis 1* (*NF1*) gene, which encodes a Ras-GTPase activating protein (38). The generation of a mouse model lacking *Nf1* specifically in osteoblasts ($Nf1_{ob-/-}$ mice) led to the demonstration that this gene plays a major physiological role in bone remodeling (**Figure 1**). The $Nf1_{ob-/-}$ mice display a high bone mass phenotype caused by increased bone turnover accompanied by an enrichment of nonmineralized osteoid (15). Further analysis of this phenotype revealed that the lack of NF1 induces an increased production of type I collagen, which is molecularly explained by a Rsk2-dependent activation of ATF4. Accordingly, transgenic mice overexpressing *ATF4* in osteoblasts display a phenotype similar to the $Nf1_{ob-/-}$ mice, and the increased type I collagen production and osteoid thickness in the $Nf1_{ob-/-}$ mice are significantly reduced by haploinsufficiency of ATF4 (15).

These molecular findings may also have therapeutic implications. Given the previously discussed function of ATF4 in amino acid import, it appeared reasonable to analyze whether the skeletal defects of the $Nf1_{ob-/-}$ mice could be affected by dietary manipulation. Indeed, the increased bone formation and osteoid thickness of $Nf1_{ob-/-}$ mice can be normalized by a low-protein diet, and the same is true for the phenotype of the transgenic mice that overexpress

ATF4 in osteoblasts (15). Conversely, the defects of osteoblast differentiation and bone formation observed in both the *ATF4*- and the *Rsk2*-deficient mice can be corrected by a high-protein diet (15). These data not only underscore the importance of ATF4 in osteoblast biology, but also demonstrate how the knowledge about its specific functions in osteoblasts can be useful for the treatment of skeletal diseases.

In summary, if we look at a cascade of cell differentiation in the skeleton we now have a fairly detailed picture of the identity and mechanisms of action of many transcription factors involved in these processes. What is remarkable, and was not necessarily anticipated at the beginning of this journey, is how many of these transcription factors are either mutated or their activities affected in human skeletal dysplasia.

SUMMARY POINTS

1. Chondrogenic differentiation of condensed mesenchymal cells, orchestrated by sex determining region Y (SRY)-related high mobility group-box gene (Sox9) is the initial event in skeletogenesis.
2. Sox9, Sox5, and Sox6 determine further differentiation in the chondrocyte lineage.
3. Runt-related 2 (Runx2) is the master gene of osteoblast differentiation.
4. Runx2 also controls chondrocyte maturation via two distinct mechanisms.
5. ATF4 is the major determinant of osteoblast function.

FUTURE ISSUES

1. How is Sox9 regulated at the transcriptional and posttranscriptional level?
2. How many more proteins interact with Runx2 in osteoblasts and chondrocytes?
3. How is Osterix (Osx) regulated at the transcriptional and posttranscriptional level?
4. How are the functions of activator protein 1 (AP1) factors connected with Runx2, Osx, and activating transcription factor-4 (ATF4) activities?
5. Do we know all the functions of ATF4?

DISCLOSURE STATEMENT

The author is not aware of any biases that might be perceived as affecting the objectivity of this review.

LITERATURE CITED

1. Bell DM, Leung KK, Wheatley SC, Ng LJ, Zhou S, et al. 1997. SOX9 directly regulates the type-II collagen gene. *Nat. Genet.* 16:174–78
2. Bi W, Deng JM, Zhang Z, Behringer RR, de Crombrugghe B. 1999. Sox9 is required for cartilage formation. *Nat. Genet.* 22:85–89
3. Bi W, Huang W, Whitworth DJ, Deng JM, Zhang Z, et al. 2001. Haploinsufficiency of Sox9 results in defective cartilage primordia and premature skeletal mineralization. *Proc. Natl. Acad. Sci. USA* 98:6698–703
4. Bialek P, Kern B, Yang X, Schrock M, Sosic D, et al. 2004. A twist code determines the onset of osteoblast differentiation. *Dev. Cell* 6:423–35

5. Bridgewater LC, Lefebvre V, de Crombrugghe B. 1998. Chondrocyte-specific enhancer elements in the *Col11a2* gene resemble the *Col2a1* tissue-specific enhancer. *J. Biol. Chem.* 273:14998–5006
6. Chinenov Y, Kerppola TK. 2001. Close encounters of many kinds: Fos-Jun interactions that mediate transcription regulatory specificity. *Oncogene* 20:2438–52
7. David JP, Mehic D, Bakiri L, Schilling AF, Mandic V, et al. 2005. Essential role of RSK2 in c-Fos-dependent osteosarcoma development. *J. Clin. Invest.* 115:664–72
8. Dobreva G, Chahrour M, Dautzenberg M, Chirivella L, Kanzler B, et al. 2006. SATB2 is a multifunctional determinant of craniofacial patterning and osteoblast differentiation. *Cell* 125:971–86
9. Ducy P. 2000. Cbfa1: a molecular switch in osteoblast biology. *Dev. Dyn.* 19:461–71
10. Ducy P, Amling M, Takeda S, Priemel M, Schilling AF, et al. 2000. Leptin inhibits bone formation through a hypothalamic relay: a central control of bone mass. *Cell* 100:197–207
11. Ducy P, Karsenty G. 1995. Two distinct osteoblast-specific cis-acting elements control expression of a mouse osteocalcin gene. *Mol. Cell Biol.* 15:1858–69
12. Ducy P, Zhang R, Geoffroy V, Ridall AL, Karsenty G. 1997. Osf2/Cbfa1: a transcriptional activator of osteoblast differentiation. *Cell* 89:747–54
13. Eferl R, Hoebertz A, Schilling AF, Rath M, Karreth F, et al. 2004. The Fos-related antigen Fra-1 is an activator of bone matrix formation. *EMBO J.* 23:2789–99
14. Elefteriou F, Ahn JD, Takeda S, Starbuck M, Yang X, et al. 2005. Leptin regulation of bone resorption by the sympathetic nervous system and CART. *Nature* 434:514–20
15. Elefteriou F, Benson MD, Sowa H, Starbuck M, Liu X, et al. 2006. ATF4 mediation of NF1 functions in osteoblast reveals a nutritional basis for congenital skeletal dysplasiae. *Cell. Metab.* 4:441–51
16. El Ghouzzi V, Le Merrer M, Perrin-Schmitt F, Lajeunie E, Benit P, et al. 1997. Mutations of the *TWIST* gene in the Saethre-Chotzen syndrome. *Nat. Genet.* 15:42–46
17. FitzPatrick DR, Carr IM, McLaren L, Leek JP, Wightman P, et al. 2003. Identification of *SATB2* as the cleft palate gene on 2q32-q33. *Hum. Mol. Genet.* 12:2491–501
18. Foster JW, Dominguez-Steglich MA, Guioli S, Kowk G, Weller PA, et al. 1994. Campomelic dysplasia and autosomal sex reversal caused by mutations in an SRY-related gene. *Nature* 372:525–30
19. Fu L, Patel MS, Bradley A, Wagner EF, Karsenty G. 2005. The molecular clock mediates leptin-regulated bone formation. *Cell* 122:803–15
20. Grigoriadis AE, Schellander K, Wang ZQ, Wagner EF. 1993. Osteoblasts are target cells for transformation in c-*fos* transgenic mice. *J. Cell Biol.* 122:685–701
21. Grigoriadis AE, Wang ZQ, Cecchini MG, Hofstetter W, Felix R, et al. 1994. c-Fos: a key regulator of osteoclast-macrophage lineage determination and bone remodeling. *Science* 266:443–48
22. Hall BK, Miyake T. 2000. All for one and one for all: condensations and the initiation of skeletal development. *Bioesssays* 22:138–47
23. Harding HP, Zhang Y, Zeng H, Novoa I, Lu PD, et al. 2003. An integrated stress response regulates amino acid metabolism and resistance to oxidative stress. *Mol. Cell* 11:619–33
24. Hinoi E, Bialek P, Chen YT, Rached MT, Groner Y, et al. 2006. Runx2 inhibits chondrocyte proliferation and hypertrophy through its expression in the perichondrium. *Genes. Dev.* 20:2937–42
25. Horton WA. 1993. Cartilage Morphology. In *Extracellular Matrix and Heritable Disorders of Connective Tissue*, ed. PM Royce, B Steinman, pp. 73–84. New York: Liss
26. Howard TD, Paznekas WA, Green ED, Chiang LC, Ma L, et al. 1997. Mutations in *TWIST*, a basic helix-loop-helix transcription factor, in Sathre-Chotzen syndrome. *Nat. Genet.* 15:36–41
27. Huang W, Chung UI, Kronenberg HM, de Crombrugghe B. 2001. The chondrogenic transcription factor Sox9 is a target of signaling by the parathyroid hormone-related peptide in the growth plate of endochondral bones. *Proc. Natl. Acad. Sci. USA* 98:160–65
28. Jabs EW, Muller U, Li X, Ma L, Luo W, et al. 1993. A mutation in the homeodomain of the human *MSX2* gene in a family affected with autosomal dominant craniosynostosis. *Cell* 75:443–50
29. Jochum W, David JP, Elliott C, Wutz A, Plenk H Jr, et al. 2000. Increased bone formation and osteosclerosis in mice overexpressing the transcription factor Fra-1. *Nat. Med.* 6:980–84
30. Jochum W, Passegue E, Wagner EF. 2001. AP-1 in mouse development and tumorigenesis. *Oncogene* 20:2401–12

31. Jones DC, Wein MN, Oukka M, Hofstaetter JG, Glimcher MJ, Glimcher LH. 2006. Regulation of adult bone mass by the zinc finger adapter protein Schnurri-3. *Science* 312:1223–27
32. Kanzler B, Kuschert SJ, Liu YH, Mallo M. 1998. Hoxa-2 restricts the chondrogenic domain and inhibits bone formation during development of the branchial area. *Development* 125:2587–97
33. Karin M, Liu Z, Zandi E. 1997. AP-1 function and regulation. *Curr. Opin. Cell Biol.* 9:240–46
34. Karsenty G. 2006. Convergence between bone and energy homeostases: Leptin regulation of bone mass. *Cell Metab.* 4:341–48
35. Kenner L, Hoebertz A, Beil T, Keon N, Karreth F, et al. 2004. Mice lacking JunB are osteopenic due to cell-autonomous osteoblast and osteoclast defects. *J. Cell Biol.* 164:613–23
36. Kim IS, Otto F, Zabel B, Mundlos S. 1999. Regulation of chondrocyte differentiation by Cbfa1. *Mech. Dev.* 80:159–70
37. Kim S, Koga T, Isobe M, Kern BE, Yokochi T, et al. 2003. Stat1 functions as a cytoplasmic attenuator of Runx2 in the transcriptional program of osteoblast differentiation. *Genes. Dev.* 17:1979–91
38. Klose A, Ahmadian MR, Schuelke M, Scheffzek K, Hoffmeyer S, et al. 1998. Selective disactivation of neurofibromin GAP activity in neurofibromatosis type 1. *Hum. Mol. Genet.* 7:1261–68
39. Koga T, Matsui Y, Asagiri M, Kodama T, de Crombrugghe B, et al. 2005. NFAT and Osterix cooperatively regulate bone formation. *Nat. Med.* 11:880–85
40. Komori T, Yagi H, Nomura S, Yamaguchi A, Sasaki K, et al. 1997. Targeted disruption of Cbfa1 results in a complete lack of bone formation owing to maturational arrest of osteoblasts. *Cell* 89:755–64
41. Lee B, Thirunavukkarasu K, Zhou L, Pastore L, Baldini A, et al. 1997. Missense mutations abolishing DNA binding of the osteoblast-specific transcription factor OSF2/CBFA1 in cleidocranial dysplasia. *Nat. Genet.* 16:307–10
42. Lee NK, Sowa H, Hinoi E, Ferron M, Ahn JD, et al. 2007. Endocrine regulation of energy metabolism by the skeleton. *Cell* 130:456–69
43. Lefebvre V, Huang W, Harley VR, Goodfellow PN, de Crombrugghe B. 1997. SOX9 is a potent activator of the chondrocyte-specific enhancer of the Proα1(II) collagen gene. *Mol. Cell Biol.* 17:2336–46
44. Lefebvre V, Ping L, de Crombrugghe B. 1998. A new long form of Sox5 (L-Sox5), Sox6 and Sox9 are coexpressed in chondrogenesis and cooperatively activate the type II collagen gene. *EMBO J.* 17:5718–33
45. Liu J, Farmer JD Jr, Lane WS, Friedman J, Weissman I, Schreiber SL. 1991. Calcineurin is a common target of cyclophilin-cyclosporin A and FKBP-FK506 complexes. *Cell* 66:807–15
46. Liu Z, Xu J, Colvin JS, Ornitz DM. 2002. Coordination of chondrogenesis and osteogenesis by fibroblast growth factor 18. *Genes Dev.* 16:859–69
47. Mundlos S, Otto F, Mundlos C, Mulliken JB, Aylsworth AS, et al. 1997. Mutations involving the transcription factor CBFA1 cause cleidocranial dysplasia. *Cell* 89:773–79
48. Nakashima K, Zhou X, Kunkel G, Zhang Z, Deng JM, et al. 2002. The novel zinc finger-containing transcription factor osterix is required for osteoblast differentiation and bone formation. *Cell* 108:17–29
49. Naski MC, Wang Q, Xu J, Ornitz DM. 1996. Graded activation of fibroblast growth factor receptor 3 by mutations causing achondroplasia and thanatophoric dysplasia. *Nat. Genet.* 13:233–37
50. Ng LJ, Wheatley S, Muscat GE, Conway-Campbell J, Bowles J, et al. 1997. SOX9 binds DNA, activates transcription, and coexpresses with type II collagen during chondrogenesis in the mouse. *Dev. Biol.* 183:108–21
51. Nishio Y, Dong Y, Paris M, O'Keefe RJ, Schwarz EM, Drissi H. 2006. Runx2-mediated regulation of the zinc finger Osterix/Sp7 gene. *Gene* 372:62–70
52. Ohbayashi N, Shibayama M, Kurotaki Y, Imanishi M, Fujimori T, et al. 2002. Fgf18 is required for osteogenesis and chondrogenesis in mice. *Genes Dev.* 16:870–79
53. Otto F, Thornell AP, Crompton T, Denzel A, Gilmour KC, et al. 1997. *Cbfa1*, a candidate gene for cleidocranial dysplasia syndrome, is essential for osteoblast differentiation and bone development. *Cell* 89:765–71
54. Rodino MA, Shane E. 1998. Osteoporosis after organ transplantation. *Am. J. Med.* 104:459–69
55. Rousseau F, Bonaventure J, Legeai-Mallet L, Pelet A, Rozet JM, et al. 1994. Mutations in the gene encoding fibroblast growth factor receptor-3 in achondroplasia. *Nature* 371:252–54
56. Ruggieri M, Pavone V, De Luca D, Franzo A, Tine A, Pavone L. 1999. Congenital bone malformations in patients with neurofibromatosis type 1 (Nf1). *J. Pediatr. Orthop.* 19:301–5

57. Sabatakos G, Sims NA, Chen J, Aoki K, Kelz MB, et al. 2000. Overexpression of ΔFosB transcription factor(s) increases bone formation and inhibits adipogenesis. *Nat. Med.* 6:985–90
58. Satokata I, Ma L, Ohshima H, Bei M, Woo I, et al. 2000. Msx2 deficiency in mice causes pleiotropic defects in bone growth and ectodermal organ formation. *Nat. Genet.* 24:391–95
59. Schinke T, Karsenty G. 1999. Characterization of Osf1, an osteoblast-specific transcription factor binding to a critical cis-acting element in the mouse osteocalcin promoters. *J. Biol. Chem.* 274:30182–89
60. Schipani E, Ryan HE, Didrickson S, Kobayashi T, Knight M, Johnson RS. 2001. Hypoxia in cartilage: HIF-1α is essential for chondrocyte growth arrest and survival. *Genes Dev.* 15:2865–76
61. Sekiya I, Tsuji K, Koopman P, Watanabe H, Yamada Y, et al. 2000. SOX9 enhances aggrecan gene promoter/enhancer activity and is up-regulated by retinoic acid in a cartilage-derived cell line, TC6. *J. Biol. Chem.* 275:10738–44
62. Shiang R, Thompson LM, Zhu YZ, Church DM, Fielder TJ, et al. 1994. Mutations in the transmembrane domain of FGFR3 cause the most common genetic form of dwarfism, achondroplasia. *Cell* 78:335–42
63. Stevenson DA, Birch PH, Friedman JM, Viskochil DH, Balestrazzi P, et al. 1999. Descriptive analysis of tibial pseudarthrosis in patients with neurofibromatosis 1. *Am. J. Med. Genet.* 84:413–19
64. Takeda S, Bonnamy JP, Owen MJ, Ducy P, Karsenty G. 2001. Continuous expression of *Cbfa1* in nonhypertrophic chondrocytes uncovers its ability to induce hypertrophic chondrocyte differentiation and partially rescues Cbfa1-deficient mice. *Genes Dev.* 15:467–81
65. Takeda S, Elefteriou F, Levasseur R, Liu X, Zhao L, et al. 2002. Leptin regulates bone formation via the sympathetic nervous system. *Cell* 111:305–17
66. Tavormina PL, Shiang R, Thompson LM, Zhu YZ, Wilkin DJ, et al. 1995. Thanatophoric dysplasia (types I and II) caused by distinct mutations in fibroblast growth factor receptor 3. *Nat. Genet.* 9:321–28
67. Teitelbaum SL, Ross FP. 2003. Genetic regulation of osteoclast development and function. *Nat. Rev. Genet.* 4:638–49
68. Tribioli C, Lufkin T. 1999. The murine *Bapx1* homeobox gene plays a critical role in embryonic development of the axial skeleton and spleen. *Development* 126:5699–711
69. Trivier E, De Cesare D, Jacquot S, Pannetier S, Zackai E, et al. 1996. Mutations in the kinase Rsk-2 associated with Coffin-Lowry syndrome. *Nature* 384:567–70
70. Wagner EF, Eferl R. 2005. Fos/AP-1 proteins in bone and the immune system. *Immunol. Rev.* 208:126–40
71. Wagner T, Wirth J, Meyer J, Zabel B, Held M, et al. 1994. Autosomal sex reversal and campomelic dysplasia are caused by mutations in and around the *SRY*-related gene *SOX9*. *Cell* 79:1111–20
72. Wilkie AOM. 1997. Craniosynostosis: genes and mechanisms. *Hum. Mol. Genet.* 6:1647–56
73. Wilkie AOM, Tang Z, Elanko N, Walsh S, Twigg SRF, et al. 2000. Functional haploinsufficiency of the human homeobox gene *MSX2* causes defects in skull ossification. *Nat. Genet.* 24:387–90
74. Wu W, Glinka A, Delius H, Niehrs C. 2000. Mutual antagonism between *dickkopf1* and *dickkopf2* regulates Wnt/β-catenin signalling. *Curr. Biol.* 10:1611–14
75. Xiao G, Jiang D, Ge C, Zhao Z, Lai Y, et al. 2005. Cooperative interactions between activating transcription factor 4 and Runx2/Cbfa1 stimulate osteoblast-specific osteocalcin gene expression. *J. Biol. Chem.* 280:30689–96
76. Xie WF, Zhang X, Sakano S, Lefebvre V, Sandell LJ. 1999. Trans-activation of the mouse cartilage-derived retinoic acid-sensitive protein gene by Sox9. *J. Bone Miner. Res.* 14:757–63
77. Yang X, Karsenty G. 2004. ATF4, the osteoblast accumulation of which is determined post-translationally, can induce osteoblast-specific gene expression in nonosteoblastic cells. *J. Biol. Chem.* 279:47109–14
78. Yang X, Matsuda K, Bialek P, Jacquot S, Masuoka HC, et al. 2004. ATF4 is a substrate of RSK2 and an essential regulator of osteoblast biology; implication for Coffin-Lowry Syndrome. *Cell* 117:387–98
79. Yoshida CA, Yamamoto H, Fujita T, Furuichi T, Ito K, et al. 2004. Runx2 and Runx3 are essential for chondrocyte maturation, and Runx2 regulates limb growth through induction of *Indian hedgehog*. *Genes Dev.* 18:952–63
80. Zhou G, Chen Y, Zhou L, Thirunavukkararsu K, Hecht J, et al. 1999. *CBFA1* mutation analysis and functional correlation with phenotypic variability in cleidocranial dysplasia. *Hum. Mol. Gen.* 8:2312–6

A Mechanistic View of Genomic Imprinting

Ky Sha

Department of Biology, Massachusetts Institute of Technology, Cambridge, Massachusetts 02139; email: ksha@mit.edu

Annu. Rev. Genomics Hum. Genet. 2008. 9:197–216

First published online as a Review in Advance on June 3, 2008

The *Annual Review of Genomics and Human Genetics* is online at genom.annualreviews.org

This article's doi:
10.1146/annurev.genom.122007.110031

1527-8204/09/0922-0197$20.00

Key Words

chromatin, epigenetics, *C. elegans*

Abstract

Genomic imprinting results in the expression of genes in a parent-of-origin-dependent manner. The mechanism and developmental consequences of genomic imprinting are most well characterized in mammals, plants, and certain insect species (e.g., sciarid flies and coccid insects). However, researchers have observed imprinting phenomena in species in which imprinting of endogenous genes is not known to exist or to be developmentally essential. In this review, I survey the known mechanisms of imprinting, focusing primarily on examples from mammals, where imprinting is relatively well characterized. Where appropriate, I draw attention to imprinting mechanisms in other organisms to compare and contrast how diverse organisms employ different strategies to perform the same process. I discuss how the various mechanisms come into play in the context of the imprint life cycle. Finally, I speculate why imprinting may be more widely prevalent than previously thought.

INTRODUCTION

In diploid organisms, every gene exists as two copies; one is contributed by the female parent (the maternal allele) and one by the male parent (the paternal allele). The vast majority of genes can be expressed from either the maternal or the paternal allele. However, for a small subset of genes (approximately 84 identified in mice as of the end of 2007; **http://www.mgu.har.mrc.ac.uk/research/imprinting/imprin-intro.html**), differences in expression levels are observed when the maternal and paternal alleles are measured separately. This biased expression of one parental allele over the other is called genomic imprinting. Genomic imprinting results in the nonequivalence of two segregating alleles of a gene and thus violates Mendelian genetics. In mammals, failure to express the proper allele according to the correct temporal and/or spatial requirement for gene function can result in mental and/or physical disorders (if the effect is mild) or abortion of the fetus in severe cases.

Geneticists recognize imprinted genes by the nonequivalence of reciprocal matings. How does an organism recognize which allele of an imprinted gene originated from which parent? The primary sequence plays only a partial role, although Luedi and coworkers (111) recently reported the successful identification of two new imprinted genes as well as more than one hundred candidates on the basis of primary sequence. Thus far, studies of imprinting in multiple organisms point to a mechanism that involves chromatin, *cis*-acting elements, and genomic context. The chromatin states of paternally derived and maternally derived alleles in the zygote can be nonidentical. In cells of the resulting animal, the differences in chromatin states can be maintained, read, and have consequences for expression. In germ cells (egg and sperm), differences in chromatin between paternal and maternal alleles are generally erased and sex-specific marks are re-established. Thus, the imprinting process cycles susceptible DNA sequences between a male state and a female state.

Here I review the state of knowledge of imprinting mechanisms, primarily using examples from mammals and plants. I also draw on examples from rare cases in which imprinting effects have been observed in organisms for which imprinting of endogenous genes was not known to exist to provide additional mechanistic insight.

KNOWN IMPRINTING MECHANISMS

Genomic Context

No single characteristic distinguishes an imprinted gene from a nonimprinted gene. Rather, imprinted and nonimprinted genes are distinguished by a list of parameters that often vary among imprinted genes. Although not universal, many imprinted genes share certain common genomic features. Hurst and colleagues (71) reported that imprinted genes contain fewer and smaller introns than do nonimprinted genes. However, many exceptions to this observation were found. Neumann and coworkers (124) observed that imprinted genes tend to exhibit some degree of repetitive sequences. This observation has been corroborated by comparative genomics studies that indicate a high content of retrotransposable elements in imprinted regions; of the approximately 80 genes known to be imprinted in mammals, more than 23 contain tandemly repeated sequences (174). Mary Lyon (115), noting the high abundance of repetitive elements on the mammalian X chromosome, suggested that perhaps repetitive elements facilitate the initiation and spreading of a heterochromatic state during dosage compensation. Other investigators have since adapted Lyon's hypothesis to imprinted loci. Although imprinted genes generally contain repetitive elements, the actual role such repetitive elements play in the imprinting process remains unclear. For example, the *Impact* gene is imprinted in mouse, rat, and rabbit (129). Tandem repeats within the mouse and rat *Impact* gene are methylated in a parent-of-origin-specific manner. The rabbit gene lacks the tandem repeats, but is apparently still imprinted (129). Other known examples of imprinted genes that do not involve nearby repetitive elements include *Medea* in *Arabidopsis* (159) and the noncoding RNA *Kcnq1ot1* in mouse (118).

In mammals, approximately 80% of imprinted genes exist in clusters or in close proximity to each other (140, 169). Differential cytosine methylation of the imprint control regions (ICRs) is believed to contribute to coordinated expression of imprinted genes (1, 141, 164, 179). For example, the *Kcnq1* gene in humans contains an imprint control region called ICR2 (117). In addition to controlling the imprinting of *Kcnq1*, Icr2 also controls the imprinting of five other genes in the cluster, including *Cdkn1c*, *Acl2*, *Phlda2*, *Tssc4*, and *Slc22a1l* (42, 69).

Further evidence linking genomic context and imprinting of genes comes from studies in *Drosophila*. Imprinting of endogenous genes has not been reported for wild-type *Drosophila*. Nevertheless, certain chromosomal translocations can confer a parent-of-origin effect to previously nonimprinted endogenous genes (104, 105). Remarkably, all genes on the translocated segment can acquire an expression pattern dependent upon parent of origin. In these *Drosophila* examples, translocations that impart imprinting properties to previously nonimprinted genes often occur in or next to heterochromatic regions of the genome.

Cis-Acting Elements

Extensive studies suggest contributions to the imprinting mechanism from diverse regulatory signals within, or in proximity to, imprinted genes. Called ICRs or differentially methylated regions/domains (DMR/D), such *cis*-acting elements are often the sites of DNA methylation in a manner dependent upon the parent of origin (68). DMRs are relatively rich in CpG dinucleotides, but the CpG content is generally less than that of CpG islands (90). Also, repetitive elements found in imprinted clusters are usually located within DMRs (174). Depending on the methylation status, the *cis*-acting sequences may recruit different regulatory factors, such as methyl-CpG binding domain (MBD) proteins, which contain a domain that recognizes and binds methylated DNA.

Certain DMRs can also function as insulators. An insulator has two properties (95): 1) When placed between an enhancer and its target promoter, the insulator prevents activation of the promoter; and 2) when placed upstream and/or downstream of a transgene, the insulator can protect the transgene from position effects. One of the better-studied paradigms of a *cis*-acting element with both DMR and insulator roles is the DMR at the *H19/Igf2* imprinted region (**Figure 1**). *Igf2* and *H19* are reciprocally imprinted genes; *Igf2* is expressed from the paternal allele and *H19* is expressed from the maternal allele (138). The DMR of the maternal allele is not methylated. As a result, a key transcription factor (CTCF) is able to bind the DMR and prevent *Igf2* promoter activation by a downstream enhancer. In the

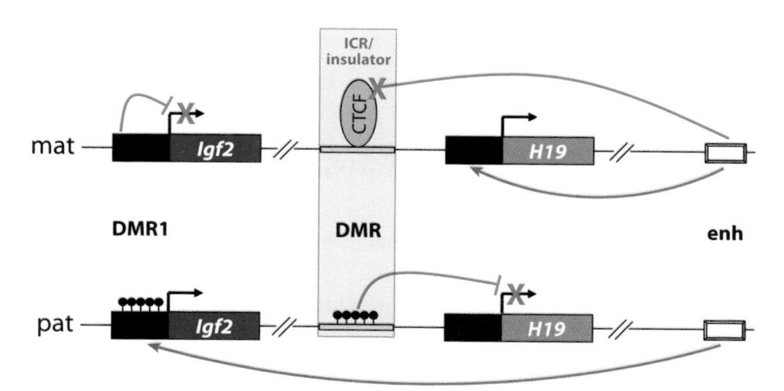

Figure 1

DMR imprint control at the *Igf2/H19* locus. The differentially methylated region (DMR) controlling allele-specific expression at the *Igf2/H19* imprinted cluster (*yellow bar*) is distinct from DMR1, another *cis*-acting element at the *Igf2/H19* locus. When bound by CTCF, the DMR acts as an insulator and blocks activation of the maternal *Igf2* promoter by the downstream enhancer. On the paternal chromosome, methylation at the DMR inhibits CTCF binding, allowing the activation of the paternal *Igf2* promoter by the enhancer. Black lollipops represent methylated cytosine residues. Black arrows indicate direction of transcription. Enh, enhancer.

paternal allele, the methylated DMR prevents CTCF binding, and as a result the downstream enhancer is able to activate transcription of *Igf2* (11, 58, 77, 161). Insulators can be swapped between different organisms and still retain buffering function, suggesting that they operate via an evolutionarily conserved mechanism.

Whatever the mechanism of imprinting control by DMRs, transgene experiments provide strong evidence that DMRs can confer imprinting capacity to nonimprinted sequences placed under their regulation. A particularly clear example of this came from the study of *RSVIgmyc*, a mouse transgene created from the Rous sarcoma virus long terminal repeats (LTRs) plus fragments from the Ig heavy chain and *c-myc* gene (160). Characterization of *RSVIgmyc* revealed that it is expressed in a parent-of-origin manner. Thus, an imprinted transgene had been created from nonimprinted sources. Reinhart and coworkers (141) created a nonimprinted derivative of *RSVIgmyc* by mapping and deleting its DMR. Imprinting of the nonimprinted derivative was restored by substitution of a DMR from the *Igf2/H19* locus (141). In a similar experiment, the *H19* DMR was able to impart imprinting status to the normally nonimprinted β-globin locus (163). This and other DMR swapping experiments suggest the presence of an imprinting signal present in the DMRs of different imprinted loci. Because DMRs are defined by divergent sequences, primary sequence is not likely to be the signal. Rather, the repetitive character, high frequency of certain sequence features such as CpG dinucleotides, and differential chromatin marks may allow formation of secondary structures that recruit regulatory factors (133).

Covalent Modifications: DNA Methylation

What additional marks, besides *cis*-acting regulatory elements and genomic context, might an organism use to distinguish maternal and paternal alleles? In model systems in which imprinting is best characterized (mammals, plants, and certain insects), imprinting is invariably associated with 5-methyl-cytosine methylation. This is especially relevant in organisms that use CpG islands as promoter elements, such as mammals and plants. Differences in regional and/or the degree of CpG island methylation may provide additional marks to distinguish between the maternal and paternal allele. For example, the ICR2 of *Kcnq1*, mentioned previously, is a CpG island. The maternal ICR2 is completely methylated whereas the paternal ICR2 is devoid of any detectable methylation (10).

DNA methylation is also involved in imprinting in insects. Insect imprinting is best characterized in the coccid insects (mealybugs). Unlike mammals, where imprinting occurs at certain sparse clusters of genes, imprinting in mealybugs occurs by heterochromatization and transcriptional silencing of the entire paternal genome (12, 83).

In mammalian development, the methylation status of an allele requires two sets of DNA methyltransferases. During gametogenesis in mammals, de novo methyltransferases *Dnmt3A* and *Dnmt3L* methylate DMRs in a gamete-specific manner (15, 78). The action of de novo methylation therefore results in distinct DNA methylation marks that differ between the two parental alleles. After fertilization, maintenance methyltransferases such as *Dnmt1* maintain the methylation state (48, 70). Methylation of an allele is generally associated with lower activity of that allele. In plants, maintenance methylation, but not de novo methylation, is required for imprinting (74, 152, 170), apparently owing to the different mechanism by which mammals and plants imprint genes (discussed below in the section Selective Inactivation Versus Selective Activation).

How might differential DNA methylation at DMRs coordinate control of multiple genes within a given cluster? Turker & Bestor (167) originally proposed a model in which DNA methylation is initially established at an ICR and subsequently spreads to surrounding regions. A more recent model posits that ICRs can modulate long-range chromatin interactions through looping (107, 123). Both models appear to be supported by experimental evidence.

Despite the ubiquity of DNA methylation in imprinting, many investigators concede that it is not known whether cytosine methylation is the cause or merely the manifestation of the imprinting process. In some cases, DNA methylation appears to be essential for proper expression of imprinted genes. For example, aberrant DNA methylation has been linked to a number of imprinting-related disorders in humans, apparently owing to improper spatial and/or temporal allele-specific expression (135, 145). In *Arabidopsis*, maintenance of DNA methylation is required for the proper imprinted expression of *FWA* and *FIS2* (74, 84, 152). However, DNA methylation appears to play only a partial role in regulating imprinting of the *Arabidopsis Medea* gene (44, 74).

Examples of a nonessential role for DNA methylation in genomic imprinting come from *Drosophila* and *Caenorhabditis elegans*. Although cytosine methylation has been detected in *Drosophila* (92, 112), its role in the ability of *Drosophila* to imprint endogenous genes in certain cases has not been established. DNA methylation has not been reported for *C. elegans* (54, 66, 157), yet this species can apparently imprint a subset of transgenes (154) and establish certain gamete-specific histone marks on the X chromosome during early embryogenesis (9). And in contrast to mammals and plants, where DNA methylation is generally associated with gene inactivation, mealybugs show the opposite relationship between DNA methylation and gene activity: The inactive heterochromatinized paternal genome is hypomethylated relative to the active, hypermethylated maternal genome (12, 14). Hence, although DNA methylation appears to be widely associated with the imprinting process, its true mechanistic role remains largely unknown.

Marking DNA with a methyl group appears to be an especially useful evolutionary invention that allows regulatory pathways to read additional information beyond the primary DNA sequence. For example, the simple addition of methyl groups to adenosine residues allows prokaryotes to distinguish not only self from nonself DNA, but also temporal information about self DNA (newly replicated DNA is not yet methylated) (6, 106). In epigenetic processes, DNA methylation, in conjunction with histone modification, gives information about the parent of origin of a DNA sequence (as in imprinting) or the transcriptional competency of a particular locus (i.e., X inactivation and paramutation). It is not surprising, then, that DNA methylation (whether cytosine or adenine methylation) is one of the ubiquitous themes in biology, used by diverse taxa for diverse biological processes.

Covalent Modifications: Histone Modifications

Cells gain another level of information by adding histone modifications to primary sequence and DNA methylation. It is now well established that histone modifications play crucial roles in gene regulation by allowing cells to regulate the temporal and spatial expression of genes reversibly. This is especially relevant to epigenetic processes such as imprinting, which requires the same DNA sequence to carry different epigenetic information depending upon its parent of origin and which requires resetting to a new state in a subsequent generation.

It is now known that DMRs are not only sites of differential DNA modification, but also sites of differential chromatin modification. This can lead to differences in chromatin structure (and hence accessibility to regulatory factors) between the two alleles of an imprinted gene. For example, the DMRs of *Igf2/H19* (81), *IC2/Kcnq1* (76), and *PWS-IC* (151) are sensitive to DNase I digestion in a parent-of-origin manner. Both activating (i.e., H3 and H4 acetylation) and deactivating histone modifications (H3 lysine 9 and lysine 27 methylation) are found at these DMRs (101). The pattern of histone modification appears to be correlated with the pattern of DNA methylation. For example, activating histone marks are found at hypomethylated DMRs, whereas deactivating marks are generally associated with highly methylated DMRs. However, histone modification at DMRs can occur independently of DNA methylation (100). Differential histone modification can also occur outside of DMRs to modulate imprinted gene expression. For example, at the *Igf2r* locus, it is allele-specific histone modification at the promoter, possibly in conjunction with DNA methylation at the ICR, that is thought to regulate imprinting (173).

The involvement of differential histone modifications in imprinted loci is also found in certain insect species. In mealybugs, differential histone modifications contribute to marking the entire paternal genome as distinct from the maternal genome (13, 32, 39). In embryos that develop into males, the paternal chromosomes are marked with deactivating histone marks and transcriptionally silenced (14, 16, 46, 83). Similar mechanisms in Sciarid flies may mark selected paternal chromosomes for elimination (33, 47) during various stages of the organism's life cycle.

Beyond covalent histone modifications, one might imagine rules for replacement of canonical histones with histone variants in the modulation of imprinting activities. The deposition of variant histones at a locus serves as a landmark that recruits regulatory factors to sites of deposition. The effect would be to allow gene expression activities to be regulated at defined regions along the chromosome (57). For example, allele-specific variant histone macroH2A1 replacement occurs at the ICRs of several known imprinted genes, including *Peg3*, *Igf2/H19*, *Gtl2/Dlk1*, and *Gnas* (30). Because substitution of variant histones can be replication independent (60), cells might achieve a dynamic level of gene regulation that is independent of the cell cycle. As an example, local deposition of histone H3.3 could be expected to make the site transcriptionally active. Such a mechanism might contribute to turn specific genes on in terminally differentiated cells.

The regulation of gene activities by variant histone replacement appears to be a ubiquitous process found in many organisms and biological processes. Some examples include macroH2A and X chromosome inactivation in mouse (21, 22, 26, 62), the involvement of H1.1 in chromatin silencing and germline development in *C. elegans* (72), AtMGH3 and sperm-specific chromatin remodeling in *Arabidopsis* (128), His1-3 and drought stress in *Arabidopsis* (3), H3.3 involvement in *Drosophila* sperm chromatin assembly (109), H2A.Z and chromosome segregation in *Xenopus* (142), and H2A1.2 in mammalian genome stability (8).

Noncoding RNAs

A characteristic feature of some imprinted clusters is the presence of one or more noncoding transcripts, often transcribed from the opposite direction as protein-coding genes in the cluster (148). In mouse, the *Air* noncoding RNA is transcribed from the paternal allele in the opposite direction to the imprinted *Igf2r* gene (113). The promoter that drives *Air* transcription is located within an intron of *Igf2r* (113). Sleutels and coworkers (158) have shown that the *Air* RNA or its transcription has a regulatory role in the imprinting of flanking genes. As in the case of *Air*, the paternally expressed *Kcnq1ot1* transcript RNA has been suggested to have a similar regulatory function at the *Ascl2/Cdkn1c/Kcnq1* imprinted cluster (118). These and other examples of regulation of an imprinted cluster by a noncoding antisense RNA are reminiscent of X chromosome inactivation in mammals.

Research into the RNA interference (RNAi) and microRNA (miRNA) pathways in the past 10 years or so has led to the discovery of a vast number of biological processes regulated by small regulatory RNAs. It is therefore not surprising to find the involvement of these classes of noncoding RNAs in imprinting processes. Two loci that provide good paradigms for investigating the involvement of small RNAs in imprinting are the Prader-Willi locus and the *Dlk1/Gtl2* (*callipyge* in sheep) clusters in humans (148). I limit the present discussion to the latter example. The *Dlk1/Gtl2* locus contains a complex set of gene arrangement and expression patterns. In this gene cluster, the paternal chromosome expresses three protein-coding genes: *Dlk1* (Delta-like 1), *Rtl1/Peg11*, and *Dio3* (45). In contrast, the maternal chromosome expresses various classes of noncoding RNAs, including miRNAs, small nucleolar RNAs (snoRNAs), and a transcript of the *Gtl2* gene. The maternal *Rtl1/Peg11* noncoding RNA is transcribed antisense to the paternally expressed *Rtl1/Peg11* transcript. This maternal *Rtl1/Peg11* transcript contains a cluster of miRNAs that mediate degradation of the paternal *Rtl1/Peg11* transcript via the RNAi pathway (34). Interestingly, *Dlk1* and *Rtl1* (Retrotransposon-like 1) appear to be remnants of retrotransposable elements (114, 183).

Polycomb and Trithorax Group Proteins

Polycomb group (PcG) and Trithorax group (TrxG) proteins maintain mitotically stable chromatin states. In *Drosophila*, these proteins assemble into multiunit complexes that recognize and bind target sequences called polycomb response elements (PREs) or trithorax response elements (TREs), respectively, to keep the target gene in the silenced (PcG-mediated) or active (TrxG-mediated) state through subsequent mitotic divisions (20). It should be noted that PREs and TREs have been identified only for *Drosophila* so far. Also, PREs and TREs are defined functionally: They are "DNA sequences to which PcG and TrxG complexes bind, directly or indirectly" (89). Thus, PcG and TrxG group proteins are certainly capable of using affiliated proteins for DNA recognition and/or binding activities.

PcG and TrxG proteins are found in many organisms and are involved in numerous developmental processes that require the maintenance of long-term gene expression states. Not surprisingly, PcG/TrxG proteins are involved in genomic imprinting. In mammals, PcG proteins are required for parent-specific allele repression (100, 116, 168). Recently, the mammalian PcG genes *Eed* and *Ezh2* and a *Drosophila* PcG gene were linked to DNA methylation (40, 116, 171). These are important findings that add to the repertoire of mechanisms by which PcG/TrxG proteins mediate epigenetic states.

PcG and TrxG proteins mediate gene activities through maintenance of cellular memory, either by directly carrying out histone modifications or by recruiting histone modification factors and/or DNA modification factors to local sites (19, 93, 94, 144, 171, 175). The simplest mechanism one can envision for PcG/TrxG maintainence of epigenetic states is physical exclusion of *trans*-acting factors from target sequences, but there is little experimental support for this model (137). Instead, experimental evidence from multiple studies seems to suggest that the mechanism is more complex, likely involving a series of recruitment and modification steps. Recognition and binding to target sites (assisted perhaps by local histone modifications) is likely the initiating step that leads to recruitment of histone and/or DNA modification factors to the region (5, 20, 63, 186). Once thought to maintain the repressed or active state of a target gene for the life of the organism, emerging evidence indicates that PcG/TrxG-mediated gene expression can be dynamic, allowing cells to switch fates during development (28, 41, 88, 99).

THE IMPRINT LIFE CYCLE

For imprinted genes, the same DNA sequence must be cycled between the male and female germline in different generations. This entails three distinct phases in the life cycle of an imprint: establishment of parent-specific imprints in the germline, maintenance and readout of the imprint in somatic cells of the progeny generation, and resetting and re-establishment of new imprints in the progeny germline (**Figure 2**). In discussing the imprint life cycle, it is useful to keep the soma and germline distinct; imprints are maintained and read in somatic cells, but erasure (of the previous generation's imprints) and establishment of new imprints occur in the germline. How these processes are thought to occur is discussed below.

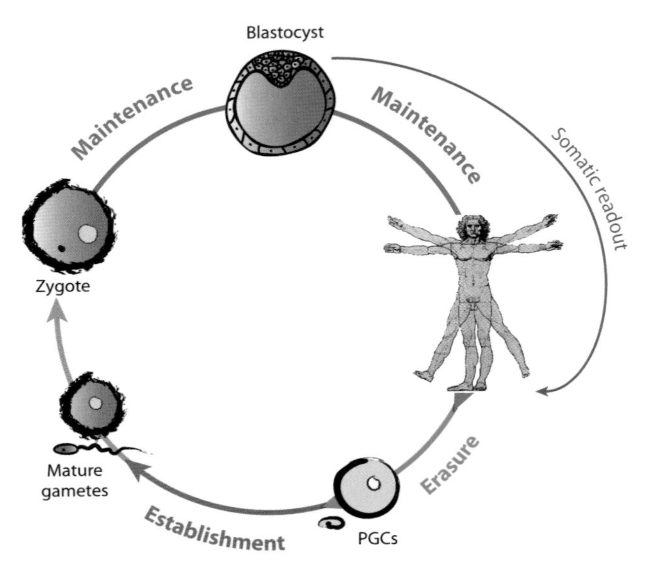

Figure 2

The imprint life cycle. Beginning at fertilization, the parental imprints are maintained in the blastocyst and all subsequent somatic lineages. In somatic cells of the embryo and adult organism, imprints are maintained and read. In the developing germline, parental imprints undergo erasure in primordial germ cells (PGCs). As germ cells mature, sex-specific imprints are re-established. Fertilization completes the imprint life cycle.

Establishment of the Imprinted State

Early studies of imprinting in mice indicated that germline passage is a requirement for the establishment of gamete-specific imprints (166). Thus, imprints are established during oogenesis and gametogenesis. Because the same DNA sequence can be made to carry different epigenetic information by alternating it between the male and female germlines, imprints likely involve gamete-specific factors that add gamete-specific marks.

Gamete-specific DNA methylation patterns have been observed since the earliest studies on genomic imprinting in mammals (23, 139, 160). Differential methylation is often targeted to DMRs, which are divided into two classes on the basis of the timing of their acquisition of methylation. Primary DMRs acquire gamete-specific methylation during gametogenesis and thus carry information about gamete origin. Primary DMRs are the targets of de novo DNA methyltransferases such as *Dnmt3A* and *Dnmt3L* (59, 78). Secondary DMRs acquire gamete-specific methylation after fertilization and may carry information about tissue specificity, for example. In mammals, secondary DMRs are targeted by the maintenance DNA methyltransferase *Dnmt1* (70).

In addition to being targets for differential DNA methylation, DMRs are also sites of gamete-specific histone modifications that occur during gametogenesis. In mice, sperm-specific marks are laid down before sperm histones are replaced by protamines (35). Researchers have found both activating and deactivating modifications that correlate with the parent-specific expression of the locus. *Snrpn*, *Igf2r*, and *U2af1-rs1* are all expressed from the paternal allele (43). Analysis of the DMRs of these imprinted loci revealed that the paternal DMRs contain activating histone modifications (i.e., H3-K4 methylation and H3-K9/K14 acetylation), whereas the corresponding maternal DMRs contain deactivating histone marks (i.e., H3-K9 methylation) (43). The maternal *U2af1-rs1* DMR is hypermethylated and is associated with the MBD proteins MeCP2, MBD1, and MBD2 (43). Thus, the deactivating histone marks on the maternal DMR are associated with heavy methylation and a deactivating chromatin mark. Numerous examples of differential histone modifications at DMRs have been found for other imprinted clusters, including PWS/AS (181) and BWS (64).

The deposition of variant histones during gametogenesis has been documented in a number of species (2, 61, 108, 120, 125, 176). Variant histone replacement during gametogenesis is believed to help with chromatin packaging, particularly in sperm. Recently, Choo and coworkers (30) reported the allele-specific deposition of histone variant macroH2A1 at some imprinted loci. Although the study does not address the question of whether deposition occurred during gametogenesis or postfertilization (the analysis was performed on mouse somatic tissue), it does reaffirm the numerous strategies cells employ to modulate epigenetic states.

PcG and TrxG group proteins are required for germline development in multiple organisms (67, 73, 80, 91, 97, 134). The activities of these proteins are detected during embryogenesis, during which they maintain epigenetic states. The maintenance function of PcG/TrxG proteins during embryogenesis is well documented in *Drosophila* (25, 50, 131, 143), but has also been reported in mammals (38, 65, 126, 155, 162, 184), plants (49, 52, 53), and possibly in *C. elegans* (24, 185). However, to my knowledge the establishment of epigenetic states by PcG proteins during gametogenesis has not been reported. PcG proteins are present during oogenesis in *Drosophila* but have not been shown to be localized to the oocyte nucleus (134).

Currently available data suggest that PcG/TrxG proteins could have a maintenance role, rather than an establishment role, in the regulation of imprinted gene activities. In principle, cells should be able to set epigenetic states during gametogenesis via the use of PcG/TrxG proteins. Two mechanisms can be envisioned; these two modes of action of PcG/TrxG proteins have been reported in the regulation of gene expression postfertilization.

Mechanism 1. PcG/TrxG proteins could function as primary recruiters that initially lay down the signals to recruit trans-acting factors (histone acetyltransferases, histone deacetylases, histone sumoylation/phosphorylation/ubiquitination factors) to target sites. In this scenario, the PcG/TrxG proteins themselves possess histone modification activities and lay down histone marks at target sites. By reading this PcG/TrxG-mediated histone code, other histone modification factors would be recruited to the site. The human homolog of PRC1(hPRC1L), for example, contains the ubiquitinase activity for histone H2A (175). PRC1-mediated ubiquitination of H2A leads to initiation of X inactivation (37).

Mechanism 2. PcG or TrxG complexes could be recruited to target sites by reading the histone code laid down by other chromatin modification factors. In this mode of action, PcG/TrxG proteins could function as bridges between the various effectors of chromatin remodeling. For example, Pc is a subunit of the *Drosophila* PRC1 complex that recognizes methylated H3-K27 (186). When present in the *Ubx* PRE, H3-K27-methyl acts as a code to attract Pc, which in turn recruits the PRC1 complex to the *Ubx* PRE to repress *Ubx* expression (186). A similar mechanism operates at the *Drosophila Pho* locus (89).

Maintenance and Readout of the Imprint

During normal fertilization events in mammals, both the paternal and maternal genomes of somatic cells undergo genome-wide demethylation shortly after fertilization (7, 121, 132, 149). An interesting aspect of imprinted genes is their resistance to this genome-wide demethylation event. What are the mechanisms of maintenance of parental imprints? DNA methylation patterns are preserved by the maintenance DNA methyltransferase *Dnmt1* (70). For example, the *H19* gene is normally expressed only from the maternal allele, but in a *Dnmt1* loss-of-function background, the paternal allele is also expressed (102). How exactly the methyltransferase recognizes which sequence to target remains unknown, but evidence suggests that primary DMRs may serve as guides.

Somatic cells must not only maintain the parental generation's imprints, but also interpret the imprints and express the proper alleles. What is the mechanism by which this occurs? Not all cells will read and respond to an imprint in the same manner. Tissue-specific regulatory elements must regulate the spatial and temporal expression of imprints. For example, the imprinted *Igf2* gene is expressed from the paternal allele in most tissues. In the liver, alternative promoter usage leads to biallelic *Igf2* expression (172). Hagège and coworkers (55) showed that additional tissue specificity of *Igf2* requires a *cis*-acting sequence located 3′ of the *H19* gene. Further studies by Ohno and coworkers (127) indicated that temporal regulation of the *Igf2* imprint leads to a transition from biallelic to gamete-origin-specific transcription during the blastocyst stage. Numerous other studies by different groups confirmed that additional imprinted loci are imprinted in a spatial and/or temporal manner, including *Neurabin*, *Pon2*, and *Pon3* (130); *Tnfrh1* (31); *Dio3* (165); and *Nesp* and *Nespas* (4). Thus, establishment of the imprint simply marks the two parental alleles as different. The subsequent generation must maintain, interpret, and express the imprints in the proper spatial and/or temporal manner.

The most logical explanation for the expression of one parental allele over another is that only one parental allele is targeted for transcription. Asynchronous DNA replication and/or subnuclear localization of one allele can target a specific allele for expression (51, 75, 86, 103). Thus, asynchronous DNA replication and subnuclear localization could be downstream events that occur in somatic cells and depend on the upstream epigenetic events that occur during gametogenesis. Surprisingly, asynchronous replication of imprinted genes has also been found to be established during gametogenesis and maintained through development (156).

Imprint Erasure and Resetting

Whereas the somatic cells must maintain, read, and express the parental imprints, the germline must reset the previous generation's imprints and lay down its own sex-specific imprints. Two models were originally put forth to account for the mechanism of imprint erasure (147). One postulates that the imprints of the same-sex parent are preserved, whereas the imprints of the opposite-sex parent are reversed. The other model posits that both imprints are erased, followed by re-establishment of gamete-specific imprints. In mammals, at least, a preponderance of data indicates the second model to be more likely.

In mammals and other organisms investigated, parental imprints are reset during early embryonic development (140), during which the primordial germ cells undergo global demethylation (i.e., erasure). Two experimental strategies have been adopted to determine the exact timing of this event. One class of experiments aims at determining the methylation status of specific imprinted loci in variously staged gametes; another class of experiments employs nuclear transfer or cloning of mice from primordial germ cells, because only properly imprinted nuclei are assumed to be competent to direct the normal development of embryos. Collectively, these two types of experiments carried out by numerous groups have led to three general conclusions about imprint erasure in the germline: 1) Erasure occurs early, approximately 9.5–10.5 days post coitum; 2) erasure is asynchronous, i.e., it does not occur simultaneously for all imprinted genes; and 3) erasure is rapid and possibly involves an active mechanism of demethylation (36, 56, 98, 110, 150, 182).

Results from experiments in *C. elegans* suggest not only that gametogenesis is the step at which parent-of-origin epigenetic states are established, but also that the stability of the marks depends on the length of time a susceptible sequence spends in a particular gametic environment. For example, *C. elegans* populations that received a GFP transgene from the male germline exhibited a population average of twofold greater expression compared with genotypically identical populations that received the same transgene from the female germline (154). The greater expression advantage of the sperm-transmitted transgene can be completely reset by a single generation transmission through the oocyte. However, transmission of the same transgene continuously through the male germline for multiple generations resulted in not only a greater than twofold expression advantage, but also a more stable epigenetic state that required multiple generations through the female germline to completely reset. Thus, the stability of the imprint can be experimentally manipulated to be both mitotically and meiotically stable. This result suggests that the stability of the epigenetic state is a function of the length of time it spends in the male germline.

SELECTIVE INACTIVATION VERSUS SELECTIVE ACTIVATION OF IMPRINTED GENES

In mammals, the imprinting process leads to selective inactivation of the nonexpressed allele. The process of selective inactivation also operates on imprinting in two insect groups: coccids, in which the imprint life cycle operates at the level of the entire paternal genome (12–14, 83); and the sciara flies, in which various paternal chromosomes are selectively inactivated and eliminated during various phases of the life cycle (17, 33, 46).

Plants exhibit very divergent biology compared with animals. It is therefore not surprising that plants have evolved a different strategy to imprint genes. In plants, the expressed allele of an imprinted gene appears to be selectively activated (152, 153). *MEDEA* and *FWA* are two well-characterized imprinted genes in *Arabidopsis*, and both are expressed only from the female gamete (27, 52, 84, 85, 87). By default both alleles are inactive and are associated with heavy DNA methylation by the maintenance methyltransferase *MET1*, a homolog of the human maintenance methyltransferase *Dnmt1* (79, 146, 180). Activation of *MEDEA* and *FWA* in female gametes requires *DEMETER*, a DNA glycosylase implicated to have DNA demethylation activity (29, 84, 180). Hence, organisms employ one of two strategies to achieve the selective expression of only one allele of an imprinted locus.

IMPRINTING: MORE COMMON THAN WE THINK?

The analysis of imprinting mechanisms has been carried out, in the majority of cases, on loci that possess phenotypic manifestations. This analysis leads to the conclusion that only a small subset of endogeneous genes (in plants and mammals) are imprinted. The present tally of imprinted genes in mammals is likely to be a gross underestimate for two reasons. First, the phenotype-driven, development-centric approach to the identification of imprinted genes neglects those loci that have a lethal phenotype or those whose phenotypes are too subtle for detection. For example, a number of loci are imprinted only temporally and/or spatially; that is, monoallelic expression occurs only at certain stages of development and/or only in a subset of tissues (55, 96, 119, 136, 177, 178). It is difficult to identify tissue and developmental stage-specific imprinting of a gene, because the researcher must not only pinpoint expression of the gene to a specific tissue and developmental time point, but also correlate the transcription as originating from only one parental chromosome. Second, parent-of-origin differences at the epigenotype level (i.e., parent-of-origin-dependent differences in chromatin marks) have not been rigorously investigated either at the chromosomal level or at the genome-wide level. Certain sequences might initially acquire parent-specific marks that are neither maintained nor read. For example, in *C. elegans* the spermatogenesis-derived X chromosome and oogenesis-derived X chromosome bear differential chromatin marks, but these are transient and persist only until the 20-cell stage of the embryo (9). Fortunately, with advances in genome-wide location analysis technologies such as chromatin immunoprecipitation coupled to high-throughput sequencing (ChIP-Seq), it is now possible to query for genome-wide parent-of-origin differences in chromatin marks. For example, Mikkelsen and coworkers (122) recently showed that ChIP-Seq can be used to assay for allele-specific chromatin marks using single nucleotide polymorphisms.

Because passage through gametogenesis is a requirement for the acquisition of parent-of-origin marks for imprinted genes, and available data indicate that the acquisition of such marks is largely independent of primary sequence, one might ask whether any DNA sequence (endogenous or exogenous) can acquire a parent-of-origin mark after having emerged from gametogenesis, or are certain *cis*-acting contexts (genomic location, repetitive character, *cis/trans*-acting elements, etc.) required to make the sequence susceptible? In other words, might nonimprinted, developmentally neutral sequences also be able to acquire parent-of-origin marks but somehow escaped maintenance and/or differential readout? This is an intriguing possibility that remains a speculation. An excellent model with which to answer this question is the mealybug. Because the entire paternally contributed genome is targeted for silencing in this group of insects, one intriguing question is whether any arbitrary DNA sequence integrated into the genome of this organism can be silenced after transmission through the male germline. Unfortunately, to my knowledge transgene manipulation has not been reported for this group of insects. Radiation-induced chromosomal fragments in male mealybugs become heterochromatinized and genetically inactive, independent of the size of the fragment (18). This observation has been interpreted to mean that the signal for heterochromatinization is chromosome wide, i.e., acts in *cis*. (82). If this interpretation is true (and if the heterochromatin can spread), arbitrary sequences integrated into the chromosomes of coccid insects would become heterochromatinized upon passage through male gametogenesis.

CONCLUSION

The study of genomic imprinting in mammalian systems and plants has revealed mechanistic insights into this pan-taxa biological phenomenon, as well as provided the basis for understanding certain human genetic disorders. However, that *C. elegans*, *Drosophila*, and zebrafish, three organisms for which imprinting of endogenous genes does not exist as a normal part of development, can be induced to imprint certain DNA sequences, and that imprinted sequences can be constructed from nonimprinted sequences (i.e., *RSVIgmyc*) demonstrate that imprinting can occur outside the context of development. The capacity to decouple imprinting from developmental pathways can allow the investigation of imprinting (as a purely gene silencing process) without the constraints of development. Advances in genomics technology will facilitate the genome-wide identification of imprinted loci in a development-neutral context. The genome-wide approaches to identifying imprinted loci on the basis of chromatin marks, along with the more traditional phenotype-driven approach, will undoubtedly lead to the identification of greater numbers of imprinted loci.

DISCLOSURE STATEMENT

The author is not aware of any biases that might be perceived as affecting the objectivity of this review.

ACKNOWLEDGMENTS

During the writing of this manuscript I was a graduate student in the Biology Department at Johns Hopkins University and the Department of Pathology at the Stanford University School of Medicine. I would like to thank Dr. Aravinda Chakravarti and my thesis advisor Dr. Andrew Fire for their critical reading of this manuscript.

LITERATURE CITED

1. Ainscough JF, Koide T, Tada M, Barton S, Surani MA. 1997. Imprinting of *Igf2* and *H19* from a 130 kb YAC transgene. *Development* 124:3621–32
2. Akhmanova A, Miedema K, Wang Y, van Bruggen M, Berden JH, et al. 1997. The localization of histone H3.3 in germ line chromatin of *Drosophila* males as established with a histone H3.3-specific antiserum. *Chromosoma* 106:335–47

3. Ascenzi R, Gantt JS. 1997. A drought-stress-inducible histone gene in *Arabidopsis thaliana* is a member of a distinct class of plant linker histone variants. *Plant Mol. Biol.* 34:629–41
4. Ball ST, Williamson CM, Hayes C, Hacker T, Peters J. 2001. The spatial and temporal expression pattern of *Nesp* and its antisense *Nespas*, in mid-gestation mouse embryos. *Mech. Dev.* 100:79–81
5. Bantignies F, Cavalli G. 2006. Cellular memory and dynamic regulation of polycomb group proteins. *Curr. Opin. Cell Biol.* 18:275–83
6. Barras F, Marinus MG. 1989. The great GATC: DNA methylation in *E. coli*. *Trends Genet.* 5:139–43
7. Barton SC, Arney KL, Shi W, Niveleau A, Fundele R, et al. 2001. Genome-wide methylation patterns in normal and uniparental early mouse embryos. *Hum. Mol. Genet.* 10:2983–87
8. Bassing CH, Chua KF, Sekiguchi J, Suh H, Whitlow SR, et al. 2002. Increased ionizing radiation sensitivity and genomic instability in the absence of histone H2AX. *Proc. Natl. Acad. Sci. USA* 99:8173–78
9. Bean CJ, Schaner CE, Kelly WG. 2004. Meiotic pairing and imprinted X chromatin assembly in *Caenorhabditis elegans*. *Nat. Genet.* 36:100–5
10. Beatty L, Weksberg R, Sadowski PD. 2006. Detailed analysis of the methylation patterns of the KvDMR1 imprinting control region of human chromosome 11. *Genomics* 87:46–56
11. Bell AC, Felsenfeld G. 2000. Methylation of a CTCF-dependent boundary controls imprinted expression of the *Igf2* gene. *Nature* 405:482–85
12. Bongiorni S, Cintio O, Prantera G. 1999. The relationship between DNA methylation and chromosome imprinting in the coccid *Planococcus citri*. *Genetics* 151:1471–78
13. Bongiorni S, Mazzuoli M, Masci S, Prantera G. 2001. Facultative heterochromatization in parahaploid male mealybugs: involvement of a heterochromatin-associated protein. *Development* 128:3809–17
14. Bongiorni S, Prantera G. 2003. Imprinted facultative heterochromatization in mealybugs. *Genetica* 117:271–79
15. Bourc'his D, Xu GL, Lin CS, Bollman B, Bestor TH. 2001. Dnmt3L and the establishment of maternal genomic imprints. *Science* 294:2536–39
16. Brown SW. 1959. Lecanoid chromosome behavior in three more families of the Coccoidea (Homoptera). *Chromosoma* 10:278–300
17. Brown SW, Chandra HS. 1977. Chromosome imprinting and the differential regulation of homologous chromosomes. In *Cell Biology: A Comprehensive Treatise*, ed. L Goldstein, DM Prescott, 1:109–89. New York: Academic.
18. Brown SW, Nelsen-Rees WA. 1961. Radiation analysis of the lecanoid genetic system. *Genetics* 42:510–23
19. Cao R, Tsukada Y, Zhang Y. 2005. Role of Bmi-1 and Ring1A in H2A ubiquitylation and Hox gene silencing. *Mol. Cell* 20:845–54
20. Cernilogar FM, Orlando V. 2005. Epigenome programming by Polycomb and Trithorax proteins. *Biochem. Cell Biol.* 83:322–31
21. Chadwick BP, Valley CM, Willard HF. 2001. Histone variant macroH2A contains two distinct macrochromatin domains capable of directing macroH2A to the inactive X chromosome. *Nucleic Acids Res.* 29:2699–705
22. Chadwick BP, Willard HF. 2003. Chromatin of the Barr body: histone and nonhistone proteins associated with or excluded from the inactive X chromosome. *Hum. Mol. Genet.* 12:2167–78
23. Chaillet JR, Vogt TF, Beier DR, Leder P. 1991. Parental-specific methylation of an imprinted transgene is established during gametogenesis and progressively changes during embryogenesis. *Cell* 66:77–83
24. Chamberlin HM, Thomas JH. 2000. The bromodomain protein LIN-49 and trithorax-related protein LIN-59 affect development and gene expression in *Caenorhabditis elegans*. *Development* 127:713–23
25. Chanas G, Maschat F. 2005. Tissue specificity of hedgehog repression by the Polycomb group during *Drosophila melanogaster* development. *Mech. Dev.* 122:975–87
26. Changolkar LN, Pehrson JR. 2006. macroH2A1 histone variants are depleted on active genes but concentrated on the inactive X chromosome. *Mol. Cell Biol.* 26:4410–20
27. Chaudhury AM, Ming L, Miller C, Craig S, Dennis ES, Peacock WJ. 1997. Fertilization-independent seed development in *Arabidopsis thaliana*. *Proc. Natl. Acad. Sci. USA* 94:4223–28
28. Chen X, Hiller M, Sancak Y, Fuller MT. 2005. Tissue-specific TAFs counteract Polycomb to turn on terminal differentiation. *Science* 310:869–72

29. Choi Y, Gehring M, Johnson L, Hannon M, Harada JJ, et al. 2002. DEMETER, a DNA glycosylase domain protein, is required for endosperm gene imprinting and seed viability in arabidopsis. *Cell* 110:33–42
30. Choo JH, Kim JD, Chung JH, Stubbs L, Kim J. 2006. Allele-specific deposition of macroH2A1 in imprinting control regions. *Hum. Mol. Genet.* 15:717–24
31. Clark L, Wei M, Cattoretti G, Mendelsohn C, Tycko B. 2002. The *Tnfrh1* (*Tnfrsf23*) gene is weakly imprinted in several organs and expressed at the trophoblast-decidua interface. *BMC Genet.* 3:11
32. Cowell IG, Aucott R, Mahadevaiah SK, Burgoyne PS, Huskisson N, et al. 2002. Heterochromatin, HP1 and methylation at lysine 9 of histone H3 in animals. *Chromosoma* 111:22–36
33. Crouse HV. 1960. The controlling element in sex chromosome behavior in Sciara. *Genetics* 45:1429–43
34. Davis E, Caiment F, Tordoir X, Cavaillé J, Ferguson-Smith A, et al. 2005. RNAi-mediated allelic trans-interaction at the imprinted *Rtl1/Peg11* locus. *Curr. Biol.* 15:743–49
35. Delaval K, Govin J, Cerqueira F, Rousseaux S, Khochbin S, Feil R. 2007. Differential histone modifications mark mouse imprinting control regions during spermatogenesis. *EMBO J.* 26:720–29
36. Durcova-Hills G, Ainscough J, McLaren A. 2001. Pluripotential stem cells derived from migrating primordial germ cells. *Differentiation* 68:220–26
37. Fang J, Chen T, Chadwick B, Li E, Zhang Y. 2004. Ring1b-mediated H2A ubiquitination associates with inactive X chromosomes and is involved in initiation of X inactivation. *J. Biol. Chem.* 279:52812–15
38. Faust C, Lawson KA, Schork NJ, Thiel B, Magnuson T. 1998. The *Polycomb*-group gene *eed* is required for normal morphogenetic movements during gastrulation in the mouse embryo. *Development* 125:4495–506
39. Ferraro M, Buglia GL, Romano F. 2001. Involvement of histone H4 acetylation in the epigenetic inheritance of different activity states of maternally and paternally derived genomes in the mealybug *Planococcus citri*. *Chromosoma* 110:93–101
40. Ferres-Marco D, Gutierrez-Garcia I, Vallejo DM, Bolivar J, Gutierrez-Aviño FJ, Dominguez M. 2006. Epigenetic silencers and notch collaborate to promote malignant tumours by Rb silencing. *Nature* 439:430–36
41. Ficz G, Heintzmann R, Arndt-Jovin DJ. 2005. Polycomb group protein complexes exchange rapidly in living *Drosophila*. *Development* 132:3963–76
42. Fitzpatrick GV, Soloway PD, Higgins MJ. 2002. Regional loss of imprinting and growth deficiency in mice with a targeted deletion of *KvDMR1*. *Nat. Genet.* 32:426–31
43. Fournier C, Goto Y, Ballestar E, Delaval K, Hever AM, et al. 2002. Allele-specific histone lysine methylation marks regulatory regions at imprinted mouse genes. *EMBO J.* 21:6560–70
44. Gehring M, Huh JH, Hsieh TF, Penterman J, Choi Y, et al. 2006. DEMETER DNA glycosylase establishes *MEDEA* polycomb gene self-imprinting by allele-specific demethylation. *Cell* 124:495–506
45. Georges M, Charlier C, Cockett N. 2003. The callipyge locus: evidence for the *trans* interaction of reciprocally imprinted genes. *Trends Genet.* 19:248–52
46. Goday C, Esteban MR. 2001. Chromosome elimination in sciarid flies. *BioEssays* 23:242–50
47. Goday C, Ruiz MF. 2002. Differential acetylation of histones H3 and H4 in paternal and maternal germline chromosomes during development of sciarid flies. *J. Cell Sci.* 115:4765–75
48. Goll MG, Bestor TH. 2005. Eukaryotic cytosine methyltransferases. *Annu. Rev. Biochem.* 74:481–514
49. Goodrich J, Puangsomlee P, Martin M, Long D, Meyerowitz EM, Coupland G. 1997. A Polycomb-group gene regulates homeotic gene expression in *Arabidopsis*. *Nature* 386:44–51
50. Gould AP, Lai RY, Green MJ, White RA. 1990. Blocking cell division does not remove the requirement for Polycomb function in *Drosophila* embryogenesis. *Development* 110:1319–25
51. Gribnau J, Hochedlinger K, Hata K, Li E, Jaenisch R. 2003. Asynchronous replication timing of imprinted loci is independent of DNA methylation, but consistent with differential subnuclear localization. *Genes Dev.* 17:759–73
52. Grossniklaus U, Vielle-Calzada JP, Hoeppner MA, Gagliano WB. 1998. Maternal control of embryogenesis by *MEDEA*, a polycomb group gene in *Arabidopsis*. *Science* 280:446–50
53. Guitton A, Berger F. 2005. Loss of function of MULTICOPY SUPPRESSOR OF IRA 1 produces nonviable parthenogenetic embryos in *Arabidopsis*. *Curr. Biol.* 15:750–54

54. Gutierrez A, Sommer RJ. 2004. Evolution of *dnmt-2* and *mbd-2*-like genes in the free-living nematodes *Pristionchus pacificus*, *Caenorhabditis elegans* and *Caenorhabditis briggsae*. *Nucleic Acids Res.* 32:6388–96
55. Hagège H, Nasser R, Weber M, Milligan L, Aptel N, et al. 2006. The 3′ portion of the mouse *H19* Imprinting-Control Region is required for proper tissue-specific expression of the *Igf2* gene. *Cytogenet. Genome Res.* 113:230–37
56. Hajkova P, Erhardt S, Lane N, Haaf T, El-Maarri O, et al. 2002. Epigenetic reprogramming in mouse primordial germ cells. *Mech. Dev.* 117:15–23
57. Hake SB, Allis CD. 2006. Histone H3 variants and their potential role in indexing mammalian genomes: The "H3 barcode hypothesis." *Proc. Natl. Acad. Sci. USA* 103:6428–35
58. Hark AT, Schoenherr CJ, Katz DJ, Ingram RS, Levorse JM, Tilghman SM. 2000. CTCF mediates methylation-sensitive enhancer-blocking activity at the *H19/Igf2* locus. *Nature* 405:486–89
59. Hata K, Okano M, Lei H, Li E. 2002. Dnmt3L cooperates with the Dnmt3 family of de novo DNA methyltransferases to establish maternal imprints in mice. *Development* 129:1983–93
60. Henikoff S, Ahmad K. 2005. Assembly of variant histones into chromatin. *Annu. Rev. Cell Dev. Biol.* 21:133–53
61. Hennig W. 2003. Chromosomal proteins in the spermatogenesis of *Drosophila*. *Chromosoma* 111:489–94
62. Hernández-Muñoz I, Lund AH, van der Stoop P, Boutsma E, Muijrers I, et al. 2005. Stable X chromosome inactivation involves the PRC1 Polycomb complex and requires histone MACROH2A1 and the CULLIN3/SPOP ubiquitin E3 ligase. *Proc. Natl. Acad. Sci. USA* 102:7635–40
63. Hernández-Muñoz I, Taghavi P, Kuijl C, Neefjes J, van Lohuizen M. 2005. Association of BMI1 with polycomb bodies is dynamic and requires PRC2/EZH2 and the maintenance DNA methyltransferase DNMT1. *Mol. Cell Biol.* 25:11047–58
64. Higashimoto K, Urano T, Sugiura K, Yatsuki H, Joh K, et al. 2003. Loss of CpG methylation is strongly correlated with loss of histone H3 lysine 9 methylation at DMR-LIT1 in patients with Beckwith-Wiedemann syndrome. *Am. J. Hum. Genet.* 73:948–56
65. Hobert O, Sures I, Ciossek T, Fuchs M, Ullrich A. 1996. Isolation and developmental expression analysis of *Enx-1*, a novel mouse Polycomb group gene. *Mech. Dev.* 55:171–84
66. Hodgkin J. 1994. Epigenetics and the maintenance of gene activity states in *Caenorhabditis elegans*. *Dev. Genet.* 15:471–77
67. Holdeman R, Nehrt S, Strome S. 1998. MES-2, a maternal protein essential for viability of the germline in *Caenorhabditis elegans*, is homologous to a *Drosophila* Polycomb group protein. *Development* 125:2457–67
68. Holmes R, Soloway PD. 2006. Regulation of imprinted DNA methylation. *Cytogenet. Genome Res.* 113:122–29
69. Horike S, Mitsuya K, Meguro M, Kotobuki N, Kashiwagi A, et al. 2000. Targeted disruption of the human *LIT1* locus defines a putative imprinting control element playing an essential role in Beckwith-Wiedemann syndrome. *Hum. Mol. Genet.* 9:2075–83
70. Howell CY, Bestor TH, Ding F, Latham KE, Mertineit C, et al. 2001. Genomic imprinting disrupted by a maternal effect mutation in the *Dnmt1* gene. *Cell* 104:829–38
71. Hurst LD, McVean G, Moore T. 1996. Imprinted genes have few and small introns. *Nat. Genet.* 12:234–37
72. Jedrusik MA, Schulze E. 2001. A single histone H1 isoform (H1.1) is essential for chromatin silencing and germline development in *Caenorhabditis elegans*. *Development* 128:1069–80
73. Johnson J, Canning Tilly J, Tilly J. 2004. *Expression of Bmi-1, a polycomb group gene family member, in postnatal mouse ovary*. Presented at Soc. Study Reproduction, Univ. Br. Columbia, Vancouver, Can.
74. Jullien PE, Kinoshita T, Ohad N, Berger F. 2006. Maintenance of DNA methylation during the *Arabidopsis* life cycle is essential for parental imprinting. *Plant. Cell* 18:1360–72
75. Kagotani K, Takebayashi S, Kohda A, Taguchi H, Paulsen M, et al. 2002. Replication timing properties within the mouse distal chromosome 7 imprinting cluster. *Biosci. Biotechnol. Biochem.* 66:1046–51
76. Kanduri C, Fitzpatrick G, Mukhopadhyay R, Kanduri M, Lobanenkov V, et al. 2002. A differentially methylated imprinting control region within the *Kcnq1* locus harbors a methylation-sensitive chromatin insulator. *J. Biol. Chem.* 277:18106–10

77. Kanduri C, Pant V, Loukinov D, Pugacheva E, Qi CF, et al. 2000. Functional association of CTCF with the insulator upstream of the *H19* gene is parent of origin-specific and methylation-sensitive. *Curr. Biol.* 10:853–56
78. Kaneda M, Okano M, Hata K, Sado T, Tsujimoto N, et al. 2004. Essential role for de novo DNA methyltransferase Dnmt3a in paternal and maternal imprinting. *Nature* 429:900–3
79. Kankel M, Ramsey DE, Stokes TL, Flowers SK, Haag JR, et al. 2003. *Arabidopsis MET1* cytosine methyltransferase mutants. *Genetics* 163:1109–22
80. Kelly WG, Fire A. 1998. Chromatin silencing and the maintenance of a functional germline in *Caenorhabditis elegans*. *Development* 125:2451–56
81. Khosla S, Aitchison A, Gregory R, Allen ND, Feil R. 1999. Parental allele-specific chromatin configuration in a boundary-imprinting-control element upstream of the mouse H19 gene. *Mol. Cell Biol.* 19:2556–66
82. Khosla S, Augustus M, Brahmachari V. 1999. Sex-specific organisation of middle repetitive DNA sequences in the mealybug *Planococcus lilacinus*. *Nucleic Acids Res.* 27:3745–51
83. Khosla S, Mendiratta G, Brahmachari V. 2006. Genomic imprinting in the mealybugs. *Cytogenet. Genome Res.* 113:41–52
84. Kinoshita T, Miura A, Choi Y, Kinoshita Y, Cao X, et al. 2004. One-way control of *FWA* imprinting in *Arabidopsis* endosperm by DNA methylation. *Science* 303:521–23
85. Kinoshita T, Yadegari R, Harada JJ, Goldberg RB, Fischer RL. 1999. Imprinting of the *MEDEA* Polycomb gene in the *Arabidopsis* endosperm. *Plant. Cell* 11:1945–52
86. Kitsberg D, Selig S, Brandeis M, Simon I, Keshet I, et al. 1993. Allele-specific replication timing of imprinted gene regions. *Nature* 364:459–63
87. Kiyosue T, Ohad N, Yadegari R, Hannon M, Dinneny J, et al. 1999. Control of fertilization-independent endosperm development by the *MEDEA* polycomb gene in *Arabidopsis*. *Proc. Natl. Acad. Sci. USA* 96:4186–91
88. Klebes A, Sustar A, Kechris K, Li H, Schubiger G, Kornberg TB. 2005. Regulation of cellular plasticity in *Drosophila* imaginal disc cells by the Polycomb group, trithorax group, and *lama* genes. *Development* 132:3753
89. Klymenko T, Papp B, Fischle W, Kocher T, Schelder M, et al. 2006. A Polycomb group protein complex with sequence-specific DNA-binding and selective methyl-lysine-binding activities. *Genes. Dev.* 20:1110–22
90. Kobayashi H, Suda C, Abe T, Kohara Y, Ikemura T, Sasaki H. 2006. Bisulfite sequencing and dinucleotide content analysis of 15 imprinted mouse differentially methylated regions (DMRs): paternally methylated DMRs contain less CpGs than maternally methylated DMRs. *Cytogenet. Genome Res.* 113:130–37
91. Korf I, Fan Y, Strome S. 1998. The Polycomb group in *Caenorhabditis elegans* and maternal control of germline development. *Development* 125:2469–78
92. Kunert N, Marhold J, Stanke J, Stach D, Lyko F. 2003. A Dnmt2-like protein mediates DNA methylation in *Drosophila*. *Development* 130:5083–90
93. Kuzmichev A, Jenuwein T, Tempst P, Reinberg D. 2004. Different EZH2-containing complexes target methylation of histone H1 or nucleosomal histone H3. *Mol. Cell* 14:183–93
94. Kuzmichev A, Margueron R, Vaquero A, Preissner TS, Scher M, et al. 2005. Composition and histone substrates of polycomb repressive group complexes change during cellular differentiation. *Proc. Natl. Acad. Sci. USA* 102:1859–64
95. Labrador M, Corces VG. 2002. Setting the boundaries of chromatin domains and nuclear organization. *Cell* 111:151–4
96. Lau JC, Hanel ML, Wevrick R. 2004. Tissue-specific and imprinted epigenetic modifications of the human *NDN* gene. *Nucleic Acids Res.* 32:3376–82
97. Lawrence PA, Johnston P, Struhl G. 1983. Different requirements for homeotic genes in the soma and germ line of *Drosophila*. *Cell* 35:27–34
98. Lee J, Inoue K, Ono R, Ogonuki N, Kohda T, et al. 2002. Erasing genomic imprinting memory in mouse clone embryos produced from day 11.5 primordial germ cells. *Development* 129:1807–17
99. Lee N, Maurange C, Ringrose L, Paro R. 2005. Suppression of Polycomb group proteins by JNK signalling induces transdetermination in *Drosophila* imaginal discs. *Nature* 438:234–37

100. Lewis A, Mitsuya K, Umlauf D, Smith P, Dean W, et al. 2004. Imprinting on distal chromosome 7 in the placenta involves repressive histone methylation independent of DNA methylation. *Nat. Genet.* 36:1291–95
101. Lewis A, Reik W. 2006. How imprinting centres work. *Cytogenet. Genome Res.* 113:81–89
102. Li E, Beard C, Jaenisch R. 1993. Role for DNA methylation in genomic imprinting. *Nature* 366:362–65
103. Lin MS, Zhang A, Fujimoto A. 1995. Asynchronous DNA replication between 15q11.2q12 homologs: cytogenetic evidence for maternal imprinting and delayed replication. *Hum. Genet.* 96:572–76
104. Lloyd V. 2000. Parental imprinting in *Drosophila*. *Genetica* 109:35–44
105. Lloyd VK, Sinclair DA, Grigliatti TA. 1999. Genomic imprinting and position-effect variegation in *Drosophila melanogaster*. *Genetics* 151:1503–16
106. Lobner-Olesen A, Skovgaard O, Marinus MG. 2005. Dam methylation: coordinating cellular processes. *Curr. Opin. Microbiol.* 8:154–60
107. Lopes S, Lewis A, Hajkova P, Dean W, Oswald J, et al. 2003. Epigenetic modifications in an imprinting cluster are controlled by a hierarchy of DMRs suggesting long-range chromatin interactions. *Hum. Mol. Genet.* 12:295–305
108. Lopez-Alanon DM, Lopez-Fernandez LA, Castaneda V, Krimer DB, Del Mazo J. 1997. H3.3 A variant histone mRNA containing an alpha-globin insertion: modulated expression during mouse gametogenesis correlates with meiotic onset. *DNA Cell Biol.* 16:639–44
109. Loppin B, Bonnefoy E, Anselme C, Laurençon A, Karr TL, Couble P. 2005. The histone H3.3 chaperone HIRA is essential for chromatin assembly in the male pronucleus. *Nature* 437:1386–90
110. Lucifero D, Mann MR, Bartolomei MS, Trasler JM. 2004. Gene-specific timing and epigenetic memory in oocyte imprinting. *Hum. Mol. Genet.* 13:839–49
111. Luedi PP, Dietrich FS, Weidman JR, Bosko JM, Jirtle RL, Hartemink AJ. 2007. Computational and experimental identification of novel human imprinted genes. *Genome Res.* 17:1723–30
112. Lyko F, Ramsahoye BH, Jaenisch R. 2000. DNA methylation in *Drosophila melanogaster*. *Nature* 408:538–40
113. Lyle R, Watanabe D, te Vruchte D, Lerchner W, Smrzka OW, et al. 2000. The imprinted antisense RNA at the *Igf2r* locus overlaps but does not imprint *Mas1*. *Nat. Genet.* 25:19–21
114. Lynch C, Tristem M. 2003. A co-opted gypsy-type LTR-retrotransposon is conserved in the genomes of humans, sheep, mice, and rats. *Curr. Biol.* 13:1518–23
115. Lyon MF. 1998. X-chromosome inactivation: a repeat hypothesis. *Cytogenet. Cell Genet.* 80:133–37
116. Mager J, Montgomery ND, de Villena FP, Magnuson T. 2003. Genome imprinting regulated by the mouse *Polycomb* group protein Eed. *Nat. Genet.* 33:502–7
117. Mancini-DiNardo D, Steele SJ, Ingram RS, Tilghman SM. 2003. A differentially methylated region within the gene *Kcnq1* functions as an imprinted promoter and silencer. *Hum. Mol. Genet.* 12:283–94
118. Mancini-DiNardo D, Steele SJ, Levorse JM, Ingram RS, Tilghman SM. 2006. Elongation of the *Kcnq1ot1* transcript is required for genomic imprinting of neighboring genes. *Genes Dev.* 20:1268–82
119. Mantovani G, Bondioni S, Locatelli M, Pedroni C, Lania AG, et al. 2004. Biallelic expression of the Gsα gene in human bone and adipose tissue. *J. Clin. Endocrinol. Metab.* 89:6316–19
120. Martianov I, Brancorsini S, Catena R, Gansmuller A, Kotaja N, et al. 2005. Polar nuclear localization of H1T2, a histone H1 variant, required for spermatid elongation and DNA condensation during spermiogenesis. *Proc. Natl. Acad. Sci. USA* 102:2808–13
121. Mayer W, Niveleau A, Walter J, Fundele R, Haaf T. 2000. Demethylation of the zygotic paternal genome. *Nature* 403:501–2
122. Mikkelsen TS, Ku M, Jaffe DB, Issac B, Lieberman E, et al. 2007. Genome-wide maps of chromatin state in pluripotent and lineage-committed cells. *Nature* 448:553–60
123. Murrell A, Heeson S, Reik W. 2004. Interaction between differentially methylated regions partitions the imprinted genes *Igf2* and *H19* into parent-specific chromatin loops. *Nat. Genet.* 36:889–93
124. Neumann B, Kubicka P, Barlow DP. 1995. Characteristics of imprinted genes. *Nat. Genet.* 9:12–13
125. Nickel BE, Roth SY, Cook RG, Allis CD, Davie JR. 1987. Changes in the histone H2A variant H2A.Z and polyubiquitinated histone species in developing trout testis. *Biochemistry* 26:4417–21
126. O'Carroll D, Erhardt S, Pagani M, Barton SC, Surani MA, Jenuwein T. 2001. The polycomb-group gene *Ezh2* is required for early mouse development. *Mol. Cell Biol.* 21:4330–36

127. Ohno M, Aoki N, Sasaki H. 2001. Allele-specific detection of nascent transcripts by fluorescence *in situ* hybridization reveals temporal and culture-induced changes in *Igf2* imprinting during preimplantation mouse development. *Genes Cells* 6:249–59
128. Okada T, Endo M, Singh MB, Bhalla PL. 2005. Analysis of the histone H3 gene family in *Arabidopsis* and identification of the male-gamete-specific variant *AtMGH3*. *Plant. J.* 44:557–68
129. Okamura K, Sakaki Y, Ito T. 2005. Comparative genomics approach toward critical determinants for the imprinting of an evolutionarily conserved gene Impact. *Biochem. Biophys. Res. Commun.* 329:824–30
130. Ono R, Shiura H, Aburatani H, Kohda T, Kaneko-Ishino T, Ishino F. 2003. Identification of a large novel imprinted gene cluster on mouse proximal chromosome 6. *Genome. Res.* 13:1696–705
131. Orlando V, Jane EP, Chinwalla V, Harte PJ, Paro R. 1998. Binding of trithorax and Polycomb proteins to the bithorax complex: dynamic changes during early *Drosophila* embryogenesis. *EMBO J.* 17:5141–50
132. Oswald J, Engemann S, Lane N, Mayer W, Olek A, et al. 2000. Active demethylation of the paternal genome in the mouse zygote. *Curr. Biol.* 10:475–78
133. Paoloni-Giacobino A, D'Aiuto L, Cirio MC, Reinhart B, Chaillet JR. 2007. Conserved features of imprinted differentially methylated domains. *Gene* 399:33–45
134. Paro R, Zink B. 1992. The Polycomb gene is differentially regulated during oogenesis and embryogenesis in *Drosophila melanogaster*. *Mech. Dev.* 40:37–46
135. Paulsen M, Ferguson-Smith AC. 2001. DNA methylation in genomic imprinting, development, and disease. *J. Pathol.* 195:97–110
136. Peters J, Holmes R, Monk D, Beechey CV, Moore GE, Williamson CM. 2006. Imprinting control within the compact *Gnas* locus. *Cytogenet. Genome Res.* 113:194–201
137. Pirrotta V. 1997. PcG complexes and chromatin silencing. *Curr. Opin. Genet. Dev.* 7:249–58
138. Rachmilewitz J, Goshen R, Ariel I, Schneider T, de Groot N, Hochberg A. 1992. Parental imprinting of the human H19 gene. *FEBS Lett.* 309:25–28
139. Reik W, Collick A, Norris ML, Barton SC, Surani MA. 1987. Genomic imprinting determines methylation of parental alleles in transgenic mice. *Nature* 328:248–51
140. Reik W, Walter J. 2001. Genomic imprinting: parental influence on the genome. *Nat. Rev. Genet.* 2:21–32
141. Reinhart B, Eljanne M, Chaillet JR. 2002. Shared role for differentially methylated domains of imprinted genes. *Mol. Cell Biol.* 22:2089–98
142. Ridgway P, Brown KD, Rangasamy D, Svensson U, Tremethick DJ. 2004. Unique residues on the H2A.Z containing nucleosome surface are important for *Xenopus laevis* development. *J. Biol. Chem.* 279:43815–20
143. Riley PD, Carroll SB, Scott MP. 1987. The expression and regulation of Sex combs reduced protein in *Drosophila* embryos. *Genes Dev.* 1:716–30
144. Ringrose L, Ehret H, Paro R. 2004. Distinct contributions of histone H3 lysine 9 and 27 methylation to locus-specific stability of polycomb complexes. *Mol. Cell* 16:641–53
145. Robertson KD. 2005. DNA methylation and human disease. *Nat. Rev. Genet.* 6:597–610
146. Ronemus MJ, Galbiati M, Ticknor C, Chen J, Dellaporta SL. 1996. Demethylation-induced developmental pleiotropy in *Arabidopsis*. *Science* 273:654–57
147. Rossant J. 1993. Immortal germ cells? *Curr. Biol.* 3:47–49
148. Royo H, Bortolin ML, Seitz H, Cavaille J. 2006. Small noncoding RNAs and genomic imprinting. *Cytogenet. Genome Res.* 113:99–108
149. Santos F, Hendrich B, Reik W, Dean W. 2002. Dynamic reprogramming of DNA methylation in the early mouse embryo. *Dev. Biol.* 241:172–82
150. Sato S, Yoshimizu T, Sato E, Matsui Y. 2003. Erasure of methylation imprinting of *Igf2r* during mouse primordial germ-cell development. *Mol. Reprod. Dev.* 65:41–50
151. Schweizer J, Zynger D, Francke U. 1999. *In vivo* nuclease hypersensitivity studies reveal multiple sites of parental origin-dependent differential chromatin conformation in the 150 kb *SNRPN* transcription unit. *Hum. Mol. Genet.* 8:555–66
152. Scott RJ, Spielman M. 2004. Epigenetics: imprinting in plants and mammals—the same but different? *Curr. Biol.* 14:R201–3
153. Scott RJ, Spielman M. 2006. Genomic imprinting in plants and mammals: how life history constrains convergence. *Cytogenet Genome Res.* 113:53–67

154. Sha K, Fire A. 2005. Imprinting capacity of gamete lineages in *Caenorhabditis elegans*. *Genetics* 170:1633–52
155. Shumacher A, Faust C, Magnuson T. 1996. Positional cloning of a global regulator of anterior-posterior patterning in mice. *Nature* 383:250–53
156. Simon I, Tenzen T, Reubinoff BE, Hillman D, McCarrey JR, Cedar H. 1999. Asynchronous replication of imprinted genes is established in the gametes and maintained during development. *Nature* 401:929–32
157. Simpson VJ, Johnson TE, Hammen RF. 1986. *Caenorhabditis elegans* DNA does not contain 5-methylcytosine at any time during development or aging. *Nucleic Acids Res.* 14:6711–19
158. Sleutels F, Zwart R, Barlow DP. 2002. The noncoding *Air* RNA is required for silencing autosomal imprinted genes. *Nature* 415:810–13
159. Spillane C, Baroux C, Escobar-Restrepo JM, Page DR, Laoueille S, Grossniklaus U. 2004. Transposons and tandem repeats are not involved in the control of genomic imprinting at the *MEDEA* locus in *Arabidopsis*. *Cold Spring Harbor Symp. Quant. Biol.* 69:465–75
160. Swain JL, Stewart TA, Leder P. 1987. Parental legacy determines methylation and expression of an autosomal transgene: a molecular mechanism for parental imprinting. *Cell* 50:719–27
161. Szabo P, Tang SH, Rentsendorj A, Pfeifer GP, Mann JR. 2000. Maternal-specific footprints at putative CTCF sites in the *H19* imprinting control region give evidence for insulator function. *Curr. Biol.* 10:607–10
162. Takihara Y, Tomotsune D, Shirai M, Katoh-Fukui Y, Nishii K, et al. 1997. Targeted disruption of the mouse homologue of the *Drosophila* polyhomeotic gene leads to altered anteroposterior patterning and neural crest defects. *Development* 124:3673–82
163. Tanimoto K, Shimotsuma M, Matsuzaki H, Omori A, Bungert J, et al. 2005. Genomic imprinting recapitulated in the human β-globin locus. *Proc. Natl. Acad. Sci. USA* 102:10250–55
164. Thorvaldsen JL, Duran KL, Bartolomei MS. 1998. Deletion of the *H19* differentially methylated domain results in loss of imprinted expression of *H19* and *Igf2*. *Genes Dev.* 12:3693–702
165. Tsai CE, Lin SP, Ito M, Takagi N, Takada S, Ferguson-Smith AC. 2002. Genomic imprinting contributes to thyroid hormone metabolism in the mouse embryo. *Curr. Biol.* 12:1221–26
166. Tucker KL, Beard C, Dausmann J, Jackson-Grusby L, Laird PW, et al. 1996. Germ-line passage is required for establishment of methylation and expression patterns of imprinted but not of nonimprinted genes. *Genes Dev.* 10:1008–20
167. Turker MS, Bestor TH. 1997. Formation of methylation patterns in the mammalian genome. *Mutat. Res.* 386:119–30
168. Umlauf D, Goto Y, Cao R, Cerqueira F, Wagschal A, et al. 2004. Imprinting along the *Kcnq1* domain on mouse chromosome 7 involves repressive histone methylation and recruitment of polycomb group complexes. *Nat. Genet.* 36:1296
169. Verona RI, Mann MR, Bartolomei MS. 2003. Genomic imprinting: intricacies of epigenetic regulation in clusters. *Annu. Rev. Cell Dev. Biol.* 19:237–59
170. Vielle-Calzada JP, Thomas J, Spillane C, Coluccio A, Hoeppner MA, Grossniklaus U. 1999. Maintenance of genomic imprinting at the *Arabidopsis medea* locus requires zygotic *DDM1* activity. *Genes Dev.* 13:2971–82
171. Vire E, Brenner C, Deplus R, Blanchon L, Fraga M, et al. 2006. The polycomb group protein EZH2 directly controls DNA methylation. *Nature* 439:871–74
172. Vu TH, Hoffman AR. 1994. Promoter-specific imprinting of the human insulin-like growth factor-II gene. *Nature* 371:714–17
173. Vu TH, Li T, Hoffman AR. 2004. Promoter-restricted histone code, not the differentially methylated DNA regions or antisense transcripts, marks the imprinting status of *IGF2R* in human and mouse. *Hum. Mol. Genet.* 13:2233–45
174. Walter J, Hutter B, Khare T, Paulsen M. 2006. Repetitive elements in imprinted genes. *Cytogenet. Genome Res.* 113:109–15
175. Wang H, Wang L, Erdjument-Bromage H, Vidal M, Tempst P, et al. 2004. Role of histone H2A ubiquitination in Polycomb silencing. *Nature* 431:873–78
176. Watson CE, Gauthier SY, Davies PL. 1999. Structure and expression of the highly repetitive histone H1-related sperm chromatin proteins from winter flounder. *Eur. J. Biochem.* 262:258–67

177. Wilkins JF. 2006. Tissue-specific reactivation of gene expression at an imprinted locus. *J. Theor. Biol.* 240:277–87
178. Williamson CM, Ball ST, Nottingham WT, Skinner JA, Plagge A, et al. 2004. A *cis*-acting control region is required exclusively for the tissue-specific imprinting of *Gnas*. *Nat. Genet.* 36:894–99
179. Wutz A, Smrzka OW, Schweifer N, Schellander K, Wagner EF, Barlow DP. 1997. Imprinted expression of the *Igf2r* gene depends on an intronic CpG island. *Nature* 389:745–49
180. Xiao W, Gehring M, Choi Y, Margossian L, Pu H, et al. 2003. Imprinting of the *MEA* Polycomb gene is controlled by antagonism between MET1 methyltransferase and DME glycosylase. *Dev. Cell* 5:891–901
181. Xin Z, Allis CD, Wagstaff J. 2001. Parent-specific complementary patterns of histone H3 lysine 9 and H3 lysine 4 methylation at the Prader-Willi syndrome imprinting center. *Am. J. Hum. Genet.* 69:1389–94
182. Yamazaki Y, Low EW, Marikawa Y, Iwahashi K, Bartolomei MS, et al. 2005. Adult mice cloned from migrating primordial germ cells. *Proc. Natl. Acad. Sci. USA* 102:11361–66
183. Youngson NA, Kocialkowski S, Peel N, Ferguson-Smith AC. 2005. A small family of sushi-class retrotransposon-derived genes in mammals and their relation to genomic imprinting. *J. Mol. Evol.* 61:481–90
184. Yu BD, Hanson RD, Hess JL, Horning SE, Korsmeyer SJ. 1998. MLL, a mammalian *trithorax*-group gene, functions as a transcriptional maintenance factor in morphogenesis. *Proc. Natl. Acad. Sci. USA* 95:10632–36
185. Zhang H, Azevedo RB, Lints R, Doyle C, Teng Y, et al. 2003. Global regulation of Hox gene expression in *C. elegans* by a SAM domain protein. *Dev. Cell* 4:903–15
186. Zhang Y, Cao R, Wang L, Jones RS. 2004. Mechanism of Polycomb group gene silencing. *Cold Spring Harbor Symp. Quant. Biol.* 69:309–17

Phylogenetic Inference Using Whole Genomes

Bruce Rannala[1] and Ziheng Yang[2]

[1]Genome Center and Department of Evolution and Ecology, University of California, Davis, California 95616; email: bhrannala@ucdavis.edu

[2]Department of Biology, University College London, London WC1E 6BT United Kingdom; Laboratory of Biometrics, Graduate School of Agriculture and Life Sciences, University of Tokyo, Tokyo, Japan; email: z.yang@ucl.ac.uk

Annu. Rev. Genomics Hum. Genet. 2008. 9:217–31

First published online as a Review in Advance on June 3, 2008

The *Annual Review of Genomics and Human Genetics* is online at genom.annualreviews.org

This article's doi:
10.1146/annurev.genom.9.081307.164407

1527-8204/08/0922-0217$20.00

Key Words

genomic sequences, coalescent process, gene tree

Abstract

The availability of genome-wide data provides unprecedented opportunities for resolving difficult phylogenetic relationships and for studying population genetic processes of mutation, selection, and recombination on a genomic scale. The use of appropriate statistical models becomes increasingly important when we are faced with very large datasets, which can lead to improved precision but not necessarily improved accuracy if the analytical methods have systematic biases. This review provides a critical examination of methods for analyzing genomic datasets from multiple loci, including concatenation, separate gene-by-gene analyses, and statistical models that accommodate heterogeneity in different aspects of the evolutionary process among data partitions. We discuss factors that may cause the gene tree to differ from the species tree, as well as strategies for estimating species phylogenies in the presence of gene tree conflicts. Genomic datasets provide computational and statistical challenges that are likely to be a focus of research for years to come.

INTRODUCTION

Species tree: the unobserved tree of genealogical relationships among the species from which the genes are sampled

Gene tree: the unobserved tree of genealogical relationships among genes through time

The genome sequencing era began three decades ago with the sequencing, in 1977, of the 5368-bp DNA genome of the bacteriophage virus ϕX174 (53). Automated sequencing technologies enabled the first bacterial genome, the 1830-kb genome of *Haemophilus influenzae* (21), to be sequenced in 1995, followed by the first eukaryotic genome, the 12.5-Mb genome of the budding yeast *Saccharomyces cerevisiae* (23), in 1997. During the current decade, the number of sequenced genomes has grown exponentially. As of October 2007, Entrez (**http://ncbi.nlm.nih.gov/**) lists 543 completed genomes for eubacterial species and 47 for archaeal species. There are 23 completed eukaryotic genomes and 129 draft genomes. These include two completed mammalian genomes (human and mouse), 21 draft assembly mammalian genomes, and 26 mammalian genomes in progress. These numbers are likely to increase by as much as tenfold by the close of the current decade.

During the last several years the potential value of comparative genomics for the identification of genes, regulatory regions, and other genome features has shifted sequencing efforts away from model organisms such as mouse and *Drosophila* to include other related species. Numerous completed genome sequences are now available for evolutionarily related species, opening up the possibility of using whole genomes to infer phylogenetic relationships and divergence times among species. Moreover, new sequencing technologies have enabled resequencing of genomes for multiple individuals of a single species, strain, or population. The newly emergent fields of phylogenomics and population genomics are one consequence of these technological advances. Although the availability of whole genome sequences is quite new, the basic principles of multilocus inference in phylogenetics and population genetics, developed and refined over the last two decades, are relatively well established.

The objective of this review is to describe how existing tools for phylogenetic inference can be applied to whole genomes. The use of robust methods of analysis is clearly extremely important when such large amounts of data are analyzed; errors induced by phylogenetic inference techniques known to be prone to large-sample problems such as statistical inconsistency can be greatly magnified by the use of whole genome data (43, 50). Moreover, new problems arise, such as how to account for the effects of recombination, gene conversion, and horizontal gene transfer. In this review, we do not deal with details of genome sequencing such as sequence assembly. Nor do we consider the important related problem of sequence alignment. Rather, we focus exclusively on the problem of accurately inferring phylogenetic trees and species divergence times using a sample of aligned orthologous sequences for regions that span an entire genome. Even this relatively focused endeavor can become quite complicated in many situations of biological interest.

SPECIES TREES AND GENE TREES

Amino acid sequences, cross-reactivity of antibodies, and other measures of evolutionary divergence among multiple proteins (and species) first became widely available during the 1960s (13). Researchers developing measures of genetic distance intended for use with such data noted early on the distinction between organismal phylogenies and molecular phylogenies. Fitch (20), for example, proposed new terms to clarify the distinction between orthologs (genes descending from a shared ancestral gene owing to a shared species divergence event) and paralogs (genes descending from a shared ancestral gene owing to a gene duplication event). Nei (41), considering the problem of dating species divergence events using immunological distances, recognized that the objective was to "reconstruct or estimate the evolutionary tree of *the organisms used* rather than that of a protein." Tateno and coworkers (62) made a similar distinction, noting that "the primary objective of molecular taxonomy or phylogenetics is to construct a species tree rather than a gene tree" and speculating that "the only way to reduce the errors involved in an estimated tree is to increase the number of genes used." In the last two decades, the importance of distinguishing between gene and species trees has been widely recognized and researchers have identified new sources of gene and species tree conflict that were previously unknown (or highly speculative). Recent genome sequencing efforts provide the opportunity for an almost limitless number of genes to be employed in a phylogenetic analysis aimed at identifying a species tree.

Sources of Gene Tree Conflict

In discussing gene tree conflict, it is essential to distinguish between an estimated gene tree and the true gene tree. The true gene tree is the unobserved tree of genealogical relationships among genes through time. The estimated gene tree is the current best estimate of that tree based on DNA sequence data. Even when true gene trees are identical between genes, estimated gene trees may differ owing to random and systematic errors in phylogenetic tree reconstruction. In analyses that use well-behaved statistical methods known to produce consistent estimates, such as maximum likelihood or Bayesian inference, such errors can be reduced by adding sites to a gene. However, underlying biological processes can cause the true gene trees to differ, in which case the estimated gene trees can differ regardless of the number of sites examined (see discussion below). Here, we use the term "gene tree conflict" to refer only to cases in which true gene trees differ. True gene trees can differ either in divergence times or tree topology. The species tree is the unobserved tree of genealogical relationships among the species from which the genes are sampled.

The two major types of biological process that can lead to conflicts among gene trees are, first, population genetic processes such as drift operating in an ancestral species, and second, genomic recombination, either within a single species (e.g., gene conversion, transposition, or meiotic crossovers) or between species (horizontal/lateral gene transfer). Because the effects of genomic recombination in causing conflicts between gene and species trees may depend on whether ancestral polymorphisms are present, we consider ancestral population effects first. We then go on to discuss the role of genomic recombination in generating conflicts and the influence of processes such as meiotic crossovers within populations on the probability of gene and species tree conflicts.

Maximum likelihood: a statistical method for estimating parameters in a statistical model by maximizing the probability of the observed data

Bayesian inference: an approach to statistical inference that uses probability distributions to describe uncertainties in model parameters

Genomic recombination: the exchange of sequence information between distinct DNA molecules

Ancestral polymorphism: polymorphism or sequence differences in an extinct ancestral species

Ancestral Polymorphisms

Species divergence times and ages of most recent common ancestors. To illustrate the effect of ancestral polymorphism, consider a homologous segment of DNA in the human and chimpanzee genomes. Suppose we sample a single human and chimpanzee sequence; we are interested in using the pattern of DNA substitutions between the pair of aligned sequences to infer the age of the human-chimpanzee speciation event, t_{HC}. Assuming that no gene flow occurs subsequent to the speciation event (which indeed may be taken as the definition of speciation), it is impossible for the sequences to share a most recent common ancestor (MRCA) that is younger than the age of the speciation event. The time until the MRCA, T_{HC}, is then determined by population genetic processes operating in the human-chimpanzee ancestral population, such as genetic drift (**Figure 1**). Coalescent theory (25, 30, 60) can be used to calculate the probability density of the discrepancy $T_{HC} - t_{HC}$:

$$f(T_{HC}) = \frac{1}{2N_e} e^{-(T_{HC} - t_{HC})/(2N_e)}, \quad \text{for } T_{HC} > t_{HC},$$

where time is measured in units of generations. This is an exponential distribution with mean $E(T_{HC} - t_{HC}) = 2N_e$ and variance $\text{var}(T_{HC}) = 4N_e^2$. The degree to which such discrepancies can be detected (and thus influence resulting estimates) when using sequence data depends on the relative accuracy of the branch length estimates (in units of expected numbers of substitutions), the mutation rate, the effective population size, and the generation time. The

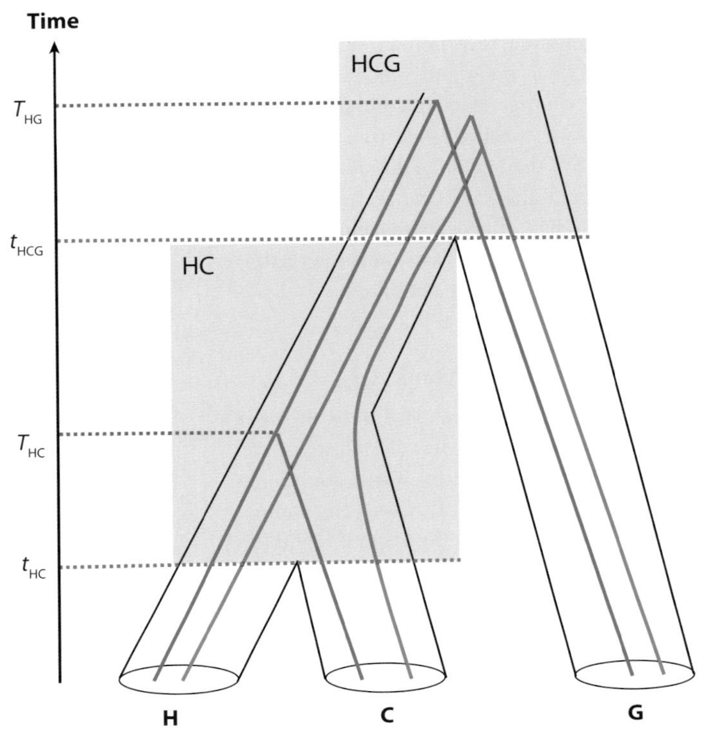

Figure 1

A species tree for three species [human (H), chimpanzee (C), and gorilla (G)] used to illustrate lineage sorting, which may generate a species tree–gene tree conflict. The species tree is ((H, C), G), with the species divergence times t_{HC} and t_{HCG}. The two ancestral species (populations) are represented as HC and HCG. Two gene trees are shown. In the blue gene tree, the H and C sequences coalesce in population HC, and the resulting gene tree matches the species tree. In the red gene tree, both coalescent events occur in population HCG, and the resulting gene tree differs from the species tree.

accuracy of branch length estimation depends on the number of sites analyzed and the substitution model used. If $v = v_1 - v_0$ is the smallest discrepancy that can be detected (in units of expected mutations) then the probability that a detectable discrepancy is observed between the branch length expected under the true species divergence time, $v_0 = t_{HC}G\mu$, and that of the MRCA, $v_1 = T_{HC}G\mu$, is

$$p = \int_{v/\mu}^{\infty} \frac{e^{-y/(2N_e G)}}{2N_e G} dy = e^{-v/(2N_e G\mu)},$$

where G is the generation time (in years) and μ is the mutation rate (per year). Considering the human-chimpanzee divergence, for example, if the human-chimpanzee ancestral effective population size is $N_e = 100{,}000$ (61), the generation time is $G = 20$ years, and the mutation rate is $\mu = 10^{-9}$ mutations per site per year, then a branch length discrepancy of $v = 0.01$ or larger will occur with probability $p = 0.08$ and a branch length discrepancy of $v = 0.005$ or larger will occur with probability $p = 0.29$. Given that potentially thousands of independent genes can be examined in a whole-genome phylogenetic analysis, discrepancies larger than $v = 0.01$ can be expected to commonly occur. If such discrepancies occur and are not accounted for it is evident that using node ages on the inferred phylogeny will tend to overestimate species divergence times. Indeed, if the human-chimpanzee speciation event occurred 5 million years ago, the average sequence divergence between the two species will be $E(T_{HC})G\mu = t_{HC}G\mu + 2N_eG\mu = 0.005 + 0.004$, so that $0.004/0.009 = 44\%$ of the divergence is due to ancestral polymorphism. To accurately estimate species divergence times in such cases population genetic parameters, such as ancestral effective population sizes, should be jointly estimated with species divergence times using sequence data under a population genetic model (e.g., see 61).

Lineage sorting and conflicting gene and species trees. If three or more species are examined, the ancestral coalescent process may generate discrepancies between species tree and gene tree topologies (26). Consider the species tree of human (H), chimpanzee (C), and gorilla (G), shown in **Figure 1**. If the H and C sequences coalesce in the ancestral species HC, the gene tree will be topologically the same as the species tree. However, if the H and C sequences do not coalesce in the HC population, they will enter the ancestral population HCG. Then all three sequences will coalesce in random order, and only one of the three possible resulting trees matches the species tree. Thus the probability that the gene tree differs from the species tree equals 2/3 the probability that the H and C sequences do not coalesce in the HC ancestral population

$$P_{SG} = \frac{2}{3} e^{-(t_{HCG} - t_{HC})/(2N_e)},$$

where N_e is the effective population size of population HC (26). Note that the mismatch probability P_{SG} is greater when the two speciation events are closer in time and when the ancestral HC population is larger.

In real data analysis, the gene tree is unknown and is inferred from sequence data at the locus. Errors in phylogeny reconstruction will then inflate the mismatch probability; that is, the probability that the species tree differs from the estimated gene tree, P_{SE}, is always greater than P_{SG} (69). A commonly used approach to estimating ancestral population sizes in the case of three species is to equate the observed mismatch probability, P_{SE}, with the expected probability, P_{SG}, ignoring errors in phylogeny reconstruction (8). This so-called tree-mismatch method can seriously overestimate ancestral population sizes. **Figure 2** illustrates the relationship between the probability of a gene and species tree mismatch for either estimated or true gene trees as a function of sequence length.

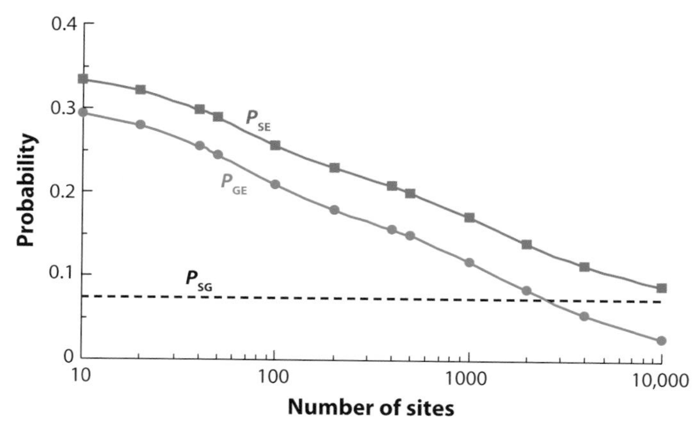

Figure 2

Tree mismatch probabilities plotted against locus size (in number of sites), calculated using computer simulation. P_{SG} is the mismatch probability between the species tree and the gene tree (0.0739 for the parameter values used). P_{SE} is the mismatch probability between the species tree and the estimated gene tree. This is much greater than P_{SG} for short loci but converges to P_{SG} with increasing sequence length. P_{GE} is the probability of a mismatch between the true gene tree and the estimated gene tree, and is the probability of error in phylogeny reconstruction. The simulation was conducted using the species tree of **Figure 1** and the following parameter values: $\theta_{HC} = 4N_{HC}G\mu = 0.0010$ for population HC, $\theta_{HCG} = 0.0031$ for population HCG, $t_{HC}G\mu = 0.0052$ for the H-C divergence, and $t_{HCG}G\mu = 0.0063$ for the HC-G divergence. Redrawn according to 69.

Similar to the case of three species, ancestral polymorphism can cause species tree and gene tree topologies to differ when the number of species is greater than three. Several authors have derived the gene and species tree mismatch probabilities for various numbers of species when the species tree is fixed (10, 25, 45).

Another question of relevance for phylogenetic inference is how often the most common gene tree differs from the species tree (9, 33). Recently, the finding that under some conditions the most common gene tree topology does not match the species tree has attracted much attention. This occurs in situations where the species tree is highly asymmetrical and arises from the fact that the coalescent process places a uniform prior on labeled histories, rather than on topologies. The labeled history (14) takes the rank ordering of the nodes in a tree into account as well as the cladogenic relationships (47). A completely asymmetrical (unbalanced) species tree is compatible with only one labeled history (because there is only one ordering of the coalescent times). However, if the species tree is not completely asymmetrical, it can correspond to several labeled histories and thus receives more weight in the prior than a completely asymmetrical tree. This is only an issue for particular tree shapes, but in such situations it could cause phylogenetic procedures based on unweighted summaries of gene trees (such as supertree methods or gene concatenation–based methods) to be inconsistent (9, 33). In practice, such situations are likely to be rare but the results do suggest that it is advisable to use a multigene phylogenetic inference method with independent gene trees under a coalescent prior conditional on the species tree (see below). The coalescent prior gives appropriate prior weight to different possible gene trees for a particular species tree, leading to a consistent estimator of the species tree and avoiding the problem identified by Degnan & Rosenberg (9) and Kubatko & Degnan (33).

Effects of Genomic Recombination

Correlated gene trees and recombination. Genes that are located on the same linear segment of DNA (e.g., on a single chromosome) will, in the absence of processes such as gene conversion or meiotic recombination, have identical gene trees. This is the case for genes of the mitochondrial genome, for example. Although the gene trees are identical in this case, they may still differ from the species tree. In a phylogenetic analysis of genes that are expected to have identical trees, it is appropriate to model the process assuming a common gene tree. Substitution rates and other parameters may still vary across genes, however, so separate parameters for these processes may still be needed (see discussion below). For regions that have undergone meiotic recombination the gene trees may differ but are often highly correlated. Hudson (25) described the joint probability distribution of the gene trees for a pair of linked loci under the ancestral recombination and coalescence process within a single population.

The probability that, in a diploid species, the first crossover event on the interval between the linked genes occurs prior to time T (in units of generations) is approximately

$$\int_0^T e^{-4N_e rt}\,dt = \frac{1}{4N_e r}\left(1 - e^{-4N_e rT}\right),$$

where r is the recombination fraction between the genes per generation (e.g., the linkage distance in units of Morgans). The probability that the first coalescence occurs prior to time T is approximately

$$\int_0^T \frac{1}{2N_e} e^{-t/2N_e}\,dt = 1 - e^{-T/2N_e}.$$

If ancestral polymorphism exists, the recombination and coalescence processes compete to determine the gene tree correlations. If a coalescence event occurs first, there is an identical

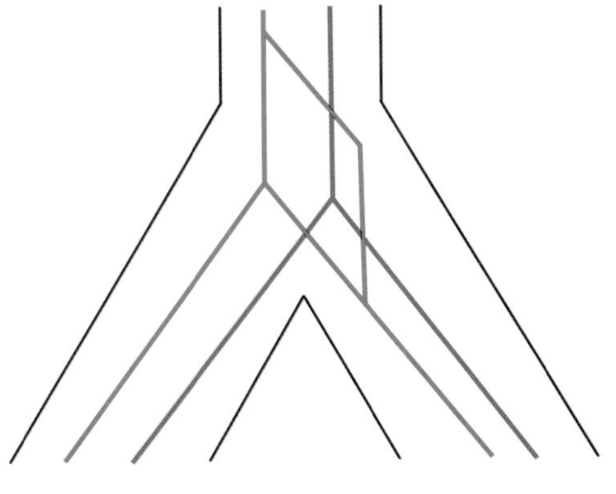

Figure 3

Node of a species tree illustrating the effects of coalescence and meiotic recombination in determining whether two genes have identical or different ages for a most recent common ancestor (MRCA) in the gene tree. The lineages in blue represent a case where a coalescence event occurs before a recombination. In that case, there is a shared MRCA for the two genes. The lineages in red represent a case where a recombination occurs before a coalescence event. Two MRCAs then exist and the two genes have independent histories at the node.

gene and species tree at the node for the linked genes. If a recombination occurs first, there are independent gene and species trees at the node (**Figure 3**).

Recombinations and coalescences have the same cumulative probability distribution if $r = 1/(4N_e)$, whereas if $r \gg 1/(4N_e)$ the recombination process will dominate (and gene trees will tend to be independent) and if $r \ll 1/(4N_e)$ the coalescent process will dominate (and gene trees will tend to be identical). For example, if the human-chimpanzee ancestor had an effective population size of $N_e = 100{,}000$ (61), then $1/(4N_e) = 1/400000 = 0.0000025$ is the critical value for r. On average, in the human genome 1 Mb $\approx$ 1 cM and the critical value therefore corresponds to a physical interval approximately 250 bp in size. Thus, gene trees for human autosomal genes separated by a distance greater than $\approx$2.5 kb might be treated as independent for phylogenetic inference. Of course, if the ancestral population size is small then all gene trees will agree with the species tree and the recombination process is irrelevant; this is probably not true for the human-chimpanzee divergence mentioned earlier.

The theory presented above provides, at best, a rough guide for choosing whether to model gene trees for linked genes as dependent or independent in a phylogenetic analysis. Both N_e and the relationship between physical distance and genetic distance can be expected to vary considerably even among closely related species. Moreover, the relationship between physical distance and genetic distance can also vary greatly across a genome even within species; this is evident from recent studies in humans and other species for which relatively precise meiotic recombination rates as well as a completed genome sequence are available (31, 37). The effective population size can also vary across a genome owing to past evolutionary forces such as directional selective sweeps affecting particular loci (which leads to younger genealogies than expected under a neutral coalescent and smaller N_e), or overdominance (which leads to older genealogies and larger N_e).

Instead of using population genetic models to predict the physical distances over which genes may be treated as independent for the purposes of a phylogenetic analysis, as we have done above, one could attempt to infer the presence of recombination using the sequence data (17, 38), so that the correlations among gene trees are inferred as part of the phylogenetic analysis. Although promising, such approaches can present significant computational challenges. A conservative approach would be to use only a small subset of the tens or hundreds of thousands of genes available for many genomes, choosing genes separated by intervals of say 1 Mb. Although this might seem wasteful, it is clear from many empirical analyses and simulation studies that species phylogenies can often be precisely inferred with far fewer genes than are available in the genome as a whole, suggesting that the gain in simplicity and reduced model complexity and assumptions may at least partially offset the loss of data in this case.

Horizontal (lateral) gene transfer. Horizontal gene transfer (HGT) between species is well documented in prokaryotes (32), and typically occurs via processes such as bacterial transformation and conjugation. HGT is also likely to occur in most eukaryotes via processes such as transduction of viral genes. Even very low rates of HGT can have a large impact on phylogeny, disrupting the usual patterns of vertical transmission of genes from parents to offspring and causing gene trees to differ from the species tree. Indeed, with high levels of HGT even the existence of a species tree may be questioned (12). In principle, one could develop a model of horizontal gene transfer among species, exploiting similarities to the population genetic process of migration among populations. However, most recent attempts to model HGT for the purposes of phylogenetic inference have used much simpler models (see below).

Horizontal gene transfer (HGT): the transfer of genetic material from one species to another species; also known as lateral gene transfer

PHYLOGENOMIC INFERENCE

Analyzing Multigene Data

Traditional approaches to phylogenetic inference make the assumption (implicitly or explicitly) that a single phylogenetic tree underlies the data. The population genetic and evolutionary processes outlined above can contradict this assumption when sequence data from multiple genes are analyzed, potentially producing erroneous conclusions in a phylogenetic analysis. Here we discuss some of the methods that have been proposed for analyzing multigene datasets, with an emphasis on parametric statistical methods.

The basic parameters of a phylogenetic analysis are the tree topology, τ, the branch lengths, ν, and the substitution model parameters, θ (e.g., nucleotide frequencies and transition/transversion rate ratios). With a single gene the maximum likelihood estimator of τ and ν is obtained by maximizing the likelihood of the sampled sequences $\mathbf{X} = \{X_i\}$ with respect to the model parameters (18) for a given tree topology τ_k,

$$L_k = \max_{\theta, \nu} \prod_{i=1}^{n} f(X_i | \theta, \nu, \tau_k),$$

and then choosing the topology τ_j with $L_j > L_k$ for all $k \neq j$. The equivalent Bayesian formulation (47) places priors on all parameters and produces a posterior distribution of phylogenetic trees,

$$f(\tau|\mathbf{X}) = \int_\theta \int_\nu f(\mathbf{X}|\theta, \nu, \tau) f(\tau, \nu) f(\theta) d\theta d\nu / f(\mathbf{X}).$$

Several strategies for analyzing data from multiple loci or genome regions are available. The simplest is to concatenate the sequences, replacing the missing data with question marks, and analyze the data as one "supergene." This approach is commonly used (e.g., see 40). In likelihood and Bayesian methods, such an analysis uses the same set of parameters for all genes and ignores possible heterogeneities among the genes. It is known that different genes may evolve at very different rates and could have different base compositions or different transition-transversion rate biases, etc., but such differences are ignored by the concatenation approach. If factors leading to gene and species tree conflicts exist, such as ancestral polymorphism or HGT, concatenation can also lead to inconsistency of the phylogenetic inference method (e.g., see 33).

An alternative method is to analyze the different genes separately, and then sum log likelihoods across genes. In a modeling framework, the concatenation approach assumes the simplest model, in which one set of parameters applies to all genes, whereas the separate analysis uses the most general model, in which each gene has its own set of parameters. Proposed methods of multigene analysis correspond to various ways of partitioning the genome, allowing model parameters to vary across genes. In this review, we distinguish between partitioning of the genome according to the substitution model and partitioning according to the underlying gene trees. Often, both types of partitions are needed for accurate phylogenetic inference.

Partitioning the genome according to substitution process. An important factor to accommodate in all multigene analyses is variation in substitution rates (expected branch lengths) and parameters of the substitution model across genes. The Bayesian formulation is

$$f(\tau, \underline{\nu}|\mathbf{X}) = \int_{\underline{\theta}} f(\mathbf{X}|\underline{\theta}, \underline{\nu}, \tau) f(\tau, \underline{\nu}) f(\underline{\theta})\, \mathrm{d}\underline{\theta} / f(\mathbf{X}),$$

where, if there are k genes, $\underline{\theta} = \{\theta_1, \ldots, \theta_k\}$, $\underline{\nu} = \{\nu_1, \ldots, \nu_k\}$, θ_j are the parameters of the substitution model for the *j*th gene, ν_j are the branch lengths for the *j*th gene, and so on.

Yang (68) discussed strategies for partitioning the data according to the substitution model and used maximum likelihood to implement a number of models that lie between the two extremes mentioned above (i.e., one model for all genes versus a separate model for each gene), which allow some aspects of the evolutionary process to be the same among genes while other aspects are different. All models implemented by Yang (68) assume that the branch lengths are proportional. Biologically, this model assumes that either a molecular clock exists (with different genes evolving at different overall rates) or lineage-specific rate changes apply similarly across all genes. Pupko and colleagues (46) implemented additional models, and used the Akaike information criterion (AIC) (1) to evaluate their fit to real data sets.

Ren and coworkers (49), Shapiro and coworkers (55), and Bofkin & Goldman (5) evaluate different strategies for analyzing protein-coding genes and conclude that it is important to account for the differences in the evolutionary process (such as rates, base compositions, and transition/transversion rate ratios) among the three codon positions. Several studies also evaluated the utility of codon models (24, 42) for phylogenetic analysis of protein-coding genes, and found that although computationally expensive, they were effective in recovering difficult phylogenies (43, 49). Nucleotide-based models that account for the differences in the three codon positions offer a computationally feasible substitute (55). Differences among the codon positions, likely a reflection of how purifying natural selection acting on the protein interacts with the genetic code, are a major feature of the evolutionary process of protein-coding genes, and it is important to accommodate them in a phylogenetic analysis of such data (6, 56, 64). Rate variation among sites within a partition (codon position) can be accommodated via the use of a random rate distribution such as the gamma (65, 66).

Programs that implement partitioned substitution models. Models for phylogenetic analysis of data from multiple genes or genomic partitions using maximum likelihood are not well developed in currently available programs. The BASEML and CODEML programs in the PAML package implement the models of Yang (68), but these programs do not include efficient tree search algorithms, and are not usable in phylogeny reconstruction with more than a dozen species. The PAUP* program (59) includes the site-specific rates model (invoked by setting up partitions and then specifying the option rates = sitespec in the lset command). This is the same as the proportional-branch model (68) and allows different partitions to have different rates—the option might thus be more appropriately called a partition-specific rates model. The program does not include models that allow other features of the evolutionary process (such as base compositions, transition/transversion rate ratio, and among-site rate variation) to differ among partitions.

The likelihood models discussed above can be used in Bayesian phylogenetic inference, as demonstrated by Suchard and colleagues (58) and Nylander and coworkers (44). MrBayes (51) is currently the only Bayesian tree-reconstruction program that has implemented a variety of models for combined analysis of multipartition datasets (invoked by setting up partitions and then using the link and unlink commands).

Partitioning the genome across local gene trees. To allow for processes such as lineage sorting and HGT that can cause gene trees and species trees to differ, one must partition the genome so that different genes may have potentially different underlying (true) gene trees. Such processes lead to different priors on the set of gene trees for the sampled genes (possibly conditional on a species tree). For example, Rannala & Yang (48) used a prior on gene trees derived on the basis of a coalescent process operating within the context of a fixed species tree topology (48). Under the molecular clock or relaxed clock models the species divergence times (rather than the branch lengths) are considered parameters of the model and may be estimated if some independent information is available concerning rates of substitution, for example from fossil-based age calibrations on one or more nodes (48). This approach effectively allows for variable gene trees owing to lineage sorting in estimating ancestral population sizes and species divergence times by integrating over the probability distribution of unobserved gene trees. Although the probability density of gene trees was not of specific interest, this information is also generated as a byproduct of the analysis. By integrating over unobserved gene trees under a coalescent prior the lineage sorting process is explicitly accounted for in the model, and the problems with inconsistency of phylogenetic inference methods described by Degnan & Rosenberg (9) are no longer an issue.

The Bayesian approach has been extended to integrate across the unobserved species tree as well as gene trees, effectively allowing a common species tree to be estimated while accounting for the effects of lineage sorting (15). The difficulty of performing the numerical calculations for this model led to the use of several ad hoc approximations, whose effect on accuracy needs to be studied further; the development of fully Bayesian methods is desirable.

Researchers have also developed phylogenetic inference methods that are intended to accommodate gene tree variations that arise

Lineage sorting: random coalescent events in ancestral populations that generate gene trees for the extant species that differ from the species tree as a result of horizontal gene transfer. These methods include a Bayesian approach that integrates over gene trees using a heuristic model of recombination based on the SPR algorithm. The SPR algorithm is commonly used to search among phylogenetic trees—however, in this case, the SPR is used as a physical model of HGT (57). The use of SPR in this context has been criticized because it does not impose the obvious time constraint that lineages involved in a horizontal gene transfer event must be contemporary (22). Approximate maximum likelihood methods have also been proposed that aim to estimate the extent of HGT (36) or to estimate phylogenetic networks (29). The likelihood method of Linz (36) is practical with only a very small number of taxa and a composite-likelihood approximation is used to deal with larger numbers of taxa. It is not currently clear how well these approximations perform. Many methods have been proposed for constructing networks from data comprised of genes with potentially different underlying gene trees. However, such networks have no clear biological interpretations. Bayesian phylogenetic models of host-parasite cospeciation (28) in which host switching occurs share many features with HGT processes and could potentially be adapted for use in modeling HGT for phylogenetic inference, with the host tree playing the same role as the species tree and the parasite trees playing the role of gene trees.

The recent likelihood method of Ané and colleagues (2) allows an independent gene tree and substitution model to underlie each gene. Genes are analyzed separately using Bayesian analysis and posterior distributions of gene trees are then combined through the use of a gene-to-tree map that is, in turn, used to estimate the proportion of genes for which any given clade is true (the sample-wide concordance factor). A drawback of this method is that the prior does not explicitly model biological processes such as lineage sorting or HGT and therefore the clades with high concordance factors are not necessarily present in the species tree.

Strategies and Difficulties in Site Partitioning

Thus far, we have used the term gene to refer operationally to a particular set of sites in a sequence alignment. Although it is often natural to partition sites according to genes, other strategies may be more appropriate for particular data sets. The main consideration in partitioning sites should be to accommodate the most important types of large-scale heterogeneity among sites. Features to be considered may include the evolutionary rate, the base composition, or the local genealogical tree topology induced by coalescent and/or HGT processes. For example, in vertebrate mitochondrial protein-coding genes, the three codon positions have very different evolutionary rates as well as different base compositions and transition/transversion rate ratios, but the differences among the genes are not so great (e.g., see 35). In this case it is better to partition the data by codon position than by gene (e.g., see 54, 67).

In the maximum likelihood method, allowing separate rates for each gene implies a great number of parameters if thousands of genes are analyzed jointly. Estimating so many parameters by maximum likelihood may pose computational problems. Furthermore, the statistical performance of the method may be affected as well, especially if some genes are small or otherwise uninformative (19). A standard statistical practice for dealing with the problem of too many parameters is to use a statistical distribution to describe the among-partition rate variation, in the same way that the gamma model is used to describe within-partition substitution rate variation among sites (65). The likelihood function then involves integrating over the among-partition distribution, and may be expensive to calculate.

In comparison, such multiparameter models are relatively easy to implement in a hierarchical Bayesian framework. The rate (or other parameters reflecting features of the evolutionary process) for a partition is assigned a prior, and integration over the prior is carried out in the Markov chain Monte Carlo algorithm in a straightforward manner. As in the case of likelihood methods, careful thought is needed to choose an appropriate partitioning of the sites.

Supermatrix and Supertree Controversies

Debate is ongoing in the literature concerning two particular strategies for phylogenetic analysis of multigene data sets, especially when some of the genes to be analyzed are not yet sequenced in some species. The supermatrix method concatenates sequences from multiple loci into a supersequence, with missing data represented as question marks, and uses the resulting data supermatrix to perform phylogenetic analysis. The supertree method instead conducts phylogenetic analysis on individual genes separately, and then uses one of several heuristic algorithms to combine the subtrees from the individual genes into a supertree for all species (for a summary, see 4). Several reviews have been published that either support the use of supermatrix methods (e.g., see 11) or instead advocate the use of supertree methods (3, 4, 52).

From a statistical modeling perspective, the debate is moot because both methods have serious drawbacks when used to analyze multigene data. The supermatrix method uses a simplistic substitution model that ignores heterogeneities in the evolutionary process among genes. Numerous simulation studies suggest that ignoring among-site or among-partition heterogeneities in the model can adversely affect phylogenetic analysis, sometimes causing systematic biases in the estimated tree (e.g., see 27, 34, 63). Moreover, as mentioned previously, simulation studies (33) show that lineage sorting may lead to inconsistent estimates of the species tree when concatenation is used, although the circumstances in which this may occur are probably rare.

The supertree method estimates an independent set of parameters for every gene and may overfit the data, inflating the variances of the estimates. Most supertree methods for constructing composite species trees use heuristic algorithms that lack a statistical basis and ignore uncertainties in the estimated subtrees (such as bootstrap support values, Bayesian posterior clade probabilities, or estimated branch lengths). Although ad hoc approaches have been suggested to remedy this (7, 39), their statistical performance has not been adequately studied. Computer simulations (e.g., see 16) tend to suggest that supertree algorithms can perform poorly even in the best-case scenario where the multiple genes are of the same length and evolve at the same rate under the same evolutionary model so that the information content in each gene is roughly equal. For example, the performance of some supertree methods can even deteriorate with the inclusion of more genes in the dataset.

In summary, although supertree methods can be useful as a tool for generating empirical summaries of the phylogenetic trees obtained in different studies from different types of characters, they are not statistically efficient for analyzing genomic data from multiple loci. Much of the theoretical research on supertree algorithms, mostly using the parsimony method of phylogenetic tree reconstruction, has emphasized combinatorial properties and computational algorithms but has neglected to examine basic statistical properties of the methods.

FUTURE ISSUES

Phylogenetic inference using whole genome data poses tremendous statistical and computational challenges. There is a profound need to develop new models for the analysis of multigene or multipartition datasets that can accommodate factors such as the heterogeneity of the evolutionary process among genes, or partitions, in whole genome phylogenetic analyses. Improved statistical methods are needed that account for genomic variation in evolutionary rates, transition/transversion rate ratios, and local gene trees. Moreover, there is an urgent need to develop efficient computer programs for combined analysis of multipartition datasets, particularly those suitable for parallel computer systems. Genome-wide phylogenetic inference will likely continue to present challenging problems for computational biologists for years to come.

SUMMARY POINTS

1. Genome-wide datasets offer opportunities for resolving difficult phylogenetic problems but also pose computational and statistical challenges for data analysis.
2. Multigene datasets should ideally be analyzed jointly, with the heterogeneity among data partitions appropriately accounted for in the model.
3. Neither supermatrix nor supertree methods are adequate for analysis of multipartition genome-wide data sets.
4. It is important to develop new statistical models and computational algorithms for efficient analysis of multigene data sets.

DISCLOSURE STATEMENT

The authors are not aware of any biases that might be perceived as affecting the objectivity of this review.

ACKNOWLEDGMENTS

This paper was written while the authors were guests of the Laboratory of Biometrics, University of Tokyo. We express our gratitude for the generous hospitality of our host, Professor Hiro Kishino. Z.Y. is supported by the Natural Environmental Sciences Research Council (NERC, UK).

LITERATURE CITED

1. Akaike H. 1974. A new look at the statistical model identification. *IEEE Trans. Autom. Contr AC* 19:716–23
2. Ané C, Larget B, Baum DA, Smith SD, Rokas A. 2007. Bayesian estimation of concordance among gene trees. *Mol. Biol. Evol.* 24:412–26
3. Bininda-Emonds ORP. 2004. *Phylogenetic Supertrees: Combining Information to Reveal the Tree of Life*. Dordrecht, the Netherlands: Kluwer Academic
4. Bininda-Emonds ORP. 2005. Supertree construction in the genomic age. *Methods Enzymol.* 395:745–57
5. Bofkin L, Goldman N. 2007. Variation in evolutionary processes at different codon positions. *Mol. Biol. Evol.* 24:513–21
6. Buckley TR, Simon C, Chambers GK. 2001. Exploring among-site rate variation models in a maximum likelihood framework using empirical data: effects of model assumptions on estimates of topology, branch lengths, and bootstrap support. *Syst. Biol.* 50:67–86
7. Burleigh JG, Driskell AC, Sanderson MJ. 2006. Supertree bootstrapping methods for assessing phylogenetic variation among genes in genome-scale data sets. *Syst. Biol.* 55:426–40
8. Chen F-C, Li W-H. 2001. Genomic divergences between humans and other hominoids and the effective population size of the common ancestor of humans and chimpanzees. *Am. J. Hum. Genet.* 68:444–56
9. Degnan JH, Rosenberg NA. 2006. Discordance of species trees with their most likely gene trees. *PLoS Genet.* 2:e68
10. Degnan JH, Salter LA. 2005. Gene tree distributions under the coalescent process. *Evolution* 59:24–37
11. de Queiroz A, Gatesy J. 2007. The supermatrix approach to systematics. *Trends Ecol. Evol.* 22:34–41
12. Doolittle WF. 1999. Phylogenetic classification and the universal tree. *Science* 284:2124–29

13. Eck RV, Dayhoff MO. 1966. *Atlas of Protein Sequence and Structure*. Silver Spring, MD: National Biomedical Research Foundation
14. Edwards AWF. 1970. Estimation of the branching points of a branching diffusion process. *J. R. Stat. Soc. B* 32:155–74
15. Edwards SV, Liu L, Pearl DK. 2007. High-resolution species trees without concatenation. *Proc. Natl. Acad. Sci. USA* 104:5936–41
16. Eulenstein O, Chen D, Burleigh JG, Fernandez-Baca D, Sanderson MJ. 2004. Performance of flip supertree construction with a heuristic algorithm. *Syst. Biol.* 53:299–308
17. Fang F, Ding J, Minin VN, Suchard MA, Dorman KS. 2007. Brother: relaxing parental tree assumptions for Bayesian recombination detection. *Bioinformatics* 23:507–8
18. Felsenstein J. 1981. Evolutionary trees from DNA sequences: A maximum likelihood approach. *J. Mol. Evol.* 17:368–76
19. Felsenstein J. 2001. Taking variation of evolutionary rates between sites into account in inferring phylogenies. *J. Mol. Evol.* 53:447–55
20. Fitch WM. 1970. Distinguishing homologous from analogous proteins. *Syst. Zool.* 19:99–113
21. Fleischmann RD, Adams MD, White O, Clayton RA, Kirkness EF, et al. 1995. Whole-genome random sequencing and assembly of haemophilus influenzae rd. *Science* 269:496–98, 507–12
22. Galtier N. 2007. A model of horizontal gene transfer and the bacterial phylogeny problem. *Syst. Biol.* 56:633–42
23. Goffeau A, Aert R, Agostini-Carbone ML, Almed A, Aigle M, et al. 1997. The yeast genome directory. *Nature* 387(suppl.):1–105
24. Goldman N, Yang Z. 1994. A codon-based model of nucleotide substitution for protein-coding DNA sequences. *Mol. Biol. Evol.* 11:725–36
25. Hudson RR. 1983. Properties of a neutral allele with intragenic recombination. *Theor. Popul. Biol.* 23:183–201
26. Hudson RR. 1983. Testing the constant-rate neutral allele model with protein sequence data. *Evolution* 37:203–17
27. Huelsenbeck JP. 1995. The robustness of two phylogenetic methods: four-taxon simulations reveal a slight superiority of maximum likelihood over neighbor joining. *Mol. Biol. Evol.* 12:843–49
28. Huelsenbeck JP, Rannala B, Larget B. 2000. A Bayesian framework for the analysis of cospeciation. *Evolution* 54:352–64
29. Jin G, Nakhleh L, Snir S, Tuller T. 2006. Maximum likelihood of phylogenetic networks. *Bioinformatics* 22:2604–11
30. Kingman JFC. 1982. The coalescent. *Stoch. Process Appl.* 13:235–48
31. Kong A, Gudbjartsson DF, Sainz J, Jonsdottir GM, Gudjonsson SA, et al. 2002. A high-resolution recombination map of the human genome. *Nat. Genet.* 31:241–47
32. Koonin E, Makarova K, Aravind L. 2001. Horizontal gene transfer in prokaryotes: Quantification and classification. *Annu. Rev. Microbiol.* 55:709–42
33. Kubatko LS, Degnan JH. 2007. Inconsistency of phylogenetic estimates from concatenated data under coalescence. *Syst. Biol.* 56:17–24
34. Kuhner MK, Felsenstein J. 1994. A simulation comparison of phylogeny algorithms under equal and unequal evolutionary rates. *Mol. Biol. Evol.* 11:459–68. Erratum. 1995. *Mol. Biol. Evol.* 12:525
35. Kumar S. 1996. Patterns of nucleotide substitution in mitochondrial protein coding genes of vertebrates. *Genetics* 143(1):537–48
36. Linz S, Radtke A, von Haeseler A. 2007. A likelihood framework to measure horizontal gene transfer. *Mol. Biol. Evol.* 24:1312–19
37. McVean GAT, Myers SR, Hunt S, Deloukas P, Bentley DR, Donnelly P. 2004. The fine-scale structure of recombination rate variation in the human genome. *Science* 304:581–84
38. Minin VN, Dorman KS, Fang F, Suchard MA. 2007. Phylogenetic mapping of recombination hotspots in human immunodeficiency virus via spatially smoothed change-point processes. *Genetics* 175:1773–85
39. Moore BR, Smith SA, Donoghue MJ. 2006. Increasing data transparency and estimating phylogenetic uncertainty in supertrees: Approaches using nonparametric bootstrapping. *Syst. Biol.* 55:662–76

40. Murphy WJ, Eizirik E, O'Brien SJ, Madsen O, Scally M, et al. 2001. Resolution of the early placental mammal radiation using bayesian phylogenetics. *Science* 294:2348–51
41. Nei M. 1977. Standard error of immunological dating of evolutionary time. *J. Mol. Evol.* 9:203–11
42. Nielsen R, Yang Z. 1998. Likelihood models for detecting positively selected amino acid sites and applications to the HIV-1 envelope gene. *Genetics* 148:929–36
43. Nishihara H, Okada N, Hasegawa M. 2007. Rooting the eutherian tree—the power and pitfalls of phylogenomics. *Genome Biol.* 8:R199
44. Nylander JAA, Ronquist F, Huelsenbeck JP, Nieves-Aldrey JL. 2004. Bayesian phylogenetic analysis of combined data. *Syst. Biol.* 53:47–67
45. Pamilo P, Nei M. 1988. Relationships between gene trees and species trees. *Mol. Biol. Evol.* 5:568–83
46. Pupko T, Huchon D, Cao Y, Okada N, Hasegawa M. 2002. Combining multiple data sets in a likelihood analysis: which models are the best? *Mol. Biol. Evol.* 19:2294–307
47. Rannala B Yang Z. 1996. Probability distribution of molecular evolutionary trees: A new method of phylogenetic inference. *J. Mol. Evol.* 43:304–11
48. Rannala B, Yang Z. 2003. Bayes estimation of species divergence times and ancestral population sizes using DNA sequences from multiple loci. *Genetics* 164:1645–56
49. Ren F, Tanaka H, Yang Z. 2005. An empirical examination of the utility of codon-substitution models in phylogeny reconstruction. *Syst. Biol.* 54:808–18
50. Rodriguez-Ezpeleta N, Brinkmann H, Roure B, Lartillot N, Lang BF, Philippe H. 2007. Detecting and overcoming systematic errors in genome-scale phylogenies. *Syst. Biol.* 56:389–99
51. Ronquist F, Huelsenbeck JP. 2003. MrBayes 3: Bayesian phylogenetic inference under mixed models. *Bioinformatics* 19:1572–74
52. Sanderson MJ. 1998. Phylogenetic supertrees: assembling the trees of life. *Trends Ecol. Evol.* 13:105–9
53. Sanger F, Air GM, Barrell BG, Brown NL, Coulson AR, et al. 1977. Nucleotide sequence of bacteriophage ϕX174. *Nature* 265:687–95
54. Sasaki T, Nikaido M, Hamilton H, Goto M, Kato H, et al. 2005. Mitochondrial phylogenetics and evolution of mysticete whales. *Syst. Biol.* 54:77–90
55. Shapiro B, Rambaut A, Drummond AJ. 2006. Choosing appropriate substitution models for the phylogenetic analysis of protein-coding sequences. *Mol. Biol. Evol.* 23:7–9
56. Simon C, Buckley TR, Frati F, Stewart JB, Beckenbach AT. 2006. Incorporating molecular evolution into phylogenetic analysis, and a new compilation of conserved polymerase chain reaction primers for animal mitochondrial DNA. *Annu. Rev. Ecol. Evol. Syst.* 37:545–79
57. Suchard MA. 2005. Stochastic models for horizontal gene transfer: Taking a random walk through tree space. *Genetics* 170:419–31
58. Suchard MA, Kitchen CM, Sinsheimer JS, Weiss RE. 2003. Hierarchical phylogenetic models for analyzing multipartite sequence data. *Syst. Biol.* 52:649–64
59. Swofford DL. 2003. *PAUP*. Phylogenetic Analysis Using Parsimony (*and Other Methods), Version 4*. Sunderland, Massachusetts: Sinauer Associates
60. Tajima F. 1983. Evolutionary relationship of DNA-sequences in finite populations. *Genetics* 105:437–60
61. Takahata N, Satta Y, Klein J. 1995. Divergence time and population size in the lineage leading to modern humans. *Theor. Popul. Biol.* 48:198–221
62. Tateno Y, Nei M, Tajima F. 1982. Accuracy of estimated phylogenetic trees from molecular data. *J. Mol. Evol.* 18:387–404
63. Tateno Y, Takezaki N, Nei M. 1994. Relative efficiencies of the maximum-likelihood, neighbor-joining, and maximum-parsimony methods when substitution rate varies with site. *Mol. Biol. Evol.* 11:261–77
64. Whelan S, de Bakker PI, Quevillon E, Rodriguez N, Goldman N. 2006. Pandit: an evolution-centric database of protein and associated nucleotide domains with inferred trees. *Nucleic Acids Res.* 34:D327–31
65. Yang Z. 1993. Maximum likelihood estimation of phylogeny from DNA sequences when substitution rates differ over sites. *Mol. Biol. Evol.* 10:1396–401
66. Yang Z. 1994. Maximum likelihood phylogenetic estimation from DNA sequences with variable rates over sites: approximate methods. *J. Mol. Evol.* 39:306–14

67. Yang Z. 1995. A space-time process model for the evolution of DNA sequences. *Genetics* 139:993–1005
68. Yang Z. 1996. Maximum-likelihood models for combined analyses of multiple sequence data. *J. Mol. Evol.* 42:587–96
69. Yang Z. 2002. Likelihood and Bayes estimation of ancestral population sizes in hominoids using data from multiple loci. *Genetics* 162:1811–23

Transgenerational Epigenetic Effects

Neil A. Youngson and Emma Whitelaw

Department of Population Studies and Human Genetics, Queensland Institute of Medical Research, Brisbane 4006, Australia; email: emma.whitelaw@qimr.edu.au

Annu. Rev. Genomics Hum. Genet. 2008. 9:233–57

First published online as a Review in Advance on June 3, 2008

The *Annual Review of Genomics and Human Genetics* is online at genom.annualreviews.org

This article's doi:
10.1146/annurev.genom.9.081307.164445

1527-8204/08/0922-0233$20.00

Key Words

non-Mendelian inheritance, soft inheritance, parental effects, resistance to reprogramming

Abstract

Transgenerational epigenetic effects include all processes that have evolved to achieve the nongenetic determination of phenotype. There has been a long-standing interest in this area from evolutionary biologists, who refer to it as non-Mendelian inheritance. Transgenerational epigenetic effects include both the physiological and behavioral (intellectual) transfer of information across generations. Although in most cases the underlying molecular mechanisms are not understood, modifications of the chromosomes that pass to the next generation through gametes are sometimes involved, which is called transgenerational epigenetic inheritance. There is a trend for those outside the field of molecular biology to assume that most cases of transgenerational epigenetic effects are the result of transgenerational epigenetic inheritance, in part because of a misunderstanding of the terms. Unfortunately, this is likely to be far from the truth.

Transgenerational epigenetic effects: phenotypes present in successive generations that are not genetically determined

Transgenerational epigenetic inheritance or gametic epigenetic inheritance: a phenotype present in successive generations that is nongenetically determined and results from epigenetic modifications passed via the gametes that escape reprogramming

Hard inheritance: essentially Mendelian; hereditary material remains constant between generations (except for rare random mutations)

Soft inheritance: the generation of a new phenotype is less rigidly determined and shows a more rapid response to environment

Parental effects: effects on the phenotype of offspring that are not determined by the offspring's own genotype but by the genotype or environmental experience of its parents

Paramutation: an interaction between two alleles of a locus, resulting in a heritable epigenetic change of one allele induced by the other allele

INTRODUCTION

Transgenerational epigenetic effects and transgenerational epigenetic inheritance are not the same, but a novice to the discipline would find this hard to understand. The situation has arisen because the word epigenetic has changed its meaning over the past fifty years. In the phrase 'transgenerational epigenetic effects,' epigenetic is being used in its broader (and original) sense to include all processes that have evolved to achieve the nongenetic determination of phenotype. Waddington, who coined the word epigenetics, was interested in how gene expression patterns are modified during differentiation and development. He was a developmental biologist, with no particular interest in transgenerational events. However, evolutionary biologists have studied the transgenerational nongenetic determination of phenotype for centuries. This has been termed soft inheritance, non-Mendelian inheritance, parental effects, and fetal programming, among others. Although in many instances these phrases refer to different phenomena, including both physiological and behavioral (intellectual) processes, they all involve a transfer of nongenetic information across generations; i.e., they are transgenerational epigenetic effects. Although in most cases the underlying molecular mechanisms are not understood, modifications to the chromosomes that pass to the next generation through the gametes are sometimes involved. The latter has to-date been called 'transgenerational epigenetic inheritance.' Here, the word epigenetic refers to mitotically and/or meiotically heritable changes in gene function that cannot be explained by changes in gene sequence. This narrower definition of epigenetics has recently become widely accepted among molecular biologists.

We have learned a considerable amount about epigenetic (in its more recent sense) modifications, including both the methylation of the cytosine residue of DNA and the modification of the chromatin proteins that package the DNA. In general these marks are established in early development and are stable through rounds of mitosis. Recent evidence shows that the establishment of epigenetic state can be influenced by environmental factors (33, 40, 129). To ensure the totipotency of the zygote and to prevent perpetuation of abnormal epigenetic states, most gene regulatory, i.e., epigenetic, information is not transferred between generations. Several mechanisms have evolved to erase the marks, including germline and somatic reprogramming of DNA methylation and chromatin proteins. However, we know that at some loci the epigenetic marks are not cleared. Examples of this include genomic imprinting in mammals, mating type switching in yeast, and paramutation in plants. Although exceptional with respect to their resistance to reprogramming, these examples can be considered part of normal development, and they are not dependent on environmental cues. Two issues that we now need to address are firstly, the extent of this resistance to transgenerational epigenetic reprogramming and secondly, whether or not epigenetic marks established in response to environmental cues are also resistant.

There is a current trend for those outside the field of molecular biology to assume that all cases of transgenerational epigenetic effects are the result of transgenerational epigenetic inheritance, in part because of a misunderstanding of the terms. This is misleading. When discussing transgenerational epigenetic effects, care must be taken not to make assumptions about the underlying mechanisms. In an attempt to improve this situation, we propose that the term 'transgenerational epigenetic inheritance' be replaced by 'gametic epigenetic inheritance,' which is a more precise description of the event. In this review we discuss the current knowledge in the broad area of non-Mendelian inheritance and attempt to highlight those cases where gametic epigenetic inheritance is known to occur.

SOFT INHERITANCE

Mendelian genetics is built on the inheritance of stable traits and the evolution of such traits occurs slowly as a result of rare genetic mutation, selection, and drift. The slow reactivity of this 'hard inheritance' is not ideal for an organism or population to thrive in a dynamic environment. Another more pliable system, which fine tunes the next generation to their future environment, would be an advantage. Ernst Mayr (1904–2005) (77, 78) first proposed the term 'soft inheritance' to describe this type of system. Soft inheritance would be especially suited to adaptation to fluctuations in nutrition, predation, or disease, which occur relatively unpredictably and may endure for more than one generation. Soft inheritance is adaptive in the sense used by evolutionary biologists, i.e., advantageous to the individual or species. The ability of epigenetic mechanisms to perpetuate gene expression patterns relatively stably, and to retain the capacity to react to environmental cues, makes them ideal for facilitating soft inheritance. The largest barrier to such a system is the resetting of epigenetic marks between generations.

The notion of soft inheritance is still viewed by some as controversial. This is mainly due to its association with the rejected evolutionary ideas of Jean-Baptiste Lamarck (1744–1829). His pre-Darwinian work proposed a mechanism for the transformation of species through the inheritance of characters that are acquired during the lifetime of an organism. According to Lamarck, both environment and behavior direct organic change in an organism's form and guide adaptation through the generations. A major problem with Lamarckian evolution, as pointed out initially by August Weismann in the nineteenth century, is the separation of germline and soma. How could environmentally induced epigenetic adaptations in somatic lineages be transmitted to the germline? The precise time point of germline separation from somatic tissues varies among species. In mammals, primordial germ cells (PGCs) are derived from the epiblast and arise in the posterior primitive streak during gastrulation. So, there is an extremely short period for epigenetic alterations to be included in the germline. In contrast, in plants there is no early separation of germline and soma and the gametes are derived from vegetative tissue after most development is complete. This may provide plants with a greater opportunity for soft inheritance than mammals.

It should be emphasized that the examples of soft inheritance described in this review, although Lamarckian in their environmental determination, involve short-term adaptations that supplement the evolutionary processes of Darwin and Mendel. Thus, they are distinct from Lamarck's proposed overall mechanism of evolution.

Adaptation: a phenotypic change that is ultimately beneficial to the reproductive success of an organism

Adaptive Parental Effects in Plants and Insects

Parental effects are defined as effects on the phenotype of offspring that are not determined by the offspring's genotype but instead are determined by the genotype or environmental experience of the parents. These effects can be paternal or maternal and have been reported in a range of multicellular organisms. Such effects fit within the confines of transgenerational epigenetic effects. Some are known to involve gametic epigenetic inheritance, others are not. Some are classified as adaptive by evolutionary biologists, others are not. Adaptive parental effects are examples of soft inheritance and are extensively reviewed elsewhere (48, 85). Here we limit our discussion of adaptive parental effects to a few of the better-understood cases.

Numerous examples of adaptive maternal effects exist in plants. A range of environmental stimuli acting on the mother, including predation (2), competition (97), soil type (106, 107), temperature (68), light (48, 108, 121), and nutrient availability (80, 108, 120, 121), has been found to induce changes to F1 phenotype. For example, offspring of *Polygonum persicaria* grown under low light allocate proportionately more resources to shoot growth than those of parents with higher light exposures (121). Conversely, offspring of limited-nutrient plants allocate proportionately more to root growth than genetically similar individuals with nutrient-rich parents (121). Wild radishes (*Raphanus raphanistrum*) produce physical spines and insect-repellent chemicals in response to predation by caterpillars (2). These adaptations provide protection against further attack. Agrawal and colleagues (2) showed that seedlings from parent plants that had been damaged by caterpillars develop phenotypic features (spines) more like their parents than seedlings from unexposed plants. These changes to the F1 are associated with reduced predation from caterpillars. The authors noticed shifts in the profiles of the defensive chemicals in the seeds from predated mothers, suggesting a possible mechanism for the transgenerational inheritance.

Paternal effects in plants have also been described, but less frequently than maternal effects (68, 106). This may be because seedlings are likely to grow up in an environment more similar to their mothers' than their fathers', because seed dispersal is limited compared with that of pollen. Further reasons could include the larger maternal (2n) than paternal (n) nuclear contribution to the endosperm (the organ providing nutrients to the developing embryo) and the maternal origin of the seed coat.

In addition to their work on wild radishes, Agrawal and colleagues (2) examined the defensive responses of the water flea, *Daphnia*, which is subject to predation by other insects. When females are exposed to chemical signals associated with the presence of predators, they develop a protective helmet, which renders them less vulnerable to attack. Females exposed to these signals lay eggs that, as neonates, develop the same defense as their mothers—even in the absence of the predator-related signals. Subsequent maternal broods, initiated after the mothers were transferred to signal-free environments, also show enhanced defenses as neonates. The effect diminishes by the second generation, though subtle grandparental effects are evident. The average helmet size of neonates whose mothers, but not grandmothers, were exposed to the predation cue is not as large as those whose mothers and grandmothers were exposed to the signal.

Another study in *Daphnia* investigated their ability to produce dormant eggs in response to cues for the forthcoming food supply (3). Alekseev and Lampert (6) manipulated the photoperiod of mothers to mimic conditions that would predict poor food availability, even though the mother had ample food. The daughters of females exposed to short days (which stimulate dormant egg production) were more likely to produce dormant eggs than daughters of mothers exposed to long days.

The relevance of the findings in *Daphnia* to events in higher animals is somewhat tempered by the fact that *Daphnia* reproduce mainly parthenogenetically. Soft inheritance may be more important to asexual reproducers that cannot adapt to environmental changes by obtaining new genetic information through sexual reproduction.

In the cases of the adaptive parental effects described above, we can be relatively confident that the information is transferred via the gametes.

Adaptive Parental Effect in Nonhuman Vertebrates

The best-characterized case of adaptive transgenerational epigenetic effects in mammals is that of the maternally transmitted responses to stress in rats. Similar to the maternal effect observed in the radish and *Daphnia*, these responses are thought to represent an inducible defense mechanism (140). In times of increased environmental stress, such as when more predators are present, there is less time for maternal care in the form of postnatal maternal licking/grooming and arched-back nursing (LG-ABN). Low levels of LG-ABN in the first week after birth cause offspring to be more fearful; the theory is that their increased watchfulness will increase their survival chances. In contrast, the offspring of high LG-ABN mothers are less fearful. These behavioral traits persist into adulthood, when a female will usually display the same behavior as her mother, thus perpetuating the trend. Cross fostering pups from one mother type to the other in the first week of

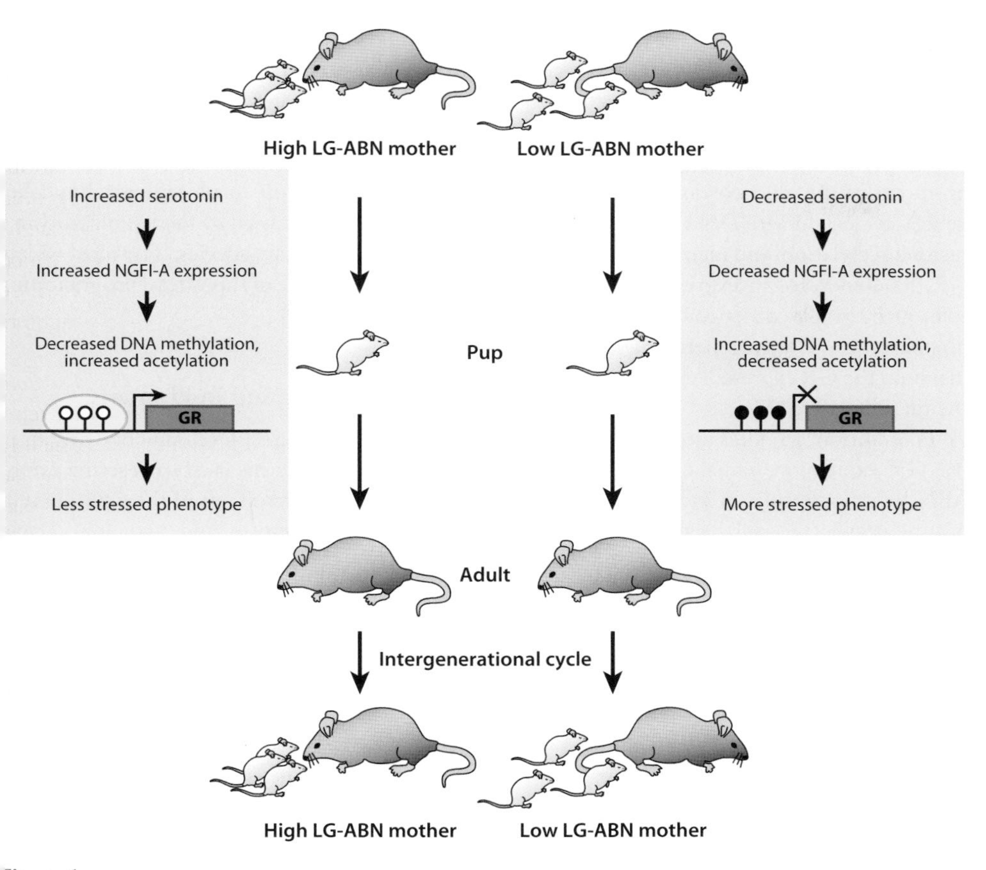

Figure 1

Transgenerational inheritance of mothering style and stress in rat. Mothering style as characterized by licking/grooming (LG) and arched-back nursing (ABN) is perpetuated across generations by a cascade of molecular events set in the the first week of life. High LG-ABN mothering results in a high serotonergic tone in the hippocampus of the pups, leading to increased expression of the transcription factor nerve growth factor inducible protein A (NGFI-A). Binding of NGFI-A to the promoter of the glucocorticoid receptor (GR) gene stimulates DNA hypomethylation, histone acetylation, and increased expression of GR. Higher glucocorticoid receptor numbers in the hippocampus are associated with reduced stress levels. The epigenetic marks maintain the GR expression state into adulthood and in females will determine the level of LG-ABN mothering, thus perpetuating the phenotype. Open circle lollipops are unmethylated CpGs, filled lollipops are methylated CpGs, and yellow ovals are nerve growth factor inducible protein A (NGFI-A) (131, 132).

life causes pups to have the stress type of their adoptive mothers (47). Therefore, in this case the adaptive maternal effect is epigenetic (in its broader sense) but not gametic.

A remarkable feature of this case is that many aspects of the mechanism have been uncovered, revealing an elegant multilevel process that involves behavioral, physiological, cellular, and molecular events (**Figure 1**). Indeed, epigenetic modifications to the regulatory elements of some relevant genes have been detected. Stress responses in mammals are mediated through the hypothalamic-pituitary-adrenal (HPA) axis and involve the action of glucocorticoid hormones. The reduced fearfulness of high LG-ABN rats is the result of an increase in the number of glucocorticoid receptors in the hippocampus. High LG-ABN mothering results

Epigenetic modifications: chromatin and DNA modifications that influence genome function but do not change the underlying DNA sequence

in a high serotonergic tone in the hippocampus of the pups, leading to activation of cAMP and increased expression of the transcription factor nerve growth factor inducible protein A (NGFI-A). Increased binding of NGFI-A to the promoter of the glucocorticoid receptor (GR) gene is associated with DNA hypomethylation, histone acetylation, and increased expression of GR. This increase in expression, in turn, results in more glucocorticoid receptors in the hippocampus. The epigenetic marks appear to maintain the GR expression state for the rest of the rat's life (131, 132).

In addition to alterations in hippocampal GR expression, enhanced maternal LG-ABN behavior results in increased hippocampal neuronal survival, synaptogenesis, and improved cognitive performance under stressful conditions (72, 73, 133). Microarray expression analysis identified over three hundred genes with differing hippocampal expression patterns in offspring of low LG-ABN compared with high LG-ABN mothers. Furthermore, intracerebroventricular infusions of the histone deacetylase inhibitor trichostatin A (TSA) or the methyl-donor L-Methionine modified the expression, consistent with epigenetic regulation (134). So, the modulation of GR is likely to be part of a larger group of adaptive effects that result from maternal nurturing. To our knowledge, this is the only study that links an adaptive maternal effect to an epigenetic change; it suggests that, at least in this case, the epigenetic marks are the molecular memory that confers persistence of the phenotype into adulthood.

Exposure to hormones in utero, in the egg, or even postnatally has been proposed to facilitate adaptation to the future environment in a number of vertebrate species. In birds, the maternally determined level of androgens deposited in egg yolk influences the offspring's embryonic development, postnatal growth, competitiveness in the nest, and dispersal distances from the nest (51). Injection of testosterone into eggs recapitulates the effects on dispersal patterns (126). Elevated in utero androgen exposure also increases dispersal distance in voles (59). However, the pleiotropic effects of changes in hormone level make it hard to know whether this process is truly adaptive, in the sense of being advantageous. The same issue has emerged over claims of adaptive significance in relation to physiological and behavioral changes caused by food availability in a number of vertebrate species. The most widely publicized example of this is fetal programming in humans.

Fetal Programming in Humans

Fetal programming or developmental origins of adult disease are terms used to describe extensive and permanent effects of the environment experienced by fetuses and neonates. David Barker (14) reported an inverse relationship between birth weight and the risk of hypertension, cardiovascular disease, and type 2 diabetes in adulthood. The effect seems to be exacerbated when the individual is well nourished postnatally (45, 46). As a consequence of these observations, Barker proposed that adverse effects in utero induce compensatory responses in the fetus. The cellular, physiological, and metabolic responses are thought to represent adaptations made by the fetus to prepare for postnatal life. This is called the thrifty phenotype hypothesis (10). According to this hypothesis, the increased levels of insulin resistance in offspring of starved mothers, rather than being an inevitable consequence of a poor early environment, is actually deliberately induced because it will confer an advantage later in life. Increased insulin resistance causes energy conservation and reduced somatic growth to allow the offspring a better chance of survival in an environment where nutrition is poor. However, with insulin resistance, the higher blood plasma levels of fatty acids, insulin, and glucose become a problem if food becomes abundant.

Some evidence in support of this hypothesis comes from the offspring of women pregnant during two civilian famines of World War II, the Dutch Hunger Winter (1944–1945) and the Siege of Leningrad (1941–44) (15). In the former case, those who were starved prenatally were found to have impaired glucose tolerance in adulthood; this was not observed in the latter case (116). So although in both situations the fetuses predicted a poor postnatal environment, it proved to be true only in the Leningrad case, where nutrition stayed poor in the subsequent years. Whether the postnatal effects of gestational undernutrition are truly adaptations, or developmental abnormalities that resemble them, remains unclear.

Fetal Programming Across More Than One Generation

Reports of multigenerational epigenetic effects in human populations are scant, in part because phenotypic records across generations have rarely been collected and in part because ruling out genetic and environmental confounders is extremely difficult. However, a few studies have been published.

Follow up work on the Dutch Hunger Winter initially suggested that mothers who were exposed to famine as fetuses delivered offspring (F2) of lower birth weight than those with no fetal exposure to famine (74). However, this study was flawed in a number of ways. In particular, birth weights in famine-exposed mothers were not measured directly, but were instead extrapolated from another group. A subsequent study by the same author found no significant effect of maternal fetal exposure to famine on the birth weights of the next generation (117).

Studies on other cohorts have revealed some association between grandparental nutrition and grandchild (referred to as the proband in this work) phenotype. Extensive records of a population in Överkalix in Sweden, including yearly crop yields over multiple generations, revealed a link between grandparental and parental periods of low or high food availability with proband mortality and disease risk (27, 63, 64, 94). The work highlighted the possible importance of food availability during the paternal grandparental prepubertal slow growth periods (SGP), between age 8–10 in girls and 9–12 in boys. If the SGP of the paternal grandfather was a period of high food availability then male probands had reduced longevity (27, 94), an effect later shown to be related to increased risk of death by cardiovascular disease or diabetes (63). Also, abundant food in the SGP of the maternal grandmother was associated with an increased mortality in female probands (94).

No molecular data exist to explain the findings, but the involvement of epigenetic marks in the form of gametic epigenetic inheritance has been suggested (94). However, these studies reveal a complex process with sex- and age-specific variations. For example, in addition to the effect of food availability during paternal grandmaternal SGP on female probands, researchers also noted an effect of the food availability during a grandmother's first five years of life—but with an opposite influence on proband mortality. Moreover, the paternal grandfather to grandson effect was not seen in all cohorts (63). Independent replication in another cohort would be helpful. The possibility of societal confounders in these studies remains high and in the absence of molecular evidence, the conclusion that this is a case of gametic epigenetic inheritance seems unwarranted.

NONADAPTIVE TRANSGENERATIONAL EPIGENETIC EFFECTS

In all the examples cited so far, the unifying factor is the concept of an adaptive response to the environment and, in general, the studies have been carried out by behavioral psychologists, evolutionary biologists, or epidemiologists. Many other examples of transgenerational epigenetic effects exist that are not necessarily adaptive, such as the gametic transgenerational inheritance of epigenetic state at paramutated alleles or transgenes. In many of these cases the inherited phenotype is actually detrimental to the organism. These cases have taught us what little we know about the underlying molecular mechanisms of gametic transgenerational inheritance.

Table 1 Transgenerationally inherited epimutations and metastable epialleles

Locus/epiallele	Organism	Mechanism[a]	Phenotype	Stability	Parent of origin effect	Reference
a locus a-m2–7991A1	Maize	Methylation levels of a Spm transposon can switch between hyper and hypo to affect expression of the nearby a gene	Pigmentation and transposition	Metastable	Yes	(11)
b1locus B' epiallele	Maize	Paramutation-induced methylation	Reduced pigmentation	Stable	No	(115)
Lcyc	*Linaria vulgaris*	Epimutation. No apparent genetic mutation	Radially symmetrical flowers	Metastable	No	(36)
A^{vy}	Mouse	IAP-LTR promoter drives ectopic expression of a nearby gene	Coat color, diabetes, obesity	Metastable	Partially inherited down female line	(83)
$Axin^{Fu}$	Mouse	IAP-LTR promoter drives eptopic expression of a nearby gene	Kinked tail	Metastable	Partially inherited down male and female lines	(100)
MLH1	Human	Presence of an epimutation in all three germ layers suggests a germline origin	Colon cancer	Unknown	N/A	(57)

[a]Abbreviations: IAP, intracisternal A particle; LTR, long terminal repeat.

Nonadaptive Transgenerational Effects in Plants

The most famous example of gametic epigenetic inheritance in plants involves the peloric variant of toadflax (*Linaria*); in this case the modified phenotype is relatively stably inherited over many generations (36) (**Table 1**). Silencing of the *Lcyc* gene causes the symmetry of the flower to change from bilateral to radial. The silencing occurs not through mutation of the DNA sequence but through methylation of the promoter. This became one of the first reported cases of an epimutation. However, although this appears to be a clear case of gametic epigenetic inheritance, whether the phenotype is perpetuated across generations by cytosine methylation or by other epigenetic factors remains unknown. Similarly, DNA methylation of a transposon at the promoter of the *a* gene in maize influences the gene's expression and is stably transmitted through meiosis (11) (**Table 1**). Another gene in maize, the *b1* locus, can become stably repressed by epigenetic modifications through paramutation; this repressed state is heritable across generations (**Table 1**). Moreover, the repressed state can silence unmethylated alleles introduced by breeding. Similarly, an epimutation at the *P* locus in maize was stably inherited over five generations, though reversions were noted (37) (**Table 2**). In all these cases, no genetic mutations have been

Epimutations: abnormal epigenetic patterns that can occur in response to a DNA mutation, but the term is generally used in cases without an underlying DNA sequence change

Table 2 Genetic mutation–induced transgenerational epigenetic inheritance

Locus/ epiallele	Organism	Manipulation	Mechanism	Stability	Phenotype	Observed parent of origin effect	Reference
pai2	Maize	Deletion of inverted repeat source of RNA-directed DNA methylation	Upon deletion of the inverted repeat homologous sequences retain methylation in successive generations	Metastable	Metabolic	Maternally inherited, paternal inheritance unknown	(19, 79)
P-pr	Maize	Strain used has increased frequencies of somatic mutation so probably a mutation in cis or in epigenetic modifier caused epimutation	Inherited epimutation (no associated genetic lesion identified)	Stable	Reduced pigmentation	No	(37)
bal	*Arabidopsis thaliana*	DDM1 SWI/ SNF-like chromatin remodeling factor mutant	DDM1 mutant–generated epimutation	Metastable	Dwarfism, elevated disease resistance	No	(119)
sup	*Arabidopsis thaliana*	Chemical mutagenesis and a variety of epigenetic modifier mutants cause the phenotype	Variety of mutants cause epimutation at the SUP gene	Metastable	Abnormal floral organ number	No	(62)
fwa	*Arabidopsis thaliana*	Possible chemical or radiation-induced mutation of epigenetic modifier; phenotype recapitulated in DDM1 mutants	Epimutation (lack of methylation) SINE retrotransposon 5′ of gene causes ectopic expression.	Metastable	Delayed flowering	No	(112)
Genome-wide	Mouse	Pronuclei transfer between different mouse strains	Epimutations in transplanted embryos are paternally transmitted to the next generation; genes in pheromone systems are particularly affected	Metastable	Reduced stature and multiple gene mis-regulation	Possibly male line–specific	(103)

(*Continued*)

Table 2 (*Continued*)

Locus/ epiallele	Organism	Manipulation	Mechanism	Stability	Phenotype	Observed parent of origin effect	Reference
Fab-7 construct	*Drosophila melanogaster*	Fab-7 PRE/TRE construct	Active state of construct is inherited through female line; H4 hyperacetylation transgenerational persistence	Metastable	Larval LacZ and adult eye color markers	No, but phenotype is influenced by sex	(28, 29)
JAK kinase	*Drosophila melanogaster*	JAK kinase overexpression mutant	Maternally-inherited JAK kinase signaling protein overexpression disrupts reprogramming in the early embryo	Metastable	Enhanced offspring tumori-genesis	Yes, parental effects	(136)
Mod(mdg4)	*Drosophila melanogaster*	Mod(mdg4) mutant	Mutation causes abnormal chromatin configuration on the Y chromosome that can be stably inherited	Stable	Enhanced position effect variegation (PEV)	Yes	(41)
Kruppel com-bined with various others	*Drosophila melanogaster*	Kruppel repetitive element insertional mutant combined with various maternal effect modifier mutations or Hsp90 chemical inhibition	Ectopic overexpression of Kruppel combined with chemical inhibition of Hsp90 causes ectopic bristles in the eye; artificial selection can either fix or remove the phenotype from a population	Metastable	Ectopic large bristle outgrowths from eye	Yes, several maternal effect modifiers	(111)

Variegation: mosaic expression of a particular phenotype among cells of the same cell type; for example, mottled coat color

identified at or near the gene of interest but causative DNA mutations *in cis* remain possible. The recent discovery of previously unnoticed copy number variants (CNVs) in vertebrates is a salient reminder of our need for caution in this regard (17).

Transgenes in plants (and animals, see below) are susceptible to silencing by epigenetic mechanisms. This silencing can be due to integration adjacent to a heterochromatic region or to a poorly understood genome defense system that recognizes the transgene as foreign (76). In many cases the silencing is probabilistic, resulting in mosaic patterns of expression called variegation or position effect variegation (PEV). The silent state is sometimes heritable across generations (76).

Seemingly nonadaptive transgenerational epigenetic effects have been reported following ionizing radiation in plants. The mutagenic properties of ionizing radiation mean that, after exposure, some inheritance of abnormal phenotype will be the result of inherited genetic lesions. However, growing evidence shows that ionizing radiation also produces nongenetic heritable effects (13). The best example comes from observations in *Arabidopsis thaliana*, where elevated rates of somatic homologous recombination in response to UV-C (UVC) persisted in untreated progeny for up to four generations (82). The phenomena are likely to be epigenetic because the whole population changes its behavior each generation. Mutation would affect only 50% of plants (those that had inherited the mutation). Furthermore, the effect acts *in trans* on a reporter transgene introduced from an untreated parent plant. The effects are independent of the sex of the transmitting parent, suggesting that the memory can be inherited through either gametophyte. The molecular marks that provide this transgenerational persistence of the response to the ionizing radiation are unknown. All the cases of nonadaptive transgenerational effects described in this section are likely to be instances of gametic epigenetic inheritance.

Nonadaptive Transgenerational Epigenetic Effects in Insects

Cases of transgenerational epigenetic effects in *Drosophila* tend to be complex with the involvement of parental effects and genetically compromised backgrounds. All cases appear to involve gametic epigenetic inheritance. Mod(mdg4) is a protein with several roles, including chromatin insulation, apoptosis, and homolog pairing in meiosis. Dorn and coworkers (41) showed that the sons of heterozygous *mod(mdg4)* mutants display enhanced PEV of a reporter locus, even if they do not inherit the mutation. The effect is thought to be caused by an abnormal chromatin configuration on the Y chromosome, which is stably inherited in wild-type males for at least 11 generations. This suggests that certain epigenetic states in flies are not reset each generation and consequently, their perturbation is not rectified.

Transgenerational persistence of polycomb/trithorax–mediated transcriptional regulation has been studied with a reporter construct containing the homeodomain regulator element Fab-7, which contains a polycomb/trithorax response element (PRE/TRE) (28, 29). After embryonic induction of trithorax-mediated expression through the transient binding of the GAL4 transcription factor, the active state is maintained in both the soma and the germline. High levels (though reduced compared with that of the F1) of reporter (lacZ in embryo, red eye color in adult) are still detectable in the F2 and F3 generations. Activated Fab-7 is marked with hyperacetylation of histone H4 and this may be the transgenerationally stable mark. However, it is not known why this reporter construct, unlike endogenous PRE/TRE elements, escapes reprogramming.

Position effect variegation (PEV): variegation caused by the inactivation of a gene in some, but not all, cells of the same cell type through its abnormal juxtaposition with heterochromatin

Xing and coworkers (136) recently reported a mutant fly in which reprogramming in the early embryo has been disrupted (by overexpression of the JAK kinase signaling protein), and that shows transgenerational inheritance of tumorigenic epimutations. How JAK signaling interferes with reprogramming is unclear, but it inhibits heterochromatin formation (110). The full extent of the epimutations is not understood.

In *Drosophila* and plants, reduction in the level of the stress response protein Hsp90, by mutation or chemical inhibition, induces unusual morphologies (99, 105). These morphologies are the result of the expression of natural variation that was previously hidden by Hsp90's chaperone function. Selection of the abnormal phenotypes can lead to their fixation in a population. Work by Sollars and colleagues (111) suggests that the phenomenon is, at least in part, epigenetic. Consistent with this hypothesis, mutations in genes encoding trithorax group proteins were commonly found in a screen carried out to identify modifiers of the process. However, transgenerational persistence of an epigenetic mark is yet to be confirmed. These actions of Hsp90 are proposed to be a form of soft inheritance (105, 111). The theory is that environmental stress

Metastable epialleles: alleles at which the epigenetic state can switch, creating different phenotypes, and where the establishment is a probabilistic event

Genomic or parental imprinting: the expression of certain genes in a parent-of-origin-specific manner; involves the sex-dependent epigenetic resetting of a germline regulatory element

diverts Hsp90 from its chaperone function of stabilizing aberrant proteins, revealing hidden phenotypes, and that advantageous ones can be selected and fixed. Importantly, an adaptive response would be made without the need to wait for the generation of novel genetic mutations. However, opponents of the theory argue that the effects of Hsp90 reduction are merely nonfunctional consequences and not an evolved evolutionary mechanism (39).

Despite the excellent genetic tractability of flies much remains to be discovered about the mechanisms of gametic epigenetic inheritance in this organism, in particular the nature of the transgenerationally resistant mark.

Nonadaptive Transgenerational Epigenetic Effects in Nonhuman Mammals

There are a number of examples of nonadaptive transgenerational epigenetic effects in mammals. Some cases involve the transgenerational persistence of environmentally induced phenotypes, some display gametic epigenetic inheritance, and a few notable cases involve both.

Transgenes and metastable epialleles. The first molecular evidence for transgenerational epigenetic inheritance (i.e., gametic epigenetic inheritance) in mammals came from studies of metastable epialleles in inbred mouse strains (83, 100) (**Figure 2*a*,*b***; **Table 1**). Inbred mouse strains provide an opportunity to study phenotype differences that occur among genetically identical individuals. Metastable epialleles are loci at which activity is dependent on the epigenetic state. A handful of such alleles have been reported, including the *agouti viable yellow* (A^{vy}) and *axin fused* ($Axin^{Fu}$) alleles, both of which contain intracisternal A particle (IAP) retrotransposons that influence expression of linked genes; this influence is dependent on the methylation status of a cryptic promoter in the IAP long terminal repeat (LTR). The transgenerational memory of these epigenetic states involves gametic epigenetic inheritance and current evidence suggests that DNA methylation is not the mark that is directly inherited (22).

Some transgenes in mice (and in plants, as mentioned previously) show gametic epigenetic inheritance and in many cases this inheritance is multigenerational (5, 52, 67, 123). Interestingly, in most cases the transgenes also show some degree of genomic imprinting (5, 52, 67, 124). In mammals there are approximately one hundred endogenous genes that undergo genomic (parental) imprinting, for which reciprocal DNA methylation patterns are set in male and female germlines. The epigenetic marks associated with imprinting are generally resistant to reprogramming in the early embryo but undergo reprogramming in the germline each generation. Importantly, the finding of long-term transgenerational effects at transgenes implies that in these cases the epigenetic marks also escape reprogramming in the germline. For example, the Tg(13HBV)E36-P transgene, when inherited paternally, is unmethylated, but maternal transmission results in silencing that cannot be reversed, even with subsequent passage through the male germline (52). These studies provided the first models to study the epigenetic transition of a single locus from expressed to permanently transgenerationally silenced. However, what actually makes these sequences resistant to reprogramming remains unclear.

In response to ionizing radiation. Transgenerational epigenetic effects following ionizing radiation, similar to those reported in plants, have been seen in mice. These studies examined germline mutation rates at expanded simple tandem repeat (ESTR) loci following irradiation (12, 43). Exposure of the F0 male with X-rays caused elevated rates of mutation in the F1 and F2 generations. As in plants, the effect can act *in trans*, i.e., ESTR alleles from unexposed mice become unstable in the germlines of progeny of exposed mice. If the effects were caused by mutations in genes that maintain ESTR stability then the effects would lessen through the generations when breeding to wild-type mice as the mutated allele(s)

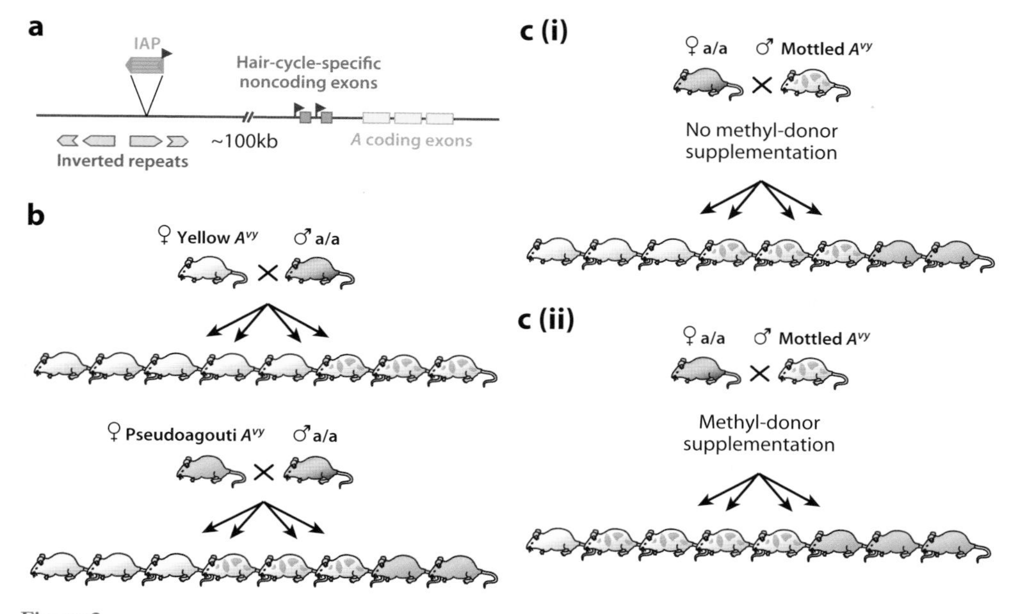

Figure 2

Intracisternal A particle (IAP)-mediated transgenerational epigenetic inheritance at the A^{vy} locus. (*a*) The agouti gene (*A*) is ectopically expressed; a transcript originates from an IAP retrotransposon upstream of the normal promoters. The expression of the cryptic IAP promoter is highly variable among isogenic mice. The agouti protein indirectly results in a yellow coat. The presence or absence of the ectopic transcript correlates with differential DNA methylation at the IAP promoter. The variable expressivity of the IAP creates a range of coat colors from yellow, through mottled, to pseudoagouti. (*b*) A^{vy}/a mice were mated to congenic a/a black mice, and the offspring scored for phenotype at weaning. The phenotype of the A^{vy}/a mother affects the phenotype of the offspring; yellow dams produce a higher proportion of yellow offspring than pseudoagouti dams. There is some memory of the epigenetic state of the maternal A^{vy} locus in the offspring. (*c*) The diets of female a/a mice were supplemented with methyl-donating substances [folic acid, choline, vitamin B_{12}, and betaine (129) or the phytoestrogen genistein (40)] two weeks before mating with male A^{vy}/a mice and throughout pregnancy and lactation. The range of coat colors was shifted toward pseudoagouti in the offspring of mothers with the supplemented diet (*i*) compared with controls (*ii*).

segregate. This is not seen. Mutation rates remain high in the F1 and F2 germline, which points to an epigenetic mechanism. As yet the molecular nature of the transgenerational memory is unknown.

Maternal exposure to changes in nutrition. Nonadaptive transgenerational epigenetic effects elicited by changes in nutrition have been reported in rodents and lagomorphs. For example, evidence shows that insulin resistance (1, 24), high blood pressure (7, 38), and elevated glucocorticoids (42, 90, 91) can increase the risk of the same condition in the next generation down the female line. Although this constitutes nongenetic perpetuation of phenotype, gametic mechanisms are not necessarily the explanation. For example, low protein diets of F0 females, while pregnant, are associated with a number of abnormalities in the F2 despite normal F1 postnatal nutrition (139). However, because the F1 experienced poor nutrition directly while in utero, the effect in what the authors call F2 is actually only a single generation after the one that experienced dietary restriction (see sidebar, Possible Explanations for Phenotypes Inherited Down the Female Line That Do Not Involve Gametic Epigenetic Inheritance). That is to say, the F2 phenotype could be due to the F1's incapacity to care for the F2 fetus; this

POSSIBLE EXPLANATIONS FOR PHENOTYPES INHERITED DOWN THE FEMALE LINE THAT DO NOT INVOLVE GAMETIC EPIGENETIC INHERITANCE

a) Postfertilization transfer of virus or toxin. This process can occur through the placenta or milk. Examples include the following: a gray mouse phenotype caused by virus in milk, ethanol transfer across the placenta, undernutrition during pregnancy.
b) Poor maternal health compromises pregnancy, which induces a similar phenotype in the next generation. Examples include insulin resistance, high blood pressure, or increased glucocorticoids in females causing complications during pregnancy, which lead indirectly to the same phenotype in the child.
c) Behavioral interactions between mother and child can perpetuate a phenotype. One example is in rats, where reduced maternal care induces a stressed phenotype in offspring and those rats become poor mothers, thus perpetuating the stressed phenotype.

would be a single generation maternal effect. Furthermore, the genome and/or epigenome of the F2 could have been directly affected by the environmental change because the specification of cells in the female germline occurs while the female is still in utero (see sidebar, Possible Explanations for Phenotypes Inherited Down the Female Line That Do Not Involve Gametic Epigenetic Inheritance; **Figure 3**). The problems in interpreting multigenerational effects down the female line also arise in other organisms. For example, in the plant *Plantago major* a collection of juvenile and adult characters in the F2 were influenced by a grandmaternal (F0) nutrient pulse (80). However, the nutrient pulse was administered during the stage of fruit maturation when the seeds that will become the F1 generation are themselves undergoing embryonic development within the F0. Thus, the F1 can be considered to have experienced the nutrient pulse and, similar to the situation in rodents, the effects in the F2 could actually be a maternal effect.

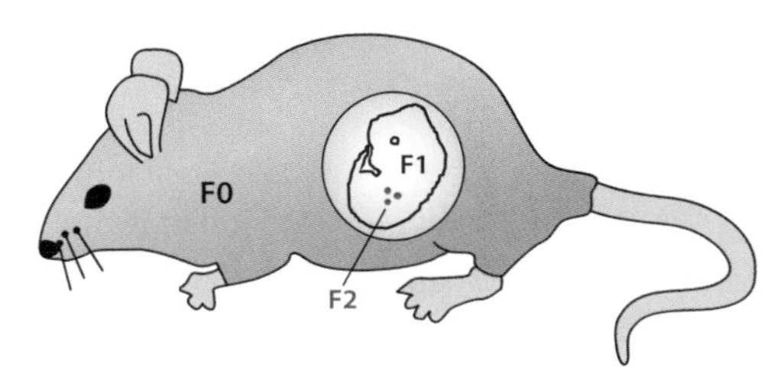

Figure 3

Three generation environmental exposure in pregnant females. In a gestating mother three generations directly experience environmental conditions. The mother (F0), embryo (F1), and the next generation (F2) in the form of the developing germline within the embryo can all be exposed to toxic chemicals, radiation, or dietary fluctuations.

Interestingly, Benyshek and coworkers (20) recently reported effects on glucose metabolism in F3 rats when pregnant (F0) females were fed a protein-restricted diet. Similarly, Stewart and colleagues (118) showed that feeding a low protein or unpalatable diet to rats for ten to twelve generations progressively reduces birth weight, which returns to control levels only three generations after reinstating a balanced diet. In the latter two cases, we can be certain that effects were seen in generations whose germline did not directly experience poor nutrition. Such examples are rare and are not necessarily the result of gametic epigenetic inheritance, because the perpetuation of effects could be mediated by postfertilization fetal-maternal or pup-dam interactions (See sidebar, Possible Explanations for Phenotypes Inherited Down the Female Line That Do Not Involve Gametic Epigenetic Inheritance; **Figure 3**). A good example of the latter is a gray mouse phenotype that is transgenerationally inherited down the female line as a result of transmission of a virus via the mother's milk (84).

Effects on the offspring of mothers fed a protein-restricted diet while pregnant can be reversed by supplementing the pregnant mother with methyl donors (61, 125). Protein restriction of pregnant F0 rats induces DNA hypomethylation and increases the expression of the *GR* and peroxisomal proliferation–activated receptor α (*PPAR*α) genes in the liver of adult F1 offspring (71). So far the molecular studies have been confined to candidate genes.

A follow-up study has found a persistence of GR and PPARα gene promoter hypomethylation in the F2 when the F1s were fed a normal diet (26). Whether this is the result of survival of epigenetic marks through the germline or de novo induction of the state in each generation through maternal-fetal interactions remains unclear.

Other studies reveal methyl donor supplementation of pregnant females with folic acid, vitamin B_{12}, choline, or betaine shifts the spectrum of coat color phenotypes in her offspring toward the repressed state, termed pseudoagouti, by increasing the level of DNA methylation at the A^{vy} allele (33, 135) (**Figure 2c**). Two studies addressed the question of whether increased DNA methylation at the locus is inherited by the next generation and they came to different conclusions. Waterland and coworkers (130) reported no cumulative effect on DNA methylation when successive generations had the supplemented diet and concluded that the acquired DNA methylation mark was not transgenerationally inherited. Using the same strain of mice but a slightly different approach, Cropley and colleagues (34) came to the opposite conclusion.

Maternal exposure to chemicals. Gestational exposure to carcinogens, endocrine disruptors, and other toxins has been shown to affect more than one generation in some cases; however, most studies do not investigate effects beyond the F2 generation and, as described previously, the F2 may have experienced the exposure directly. Chemicals known to induce phenotypic effects in unexposed generations include alloxan (113), cyclophosphamide (53), orthoaminoasotoluol (98), benzpyrene (35), diethylstilbestrol (DES) (89), and vinclozolin (8). The underlying mechanisms are not known and in most cases inherited genetic lesions cannot be ruled out, especially in cases of mutagens (e.g., cyclophosphamide).

In particular, the reports in rats of transgenerational epigenetic effects following exposure to the endocrine disruptor vinclozolin have raised considerable public interest. These studies showed that exposure to the fungicide vinclozolin at the time of gonadal sex determination causes a variety of abnormalities in offspring (8, 9). The effects are transmitted down the male line for at least three generations. The high incidence of the defects (approximately 90% of all males in all generations) and the absence of abnormalities when passed down the female line suggest gametic epigenetic inheritance. Importantly, increased DNA methylation was seen in sperm from vinclozolin-exposed males, and these abnormal methylation patterns (epimutations) were inherited. The most important feature of this work is the suggestion that environmentally induced epigenetic marks can survive reprogramming events over multiple generations; this work also highlights potential dangers of current environmental exposures on the health of future generations. However, many unresolved issues remain. How does the induced DNA methylation resist transgenerational reprogramming? How does the passage through the female germline terminate the transmission of the defects? How extensive are the DNA methylation changes? Could genetic changes to the Y chromosome be involved?

Nonadaptive Transgenerational Epigenetic Effects in Humans

For almost 30 years, clinicians prescribed the synthetic estrogen DES to women to prevent miscarriages. Women exposed to DES before birth were later shown to have a greatly increased risk of vaginal adenocarcinoma (55). Animal experiments showed increased tumor risk in the F2 and there is some evidence that this may also be the case in humans (21, 89). Pre- and neonatal DES exposure causes a wide range of gene expression changes and some DNA hypomethylation (70). However, survival of DES-induced epigenetic marks through transgenerational reprogramming is not necessarily the explanation. Estrogenic compounds can induce DNA damage, so the transgenerational incidence of DES-induced tumors could involve DNA mutation (88). Considering society's increased concern about environmental pollutants, it is likely that this area of research will grow.

Two families with an increased risk of colorectal cancer resulting from heterozygosity for epimutations at tumor suppressor genes provided some evidence that epimutations can be inherited transgenerationally. The best evidence comes from an individual with hereditary nonpolyposis colorectal cancer (HNPCC) (57). The subject had abnormal DNA methylation and silencing of one allele of the DNA mismatch repair/tumor suppressor gene MLH1. The presence of the epimutation in all three germ layers suggests it arose in the parental (in this case maternal) germline or in the zygote and resisted postfertilization reprogramming. No novel DNA mutations were identified in the region and some siblings inherited the same allele [as determined by single nucleotide polymorphism (SNP) analysis] in an unmethylated state. These facts argue against the theory that the epimutation is secondary to a mutation. This is in contrast to an epimutation in another case of HNPCC associated with an epimutation of another tumor suppressor gene (MSH2) (31). In this case a haplotype associated with the epimutation segregated in a Mendelian manner, and no revertants (unmethylated copies) were found. This result suggests that the epimutation was caused by a linked DNA mutation. Unfortunately, in humans it is almost impossible to prove that an epimutation was inherited because of a failure to reprogram in the germline. This is because we are outbred, so even if the DNA in the region of the epimutation has no mutation, the epigenetic status of the epimutation may actually be dependent on the unique genetic background of an individual. Several reports of HNPCC with MLH1 epimutations now exist, suggesting that this may be a hotspot for abnormal DNA methylation (49, 56, 57, 81, 122).

In summary, despite a growing amount of observational and molecular data on induced transgenerational epigenetic effects, incontrovertible evidence for gametic epigenetic inheritance of acquired marks remains scant in vertebrates. However, the recent advances in molecular technologies and high levels of interest in the topic suggest that a resolution is not far away.

GENETIC ELEMENTS NATURALLY RESISTANT TO REPROGRAMMING

Evidence exists that at some regions of the genome, such as centromeres and telomeres, gametic inheritance of epigenetic state is routine; however, these regions do not appear to be prone to acquisition of marks in response to environment and contain few if any genes. Indeed, the inheritance of the epigenetic state at such regions is probably necessary for normal chromosome structure, pairing, and segregation. The study of these genetic elements provides molecular evidence of gametic epigenetic inheritance and guides efforts to unravel events at the biochemical level.

Sequences Naturally Resistant to Reprogramming in Plants

DNA methylation in plants is more complex than in animals and occurs at both CG and CNG sites. Mathieu and colleagues (75) backcrossed mutants with no CG methylation (*met1-3* homozygotes) to wild-type (MET1/MET1) plants and selected MET1/MET1 progeny in the F2 generation. These progeny were then inbred for seven generations. Surprisingly, unmethylated sequences were still present in the F7 generation, showing robust transgenerational inheritance of the unmethylated state. This work built on the results of numerous previous studies (6, 44, 65, 102, 128), suggesting that methylation patterns in plants are resistant to reprogramming. Several studies show that after DNA methylation patterns have been disturbed it can take a number of generations before the patterns revert to normal (**Table 2**). Furthermore, phenotypic consequences of hypomethylation (transposon activation in particular) are known (141). Less is known about the transgenerational dynamics of histones and PcG proteins in plants (60, 86).

Sequences Naturally Resistant to Reprogramming in Mice

As discussed above, the epigenetic marks that govern transcriptional regulation at imprinted loci are resistant to postfertilization reprogramming and some transgenes are resistant to both postfertilization and gametic reprogramming. Some classes of retrotransposons appear to behave like transgenes and are resistant to reprogramming at both stages, at least with respect to DNA methylation (69). Constitutive heterochromatin at centromeres is resistant to postfertilization demethylation (104). During germline reprogramming, although DNA methylation is erased from centromeric regions, their heterochromatic state is maintained by the continued presence of the repressive histone modification (109).

The removal of histones and replacement with protamines during mammalian spermatogenesis is another point at which epigenetic marks are cleared and replaced. In this case we also know that the clearing is incomplete. In human sperm approximately 15% of DNA remains nucleosome-bound and in the mouse this figure is approximately 2% (18). Nucleosome-bound DNA in sperm localizes to the telomeres and centromeres, consistent with the idea that these regions enter the oocyte already marked for their subsequent heterochromatic structure (93, 127, 138). Modified histones incorporated into the sex chromosomes during spermatogenesis in *Caenorhabditis elegans* and mouse probably persist for several cell divisions postfertilization (16, 87).

Chong and coworkers (32) recently showed that heterozygosity for mutations in epigenetic modifiers can induce phenotypes in the next generation in mice, presumably owing to the retention of abnormal epigenetic states established in the gametes. Roemer and colleagues (103) had previously shown that multiple epimutations caused by pronuclear transplantation can be passed on to the next generation. Therefore it seems likely that in mammals perturbation of global epigenetic patterns established in one generation can be passed on to the next, but this is rarely associated with single copy genes.

The Molecular Nature of Gametic Epigenetic Inheritance

The normal developmental program of an organism requires that more than simply DNA is transferred to the next generation, because the zygote must have the capacity to initiate transcription. Transcription requires proteins and RNA, which must have originated in the gametes. This requirement creates a molecular memory of the genotype of the parent. In those cases where gametic epigenetic inheritance occurs, the underlying molecular mark could take the form of DNA methylation, chromatin proteins, or RNA. As described above, cytosine methylation has been shown to be involved in gametic epigenetic inheritance in plants; however, little direct evidence exists about the inheritance of chromatin protein. There is increasing interest in the idea that RNA may have a role in this process.

In plants, gametic epigenetic inheritance in the form of RNA is an attractive idea because of the widespread RNA-directed epigenetic pathways that have been uncovered (54). RNA is present in considerable quantities in pollen, where much appears to be dedicated to the growth of the pollen tube (58, 96). It is exciting to speculate that this cache of RNAs could also act as a source of inherited memory to initiate the silencing of relevant transposon classes or genes in the next generation. Consistent with this idea, a genetic screen to identify modifiers of paramutation (which involves gametic epigenetic inheritance) at the aforementioned *b1* locus identified a gene encoding a protein that acts as an RNA-dependent RNA polymerase (4). Alleman and coworkers (4) propose that the polymerase is required to establish and maintain the heritable chromatin state associated with paramutation. Some information is available on RNA transcripts unique to the female *A. thaliana* gametophyte (137), but the specific functions of these transcripts are unknown.

In animals, the RNA stores in the female gamete are vital for early development. These stores are produced by a set of genes, called maternal effect genes, that are transcribed before the completion of meiosis and originate from alleles present in the mother but not necessarily present in the haploid genetic complement of the oocyte. In insects, RNA stores can also be produced by adjacent diploid nurse cells, which are connected to the oocyte by cytoplasmic bridges (114); the same process may also occur in mammals (95). Similar mechanisms are present during male gametogenesis. Human sperm has 5–10 femtograms of RNA (23), consisting of around 2700 different transcripts (92, 142). There is certainly ample opportunity during male gametogenesis for paternal effects resulting from either RNA or proteins made prior to meiosis or shared between spermatids (25).

In *C. elegans*, microinjection of small RNAs that target genes expressed in the maternal germline can induce phenotypes that last up to three generations (50). Furthermore, in the mouse, a white tail phenotype generally caused by a mutation at the *Kit* gene has been detected in offspring that do not inherit the mutation (101). The phenotype is weaker but still present in F2 offspring from crosses between affected 'wild-type' F1 mice. The authors argue that this phenomenon is the result of the inheritance of abnormal *Kit* RNA from sperm.

SUMMARY

Transgenerational reprogramming is important to ensure that the correct gene expression program is set at the start of embryonic development. The discovery of abnormal phenotypes, including cancer in humans, that are caused by epimutations emphasizes the dangers of abnormal resetting. The evidence at present suggests that for a mark to be resistant requires either a failure of the system (owing to mutation in genes encoding proteins involved in epigenetic reprogramming, abnormal nutritional availability, radiation exposure, or chemical treatment) or the locus to be part of an element involved in maintaining genome integrity (e.g., telomeres or centromeres) or susceptible to genomic imprinting. In the mouse, the IAP retrotransposons are exceptional in their resistance to DNA demethylation. These are the most active transposable elements in the mouse genome, and as such they may attract extra attention from genome defense systems (66). The resistance of some trangenes to reprogramming may occur through a similar mechanism.

Owing to their sessile nature, plants have evolved an enhanced capacity to respond to changes in their environment. Soft inheritance is therefore likely to be of greater use to plants than to animals. Unlike animals, there is no early separation of germline and soma in plants, allowing for epigenetic marks acquired throughout their lifetime to be included in the gametes. Indeed, some epialleles in plants are resistant to reprogramming for many generations, e.g., *Lcyc*. Nevertheless, it seems likely that such examples will be rare. The epimutation at *Lcyc* involves CG methylation, which is known to be resistant to reprogramming. However, most developmentally regulated genes are controlled by non-CG methylation, which requires a continuous remethylation cue and as such is continually reprogrammed (30, 141). Therefore, gain or loss of non-CG methylation at these genes is unlikely to be transferred to the next generation.

In animals, both adaptive and nonadaptive transgenerational epigenetic effects do occur. Many, perhaps most, of these effects are not the result of the direct transfer of information via the gametes. The advantage of epigenetically preadapting offspring to their future environment via the gametes appears to have been mostly outweighed by the desire to prevent inherited epimutations and safeguard the pluripotency of the epigenetic program of early development. The observations of transgenerational effects of nutritional availability, chemical exposure, and inherited epimutations are generally limited to one generation. Indeed, the theoretical arguments posited for the existence of soft inheritance emphasize the value of flexibility, so multigenerational inheritance of adaptive changes would be counterproductive.

DISCLOSURE STATEMENT

The authors are not aware of any biases that might be perceived as affecting the objectivity of this review.

ACKNOWLEDGMENTS

The authors wish to apologize to the authors of all the studies that we did not have space to include.

LITERATURE CITED

1. Aerts L, Van Assche FA. 2006. Animal evidence for the transgenerational development of diabetes mellitus. *Int. J. Biochem. Cell Biol.* 38:894–903
2. Agrawal AA, Laforsch C, Tollrian R. 1999. Transgenerational induction of defences in animals and plants. *Nature* 401:60–63
3. Alekseev V, Lampert W. 2001. Maternal control of resting-egg production in *Daphnia*. *Nature* 414:899–901
4. Alleman M, Sidorenko L, McGinnis K, Seshadri V, Dorweiler JE, et al. 2006. An RNA-dependent RNA polymerase is required for paramutation in maize. *Nature* 442:295–8
5. Allen ND, Norris ML, Surani MA. 1990. Epigenetic control of transgene expression and imprinting by genotype-specific modifiers. *Cell* 61:853–61
6. Amado L, Abranches R, Neves N, Viegas W. 1997. Development-dependent inheritance of 5-azacytidine-induced epimutations in triticale: analysis of rDNA expression patterns. *Chromosome Res.* 5:445–50
7. Anderson CM, Lopez F, Zimmer A, Benoit JN. 2006. Placental insufficiency leads to developmental hypertension and mesenteric artery dysfunction in two generations of Sprague-Dawley rat offspring. *Biol. Reprod.* 74:538–44
8. Anway MD, Cupp AS, Uzumcu M, Skinner MK. 2005. Epigenetic transgenerational actions of endocrine disruptors and male fertility. *Science* 308:1466–69
9. Anway MD, Skinner MK. 2006. Epigenetic transgenerational actions of endocrine disruptors. *Endocrinology* 147:S43–49
10. Armitage JA, Taylor PD, Poston L. 2005. Experimental models of developmental programming: consequences of exposure to an energy rich diet during development. *J. Physiol.* 565:3–8
11. Banks JA, Masson P, Fedoroff N. 1988. Molecular mechanisms in the developmental regulation of the maize *Suppressor-mutator* transposable element. *Genes Dev.* 2:1364–80
12. Barber R, Plumb MA, Boulton E, Roux I, Dubrova YE. 2002. Elevated mutation rates in the germ line of first- and second-generation offspring of irradiated male mice. *Proc. Natl. Acad. Sci. USA* 99:6877–82
13. Barber RC, Dubrova YE. 2006. The offspring of irradiated parents, are they stable? *Mutat. Res.* 598:50–60
14. Barker DJP. 1998. *Mothers, Babies and Health in Later Life*. Edinburgh: Churchill Livingstone. ix, 217 pp. 2nd ed.
15. Bateson P. 2001. Fetal experience and good adult design. *Int. J. Epidemiol.* 30:928–34
16. Bean CJ, Schaner CE, Kelly WG. 2004. Meiotic pairing and imprinted X chromatin assembly in *Caenorhabditis elegans*. *Nat. Genet.* 36:100–5
17. Beckmann JS, Estivill X, Antonarakis SE. 2007. Copy number variants and genetic traits: closer to the resolution of phenotypic to genotypic variability. *Nat. Rev. Genet.* 8:639–46
18. Bench GS, Friz AM, Corzett MH, Morse DH, Balhorn R. 1996. DNA and total protamine masses in individual sperm from fertile mammalian subjects. *Cytometry* 23:263–71
19. Bender J, Fink GR. 1995. Epigenetic control of an endogenous gene family is revealed by a novel blue fluorescent mutant of *Arabidopsis*. *Cell* 83:725–34

20. Benyshek DC, Johnston CS, Martin JF. 2006. Glucose metabolism is altered in the adequately-nourished grand-offspring (F3 generation) of rats malnourished during gestation and perinatal life. *Diabetologia* 49:1117–19
21. Blatt J, Van Le L, Weiner T, Sailer S. 2003. Ovarian carcinoma in an adolescent with transgenerational exposure to diethylstilbestrol. *J. Pediatr. Hematol. Oncol.* 25:635–36
22. Blewitt ME, Vickaryous NK, Paldi A, Koseki H, Whitelaw E. 2006. Dynamic reprogramming of DNA methylation at an epigenetically sensitive allele in mice. *PLoS Genet.* 2:e49
23. Boerke A, Dieleman SJ, Gadella BM. 2007. A possible role for sperm RNA in early embryo development. *Theriogenology* 68(Suppl 1):S147–55
24. Boloker J, Gertz SJ, Simmons RA. 2002. Gestational diabetes leads to the development of diabetes in adulthood in the rat. *Diabetes* 51:1499–506
25. Braun RE, Behringer RR, Peschon JJ, Brinster RL, Palmiter RD. 1989. Genetically haploid spermatids are phenotypically diploid. *Nature* 337:373–76
26. Burge GC, Slater-Jeffries J, Torrrens C, Phillips ES, Hanson MA, Lilliycrop KA. 2007. Dietary protein restriction of pregnant rats in the F_0 generation induces altered methylation of hepatic gene promoters in the adult male offspring in the F_1 and F_2 generations. *Br. J. Nutr.* 97:435–9
27. Bygren LO, Kaati G, Edvinsson S. 2001. Longevity determined by paternal ancestors' nutrition during their slow growth period. *Acta Biotheor.* 49:53–59
28. Cavalli G, Paro R. 1998. The *Drosophila Fab-7* chromosomal element conveys epigenetic inheritance during mitosis and meiosis. *Cell* 93:505–18
29. Cavalli G, Paro R. 1999. Epigenetic inheritance of active chromatin after removal of the main transactivator. *Science* 286:955–58
30. Chan SW, Henderson IR, Zhang X, Shah G, Chien JS, Jacobsen SE. 2006. RNAi, DRD1, and histone methylation actively target developmentally important non-CG DNA methylation in Arabidopsis. *PLoS Genet.* 2:e83
31. Chan TL, Yuen ST, Kong CK, Chan YW, Chan AS, et al. 2006. Heritable germline epimutation of *MSH2* in a family with hereditary nonpolyposis colorectal cancer. *Nat. Genet.* 38:1178–83
32. Chong S, Vickaryous N, Ashe A, Zamudio N, Youngson N, et al. 2007. Modifiers of epigenetic reprogramming show paternal effects in the mouse. *Nat. Genet.* 39:614–22
33. Cooney CA, Dave AA, Wolff GL. 2002. Maternal methyl supplements in mice affect epigenetic variation and DNA methylation of offspring. *J. Nutr.* 132:2393–400
34. Cropley JE, Suter CM, Beckman KB, Martin DI. 2006. Germ-line epigenetic modification of the murine A^{vy} allele by nutritional supplementation. *Proc. Natl. Acad. Sci. USA* 103:17308–12
35. Csaba G, Inczefi-Gonda A. 1998. Transgenerational effect of a single neonatal benzpyrene treatment on the glucocorticoid receptor of the rat thymus. *Hum. Exp. Toxicol.* 17:88–92
36. Cubas P, Vincent C, Coen E. 1999. An epigenetic mutation responsible for natural variation in floral symmetry. *Nature* 401:157–61
37. Das OP, Messing J. 1994. Variegated phenotype and developmental methylation changes of a maize allele originating from epimutation. *Genetics* 136:1121–41
38. Denton KM, Flower RL, Stevenson KM, Anderson WP. 2003. Adult rabbit offspring of mothers with secondary hypertension have increased blood pressure. *Hypertension* 41:634–39
39. Dickinson WJ, Seger J. 1999. Cause and effect in evolution. *Nature* 399:30
40. Dolinoy DC, Weidman JR, Waterland RA, Jirtle RL. 2006. Maternal genistein alters coat color and protects A^{vy} mouse offspring from obesity by modifying the fetal epigenome. *Environ. Health Perspect.* 114:567–72
41. Dorn R, Krauss V, Reuter G, Saumweber H. 1993. The enhancer of position-effect variegation of *Drosophila*, *E(var)3-93D*, codes for a chromatin protein containing a conserved domain common to several transcriptional regulators. *Proc. Natl. Acad. Sci. USA* 90:11376–80
42. Drake AJ, Walker BR, Seckl JR. 2005. Intergenerational consequences of fetal programming by in utero exposure to glucocorticoids in rats. *Am J. Physiol. Regul. Integr. Comp. Physiol.* 288:R34–38
43. Dubrova YE, Plumb M, Gutierrez B, Boulton E, Jeffreys AJ. 2000. Transgenerational mutation by radiation. *Nature* 405:37

44. Finnegan EJ, Peacock WJ, Dennis ES. 1996. Reduced DNA methylation in *Arabidopsis thaliana* results in abnormal plant development. *Proc. Natl. Acad. Sci. USA* 93:8449–54
45. Forsen T, Eriksson J, Tuomilehto J, Reunanen A, Osmond C, Barker D. 2000. The fetal and childhood growth of persons who develop type 2 diabetes. *Ann. Intern. Med.* 133:176–82
46. Forsen T, Eriksson JG, Tuomilehto J, Osmond C, Barker DJ. 1999. Growth in utero and during childhood among women who develop coronary heart disease: longitudinal study. *BMJ* 319:1403–7
47. Francis D, Diorio J, Liu D, Meaney MJ. 1999. Nongenomic transmission across generations of maternal behavior and stress responses in the rat. *Science* 286:1155–58
48. Galloway LF. 2005. Maternal effects provide phenotypic adaptation to local environmental conditions. *New Phytol.* 166:93–99
49. Gazzoli I, Loda M, Garber J, Syngal S, Kolodner RD. 2002. A hereditary nonpolyposis colorectal carcinoma case associated with hypermethylation of the *MLH1* gene in normal tissue and loss of heterozygosity of the unmethylated allele in the resulting microsatellite instability-high tumor. *Cancer Res.* 62:3925–28
50. Grishok A, Tabara H, Mello CC. 2000. Genetic requirements for inheritance of RNAi in *C. elegans*. *Science* 287:2494–97
51. Groothuis TG, Muller W, von Engelhardt N, Carere C, Eising C. 2005. Maternal hormones as a tool to adjust offspring phenotype in avian species. *Neurosci. Biobehav. Rev.* 29:329–52
52. Hadchouel M, Farza H, Simon D, Tiollais P, Pourcel C. 1987. Maternal inhibition of hepatitis B surface antigen gene expression in transgenic mice correlates with de novo methylation. *Nature* 329:454–56
53. Hales BF, Crosman K, Robaire B. 1992. Increased postimplantation loss and malformations among the F2 progeny of male rats chronically treated with cyclophosphamide. *Teratology* 45:671–78
54. Henderson IR, Jacobsen SE. 2007. Epigenetic inheritance in plants. *Nature* 447:418–24
55. Herbst AL, Ulfelder H, Poskanzer DC. 1971. Adenocarcinoma of the vagina. Association of maternal stilbestrol therapy with tumor appearance in young women. *N. Engl. J. Med.* 284:878–81
56. Hitchins M, Williams R, Cheong K, Halani N, Lin VA, et al. 2005. *MLH1* germline epimutations as a factor in hereditary nonpolyposis colorectal cancer. *Gastroenterology* 129:1392–99
57. Hitchins MP, Wong JJ, Suthers G, Suter CM, Martin DI, et al. 2007. Inheritance of a cancer-associated *MLH1* germ-line epimutation. *N. Engl. J. Med.* 356:697–705
58. Honys D, Twell D. 2004. Transcriptome analysis of haploid male gametophyte development in *Arabidopsis*. *Genome Biol.* 5:R85
59. Ims RA. 1990. Determinants of natal dispersal and space use in gray-sided voles, *Clethrionomys rufocanus*: A combined field and laboratory experiment. *Oikos* 57:106–13
60. Ingouff M, Hamamura Y, Gourgues M, Higashiyama T, Berger F. 2007. Distinct dynamics of HISTONE3 variants between the two fertilization products in plants. *Curr. Biol.* 17:1032–37
61. Jackson AA, Dunn RL, Marchand MC, Langley-Evans SC. 2002. Increased systolic blood pressure in rats induced by a maternal low-protein diet is reversed by dietary supplementation with glycine. *Clin. Sci.* 103:633–39
62. Jacobsen SE, Meyerowitz EM. 1997. Hypermethylated *SUPERMAN* epigenetic alleles in *Arabidopsis*. *Science* 277:1100–3
63. Kaati G, Bygren LO, Edvinsson S. 2002. Cardiovascular and diabetes mortality determined by nutrition during parents' and grandparents' slow growth period. *Eur. J. Hum. Genet.* 10:682–88
64. Kaati G, Bygren LO, Pembrey M, Sjostrom M. 2007. Transgenerational response to nutrition, early life circumstances and longevity. *Eur. J. Hum. Genet.* 15:784–90
65. Kakutani T, Munakata K, Richards EJ, Hirochika H. 1999. Meiotically and mitotically stable inheritance of DNA hypomethylation induced by *ddm1* mutation of *Arabidopsis thaliana*. *Genetics* 151:831–38
66. Kantheti P, Diaz ME, Peden AE, Seong EE, Dolan DF, et al. 2003. Genetic and phenotypic analysis of the mouse mutant mh2J, an Ap3d allele caused by IAP element insertion. *Mamm. Genome* 14:157–67
67. Kearns M, Preis J, McDonald M, Morris C, Whitelaw E. 2000. Complex patterns of inheritance of an imprinted murine transgene suggest incomplete germline erasure. *Nucleic Acids Res.* 28:3301–9
68. Lacey EP. 1996. Parental effects in *Plantago lanceolata* L. I. A growth chamber experiment to examine pre- and postzygotic temperature effects. *Evolution* 50:865–78
69. Lane N, Dean W, Erhardt S, Hajkova P, Surani A, et al. 2003. Resistance of IAPs to methylation reprogramming may provide a mechanism for epigenetic inheritance in the mouse. *Genesis* 35:88–93

70. Li S, Washburn KA, Moore R, Uno T, Teng C, et al. 1997. Developmental exposure to diethylstilbestrol elicits demethylation of estrogen-responsive lactoferrin gene in mouse uterus. *Cancer Res.* 57:4356–59
71. Lillycrop KA, Phillips ES, Jackson AA, Hanson MA, Burdge GC. 2005. Dietary protein restriction of pregnant rats induces and folic acid supplementation prevents epigenetic modification of hepatic gene expression in the offspring. *J. Nutr.* 135:1382–86
72. Liu D, Caldji C, Sharma S, Plotsky PM, Meaney MJ. 2000. Influence of neonatal rearing conditions on stress-induced adrenocorticotropin responses and norepinepherine release in the hypothalamic paraventricular nucleus. *J. Neuroendocrinol.* 12:5–12
73. Liu D, Diorio J, Tannenbaum B, Caldji C, Francis D, et al. 1997. Maternal care, hippocampal glucocorticoid receptors, and hypothalamic-pituitary-adrenal responses to stress. *Science* 277:1659–62
74. Lumey LH. 1992. Decreased birthweights in infants after maternal in utero exposure to the Dutch famine of 1944–1945. *Paediatr. Perinat. Epidemiol.* 6:240–53
75. Mathieu O, Reinders J, Caikovski M, Smathajitt C, Paszkowski J. 2007. Transgenerational stability of the *Arabidopsis* epigenome is coordinated by CG methylation. *Cell* 130:851–62
76. Matzke AJ, Matzke MA. 1998. Position effects and epigenetic silencing of plant transgenes. *Curr. Opin. Plant Biol.* 1:142–48
77. Mayr E. 1982. *The Growth of Biological Thought: Diversity, Evolution, and Inheritance*. Cambridge, MA/London: Belknap Press of Harvard Univ. Press. 974 pp.
78. Mayr E, Provine WB. 1980. *The Evolutionary Synthesis: Perspectives on the Unification of Biology*. Cambridge, MA/ London: Harvard Univ. Press. 487 pp.
79. Melquist S, Bender J. 2003. Transcription from an upstream promoter controls methylation signaling from an inverted repeat of endogenous genes in *Arabidopsis*. *Genes Dev.* 17:2036–47
80. Miao SL, Bazzaz FA, Primack RB. 1991. Persistence of maternal nutrient effects in *Plantago major*: The third generation. *Ecology* 72:1634–42
81. Miyakura Y, Sugano K, Konishi F, Fukayama N, Igarashi S, et al. 2003. Methylation profile of the *MLH1* promoter region and their relationship to colorectal carcinogenesis. *Genes Chromosomes Cancer* 36:17–25
82. Molinier J, Ries G, Zipfel C, Hohn B. 2006. Transgeneration memory of stress in plants. *Nature* 442:1046–49
83. Morgan HD, Sutherland HG, Martin DI, Whitelaw E. 1999. Epigenetic inheritance at the agouti locus in the mouse. *Nat. Genet.* 23:314–18
84. Morse HC 3rd, Yetter RA, Stimpfling JH, Pitts OM, Fredrickson TN, Hartley JW. 1985. Greying with age in mice: Relation to expression of murine leukemia viruses. *Cell* 41:439–48
85. Mousseau TA, Fox CW, eds. 1998. *Maternal Effects as Adaptations*. New York/Oxford: Oxford Univ. Press. 375 pp.
86. Mylne JS, Barrett L, Tessadori F, Mesnage S, Johnson L, et al. 2006. LHP1, the *Arabidopsis* homologue of HETEROCHROMATIN PROTEIN1, is required for epigenetic silencing of *FLC*. *Proc. Natl. Acad. Sci. USA* 103:5012–17
87. Namekawa SH, Park PJ, Zhang LF, Shima JE, McCarrey JR, et al. 2006. Postmeiotic sex chromatin in the male germline of mice. *Curr. Biol.* 16:660–67
88. Newbold RR, Liehr JG. 2000. Induction of uterine adenocarcinoma in CD-1 mice by catechol estrogens. *Cancer Res.* 60:235–37
89. Newbold RR, Padilla-Banks E, Jefferson WN. 2006. Adverse effects of the model environmental estrogen diethylstilbestrol are transmitted to subsequent generations. *Endocrinology* 147:S11–17
90. Newnham JP, Evans SF, Godfrey M, Huang W, Ikegami M, Jobe A. 1999. Maternal, but not fetal, administration of corticosteroids restricts fetal growth. *J. Matern. Fetal Med.* 8:81–87
91. Nyirenda MJ, Lindsay RS, Kenyon CJ, Burchell A, Seckl JR. 1998. Glucocorticoid exposure in late gestation permanently programs rat hepatic phosphoenolpyruvate carboxykinase and glucocorticoid receptor expression and causes glucose intolerance in adult offspring. *J. Clin. Invest.* 101:2174–81
92. Ostermeier GC, Dix DJ, Miller D, Khatri P, Krawetz SA. 2002. Spermatozoal RNA profiles of normal fertile men. *Lancet* 360:772–77

93. Palmer DK, O'Day K, Margolis RL. 1990. The centromere specific histone CENP-A is selectively retained in discrete foci in mammalian sperm nuclei. *Chromosoma* 100:32–36
94. Pembrey ME, Bygren LO, Kaati G, Edvinsson S, Northstone K, et al. 2006. Sex-specific, male-line transgenerational responses in humans. *Eur. J. Hum. Genet.* 14:159–66
95. Pepling ME, Spradling AC. 1998. Female mouse germ cells form synchronously dividing cysts. *Development* 125:3323–28
96. Pina C, Pinto F, Feijo JA, Becker JD. 2005. Gene family analysis of the *Arabidopsis* pollen transcriptome reveals biological implications for cell growth, division control, and gene expression regulation. *Plant Physiol.* 138:744–56
97. Platenkamp GJ, Shaw RG. 1993. Environmental and genetic maternal effects on seed characters in *Nemophila menziesii*. *Evolution* 47:540–55
98. Popova NV. 1989. Transgenerational effect of orthoaminoasotoluol in mice. *Cancer Lett.* 46:203–6
99. Queitsch C, Sangster TA, Lindquist S. 2002. Hsp90 as a capacitor of phenotypic variation. *Nature* 417:618–24
100. Rakyan VK, Chong S, Champ ME, Cuthbert PC, Morgan HD, et al. 2003. Transgenerational inheritance of epigenetic states at the murine $Axin^{Fu}$ allele occurs after maternal and paternal transmission. *Proc. Natl. Acad. Sci. USA* 100:2538–43
101. Rassoulzadegan M, Grandjean V, Gounon P, Vincent S, Gillot I, Cuzin F. 2006. RNA-mediated non-mendelian inheritance of an epigenetic change in the mouse. *Nature* 441:469–74
102. Riddle NC, Richards EJ. 2005. Genetic variation in epigenetic inheritance of ribosomal RNA gene methylation in *Arabidopsis*. *Plant J.* 41:524–32
103. Roemer I, Reik W, Dean W, Klose J. 1997. Epigenetic inheritance in the mouse. *Curr. Biol.* 7:277–80
104. Rougier N, Bourc'his D, Gomes DM, Niveleau A, Plachot M, et al. 1998. Chromosome methylation patterns during mammalian preimplantation development. *Genes. Dev.* 12:2108–13
105. Rutherford SL, Lindquist S. 1998. Hsp90 as a capacitor for morphological evolution. *Nature* 396:336–42
106. Schmid B, Dolt C. 1994. Effects of maternal and paternal environment and genotype on offspring phenotype in *Solidago altissima*. *Evolution* 48:1525–49
107. Schmitt J, Gamble SE. 1990. The effect of distance from the parental site on offspring performance and inbreeding depression in *Impatiens capensis*: A test of the local adaptation hypothesis. *Evolution* 44:2022–30
108. Schmitt J, Niles J, Wulff R. 1992. Norms of reaction of seed traits to maternal environments in *Plantago lanceolata*. *Am. Nat.* 139:451–66
109. Seki Y, Yamaji M, Yabuta Y, Sano M, Shigeta M, et al. 2007. Cellular dynamics associated with the genome-wide epigenetic reprogramming in migrating primordial germ cells in mice. *Development* 134:2627–38
110. Shi S, Calhoun HC, Xia F, Li J, Le L, Li WX. 2006. JAK signaling globally counteracts heterochromatic gene silencing. *Nat. Genet.* 38:1071–76
111. Sollars V, Lu X, Xiao L, Wang X, Garfinkel MD, Ruden DM. 2003. Evidence for an epigenetic mechanism by which Hsp90 acts as a capacitor for morphological evolution. *Nat. Genet.* 33:70–74
112. Soppe WJ, Jacobsen SE, Alonso-Blanco C, Jackson JP, Kakutani T, et al. 2000. The late flowering phenotype of *fwa* mutants is caused by gain-of-function epigenetic alleles of a homeodomain gene. *Mol. Cell* 6:791–802
113. Spergel G, Levy LJ, Goldner MG. 1971. Glucose intolerance in the progeny of rats treated with single subdiabetogenic dose of alloxan. *Metabolism* 20:401–13
114. Spradling AC. 1993. Developmental genetics of oogenesis. In *The Development of Drosophila melanogaster*, ed. M Bates, A Martinez-Arias, pp. 1–70. New York: Cold Spring Harbor Lab. Press
115. Stam M, Belele C, Dorweiler JE, Chandler VL. 2002. Differential chromatin structure within a tandem array 100 kb upstream of the maize *b1* locus is associated with paramutation. *Genes Dev.* 16:1906–18
116. Stanner SA, Bulmer K, Andres C, Lantseva OE, Borodina V, et al. 1997. Does malnutrition in utero determine diabetes and coronary heart disease in adulthood? Results from the Leningrad siege study, a cross sectional study. *Br. Med. J.* 315:1342–48

117. Stein AD, Lumey LH. 2000. The relationship between maternal and offspring birth weights after maternal prenatal famine exposure: The Dutch Famine Birth Cohort Study. *Hum. Biol.* 72:641–54
118. Stewart RJ, Sheppard H, Preece R, Waterlow JC. 1980. The effect of rehabilitation at different stages of development of rats marginally malnourished for ten to twelve generations. *Br. J. Nutr.* 43:403–12
119. Stokes TL, Kunkel BN, Richards EJ. 2002. Epigenetic variation in *Arabidopsis* disease resistance. *Genes Dev.* 16:171–82
120. Stratton DA. 1989. Competition prolongs expression of maternal effects in seedlings of *Erigeron annuus* (Asteraceae). *Am. J. Bot.* 76:1646–53
121. Sultan SE. 1996. Phenotypic plasticity for offspring traits in *Polygonum persicaria*. *Ecology* 77:1791–807
122. Suter CM, Martin DI, Ward RL. 2004. Germline epimutation of *MLH1* in individuals with multiple cancers. *Nat. Genet.* 36:497–501
123. Sutherland HG, Kearns M, Morgan HD, Headley AP, Morris C, et al. 2000. Reactivation of heritably silenced gene expression in mice. *Mamm. Genome* 11:347–55
124. Swain JL, Stewart TA, Leder P. 1987. Parental legacy determines methylation and expression of an autosomal transgene: a molecular mechanism for parental imprinting. *Cell* 50:719–27
125. Torrens C, Brawley L, Anthony FW, Dance CS, Dunn R, et al. 2006. Folate supplementation during pregnancy improves offspring cardiovascular dysfunction induced by protein restriction. *Hypertension* 47:982–87
126. Tschirren B, Fitze PS, Richner H. 2007. Maternal modulation of natal dispersal in a passerine bird: An adaptive strategy to cope with parasitism? *Am. Nat.* 169:87–93
127. van der Heijden GW, Derijck AA, Ramos L, Giele M, van der Vlag J, de Boer P. 2006. Transmission of modified nucleosomes from the mouse male germline to the zygote and subsequent remodeling of paternal chromatin. *Dev. Biol.* 298:458–69
128. Vongs A, Kakutani T, Martienssen RA, Richards EJ. 1993. *Arabidopsis thaliana* DNA methylation mutants. *Science* 260:1926–28
129. Waterland RA, Jirtle RL. 2003. Transposable elements: targets for early nutritional effects on epigenetic gene regulation. *Mol. Cell. Biol.* 23:5293–300
130. Waterland RA, Travisano M, Tahiliani KG. 2007. Diet-induced hypermethylation at *agouti viable yellow* is not inherited transgenerationally through the female. *FASEB J.* 21:3380–5
131. Weaver IC, Cervoni N, Champagne FA, D'Alessio AC, Sharma S, et al. 2004. Epigenetic programming by maternal behavior. *Nat. Neurosci.* 7:847–54
132. Weaver IC, D'Alessio AC, Brown SE, Hellstrom IC, Dymov S, et al. 2007. The transcription factor nerve growth factor-inducible protein a mediates epigenetic programming: Altering epigenetic marks by immediate-early genes. *J. Neurosci.* 27:1756–68
133. Weaver IC, Grant RJ, Meaney MJ. 2002. Maternal behavior regulates long-term hippocampal expression of BAX and apoptosis in the offspring. *J. Neurochem.* 82:998–1002
134. Weaver IC, Meaney MJ, Szyf M. 2006. Maternal care effects on the hippocampal transcriptome and anxiety-mediated behaviors in the offspring that are reversible in adulthood. *Proc. Natl. Acad. Sci. USA* 103:3480–85
135. Wolff GL, Kodell RL, Moore SR, Cooney CA. 1998. Maternal epigenetics and methyl supplements affect *agouti* gene expression in A^{vy}/a mice. *FASEB J.* 12:949–57
136. Xing Y, Shi S, Le L, Lee CA, Silver-Morse L, Li WX. 2007. Evidence for transgenerational transmission of epigenetic tumor susceptibility in *Drosophila*. *PLoS Genet.* 3:1598–606
137. Yu HJ, Hogan P, Sundaresan V. 2005. Analysis of the female gametophyte transcriptome of *Arabidopsis* by comparative expression profiling. *Plant Physiol.* 139:1853–69
138. Zalenskaya IA, Bradbury EM, Zalensky AO. 2000. Chromatin structure of telomere domain in human sperm. *Biochem. Biophys. Res. Commun.* 279:213–18
139. Zambrano E, Martinez-Samayoa PM, Bautista CJ, Deas M, Guillen L, et al. 2005. Sex differences in transgenerational alterations of growth and metabolism in progeny (F2) of female offspring (F1) of rats fed a low protein diet during pregnancy and lactation. *J. Physiol.* 566:225–36

140. Zhang TY, Bagot R, Parent C, Nesbitt C, Bredy TW, et al. 2006. Maternal programming of defensive responses through sustained effects on gene expression. *Biol. Psychol.* 73:72–89
141. Zhang X, Yazaki J, Sundaresan A, Cokus S, Chan SW, et al. 2006. Genome-wide high-resolution mapping and functional analysis of DNA methylation in *Arabidopsis*. *Cell* 126:1189–201
142. Zhao Y, Li Q, Yao C, Wang Z, Zhou Y, et al. 2006. Characterization and quantification of mRNA transcripts in ejaculated spermatozoa of fertile men by serial analysis of gene expression. *Hum. Reprod.* 21:1583–90

Evolution of Dim-Light and Color Vision Pigments

Shozo Yokoyama

Department of Biology, Emory University, Atlanta, Georgia 30322; email: syokoya@emory.edu

Annu. Rev. Genomics Hum. Genet. 2008. 9:259–82

First published online as a Review in Advance on June 10, 2008

The *Annual Review of Genomics and Human Genetics* is online at genom.annualreviews.org

This article's doi:
10.1146/annurev.genom.9.081307.164228

1527-8204/08/0922-0259$20.00

Key Words

spectral tuning, adaptive evolution, ancestral pigments, mutagenesis, protein engineering

Abstract

A striking level of diversity of visual systems in different species reflects their adaptive responses to various light environments. To study the adaptive evolution of visual systems, we need to understand how visual pigments, the light-sensitive molecules, have tuned their wavelengths of light absorption. The molecular basis of spectral tuning in visual pigments, a central unsolved problem in phototransduction, can be understood only by studying how different species have adapted to various light environments. Certain amino acid replacements at 30 residues explain some dim-light and color vision in vertebrates. To better understand the molecular and functional adaptations of visual pigments, we must identify all critical amino acid replacements that are involved in the spectral tuning and elucidate the effects of their interactions on the spectral shifts.

INTRODUCTION

Molecular adaptation: the process by which molecules are positively selected by natural selection

TM: transmembrane

λ_{max}: the wavelength of maximal absorption of a visual pigment

Spectral tuning: the phenomenon in which a chromophore attains different absorption spectra when attached to different opsins

Visual pigments: light-sensitive molecules that consist of a chromophore (either 11-*cis*-retinal or 11-*cis*-3,4-dehydroretinal) and an opsin

In vitro assay: a cell culture–based method of measuring the absorption spectra of visual pigments

Functional adaptation: the process by which a protein's function is positively selected by natural selection

Many vertebrates use vision as a principal means to interpret their environments and have consequently evolved diverse visual systems (8, 40, 71). The extensive data collected by vision scientists suggest strongly that this diversity is a result of organisms' adaptations to various photic environments and to their new lifestyles (8, 40, 71, 81, 83, 94). Did animals really modify their visual systems to adapt to different environments? If so, how did they do it? These questions touch on a remarkably difficult problem of molecular adaptation in evolutionary biology as well as on a central unsolved problem of phototransduction in vision science.

In most vertebrates, rod photoreceptors are responsible for highly sensitive dim-light vision, whereas cone photoreceptors mediate color discrimination and high visual acuity at higher light intensities (6, 60, 75). The nocturnal Tokay gecko (*Gekko gekko*) and the diurnal American chameleon (*Anolis carolinensis*) provide interesting oddities in that they have pure-rod retinas and pure-cone retinas, respectively (81).

The light-sensitive molecules, visual pigments, in these photoreceptor cells consist of an integral transmembrane (TM) protein, opsin, and a chromophore, either 11-*cis*-retinal or 11-*cis*-3,4-dehydroretinal. The chromophore is covalently bound to an opsin via a Schiff base linkage to a conserved lysine at residue 296 (K296) (70). The 11-*cis*-retinal in solution absorbs light maximally (λ_{max}) at 440 nm (36); however, by interacting with various opsins, it detects a wide range of λ_{max}s between 360 and 560 nm, which is known as spectral tuning in visual pigments (37).

Animals live in diverse light environments that range from the darkness at the bottom of the ocean to bright light on land. A strong association exists between the types of visual pigments animals possess and the environments they live in. For example, humans have a total of four types of visual pigments and their λ_{max}s range from 414 nm to 560 nm (51); zebrafish, living in shallow water, have a total of nine types of visual pigments and their λ_{max}s range from 360 to 560 nm (14), which corresponds to the wide range of light available to them. Compared with these species, coelacanths, living at a depth of 200 m, have only two types of visual pigments whose λ_{max}s are very close to the maximal wavelength of downwelling sunlight at 480 nm (94) (**Figure 1**).

Molecular genetic analyses of spectral tuning in visual pigments became feasible when the bovine rhodopsin gene was cloned (49) and when an in vitro assay (**Figure 2**) became available (46, 47, 54). Thanks to these developments, virtually any opsins in vertebrates can now be manipulated, expressed in cultured cells, reconstituted with 11-*cis*-retinal, and the λ_{max} of the resulting visual pigments can be measured (82). The recent crystal structural analyses of bovine rhodopsin (**Figure 3**) also lay a solid foundation for studying the chemical basis of spectral tuning (53, 56). Despite these developments, the molecular basis of spectral tuning in visual pigments is still not well understood. Analyses of molecular adaptations in vertebrates are also fraught with major problems because it is remarkably difficult not only to detect minute selective advantages caused by molecular changes in nature (38), but also to find genetic systems where evolutionary hypotheses can be tested (79).

To study the possible functional adaptation of visual pigments, we must understand how the spectral tuning in visual pigments works. To understand the molecular basis of spectral tuning, we must identify amino acid changes that shift the absorption spectra of visual pigments. To identify such amino acid changes, we must know how animals modified their visual pigments in the past (79). Hence, the evolutionary biology and vision science approaches are closely intertwined and share an important common goal of elucidating why and how organisms modified their visual pigments to live in their new environments (79, 81, 83).

With the possibility of many adaptive events, the availability of the in vitro assay, and the crystal structure of the bovine rhodopsin, dim-light and color vision provide an opportunity

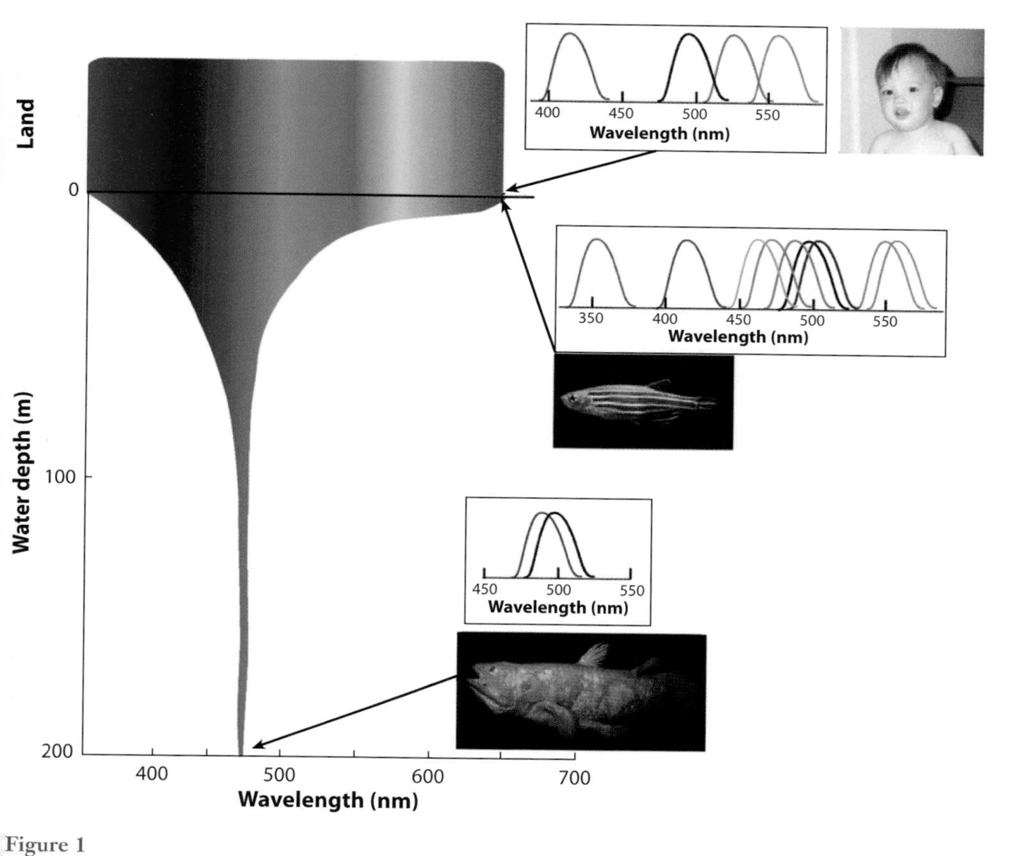

Figure 1

Photic environments and the visual pigments of human (51), zebrafish (14), and coelacanth (94). The pictures of zebrafish and coelacanth were taken by M. Noren and JJ Photo, respectively. See **http://www.fshbase.org**, version 09/2006.

to explore not only why and how molecular and functional adaptations of visual pigments occurred, but also how various visual pigments modulate their light sensitivities. Mutagenesis analyses of visual pigments have improved our understanding of the molecular bases of spectral tuning in visual pigments (25, 81). Consequently, visual pigments became one of a small number of model systems in the exploration of molecular adaptations in vertebrates (19, 29, 79, 81, 83, 93). In fact, the molecular analyses of the origin and evolution of color vision produced arguably "the deepest body of knowledge linking differences in specific genes to differences in ecology and to the evolution of species" (10).

Here I review recent developments in the functional differentiations of visual pigments that have generated red-green color vision, UV-violet vision, and dim-light vision in ancestral as well as contemporary species. The key ingredient in these analyses is mutagenesis experiments. A survey of mutagenesis results of visual pigments reveals that the direction and magnitude of the spectral shift caused by a certain amino acid change can be affected strongly by the amino acid composition of a specific visual pigment under study. Consequently, the specific mutagenesis result may not apply even to the identical mutation in other pigments, making it difficult not only to derive a general principle of the spectral tuning of visual pigments, but also to elucidate the molecular mechanism of functional adaptation of visual pigments. The seemingly conflicting mutagenesis results can

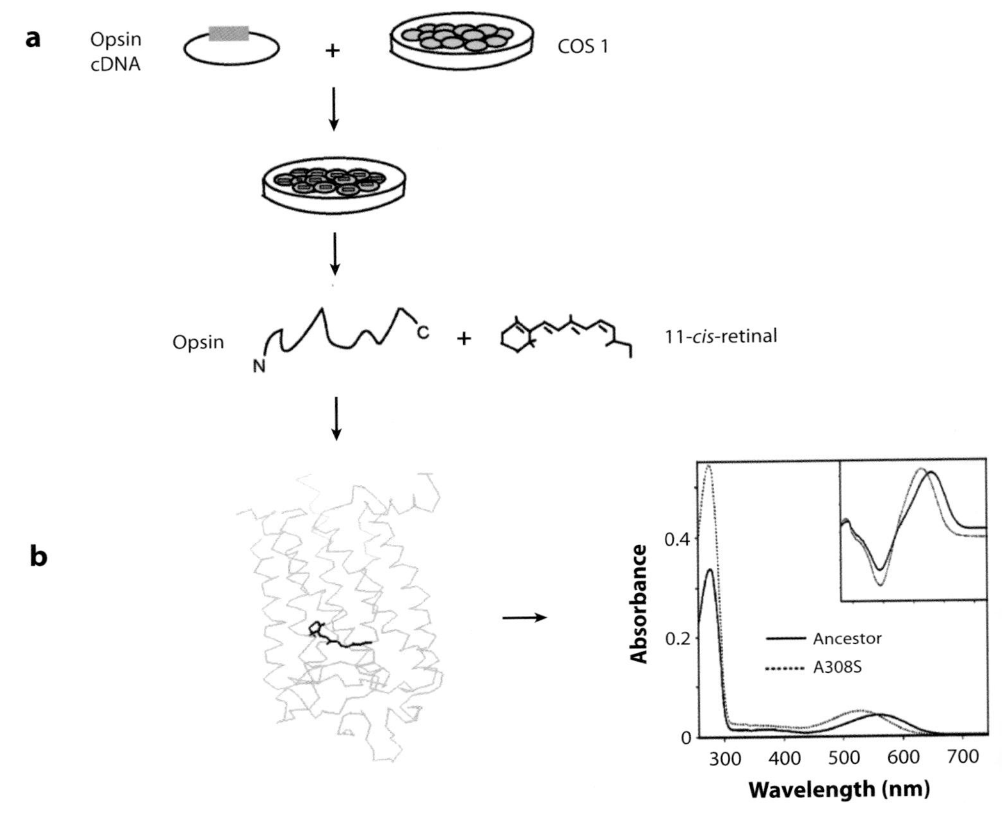

Figure 2

In vitro assay of the absorption spectra of the ancestral mammalian middle and long wavelength-sensitive (M/LWS) pigment and its mutant, containing A308S. (*a*) The opsin cDNAs are expressed in an expression vector, pMT5, in COS1 cells after transient transfection. The opsins are then reconstituted with 11-*cis*-retinal. The resulting visual pigments are purified by immunoaffinity chromatography with the monoclonal antibody 1D4 Sepharose 4B. (*b*) The absorption spectra (dark spectra) of the visual pigments are recorded in the dark using a spectrophotometer. The amino acid change A308S was made by site-directed mutagenesis. The absorption spectra in the inset show the difference spectra by subtracting a spectrum measured after photobleaching from a spectrum evaluated before light exposure.

"make sense only in the light of evolution" (21) and the way we conduct mutagenesis experiments must be re-evaluated.

Rhodopsins (RH1): visual pigments that are expressed typically in rods

RH1-like (RH2) pigments: visual pigments whose amino acid sequences are most closely related to those of RH1 pigments

EVOLUTION OF VISUAL PIGMENTS

Since the first molecular clonings of the bovine and human opsin genes (49–51), more than 470 opsin genes from ~180 vertebrate species have been characterized. For 126 of these genes, not only have the complete nucleotide sequences of the coding regions been determined but also the λ_{max}s of the corresponding visual pigments have been measured with in vitro assays. The phylogenetic analyses of these pigments show that they are divided into five groups: rhodopsins (RH1 pigments), RH1-like (RH2) pigments, short wavelength–sensitive type 1 (or SWS1) pigments, SWS type 2 (or SWS2) pigments, and middle and long wavelength–sensitive (M/LWS) pigments (79, 81, 83). These pigment groups have a tree topology of (M/LWS, (SWS1, (SWS2, (RH2, RH1)))), and

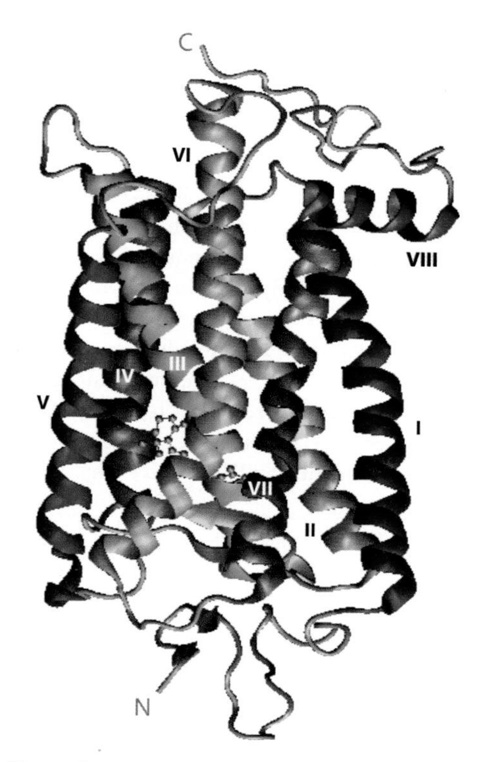

Figure 3

The seven α helices and chromophore of bovine rhodopsin (Protein Data Bank ID 1U19).

reveal two major characteristics: First, the RH1 group includes pigments from a wide range of vertebrate species, from fish to mammals, showing that the early vertebrate ancestor already possessed all five types of visual pigments. Second, humans lack RH2 and SWS2 pigments. In fact, no placental mammals have RH2 and SWS2 pigments, but platypus has SWS2 pigment (18). Hence, the RH2 gene seems to have become nonfunctional in the mammalian ancestor, whereas the SWS2 gene in the placental mammals became nonfunctional after the divergence between placental mammals and marsupials. The λ_{max}s of RH1 pigments (~480–510 nm), RH2 pigments (~450–530 nm), SWS1 pigments (~360–440 nm), SWS2 pigments (~400–450 nm), and M/LWS pigments (~510–560 nm) can overlap among different groups. These λ_{max}s have been measured in two different ways: dark spectra, which were measured in the dark, and/or difference spectra, which were measured by subtracting a spectrum measured after photobleaching from a spectrum evaluated before light exposure (dark spectrum) (**Figure 2**).

Considering representative visual pigments of the five pigment groups, I now discuss what types of amino acid replacements generated the variable λ_{max}s in contemporary pigments. In these analyses, the amino acid sequences of ancestral visual pigments were inferred using a computer program, PAML (74). Unless stated otherwise, the amino acid residues of RH1, RH2, SWS1, and SWS2 pigments are standardized by the corresponding amino acid positions of the bovine RH1 pigment and the residues of the M/LWS pigments are standardized by the corresponding amino acid positions of the human M/LWS pigments.

SWS1 pigments: short wavelength–sensitive type 1 pigments

SWS2 pigments: short wavelength–sensitive type 2 pigments

M/LWS pigments: middle and long wavelength–sensitive pigments

Several ancestral RH1 pigments were engineered by introducing all necessary amino acid changes into contemporary pigments or into engineered ancestral pigments (S. Yokoyama, T. Tada, N. Takenaka, H. Zhang & L. Britt, unpublished data). The in vitro assays of these pigments show that the RH1 pigments of early ancestors had λ_{max}s of ~500 nm (**Figure 4*a***). The ancestral RH1 pigment in the archosaur, the major branch of the diapsid reptiles, was also engineered and its λ_{max} was estimated to be 508 nm (13). Studies with various mutations introduced into the engineered ancestral pigments show that a total of nine amino acid replacements explain the λ_{max}s of most ancestral and contemporary RH1 pigments in **Figure 4*a*** (S. Yokoyama, T. Tada, N. Takenaka, H. Zhang & L. Britt, unpublished data). Interestingly, D83N, E122Q, F261Y, and A292S occurred seven, two, two, and seven times, respectively (**Table 1**); in particular, D83N/A292S occurred on five separate occasions and caused similar functional changes (**Figure 4*a***).

A limited number of ancestral RH2 pigments has been engineered and the evolutionary mechanisms of λ_{max} shifts of many RH2 pigments are still largely unknown. However, molecular bases of spectral tuning in the four zebrafish pigments (15), coelacanth 2 (P478) (94), chameleon 2 (P478), gecko 2 (P467), and medaka 2-A (P452) (68) have been examined by mutagenesis experiments (**Figure 4*b***). E122Q occurred on four separate occasions and seems to be the most common amino acid replacement that caused the major λ_{max} shifts of RH2 pigments. As in the case of the coelacanth RH1 pigment, E122Q/A292S explains the λ_{max} shift in medaka 2-A (P452) (68). However, the amino acid replacements at residues 49, 52, 83, 86, 97, and 164 explain only 65% of the λ_{max} difference between chameleon 2 (P495) and gecko 2 (P467) (68).

The λ_{max}s of engineered SWS1 pigments at various stages of vertebrate evolution (62) show that those in early ancestors had λ_{max}s of ~360 nm and were UV sensitive (**Figure 4*c***). Hence, most UV pigments in contemporary species, including fish and mouse, have inherited UV vision directly from the early ancestors. An interesting exception is the avian lineage, where the common ancestor developed violet vision with a λ_{max} of 393 nm, but a lineage of descendants, such as zebra finch and budgerigar, reinvented UV vision with a single amino acid replacement, S90C (**Figure 4*c***). This

a

P194R/N195A/A292S
486 Conger 1A (P486)
501
500 Eel 1A (P500)
501
479 Eel 1B (P479)
482
A292S
485 Conger 1B (P485)
D83N
501
E122I/F261Y
503 Cavefish 1 (P503)
501 Zebrafish 1 (P501)
502 Medaka 1 (P502)
501
483 Thornyhead 1 (P483)
E122Q D83N/A292S
501
502
492 Lampfish 1 (P492)*
F261Y
507 Scabbard 1A (P507)
Y96V/Y102F
496
481 Scabbard 1B (P481)
501
D83N/A292S
489 Viperfish 1 (P489)*
E122Q/A292S D83N/A292S
482 Coelacanth 1 (P482)*
502 Clawed frog 1 (P502)
D83N
491 Chameleon 1 (P491)
502 Pigeon 1 (P502)
508
501 Zebra finch 1 (P501)
500 Bovine 1 (P500)
488 Dolphin 1 (P488)*
D83N/A292S
496 Elephant 1 (P496)
D83N

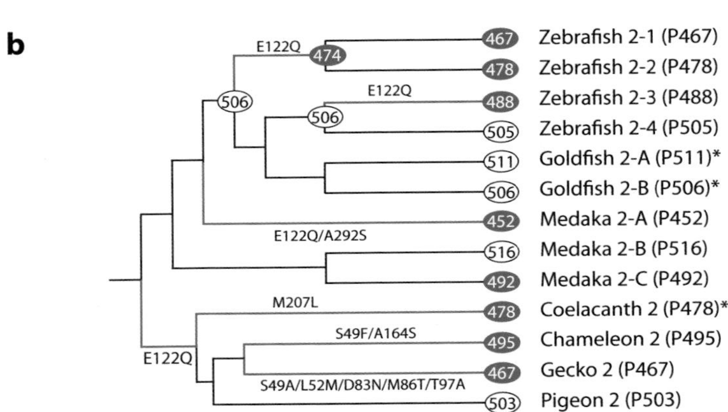

Figure 4

Phylogenetic trees of visual pigments and amino acid replacements that caused λ_{max} shifts. The numbers after P refer to dark and difference (*) spectra. The decrease (*blue line*) and increase (*red line*) of λ_{max}s are shown. (*a*) Rhodopsin (RH1) pigments: conger, eel, lampfish, scabbard fish pigments (S. Yokoyama, T. Tada, N. Takenaka, H. Zhang & L. Britt, unpublished data); medaka pigment (41); thornyhead pigment (88); elephant pigment (91); and others (82). The ancestral pigment of the pigeon and zebra finch pigments and others are from 11 and S. Yokoyama, T. Tada, N. Takenaka, H. Zhang & L. Britt (unpublished data), respectively. The ovals denote surface (*white*), intermediate (*gray*), and deep-sea (*black*) pigments. (*b*) RH1-like (RH2) pigments: zebrafish pigments (14), goldfish pigments (34), medaka pigments (41), and others (82). The ovals indicate the ancestral pigment (*white*) and its descendant pigment with a blue-shifted λ_{max} (*blue*) (15). (*c*) Short wavelength–sensitive type 1 (SWS1) pigments: bluefin killifish pigment (92), bovine pigment (28), elephant pigment (91), wallaby pigment (20), and others (82). The ovals indicate UV pigments (*purple*) and violet pigments (*blue*) (62). (*d*) SWS type 2 (SWS2) pigments: bluefin killifish pigments (92), medaka pigments (41), frog and newt pigments (66), platypus pigment (18), and others (82). The white oval indicates the ancestor of the goldfish and zebrafish pigments (16). (*e*) Middle and long wavelength–sensitive (M/LWS) pigments: zebrafish pigments (14), medaka pigments (41), wallaby pigment (20), platypus pigment (18), and others (82). The ovals indicate LWS (*red*) and MWS pigments (*green*). Data for the ancestral pigments are taken from 83.

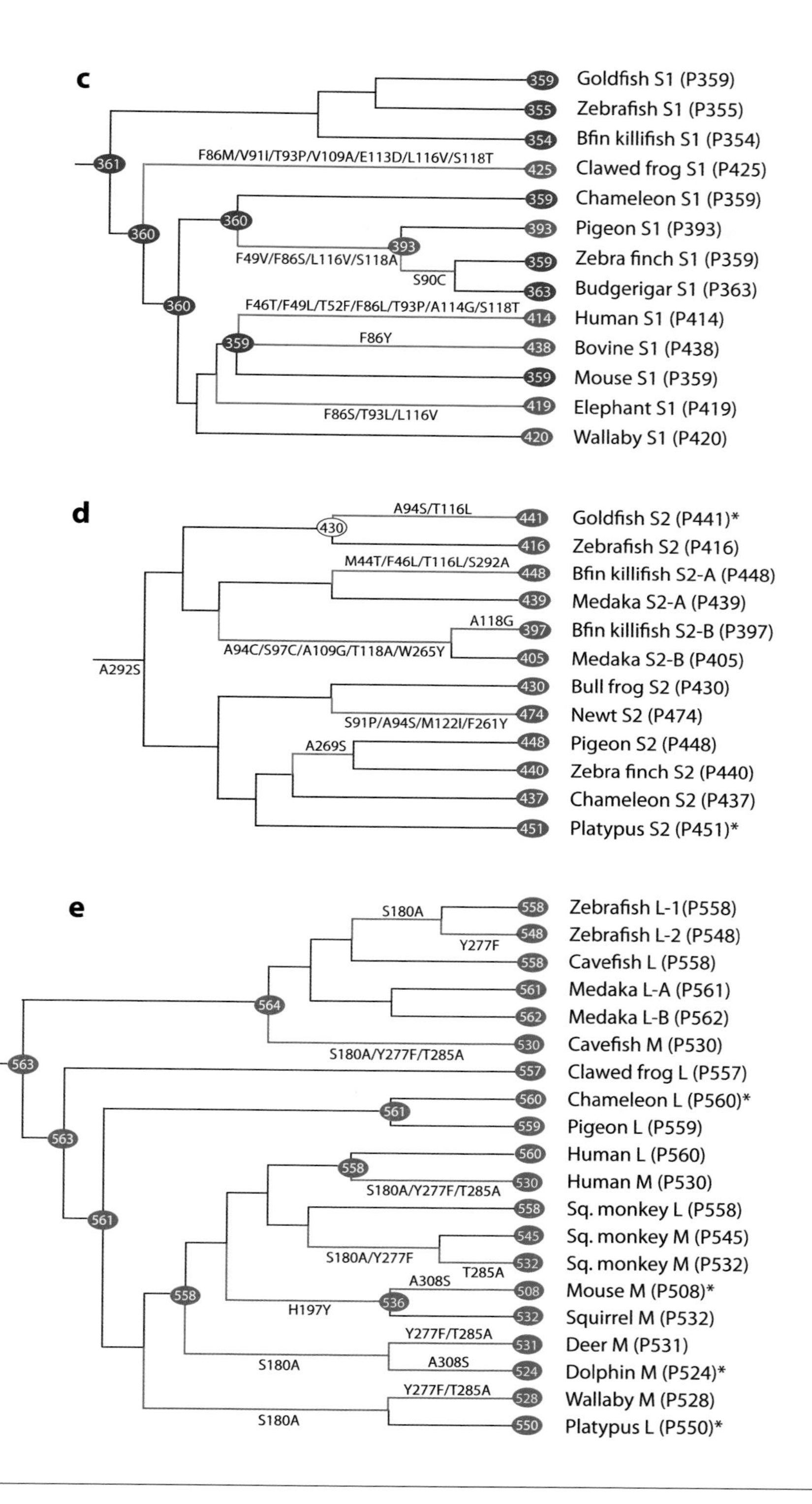

Figure 4

(Continued)

Table 1 Critical amino acid changes that shift λ_{max}s of visual pigments

Site[1]	Mutation	RH1	RH2	SWS1	SWS2	M/LWS	All
44	M44T				1		1
46	F46T			1			1
	F46L				1		1
49	S49F		1				1
	S49A		1				1
	F49V			1			1
	F49L			1			1
52	L52M		1				1
	T52F			1			1
83	D83N	7	1				8
86	M86T		1				1
	F86M			1			1
	F86S			2			2
	F86L			1			1
	F86Y			1			1
90	S90C			1			1
91	V91I			1			1
	S91P				1		1
93	T93P			2			2
	T93L			1			1
94	A94S				2		2
	A94C				1		1
96[2]	Y96V	1					1
97	T97A		1				1
	S97C			1			1
102[2]	Y102F	1					1
109	V109A			1			1
	A109G				1		1
113	E113D			1			1
114	A114G			1			1
116	L116V			3			3
	T116L				2		2
118	S118T			2			2
	S118A			1			1
	T118A				1		1
	A118G				1		1
122	E122I	1					1
	E122Q	2	4				6
	M122I				1		1
164	S164A					6	6
	A164S		1				1

(*Continued*)

Table 1 (*Continued*)

Site[1]	Mutation	RH1	RH2	SWS1	SWS2	M/LWS	All
181	H181Y					1	1
194[2]	P194R	1					1
195[2]	N195A	1					1
207	M207L		1				1
261	F261Y	2			1		3
	Y261F					6	6
265	W265Y				1		1
269	A269S				1		1
	T269A					5	5
292	A292S	7	2		1	2	12
	S292A				1		1

[1]Sites 164, 181, 261, 269, and 292 correspond to 180, 197, 277, 285, and 308 of middle and long wavelength–sensitive pigments (M/LWS) pigments, respectively. H211C and A295S in bovine 1 (P500) also cause λ_{max} shifts (32, 39, 46).
[2]S. Yokoyama, T. Tada, N. Takenaka, H. Zhang & L. Britt (unpublished data).

interpretation is also supported by mutagenesis results showing that the zebra finch and budgerigar UV pigments become violet-sensitive by the mutation C90S (73, 86) and the chicken and pigeon violet pigments become UV-sensitive by S90C (86).

Despite extensive mutagenesis analyses (27, 61, 62, 67), the molecular basis of spectral tuning in the SWS1 pigments is still poorly understood. This is simply because, with the exception of some amino acid changes at residues 86 and 90, extremely strong interactions exist among different amino acid residues. The functional differentiations of various violet pigments in some species from the ancestral UV pigment were caused by different sets of amino acid replacements. The λ_{max}s of human S1 (P414) (61) and clawed frog S1 (P425) (67) each can be explained by seven amino acid replacements, in which T93P and S118T were shared (**Figure 4*c***). Intriguingly, all these 12 amino acid changes cause no λ_{max} shift individually and the λ_{max}s of two pigments have been modified only through synergistic interactions of the seven amino acid replacements in each lineage (61, 67, Yokoyama & N. Takenaka, unpublished result). Conversely, F86Y in bovine S1 (P438) (27; see also 11, 17) and F86S in the ancestral avian pigment and in elephant S1 (P419) (61, 91) increased the λ_{max} significantly individually. In the SWS1 pigments, F86S, T93P, L113V, and S118T occurred more than once (**Table 1**).

For the SWS2 pigments, only the common ancestor of goldfish S2 (P441) and zebrafish S2 (P416) has been engineered (16). Because this ancestral pigment had a λ_{max} of 430 nm and bluefin killifish S2-B (P397) decreased its λ_{max} to 397 nm (**Figure 4*d***), the ancestral SWS2 pigment probably had a λ_{max} of 400–430 nm. This relatively low λ_{max} seems to have been caused by A292S that occurred in the ancestral SWS2 pigment (**Figure 4*d***). Mutagenesis analyses suggest that goldfish S2 (P441) (16), bluefin killifish S2-A (P448) (92), and newt S2 (P474) (66) increased their λ_{max}s, whereas bluefin killifish S2-B (P397) decreased it (**Figure 4*d***). The mutagenesis results of goldfish S2 (P441) (87) also suggest that A269S increased the λ_{max} of the common ancestor of pigeon S2 (P448) and zebra finch S2 (P440). In this group, A94S and T116L occurred twice (**Table 1**).

The engineered ancestral M/LWS pigments reveal that early vertebrate ancestors used LWS pigments with λ_{max}s of ~560 nm (85). A wide range of species, including fish, amphibians, reptiles, birds, and mammals have kept LWS pigments, whereas others have switched from the ancestral LWS pigments to MWS pigments (**Figure 4*e***). The ancestral LWS pigments had a specific amino acid composition of S180, H197, Y277, T285, and A308, from which variable λ_{max}s of contemporary M/LWS pigments have been created by certain combinations of S180A, H197Y, Y277F, T285A, and A308S. The ancestral rodent pigment is unique and reduced its λ_{max} to 536 nm by H197Y (63). One striking aspect of these amino acid replacements is that the identical changes S180A/Y277F/T285A occurred independently in cavefish M (P530), human M (P530), squirrel monkey M (P532), deer M (P531), and wallaby M (P528). In addition, the gecko MWS pigment with a λ_{max} of 527 nm also incorporated S180A/Y277F/T285A (N. S. Blow & S. Yokoyama, unpublished data). The probability that these six sets of S180A/Y277F/T285A occurred by chance alone is on the order of 10^{-28}. In addition, H198Y occurred once and S180A/A308S three times during vertebrate evolution, and the chance of the parallel replacements under neutral evolution is on the order of 10^{-9}. These rare parallel changes strongly suggest that the switches from the ancestral LWS pigment to the six MWS pigments were caused by adaptive evolution (90).

When the five groups of visual pigments are considered together, a certain pattern of amino acid replacements emerge. That is, amino acid replacements, such as D83N and A292S, occurred repeatedly in different pigment groups. In particular, A292S occurred on at least 12 separate occasions (**Table 1**). A292S offers an important lesson in understanding the molecular basis of spectral tuning in various pigments. A292S often decreases the λ_{max} of visual pigments by ~10 nm (**Figure 4*a*,*b*,*e***). However, A292S that occurred in the ancestral pigment of conger 1A (P486) does not decrease the λ_{max}; conversely, the reverse mutation, S292A, in conger 1A (P486) increases the λ_{max} by 12 nm (S. Yokoyama, T. Tada, N. Takenaka, H. Zhang & L. Britt, unpublished data). Furthermore, the reverse mutation, S292A, in human S1 (P414) does not increase the λ_{max} at all (26). These mutagenesis results suggest that synergistic interactions can occur among different amino acid residues. Such synergistic effects of different amino acids have significant implications in the analyses of spectral tuning, but they have been paid little attention.

As discussed below, critical amino acid changes that cause significant λ_{max} shifts are localized to a total of 30 residues, most of which are located near the N terminus of the TM segments. Because the chromophore is also located near the N-terminal tail (luminal face) (**Figure 3**), these amino acid changes are expected to interact with the chromophore and modify the λ_{max} of various visual pigments. However, residues 102, 194, and 195 in RH1 pigments and 197 in M/LWS pigments, which is equivalent to site 181 in the bovine RH1 pigment, are located in the luminal face, which is outside of TM segments. In particular, residues 194 and 195 are ~20 Å away from the chromophore (S. Yokoyama, T. Tada, N. Takenaka, H. Zhang & L. Britt, unpublished data). At present, the molecular structural bases of such amino acid interactions at long distance are not known.

EVOLUTION OF DIM-LIGHT AND COLOR VISION

Dim-Light Vision

One of the critical times for the survival of animals in shallow water and on land is at twilight, during which the most abundant light wavelengths are 400–500 nm (45). In this environment, a majority of animals use RH1 pigments (referred to as surface rhodopsins) with λ_{max}s of 500–507 nm. In contrast, in deep water, the distribution of downwelling sunlight is narrower at ~480 nm (33), and many deep-sea fish use RH1 pigments (deep-sea rhodopsins) with λ_{max}s of ~480–485 nm. The other RH1 pigments with λ_{max}s of ~490–495 nm can be classified as intermediate rhodopsins. On the basis of considerations of their light environments, lifestyles, and the types of RH1 pigments they use, vertebrate dim-light visions are classified mainly as surface, intermediate, and deep-sea vision (S. Yokoyama, T. Tada, N. Takenaka, H. Zhang & L. Britt, unpublished data). The transitions among the three types of dim-light visions have occurred on 12 separate occasions (**Figure 4*a***), strongly suggesting that dim-light vision has undergone adaptive evolution. The evolution of dim-light vision reveals two characteristics. First, natural selection can be subtle and selective force can differentiate even 5 nm of λ_{max} differences of RH1 pigments. Second, many transitions show the ancestral surface vision → intermediate vision (represented by lampfish and viperfish) or surface vision → intermediate vision → deep-sea vision [represented by scabbard 1B (P481)]. However, the lineage of scabbard 1A (P507) shows the transition of surface vision → intermediate vision → surface vision. Similarly, certain lineages of squirrelfish have switched back to their ancestral surface and intermediate vision from intermediate and deep-sea vision, respectively (89). In addition, as described below, a certain lineage of birds has experienced a UV vision → violet vision → UV vision transition. These observations show that evolution of dim-light vision and color vision can be reversible. To detect such reverse changes correctly, we must engineer ancestral pigments at different stages of vertebrate evolution.

Red-Green Color Vision

Vertebrates achieve red-green color vision using not only M/LWS pigments but also RH2 pigments. Some fish and primates, including humans, use LWS and MWS pigments with typical λ_{max}s of ~560 and ~530 nm, respectively, for their red-green color vision. To achieve red-green color vision, other species have modified their visual pigments and photoreceptor cells. That is, many fish species, birds, and reptiles do not have typical MWS pigments, but they can still achieve red-green color vision using 11-*cis*-3,4-dehydroretinal or colored oil droplets. For example, goldfish 2-A (P511) with 11-*cis*-retinal has a λ_{max} of 511 nm; however, when its 11-*cis*-retinal is replaced by 11-*cis*-3,4-dehydroretinal, the pigment achieves a λ_{max} of 537 nm (55). Conversely, the chicken RH2 pigment with 11-*cis*-retinal has a λ_{max} of 508 nm (52), but because it has a green oil-droplet in its photoreceptor cell, the chicken cones with the RH2 pigments actually have λ_{max}s of 533 nm (9).

Having neither 11-*cis*-3,4-dehydroretinal nor colored oil droplets, the red-green color vision of mammals is mediated solely by their M/LWS pigments. In higher primates, red-green color vision evolved in two separate ways. Hominoids and Old World monkeys use LWS and MWS opsins, encoded by two duplicated X-linked loci (51). Most New World monkeys have one M/LWS locus with three alleles (8, 31); for example, the squirrel monkey has three M/LWS alleles with λ_{max}s of 532, 545, and 558 nm (81). In such species, all males are red-green color blind, whereas females are either color blind or have red-green color vision, depending on their genotypes.

Because the molecular basis of spectral tuning in RH2 pigments is not well understood, I consider the subgroup of red-green color vision that is based only on M/LWS pigments with 11-*cis*-retinal. As noted earlier, cavefish, gecko, human, squirrel monkey, deer, and wallaby switched their LWS pigment into an MWS pigment independently. Furthermore, because of the extremely low chance of the occurrence of S180A/Y277F/T285A in the six lineages, the switches from the ancestral LWS pigment to MWS pigments in these species seem to have undergone adaptive evolution (90). This conclusion comprises one surprise; that is, the positively selected MWS pigments are found in animals with red-green color vision (cavefish, human, and squirrel monkey) and also in red-green color blind animals (gecko, deer, and wallaby). This finding contradicts a widely accepted notion that animals with red-green color vision have a selective advantage over those with color blindness (64), but it is compatible with the observation that the majority of mammalian species and many other species are red-green color blind (31, 71).

Evidence is rather scant and is sometimes controversial, but at least two observations suggest that animals with red-green color blindness can have a selective advantage over those with red-green color vision. First, colorblind people can detect color-camouflaged objects much better than those with red-green color vision (44, 57). Second, color-blind individuals of capuchin monkey, crab-eating monkeys, and chimpanzees are capable of discriminating color-camouflaged stimuli, but those with red-green color vision failed to do so (58). However, in another survey no advantage was detected between female tamarins with red-green color vision and males without red-green color vision (23). Clearly, more analyses are needed to determine whether the ability of decoding color camouflage gives a selective advantage to color blind individuals over those with red-green color vision. The decoding of color-camouflage may be only one facet of selective advantage of red-green color blindness over red-green color vision. In the future, the other causes for the selective advantage of red-green color blindness over red-green color vision may be discovered.

UV and Violet Vision

UV and violet (or blue) vision are mediated by SWS1 and SWS2 pigments, which have λ_{max}s of 360–440 and 400–450 nm, respectively. Hence, with the exception of UV pigments in the SWS1 group, the λ_{max}s of the two groups of visual pigments are indistinguishable, but the molecular mechanisms of functional differentiation of the two groups of pigments are very different (**Figure 4*c*,*d***; **Table 1**). At present, the molecular basis of spectral tuning in the SWS1 pigments is better understood than that in the SWS2 pigments. Therefore, I consider the subgroup of UV-violet vision that is based on SWS1 pigments.

The engineered ancestral pigments show that early vertebrate ancestors had UV vision (**Figure 4*c***). Because UV vision works under UV light, organisms are expected to switch from UV vision to violet vision or simply shut off the function of the *SWS1* gene in the absence of UV light. However, given abundant UV light in their environments, many organisms also switched from UV vision to violet vision. Two major causes for these changes can be considered (30). First, UV light can damage retinal tissues, and the yellow pigments in the lenses or corneas in many species, including humans, are devised to obviate most UV light from reaching the retina. In such cases, UV pigments are of no use. Second, by achieving violet vision, organisms can improve visual resolution and subtle contrast detection.

In the avian lineage, the ancestor lost UV vision, but some of its descendants restored it (**Figure 4*c***). The reinvention of UV vision seems to have been related to avian migration. For migratory birds, the pineal gland senses changes in day length and releases hormones that initiate migration (1). UV vision is also essential in orientation based on the sun (7). Surprisingly, the mouse, a nocturnal animal, also uses UV vision (**Figure 4*c***). Voles mark their runway with urine and feces, which reflect UV light and are used as a method of communication (69). Furthermore, UV pigments are the major visual pigments expressed in the third eye (or parietal eye) of chameleon (35). Clearly, UV detection through this organ is important in addition to UV vision.

Thus, the use of UV pigments and UV vision by organisms is strongly associated with their light environments and behaviors. Compared with organisms with violet vision, those with UV vision have an advantage of recognizing certain UV-reflecting objects much more quickly, but they lack precision in viewing their surroundings and are subjected to a higher chance of developing retinal damage caused by UV light. Whether or not organisms use UV vision or violet vision must depend on the relative importance of these and other conflicting characteristics associated with UV vision to them (61). To appreciate the evolution of UV-violet vision in nature, we must study the roles of UV and violet pigments of many species in various light environments.

SPECTRAL TUNING

The Problem

Certain amino acid changes at a total of 26 residues were known to have generated variable

Table 2 Forward and reverse mutations that shift λ_{max}s of visual pigments

Site	RH1	RH2	SWS1	SWS2	M/LWS
83	D83N (−6)[a]	–	–	–	–
	N83D (2)[b]	–	–	–	–
86	–	–	F86Y (66)[c]	–	–
	–	–	Y86F (−75)[d]	–	–
	–	–	F86S (17)[e]	–	–
	–	–	S86F (−52)[b]	–	–
90	G90S (−13)[f]	–	S90G (−7)[g]	–	–
			S90C (−7)[c]		
			C90S (38)[h]		
93	–	–	T93I (0)[c]	–	–
	–	–	I93T (−6)[b]	–	–
113	E113D (7)[i]	–	E113D (−4)[j]	–	–
	–	–	D113E (−12)[k]	–	–
116	–	–	L116V (0)[j]	–	–
	–	–	V116L(−3)[l]	–	–
118	T118A (−16)[f]	–	A118T (3)[m]	–	–
122	E122Q (−20)[i]	Q122E (10)[n]	–	–	–
	Q122E (10)[n]	–	–	–	–
164	A164S (2)[o]	–	–	–	S164A (−7)[p]
	–	–	–	–	A164S (6)[b]
261	F261Y (10)[o]	–	–	Y261F (−5)[q]	Y261F (−10)[p]
	Y261F (−8)[r]	–	–	–	F261Y (6)[p]
265	W265Y (−15)[s]	–	Y265W (10)[g]	–	–
269	A269T (14)[o]	–	–	A269T (6)[t]	A269T (10)[p]
					T269A (−16)[p]
292	A292S (−10)[f]	–	S292A (0)[g]	A292S (−8)[q]	S292A (28)[u]
	S292A (8)[n]				

[a]46; [b]91; [c]28; [d]17; [e]62; [f]32; [g]26; [h]86; [i]96; [j]67; [k]5; [l]90; [m]73; [n]94; [o]12; [p]4; [q]66; [r]76; [s]39; [t]87; [u]27.

λ_{max}s of visual pigments in vertebrates (92). Amino acid replacements in **Figure 4*a–e*** cover changes at 24 residues. **Table 1** also lists amino acid changes at four additional residues that are involved in the spectral tuning of RH1 pigments. Therefore, amino acid changes at a total of 30 residues are now known to cause significant λ_{max} shifts individually and synergistically.

Mutagenesis results reveal three characteristics of spectral tuning of visual pigments (**Table 2**). First, mutations in opposite directions do not necessarily shift the λ_{max} to opposite directions. For example, G90S in a RH1 pigment decreases the λ_{max} by 13 nm, but the reverse change, S90G, in a SWS1 pigment also decreases the λ_{max} by 7 nm. Similarly, E113D and D113E in two different SWS1 pigments both decrease the λ_{max}s. Second, identical amino acid changes may cause different magnitudes of λ_{max} shifts. For example, S292A in a SWS1 pigment does not shift the λ_{max}, but the same mutation in a MWS pigment increases the λ_{max} by 28 nm. Although it is not clear from **Table 2**, the λ_{max} shifts caused by S90C in different SWS1 pigments range between −46 and 0 nm (24, 27, 61, 62, 86). Third, even when the forward and reverse mutations shift the λ_{max} to opposite directions, the magnitudes of λ_{max}

shifts can differ significantly. For example, pairs of F86Y and Y86F, F86S and S86F, S90C and C90S, T118A and A118T, E122Q and Q122E, A269T and T269A, and A292S and S292A shift λ_{max}s to opposite directions, but the difference in the magnitudes of λ_{max} shifts for each pair is more than 10 nm. As more mutagenesis results accumulate, the list of these examples is expected to grow.

Hence, λ_{max} shifts caused by forward mutations that actually occurred in nature should not be inferred from those of the identical amino acid changes or corresponding reverse mutations in contemporary pigments. As the next two examples illustrate, even if we are interested in understanding the molecular basis of spectral tuning only, the actual evolutionary process cannot be ignored.

The Human M/LWS Pigments

An extensive mutagenesis analysis has been conducted using human L (P560) and human M (P530), whose difference spectra are given by 563 nm and 531 nm, respectively (4). S180A/Y277F/T285A in human L (P560) decrease the λ_{max} by 33 nm and explain fully the λ_{max} difference between the two pigments. However, the reverse changes A180S/F277Y/A285T in human M (P530) increase the λ_{max} only by 23 nm and do not explain the λ_{max} of human L (P560). In this case, not only A180S/F277Y/A285T but also Y116S/T230I/S233A/F309Y are needed to explain the λ_{max} difference between the two pigments (4). Therefore, depending on which pigment we choose to mutate, we end up with two different molecular mechanisms of spectral tuning! If we are not satisfied with two different answers, then how can we resolve the problem? One natural way is to try to understand the molecular mechanism of spectral tuning that actually occurred in the past (78).

We have seen that the engineered ancestral pigment of human L (P560) and human M (P530) had a λ_{max} of ~560 nm (**Figure 4*e***). The ancestral LWS pigment had the amino acid composition of S180/Y277/T285, and S180A, Y277F, and T285A occurred in the past. With a possible exception of S233A, it is highly unlikely that any of Y116S, A180S, T230I, F277Y, A285T, and F309Y occurred in the ancestral pigment (85). S233A decreases the λ_{max} of human L (P560) by 3 nm (4), but its actual effect on the λ_{max} shift in the ancestral pigment is unknown. In fact, when 180A/Y277F/T285A were introduced into the ancestral mammalian LWS pigment that was engineered previously (pigment d in 83), the mutant pigment had a λ_{max} of 532 nm (S. Yokoyama & H. Yang, unpublished data). Hence, the three forward mutations explain fully the spectral tuning in the human M (P530) and the effect of S233A on the λ_{max} shift is negligible. Therefore, the evolutionary interpretation of the mutagenesis results is simple: The λ_{max} of human M (P530) was achieved by S180A/Y277F/T285A, whereas human L (P560) inherited its λ_{max} directly from the ancestral pigment without any critical amino acid changes. Hence, the seven reverse amino acid changes in human M (P530) describe a mostly hypothetical situation and are unrealistic.

The Clawed Frog SWS1 Pigment and Its Ancestor

Two sets of chimeras of different SWS1 pigments (27, 62) suggested that the λ_{max} differences between pairs of UV and violet pigments were generated by amino acid differences at residues in TM I–III. Consequently, the search for amino acids that caused variable λ_{max}s among SWS1 pigments has been focused in that region. To date, a total of 13 amino acid residues in that region have been shown to be involved in the λ_{max} shift of SWS1 pigments (**Table 1**). However, considering the chimeric pigments between clawed frog S1 (P425) [or simply, frog S1 (P425)] and its ancestral amphibian pigment with a λ_{max} of 359 nm [pigment (P359)], an entirely different picture of the molecular basis of spectral tuning of SWS1 pigments has emerged (67).

The regions of interest in the two pigments were distinguished into four segments:

Table 3 The effects of transmembrane domain (TM) exchanges on the λ_{max} shift in frog S1 (P423) and its ancestor, pigment (P359)[a]

TM	Forward	Reverse
I	0	−5
II	24	−19
III	51	−15
IV–VII	1	−1
I × II	6	5
I × III	7	1
I × IV–VII	1	4
II × III	−13	−28
II × IV–VII	20	14
III × IV–VII	−7	−18
I × II × III	−12	−2
I × II × IV–VII	−3	−11
I × III × IV–VII	−8	−6
II × III × IV–VII	−17	3
I × II × III × IV–VII	14	14
Total	64	−64

[a]Data from 65.

TM I (residues 31–66), TM II (residues 67–98), TM III (residues 99–151), and TM IV–VII (residues 152–311). The amino acids at the N and C termini of the two pigments were replaced by those of chameleon S1 (P359). Then, all single and multiple combinations of these four segments were constructed (67). Considering the evolution of frog S1 (P425) from pigment (P359), the magnitudes of the λ_{max} shift caused by replacing the TM I (θ_{I}), TM II (θ_{II}), TM III (θ_{III}), and TM IV–VII (θ_{IV-VII}) of pigment (P359) by the corresponding segments of frog S1 (P425) and those of their synergistic effects $\theta_{I \times II}$, $\theta_{I \times III}$, $\theta_{I \times IV-VII}, \ldots,$ and $\theta_{I \times II \times III \times IV-VII}$ on the λ_{max} shift were evaluated (**Table 3**). The results show that TM II and TM III have significant individual effects in the spectral tuning of frog S1 (P425) and, at the same time, TM IV–VII reveal significant interactions with the other TM segments. However, the overall effect of TM IV–VII on the λ_{max} shift ($\theta_{IV-VII} + \theta_{I \times IV-VII} + \theta_{II \times IV-VII} + \theta_{III \times IV-VII} + \theta_{I \times II \times IV-VII} + \theta_{I \times III \times IV-VII} + \theta_{II \times III \times IV-VII} + \theta_{I \times II \times III \times IV-VII}$) is only 1 nm. This negligible overall effect and negligible θ_{IV-VII} give an impression that the spectral tuning in frog S1 (P425) is determined exclusively by amino acid changes in TM I–III.

By considering the change from frog S1 (P425) to pigment (P359), we can also evaluate the effects of amino acid changes in the opposite direction (**Table 3**). In this case, TM II and TM III also cause significant λ_{max} shifts, but their impacts are much smaller than those of the forward changes; in particular, the decrease in the λ_{max} caused by TM III of pigment (P359) is 36 nm smaller than the expected value from the λ_{max} shift caused by that of frog S1 (P425). In fact, the absolute values of the corresponding θ_{III}, $\theta_{II \times III}$, $\theta_{III \times IV-VII}$, $\theta_{I \times II \times III}$, and $\theta_{II \times III \times IV-VII}$ values between the forward and reverse TM exchanges differ by 10 nm or more. For the reverse changes, the overall effect of TM IV–VII on the λ_{max} shift is −1 nm and is again negligible.

The analyses of the chimeric pigments reveal three main features of spectral tuning of SWS1 pigments. First, amino acid changes not only in TM I–III but also in TM IV–VII are involved in the spectral tuning in clawed frog S1 (P425), where the critical amino acids in TM IV–VII remain to be discovered. Second, the effects of forward and reverse TM changes and amino acid changes on the λ_{max} shift can be very different. Third, despite a significant amount of interaction between TM I–III and TM IV–VII, the overall effect of amino acid changes in TM IV–VII on the λ_{max} shift is negligible. The cause and implications of the last observation are not immediately clear.

A Solution

For RH1, RH2, SWS1, and SWS2 pigment groups, we do not have sufficient information on the effects of forward amino acid changes and their interactions. However, we have a significant amount of data to study the molecular basis of spectral tuning in the M/LWS pigments. In 2001, applying multiple regression analysis to all M/LWS pigments that were known at that time, various combinations of

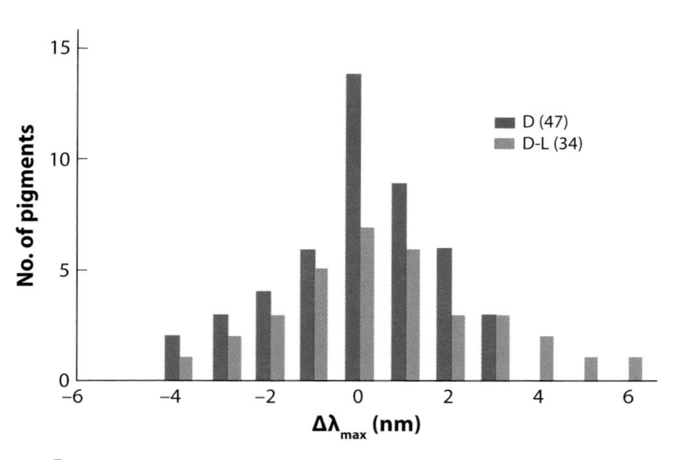

Figure 5

The distribution of the differences ($\Delta\lambda_{max}$) between the expected λ_{max} based on the five-sites rule and actual λ_{max} of middle and long wavelength–sensitive pigments (M/LWS) pigments; dark (D) (47 pigments) and difference (D-L) (34 pigments) spectra were evaluated separately.

λ_{max} shifts caused by S180A (−7 nm), H197Y (−28 nm), Y277F (−8 nm), T285A (−15 nm), A308S (−27 nm), and S180A/H197Y (11 nm) were suggested to have generated the variable λ_{max}s of M/LWS pigments (85). In the analyses, the effects of forward amino acid replacements S180A, H197Y, Y277F, and T285A on the λ_{max} shift were evaluated experimentally, but those of A308S and S180A/H197Y were not. Recently, A308S, S180A/A308S, and H197Y/A308S were introduced into the ancestral mammalian LWS pigment. The results show that the respective mutant pigments have λ_{max}s of 527, 525, and 516 nm (S. Yokoyama & H. Yang, unpublished data).

At present, the dark spectra of a total of 9 ancestral and 38 contemporary M/LWS pigments and the difference spectra of 9 ancestral and 25 contemporary pigments are available (4, 42, 43, 64, 83–85) (**Figure 5**). Applying multiple regression analysis to the λ_{max}s and the amino acid compositions at residues 180, 197, 277, 285, and 308 of these pigments, the effects of the individual and synergistic effects of the five amino acid replacements on the λ_{max} shift were evaluated. The results show that the λ_{max}s of M/LWS pigments are determined mainly by λ_{max} shifts caused by S180A (−6 nm), H197Y (−26 nm), Y277F (−10 nm), T285A (−16 nm), A308S (−33 nm), H197Y × A308S (15 nm), and S180A × H197Y × A308S (−8 nm). However, the effects of S180A × H197Y (2 nm), S180A × Y277F (2 nm), S180A × T285A (1 nm), S180A × A308S (3 nm), H197Y × T285A (−2 nm), Y277F × T285A (−1 nm), and S180A × Y277F × T285A (0 nm) on the λ_{max} shift are much smaller and are negligible. Hence, the effect of S180A × H197Y is now negligible; instead, the effects of interactions H197Y × S308A and S180A × H197Y × S308A become important. As suspected, the results depend strongly on the data set used.

Only the rodent and dolphin pigments have incorporated H197Y and A308S (**Figure 4*e***). If we exclude them, the absorption spectra of M/LWS pigments in a wide range of vertebrate species are explained mostly by the additive effects of S180A, Y277F, and T285A, and a so-called three-sites rule holds (77, 80). If we exclude only the rodent pigments from consideration, then the λ_{max}s of M/LWS pigments are modulated mostly by the additive effects of S180A, Y277F, T285A, and A308S.

The ancestral pigments with S180, H197, Y277, T285, and A308 have dark and difference spectra of 560 ± 2 and 561 ± 2 nm, respectively (S. Yokoyama & H. Yang, unpublished data). Theoretically, the λ_{max}s of all visual pigments can be evaluated by the λ_{max}s of the ancestral pigment and θs. Hence, the expected λ_{max}s based on the new five-sites rule can be compared with the corresponding observed values of M/LWS pigments. The differences between the expected and observed λ_{max}s of M/LWS pigments were evaluated for the dark and difference spectra separately (**Figure 5**). For the dark spectra, the λ_{max} differences are within 4 nm. Because the standard deviation of λ_{max}s of ancestral pigments is 2 nm, these λ_{max} differences are within the margin of experimental error. The majority of λ_{max} differences for the difference spectra are also within 4 nm, but the differences of two pairs of pigments are larger than 4 nm. Because even when they are not reliable, dark spectra are used in evaluating difference spectra, the deviations might have been caused by inaccurate estimates of the difference spectra. Overall, therefore, the variable λ_{max}s of the currently known M/LWS pigments can be explained reasonably well by the new five-sites rule (S. Yokoyama & H. Yang, unpublished result).

ADAPTIVE EVOLUTION

By studying functional differentiations of ancestral RH1 and SWS1 pigments and relating them to the associated environmental changes of organisms' habitats and to new lifestyles, we have established that dim-light and UV-violet vision have undergone adaptive evolution. For the adaptive evolution of red-green color vision, a more probabilistic argument of parallel amino acid replacements in M/LWS pigments was used. Thus, surveying the amino acid sequences and λ_{max}s of various visual pigments, followed by mutagenesis analyses, amino acid replacements that generated a wide range of λ_{max}s in nature have been uncovered (81).

Without an available functional assay, molecular adaptations have often been inferred by identifying amino acid changes using statistical methods (65, 74, 95). The mutagenesis analyses of visual pigments establish five fundamental features of molecular evolution that cannot be learned from the standard statistical analyses of protein sequence data. First of all, mutagenesis experiments can offer critical and decisive tests of whether or not amino acid changes that are inferred as adaptive actually cause any functional changes (19). Second, as exemplified by several sets of mutations (**Table 2**), the same amino acid replacements do not always produce the same functional change but instead the change can be affected by the background amino acids of the opsin. Therefore, the probability of parallel amino acid replacements, which may or may not result in any functional change, can overestimate the actual chance that functional adaptive events occur. Third, similar functional changes can be achieved by different amino acid replacements. For example, D83N/A292S, P194R/N195A/A292S, and E122Q all decrease the λ_{max} by 14–20 nm (**Figure 4*a***). Thus, by simply looking for parallel replacements of specific amino acids, one can miss other amino acid changes that generate the same functional change, thereby underestimating the chance of finding functional adaptations.

Fourth, as stressed already, not only can the identical mutations in different pigments cause different magnitudes of λ_{max} shift, but also the effects of forward and reverse amino acid changes on the λ_{max} shift can differ significantly. Hence, if we are interested in elucidating the evolutionary mechanisms of functional and phenotypic changes, we must study the effects of forward mutations, not reverse mutations. As noted earlier, this evolutionary approach also simplifies our understanding of the molecular basis of spectral tuning.

Fifth, even when the phylogenetic position of a molecule is uncertain, its functional assay can clarify the molecular evolution of functional adaptation. For example, the phylogenetic position of lampfish 1 (P492) is uncertain (**Figure 4*a***). However, because the E122Q mutation that generated its λ_{max} is different from the other critical amino acid replacements in the closely related thornyhead 1 (P483), scabbard 1B (P481), and viperfish 1 (P489) proteins (**Figure 4*a***), we can easily establish an independent origin of the functional change in lampfish 1 (P492). Therefore, to explore the adaptive evolution of certain traits, both functional and molecular analyses of such traits are valuable (19). Analyses of functional adaption of visual pigments also demonstrate the importance of relating the functional changes to the environmental or behavioral changes that presumably caused the functional and phenotypic changes in the first place.

To fully appreciate how adaptive evolution of dim-light and color vision occurred, we must study the effects of critical forward amino acid replacements on the λ_{max} shift at the chemical level as well. For example, quantum chemical analyses of the effects of forward amino acid changes on the λ_{max} shift will improve significantly our understanding of the molecular basis of spectral tuning in visual pigments (2, 3). The same analyses will, in turn, improve significantly our understanding of the chemical basis of the functional adaptations of dim-light vision and color vision.

CONCLUSIONS

Studies on the structure and function of bovine RH1 pigment by Doi and coworkers (22) and a series of subsequent papers by H. G. Khorana and his colleagues as well as other vision researchers (46, 48, 59, 72, 96) have improved dramatically our understanding of how key amino acids in visual pigments work. Unfortunately, most mutations considered in these biochemical studies are not found in nature, so their roles in the actual spectral tuning in various visual pigments are not immediately clear (79, 81). If we want to elucidate the mechanisms of spectral tuning that generated the λ_{max}s of contemporary visual pigments, then we must consider amino acid replacements that actually occurred in nature. Such changes can be inferred only by comparing the amino acid sequences of contemporary visual pigments, and the actual functional changes caused by the predicted amino acid changes can be evaluated using in vitro assays. To date, using this approach, certain amino acid replacements at a total of 30 residues have been shown to be involved in the spectral tuning of different visual pigments (**Table 1**).

Phylogenetic analyses of contemporary visual pigments show that early vertebrate ancestors already had RH1, RH2, SWS1, SWS2, and M/LWS pigments (25, 81, 93). Many contemporary species still use all five sets of visual pigments, and more recent gene duplications in some species generated additional variations in the λ_{max}s of visual pigments, whereas RH2 and SWS2 pigments have become nonfunctional in some lineages, including placental mammals, and their color vision has become more specialized (**Figure 4*a–e***). The engineered visual pigments show that the RH1, SWS1, and M/LWS in early ancestors had λ_{max}s of ~500, ~360, and ~560 nm, respectively.

Depending on the organisms' light environments, lifestyles, and the λ_{max}s of their RH1 pigments, dim-light vision of organisms can be distinguished into deep-sea, intermediate, and surface vision. The RH1 pigments of the respective groups have λ_{max}s of 480–485, 490–495, and 500–510 nm. Some species inherited the ancestral surface vision directly from the vertebrate ancestor, whereas others have switched to different types of dim-light vision. During vertebrate evolution, such transitions occurred on 12 separate occasions. As the λ_{max}s of the three types of dim-light vision indicate, natural selection can be subtle and selective force may differentiate even 5 nm of λ_{max} difference. These adaptive events were accomplished mostly by amino acid changes at nine residues, where D83N/A292S occurred seven times independently.

Many contemporary LWS pigments have maintained the ancestral λ_{max} of ~560 nm, whereas others have decreased their λ_{max}s by using various combinations of S180A, H197Y, Y277F, T285A, and A308S. In particular, identical amino acid replacements (S180A/Y277F/T285A) occurred on six separate occasions and shifted the λ_{max}s of M/LWS pigments in an additive fashion. In the lineage of rodent M/LWS pigments, H197Y occurred in their ancestral pigment, followed by A308S in some MWS pigments, and they decreased λ_{max}s individually and synergistically. Similarly, many contemporary SWS1 pigments inherited their UV sensitivities from the common ancestor, and others developed violet sensitivities using different sets of amino acid replacements, many of which remain to be discovered. Most of the currently known critical amino acid replacements modify the λ_{max} mainly through their synergistic effects, but some amino acid replacements at residues 86 and 90, including F86Y, F86S, and S90C, can cause significant λ_{max} shifts individually as well as synergistically (91).

Despite these advances, our understanding of the molecular bases of adaptive evolution and spectral tuning of visual pigments is still fragmental. This is because we still don't have much information on how the chromophore and different amino acids interact with each other. In particular, amino acid changes in opposite directions do not shift the λ_{max} in the opposite direction by the same magnitudes. Or, when introduced into different pigments, even the identical amino acid replacements can cause different magnitudes of λ_{max} shifts. The most reasonable approach in resolving these seemingly contradictory observations is to consider amino acid changes that actually generated the variable λ_{max}s of contemporary visual pigments. Then, the functional adaptation and spectral tuning of visual pigments can be understood together by studying the mechanisms of adaptive evolution of visual pigments at the molecular and phenotypic levels.

To solve the problem, we must engineer ancestral pigments for the five groups of visual pigments at various stages of vertebrate evolution and introduce mutations into them. Such ancestral pigments at various stages of vertebrate evolution have been engineered for RH1, SWS1, and M/LWS pigments, but those for RH2 and SWS2 pigments remain to be engineered. By dissecting these and contemporary visual pigments at the molecular level and relating their λ_{max}s to organisms' light environments and lifestyles, we can start to learn why and how organisms adapted to their light environments.

SUMMARY POINTS

1. Visual pigments in vertebrates are classified into rhodopsins (RH1), RH1-like (RH2), short wavelength-sensitive type 1 (SWS1), SWS type 2 (SWS2), and middle and long wavelength–sensitive (M/LWS) groups with λ_{max}s of 480–510, 450–530, 360–440, 400–450, and 510–560 nm, respectively.
2. Dim-light vision is mediated by RH1 pigments and can be classified into three different types; the evolutionary switches among them occurred on 12 separate occasions.
3. Red-green color vision and color blindness mediated by M/LWS pigments were generated by certain combinations of amino acid changes (S180A, H197Y, Y277F, T285A, and A308S); S180A/Y277F/T285 occurred on six separate occasions.
4. The parallel replacements of S180A/Y277F/T285A in various vertebrate species suggest that both red-green color vision and color blindness have undergone adaptive evolution, but the selective advantage of color blindness over red-green color vision is still not well understood.
5. Many fish, reptile, and mammalian species inherited their UV vision from the vertebrate ancestor, but the bird ancestor achieved violet vision by F4V/F86S/L116V/S118A, and some of its descendants reinvented UV vision by S90C.
6. With the exception of some amino acid changes at residues 86 and 90, the molecular basis of spectral tuning in SWS1 pigments is characterized by strong interactions among amino acid residues.
7. Mutagenesis data show that mutations in opposite directions do not necessarily cause λ_{max} shifts to the opposite directions by the same magnitudes, implying that the molecular basis of spectral tuning in visual pigments should be understood by considering forward amino acid changes that actually generated the variable λ_{max} shifts of contemporary pigments.

8. Among the five pigment groups, the molecular basis of spectral tuning is best understood for the M/LWS group and the most recent data show that the ancestral M/LWS pigment had a λ_{max} of 560 nm. Significant λ_{max} shifts have been caused mostly by S180A (−6 nm), H197Y (−26 nm), Y277F (−10 nm), T285A (−16 nm), A308S (−33 nm), H197Y/A308S (15 nm), and S180A/H197Y/A308S (−8 nm).

FUTURE ISSUES

1. All amino acid replacements that generated the variable λ_{max}s of the five groups of contemporary visual pigments need to be identified.
2. Individual and synergistic effects of these forward amino acid changes on the λ_{max} shifts need to be evaluated.
3. The molecular bases of spectral tuning in various visual pigments need to be understood in terms of the individual and synergistic effects of the forward amino acid changes on the λ_{max} shifts.
4. The spectral tuning in visual pigments need to be understood at the chemical structural level, where quantum chemical computations of visual pigments at various stages of vertebrate evolution should be performed.
5. The molecular bases of functional adaptation of visual pigments need to be understood not only by studying the molecular basis of differentiation of visual pigments, but also by relating them to organisms' move to new photic environments or to new lifestyles.

DISCLOSURE STATEMENT

The author is not aware of any biases that might be perceived as affecting the objectivity of this review.

ACKNOWLEDGMENTS

The author's research was supported by a grant from the National Institutes of Health.

LITERATURE CITED

1. Alcock J. 1997. *Animal Behavior*. Sunderland, MA: Sinauer
2. Altun A, Yokoyama S, Morokuma K. 2008. Spectral tuning in visual pigments: An ONIOM (QM:MM) study on bovine rhodopsin and its mutants. *J. Phys. Chem. B* 112:6814–27
3. Altun A, Yokoyama S, Morokuma K. 2008. Quantum mechanical/molecular mechanical studies on spectral tuning mechanisms of visual pigments and other photoactive proteins. *Photochem. Photobiol.* In press
4. Asenjo AB, Rim J, Oprian DD. 1994. Molecular determination of human red/green color discrimination. *Neuron* 12:1131–38
5. Babu KR, Dukkipati A, Birge RR, Knox BE. 2001. Regulation of phototransduction in short-wavelength cone visual pigments via the retinylidene Schiff base counterion. *Biochemistry* 40:13760–66
6. Baylor D. 1996. How photons start vision. *Proc. Natl. Acad. Sci. USA* 93:560–65
7. Bennett ATD, Cuthill IC. 1994. Ultraviolet vision in birds: What is its function? *Vis. Res.* 34:1471–78

8. Bowmaker JK. 1991. Evolution of photoreceptors and visual pigments. In *Evolution of the Eye and Visual Pigments*, ed. JR Cronly-Dillon, RJ Gregory, pp. 63–81. Boca Raton, FL: CRC Press
9. Bowmaker JK, Knowles A. 1977. The visual pigments and oil droplets of the chicken retina. *Vis. Res.* 17:755–64
10. Carroll SB. 2006. *The Making of The Fittest*. New York: Norton
11. Carvalho LD, Cowing JA, Wilkie SE, Bowmaker JK, Hunt DM. 2006. Shortwave visual sensitivity in tree and flying squirrels reflects changes in lifestyle. *Curr. Biol.* 16:R81–83
12. Chan T, Lee M, Sakmar TP. 1992. Introduction of hydroxyl-bearing amino acids causes bathochromic spectral shifts in rhodopsin: amino acid substitutions responsible for red-green color pigment spectral tuning. *J. Biol. Chem.* 267:9478–80
13. Chang BSW, Jonsson K, Kazmi MA, Donoghue MJ, Sakmar TP. 2002. Recreating a functional ancestral archosaur visual pigment. *Mol. Biol. Evol.* 19:1483–89
14. Chinen A, Hamaoka T, Yamada Y, Kawamura S. 2003. Gene duplication and spectral diversification of cone visual pigments of zebrafish. *Genetics* 163:663–75
15. Chinen A, Matsumoto Y, Kawamura S. 2005. Reconstitution of ancestral green visual pigments of zebrafish and molecular mechanism of their spectral differentiation. *Mol. Biol. Evol.* 22:1001–10
16. Chinen A, Matsumoto Y, Kawamura S. 2005. Spectral differentiation of blue opsins between phylogenetically close but ecologically distant goldfish and zebrafish. *J. Biol. Chem.* 280:9460–66
17. Cowing JA, Poopalasundaram S, Wilkie SE, Robinson PR, Bowmaker JK, et al. 2002. The molecular mechanism for the spectral shifts between vertebrate UV- and violet-sensitive cone visual pigments. *Biochem. J.* 367:129–35
18. Davies WL, Carvalho LS, Cowing JA, Beazley LD, Hunt DM, et al. 2007. Visual pigments of the platypus: A novel route to mammalian colour vision. *Curr. Biol.* 17:R161–63
19. Dean AM, Thornton JW. 2007. Mechanistic approaches to the study of evolution: the functional synthesis. *Nat. Rev. Genet.* 8:675–88
20. Deeb SS, Wakefield MJ, Tada T, Marotte L, Yokoyama S, Graves JAM. 2003. The cone visual pigments of an Australian marsupial, the tammar wallaby (*Macropus eugenii*): sequence, spectral tuning and evolution. *Mol. Biol. Evol.* 20:1642–49
21. Dobzhansky T. 1973. Nothing in biology makes sense except in the light of evolution. *Am. Biol. Teach.* 35:125–29
22. Doi T, Molday RS, Khorana HG. 1990. Role of the intradiscal domain in rhodopsin assembly and function. *Proc. Natl. Acad. Sci. USA* 87:4991–95
23. Dominy NJ, Garber PA, Bicca-Marques JC, Azevedo-Lopes MADO. 2003. Do female tamarins use visual cues to detect fruit rewards more successfully than do males? *Anim. Behav.* 66:829–37
24. Dukkipati A, Vought BW, Singh D, Birge RR, Knox BE. 2001. Serine 85 in transmembrane helix 2 of short-wavelength visual pigments interacts with the retinylidene Schiff base counterion. *Biochemistry* 40:15098–108
25. Ebrey TG, Takahashi Y. 2002. Photobiology of retinal proteins. In *Photobiology for the 21st Century*, ed. TP Coohill, DP Valenzeno, pp. 101–33. Overland Park, KS: Valdenmar
26. Fasick JI, Lee N, Oprian DD. 1999. Spectral tuning in the human blue cone pigment. *Biochemistry* 38:11593–96
27. Fasick JI, Applebury ML, Oprian DD. 2002. Spectral tuning in the mammalian short wavelength sensitive cone pigments. *Biochemistry* 41:6860–65
28. Fasick JI, Robinson PR. 1998. Mechanism of spectral tuning in the dolphin visual pigments. *Biochemistry* 37:433–38
29. Golding GB, Dean AM. 1998. The structural basis of molecular adaptation. *Mol. Biol. Evol.* 15:355–69
30. Jacobs GH. 1992. Ultraviolet vision in vertebrates. *Am. Zool.* 342:544–54
31. Jacobs GH. 1993. The distribution and nature of color vision among the mammals. *Biol. Rev.* 68:413–71
32. Janz JM, Farrens DL. 2001. Engineering a functional blue-wavelength-shifted rhodopsin mutant. *Biochemistry* 40:7219–27
33. Jerlov NG. 1976. *Marine Optics*. Amsterdam: Elsevier
34. Johnson R, Grant KB, Zankel TC, Boehm MF, Merbs SL, et al. 1993. Cloning and expression of goldfish opsin sequences. *Biochemistry* 32:208–14

35. Kawamura S, Yokoyama S. 1997. Expression of visual and nonvisual opsins in American chameleon. *Vis. Res.* 37:1867–71
36. Kito Y, Suzuki T, Azuma M, Sekoguchi Y. 1968. Absorption spectrum of rhodopsin denatured with acid. *Nature* 218:955–57
37. Kochendoerfer GG, Lin S, Sakmar TM, Mathies RA. 1999. How color visual pigments are tuned. *Trends Biochem. Sci.* 24:300–5
38. Lewontin RC. 1979. Adaptation. *Sci. Am.* 239:156–69
39. Lin SW, Kochendoerfer CKS, Wang D, Mathies RA, Sakmar TP. 1998. Mechanisms of spectral tuning in blue cone visual pigments: visible and Raman spectroscopy of blue-shifted rhodopsin mutants. *J. Biol. Chem.* 273:24583–91
40. Lythgoe JN. 1979. *The Ecology of Vision*. Oxford: Clarendon Press
41. Matsumoto Y, Fukamachi S, Mitani H, Kawamura S. 2006. Functional characterization of visual opsin repertoire in Medaka (*Oryzias latipes*). *Gene* 371:268–78
42. Merbs SL, Nathans J. 1992. Absorption spectrum of human cone pigments. *Nature* 356:433–35
43. Merbs SL, Nathans J. 1993. Role of hydroxyl-bearing amino-acids in differentially tuning the absorption-spectra of the human red and green cone pigments. *Photochem. Photobiol.* 58:706–10
44. Morgan MJ, Adam A, Mollon JD. 1992. Dichromats detect colour-camouflaged objects that are not detected by trichromats. *Proc. R. Soc. London Ser. B* 248:291–95
45. Muntz FW, McFarland WN. 1977. Evolutionary adaptations of fishes to the photic environment. In *Handbook of Sensory Physiology, VII/5*, ed. F Crescitelli, pp. 193–274. Berlin: Springer-Verlag
46. Nathans J. 1990. Determinations of visual pigment absorbance: role of charged amino acids in the putative transmembrane segments. *Biochemistry* 29:937–42
47. Nathans J. 1990. Determinants of visual pigment absorbance: identification of the retinylidene Schiff's base counterion in bovine rhodopsin. *Biochemistry* 29:9746–52
48. Nathans J, Davenport CM, Maumenee IH, Lewis RA, Hejtmancik JF, et al. 1998. Molecular genetics of human blue cone monochromacy. *Science* 245:831–38
49. Nathans J, Hogness DS. 1983. Isolation, sequence analysis, and intron-exon arrangement of the gene encoding bovine rhodopsin. *Cell* 34:807–14

49. Presents the first cloning of a vertebrate opsin gene.

50. Nathans J, Hogness DS. 1984. Isolation and nucleotide sequence of the gene encoding human rhodopsin. *Proc. Natl. Acad. Sci. USA* 81:4851–55
51. Nathans J, Thomas D, Hogness DS. 1986. Molecular genetics of human color vision: The genes encoding blue, green, and red pigments. *Science* 232:193–201

51. Presents the first cloning of cone opsin genes.

52. Okano T, Kojima D, Fukada Y, Shichida Y, Yoshizawa T. 1992. Primary structures of chicken cone visual pigments; Vertebrate rhodopsins have evolved out of cone visual pigments. *Proc. Natl. Acad. Sci. USA* 89:5932–36
53. Okada T, Sugihara M, Bondar AN, Elstner M, Entel P, et al. 2004. The retinal conformation and its environment in rhodopsin in light of a new 2.2 Å crystal structure. *J. Mol. Biol.* 342:571–83
54. Oprian DD, Molday RS, Kaufman RJ, Khorana HG. 1987. Expression of a synthetic bovine rhodopsin gene in monkey kidney cells. *Proc. Natl. Acad. Sci. USA* 84:8874–78

54. Presents the first in vitro assay of bovine rhodopsin.

55. Palacios AG, Varela FJ, Srivastava R, Goldsmith TJ. 1998. Spectral sensitivity of cones in the goldfish, *Carassius auratus*. *Vis. Res.* 38:2135–46
56. Palczewski K, Kumasaka T, Hori T, Behnke CA, Motoshima H, et al. 2000. Crystal structure of rhodopsin: A G protein-coupled receptor. *Science* 289:739–45

56. Presents the first crystal structure of a G protein–coupled receptor.

57. Saitou A, Mikami A, Hosokawa T, Hasegawa T. 2006. Advantage of dichromats over trichromats in discrimination of color-camouflaged stimuli in humans. *Percept. Motor Skills* 102:3–12
58. Saitou A, Mikami A, Kawamura S, Ueno Y, Hiramatsu C, et al. 2005. Advantage of dichromats over trichromats in discrimination of colour-camouflaged stimuli in nonhuman primates. *Am. J. Primatol.* 67:425–36
59. Sakmar TP, Franke RR, Khorana HG. 1989. Glutamic acid-113 serves as the retinylidene Schiff base counterion in bovine rhodopsin. *Proc. Natl. Acad. Sci. USA* 86:8309–13
60. Schnapf JL, Baylor DA. 1987. How photoreceptor cells respond to light. *Sci. Am.* 256:32–39
61. Shi Y, Yokoyama S. 2003. Molecular analysis of the evolutionary significance of UV vision in vertebrates. *Proc. Natl. Acad. Sci. USA* 100:8308–13

62. Shi YS, Radlwimmer FB, Yokoyama S. 2001. Molecular genetics and the evolution of UV vision in vertebrates. *Proc. Natl. Acad. Sci. USA* 98:11731–36
63. Sun H, Macke JP, Nathans J. 1997. Mechanisms of spectra tuning in the mouse green cone pigment. *Proc. Natl. Acad. Sci. USA* 94:8860–65
64. Surridge AK, Osorio D, Mundy NI. 2003. Evolution and selection of trichromatic vision in primates. *Trends Ecol. Evol.* 18:198–205
65. Suzuki Y, Gojobori T. 1999. A method for detecting positive selection at single amino acid sites. *Mol. Biol. Evol.* 16:1315–28
66. Takahashi Y, Ebrey TG. 2003. Molecular basis of spectral tuning in the newt short wavelength sensitive visual pigment. *Biochemistry* 42:6025–34
67. Takahashi Y, Yokoyama S. 2005. Genetic basis of spectral tuning in the violet-sensitive visual pigments of African clawed frog, *Xenopus laevis*. *Genetics* 171:1153–60
68. Takenaka N, Yokoyama S. 2007. Mechanisms of spectral tuning in the RH2 pigments of Tokay gecko and American chameleon. *Gene* 399:26–32
69. Viitala J, Korpimaki E, Palokangas P, Koivula M. 1995. Attraction of kestrels to vole scent marks visible in UV light. *Nature* 373:425–27
70. Wald G. 1968. Molecular basis of visual excitation. *Science* 162:230–39
71. Walls GL. 1942. *The Vertebrate Eye and Its Adaptive Radiation*. New York: Hafner
72. Weitz CJ, Nathans J. 1993. Rhodopsin activation: effect of the metarhodopsin I-metarhodopsin II equilibrium of neutralization or introduction of charged amino acids within putative transmembrane segments. *Biochemistry* 32:14176–82
73. Wilkie SE, Robinson PR, Cronin TW, Poopalasundaram S, Bowmaker JK, et al. 2000. Spectral tuning of avian violet- and UV-sensitive visual pigments. *Biochemistry* 39:7895–901
74. Yang Z. 1997. PAML: a program package for phylogenetic analysis by maximum likelihood. *Comput. Appl. Biosci.* 13:555–56
75. Yau KW. 1994. Phototransduction mechanism in retinal rods and cones. *Invest. Ophthalmol. Vis. Sci.* 35:9–32
76. Yokoyama R, Knox BE, Yokoyama S. 1995. Rhodopsin from the fish, Astyanax: Role of tyrosine 261 in the red shift. *Invest. Ophthalmol. Vis. Sci.* 36:939–45
77. **Yokoyama R, Yokoyama S. 1990. Convergent evolution of the red- and green-like visual pigment genes in fish, *Astyanax fasciatus*, and human. *Proc. Natl. Acad. Sci. USA* 87:9315–18**
78. Yokoyama S. 1995. Amino acid replacements and wavelength absorption of visual pigments in vertebrates. *Mol. Biol. Evol.* 12:53–61
79. Yokoyama S. 1997. Molecular genetic basis of adaptive selection: Examples from color vision in vertebrates. *Annu. Rev. Genet.* 31:315–36
80. Yokoyama S. 1998. The "five-sites" rule and the evolution of red and green color vision in mammals. *Mol. Biol. Evol.* 15:560–67
81. Yokoyama S. 2000. Molecular evolution of vertebrate visual pigments. *Prog. Retin. Eye Res.* 19:385–419
82. Yokoyama S. 2000. Phylogenetic analysis and experimental approaches to study color vision in vertebrates. *Methods Enzymol.* 315:312–25
83. Yokoyama S. 2002. Molecular evolution of color vision in vertebrates. *Gene* 300:69–78
84. Yokoyama S, Radlwimmer FB. 1999. The molecular genetics of red and green color vision in mammals. *Genetics* 153:919–32
85. **Yokoyama S, Radlwimmer FB. 2001. The molecular genetics of red and green color vision in vertebrates. *Genetics* 158:1697–710**
86. Yokoyama S, Radlwimmer FB, Blow NS. 2000. Ultraviolet pigments in birds evolved from violet pigments by a single amino acid change. *Proc. Natl. Acad. Sci. USA* 97:7366–71
87. Yokoyama S, Tada T. 2003. The spectral tuning in the short wavelength-sensitive type 2 pigments. *Gene* 306:91–98
88. Yokoyama S, Tada T, Yamato T. 2007. Modulation of the absorption maximum of rhodopsin by amino acids in the C-terminus. *Photochem. Photobiol.* 83:236–41
89. Yokoyama S, Takenaka N. 2004. The molecular basis of rhodopsin evolution of squirrelfish and soldierfish (Holocentridae). *Mol. Biol. Evol.* 21:2071–78

77. Presents the first prediction that amino acid differences at sites 180, 277, and 285 caused the λ_{max} difference between LWS (red) and MWS (green) pigments.

85. Presents the first genetically engineered ancestral visual pigments.

90. Yokoyama S, Takenaka N. 2005. Statistical and molecular analyses of evolutionary significance of red-green color vision and color blindness in vertebrates. *Mol. Biol. Evol.* 22:968–75
91. Yokoyama S, Takenaka N, Agnew DW, Shoshani J. 2005. Elephants and human color-blind deuteranopes have identical sets of visual pigments. *Genetics* 170:335–44
92. Yokoyama S, Takenaka N, Blow N. 2007. A novel spectral tuning in the short wavelength-sensitive (SWS1 and SWS2) pigments of bluefin killifish (*Lucania goodei*). *Gene* 396:196–202
93. Yokoyama S, Yokoyama R. 1996. Adaptive evolution of photoreceptors and visual pigments in vertebrates. *Annu. Rev. Ecol. Syst.* 27:543–67
94. Yokoyama S, Zhang H, Radlwimmer FB, Blow NS. 1999. Adaptive evolution of color vision of the Comoran coelacanth (*Latimeria chalumnae*). *Proc. Natl. Acad. Sci. USA* 96:6279–84
95. Zhang J. 2006. Parallel adaptive origins of digestive RNases in Asian and African leaf monkeys. *Nat. Genet.* 38:819–23
96. Zhukovsky EA, Oprian DD. 1989. Effect of carboxylic acid side chains on the absorption maximum of visual pigments. *Science* 246:928–30

Genetic Basis of Thoracic Aortic Aneurysms and Dissections: Focus on Smooth Muscle Cell Contractile Dysfunction

Dianna M. Milewicz, Dong-Chuan Guo, Van Tran-Fadulu, Andrea L. Lafont, Christina L. Papke, Sakiko Inamoto, Carrie S. Kwartler, and Hariyadarshi Pannu

Department of Internal Medicine, University of Texas, Houston, Texas 77030; email: Dianna.M.Milewicz@uth.tmc.edu

Annu. Rev. Genomics Hum. Genet. 2008. 9:283–302

First published online as a Review in Advance on June 10, 2008

The *Annual Review of Genomics and Human Genetics* is online at genom.annualreviews.org

This article's doi: 10.1146/annurev.genom.8.080706.092303

1527-8204/08/0922-0283$20.00

Key Words

α-actin, β-myosin, fibrillin, TGFBR1, TGFBR2

Abstract

Thoracic aortic aneurysms leading to type A dissections (TAAD) can be inherited in isolation or in association with genetic syndromes, such as Marfan syndrome and Loeys-Dietz syndrome. When TAAD occurs in the absence of syndromic features, it is inherited in an autosomal dominant manner with decreased penetrance and variable expression, the disease is referred to as familial TAAD. Familial TAAD exhibits significant clinical and genetic heterogeneity. The first genes identified to cause TAAD were *FBN1*, *TGFBR2*, and *TGFBR1*. The identification and characterization of these genes suggested that increased TGF-β signaling plays a role in pathogenesis. The recent discovery that mutations in the vascular smooth muscle cell (SMC)-specific β-myosin (*MYH11*) and α-actin (*ACTA2*) can also cause this disorder has focused attention on the importance of the maintenance of SMC contractile function in preserving aortic structure and preventing TAAD.

INTRODUCTION

Aneurysms and dissections of the thoracic aorta are the fifteenth leading cause of death in the United States. Aortic aneurysms are typically classified in terms of their anatomical location, and thoracic aortic aneurysms primarily involve the ascending thoracic aorta (**Figure 1*a***). Thoracic aortic aneurysms tend to be asymptomatic and often are not diagnosed before an aortic dissection occurs. During an acute dissection, blood in the aortic lumen enters the aortic wall through an intimal tear and dissects along the plane of the wall, leading to the establishment of a false lumen. Aortic dissections originate primarily in the ascending aorta just above the aortic valve (type A dissections) but can also occur in the descending thoracic aorta just distal to the origin of the left subclavian artery (type B) (**Figure 1*a*** and ***b***). Rupture and dissection of an aneurysm are associated with a high degree of morbidity and mortality, despite continued improvements in surgical techniques. In contrast, prophylactic repair of an ascending aortic aneurysm prior to rupture or dissection is associated with very low morbidity and mortality. Thus, current recommendations are that the ascending aorta should be surgically repaired when the aortic aneurysm enlarges to a diameter of 5.0–5.5 cm (56). Prevention of untimely death from thoracic aortic dissections depends on early identification of individuals predisposed to thoracic aortic aneurysms leading to type A dissections (TAAD) by carefully monitoring the diameter of the ascending aorta and performing surgical repair of diseased segments in a timely fashion (18, 80).

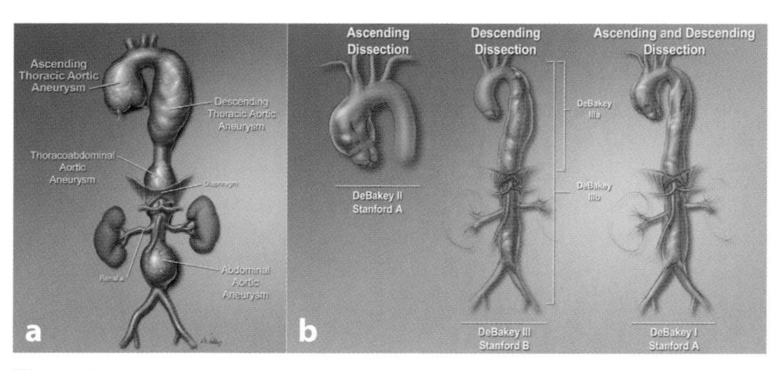

Figure 1

Anatomical location and classification of aortic aneurysms and dissections. (*a*) Anatomical location of aortic aneurysms. Ascending thoracic aortic aneurysms (*highlighted in yellow*) are the focus of this review. (*b*) Classification of aortic dissections initiating in the thoracic aorta.

The aorta is composed of three layers: a thin inner layer, the tunica intima; a thick middle layer, the tunica media; and a thin outer layer, the tunica adventitia (**Figure 2*a*** and ***c***). The tensile strength and elasticity of the aorta reside in the medial layer, which is composed of concentrically arranged elastic fibers and smooth muscle cells (SMCs). The SMCs are longitudinally oriented and dispersed among the circular elastic fibers. By contracting in response to pulsatile blood flow, the SMCs regulate blood flow and pulse pressure. The pathological hallmark of TAAD is medial degeneration, which is characterized by loss and fragmentation of elastic fibers and accumulation of proteoglycans in the aortic media (**Figure 2*b*** and ***d***). Debate persists as to whether SMC loss or hyperplasia occurs in the aortic media associated with medial degeneration extends into the aortic wall (84). Inflammatory cells often accompany medial degeneration, but the role of inflammation in disease progression remains to be defined (28, 85).

Numerous genetic syndromes predispose individuals to TAAD. Most prominent among these syndromes is Marfan syndrome (MFS), where virtually every patient develops ascending aortic disease. Other genetic syndromes and disorders associated with TAAD include Loeys-Dietz syndrome, Ehlers-Danlos syndrome (vascular form), and filamin A mutations. However, most patients with TAAD do not have a genetic syndrome, although many have an inherited genetic predisposition for TAAD. This review focuses on the genes that predispose to TAAD and describes how the identification of these genes has provided insights into the pathogenesis of this deadly disease.

GENETIC SYNDROMES THAT PREDISPOSE INDIVIDUALS TO TAAD

It has long been recognized that patients with MFS are predisposed to thoracic aortic disease. Typically, these patients have an acute aortic dissection in early adulthood unless they undergo prophylactic aortic repair. A similar predisposition for TAAD is seen in patients with Loeys-Dietz syndrome (LDS), vascular Ehlers-Danlos syndrome (EDS), and EDS caused by filamin A mutations. Other genetic syndromes associated with a slightly increased risk for thoracic aortic disease include Turner syndrome, Noonan syndrome, adult polycystic kidney disease, osteogenesis imperfecta, and Alagille syndrome (1, 16, 34, 41, 45, 47, 48, 74).

Marfan syndrome is an autosomal dominant disorder with pleiotropic manifestations that involve the skeletal (pectus deformities, kyphoscoliosis, dolichostenomelia, and arachnodactyly), ocular (ectopia lentis), pulmonary (apical lung blebs and pneumothoraxes), integumentary (striae), and cardiovascular system (TAAD and valvular insufficiencies). The progressive dilation of the aortic root culminates in aortic dissection, which is the major cause of morbidity and mortality in this disorder (18, 79, 91). MFS is caused by heterozygous mutations in the fibrillin-1 gene (*FBN1*) on chromosome 15. Fibrillin-1 is a large, 350-kDa, 2871-amino-acid cysteine-rich glycoprotein that is a major structural component of extracellular matrix (ECM) microfibrils, which are found at the periphery of elastic fibers. More than 600 *FBN1* mutations have been identified in patients with MFS (**http://www.umd.be/**). The majority of *FBN1* mutations are missense mutations. In addition to MFS, *FBN1* mutations may also cause isolated clinical manifestations of the disease, including ectopia lentis, isolated skeletal features, and familial TAAD (2, 20, 35, 57, 58).

Immunohistochemical and biochemical studies have shown that fibroblast cells explanted from patients with MFS have a significant reduction in fibrillin-1-containing microfibrils in the ECM that is less than the 50% predicted by a heterozygous mutation, which suggests that these mutations have a dominant negative effect on microfibril formation (4, 31, 59). Analysis of the fibrillin-1-deficient and knockin missense mutation mouse models of MFS indicated that a connection between the decreased formation of microfibrils and the manifestations of Marfan syndrome may cause this disorder (25, 61, 62, 68). Fibrillin-1-deficient mice have an increase in active transforming growth factor-β (TGF-β) in tissues where fibrillin-1 is expressed when compared with wild-type mice. These results suggest that a reduction in fibrillin-1-containing microfibrils increases the bioavailability of active TGF-β in tissues. Furthermore, antagonism of TGF-β signaling prevents the pulmonary parenchymal abnormalities, mitral valve abnormalities, and aortic dilatation observed in mouse models of MFS, suggesting a critical role for TGF-β signaling in MFS (25, 61, 62).

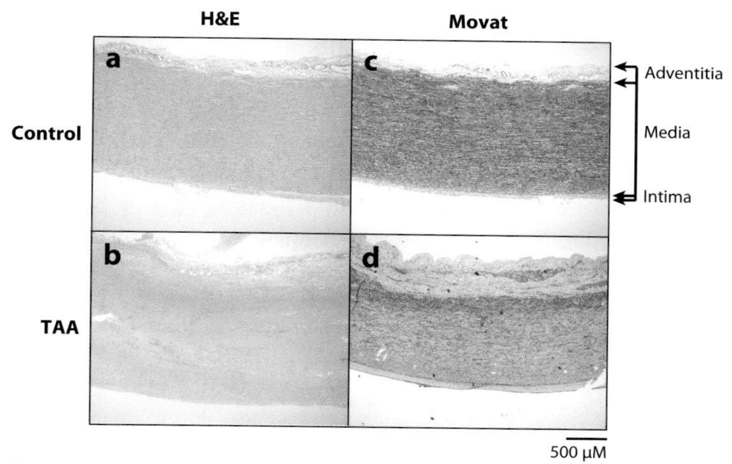

Figure 2

Aortic pathology associated with thoracic aortic aneurysms leading to type A dissections (TAAD). All panels are oriented with the adventitia at the top and the intima at the bottom. Hematoxylin and eosin (H&E) staining of aortic sections from (*a*) a normal subject and (*b*) a patient with TAAD, illustrating medial degeneration with the fragmentation of elastic fibers and loss of smooth muscle cells (SMCs). Movat staining of aortic sections from (*c*) a normal subject and (*d*) TAAD patients shows fragmentation of elastic fibers (*stained black*), loss of SMCs (*cells stained red and nuclei stained violet*), and accumulation of proteoglycans (*stained blue*) in the medial layer. 40× magnification; scale bar represents 500 μM for all panels.

More than a decade ago, genetic heterogeneity for MFS was proposed and a second locus for MFS, termed the MFS2 locus, was mapped to chromosome 3p24-25. This locus was identified using a single large family, although it remained controversial whether the family met the diagnostic criteria for MFS. More recently, a heterozygous mutation in the gene encoding the transforming growth factor-β type 2 receptor (*TGFBR2*) was identified in this family. The mutation disrupts splicing in the kinase domain of *TGFBR2*. *TGFBR2* missense mutations in the intracellular domain were also found in three of nine families with MFS and no mutation in FBN1.

More recently, mutations in either *TGFBR2* or the transforming growth factor-β type 1 receptor gene (*TGFBR1*) have been shown to cause a syndrome called Loeys-Dietz syndrome (LDS), which is characterized by TAAD, arterial aneurysms and tortuousity, craniosynostosis, cleft palate, congenital heart disease, and thin, translucent skin. Rupture or dissection of the aorta occurs at a young age and often when the aortic diameters are less than 5.0 cm, which is the usual cut off for surgical intervention to avoid aortic dissection (50).

Most mutations that cause LDS are missense mutations that affect amino acids in the intracellular kinase domain of the receptors. Many of the disease-causing *TGFBR2* missense mutations reduce receptor signaling activity in response to TGF-β binding. However, cells in the medial layer of aortic tissue from individuals with LDS showed increased expression of collagen and connective tissue growth factor, as well as enrichment of nuclear phosphorylated Smad2; both observations suggest increased TGF-β cellular signaling rather than decreased signaling (49, 50).

Thus, data from both the MFS mouse model and tissues from individuals carrying *TGFBR2* and *TGFBR1* mutations suggest a common theme of increased TGF-β signaling as a cause of the thoracic aortic disease associated with MFS and LDS. Interestingly, pharmacologic studies in the MFS mouse model designed to evaluate the effectiveness of losartan, an angiotensin type I receptor blocker that also inhibits TGF-β signaling, concluded that losartan treatment is more effective than β-adrenergic blockade in successfully treating the lung and aortic abnormalities observed in these mice (25). A randomized clinical trial is currently in progress to assess the comparative effectiveness of losartan and β-adrenergic blockade treatment in the prevention of aortic root dilatation and improvement of other cardiovascular outcomes in individuals with MFS (37).

Ehlers-Danlos syndrome type IV, the vascular type, results from mutations in the gene for type III procollagen (*COL3A1*) (83). Affected patients are at risk for aneurysms and rupture or dissection, especially of medium-sized arteries (67). Although vascular EDS patients can present with aortic aneurysms and dissections, the frequency of aortic disease versus other arteries in this population is not well established. Attempts to surgically repair arteries are often complicated owing to the presence of friable tissue that does not heal well. The diffuse vascular disease and the difficulties encountered during repair lead to premature deaths; vascular EDS patients have a median life span of 48 years. These patients also have thin, translucent skin, wounds that heal poorly and cause atrophic scars, and a high risk for bowel and uterine rupture. Interestingly, these complications are similar to complications observed in patients with LDS (50).

Filamin A (*FLNA*) mutations result in X-linked inheritance of a brain malformation known as periventricular heterotopia (19). The disorder occurs mostly in females and affected women have an increased number of miscarriages of male fetuses, suggesting that hemizygous males die perinatally. In addition, *FLNA* mutations also cause an Ehlers-Danlos syndrome with joint and skin hyperextensibility and aortic dissections (75). *FLNA* encodes a nonmuscle actin binding protein that plays an important role in cross-linking cortical actin filaments into a dynamic three-dimensional structure (81).

FAMILIAL THORACIC AORTIC ANEURYSMS AND DISSECTIONS

Although it has been well appreciated for more than 40 years that patients with MFS are predisposed to TAAD, only in the past 10 years did it become clear that genetic factors contribute importantly to isolated TAAD. The first family with nonsyndromic TAAD in which both MFS and vascular EDS were excluded was reported in 1989 (63). More recent family studies revealed that up to 19% of TAAD patients had a first degree relative with TAAD (7, 12). These studies also showed that individuals with a family history presented with disease at a significantly younger age than those with sporadic aneurysms.

In most of the families with TAAD the disorder segregates as an autosomal dominant trait with decreased penetrance and variable expression with respect to the age of onset of the aortic disease, the location of the aneurysm, and the degree of aortic dilatation prior to dissection (55) (**Figure 3**). Imaging the aorta of family members can reveal many asymptomatic aneurysms in family members and confirm autosomal dominant inheritance of the genetic predisposition. The incidence of aortic disease increases with advancing age in these families. There is more variability in the age of onset of aortic disease in these families than is seen in MFS. Another feature that complicates the genetic analysis of these families is incomplete penetrance, especially among female carriers.

The variability of the vascular disease presentation and associated features highlights a significant amount of clinical heterogeneity in families with multiple members with TAAD. In a small subset of TAAD families, family members experience aortic dissection with no aortic dilatation. Some families include individuals who have inherited the defective gene on the basis of their location in the pedigree but they have had cerebral aneurysms rather than TAAD (**Figure 3*e***). Other families have TAAD associated with a bicuspid aortic valve (BAV). The association of BAV and TAAD as a familial condition has been reported in the literature frequently, leading to the suggestion that BAV and TAAD can be manifestations of a single gene defect (11, 21, 52, 54). Family members can have BAV, TAAD, or both cardiovascular conditions. Similar to other TAAD families, the inheritance of the predisposition is autosomal dominant with incomplete penetrance (52). Finally, a rare association found in TAAD is families with aortic disease and a patent ductus arteriosus (PDA). The region of aneurysmal involvement in the ascending aorta is another feature of the disease that varies between TAAD families. Some families have ascending aneurysms that initially involve the sinuses of Valsalva, similar to patients with MFS and LDS. Other families have members who present with aneurysms that spare the sinuses of Valsalva and instead the dilatation occurs in the ascending aorta and often extends into the arch of the aorta (**Figure 4**). The ascending aorta is the most common location for thoracic aortic aneurysms (TAAs) and this is also the location for aneurysms associated with long standing hypertension.

The clinical heterogeneity observed in TAAD families predicts an underlying genetic heterogeneity for the disease. Significant genetic heterogeneity was known to be associated with other adult-onset cardiovascular diseases caused by single gene mutations, including hypertrophic cardiomyopathy and long Q-T syndrome (3, 93). Therefore, our approach to map loci for familial TAAD was to use individual families with multiple affected members such that a LOD score greater than three could be generated from the single family alone. We addressed the documented decreased penetrance of the disease by not scoring family members at risk for inheriting the defective gene as unaffected for linkage analysis even if they had a normal aorta by imaging, and using an age-related penetrance model for linkage. We and other investigators have had significant success in mapping loci for TAAD using this approach and have mapped six chromosomal loci: *TAAD1* at 5q13-14, *FAA1* at 11q23-24, *TAAD2* at 3p24-25, *TAAD3* at 15q24-26, *TAAD4* at 10q23-24, and a gene on 16p12-13 (5, 22, 23, 26, 94)

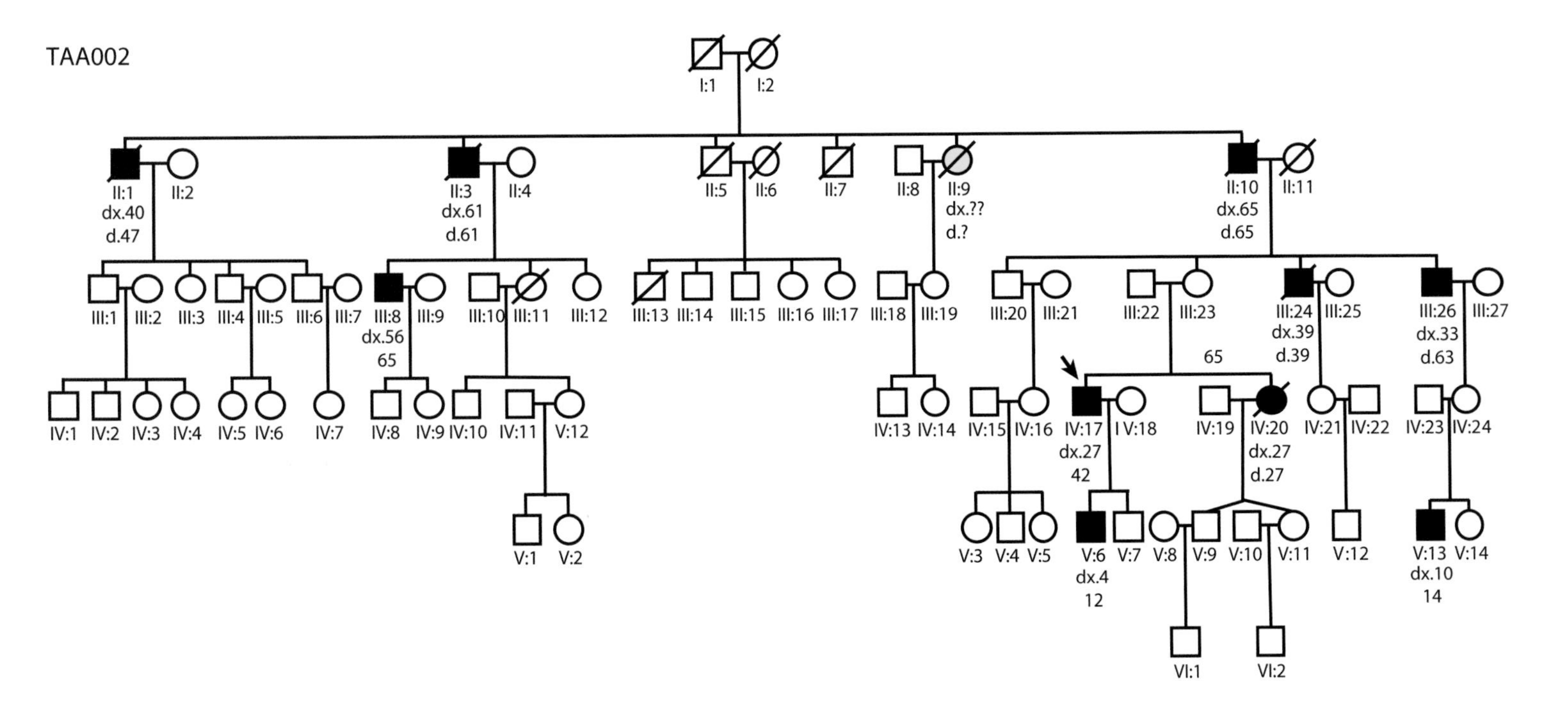

Figure 3

Genetic and phenotypic heterogeneity in familial thoracic aortic aneurysms and dissections (TAAD). Pedigrees indicate males as squares and females as circles, a line through an individual indicates death, solid black symbols represent affected status due to TAAD, the yellow symbol indicates ICA, and clear symbols indicate unaffected status. Numbers represent ages, "dx." indicates age of diagnosis, and "d." indicates age of death. Family TAA002 illustrates a four-generation pedigree with TAAD exhibiting an autosomal dominant inheritance pattern with decreased penetrance and variable expression. The proband (IV:17) was a 30-year-old Caucasian man, who was evaluated via aortic imaging due to the family history of aortic aneurysms and dissections. He was found to have a 6.5 cm aortic root that rapidly expanded to 8 cm. He underwent composite aortic root surgical repair. At 38 years of age he was found to have a Type B dissection, which is currently being managed medically. His family history was positive for a sister (IV:20) who died suddenly of a Type A dissection with hemopericardium at 27 years of age. His mother (III:23) at 63 years of age has had annual echocardiograms that show a normal diameter of her aortic root, demonstrating decreased penetrance of the disease gene in this individual. The proband's maternal grandfather (II:10) died suddenly at 65 years of age and the autopsy revealed the death was due to an aortic rupture. One of the proband's maternal uncles (III:24) died suddenly at 39 years of age and the autopsy revealed an aortic aneurysm of 13.5 cm and a Type A dissection with hemopericardium. The proband's youngest uncle (III:26) underwent surgical repair of his descending aorta at 33 years of age and at 48 years of age underwent ascending aortic aneurysm repair. At 54 years of age, he was found to have an abdominal aortic aneurysm. His daughter (IV:24) currently has a normal size aorta but her 14-year-old son (V:13) has a dilated ascending aorta measuring 4.0 cm. Two of the proband's great uncles died of Type A dissections, one at 47 years of age (II:1), and the other at age 61 years (II:3).

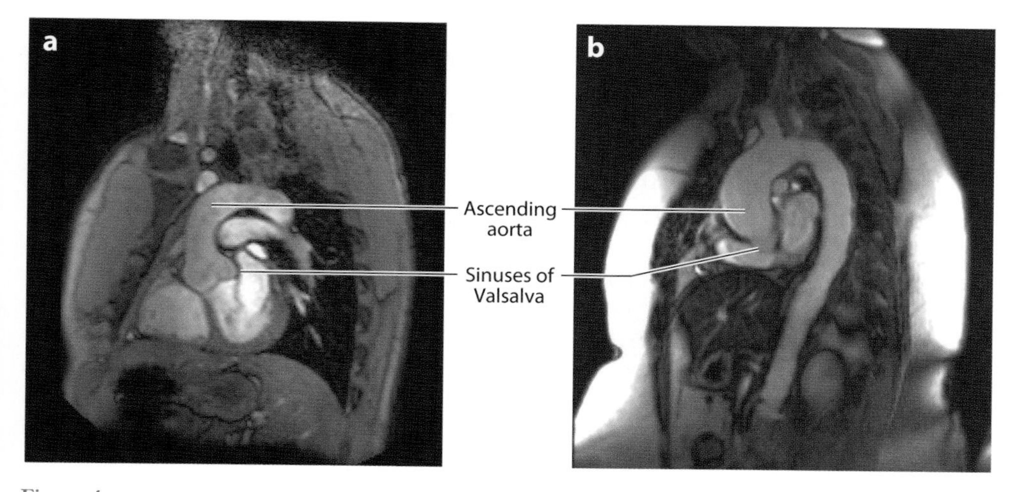

Figure 4

Location of aneurysms of the ascending thoracic aorta. (*a*) Aneurysm in a patient with familial thoracic aortic aneurysms leading to type A dissections (TAAD) that involves the sinuses of Valsalva and spares the ascending aorta. (*b*) Aneurysm that involves the ascending aorta and aortic arch and spares the sinuses of Valsalva.

(**Figure 5**). These mapping studies have firmly established that genetic heterogeneity for familial TAAD exists. Many families with TAAD are not linked to these known loci, which indicates that additional genes remain to be mapped (D.M. Milewicz and D.C. Guo, unpublished data). The causative genes have been identified at the following three loci: *TGFBR2* at the *TAAD2* locus, the gene encoding smooth muscle myosin heavy chain (*MYH11*) at the 16p locus, and the gene encoding smooth muscle α-actin (*ACTA2*) at the *TAAD4* locus (23, 65, 96).

The defective gene at the *TAAD2* locus was identified as *TGFBR2*, a gene that is mutated in a small subset of MFS patients and approximately two-thirds of LDS patients. Sequencing of 80 unrelated families with familial TAAD determined that *TGFBR2* mutations were responsible for disease in only four families (5%) (65). Strikingly, all four families had *TGFBR2* mutations that altered arginine 460 in the intracellular domain of the receptor, suggesting that this genotype is associated with familial TAAD. A subsequent report of a large TAAD family also with a *TGFBR2* mutation that disrupts arginine 460 further validates this hypothesis (40). On physical examination, these families do not have features of LDS, i.e., hypertelorism, bifid uvula/cleft palate, and craniosynostosis. Arterial tortuousity and translucent skin are absent in the subset of family members that have been evaluated for these features. Although the

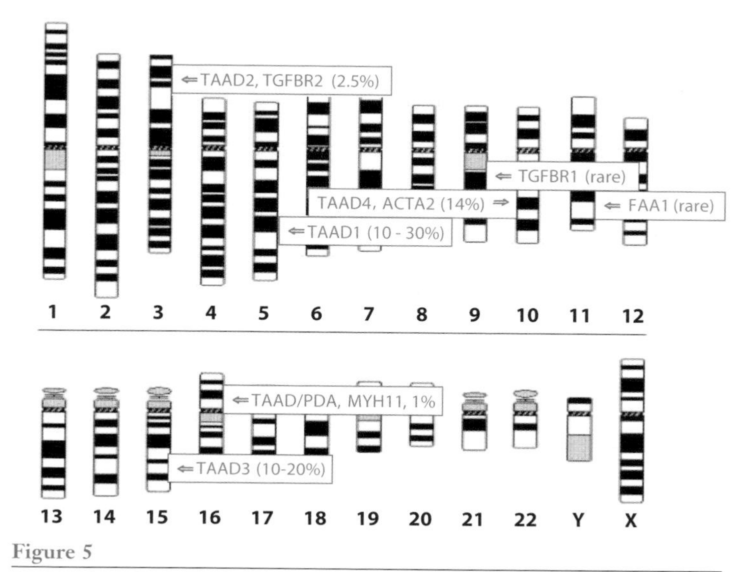

Figure 5

Genetic loci identified for familial thoracic aortic aneurysms leading to type A dissections (TAAD). Chromosomal locations, defective genes (when identified), and the estimated percentage of familial TAAD patients with involvement of the locus are indicated on a normal male karyotype.

vast majority of affected individuals in these families presents with TAAD, some have subsequently developed descending aortic disease and aneurysms of other arteries, including cerebral, carotid, and popliteal aneurysms. Three individuals with the *TGFBR2* R460 mutation have had dissections prior to reaching an aortic diameter of 5.0 cm. Other affected family members have had dissections with aortic diameters as small as 4.2 cm. These dissections with minimal dilatation have led physicians to recommend prophylactic repair when an individual who carries the mutation reaches an aortic diameter of 4.0–4.2 cm (43).

We have identified four multigeneration TAAD families with *TGFBR1* mutations (out of 150 families screened for mutations) and these families also do not have features of LDS (D.M. Milewicz and V. Tran Fadulu, manuscript submitted). In fact, the first family reported with familial TAAD in which MFS and EDS were excluded has a *TGFBR1* mutation as the cause of the disease (63). Therefore, *TGFBR1* mutations can also lead to familial TAAD but are a rare cause of the disease.

A large French family with TAAD associated with PDA was used to map the defective gene responsible for this phenotype to 16p (92). The defective gene at this locus was identified as the *MYH11* gene, which encodes the SMC-specific myosin heavy chain, a major component of the contractile unit in SMCs. Subsequent analysis of 93 unrelated families with TAAD failed to identify any *MYH11* mutations. Sequencing DNA from three unrelated families with TAAD associated with PDA identified *MYH11* mutations in two of these families (66); the remaining family had a *TGFBR2* mutation as the cause of the TAAD and PDA (D.M. Milewicz and H. Pannu, unpublished data). The spectrum of *MYH11* mutations identified for the familial TAAD/PDA phenotype is limited to four mutations: a small deletion, a splice site mutation, and two missense mutations. Therefore, *MYH11* mutations can cause the disease in the small subset of families with TAAD and PDA.

Recently we mapped a novel locus for familial TAAD (*TAAD4*) to 10q23-24 and identified the gene defective in one large family as *ACTA2*, which encodes the SMC-specific α-actin, a component of the contractile complex and the most abundant protein in vascular SMCs (17, 23). Out of 98 unrelated TAAD families, 14 families (14%) had mutations in *ACTA2*; thus mutations in *ACTA2* are the most common genetic defect causing TAAD (23). Several different missense mutations, including two recurrent mutations found in unrelated families, have been identified. In the family in which the disease was mapped, the mutation in *ACTA2* (R149C) segregates invariantly with a skin rash caused by dermal capillary and small artery occlusion referred to as livedo reticularis. Other features associated in a subset of families with *ACTA2* mutations include iris flocculi, PDA, and BAV. The penetrance of TAAD in family members heterozygous for *ACTA2* mutations is low (0.48) and does not increase with age, thus distinguishing these patients from those with other identified forms of familial TAAD. The majority of affected individuals presented with acute type A dissections or type B dissections, and 16 of 24 deaths occured due to type A dissections. Two of 13 individuals experienced type A dissections with a documented ascending aortic diameter less than 5.0 cm. Aortic dissections occurred in three individuals under 20 years of age and two women died of dissections postpartum. Finally, three young men had type B dissections complicated by rupture or aneurysm formation at the ages of 13, 16, and 21 years.

IDENTIFICATION OF DEFECTIVE GENES CAUSING THORACIC AORTIC ANEURYSMS AND DISSECTIONS: FOCUS ON SMOOTH MUSCLE CELL CONTRACTION

The identification of mutations in SMC-specific contractile proteins as causes of familial TAAD provides insight into the pathogenesis of these aortic diseases. The SMC contractile unit is composed of thin and thick filaments that contain SMC-specific α-actin and β-myosin

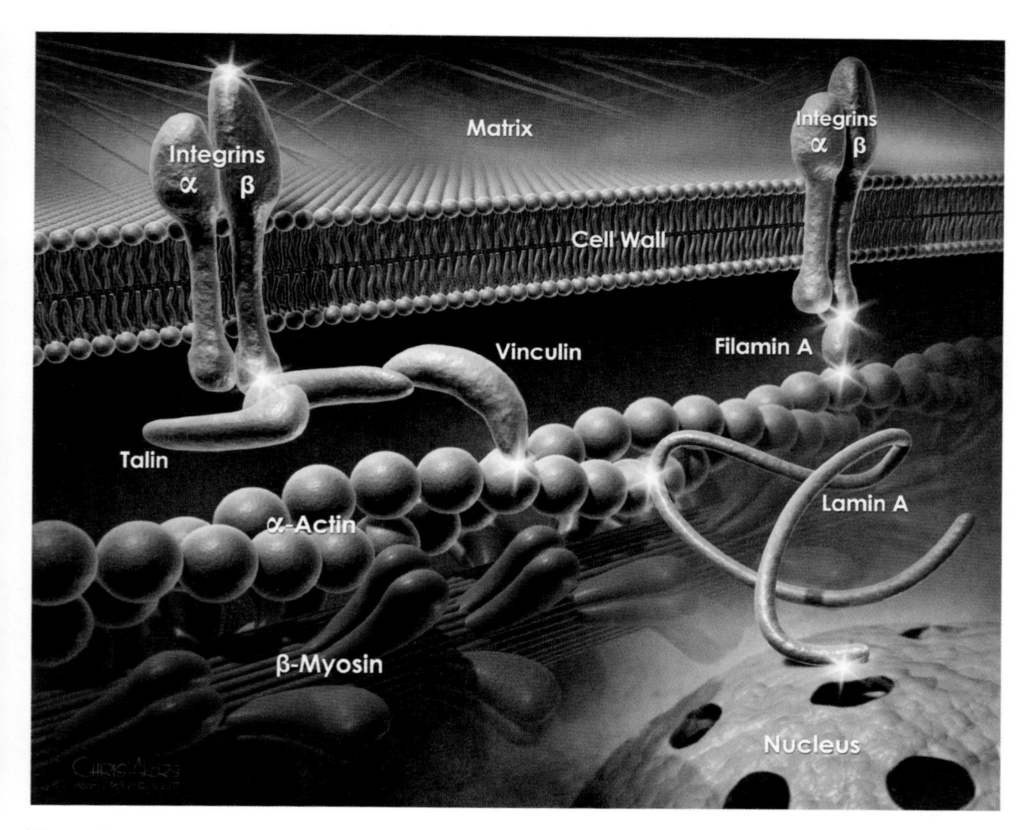

Figure 6

The mechanotransduction complex in smooth muscle cells (SMCs). Contractile and cytoskeletal filaments containing actin and myosin link to the fibrillin microfibrils in the extracellular matrix through heterodimeric integrin receptors composed of α and β subunits. This association is reversible and is mediated by several proteins, including vinculin, talin, and filamin A. Contractile and cytoskeletal filaments are anchored to the nuclear membrane through an interaction between actin and lamin A.

heavy chain, respectively (**Figure 6**). The structural lattice in SMCs is composed of the contractile unit, along with the cytoskeleton (comprised of nonmuscle actin and intermediate filaments), which links to the cell surface through filamin A (81). The actin filaments of the contractile unit interface with the cytoskeleton at the cytoplasmic dense bodies and at dense plaques on the cell surface, which contain integrin receptors. Integrin receptors are the principal receptors for the ECM and serve as a transmembrane link between the matrix, the actin cytoskeleton, and contractile units (60) (**Figure 6**). Integrins are composed of α and β subunits, and each αβ combination has its own ligand specificity. This cellular complex, termed the mechanotransduction complex, provides the interface between the contractile machinery on the interior of the cell and the ECM on the exterior, to which force is transmitted.

Detailed electron microscopy studies of the developing mouse aorta reveal that SMC contractile filaments link to microfibrils in the ECM early in development and these links are maintained through adulthood (13) (**Figure 7**). Fibrillin-1 is the major protein component of these microfibrils, raising the possibility that the integrin receptors in the dense plaques bind to fibrillin-1. Further support of this binding is provided by the fact that fibrillin contains a recognition sequence for integrin receptor binding [RGD (arginine-glycine-aspartic acid)]. Furthermore, fibroblasts adhere to purified microfibrils and recombinant fibrillin fragments containing the RGD sequence through the $\alpha5\beta1$ and $\alpha V\beta3$ receptors. This would suggest that fibrillin-1-containing microfibrils are a component of the mechanotransduction system of SMCs, linking fibrillin-1 in the matrix to the intracellular actin filaments.

Cyclic interaction of the myosin motor with actin filaments, fueled by ATP hydrolysis, leads to SMC shortening and contractile force generation (14, 90). Smooth muscle myosin consists of two heavy chains and four light chains. The long C-terminal stalks of the heavy chains dimerize to yield an α-helical coiled-coil. The N terminus harbors the motor domain (MD), containing both the ATP- and actin-binding sites, which together with the light chains form two globular heads (cross bridges). The tail portion functions as a lever arm that translates conformational changes within the MD into rowing motions by which myosin moves actin (69). The SH1 α-helix, which functions as a joint between the MD and the converter/lever arm unit, is key to force development. Both biochemical and structural studies illustrate that the SH1 helix undergoes conformational changes during ATP hydrolysis (32). A *MYH11* missense mutation causing familial TAAD that alters arginine 712, an invariant amino acid in all type II myosins, is part of this α-helix. Other *MYH11* mutations are predicted to disrupt the

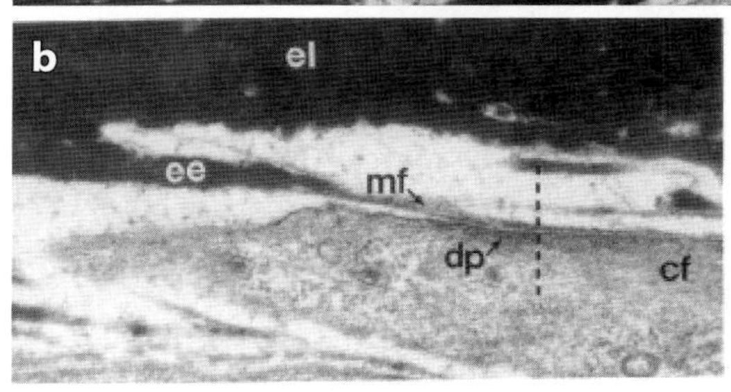

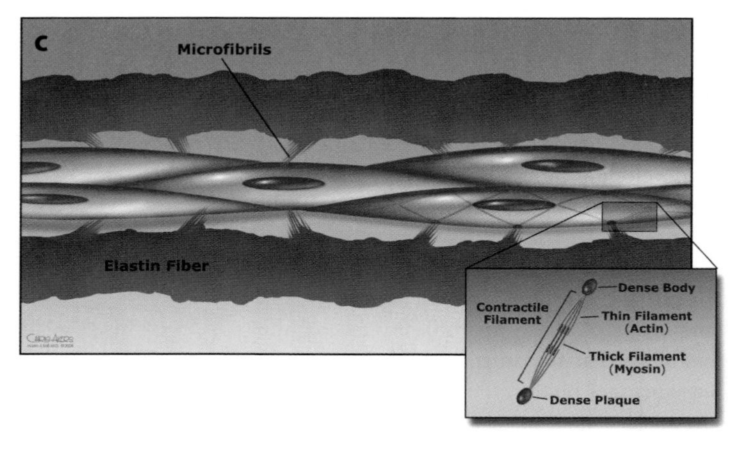

Figure 7

Electron microscopy images of rat aortic smooth muscle cells (SMCs) that demonstrate the connection between the contractile units inside the cells, the dense plaques, and the microfibrils in the extracellular matrix. (*a*) Longitudinal section of a SMC in the aortic media. A bundle of contractile filaments (cf) can be seen oriented oblique to the long axis of the cell. The contractile filaments anchor on either side in membrane-associated dense plaques (dp). (*b*) SMC in longitudinal section showing association of the cell with the microfibrils, forming a lateral association with the cell surface in a region overlying a membrane-associated dense plaque. Contractile filaments, anchored to dense plaques, have the same orientation as elastin extension (ee) and microfibrils (mf). (*c*) The connections between the elastic fibers, fibrillin-1-containing microfibrils, and the contractile unit in the aortic SMCs. Reproduced with permission from Reference 13.

coiled coil domain of the long C-terminal domain (66, 96). Therefore, disruption of either the motor domain or coiled-coil domain of myosin is predicted to disrupt the structure or function of myosin.

Similarly, the missense *ACTA2* mutations identified in familial TAAD are predicted to alter the dynamics of actin assembly into filaments, by either disrupting the actin-actin interaction sites or interfering with ATP hydrolysis. Analysis of SMCs explanted from patients heterozygous for *ACTA2* mutations demonstrated reduced ACTA2-containing fibers and therefore confirmed that *ACTA2* missense mutations disrupt actin fiber assembly or stability (23).

Therefore, the various *MYH11* and *ACTA2* mutations identified are predicted to disrupt SMC contractile function, leading to the hypothesis that decreased SMC contractile function may be the underlying cause of the disease. As described previously, contractile filaments link to the ECM through heterodimeric integrin receptors. Structural and biochemical studies suggest that $\alpha V\beta 3$ and $\alpha 5\beta 1$ integrin receptors bind to fibrillin-containing microfibrils at the periphery of elastic fibers. Hence, *FBN1* mutations may also disrupt SMC contraction through disruption of the SMC mechanotransduction complex. *FBN1* missense mutations disrupt the polymerization of fibrillin-1 monomers into microfibrils, resulting in a decrease in the amount of fibrillin-containing microfibrils. Interestingly, the first ultrastructural abnormality noted in the mouse model for MFS is an unusually smooth surface of the elastic laminae, likely due to the loss of cell attachments that are normally mediated by fibrillin-1 (9).

Filamin A is a component of the intermediate filaments in SMCs and mutations in *FLNA* lead to a syndrome of joint laxity and aortic dissections. Despite the fact that the intermediate filaments do not have an established role in mechanotransduction, it is interesting to speculate that these mutations also disrupt the connections between the contractile filaments and the integrin receptors on the cell surface, a possibility that has been discussed in prior reviews (75, 76).

Mutations in type III collagen (*COL3A1*) that cause vascular EDS also may disrupt SMC contractility by binding SMCs to type III collagen in the matrix. Type III collagen interacts with integrin receptor $\alpha 2\beta 1$ (70). Interestingly, this binding is not through an RGD recognition sequence, but rather through GxxGER motifs or GxxGEN motifs located in the triple-helical domain of the protein. In addition, fibroblasts with *COL3A1* mutations decrease the expression of $\alpha 2\beta 1$ receptors (97). Because type III collagen is a major constituent of the medial layer of arteries, along with type I collagen, SMCs may also anchor to type III collagens in the matrix (77).

Thus, the majority of gene mutations identified in familial TAAD could potentially disrupt proteins involved in mechanotransduction in SMCs and lead to decreased contractility. The notable exceptions to this observation are mutations in the *TGFBR1* and *TGFBR2* genes. These encode the cell surface receptors for TGF-β and ligand binding to these receptors initiates a downstream signaling cascade. Although these receptors are not part of the contractile complex in SMCs, TGF-β signaling does play a major role in the differentiation of SMCs, including the expression of contractile proteins. Vascular SMCs originate from two sources during vascular development: The SMCs in the ascending aorta and arch of the aorta are derived from the neural crest, whereas SMCs in the descending aorta and other arteries are derived from the mesoderm (29, 87). TGF-β induces neural crest stem cells to differentiate into SMCs that specifically express SMC contractile proteins such as α-actin and β-myosin (10). Mesenchymal cells also differentiate into SMCs upon exposure to TGF-β (30). Almost all the *TGFBR2* and *TGFBR1* missense mutations leading to LDS, MFS, and familial TAAD alter amino acids in the intracellular domain of the receptor and are predicted to disrupt kinase activity of the receptors, and thus likely prevent proper signaling, thereby disrupting the differentiation of neural crest and mesenchymal cells into vascular SMCs. The lack of contractile units in the vascular SMCs owing to disruption of differentiation would be predicted to lead to altered SMC contractility, similar to the predicted effect of *ACTA2* and *MYH11* mutations.

The genetic mutations in the SMC-specific isoforms of contractile proteins can be correlated with mutations in cardiac- and skeletal muscle–specific isoforms of these proteins, which lead to hypertrophic cardiomyopathy (HCM) or dilated cardiomyopathy and muscular dystrophy, respectively. Mutations in 11 genes that encode cardiac sarcomeric proteins involved in contractile function have been identified for HCM, including the following: *MYH7*, *MYBPC3*, *TNNI3*, *TNNT2*, *TPM1*, *ACTC*, *TNNC1*, *MYL3*, *MYL2*, *MYH6*, and *TTN* (86). Despite the significant genetic heterogeneity of HCM, the genes mutated in this disorder involve components of the cardiac contractile unit. Dilated cardiomyopathy is also caused by mutations in proteins involved in cardiac contraction, specifically in cytoskeletal and sarcomeric proteins (88). Many of the mutated genes in this disorder are predicted to disrupt the connection of the cytoskeleton to the sarcomere. Therefore, mutations in proteins that disrupt the cytoskeleton/contractile complex are the cause of both HCM and dilated cardiomyopathy.

Mutations in the skeletal muscle–specific isoforms of the cardiac contractile genes lead to muscular dystrophies (39) by disrupting the physical link between the ECM and the cytoskeleton/contractile apparatus. The dystrophin-glycoprotein complex links the ECM to the cytoskeleton, and mutations in components of this complex cause a variety of muscular dystrophies. Other myopathies, such as nemaline and actin myopathies, are caused by mutations in genes encoding other proteins in the contractile apparatus, including actin (*ACTA1*) and troponin T (*TNNT1*) (36). Finally, mutations in ECM proteins, such as laminin or collagen VI, also cause muscular dystrophies. As previously speculated, nearly all the genes identified for inherited dilated cardiomyopathy are also known to cause skeletal myopathy in human or mouse models, and mutations in some genes cause both diseases (89). It is interesting to speculate that a similar pattern of mutations is emerging for TAAD, in which mutations that lead to SMC myopathy may cause a genetic predisposition to familial TAAD.

GENETIC MUTATIONS LEAD TO ACTIVATION OF CELLULAR PATHWAYS

The identification of *ACTA2* and *MYH11* mutations as causes of familial TAAD suggests that SMC contractile function plays a key role in maintaining the structural integrity of the ascending aorta and preventing the development of aortic aneurysms. Despite accumulating evidence supporting this hypothesis, no clear link exists between the defective gene and the pathology observed in the aortas in TAAD patients. The only exception is the data showing that increased TGF-β signaling occurs in the mouse model of MFS.

The observed physiological effects of cardiac β-myosin (*MYH7*) mutations causing HCM provide insights into the effects of mutations in *MYH11* on SMCs. The incorporation of mutant β-myosin into myofibrils is proposed to lead to decreased cardiomyocyte contractility (53). The cardiac cell responds to the defective contractile apparatus by upregulating mitotic and trophic factors, including IGF-1, TGF-β1, and angiotensin II (8, 44, 46), which in turn promotes hypertrophy of the cardiomyocytes.

A similar pathogenic sequence occurs in the aortic wall of patients with *MYH11* and *ACTA2* mutations. Analysis of the aortic tissue from these patients demonstrates proteoglycan accumulation, fragmentation of elastic fibers, and a decreased number of SMCs, which are all features of medial degeneration. Atypical findings include focal areas of increased numbers of SMCs that are in apparent disarray and not oriented parallel to the lumen of the aorta (23, 66). The aortas are notable for marked medial SMC proliferation, causing stenosis or occlusion of the vessel in focal areas of the vasa vasorum. The blood vessels in the vasa vasorum of the diseased aorta are of increased size compared with controls (23, 66).

Explanted SMCs from a patient with a *MYH11* mutation who had a rapidly expanding ascending aortic aneurysm demonstrated a dramatic increase in the level of expression of *IGF-1*, with no increase in the expression of *TGF-β1* or platelet derived growth factor-β (*PDGF-B*) (66). Increased IGF-1 immunostaining was also observed in the patient's aortic tissues. Therefore, the focal SMC proliferation observed in the aortic media and the vasa vasorum could be caused by increased IGF-1 secreted by the SMCs. Furthermore, the SMCs demonstrated increased expression of angiotensin converting enzyme (ACE); in conjunction with the increased vascularity of the aneurysms, this increase could potentially lead to increased tissue production of angiotensin II from circulating precursors, further driving SMC proliferation. Finally, our studies documented increased MIP-α and MIP-β expression and protein production by the SMCs, which could also contribute to the increased vascularity of the tissue. A similar increase in IGF-1 expression and production has been observed in SMCs that harbor heterozygous *ACTA2* mutations (D.M. Milewicz and C. Papke, unpublished data).

In summary, our studies suggest similarities between the proposed pathogenesis of HCM and TAAD. The underlying *MYH11* mutations are predicted to alter contractility of the SMCs, increasing SMC stress and initiating adaptive pathways to repair the defect in the SMCs, which leads to the production of trophic factors that include IGF-1, ACE, MIP-1α, and MIP-1β. These trophic factors most likely lead to many aspects of the observed pathology, including SMC proliferation, disarray, and increased size and extent of the vasa vasorum (66).

Although the increased expression of secreted growth factors and chemokines may be the basis of some of the pathology observed with *MYH11* and *ACTA2* mutations, the production of these factors does not explain all the features of TAAD, including the increased levels of proteoglycans and loss and fragmentation of the elastic fibers seen in the aortas. Studies have demonstrated increased levels of metalloproteinases, specifically matrix metalloproteinase 2 and 9 (MMP2 and MMP9), in the aortic tissues of patients with both sporadic and genetic forms of TAAD (73). In the context where SMC contractile dysfunction may be the underlying defect leading to TAAD, it is notable that cyclic stretch of SMCs leads to increased production of MMP2 and MMP9 by SMCs. Furthermore, low levels of mechanical strain increase the production of proteoglycans by SMCs, which is another feature of medial degeneration and TAAD (24, 42). The decreased SMC contraction due to gene mutations could lead to upregulation of stretch pathways in SMCs, resulting in increased production of MMPs and proteoglycans by these cells.

Therefore, a stepwise disease pathogenesis is proposed for mutations that cause TAAD (**Figure 8**). The mutations disrupt SMC contractile function, leading to the activation of both stress and stretch pathways in SMCs. The activation of these SMC pathways leads to medial degeneration and the clinical disease. With *MYH11* and *ACTA2* mutations, increased IGF-1 expressed by aortic SMCs increases the expression of the contractile proteins in the SMCs and may also lead to hyperplasia of the SMCs. Therefore, the expression of IGF-1 may be a compensatory action of the aortic SMCs in an attempt to correct the contractile defect due to the *MYH11* mutation. SMCs from MFS and LDS may instead increase TGF-β in an attempt to drive SMC differentiation or increase ECM deposition. Therefore, the increased TGF-β in the tissues of the Marfan mouse model and LDS patients may also result from the increased production of TGF-β by the aortic SMCs.

NONGENETIC CAUSES AND MEDICAL THERAPY OF THORACIC AORTIC ANEURYSMS AND DISSECTIONS

Although single gene mutations cause up to 20% of TAAD, the majority of patients do not have a family history of disease. The

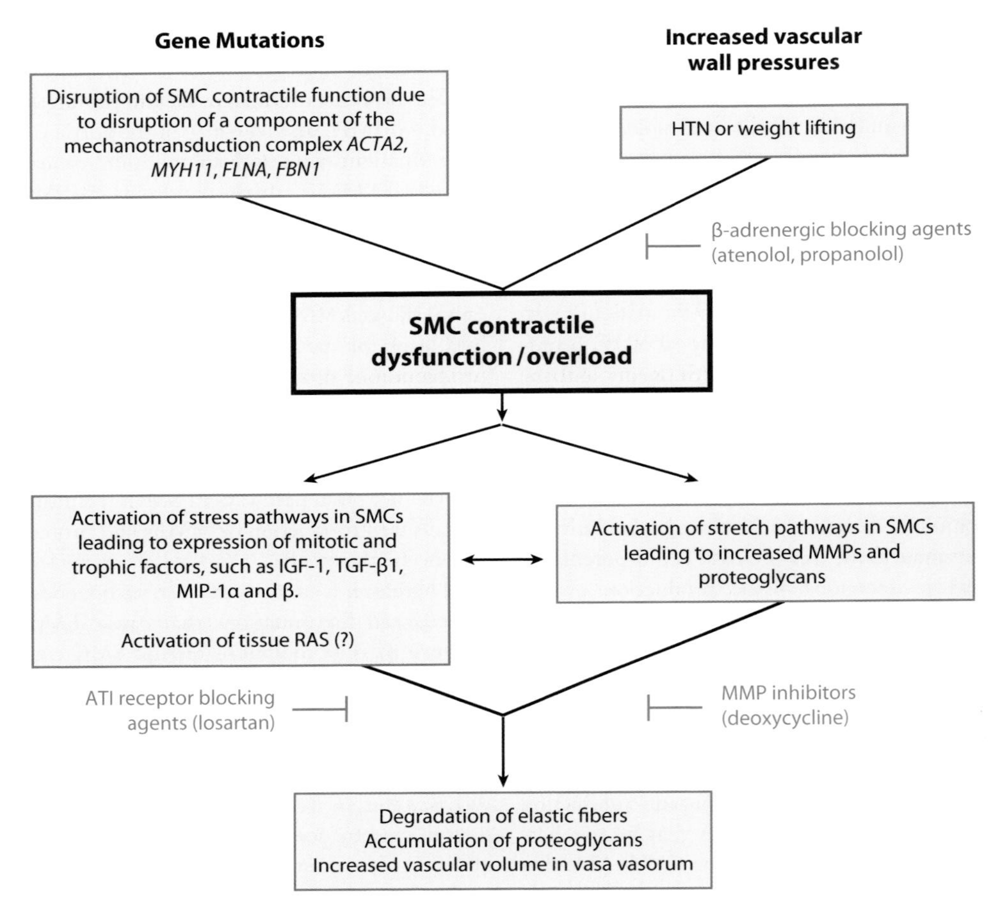

Figure 8

The pathways involved in the pathogenesis of thoracic aortic aneurysms leading to type A dissections (TAAD). Gene mutations and/or environmental factors such as hypertension that cause increased vascular pressure can lead to activation of stress- and stretch-induced pathways in the aortic smooth muscle cells (SMCs). Pathways activated in the SMCs lead to increased production of multiple trophic factors, proteoglycans, and metalloproteinases. Persistent stress or stretch abnormalities result in continued upregulation of these pathways, and ultimately lead to proteoglycan accumulation and the degradation of elastic fibers. HTN, hypertension; RAS, renin angiotensin system; MMP, metalloproteinase.

identification of the defective genes that cause familial TAAD, along with the understanding of the pathogenesis of TAAD, has implications for the risk factors associated with sporadic TAAD, which include the presence of a BAV, hypertension, and weight lifting.

The association of a BAV with TAAD is well established (6, 15, 71). Although the prevalence of BAV is 0.5% in the population, the prevalence of BAV among patients with aortic dissections is 9% (15). As previously indicated, the inheritance of the combined phenotype of BAV/TAAD suggests a common genetic defect underlies this association. It is interesting to note that 3% of our patients with *ACTA2* mutations (4 out of 119 patients) and 2.5% of our patients with *TGFBR2* mutations (2 out of 81) have BAV, implying that these genetic defects increase the risk of BAV. This observation also suggests that increased risk of BAV may be associated with many of the genes that cause a predisposition to TAAD.

Hypertension leads to increased risk and progression of TAAD (33). The pathology associated with hypertensive aortic disease is indistinguishable from the pathology found in patients with familial TAAD. Elevated blood pressure imposes increased mechanical stresses on the vascular wall and mechanical strain is a mitogenic stimulus for SMCs, increasing the production of IGF-1 and TGF-β1 and increasing stretch-induced pathways (64, 82). In animal models of essential hypertension [spontaneously hypertensive rats (SHR) and stroke-prone spontaneously hypertensive rats (SHR-SP)], structural modifications of the arterial wall include an increase in the number of elastin/SMC connections, leading to the suggestion that these changes result in a redistribution of the mechanical load to elastic fibers (38). Therefore, many of the same SMC pathways that are activated because of genetic defects in the contractile complex in SMCs are involved in vascular disease because of hypertension and may be the basis for the increased risk for the development and progression of TAAD (**Figure 8**).

Acute aortic dissections can occur with heavy weight lifting or other extreme exertions. A recent report of 31 cases of acute aortic dissections that occurred in individuals during heavy exertion, during exercise, or as a part of their job provides further evidence of this association (27). Weight lifting and other high-resistance activities lead to significant increases in blood pressure, heart rate, myocardial contractility and cardiac output (51). Blood pressure is particularly increased in part to provide blood flow into muscles that have compressed their arterial inflow as a result of increases in intramuscular pressure created by the skeletal muscle contraction; blood pressures as high as 380 mm Hg have been recorded in competitive weight lifters (27). These dramatic increases in blood pressure lead to increased wall stress and could potentially activate stress and stretch pathways even in the absence of a defect in SMC contraction.

The proposed pathogenesis also has implications for the treatment to prevent TAAD. Current medical management of TAAD is limited to treatment with β-adrenergic blocking agents. Treatment of MFS patients with β-blockers has indicated that such therapy is successful in some patients (72, 78). Treated patients show slower aortic root growth, fewer cardiovascular complications (such as aortic regurgitation, dissection, or surgical repair), and improved survival. The protective effect of β-blockers is derived from the negative chronotropy and decreased rate of blood volume and pressure in the ascending aorta that results from negative inotropy. On the basis of the proposed mechanism of disease progression in TAAD, medications that reduce blood flow and pressure on the aortic wall should decrease the activation of SMC cellular pathways that cause the associated pathology and ultimately the clinical disease.

As previously discussed, losartan has proven to be an effective therapy for preventing the formation of aortic aneurysms in the mouse model of MFS (25). Losartan not only blocks TGF-β signaling but also would prevent angiotensin II signaling through blockade of the AT1 receptor, which may be activated in some forms of TAAD (66). Finally, recent studies have indicated that doxycycline, a nonspecific MMP inhibitor, significantly delays aneurysm rupture in a mouse model of MFS by inhibiting activity of tissue MMP2 and MMP9 (95). Therefore, all these therapies have the potential to block the activation of stress and stretch pathways in SMCs activated by the underlying genetic defect (**Figure 8**). Although these drugs are currently being tested as single therapies, combinations of these agents may prove to be of greater benefit if these agents do block different pathways.

DISCLOSURE STATEMENT

The authors are not aware of any biases that might be perceived as affecting the objectivity of this review.

ACKNOWLEDGMENTS

The authors are grateful to the families participating in our research studies and to Chris Akers for excellent graphic assistance. The following sources provided funding for these studies: RO1 HL62594 (D.M.M.), P50HL083794-01 (D.M.M.), UL1 RR024148 (CTSA), and TexGen Foundation. D.M.M. is a Doris Duke Distinguished Clinical Scientist.

LITERATURE CITED

1. Adeola T, Adeleye O, Potts JL, Faulkner M, Oso A. 2001. Thoracic aortic dissection in a patient with autosomal dominant polycystic kidney disease. *J. Natl. Med. Assoc.* 93:282–87
2. Ades LC, Sreetharan D, Onikul E, Stockton V, Watson KC, Holman KJ. 2002. Segregation of a novel *FBN1* gene mutation, G1796E, with kyphoscoliosis and radiographic evidence of vertebral dysplasia in three generations. *Am. J. Med. Genet.* 109:261–70
3. Ahmad F, Seidman JG, Seidman CE. 2005. The genetic basis for cardiac remodeling. *Annu. Rev. Genomics Hum. Genet.* 6:185–216
4. Aoyama T, Francke U, Dietz HC, Furthmayr H. 1994. Quantitative differences in biosynthesis and extracellular deposition of fibrillin in cultured fibroblasts distinguish five groups of Marfan syndrome patients and suggest distinct pathogenetic mechanisms. *J. Clin. Invest.* 94:130–37
5. Avidan N, Tran-Fadulu V, Chen J, Yuan J, Yu R, et al. 2005. Submitted. Mapping a third locus for familial TAAD (TAAD3) using samples from a single family with multiple affected individuals and determining the contribution of this locus to familial disease. Poster 1515 presented at Am. Soc. Hum. Genet. Meet.
6. Basso C, Boschello M, Perrone C, Mecenero A, Cera A, et al. 2004. An echocardiographic survey of primary school children for bicuspid aortic valve. *Am. J. Cardiol.* 93:661–63
7. Biddinger A, Rocklin M, Coselli J, Milewicz DM. 1997. Familial thoracic aortic dilatations and dissections: a case control study. *J. Vasc. Surg.* 25:506–11
8. Broglio F, Fubini A, Morello M, Arvat E, Aimaretti G, et al. 1999. Activity of GH/IGF-I axis in patients with dilated cardiomyopathy. *Clin. Endocrinol.* 50:417–30
9. Bunton TE, Biery NJ, Myers L, Gayraud B, Ramirez F, Dietz HC. 2001. Phenotypic alteration of vascular smooth muscle cells precedes elastolysis in a mouse model of Marfan syndrome. *Circ. Res.* 88:37–43
10. Chen S, Lechleider RJ. 2004. Transforming growth factor-β-induced differentiation of smooth muscle from a neural crest stem cell line. *Circ. Res.* 94:1195–202
11. Clementi M, Notari L, Borghi A, Tenconi R. 1996. Familial congenital bicuspid aortic valve: a disorder of uncertain inheritance. *Am. J. Med. Genet.* 62:336–38
12. Coady MA, Davies RR, Roberts M, Goldstein LJ, Rogalski MJ, et al. 1999. Familial patterns of thoracic aortic aneurysms. *Arch. Surg.* 134:361–67
13. Davis EC. 1993. Smooth muscle cell to elastic lamina connections in developing mouse aorta. Role in aortic medial organization. *Lab. Invest.* 68:89–99
14. Dillon PF, Aksoy MO, Driska SP, Murphy RA. 1981. Myosin phosphorylation and the cross-bridge cycle in arterial smooth muscle. *Science* 211:495–97
15. Edwards WD, Leaf DS, Edwards JE. 1978. Dissecting aortic aneurysm associated with congenital bicuspid aortic valve. *Circulation* 57:1022–25
16. Elsheikh M, Dunger DB, Conway GS, Wass JA. 2002. Turner's syndrome in adulthood. *Endocr. Rev.* 23:120–40
17. Fatigati V, Murphy RA. 1984. Actin and tropomyosin variants in smooth muscles. Dependence on tissue type. *J. Biol. Chem.* 259:14383–88
18. Finkbohner R, Johnston D, Crawford ES, Coselli J, Milewicz DM. 1995. Marfan syndrome. Long-term survival and complications after aortic aneurysm repair. *Circulation* 91:728–33
19. Fox JW, Lamperti ED, Eksioglu YZ, Hong SE, Feng Y, et al. 1998. Mutations in filamin 1 prevent migration of cerebral cortical neurons in human periventricular heterotopia. *Neuron* 21:1315–25
20. Francke U, Berg MA, Tynan K, Brenn T, Liu WG, et al. 1995. A Gly1127Ser mutation in an Egf-like domain of the fibrillin-1 gene is a risk factor for ascending aortic-aneurysm and dissection. *Am. J. Hum. Genet.* 56:1287–96

21. Glick BN, Roberts WC. 1994. Congenitally bicuspid aortic valve in multiple family members. *Am. J. Cardiol.* 73:400–4
22. Guo D, Hasham S, Kuang SQ, Vaughan CJ, Boerwinkle E, et al. 2001. Familial thoracic aortic aneurysms and dissections: genetic heterogeneity with a major locus mapping to 5q13–14. *Circulation* 103:2461–88
23. Guo DC, Pannu H, Papke CL, Yu RK, Avidan N, et al. 2007. Mutations in smooth muscle α-actin (*ACTA2*) lead to thoracic aortic aneurysms and dissections. *Nat. Genet.* 39:1488–93
24. Gupta V, Grande-Allen KJ. 2006. Effects of static and cyclic loading in regulating extracellular matrix synthesis by cardiovascular cells. *Cardiovasc. Res.* 72:375–83
25. Habashi JP, Judge DP, Holm TM, Cohn RD, Loeys BL, et al. 2006. Losartan, an AT1 antagonist, prevents aortic aneurysm in a mouse model of Marfan syndrome. *Science* 312:117–21
26. Hasham SN, Willing MC, Guo DC, Muilenburg A, He RM, et al. 2003. Mapping a locus for familial thoracic aortic aneurysms and dissections (*TAAD2*) to 3p24-25. *Circulation* 107:3184–90
27. Hatzaras I, Tranquilli M, Coady M, Barrett PM, Bible J, Elefteriades JA. 2007. Weight lifting and aortic dissection: more evidence for a connection. *Cardiology* 107:103–6
28. He R, Guo DC, Estrera AL, Safi HJ, Huynh TT, et al. 2006. Characterization of the inflammatory and apoptotic cells in the aortas of patients with ascending thoracic aortic aneurysms and dissections. *J. Thorac. Cardiovasc. Surg.* 131:671–78
29. Hellstrand P, Albinsson S. 2005. Stretch-dependent growth and differentiation in vascular smooth muscle: role of the actin cytoskeleton. *Can. J. Physiol. Pharmacol.* 83:869–75
30. Hirschi KK, Rohovsky SA, D'Amore PA. 1998. PDGF, TGF-β, and heterotypic cell-cell interactions mediate endothelial cell-induced recruitment of 10T1/2 cells and their differentiation to a smooth muscle fate. *J. Cell Biol.* 141:805–14
31. Hollister DW, Godfrey M, Sakai LY, Pyeritz RE. 1990. Immunohistologic abnormalities of the microfibrillar-fiber system in the Marfan syndrome. *N. Engl. J. Med.* 323:152–59
32. Huston EE, Grammer JC, Yount RG. 1988. Flexibility of the myosin heavy chain: direct evidence that the region containing SH1 and SH2 can move 10 Å under the influence of nucleotide binding. *Biochemistry* 27:8945–52
33. Ince H, Nienaber CA. 2007. Etiology, pathogenesis and management of thoracic aortic aneurysm. *Nat. Clin. Pract. Cardiovasc. Med.* 4:418–27
34. Kamath BM, Spinner NB, Emerick KM, Chudley AE, Booth C, et al. 2004. Vascular anomalies in Alagille syndrome: a significant cause of morbidity and mortality. *Circulation* 109:1354–58
35. Katzke S, Booms P, Tiecke F, Palz M, Pletschacher A, et al. 2002. TGGE screening of the entire *FBN1* coding sequence in 126 individuals with Marfan syndrome and related fibrillinopathies. *Hum. Mutat.* 20:197–208
36. Kimura A, Harada H, Park JE, Nishi H, Satoh M, et al. 1997. Mutations in the cardiac troponin I gene associated with hypertrophic cardiomyopathy. *Nat. Genet.* 16:379–82
37. Lacro RV, Dietz HC, Wruck LM, Bradley TJ, Colan SD, et al. 2007. Rationale and design of a randomized clinical trial of β-blocker therapy (atenolol) versus angiotensin II receptor blocker therapy (losartan) in individuals with Marfan syndrome. *Am. Heart J.* 154:624–31
38. Laurent S, Boutouyrie P, Lacolley P. 2005. Structural and genetic bases of arterial stiffness. *Hypertension* 45:1050–55
39. Laval SH, Bushby KM. 2004. Limb-girdle muscular dystrophies–from genetics to molecular pathology. *Neuropathol. Appl. Neurobiol.* 30:91–105
40. Law C, Bunyan D, Castle B, Day L, Simpson I, et al. 2006. Clinical features in a family with an R460H mutation in transforming growth factor βreceptor 2 gene. *J. Med. Genet.* 43:908–16
41. Lee CC, Chang WT, Fang CC, Tsai IL, Chen WJ. 2004. Sudden death caused by dissecting thoracic aortic aneurysm in a patient with autosomal dominant polycystic kidney disease. *Resuscitation* 63:93–96
42. Lee RT, Yamamoto C, Feng Y, Potter-Perigo S, Briggs WH, et al. 2001. Mechanical strain induces specific changes in the synthesis and organization of proteoglycans by vascular smooth muscle cells. *J. Biol. Chem.* 276:13847–51
43. LeMaire SA, Pannu H, Tran-Fadulu V, Carter SA, Coselli JS, Milewicz DM. 2007. Severe aortic and arterial aneurysms associated with a *TGFBR2* mutation. *Nat. Clin. Pract. Cardiovasc. Med.* 4:167–71

44. Li G, Borger MA, Williams WG, Weisel RD, Mickle DA, et al. 2002. Regional overexpression of insulin-like growth factor-I and transforming growth factor-β1 in the myocardium of patients with hypertrophic obstructive cardiomyopathy. *J. Thorac. Cardiovasc. Surg.* 123:89–95
45. Lie JT. 1982. Aortic dissection in Turner's syndrome. *Am. Heart J.* 103:1077–80
46. Lim DS, Lutucuta S, Bachireddy P, Youker K, Evans A, et al. 2001. Angiotensin II blockade reverses myocardial fibrosis in a transgenic mouse model of human hypertrophic cardiomyopathy. *Circulation* 103:789–91
47. Lin AE, Garver KL, Allanson J. 1987. Aortic-root dilatation in Noonan's syndrome. *N. Engl. J. Med.* 317:1668–69
48. Lin SH, Bichet DG, Sasaki S, Kuwahara M, Arthus MF, et al. 2002. Two novel aquaporin-2 mutations responsible for congenital nephrogenic diabetes insipidus in Chinese families. *J. Clin. Endocrinol. Metab.* 87:2694–700
49. Loeys BL, Chen J, Neptune ER, Judge DP, Podowski M, et al. 2005. A syndrome of altered cardiovascular, craniofacial, neurocognitive and skeletal development caused by mutations in *TGFBR1* or *TGFBR2*. *Nat. Genet.* 37:275–81
50. Loeys BL, Schwarze U, Holm T, Callewaert BL, Thomas GH, et al. 2006. Aneurysm syndromes caused by mutations in the TGF-βreceptor. *N. Engl. J. Med.* 355:788–98
51. Longhurst JC, Stebbins CL. 1992. The isometric athlete. *Cardiol. Clin.* 10:281–94
52. Loscalzo ML, Goh DL, Loeys B, Kent KC, Spevak PJ, Dietz HC. 2007. Familial thoracic aortic dilation and bicommissural aortic valve: a prospective analysis of natural history and inheritance. *Am. J. Med. Genet. A* 143:1960–67
53. Marian AJ. 2000. Pathogenesis of diverse clinical and pathological phenotypes in hypertrophic cardiomyopathy. *Lancet* 355:58–60
54. McKusick VA. 1972. Association of congenital bicuspid aortic valve and Erdheim's cystic medial necrosis. *Lancet* 1:1026–27
55. Milewicz DM, Chen H, Park ES, Petty EM, Zaghi H, et al. 1998. Reduced penetrance and variable expressivity of familial thoracic aortic aneurysms/dissections. *Am. J. Cardiol.* 82:474–79
56. Milewicz DM, Dietz HC, Miller DC. 2005. Treatment of aortic disease in patients with Marfan syndrome. *Circulation* 111:e150–e157
57. Milewicz DM, Grossfield J, Cao SN, Kielty C, Covitz W, Jewett T. 1995. A mutation in *FBN1* disrupts profibrillin processing and results in isolated skeletal features of the Marfan syndrome. *J. Clin. Invest.* 95:2373–78
58. Milewicz DM, Michael K, Fisher N, Coselli JS, Markello T, Biddinger A. 1996. Fibrillin-1 (FBN1) mutations in patients with thoracic aortic aneurysms. *Circulation* 94:2708–11
59. Milewicz DM, Pyeritz RE, Crawford ES, Byers PH. 1992. Marfan syndrome: defective synthesis, secretion, and extracellular matrix formation of fibrillin by cultured dermal fibroblasts. *J. Clin. Invest.* 89:79–86
60. Moiseeva EP. 2001. Adhesion receptors of vascular smooth muscle cells and their functions. *Cardiovasc. Res.* 52:372–86
61. Neptune ER, Frischmeyer PA, Arking DE, Myers L, Bunton TE, et al. 2003. Dysregulation of TGF-β activation contributes to pathogenesis in Marfan syndrome. *Nat. Genet.* 33:407–11
62. Ng CM, Cheng A, Myers LA, Martinez-Murillo F, Jie C, et al. 2004. TGF-β-dependent pathogenesis of mitral valve prolapse in a mouse model of Marfan syndrome. *J. Clin. Invest.* 114:1586–92
63. Nicod P, Bloor C, Godfrey M, Hollister D, Pyeritz RE, et al. 1989. Familial aortic dissecting aneurysm. *J. Am. Coll. Cardiol.* 13:811–19
64. O'Callaghan CJ, Williams B. 2000. Mechanical strain-induced extracellular matrix production by human vascular smooth muscle cells: role of TGF-β_1. *Hypertension* 36:319–24
65. Pannu H, Fadulu V, Chang J, Lafont A, Hasham SN, et al. 2005. Mutations in transforming growth factor-β receptor type II cause familial thoracic aortic aneurysms and dissections. *Circulation* 112:513–20
66. Pannu H, Tran-Fadulu V, Papke CL, Scherer S, Liu Y, et al. 2007. *MYH11* mutations result in a distinct vascular pathology driven by insulin-like growth factor 1 and angiotensin II. *Hum. Mol. Genet.* 16:3453–62
67. Pepin M, Schwarze U, Superti-Furga A, Byers PH. 2000. Clinical and genetic features of Ehlers-Danlos syndrome type IV, the vascular type. *N. Engl. J. Med.* 342:673–80

68. Pereira L, Lee SY, Gayraud B, Andrikopoulos K, Shapiro SD, et al. 1999. Pathogenetic sequence for aneurysm revealed in mice underexpressing fibrillin-1. *Proc. Natl. Acad. Sci. USA* 96:3819–23
69. Rayment I, Rypniewski WR, Schmidt-Base K, Smith R, Tomchick DR, et al. 1993. Three-dimensional structure of myosin subfragment-1: a molecular motor. *Science* 261:50–58
70. Raynal N, Hamaia SW, Siljander PR, Maddox B, Peachey AR, et al. 2006. Use of synthetic peptides to locate novel integrin α2β1-binding motifs in human collagen III. *J. Biol. Chem.* 281:3821–31
71. Roberts CS, Roberts WC. 1991. Aortic dissection with the entrance tear in the descending thoracic aorta. Analysis of 40 necropsy patients. *Ann. Surg.* 213:356–68
72. Rossi-Foulkes R, Roman MJ, Rosen SE, Kramer-Fox R, Ehlers KH, et al. 1999. Phenotypic features and impact of beta blocker or calcium antagonist therapy on aortic lumen size in the Marfan syndrome. *Am. J. Cardiol.* 83:1364–68
73. Segura AM, Luna RE, Horiba K, Stetler-Stevenson WG, McAllister HA, et al. 1998. Immunohistochemistry of matrix metalloproteinases and their inhibitors in thoracic aortic aneurysms and aortic valves of patients with Marfan's syndrome 2. *Circulation* 98:II331–II337
74. Shachter N, Perloff JK, Mulder DG. 1984. Aortic dissection in Noonan's syndrome (46 XY Turner). *Am. J. Cardiol.* 54:464–65
75. Sheen VL, Jansen A, Chen MH, Parrini E, Morgan T, et al. 2005. Filamin A mutations cause periventricular heterotopia with Ehlers-Danlos syndrome. *Neurology* 64:254–62
76. Sheen VL, Walsh CA. 2005. Periventricular heterotopia: new insights into Ehlers-Danlos syndrome. *Clin. Med. Res.* 3:229–33
77. Shekhonin BV, Domogatsky SP, Muzykantov VR, Idelson GL, Rukosuev VS. 1985. Distribution of type I, III, IV and V collagen in normal and atherosclerotic human arterial wall: immunomorphological characteristics. *Coll. Relat. Res.* 5:355–68
78. Shores J, Berger KR, Murphy EA, Pyeritz RE. 1994. Progression of aortic dilatation and the benefit of long-term β-adrenergic blockade in Marfan's syndrome. *N. Engl. J. Med.* 330:1335–41
79. Silverman DI, Burton KJ, Gray J, Bosner MS, Kouchoukos NT, et al. 1995. Life expectancy in the Marfan syndrome. *Am. J. Cardiol.* 75:157–60
80. Silverman DI, Gray J, Roman MJ, Bridges A, Burton K, et al. 1995. Family history of severe cardiovascular disease in Marfan syndrome is associated with increased aortic diameter and decreased survival. *J. Am. Coll. Cardiol.* 26:1062–67
81. Small JV, Gimona M. 1998. The cytoskeleton of the vertebrate smooth muscle cell. *Acta Physiol. Scand.* 164:341–48
82. Standley PR, Obards TJ, Martina CL. 1999. Cyclic stretch regulates autocrine IGF-I in vascular smooth muscle cells: implications in vascular hyperplasia. *Am. J. Physiol.* 276:E697–E705
83. Superti-Furga A, Gugler E, Gitzelmann R, Steinmann B. 1988. Ehlers-Danlos syndrome type IV: a multi-exon deletion in one of the two *COL3A1* alleles affecting structure, stability, and processing of type III procollagen. *J. Biol. Chem.* 263:6226–32
84. Tang PC, Coady MA, Lovoulos C, Dardik A, Aslan M, et al. 2005. Hyperplastic cellular remodeling of the media in ascending thoracic aortic aneurysms. *Circulation* 112:1098–105
85. Tang PC, Yakimov AO, Teesdale MA, Coady MA, Dardik A, et al. 2005. Transmural inflammation by interferon-gamma-producing T cells correlates with outward vascular remodeling and intimal expansion of ascending thoracic aortic aneurysms. *FASEB J.* 19:1528–30
86. Tardiff JC. 2005. Sarcomeric proteins and familial hypertrophic cardiomyopathy: linking mutations in structural proteins to complex cardiovascular phenotypes. *Heart Fail. Rev.* 10:237–48
87. Topouzis S, Majesky MW. 1996. Smooth muscle lineage diversity in the chick embryo. Two types of aortic smooth muscle cell differ in growth and receptor-mediated transcriptional responses to transforming growth factor-beta 3. *Dev. Biol.* 178:430–45
88. Towbin JA, Bowles NE. 2002. The failing heart. *Nature* 415:227–33
89. Towbin JA, Bowles NE. 2006. Dilated cardiomyopathy: a tale of cytoskeletal proteins and beyond. *J. Cardiovasc. Electrophysiol.* 17:919–26
90. Vale RD, Milligan RA. 2000. The way things move: looking under the hood of molecular motor proteins. *Science* 288:88–95

91. van Karnebeek CD, Naeff MS, Mulder BJ, Hennekam RC, Offringa M. 2001. Natural history of cardiovascular manifestations in Marfan syndrome. *Arch. Dis. Child* 84:129–37
92. Van Kien PK, Mathieu F, Zhu L, Lalande A, Betard C, et al. 2005. Mapping of familial thoracic aortic aneurysm/dissection with patent ductus arteriosus to 16p12.2-p13.13. *Circulation* 112:200–6
93. Vatta M, Li H, Towbin JA. 2000. Molecular biology of arrhythmic syndromes. *Curr. Opin. Cardiol.* 15:12–22
94. Vaughan CJ, Casey M, He J, Veugelers M, Henderson K, et al. 2001. Identification of a chromosome 11q23.2-q24 locus for familial aortic aneurysm disease, a genetically heterogeneous disorder. *Circulation* 103:2469–75
95. Xiong W, Knispel RA, Dietz HC, Ramirez F, Baxter BT. 2008. Doxycycline delays aneurysm rupture in a mouse model of Marfan syndrome. *J. Vasc. Surg.* 47:166–72
96. Zhu L, Vranckx R, Khau Van KP, Lalande A, Boisset N, et al. 2006. Mutations in myosin heavy chain 11 cause a syndrome associating thoracic aortic aneurysm/aortic dissection and patent ductus arteriosus. *Nat. Genet.* 38:343–49
97. Zoppi N, Gardella R, De PA, Barlati S, Colombi M. 2004. Human fibroblasts with mutations in *COL5A1* and *COL3A1* genes do not organize collagens and fibronectin in the extracellular matrix, down-regulate $\alpha 2\beta 1$ integrin, and recruit $\alpha V\beta 3$ instead of $\alpha 5\beta 1$ integrin. *J. Biol. Chem.* 279:18157–68

Cohesin and Human Disease

Jinglan Liu[1] and Ian D. Krantz[1,2]

[1]Division of Human Genetics, The Children's Hospital of Philadelphia and [2]The University of Pennsylvania School of Medicine, Philadelphia, Pennsylvania 19104;
email: ian2@mail.med.upenn.edu, liujin@email.chop.edu

Annu. Rev. Genomics Hum. Genet. 2008. 9:303–20

The *Annual Review of Genomics and Human Genetics* is online at genom.annualreviews.org

This article's doi:
10.1146/annurev.genom.9.081307.164211

1527-8204/08/0922-0303$20.00

Key Words

Cornelia de Lange syndrome, Roberts-SC phocomelia, NIPBL, SMC1A, SMC3, ESCO2

Abstract

Cornelia de Lange syndrome (CdLS) is a dominant multisystem disorder caused by a disruption of cohesin function. The cohesin ring complex is composed of four protein subunits and more than 25 additional proteins involved in its regulation. The discovery that this complex also has a fundamental role in long-range regulation of transcription in *Drosophila* has shed light on the mechanism likely responsible for its role in development. In addition to the three cohesin proteins involved in CdLS, a second multisystem, recessively inherited, developmental disorder, Roberts-SC phocomelia, is caused by mutations in another regulator of the cohesin complex, ESCO2. Here we review the phenotypes of these disorders, collectively termed cohesinopathies, as well as the mechanism by which cohesin disruption likely causes these diseases.

CORNELIA DE LANGE SYNDROME

CdLS: Cornelia de Lange syndrome

Cornelia de Lange syndrome (CdLS) (OMIM #122470 and #300590) was first described in 1849 by Vrolik, who reported a severe example of oligodactyly (72). In 1916, Brachmann (10) provided a detailed account of an individual with symmetrical monodactyly, antecubital webbing, dwarfism, cervical ribs, and hirsutism (64). In the 1930s Cornelia de Lange, a Dutch pediatrician, reported two unrelated girls with strikingly similar features and named the condition after the city in which she worked: degeneration typus amstelodamensis (17, 18). Although some literature refers to the disorder as Brachmann-de Lange syndrome (BDLS), the disorder is widely referred to as Cornelia de Lange syndrome in honor of Dr. de Lange's contributions to the formal characterization of this disorder.

CdLS is a dominantly inherited, genetically heterogeneous, multisystem developmental disorder. The phenotype consists of characteristic facial features (including synophrys, long eyelashes, depressed nasal root with an uptilted tip of the nose and anteverted nares, long philtrum, thin upper lip, small widely spaced teeth, small brachycephalic head, and low-set, posteriorly angulated ears), hirsutism, various ophthalmologic abnormalities, abnormalities of the upper extremities that range from small hands with single palmar creases and subtle changes in the phalanges and metacarpal bones to severe forms of oligodactyly and truncation of the forearms that primarily involves the ulnar structures, gastroesophageal dysfunction, growth retardation, and neurodevelopmental delay (42, 48) (**Figure 1**). Other frequently seen findings include ptosis, myopia, intestinal malrotation, cryptorchidism, hypospadias,

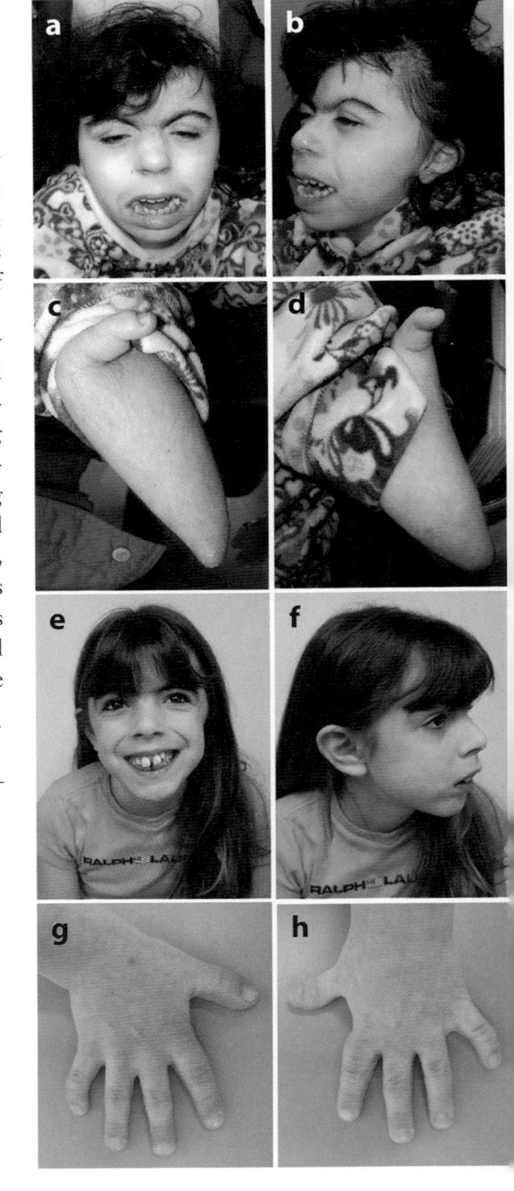

Figure 1

Typical phenotypic characteristics of Cornelia de Lange syndrome (CdLS). (*a–d*) Manifestations in a severely affected 17-year-old girl. (*a–b*) Note characteristic facial features such as arched eyebrows, synophrys, long eyelashes, ptosis, short nose, long philtrum, thin upper lip, and posteriorly angulated ears. (*c–d*) Note severe oligodactyly with single radial digit, hypoplasia of ulnar structures, and pterygia of antecubital region bilaterally. (*e–h*) The characteristic mild phenotype in an 8-year-old girl. (*e–f*) Note the characteristic facial features described above (without the ptosis) but with much milder expression. (*g–h*) Note the small hands with fifth finger clinodactyly and proximally placed thumbs bilaterally.

pyloric stenosis, congenital diaphragmatic hernias, cardiac septal defects, seizures, and hearing loss. The mental retardation seen in CdLS is severe; IQs range from less than 30 to 102 with an average of 53. Many patients also demonstrate autistic behavior, including self-destructive tendencies, and may avoid or reject social interactions and physical contact (48). The prevalence of this syndrome has been estimated to be approximately 1 in 10,000 (73), although this may be an underestimate because gene testing has allowed for an even broader expansion of our clinical understanding of this disorder; the mildest end of the spectrum approaches apparent isolated mild mental retardation (19).

The facial features seen in individuals with classical CdLS are striking and easily recognizable, and may be one of the most useful diagnostic signs. However, marked variability exists and a milder phenotype has been consistently described (1, 42, 83, 89, 101) (**Figure 1**). In fact, even the first descriptions of CdLS by Brachmann in 1916 (10) and de Lange in 1933 (17, 18) were discrepant in the lack of major limb abnormalities in de Lange's cases. With increasing recognition of a milder CdLS phenotype, both isolated and familial cases have been diagnosed and reported (14, 32, 82). The mild phenotype is distinguished by less significant psychomotor and growth retardation, a lower incidence of major malformations, and milder limb anomalies as compared with the more severe phenotype (1). The milder phenotype has been estimated to account for approximately 20–30% of the CdLS population (1). Again, this is most likely an underrepresentation because many of these children are unlikely to be diagnosed with CdLS. Standard growth charts for height, weight, and head circumference as well as specific psychomotor developmental milestone charts have been developed for CdLS (51, 52).

GENE IDENTIFICATION

Most individuals with CdLS have normal chromosomes; however, a number of chromosomal rearrangements in individuals with CdLS have been reported over the years (21, 54). In some of the earlier described cases the diagnosis is now considered insecure and in several cases the same type of abnormality has been reported in unaffected individuals. However, in some CdLS individuals the reported chromosomal abnormality may be specific. Researchers have reported three affected sporadic children with CdLS with apparently balanced de novo translocations: t(14;21)(q32;q11), t(X;8)(p11.2;q24.3), and t(3;17)(q26.3;q23.1) (26, 43, 110). Recently a 0.6-Mb de novo 9p duplication was found in a Swedish boy with a CdLS-like phenotype after array comparative genomic hybridization analysis (84). The 3q26.3 breakpoint had been considered to be of particular interest because of phenotypic overlap between CdLS individuals and individuals with duplication 3q syndrome. A critical region has been defined at 3q26.3-q27 (3, 44, 78). The dup 3q syndrome has long been considered to be a phenocopy of CdLS because of the clinical overlap (39, 63, 65, 87, 109), but the two entities can be differentiated easily on clinical exam.

Several genes that map to these previously reported candidate regions were screened as potential CdLS disease genes, including *CHRD* and *SOX2* (3q27) (93), *SHOT* (3q25-q26) (7, 74), *GSC* (14q32) (93), *NAALADL2* (3q26.3) [found to be disrupted by the t(3;17) (98)], and *NLGN1*, a gene involved in synaptogenesis in the central nervous system, the gene dosage of which has been implicated in the mental retardation associated with the dup(3q) syndrome (65). Unfortunately, none of these genes were found to be mutated among other CdLS patients and the *CDL1* locus at 3q26.3 did not segregate in at least half of familial cases studied, questioning the veracity of this region as a CdLS locus (56).

A genome-wide linkage exclusion analysis performed on 12 families was used to narrow candidate gene loci to five genomic regions. One region at 5p13.1 corresponded to an additional patient with a de novo balanced t(5;13)(p13.1;q11.2). Candidate gene screening

NIPBL: nipped-B-like

SMC1A: structural maintenance of chromosome 1A

SMC3: structural maintenance of chromosome 3

was used to identify a novel gene named *Nipped-B-Like* (*NIPBL*) after its *Drosophila* homolog, *nipped-b*; heterozygous mutations in *NIPBL* were found to cause CdLS (55). NIPBL is a regulator of the cohesin complex. This finding was the first implication of the cohesin complex and its regulators with a developmental disorder. Shortly afterwards mutations in other cohesin components, structural maintenance of chromosomes 1A and 3 (*SMC1A* and *SMC3*), were also found to cause CdLS (19, 69).

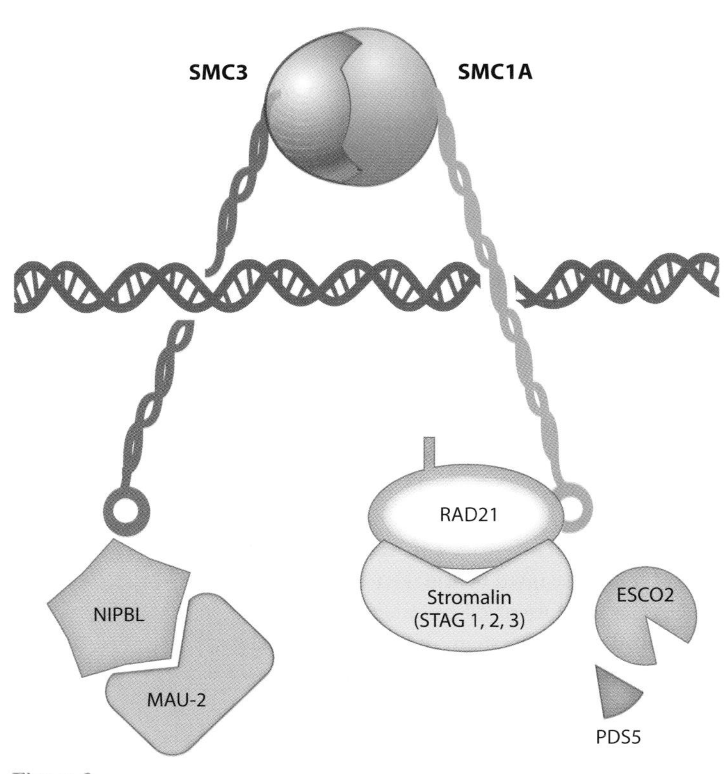

Figure 2

Schematic representation of a single open cohesin molecule on a DNA strand. Structural maintenance of chromosomes 1A and 3 (SMC1A and SMC3), mutations in which have been identified in Cornelia de Lange syndrome (CdLS), are attached at their hinge domains. Coiled-coil arms connect the hinge domain to the head domain. The head domains contain ATP-binding cassettes crucial in the dynamic opening and closing of the ring structure, which is mediated by the RAD21 and stromalin proteins. Also shown are the cohesin regulatory proteins Nipped-B-Like (NIPBL) (and interacting protein MAU-2) and establishment of cohesion 1 homolog 2 (ESCO2) (and putative interacting factor PDS5), involved in CdLS and Roberts/SC phocomelia syndrome (RBS-SC), respectively.

THE COHESIN COMPLEX AND ITS REGULATORS

During mitosis and meiosis, sister chromatid cohesion is established when the DNA replication products are held together by a multisubunit complex called cohesin. In yeast, the cohesin complex consists of an Smc1 and Smc3 heterodimer and at least two non-SMC subunits [Scc1/Mcd1/RAD21 and Scc3/Stromalins (SAs)] (**Figure 2**). The Smc1 and Smc3 subunits are rod-shaped molecules with globular ATPase domains at one end and dimerization domains at the other end. The ATPase head domains of Smc1 and Smc3 are formed by their N and C termini, and the intervening sequence forms antiparallel coiled-coil structures that fold back on themselves in between. Smc1 and Smc3 interact between their hinge regions and form a V-shaped heterodimer. The two ATPase head domains further interact with the N and C termini of Scc1/Rad21, respectively, creating a triangular structure that could trap sister chromatids (2, 33) (**Figure 2**). During interphase, cohesin binds along chromosome arms roughly every 10 or 20 kb in yeast or higher organisms (25, 31). Centromeres recruit more cohesin than other chromosome sites do; protein complexes constructed by cohesin and heterochromatin proteins appear to regulate several cellular functions (4, 76). Most cohesin bound to chromosome arms is removed at mitotic prophase (97, 106). Centromere-associated cohesin is regulated differently and involves the shugosin protein (107). Cohesin dissociates completely from chromosomes at anaphase owing to separase cleavage of Scc1/Rad21 and sister chromatid disjunction is triggered afterwards (70) (**Figure 3**). Cohesin starts to be loaded onto chromosomes in telophase, but in *Saccharomyces cerevisiae* the loading starts at the G_1/S phase transition (66). The mechanisms by which sister chromatid cohesion is established remain unclear, although it appears that the cohesin loading (or adherin) complex Scc2/Scc4 and the acetyl-transferase Eco1/Ctf7 (chromosome transmission fidelity) are essential for this

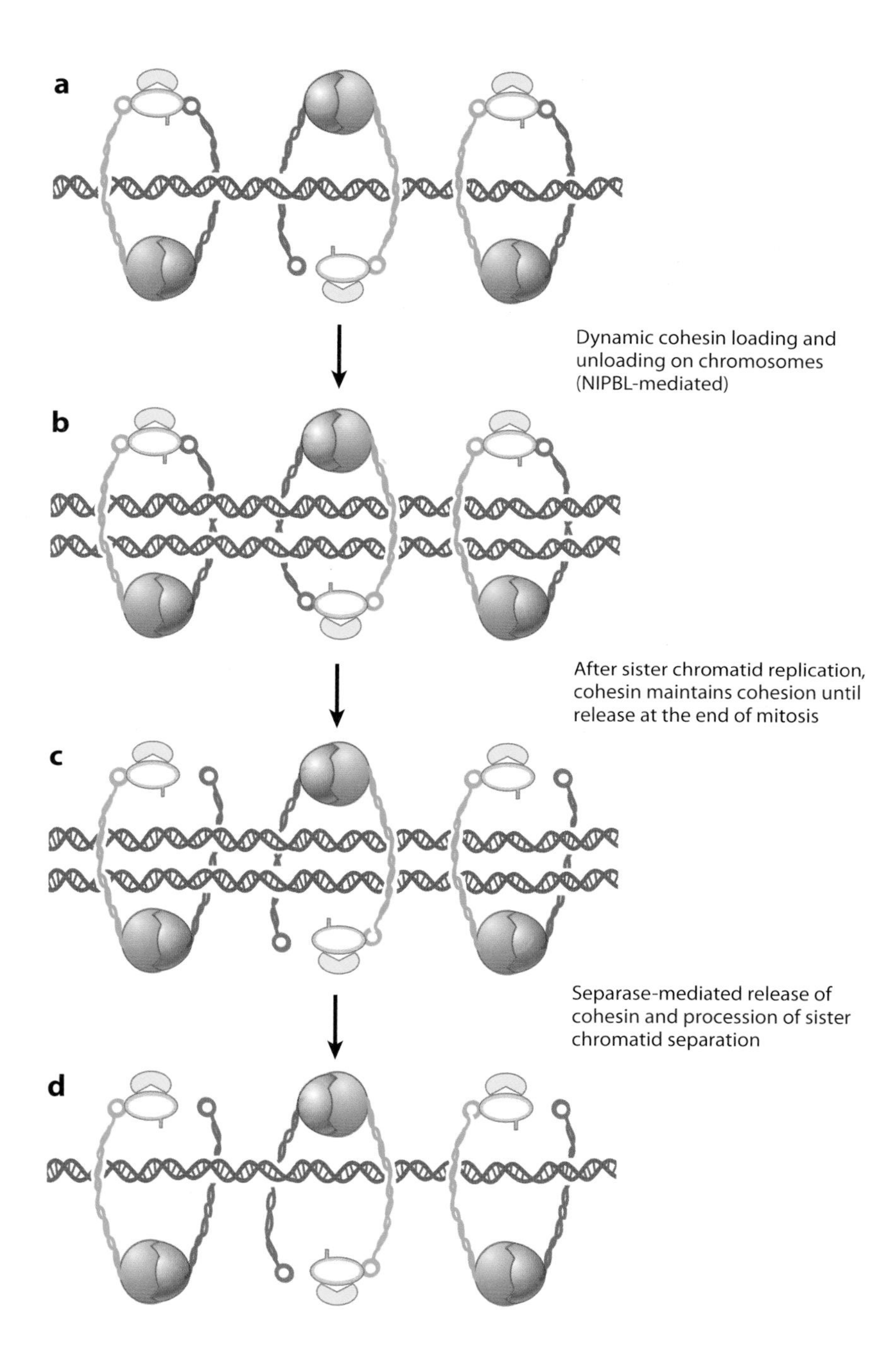
a
Dynamic cohesin loading and unloading on chromosomes (NIPBL-mediated)
b
After sister chromatid replication, cohesin maintains cohesion until release at the end of mitosis
c
Separase-mediated release of cohesin and procession of sister chromatid separation
d

Figure 3

The sister chromatid "embracing" model of cohesin function, which depicts the dynamic way in which cohesin is opened and closed and loaded on and off of chromosomes at precise times during the cell cycle to correctly release sister chromatids for appropriate segregation to daughter cells after mitosis. NIPBL, Nipped-B-Like.

process (see 71 and 38 for reviews). Mutations in the human homolog of Eco1/Ctf7, establishment of cohesion 1 homolog 2 (ESCO2) result in Roberts syndrome (see below). During or after DNA replication, replication factor C and DNA helicase are also required to establish cohesion (see 91 for a review).

Four topological models of sister chromatid cohesion have been proposed, differing in whether one ring (embracing model) or two

ESCO2: establishment of cohesion 1 homolog 2

rings act in concert to secure the chromatids, and whether cohesins bind directly to chromatin or not (61). In most eukaryotic cells, the cohesin cleaved by separase at the onset of anaphase composes only a small portion of the cohesin population (106), whereas the majority of cohesin dissociates from chromosome arms owing to separase-independent mechanisms during prophase (28, 57). Therefore, a large amount of intact and free cohesin is available to reassociate with chromosomes near the beginning of the G_1 phase in the following cell cycle. Hence, cohesin steadily associates with chromosomes for most of the cell division cycle and has expected functions linked to replication. Many recent lines of evidence implicate cohesin as a key regulator of gene expression. Because sister chromatid cohesion seems to be minimally affected or not affected at all by dosage changes of cohesin and its associated proteins (as evidenced by the rare presence of sister chromatid cohesion defects), the clinical phenotype of CdLS is likely the result of cohesin-mediated gene dysregulation during embryonic development (reviewed in 24). Defects in the establishment or maintenance of sister chromatid cohesion may contribute to the phenotypic manifestations in CdLS because a mild effect on sister chromatids, named precocious sister chromatid separation (PSCS), has been identified in CdLS probands with heterozygous *NIPBL* mutations (49).

HETEROZYGOUS MUTATIONS OF THE COHESIN LOADING PROTEIN NIPBL AND COHESIN SUBUNITS SMC1A AND SMC3 LEAD TO CORNELIA DE LANGE SYNDROME

Human *NIPBL* is located on chromosome 5q13 and the full-length transcript is approximately 10 kb in length. Various smaller alternative transcripts exist, and two protein isoforms are conserved in vertebrates (55, 99). The NIPBL protein is also referred to as delangin. *SMC1A* is located on Xp11.22-p11.21 and escapes X inactivation in humans (12). *SMC3* is located on 10q25. *NIPBL* transcripts are ubiquitously expressed in humans but obvious tissue diversities exist, with the highest expression found in heart and skeletal tissue (55, 99). In situ hybridization studies on whole-mount mouse embryos showed *NIPBL* expression in developing limbs, craniofacial bones and muscles, the craniofacial mesenchyme surrounding the cochlear canal, the respiratory and gastrointestinal systems, heart, renal tubules, and genitourinary system, concordant with the clinically involved systems in CdLS (55). Mouse *Nipbl* is robustly expressed in limb buds, branchial arch, and craniofacial mesenchyme, although other tissue and organs also show expression at various levels (99). *Xenopus tropicalis* delangin is found throughout the ectoderm and in the mesoderm during gastrulation (95). UniGene cluster and expressed sequence tag (EST) studies showed that both SMC1A and SMC3 are universally expressed and there is some increased expression in certain tissues (**http://source.stanford.edu/cgi-bin/source/sourceSearch**).

NIPBL MUTATION SCREENING AND GENOTYPE-PHENOTYPE CORRELATIONS IN CORNELIA DE LANGE SYNDROME

Mutational analysis by sequencing of *NIPBL* exons and flanking intronic regions has been performed among various ethnic groups (5, 29, 84, 90, 111), as has *SMC1A* mutation analysis (8, 19). A wide variety of pathogenic mutations have been identified. Overall only approximately 50% of CdLS probands are found to carry an *NIPBL* mutation, most of which are point mutations or small insertions and deletions in coding regions or splice junctions. These mutations are assumed to produce either malfunctioning full-length or truncated NIPBL proteins, consistent with haploinsufficiency (29). In rare cases, researchers have found large genomic rearrangements as well as alterations in the upstream noncoding region of the gene. For example, one affected girl and her mildly affected father have a heterozygous deletion-insertion mutation in the *NIPBL* 5′-untranslated region (UTR) (9), and in another patient multiplex ligation-dependent probe amplification (MLPA) revealed a 5.2-kb deletion that encompasses exons 41–42 of *NIPBL* (6). Among *NIPBL* mutation–negative CdLS probands approximately 8% (4% of the total CdLS population) carry an *SMC1A* mutation (19). One mildly affected adult male with CdLS was found to carry a de novo *SMC3* mutation. At least 40% of clinically diagnosed patients with CdLS do not carry a mutation in any of these genes, suggesting the existence of other CdLS genes or that potential alternative mechanisms of alteration in these known genes may be involved.

Genotype-phenotype correlations are beginning to emerge but mostly involve gross observations that individuals with missense mutations or no identifiable mutations tend to present a milder phenotype than those with truncating mutations; individuals with *SMC1A* mutations also tend to have a milder phenotype and rarely, if ever, manifest structural organ or limb defects (19, 29). At the cellular level, B-lymphoblastoid cells from CdLS patients also manifest some phenotypic features: (*a*) 41% of metaphase spreads from *NIPBL* mutation–positive probands show PSCS, whereas this phenotype appears in only 9% of control samples (49); and (*b*) both fibroblast and B-lymphoblastoid cells derived from CdLS patients, either with or without detectable *NIPBL* mutations, are more sensitive to DNA crosslinker mitomycin C (MMC) than wild type and demonstrate a decreased ability to repair double-strand DNA breaks at G_2 phase after X-ray exposure (105). Rare CdLS-associated malignant cases such as Wilm's tumor have been reported, but in general cancer is not a predominant manifestation (13).

COHESIN BINDING IMPACTS GENE EXPRESSION

In both yeast and *Xenopus* the loading of cohesin onto chromatin in G_1 phase is functionally separable from the establishment of sister chromatid cohesion at S phase (62, 100). In *S. cerevisiae*, cohesin loading requires a protein complex formed by Scc2 and Scc4 (15) and cohesion establishment and maintenance require Pds5 (36, 75). Scc2, Scc4, and Pds5 all contain HEAT (Huntington, elongation factor 3, PR65/A, TOR) domains.

It is becoming increasingly clear that different factors manage the binding and dissociation of cohesin for different chromosomal regions [e.g., arms, centromeres, telomeres, ribosomal DNA (rDNA)], raising the possibility that different chromosomal regions use different pathways for cohesin-mediated sister chromatid cohesion. In addition, not all cohesion is due to cohesin. Other mechanisms independent of cohesin, such as phosphatase Cdc14- and condensin-mediated cohesion, have also been proposed in budding yeast (16, 96). The presence of cohesin in chromatin may serve functions other than cohesion, e.g., the establishment of silencing or regulation of transcription by blocking the communication between enhancers and promoters (24).

The cohesin apparatus is composed of a ring-like structure that contains four subunits: Smc1, Smc3, Scc1/Rad21, and Scc3/SA (**Figure 2**). During the mitotic cell cycle, cohesin is loaded onto chromosomes at telophase and removed from the arms at the subsequent prophase. Several mechanisms have been proposed for the formation of the ring structure and the corresponding chromatid encircling models (38, 45, 46). Nipped-B, the homolog of cohesin loading protein Scc2, was identified in a genetic screen in *Drosophila* that measured the activation of *cut* and *Ultrabithorax* gene expression. Nipped-B alleviates the *gypsy* insulator function by assisting long distance promoter-enhancer interactions (81). *Cut* regulates *Drosophila* wing and limb development. *Nipped-B* null mutations are lethal to the fly, whereas heterozygous mutations in *Nipped-B* result in lower expression of *cut* protein than wild type and notch wing phenotypes (81). In *Drosophila*, heterozygous *Nipped-B* mutants do not show cohesion defects, whereas homozygous *Nipped-B* mutants only show the defects right before death at the second instar stage, indicating sister chromatid cohesion is independent from cohesin-regulated gene expression. Cohesin binds to chromosomes throughout interphase, when gene expression also takes place; cohesin binds to the *cut* regulatory sequences in cultured cells and to the *cut* locus in *Drosophila* salivary gland chromosomes (25). Reducing cohesin levels boosts *cut* expression, whereas reducing Nipped-B levels diminishes *cut* expression in the emergent wing margin (25, 80). Cohesin binding may interfere with enhancer-promoter communication at the *cut* gene locus to inhibit gene expression. Nipped-B is able to reverse these suppression effects by dynamically removing the bound cohesin (23).

The results of genome-wide mapping of the cohesin and cohesin loading protein binding sites in yeast and *Drosophila* are conflicting. Cohesin binds to intergenic regions between convergent transcription units almost exclusively and does not reside on the same sequences as Scc2 in *S. cerevisiae*, suggesting that Scc2 loads cohesin first, and then other factors, such as RNA polymerase, push cohesin to its binding places (31, 59). In *Drosophila*, on the contrary, salivary polytene chromosome immunostaining and the *cut* locus chromatin immunoprecipitation (ChIP) assay have demonstrated the colocalization of Nipped-B and cohesin. Misulovin and coworkers (67) reported the mapping of Nipped-B and cohesin binding sites by ChIP/microarray (ChIP-on-chip) assays in the entire nonrepetitive *Drosophila* genome and found that not only do Nipped-B and cohesin colocalize, but they also both bind preferentially to active transcription units. The reason for this discrepancy between yeast and *Drosophila* is unknown, but colocalization of Nipped-B with cohesin may be critical for Nipped-B to relieve cohesin from blocking the long-range communication between promoters and enhancers, thus permitting gene activation. Because gene regulation through distant DNA elements is not a major regulatory mechanism in yeast, higher organisms may possess additional mechanisms for cohesin to bind in the genome and to regulate gene expression (67).

Scc4 (mau-2 in other species) interacts with Scc2 and is also essential for sister chromatid cohesion. In *Drosophila*, Nipped-B interacts with Mau-2 (30, 88). In *Caenorhabditis elegans* cells with RNAi knockdown of *mau-2* have abnormal axonal migration without sister chromatid cohesion defects, whereas in human cultured cells knockdown of human *MAU-2* by RNAi does cause cohesion defects (88). *Mau-2* mutations in *C. elegans* may have a similar effect as heterozygous *Nipped-B* mutations in *Drosophila*, which may reduce gene expression and activity without obvious effect on cohesion (80, 81). Compared with Nipped-B, Mau-2 expression changes seem less critical because partial knockdown of *Mau-2* by RNAi in *Drosophila* has no effects on either *cut* gene expression or sister chromatid cohesion (80, 88). These observations suggest that the sister chromatid cohesion apparatus can influence both gene expression and cohesion during development, but the latter may require more dramatic protein changes. Thus, gene expression appears to be more sensitive to subtle dosage alterations of the cohesin apparatus. Another cohesion factor, the *Pds5B* gene, is conserved from fungi to man, dynamically interacts with cohesin, and is involved in establishment and/or maintenance of sister chromatid cohesion (25). A single *pds5* gene exists in *Drosophila*, and heterozygous *pds5* mutations alter *cut* gene expression without effects on cohesion. Homozygous mutants show both gene expression and cohesion defects (25). Mammals have two Pds5 proteins. Homozygous *Pds5B* knockout mice show CdLS-like developmental abnormalities without cohesion defects, suggesting that changes in gene expression likely also underlie Pds5B function in mouse development (112).

In *Drosophila*, Nipped-B might facilitate the interaction between promoter and remote enhancers for target genes such as *cut* and *Ultrabithorax*, possibly by regulating chromosome structure (81). Nipped-B may operate at various regulating levels and directly control the expression of target genes. In humans, CdLS is caused by heterozygous loss-of-function mutations in the *Nipped-B-Like* (*NIPBL*) ortholog of *Nipped-B* and, in fewer cases, by mutations that all presumably maintain the open reading frame in the *SMC1A* or *SMC3* cohesin subunit genes (19, 55, 69, 99). The maintenance of the open reading frame and production of a protein product with residual, albeit decreased, function would be critical for *SMC1A* mutant viability. Because this gene is on the X chromosome and escapes inactivation in humans, loss-of-function mutations would likely be incompatible with life in males because they would have no functional SMC1A protein and therefore no functional cohesin. In females, loss-of-function mutations would likely result in no phenotype as long as the alternate allele was normal. Given the constellation of developmental abnormalities present in individuals with CdLS, and that the majority show no significant evidence of a cohesion defect (49, 105), the alterations of cohesin regulation and structure seen in these individuals most likely results in gene expression dysregulation similar to what has been observed in *Drosophila*.

IMPLICATION OF COHESIN APPARATUS–MEDIATED GENE REGULATION THROUGH LONG-RANGE REGULATORY ELEMENT-PROMOTER INTERACTIONS

The Nipped-B protein, an Scc2 ortholog, was identified in *Drosophila* through a genetic screen for proteins that facilitated expression of the *cut* homeobox gene in the developing wing margin. Expression of *cut* is driven by a distant transcriptional enhancer located more than 80 kb upstream of the transcription start site. Nipped-B is an essential protein; homozygous *Nipped-B* mutants die as second instar larvae (upon depletion of maternal Nipped-B protein). Heterozygous mutants reduce the ability of the enhancer to overcome the *gypsy* insulator inserted between the distal enhancer and the *cut* gene promoter and thereby enhance the phenotype caused by the *gypsy* insertion (34, 81). Nipped-B is highly dosage sensitive: Heterozygous *Nipped-B* mutations reduce *Nipped-B* messenger RNA (mRNA) levels by only 25% and a 50% reduction induced by RNAi is lethal to the fly (80).

Therefore, the critical means by which NIPBL and cohesin affect transcriptional control appears to be through long-range enhancer-promoter interactions (**Figure 4**). For most genes the region immediately upstream of the minimal promoter contains the

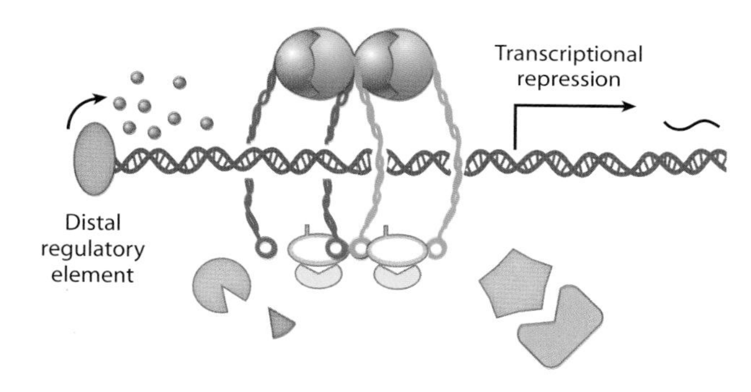

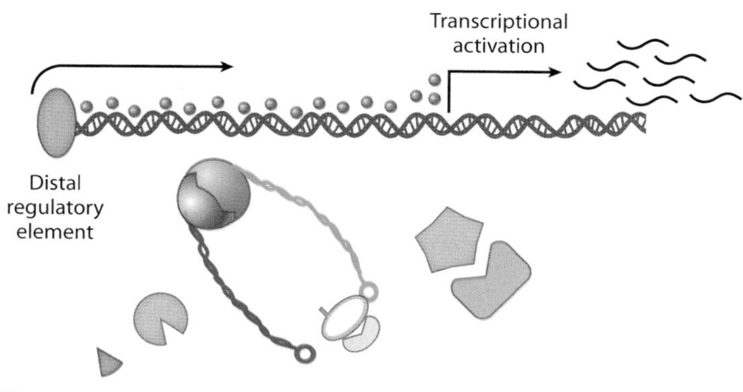

Figure 4

Putative model by which cohesin can hinder and relieve transcriptional control (in this case transcriptional repression) of genes regulated by distal elements. (*a*) Cohesin loaded on the chromosome inhibits the propagation of a transcriptional control signal from a distal enhancer to increase transcription via the target promoter. (*b*) With release of cohesin from the chromosome, the transcription activation signal from the distal enhancer is free to propagate to the promoter and increase transcription of the target gene.

requisite transcription factor binding sites to regulate expression of the gene. However, for many genes multiple *cis*-acting distal elements [most commonly enhancers, but also silencers, insulators, and locus control regions (LCRs)] are required for correct spatiotemporal expression (50). These elements may be located upstream, downstream, or within introns and can reside greater than 1 Mb from the target gene (104). Disruption of these long-range regulatory interactions can result in human disease phenotypes either through global or partial tissue-specific loss or gain of expression (reviewed in 50). The majority of the genes responsible for these disorders when disrupted are transcription factors that in turn regulate the expression of tissue-specific targets critical for normal morphogenesis of the organ systems involved. An additional layer of complexity in understanding the role of long-range enhancers in human disease is that the phenotype that results from disruption of a regulatory element can be quite different from that caused by a mutation in the coding region of the gene. This phenomenon is likely due to disruption in a subset of tissues or during a specific developmental window that results from a regulatory mutation versus an effect on all tissues and developmental stages where the gene product is affected by a coding mutation. An example of this is the varying phenotypes caused by mutations in the coding region versus the regulatory components of the *sonic hedgehog* (*SHH*) gene. Mutations in the coding region of *SHH* cause holoprosencephaly (HPE3), a defect of the midline structures of the face and brain (79), whereas disruption of a regulatory element 1 Mb upstream of the *SHH* promoter results in the limb-specific phenotype of preaxial polydactyly (60). This regulatory element mutation results in loss of the normal restriction of SHH expression to the posterior margin of the anterior limb bud, thereby allowing development of additional digits on the opposite axis.

This finding demonstrates that dysregulation of distal regulatory elements can result in not only loss of function of the target gene in a spatiotemporal manner but also gains of function when the normal suppression of expression is lost. In the study that first identified *Nipped-B* in *Drosophila*, researchers noted that the alteration of this gene and the resulting disruption of the long-range enhancer-promoter interaction affected the regulation of the homeobox gene *Ultrabithorax* (81). Homeobox gene clusters in humans, such as the *HOX* genes, that play a key role in body patterning are also regulated by distal enhancers that are capable of controlling their sequential activation. The identification of a global control region approximately 240 kb upstream of the *HOXD* cluster that regulates downstream expression of at least six genes in a tissue- and temporal-specific manner makes this an attractive candidate for some of the structural defects seen in CdLS (20, 37, 94).

The results of genome-wide ChIP-on-chip assays designed to examine the binding sites of Nipped-B, RNA polymerase II (Pol II), and cohesin subunit SMC1A in three different *Drosophila* cell lines showed that Nipped-B colocalizes with cohesin, which supports the idea that Nipped-B dynamically regulates cohesin binding (67). The preferential association of cohesin with transcribed regions suggests additional mechanisms by which cohesin binding might affect transcription, and vice versa. Results from the same study have also suggested that cohesin could interfere with both transcriptional activation and elongation because cohesin binds to the active *Abd-B* gene in Sg4 cells and some cohesin and Pol II peaks overlap within both the *Abd-B* transcriptional unit and the regulatory region. Genes with distant regulatory elements, such as *cut* and *Ubx*, may be more sensitive to Nipped-B dosage because of the combined effects on activation and elongation.

Other studies suggest that cohesin might have positive effects on gene expression as well. Inhibiting Rad21 expression results in reduced *runx* gene expression during early zebrafish development, although it is unknown if the effect is direct (40). Possible ways in which cohesin could directly facilitate gene expression include (*a*) preserving the targeted chromatin in an open state that is more accessible to transcription or (*b*) upholding chromatin boundaries, separating active and inactive chromatin regions, and preventing the distribution of silencing influence across domains. To support the second idea, in *S. cerevisiae* mutations in Smc1 cause the loss of the boundaries surrounding the *HMR* silent-mating-type loci and the spread of SIR (silent information regulator) complexes into neighboring regions (22); in *Drosophila* a cohesin/Nipped-B peak was found at the Fab-7 boundary element flanking the active *Abd-B* domain (67).

COHESIN AND CHROMOSOME REMODELING

Cohesin forms stable associations with chromatin remodeling complexes in vivo (35). A human ISWI (SNF2h)-containing complex was copurified with cohesin and NuRD. The hRAD21 subunit of the cohesin complex directly interacts with the ATPase subunit of SNF2h. Loading hRAD21 onto chromatin might involve ATPase activity of SNF2h (35). Alu sequences specifically bind to hRAD21, SNF2h, and Mi2, demonstrating that Alu repeats may also act as cohesin binding sites. Modifications of histone tails may be associated with the SNF2h/cohesin complex as well. Cohesin binds to AT-rich noncoding regions, which are the bases of chromosome loops attached by chromatin fibers and flanked by genes transcribed convergently in yeast (27). The yeast Sir2 protein recruits cohesin to multiple tandem ribosomal DNA arrays and suppresses chromatin recombination (53). Additional support for the role of NIPBL and cohesin in chromatin remodeling and epigenetic regulation came from the demonstration that NIPBL binds directly to the chromoshadow domain (CSD) of heterochromatin protein 1α (HP1α) with high affinity. The HP1α protein regulates epigenetic gene silencing by promoting and maintaining chromatin condensation. The HP1α chromodomain also binds to the methylated histone H3 (58). (See **Supplemental Material**. Follow the Supplemental Material link from the Annual Reviews home page at **http://www.annualreviews.org**.)

OTHER COHESINOPATHIES

Roberts Syndrome and SC Phocomelia

PCNA: proliferating cell nuclear antigen

The identification of a developmental disorder caused by mutation in the cohesin regulator ESCO2, the Roberts-SC phocomelia syndrome (RBS) (OMIM #268300; SC OMIM #269000), further implicates this pathway as a critical regulator of normal human development (103). *ESCO2* is homologous to yeast *Eco1/Ctf7*, which encodes an acetyltransferase that interacts with proliferating cell nuclear antigen (PCNA) and is essential for establishment of sister chromatid cohesion at S phase (68).

Despite the fact that the clinical presentations have some overlap and the molecular mechanisms are similar, these two congenital diseases are quite distinct. CdLS is a dominant disorder: 50% of probands carry heterozygous mutations in *NIPBL*, 5% carry heterozygous mutations in *SMC1A*, and one proband carries an *SMC3* mutation (19). RBS-SC phocomelia is an autosomal recessive disorder: All probands have either homozygous or compound heterozygous mutations of *ESCO2*. RBS-SC phocomelia is characterized by symmetric hypomelia, which varies from tetraphocomelia to a lesser degree of limb deficiency that is more severe in the upper limbs. Only ~25% of CdLS probands have limb reduction defects and these are usually asymmetric and involve the ulnar structures of upper extremities. RBS-SC phocomelia probands also have pre- and postnatal growth delay, as well as mild to severe mental retardation; both disorders have craniofacial anomalies, but each has distinct characteristic facial features (29, 102) (**Figure 5**). Cytogenetically, heterochromatin repulsion (HR) appears in 100% of RBS-SC phocomelia probands, is highly correlated with the phenotype and *ESCO2* mutations (85), and has been used for prenatal diagnosis (86). Genotype-phenotype correlations exist among individuals with CdLS in that probands with

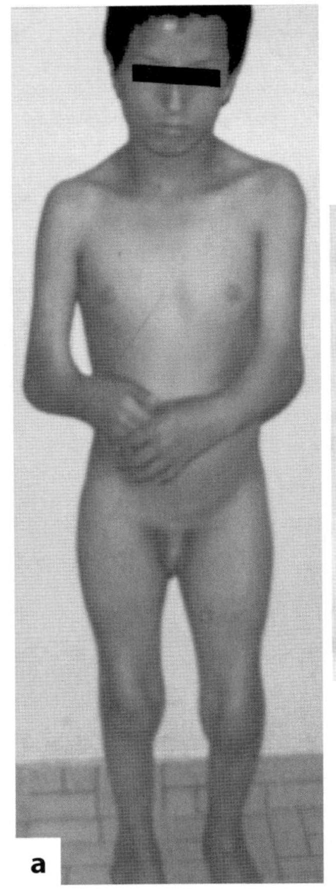

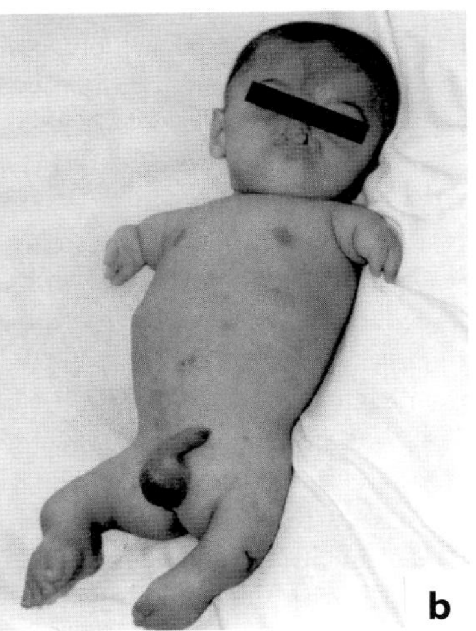

Figure 5

Phenotypic features in Roberts-SC phocomelia syndrome. Both patients have the same causative mutations in *ESCO2* (establishment of cohesion 1 homolog 2). (*a*) Typical features of SC phocomelia, including upper limb long bone shortening (radial hypoplasia > ulnar hypoplasia) and contractures with relative sparing of lower extremities. (*b*) Roberts syndrome patient with more severe phenotype, including tetraphocomelia and craniofacial abnormalities. Photographs courtesy of Dr. Hugo Vega from the Mount Sinai School of Medicine and Universidad Nacional de Colombia.

protein-truncating mutations in *NIPBL* tend to be severely affected, whereas those with missense mutations have milder phenotypes; all probands carrying an *SMC1A* mutation are more mildly affected and display almost no structural anomalies but do show mental retardation (19, 29). By contrast, there is no genotype-phenotype correlation among RBS-SC phocomelia patients (85). SC phocomelia syndrome is now known to be a milder form of RBS because it is also due to recessive mutations in *ESCO2* (the same mutations cause both RBS and SC phocomelia) and all have HR (85).

ESCO2 maps to 8p21.1 and is the human homolog of the budding yeast gene *Eco1/Ctf7*, which is involved in triggering cohesion initiation (92). *Eco1* has a C2H2 zinc finger domain at the N terminus and an acetyltransferase domain at the C terminus; Eco1 is able to acetylate itself and cohesin subunits in vitro (47), but no in vivo enzyme activity has been demonstrated (11). The protein structure of Eco1/Ctf7 is highly conserved from yeast to human. Functional Eco1/Cft7 proteins are important in all studied organisms; for example, sister chromatid cohesion is disrupted at the centromere and chromosome alignment is delayed in *Drosophila* mutants, although cohesion along sister chromatid arms seems unaffected (108). In human cell lines, ESCO2 binds to chromatin, possesses acetyltransferase activity, and is required for stable sister chromatid cohesion. Defective chromosome congregation or segregation is induced by RNAi knockdown of ESCO2 protein expression (41). The replication protein PCNA bridges cohesion establishment with DNA replication, because direct physical interaction between Eco1 and PCNA is essential for the engagement of Eco1 to chromatin and the initiation of sister chromatid cohesion in both yeast and human cells (68). The understanding of the relationship between ESCO2 and other cohesin subunits or regulatory factors is still very limited at this time.

α-Thalassemia/Mental Retardation Syndrome, X-Linked

α-Thalassemia/mental retardation syndrome, X-linked (ATRX) (OMIM #301040), caused by mutations in the *ATRX* gene on the X chromosome, was recently found to also involve a chromosome cohesion defect (77). ATRX is a multisystem disorder of postnatal growth deficiency, mental retardation, microcephaly, dysmorphic craniofacial features (midface hypoplasia; small, low-set ears; hypertelorism;

anteverted nares; full lips with protruding tongue), genital abnormalities in males, expressive speech absence, seizures, mild hypochromic microcytic anemia, and a mild form of hemoglobin H (Hb H) disease. The *ATRX* gene encodes a chromatin remodeling enzyme that is highly enriched at pericentromeric heterochromatin in mouse and human cells and associates with the chromoshadow domain of HP1α (as does NIPBL) (77). Defective sister chromatid cohesion, chromosome congression at the metaphase plate, and mitotic defects have recently been described in various mammalian cells (77). Defects in the *ATRX* gene that lead to the ATRX syndrome are postulated to be a result of perturbed cohesin targeting or loading/unloading; the phenotype is thought to be wholly or partially caused by the resultant gene dysregulation or mitotic defects.

SUMMARY POINTS

1. Studies of mutations in structural components of cohesin [structural maintenance of chromosomes 1A and 3 (*SMC1A* and *SMC3*) in Cornelia de Lange syndrome (CdLS)] and in regulators of the cohesin complex [Nipped-B-Like (*NIPBL*) in CdLS and establishment of cohesion 1 homolog 2 (*ESCO2*) in Roberts-SC phocomelia] directly implicate the role of this complex in normal human development.
2. Several other developmental disorders are likely due to a disruption of the cohesin complex and its regulation given that more than 25 genes have been primarily implicated in this cellular event. The exact manner by which cohesin mediates its control in developmental processes is unclear but is slowly being revealed and multiple molecular mechanisms are involved.
3. The regulation of gene expression through long-range regulatory elements, as well as possible roles in chromatin remodeling and epigenetic regulation of transcription, is likely the more critical underlying mechanism by which developmental control is regulated rather than by cohesin's canonical role in sister chromatid cohesion and segregation.

FUTURE ISSUES

1. Mammalian animal models, such as a CdLS mouse model, will be invaluable tools to resolve the conflicts currently being uncovered in nonmammalian species and to identify target genes regulated by cohesin at different embryonic stages and within each organ system involved in these disorders.
2. Genome-wide expression studies and the identification of cohesin and NIPBL binding sites in human cells will also be necessary to correlate cohesin binding with specific gene transcription.
3. Potential inter/intrachromosomal association mediated by NIPBL/cohesin and the corresponding impact on gene regulation need more attention.
4. Novel areas of transcriptional control (i.e., noncoding RNAs) may also be affected by cohesin dysfunction and these areas will need to be investigated on a genome-wide and spatiotemporal level in embryogenesis as well to appreciate their contribution to the phenotypes seen in the cohesinopathies.

DISCLOSURE STATEMENT

The authors are not aware of any biases that might be perceived as affecting the objectivity of this review.

ACKNOWLEDGMENTS

The authors would like to acknowledge the support of the NICHD (PO1HD052860) (I.D.K.) and the CdLS Foundation Fellowship Grant (J.L.).

LITERATURE CITED

1. Allanson JE, Hennekam RC, Ireland M. 1997. De Lange syndrome: subjective and objective comparison of the classical and mild phenotypes. *J. Med. Genet.* 34:645–50
2. Anderson DE, Losada A, Erickson HP, Hirano T. 2002. Condensin and cohesin display different arm conformations with characteristic hinge angles. *J. Cell Biol.* 156:419–24
3. Aqua MS, Rizzu P, Lindsay EA, Shaffer LG, Zackai EH, et al. 1995. Duplication 3q syndrome: molecular delineation of the critical region. *Am J. Med. Genet.* 55:33–37
4. Bernard P, Maure JF, Partridge JF, Genier S, Javerzat JP, Allshire RC. 2001. Requirement of heterochromatin for cohesion at centromeres. *Science* 294:2539–42
5. Bhuiyan ZA, Klein M, Hammond P, van Haeringen A, Mannens MM, et al. 2006. Genotype-phenotype correlations of 39 patients with Cornelia De Lange syndrome: the Dutch experience. *J. Med. Genet.* 43:568–75
6. Bhuiyan ZA, Stewart H, Redeker EJ, Mannens MM, Hennekam RC. 2007. Large genomic rearrangements in *NIPBL* are infrequent in Cornelia de Lange syndrome. *Eur. J. Hum. Genet.* 15:505–8
7. Blaschke RJ, Monaghan AP, Schiller S, Schechinger B, Rao E, et al. 1998. SHOT, a SHOX-related homeobox gene, is implicated in craniofacial, brain, heart, and limb development. *Proc. Natl. Acad. Sci. USA* 95:2406–11
8. Borck G, Zarhrate M, Bonnefont JP, Munnich A, Cormier-Daire V, Colleaux L. 2007. Incidence and clinical features of X-linked Cornelia de Lange syndrome due to *SMC1L1* mutations. *Hum. Mutat.* 28:205–6
9. Borck G, Zarhrate M, Cluzeau C, Bal E, Bonnefont JP, et al. 2006. Father-to-daughter transmission of Cornelia de Lange syndrome caused by a mutation in the 5′ untranslated region of the *NIPBL* gene. *Hum. Mutat.* 27:731–35
10. Brachmann W. 1916. Ein fall von symmetrischer Monodactylie durch Ulnadefekt, mit symmetrischer flughautbildung in den Ellenbeugen, sowie anderen Abnormalitaten (Zwerghaftigkeit, Halsrippen, Behaarung). *Jahrb. Kinderheilkd* 84:225–35
11. Brands A, Skibbens RV. 2005. Ctf7p/Eco1p exhibits acetyltransferase activity–but does it matter? *Curr. Biol.* 15:R50–51
12. Brown CJ, Miller AP, Carrel L, Rupert JL, Davies KE, Willard HF. 1995. The DXS423E gene in Xp11.21 escapes X chromosome inactivation. *Hum. Mol. Genet.* 4:251–55
13. Charles AK, Porter HJ, Sams V, Lunt P. 1997. Nephrogenic rests and renal abnormalities in Brachmann-de Lange syndrome. *Pediatr. Pathol. Lab. Med.* 17:209–19
14. Chodirker BN, Chudley AE. 1994. Male-to-male transmission of mild Brachmann-de Lange syndrome. *Am. J. Med. Genet.* 52:331–33
15. Ciosk R, Shirayama M, Shevchenko A, Tanaka T, Toth A, et al. 2000. Cohesin's binding to chromosomes depends on a separate complex consisting of Scc2 and Scc4 proteins. *Mol. Cell* 5:243–54
16. D'Amours D, Stegmeier F, Amon A. 2004. Cdc14 and condensin control the dissolution of cohesin-independent chromosome linkages at repeated DNA. *Cell* 117:455–69
17. de Knecht-van Eekelen A, Hennekam RC. 1994. Historical study: Cornelia C. de Lange (1871–1950)–a pioneer in clinical genetics. *Am. J. Med. Genet.* 52:257–66

18. de Lange C. 1933. Sur un type nouveau de dégénération (typus Amstelodamensis) [On a new type of degeneration (type Amsterdam)]. *Arch. Méd. Enfants* 36:715–19
19. Deardorff MA, Kaur M, Yaeger D, Rampuria A, Korolev S, et al. 2007. Mutations in cohesin complex members SMC3 and SMC1A cause a mild variant of Cornelia de Lange syndrome with predominant mental retardation. *Am. J. Hum. Genet.* 80:485–94

19. Identification of *SMC3* as a CdLS gene and genotype-phenotype correlation for SMC1A mutation-positive probands.

20. Del Campo M, Jones MC, Veraksa AN, Curry CJ, Jones KL, et al. 1999. Monodactylous limbs and abnormal genitalia are associated with hemizygosity for the human 2q31 region that includes the *HOXD* cluster. *Am. J. Hum. Genet.* 65:104–10
21. DeScipio C, Kaur M, Yaeger D, Innis JW, Spinner NB, et al. 2005. Chromosome rearrangements in Cornelia de Lange syndrome (CdLS): report of a der(3)t(3;12)(p25.3;p13.3) in two half sibs with features of CdLS and review of reported CdLS cases with chromosome rearrangements. *Am J. Med. Genet. A* 137:276–82
22. Donze D, Adams CR, Rine J, Kamakaka RT. 1999. The boundaries of the silenced HMR domain in *Saccharomyces cerevisiae*. *Genes Dev.* 13:698–708
23. Dorsett D. 2004. Adherin: key to the cohesin ring and Cornelia de Lange syndrome. *Curr. Biol.* 14:R834–36
24. Dorsett D. 2007. Roles of the sister chromatid cohesion apparatus in gene expression, development, and human syndromes. *Chromosoma* 116:1–13
25. Dorsett D, Eissenberg JC, Misulovin Z, Martens A, Redding B, McKim K. 2005. Effects of sister chromatid cohesion proteins on *cut* gene expression during wing development in *Drosophila*. *Development* 132:4743–53
26. Egemen A, Ulger Z, Ozkinay F, Gulen F, Cogulu O. 2005. A de novo t (X;8)(p11.2;q24.3) demonstrating Cornelia de Lange syndrome phenotype. *Genet. Couns.* 16:27–30
27. Filipski J, Mucha M. 2002. Structure, function and DNA composition of *Saccharomyces cerevisiae* chromatin loops. *Gene* 300:63–68
28. Gandhi R, Gillespie PJ, Hirano T. 2006. Human Wapl is a cohesin-binding protein that promotes sister-chromatid resolution in mitotic prophase. *Curr. Biol.* 16:2406–17
29. Gillis LA, McCallum J, Kaur M, DeScipio C, Yaeger D, et al. 2004. *NIPBL* mutational analysis in 120 individuals with Cornelia de Lange syndrome and evaluation of genotype-phenotype correlations. *Am. J. Hum. Genet.* 75:610–23
30. Giot L, Bader JS, Brouwer C, Chaudhuri A, Kuang B, et al. 2003. A protein interaction map of *Drosophila melanogaster*. *Science* 302:1727–36
31. Glynn EF, Megee PC, Yu HG, Mistrot C, Unal E, et al. 2004. Genome-wide mapping of the cohesin complex in the yeast *Saccharomyces cerevisiae*. *PLoS Biol.* 2:E259
32. Greenberg F, Robinson LK. 1989. Mild Brachmann-de Lange syndrome: changes of phenotype with age. *Am J. Med. Genet.* 32:90–92
33. Gruber S, Haering CH, Nasmyth K. 2003. Chromosomal cohesin forms a ring. *Cell* 112:765–77
34. Hagstrom KA, Meyer BJ. 2003. Condensin and cohesin: more than chromosome compactor and glue. *Nat. Rev. Genet.* 4:520–34
35. Hakimi MA, Bochar DA, Schmiesing JA, Dong Y, Barak OG, et al. 2002. A chromatin remodelling complex that loads cohesin onto human chromosomes. *Nature* 418:994–98
36. Hartman T, Stead K, Koshland D, Guacci V. 2000. Pds5p is an essential chromosomal protein required for both sister chromatid cohesion and condensation in *Saccharomyces cerevisiae*. *J. Cell Biol.* 151:613–26
37. Herault Y, Fraudeau N, Zakany J, Duboule D. 1997. *Ulnaless* (*Ul*), a regulatory mutation inducing both loss-of-function and gain-of-function of posterior *Hoxd* genes. *Development* 124:3493–500
38. Hirano T. 2006. At the heart of the chromosome: SMC proteins in action. *Nat. Rev. Mol. Cell Biol.* 7:311–22
39. Holder SE, Grimsley LM, Palmer RW, Butler LJ, Baraitser M. 1994. Partial trisomy 3q causing mild Cornelia de Lange phenotype. *J. Med. Genet.* 31:150–52
40. Horsfield JA, Anagnostou SH, Hu JK, Cho KH, Geisler R, et al. 2007. Cohesin-dependent regulation of Runx genes. *Development* 134:2639–49
41. Hou F, Zou H. 2005. Two human orthologues of Eco1/Ctf7 acetyltransferases are both required for proper sister-chromatid cohesion. *Mol. Biol. Cell* 16:3908–18

42. Ireland M, Donnai D, Burn J. 1993. Brachmann-de Lange syndrome. Delineation of the clinical phenotype. *Am. J. Med. Genet.* 47:959–64
43. Ireland M, English C, Cross I, Houlsby WT, Burn J. 1991. A de novo translocation t(3;17)(q26.3;q23.1) in a child with Cornelia de Lange syndrome. *J. Med. Genet.* 28:639–40
44. Ireland M, English C, Cross I, Lindsay S, Strachan T. 1995. Partial trisomy 3q and the mild Cornelia de Lange syndrome phenotype. *J. Med. Genet.* 32:837–38
45. Ivanov D, Nasmyth K. 2005. A topological interaction between cohesin rings and a circular minichromosome. *Cell* 122:849–60
46. Ivanov D, Nasmyth K. 2007. A physical assay for sister chromatid cohesion in vitro. *Mol. Cell* 27:300–10
47. Ivanov D, Schleiffer A, Eisenhaber F, Mechtler K, Haering CH, Nasmyth K. 2002. Eco1 is a novel acetyltransferase that can acetylate proteins involved in cohesion. *Curr. Biol.* 12:323–28
48. **Jackson L, Kline AD, Barr MA, Koch S. 1993. de Lange syndrome: a clinical review of 310 individuals. *Am. J. Med. Genet.* 47:940–46**

48. The first clinical review of a large cohort of CdLS patients in the United States.

49. Kaur M, DeScipio C, McCallum J, Yaeger D, Devoto M, et al. 2005. Precocious sister chromatid separation (PSCS) in Cornelia de Lange syndrome. *Am. J. Med. Genet. A* 138:27–31
50. Kleinjan DA, van Heyningen V. 2005. Long-range control of gene expression: emerging mechanisms and disruption in disease. *Am. J. Hum. Genet.* 76:8–32
51. Kline AD, Barr M, Jackson LG. 1993. Growth manifestations in the Brachmann-de Lange syndrome. *Am. J. Med. Genet.* 47:1042–49
52. Kline AD, Stanley C, Belevich J, Brodsky K, Barr M, Jackson LG. 1993. Developmental data on individuals with the Brachmann-de Lange syndrome. *Am. J. Med. Genet.* 47:1053–58
53. Kobayashi T, Horiuchi T, Tongaonkar P, Vu L, Nomura M. 2004. *SIR2* regulates recombination between different rDNA repeats, but not recombination within individual rRNA genes in yeast. *Cell* 117:441–53
54. Kousseff BG, Newkirk P, Root AW. 1994. Brachmann-de Lange syndrome. 1994 update. *Arch. Pediatr. Adolesc. Med.* 148:749–55
55. **Krantz ID, McCallum J, DeScipio C, Kaur M, Gillis LA, et al. 2004. Cornelia de Lange syndrome is caused by mutations in *NIPBL*, the human homolog of *Drosophila melanogaster Nipped-B*. *Nat. Genet.* 36:631–35**

55. Identification of *NIPBL* as CdLS gene.

56. Krantz ID, Tonkin E, Smith M, Devoto M, Bottani A, et al. 2001. Exclusion of linkage to the *CDL1* gene region on chromosome 3q26.3 in some familial cases of Cornelia de Lange syndrome. *Am. J. Med. Genet.* 101:120–29
57. Kueng S, Hegemann B, Peters BH, Lipp JJ, Schleiffer A, et al. 2006. Wapl controls the dynamic association of cohesin with chromatin. *Cell* 127:955–67
58. Lechner MS, Schultz DC, Negorev D, Maul GG, Rauscher FJ 3rd. 2005. The mammalian heterochromatin protein 1 binds diverse nuclear proteins through a common motif that targets the chromoshadow domain. *Biochem. Biophys. Res. Commun.* 331:929–37
59. Lengronne A, Katou Y, Mori S, Yokobayashi S, Kelly GP, et al. 2004. Cohesin relocation from sites of chromosomal loading to places of convergent transcription. *Nature* 430:573–78
60. Lettice LA, Heaney SJ, Purdie LA, Li L, de Beer P, et al. 2003. A long-range Shh enhancer regulates expression in the developing limb and fin and is associated with preaxial polydactyly. *Hum. Mol. Genet.* 12:1725–35
61. Losada A. 2007. Cohesin regulation: fashionable ways to wear a ring. *Chromosoma* 116:321–29
62. Losada A, Hirano M, Hirano T. 1998. Identification of *Xenopus* SMC protein complexes required for sister chromatid cohesion. *Genes Dev.* 12:1986–97
63. McKenney RR, Elder FF, Garcia J, Northrup H. 1996. Brachmann-de Lange syndrome: autosomal dominant inheritance and male-to-male transmission. *Am. J. Med. Genet.* 66:449–52
64. Meinecke P, Hayek H. 1990. Brief historical note on the Brachmann-de Lange syndrome: a patient closely resembling the case described by Brachmann in 1916. *Am. J. Med. Genet.* 35:449–50
65. Meins M, Hagh JK, Gerresheim F, Einhoff E, Olschewski H, et al. 2005. Novel case of dup(3q) syndrome due to a de novo interstitial duplication 3q24-q26.31 with minimal overlap to the dup(3q) critical region. *Am. J. Med. Genet. A* 132:84–89
66. Michaelis C, Ciosk R, Nasmyth K. 1997. Cohesins: chromosomal proteins that prevent premature separation of sister chromatids. *Cell* 91:35–45

67. Misulovin Z, Schwartz YB, Li XY, Kahn TG, Gause M, et al. 2007. Association of cohesin and Nipped-B with transcriptionally active regions of the *Drosophila melanogaster* genome. *Chromosoma* 117:89–102
68. Moldovan GL, Pfander B, Jentsch S. 2006. PCNA controls establishment of sister chromatid cohesion during S phase. *Mol. Cell* 23:723–32
69. **Musio A, Selicorni A, Focarelli ML, Gervasini C, Milani D, et al. 2006. X-linked Cornelia de Lange syndrome owing to *SMC1L1* mutations. *Nat. Genet.* 38:528–30**
70. Nakajima M, Kumada K, Hatakeyama K, Noda T, Peters JM, Hirota T. 2007. The complete removal of cohesin from chromosome arms depends on separase. *J. Cell Sci.* 120:4188–96
71. **Nasmyth K, Haering CH. 2005. The structure and function of SMC and kleisin complexes. *Annu. Rev. Biochem.* 74:595–648**
72. Oostra RJ, Baljet B, Hennekam RC. 1994. Brachmann-de Lange syndrome "avant la lettre." *Am. J. Med. Genet.* 52:267–68
73. Opitz JM. 1985. The Brachmann-de Lange syndrome. *Am. J. Med. Genet.* 22:89–102
74. Ozkinay F, Cogulu O, Gunduz C, Levent E, Ozkinay C. 1998. A case of Brachman de Lange syndrome with cerebellar vermis hypoplasia. *Clin. Dysmorphol.* 7:303–5
75. Panizza S, Tanaka T, Hochwagen A, Eisenhaber F, Nasmyth K. 2000. Pds5 cooperates with cohesin in maintaining sister chromatid cohesion. *Curr. Biol.* 10:1557–64
76. Partridge JF, Scott KS, Bannister AJ, Kouzarides T, Allshire RC. 2002. Cis-acting DNA from fission yeast centromeres mediates histone H3 methylation and recruitment of silencing factors and cohesin to an ectopic site. *Curr. Biol.* 12:1652–60
77. Ritchie K, Seah C, Moulin J, Isaac C, Dick F, Bérubé NG. 2008. Loss of ATRX leads to chromosome cohesion and congression defects. *J. Cell Biol.* 180:315–24
78. Rizzu P, Haddad BR, Vallcorba I, Alonso A, Ferro MT, et al. 1997. Delineation of a duplication map of chromosome 3q: a new case confirms the exclusion of 3q25-q26.2 from the duplication 3q syndrome critical region. *Am. J. Med. Genet.* 68:428–32
79. Roessler E, Belloni E, Gaudenz K, Jay P, Berta P, et al. 1996. Mutations in the human *Sonic Hedgehog* gene cause holoprosencephaly. *Nat. Genet.* 14:357–60
80. Rollins RA, Korom M, Aulner N, Martens A, Dorsett D. 2004. *Drosophila* nipped-B protein supports sister chromatid cohesion and opposes the stromalin/Scc3 cohesion factor to facilitate long-range activation of the *cut* gene. *Mol. Cell Biol.* 24:3100–11
81. **Rollins RA, Morcillo P, Dorsett D. 1999. Nipped-B, a *Drosophila* homologue of chromosomal adherins, participates in activation by remote enhancers in the *cut* and *Ultrabithorax* genes. *Genetics* 152:577–93**
82. Russell KL, Ming JE, Patel K, Jukofsky L, Magnusson M, Krantz ID. 2001. Dominant paternal transmission of Cornelia de Lange syndrome: a new case and review of 25 previously reported familial recurrences. *Am. J. Med. Genet.* 104:267–76
83. Saul RA, Rogers RC, Phelan MC, Stevenson RE. 1993. Brachmann-de Lange syndrome: diagnostic difficulties posed by the mild phenotype. *Am. J. Med. Genet.* 47:999–1002
84. Schoumans J, Wincent J, Barbaro M, Djureinovic T, Maguire P, et al. 2007. Comprehensive mutational analysis of a cohort of Swedish Cornelia de Lange syndrome patients. *Eur. J. Hum. Genet.* 15:143–49
85. Schule B, Oviedo A, Johnston K, Pai S, Francke U. 2005. Inactivating mutations in *ESCO2* cause SC phocomelia and Roberts syndrome: no phenotype-genotype correlation. *Am. J. Hum. Genet.* 77:1117–28
86. Schulz S, Gerloff C, Ledig S, Langer D, Volleth M, et al. 2008. Prenatal diagnosis of Roberts syndrome and detection of an *ESCO2* frameshift mutation in a Pakistani family. *Prenat. Diagn.* 28:42–45
87. Sciorra LJ, Bahng K, Lee ML. 1979. Trisomy in the distal end of the long arm of chromosome 3. A condition clinically similar to the Cornelia de Lange syndrome. *Am. J. Dis. Child* 133:727–30
88. Seitan VC, Banks P, Laval S, Majid NA, Dorsett D, et al. 2006. Metazoan Scc4 homologs link sister chromatid cohesion to cell and axon migration guidance. *PLoS Biol.* 4:e242
89. Selicorni A, Lalatta F, Livini E, Briscioli V, Piguzzi T, et al. 1993. Variability of the Brachmann-de Lange syndrome. *Am. J. Med. Genet.* 47:977–82
90. Selicorni A, Russo S, Gervasini C, Castronovo P, Milani D, et al. 2007. Clinical score of 62 Italian patients with Cornelia de Lange syndrome and correlations with the presence and type of *NIPBL* mutation. *Clin. Genet.* 72:98–108

69. Identification of *SMC1A* as a CdLS gene.

71. A thorough review of cohesin biology.

81. Identification of *Nipped-B* and its role in long-range enhancer promoter interactions in *Drosophila*.

91. Skibbens RV. 2005. Unzipped and loaded: the role of DNA helicases and RFC clamp-loading complexes in sister chromatid cohesion. *J. Cell Biol.* 169:841–46
92. Skibbens RV, Corson LB, Koshland D, Hieter P. 1999. Ctf7p is essential for sister chromatid cohesion and links mitotic chromosome structure to the DNA replication machinery. *Genes Dev.* 13:307–19
93. Smith M, Herrell S, Lusher M, Lako L, Simpson C, et al. 1999. Genomic organisation of the human chordin gene and mutation screening of candidate Cornelia de Lange syndrome genes. *Hum. Genet.* 105:104–11
94. Spitz F, Gonzalez F, Duboule D. 2003. A global control region defines a chromosomal regulatory landscape containing the *HoxD* cluster. *Cell* 113:405–17
95. Strachan T. 2005. Cornelia de Lange Syndrome and the link between chromosomal function, DNA repair and developmental gene regulation. *Curr. Opin. Genet. Dev.* 15:258–64
96. Sullivan M, Higuchi T, Katis VL, Uhlmann F. 2004. Cdc14 phosphatase induces rDNA condensation and resolves cohesin-independent cohesion during budding yeast anaphase. *Cell* 117:471–82
97. Sumara I, Vorlaufer E, Gieffers C, Peters BH, Peters JM. 2000. Characterization of vertebrate cohesin complexes and their regulation in prophase. *J. Cell Biol.* 151:749–62
98. Tonkin ET, Smith M, Eichhorn P, Jones S, Imamwerdi B, et al. 2004. A giant novel gene undergoing extensive alternative splicing is severed by a Cornelia de Lange-associated translocation breakpoint at 3q26.3. *Hum. Genet.* 115:139–48
99. Tonkin ET, Wang TJ, Lisgo S, Bamshad MJ, Strachan T. 2004. *NIPBL*, encoding a homolog of fungal Scc2-type sister chromatid cohesion proteins and fly Nipped-B, is mutated in Cornelia de Lange syndrome. *Nat. Genet.* 36:636–41

99. An independent study that also identified *NIPBL* as the CdLS gene.

100. Uhlmann F, Nasmyth K. 1998. Cohesion between sister chromatids must be established during DNA replication. *Curr. Biol.* 8:1095–101
101. Van Allen MI, Filippi G, Siegel-Bartelt J, Yong SL, McGillivray B, et al. 1993. Clinical variability within Brachmann-de Lange syndrome: a proposed classification system. *Am. J. Med. Genet.* 47:947–58
102. Van Den Berg DJ, Francke U. 1993. Roberts syndrome: a review of 100 cases and a new rating system for severity. *Am. J. Med. Genet.* 47:1104–23
103. Vega H, Waisfisz Q, Gordillo M, Sakai N, Yanagihara I, et al. 2005. Roberts syndrome is caused by mutations in *ESCO2*, a human homolog of yeast *ECO1* that is essential for the establishment of sister chromatid cohesion. *Nat. Genet.* 37:468–70

103. Identification of *ESCO2* as the Roberts-SC phocomelia gene.

104. Velagaleti GV, Bien-Willner GA, Northup JK, Lockhart LH, Hawkins JC, et al. 2005. Position effects due to chromosome breakpoints that map approximately 900 kb upstream and approximately 1.3 Mb downstream of *SOX9* in two patients with campomelic dysplasia. *Am. J. Hum. Genet.* 76:652–62
105. Vrouwe MG, Elghalbzouri-Maghrani E, Meijers M, Schouten P, Godthelp BC, et al. 2007. Increased DNA damage sensitivity of Cornelia de Lange syndrome cells: evidence for impaired recombinational repair. *Hum. Mol. Genet.* 16:1478–87
106. Waizenegger IC, Hauf S, Meinke A, Peters JM. 2000. Two distinct pathways remove mammalian cohesin from chromosome arms in prophase and from centromeres in anaphase. *Cell* 103:399–410
107. Watanabe Y, Kitajima TS. 2005. Shugoshin protects cohesin complexes at centromeres. *Philos. Trans. R. Soc. London Ser. B* 360:515–21
108. Williams BC, Garrett-Engele CM, Li Z, Williams EV, Rosenman ED, Goldberg ML. 2003. Two putative acetyltransferases, san and deco, are required for establishing sister chromatid cohesion in *Drosophila*. *Curr. Biol.* 13:2025–36
109. Wilson GN, Hieber VC, Schmickel RD. 1978. The association of chromosome 3 duplication and the Cornelia de Lange syndrome. *J. Pediatr.* 93:783–88
110. Wilson WG, Kennaugh JM, Kugler JP, Wyandt HE. 1983. Reciprocal translocation 14q;21q in a patient with the Brachmann-de Lange syndrome. *J. Med. Genet.* 20:469–71
111. Yan J, Saifi GM, Wierzba TH, Withers M, Bien-Willner GA, et al. 2006. Mutational and genotype-phenotype correlation analyses in 28 Polish patients with Cornelia de Lange syndrome. *Am. J. Med. Genet. A* 140:1531–41
112. Zhang B, Jain S, Song H, Fu M, Heuckeroth RO, et al. 2007. Mice lacking sister chromatid cohesion protein PDS5B exhibit developmental abnormalities reminiscent of Cornelia de Lange syndrome. *Development* 134:3191–201

Genetic Predisposition to Breast Cancer: Past, Present, and Future

Clare Turnbull and Nazneen Rahman

Section of Cancer Genetics, Institute of Cancer Research, Sutton, SM2 5NG, United Kingdom; email: clare.turnbull@icr.ac.uk, nazneen.rahman@icr.ac.uk

Annu. Rev. Genomics Hum. Genet. 2008. 9:321–45

First published online as a Review in Advance on June 10, 2008

The *Annual Review of Genomics and Human Genetics* is online at genom.annualreviews.org

This article's doi:
10.1146/annurev.genom.9.081307.164339

1527-8204/08/0922-0321$20.00

Key Words

BRCA*, allele, familial, susceptibility, penetrance

Abstract

In recent years, our understanding of genetic predisposition to breast cancer has advanced significantly. Three classes of predisposition factors, categorized by their associated risks of breast cancer, are currently known. *BRCA1* and *BRCA2* are high-penetrance breast cancer predisposition genes identified by genome-wide linkage analysis and positional cloning. Mutational screening of genes functionally related to *BRCA1* and/or *BRCA2* has revealed four genes, *CHEK2*, *ATM*, *BRIP1*, and *PALB2*; mutations in these genes are rare and confer an intermediate risk of breast cancer. Association studies have further identified eight common variants associated with low-penetrance breast cancer predisposition. Despite these discoveries, most of the familial risk of breast cancer remains unexplained. In this review, we describe the known genetic predisposition factors, expound on the methods by which they were identified, and consider how further technological and intellectual advances may assist in identifying the remaining genetic factors underlying breast cancer susceptibility.

INTRODUCTION

Linkage analysis: a statistical method for evaluating linkage between disease and markers of known location by following their inheritance in families

Penetrance: the probability that a particular phenotype/disease is expressed in an individual with a particular genotype

Through the dynamic interplay of multiple approaches, the past two decades have witnessed the gradual emergence of a clearer understanding of genetic susceptibility to breast cancer. Clinical observation underpinned these advances through recognition of unusual familial clustering and phenotypes associated with breast cancer. Observational epidemiology has also provided essential foundations to our understanding, both in quantifying the contribution of genetic factors to breast cancer and in predicting how these factors will act individually and interact. Genetic modeling has been performed to predict the profiles of the genes involved, which in turn facilitates the selection of molecular methods appropriate for identifying them. Linkage analysis, mutational screening of candidate genes, and association studies have been used to identify predisposition factors of three distinct risk-prevalence profiles: rare high-penetrance alleles, rare intermediate-penetrance alleles, and common low-penetrance alleles.

OBSERVATIONAL EPIDEMIOLOGY AND SEGREGATION ANALYSIS

Epidemiological observation of the clustering of breast cancer within families was made in Roman times and reiterated by other early socio-scientific commentators (22, 77). More than 50 studies have explored this familial aggregation of breast cancer using predominantly case-control and cohort designs. Meta-analyses of these data conclude that, overall, breast cancer is twice as common in women with an affected first-degree relative. Simulation studies suggest that an environmental factor would have to confer at least a tenfold increase in risk for shared exposure to result in even a modest increase to the familial risk (65). Because no environmental risk factors of this magnitude have been identified for breast cancer, it seems unlikely that shared environment accounts for much of the familial aggregation. Heritability studies can assist in clarifying the relative contribution of genetic factors and shared environment. Via demonstration that the risk to a monozygotic twin is substantially higher than to a dizygotic twin of an affected individual, heritability studies using twins confirmed that the predominant component of the familial aggregation in breast cancer is genetic (80, 93). Patterns of disease aggregation in twins provide evidence not just for a genetic basis for breast cancer but demonstrate a markedly skewed distribution of genetic liability, suggesting that the majority of genetic risk may lie within a genetically predisposed minority (93).

Segregation analysis (genetic modeling) involves the simulation of various scenarios whereby disease occurs owing to combinations of known or hypothetical genes of specified risk, prevalence, and mode of inheritance. The patterns of cancers predicted by the different models are compared with observed disease frequencies and the best-fitting model is favored. Segregation analyses have strongly influenced the molecular approaches with which genetic predisposition factors have been investigated. The Cancer and Steroid Hormone Study data and other early segregation analyses of families with breast cancer favored a highly penetrant autosomal dominant genetic model (16, 29, 68, 121). This prediction was validated through linkage analysis and the identification of the highly penetrant autosomal dominant breast cancer predisposition genes, *BRCA1* and *BRCA2*. Subsequent complex segregation analyses have incorporated the effects of *BRCA1* and *BRCA2* and explored the evidence for additional genes (8, 9, 35, 89). Probably the most comprehensive of these analyses, based upon an extensive series comprising both population-based cases of breast cancer and large familial clusters, has strongly favored the polygenic model. The polygenic component of risk in this model is equivalent to multiple genetic factors acting independently: This component is log-normally distributed, has a variance that declines linearly with age, and applies similarly to *BRCA* mutation carriers and noncarriers (8–10). The polygenic model is also consistent with multiple observations that the excess of familial breast cancer is distributed across many families, each typically comprising a modest number of cases, rather than just a few very extensive families (9, 35). The polygenic model has largely been accepted as an apposite explanation for the residual breast cancer predisposition and has been validated in part by identification of a number of the incumbent polygenes.

STRATEGIES FOR IDENTIFYING BREAST CANCER PREDISPOSITION FACTORS

Three principal experimental designs have been used in the molecular identification of genetic breast cancer predisposition factors: genome-wide linkage analysis, mutational screening of candidate genes, and association studies.

Linkage Analysis and Positional Cloning

Linkage studies are used to map a disease locus via analysis for cosegregation of genomic markers with a specified disease phenotype using samples from multiple members of large families. The region of the genome surrounding the linked markers is then interrogated for likely causative genes (positional cloning). Linkage analysis is suitable for mapping only high-penetrance breast cancer predisposition genes, whereby mutations result in prominent cancer families in which the majority of affected individuals carry the mutation. If a gene is of lower penetrance, the correlation between breast cancer and mutation status in mutation-positive families may be insufficient to generate a linkage signal. Nor will a significant signal be detectable if the breast cancer is linked to a particular locus in only a small proportion of the families analyzed.

Linkage maps and sets of markers that adequately cover the genome became available in the late 1980s. Using large breast cancer pedigrees, genome-wide linkage analyses were undertaken to map the high-penetrance breast cancer susceptibility genes proposed by the segregation analyses, *BRCA1* and *BRCA2* (61, 123). Linkage analysis has also been performed in groups of phenotypically distinct families to map the loci underlying specific syndromes that include an elevated risk of breast cancer. Positional cloning was then used to identify the causative genes, such as *PTEN*, *STK11*, and *CDH1* (58, 63, 64, 70, 87, 88).

Resequencing Studies: Mutational Screening of Candidate Genes

During the 1990s, emerging understanding of the molecular pathogenesis of breast cancer offered insights into possible candidate genes that predispose to breast cancer. It seemed biologically plausible that proteins interacting with BRCA1 and BRCA2 or acting in similar DNA repair pathways may also be involved in breast cancer susceptibility. A few deleterious mutations were found at frequencies amenable to evaluation by association studies. However, in the majority of DNA repair genes studied, disease-causing mutations (primarily leading to premature protein truncation or nonsense-mediated RNA decay) have been individually very rare. A robust demonstration that such genes confer predisposition to breast cancer has therefore required mutational screening of the entire coding sequence of the gene through large numbers of cases and controls to meaningfully compare the total number of pathogenic mutations. Experiments of this magnitude have been published for only a handful of DNA repair genes. Other postulated candidate genes, such as those implicated in cell cycle regulation, checkpoint control, apoptosis, and steroid hormone metabolism, have rarely been evaluated to this level.

Association Studies

In an association study, the frequency of a specified variant is compared between breast cancer cases and controls. Statistically significant differences in allele frequency between cases and controls are more readily demonstrable for

Single nucleotide polymorphism (SNP): a single-base sequence variation that commonly occurs at a particular position within the genome

Linkage disequilibrium: the non-random association of alleles at two or more loci, which suggests that they may be physically close and thus linked

variants of population frequency >5%; hence association studies are most suitable for identification of common breast cancer predisposition variants. Initial studies focused on genes proposed by function and examined the association with breast cancer of numerous common variants from recognized and candidate predisposition genes. However, experiments were often too small, subject to bias, and utilized too lenient levels of significance, resulting in inconsistency and lack of replication of findings. The majority of putative associations likely represented false positives (type 1 errors) (40, 95). To increase power and rationalize choices of candidates, collaborations have formed that undertake association studies across tens of thousands of samples (19).

The emergence of comprehensive high-density maps of single nucleotide polymorphisms (SNPs) and affordable genotyping platforms has allowed the graduation of association studies from the limitations of preconceived notions of candidacy to the agnosticism of the genome-wide approach. On account of linkage disequilibrium, a panel of a few hundred thousand reporter SNPs can be used as tags for the majority of the millions of common variants in the genome. Accordingly, owing to the degree of multiple testing, a genome-wide scan must be sufficiently well-powered to ensure that the true associations are detected. The staged experimental design is a means of optimizing the statistical power afforded by a given sample size and has been useful in studies to date because of the current high costs of whole genome tag SNP panels.

BREAST CANCER PREDISPOSITION GENES AND VARIANTS

The breast cancer predisposition factors identified to date can be stratified by risk profile into three tiers: high-penetrance genes, intermediate-penetrance genes, and low-penetrance alleles. Three further genes are associated with syndromes in which the incidence of breast cancer is elevated but the actual risk remains unclear (**Table 1**).

High-Penetrance Breast Cancer Predisposition Genes

Mutations in three high-penetrance breast cancer predisposition genes confer a greater than tenfold relative risk of breast cancer. *BRCA1* and *BRCA2* were identified through linkage analysis and positional cloning. *TP53* was deemed a plausible candidate and identified as a high-risk breast cancer gene through mutational screening.

BRCA1 and BRCA2. The first convincing report of linkage of breast cancer to 17q21 was published in 1990 (61). In 1994, positional cloning revealed the causative gene, accordingly named *BRCA1* (86). Linkage analysis and positional cloning led to the mapping and identification of *BRCA2* in 1994 and 1995, respectively (122, 123). BRCA1 and BRCA2 have important roles in the maintenance of genomic stability by facilitating repair of DNA double-strand breaks. The cellular roles of BRCA1 and BRCA2 were reviewed by Gudmundsdottir & Ashworth (56).

BRCA1 and *BRCA2* are large genes in which multiple different loss-of-function mutations have been detected. Some founder mutations are relatively frequent in particular ethnic groups, such as *BRCA1*_185delAG, *BRCA1*_5382insC, and *BRCA2*_6174delT in the Ashkenazim and *BRCA2*_999del5 in Icelanders. However, the majority of mutations are individually rare and many have been reported only in single families. Most recognized disease-associated mutations result in premature protein truncation and include nonsense mutations, deletions/insertions that result in translational frameshifts, and mutations that affect splice sites. More recently a number of exonic deletions/duplications have also been identified, especially in *BRCA1*. In addition, a large number of amino acid substitutions and synonymous nucleotide substitutions in *BRCA1* and *BRCA2* have been

Table 1 Summary of known breast cancer predisposition factors

	Gene/Locus	Relative Risk of breast cancer	Carrier Frequency†	Breast cancer subtype	Other cancers in monoallelic carriers	Syndrome in biallelic carriers	Method of identification
High penetrance	*BRCA1*	>10	0.1%	Basal (ER-negative)	Ovarian		Linkage study
	BRCA2	>10	0.1%		Ovarian prostate	Fanconi anaemia D1	Linkage study
	TP53	>10	rare		Sarcomas adrenal brain		Candidate resequencing study
Uncertain penetrance	*PTEN*	2–10	rare		Thyroid endometrium		Linkage study
	STK11	2–10	rare		Gasto-intestinal		Linkage study
	CDH1	2–10	rare	lobular	Gastric (diffuse)		Linkage study
Intermediate penetrance	*ATM*	2–3	0.4%			Ataxia telangiectasia	Epidemiology; Candidate resequencing study
	CHEK2	2–3	0.4%				Candidate resequencing study
	BRIP1	2–3	0.1%			Fanconi anaemia J	Candidate resequencing study
	PALB2	2–4	rare			Fanconi anaemia N	Candidate resequencing study
Low penetrance	10q26, 16q12, 2q35, 8q24, 5p12	1.08–1.26	24–50%	ER-positive			Genome-wide association studies
	11p15, 5q11	1.07–1.13	28–30%				Genome-wide association study
	2q33	1.13	0.87				Candidate association study

† estimated carrier frequency of mutations/risk allele in the UK; where 'rare', the carrier frequency is unlikely to be >0.1%.

detected, the majority of which are innocuous. A few missense mutations, for example those that target cysteine residues in the BRCA1 RING domain, abrogate function and are presumed to confer risks comparable to truncating mutations (43, 106).

There is evidence of genotype-phenotype correlations for mutations in *BRCA1* and *BRCA2*. In both genes, the ratio of ovarian cancers to breast cancers conferred by mutations in the central region of the gene is higher than that for mutations in the 5′ or 3′ end. In *BRCA1* this central region is bounded by nucleotides 2401 and 4191; in *BRCA2* the ovarian cancer cluster region lies between nucleotides 3035 and 6629 (114, 115). The biological mechanisms underlying these observations remain opaque (120).

BRCA1 and *BRCA2* are high penetrance breast cancer genes. Estimates of the risks of cancer conferred by mutations in these genes vary according to the ascertainment of the cases studied. Early studies of large cancer families suggested that the risk of breast cancer by age 70 may be as high as 87% [95% confidence interval (CI) = 72%–95%] for *BRCA1* and 84% (43%–95%) for *BRCA2* mutation carriers, although the upper estimates were based on relatively small numbers of families (44, 51). In population-based studies of breast cancer cases, unselected for family history, the risks are lower: 65% (51%–75%) for *BRCA1* and 45% (33%–54%) for *BRCA2* (5). The pattern of age-related risk in *BRCA2* mutation carriers resembles that of the general population (only higher). By contrast, the relative risk of breast cancer is markedly elevated in *BRCA1* mutation carriers under 40 and becomes less dramatic with advancing age (5). Linkage analysis suggests that mutations in *BRCA1* and *BRCA2* are responsible for disease in approximately two-thirds of large families with site-specific female breast cancer (≥ four cases) but the attribution diminishes sharply for smaller family clusters (51). The estimated population frequency of mutations in these genes is approximately 1/1000 per gene in the United Kingdom. Overall this equates to 15%–20% of the excess familial risk of breast cancer (4, 7, 9, 42, 91).

BRCA1 and *BRCA2* are also high-penetrance ovarian cancer genes: Mutations in *BRCA1* confer a higher risk of ovarian cancer than those in *BRCA2*, particularly for carriers below 50 years of age (5, 7, 44, 51). *BRCA1* and *BRCA2* mutations account for most of the epidemiologically observed familial coaggregation of breast and ovarian cancer (45, 92). 95% of families containing four or more breast cancers and an ovarian cancer were linked to *BRCA1* or *BRCA2*, whereas recent mutational screening detected mutations in 83% of families containing at least two cases of each cancer (51, 99).

The relative risk of male breast cancer is elevated for both genes, particularly *BRCA2*. An elevated risk of prostate cancer has also been demonstrated in *BRCA2* carriers, particularly in men aged <65 years (21, 116). Small excesses of a number of other cancers have been observed in monoallelic (heterozygous) *BRCA1* and *BRCA2* mutation carriers but larger studies are required to clarify whether these findings reflect truly elevated risks of these cancers (21, 116). Biallelic mutations in *BRCA2* result in Fanconi anemia, subtype D1. Fanconi anemia is a rare, recessive chromosomal instability syndrome characterized by skeletal abnormalities, bone marrow failure, and cancer predisposition. Subtype D1 has a distinctive phenotype that includes a high risk of childhood solid tumors (66). Biallelic *BRCA1* mutations have never been reported convincingly in humans and are presumed to be embryonically lethal (49).

BRCA1 tumors are typically high grade, invasive ductal carcinomas in which there is a high incidence of triple negative phenotype: negative staining for ER (estrogen receptor), PR (progesterone receptor), and HER2 (ERBB2) (20, 73, 75). These tumors also frequently stain positively for a subset of basal keratins characteristically expressed in the normal basal myoepithelium of the breast. This basal phenotype is distinctive and largely encompasses the previous histological descriptions of *BRCA1* tumors as medullary (69, 74). No distinctive histopathological features have been described in *BRCA2* tumors.

TP53. Li-Fraumeni syndrome is a cancer predisposition syndrome in which there is a high frequency of early onset breast cancer found in association with sarcomas and childhood cancers of the adrenal cortex, brain, and other sites. Although of a penetrance sufficient for mapping through linkage analysis, the associated early mortality and rarity of the condition impeded collection of sufficient familial samples. p53 was recognized early as a prominent transcription factor central to multiple cellular pathways and is frequently somatically mutated in tumors. These observations recommended *TP53* as a plausible candidate gene for Li-Fraumeni syndrome and in 1990 mutational screening of the gene revealed causative mutations in the five families studied (82). The overall lifetime cancer risk for women with Li-Fraumeni syndrome is grossly elevated, predominantly on account of their high risk of breast cancer (15, 28). Li-Fraumeni syndrome is rare and mutations in *TP53* are uncommon in non-Li-Fraumeni breast cancer families (18, 48, 76). Thus, the attributable risk of *TP53* mutations to familial breast cancer is very low.

Breast Cancer Predisposition Genes of Uncertain Penetrance

Three syndromes are currently clearly associated with an increased risk of breast cancer, but the magnitude of the associated risk for each remains uncertain: Cowden syndrome (caused by *PTEN* mutations), Peutz-Jeghers syndrome (caused by *STK11* mutations), and Hereditary diffuse gastric cancer syndrome (caused by *CDH1* mutations).

PTEN. Cowden syndrome is a multiple hamartoma syndrome that includes increased risk of benign and malignant tumors of the breast, thyroid, and endometrium; distinctive higher-penetrance features include mucocutaneous lesions, macrocephaly, and hamartomatous intestinal polyps. A study of 12 Cowden syndrome families revealed linkage to chromosome 10q, leading to identification of *PTEN* as the causative gene (87, 88). *PTEN* encodes a lipid phosphatase that functions as a tumor suppressor through negative regulation of a cell-survival signaling pathway. There is cross talk between this PTEN-related pathway and other pathways, including those involving Ras, p53, and TOR (36).

STK11. Peutz-Jeghers syndrome is characterized by hamartomatous intestinal polyps, mucocutaneous pigmentation, and increased incidence of several malignancies, including breast cancer. Directed by patterns of loss of heterozygosity in polyps of affected individuals, studies in 12 Peutz-Jeghers families established linkage to chromosome 19p and positional cloning led to the identification of *STK11* (*LKB1*) as the causative gene. STK11 is a serine/threonine kinase that inhibits cellular proliferation, controls cell polarity, and interacts with the TOR pathway (1, 63, 64, 70).

CDH1. Linkage analysis in a single New Zealand Maori family with multiple cases of diffuse gastric cancer enabled the identification of *CDH1* (*ECAD*) as the responsible gene (58). Many additional families have since been reported and an elevated frequency of lobular breast carcinoma has been observed (71, 96). Occasional *CDH1* mutations in families with cases of lobular breast cancer but no gastric cancer have also been reported (83). *CDH1* encodes E-cadherin, a transmembrane protein important in the maintenance of cell polarity.

These conditions were identified because the non-breast-cancer-related features were sufficiently distinctive and penetrant to reliably ascertain families and assign affection status, thus facilitating mapping of the causative genes by linkage analysis. The true breast cancer risks associated with mutations are unclear and it is possible that the published risks are inflated through the bias of studying families with prominent phenotypes. However, the relative risks are likely to be intermediate and in the range of 2 to 10. There is no current evidence that mutations in *PTEN*, *STK11*, or *CDH1* account for a substantial proportion of familial or sporadic breast cancer in the absence of their respective syndromes; because these syndromes are very rare, the attributable risk of mutations in these genes to familial breast cancer is low (14, 27, 57, 83, 107).

Intermediate-Penetrance Breast Cancer Predisposition Genes

Four intermediate-penetrance breast cancer genes have been identified via mutational screening: *CHEK2*, *ATM*, *BRIP1*, and *PALB2*. Mutations in these genes are rare and confer a relative risk of breast cancer of 2 to 4. *RAD50* may also be an intermediate-penetrance breast cancer predisposition gene, but convincing results to date pertain only to a founder mutation detected in the Finnish population. The rarity of mutations and modest associated risks are such that the attributable risk of mutations in these genes is low: Together they account for approximately 2.3% of excess familial risk (98).

CHEK2. *CHEK2* encodes CHK2, a central mediator of cellular response to DNA damage that phosphorylates both p53 and BRCA1 to regulate repair of DNA double-strand breaks. The 1100delC mutation in *CHEK2* was first reported after mutational screening of this eligible candidate gene in a single family with features of Li-Fraumeni syndrome (13). However, the population frequency of this mutation was shown to be approximately 1% (18/1620), demonstrating that *CHEK2* could not be a high-risk Li-Fraumeni syndrome gene. The identification of the *CHEK2*_1100delC mutation in 4.2% (30/718) of breast cancer families demonstrated, instead, that *CHEK2* was an intermediate-penetrance breast cancer predisposition gene ($P = 5 \times 10^{-6}$) (85). Segregation analysis in these families provided an indirect estimate of 1.70 (95% CI = 1.32–2.20) for the relative risk of breast cancer conferred by the mutation. Combining data from ten case control studies, the *CHEK2* consortium demonstrated the frequency of *CHEK2*_1100delC in population-based breast cancer cases to be 1.9% (201/10,860) compared with 0.7% in controls (64/9065) ($P = 1 \times 10^{-7}$). These frequencies equate to a direct odds ratio of breast cancer of 2.34 (95% CI = 1.72–3.20) (26). It has been proposed that *CHEK2*_1100delC may confer risks of prostate and other cancers but evidence is conflicting (37, 38, 85, 105, 117). There are also reports of other rare truncating mutations in *CHEK2*, which would be anticipated to have similar risks to *CHEK2*_1100delC (17, 39, 55, 103, 119). A number of rare *CHEK2* missense variants, such as 470T/C (I157T), have been reported in breast cancer cases but their significance has yet to be clarified (2, 17, 55, 72, 103).

ATM. The proposal of *ATM* as a breast cancer predisposition gene first came in 1976 from an epidemiological study that reported an excess of breast cancer in female relatives of patients with ataxia telangiectasia, an autosomal recessive syndrome characterized by progressive cerebellar ataxia, immune deficiency, and cancer predisposition. This astute observation preceded the mapping of the gene by almost two decades and has been subsequently replicated in a number of large epidemiological studies (41, 102, 112, 113). The candidacy of *ATM* as a breast cancer predisposition gene was further enhanced as the function of the encoded protein became apparent: ATM occupies a central role in the response to double-strand DNA breaks through initiation of a signaling cascade that involves phosphorylation of multiple proteins including p53, BRCA1, and CHK2. Results from initial mutational screening of *ATM* in breast cancer were inconclusive and/or inconsistent but convincing proof was finally published in 2006. Mutations were found in 12/443 familial cases negative for mutations in *BRCA1* and *BRCA2* and 2/521 controls ($P = 0.0047$). These mutations comprised truncations, splice-site abnormalities, and two missense mutations that were known to affect protein function and cause ataxia telangiectasia. The relative risk of breast cancer conferred by *ATM* mutations was estimated from segregation analysis to be 2.37 (95% CI = 1.51–3.78, $P = 0.0003$) (101). This indirect estimate is very similar to that derived from large epidemiological studies (113). There is some epidemiological evidence that mutations in *ATM* may predispose to other cancers but this has yet to be confirmed or corroborated by molecular data (113).

BRIP1. *BRIP1* (*BACH1*) encodes a DEAH helicase that interacts with BRCA1 and has BRCA1-dependent roles in DNA repair and checkpoint control (24, 90). In 2006, truncating *BRIP1* mutations were reported in 9/1212 index breast cancer cases from families negative for mutations in *BRCA1* and *BRCA2* compared with 2/2081 controls, thus providing convincing evidence that *BRIP1* is a breast cancer predisposition gene ($P = 0.0030$). Segregation analysis found the relative risk conferred by *BRIP1* mutations to be 2.0 (95% CI = 1.2–3.2, $P = 0.012$) (104). Of the mutations reported in *BRIP1*, approximately half are a nonsense mutation, 2392C/T (R798X), whereas the remainder include disparate small insertions and deletions. Concurrently, it emerged that biallelic mutations in *BRIP1* result in Fanconi anemia, although the Fanconi anemia subtype associated with *BRIP1* (subtype J) is phenotypically distinct from that associated with *BRCA2* (subtype D1) and has not been associated with childhood solid tumors (78, 79, 81).

PALB2. Precipitation of BRCA2-containing complexes revealed a novel protein that was shown to promote the localization and stability of BRCA2, thus facilitating BRCA2-mediated DNA repair. Further study of *PALB2* (partner and localizer of BRCA2) revealed that knockdown of the gene resulted in sensitization of cells to chromosomal damage by mitomycin C, the hallmark of Fanconi anemia (125). Biallelic truncating mutations in *PALB2* were detected in families with Fanconi anemia not caused by mutations in other genes and this subtype was designated FA-N (100, 124). The phenotypes of FA-N and Fanconi anemia caused by *BRCA2* mutations (FA-D1) are very similar and can be distinguished from classical Fanconi anemia on account of a markedly elevated frequency of childhood solid tumors such as Wilms tumor and medulloblastoma. With the recent precedents of both *BRIP1* and *BRCA2* causing Fanconi anemia in biallelic mutation carriers and conferring susceptibility to breast cancer in monoallelic carriers, the identification of *PALB2* as a Fanconi anemia gene further recommended *PALB2* as an attractive candidate breast cancer predisposition gene.

Sequencing of the gene revealed truncating mutations in 10/923 index breast cancer cases from families negative for mutations in *BRCA1* and *BRCA2* compared with 0/1084 in controls ($P = 0.0004$). Segregation analysis from these families estimated the relative risk of *PALB2* mutations to be 2.3 (95% CI = 1.4–3.9, $P = 0.0025$) (98). Mutational screening of *PALB2* in a Finnish series detected one truncating mutation: 1592delT. This mutation was identified in 3/113 (2.7%) familial cases, 18/1918 (0.9%) unselected breast cancer cases, and 6/2501 (0.2%) controls. The directly derived odds ratio of breast cancer of this mutation is therefore 3.94 (95% CI = 1.5–12.1) (47). This mutation has not been detected in other series and may represent a Finnish founder. A French Canadian founder mutation, 2323 C/T (Q775X), has also been reported (52).

RAD50. The highly conserved MRN complex (*MRE11*, *RAD50*, and *NBS1*) interacts with BRCA1 and plays a central role in DNA repair. A truncating mutation in *RAD50*, 657delT, was demonstrated in 8/317 consecutively ascertained breast cancer cases and 6/1000 controls from Finland [$P = 0.008$, odds ratio (OR) = 4.3, 95% CI = 1.5–12.5]. Other rare truncating mutations have been reported in *RAD50*, but the contribution of *RAD50* to breast cancer predisposition outside of Finland requires further clarification (62, 118).

Distinctive features of intermediate-penetrance breast cancer predisposition genes. As in *BRCA1* and *BRCA2*, multiple different truncating mutations occur in these intermediate-penetrance genes, the frequencies of which differ between populations on account of founder and migrational effects. Monoallelic mutations in these genes confer an approximately twofold increase in the risk of breast cancer. Although there is some imprecision in the indirect risk estimates derived from segregation analysis, population-based series and epidemiological data have provided some corroboration for these figures. There is currently no strong evidence that monoallelic mutations in these genes confer a phenotype beyond breast cancer predisposition. Biallelic mutations in *ATM*, *BRIP1*, and *PALB2* result in severe childhood disorders: ataxia telangiectasia and Fanconi anemia types J and N, respectively (97). Because these genes function in the same pathways as BRCA1 and BRCA2, it remains unclear why abrogation of their functions confers more modest risks of breast cancer. However, the lower penetrance of these genes means that they deviate from some of the hallmarks of pathogenicity associated with mutations of *BRCA1* and *BRCA2*. Mutations in these genes do not show the classical, near-complete pattern of segregation of disease with the family mutation. By contrast, because breast cancer is a common disease, in a typical pedigree an appreciable proportion of breast cancers arise by chance and thus may manifest in either mutation carriers or noncarriers. Likewise, one should not anticipate finding the pattern of loss of the wild-type allele detected at high frequency in tumors in *BRCA1* and *BRCA2* mutation-positive individuals. Firstly, the mechanism of tumorigenesis of these genes is unknown and may not require inactivation of the wild-type allele. Secondly, even if the classical, two-hit model of a tumor suppressor gene does apply, one could not expect to demonstrate loss of heterozygosity in all tumors occurring in mutation carriers because the mutation is not causally implicated in a significant portion of them.

Low-Penetrance Breast Cancer Predisposition Alleles

There is currently strong evidence for the association with breast cancer of eight common alleles, which each confer a relative risk of breast cancer of <1.5. A nonsynonymous variant in *CASP8* was identified through a candidate-based approach in a consortium comprising 14 studies that compared 17,109 cases and 16,423 controls (34). Seven further variants have been detected in three recent genome-wide association studies (**Table 2**). Easton and coworkers (46) performed a three-stage genome-wide experiment. In the first stage a Perlegen platform was used to study the association of 227,876 SNPs in 390 cases with a family history of breast cancer and/or bilateral disease and 364 controls from the United Kingdom. In the second stage 12,711 SNPs (the most significant 5% of SNPs from stage 1) were genotyped in 3990 cases and 3916 UK controls using a custom-designed array. In the third stage the 30 most significant SNPs were tested for confirmation in 21,860 cases and 22,578 controls from 22 international studies (46). The study identified five variants associated with breast cancer at a significance level of $P < 10^{-7}$. Stacey and colleagues (110) conducted a genome-wide scan in 1600 Icelandic cases and 11,563 controls using the Illumina HumanHap300 platform; the ten best associated SNPs were studied further in five international replication sets comprising 2954 cases and 6014 controls. This study identified one of the variants reported by Easton and coworkers (rs3803662) and a novel region of association on 2q35. As part of the National Cancer Institute Cancer Genetic Markers of Susceptibility (CGEMS) Project, Hunter and colleagues (67) used the Illumina HumanHap500 array to genotype 1145 postmenopausal invasive breast cancer cases and 1142 North American controls of European ancestry. The six best-associated signals were

Table 2 Summary of known low-penetrance breast cancer predisposition variants*

Locus	Estimated size of LD block (kb)	Genes within LD block	SNP	MAF	Heterozygote OR (95% CI)	Homozygote OR (95% CI)	Per allele OR (95% CI)	*P*- trend	Ascertain-ment	Study
10q26	25	*FGFR2*	rs2981582	0.38	1.23 (1.18–1.28)	1.63 (1.53–1.72)	1.26 (1.23–1.30)	2×10^{-76}	GW	Easton et al. (UK) (46)
			rs1219648	0.39	1.20 (1.07–1.42)	1.64 (1.42–1.90)		1.1×10^{-10}	GW	Hunter et al. (USA) (67)
16q12	160	*TNRC9* *LOC643714*	rs3803662	0.25	1.23 (1.18–1.29)	1.39 (1.26–1.45)	1.20 (1.16–1.24)	1×10^{-36}	GW	Easton et al. (UK)
			rs3803662	0.27	1.27 (1.19–1.36)	1.64 (1.45–1.85)	1.28 (1.21–1.35)	5.9×10^{-19}	GW	Stacey et al. (Iceland) (110)
2q35	117	–	rs13387042	0.50	1.11 (1.03–1.20)	1.44 (1.30–1.58)	1.20 (1.14–1.26)	1.3×10^{-13}	GW	Stacey et al. (Iceland) (110)
5p12	310	*MRPS30*	rs10941679	0.24			1.19 (1.13–1.26)	2.9×10^{-11}	GW	Stacey et al. (Iceland) (111); Easton et al. (UK) (46)
5q11	280	*MAP3K1* *MGC33648* *MIER3*	rs889312	0.28	1.13 (1.09–1.18)	1.27 (1.19–1.36)	1.13 (1.10–1.16)	7×10^{-20}	GW	Easton et al. (UK) (46)
2q33	290	*CASP8* *TRAK2* *ALS2CR12* *ALS2CR2* *ALS2CR11* *LOC389286* *LOC729191*	rs1045485	0.13	0.89 (0.85–0.94)	0.74 (0.62–0.87)	0.88 (0.84–0.92)	1.1×10^{-7}	Candidate	Cox et al. (UK) (34)
8q24	110	–	rs13281615	0.40	1.06 (1.01–1.11)	1.18 (1.1–1.25)	1.08 (1.05–1.11)	5×10^{-12}	GW	Easton et al. (UK) (46)
11p15	180	*LSP1* *TNNT3* *MRPL23* *H19* *LOC728008*	rs3817198	0.30	1.06 (1.02–1.11)	1.17 (1.08–1.25)	1.07 (1.04–1.11)	3×10^{-9}	GW	Easton et al. (UK) (46)

*Abbreviations used: LD, linkage disequilibrium; MAF, minor allele frequency in control population; OR, odds ratio (measured relative to common homozygotes); CI, confidence interval; *P*-trend, *P* value for per allele effect under multiplicative model; GW, genome-wide association study; Candidate, association studies of single nucleotide polymorphisms (SNPs) in candidate genes.

further studied in 1776 cases and 2072 controls from three North American replication sets. The results of this study confirmed the signal in intron 2 of *FGFR2* reported by Easton and coworkers (67). The final predisposition SNP on 5p12 was identified through mining and cross-referencing of associations of borderline significance from these three genome-wide association studies; this signal was verified by Stacey and coworkers (111) in a replication series of 5,028 cases and 32,090 controls.

10q26 (FGFR2 intron 2). The strongest evidence for association was for a signal within intron 2 of the gene encoding fibroblast growth factor receptor 2 (*FGFR2*) [per allele OR = 1.26 (1.23–1.30) $P = 2 \times 10^{-76}$] (46, 67, 111). Fine association mapping of the region of linkage disequilibrium has been undertaken using UK samples and refined using an Asian series (in which linkage disequilibrium is weaker). These studies have reduced the association to a minimum set of six variants within intron 2, which are too strongly correlated for their individual effects to be discerned using further genetic epidemiologic approaches (46).

Somatic *FGFR2* mutations have been reported in several cancers and result in overactivity of the protein. It therefore seems plausible that the elevated breast cancer risk is somehow mediated through increased/altered activity of FGFR2, but the molecular basis for the association is currently unknown.

The risk-prevalence profile of this allele means that the power to detect it in the genome-wide scans is high. The corollary of this is that if other alleles of a similar risk-prevalence profile exist, they should most likely have been detected. Thus, it is possible that this may be the allele of most substantial effect to exist within this category of risk alleles. The differences in the risks conferred are nevertheless modest: The risk of breast cancer by age 70 in individuals homozygous for the risk allele is 10.5% compared with 6.7% for heterozygotes and 5.5% for nonrisk allele homozygotes (based upon UK breast cancer incidence figures). However, the risk allele is common: 14% of the UK population and 19% of UK breast cancer cases are homozygous for the risk allele. Hence, this allele may account for approximately 1.9% of the excess familial risk of breast cancer.

16q12. Strong evidence for association of rs3803662 with breast cancer was reported in two studies [per allele OR = 1.20 (1.16–1.24), $P = 1 \times 10^{-36}$ in the UK study and per allele OR = 1.28 (1.21–1.35), $P = 5.9 \times 10^{-19}$ in the Icelandic study] (46, 110). This variant lies on 16q12 and tags a region of linkage disequilibrium containing the 5′ end of *TNRC9* (*TOX3*) and a hypothetical gene *LOC643714*. *TNRC9* is a gene of uncertain function that contains a putative high mobility group box motif, suggesting that it may act as a transcription factor. Results from a previous study suggested that *TNRC9* expression is predictive of metastasis of breast cancer to bone (108).

2q35. The Icelandic study found evidence for association with breast cancer of a variant on 2q35 [per allele OR = 1.20 (1.14–1.26), $P = 1.3 \times 10^{-13}$]. The region of linkage disequilibrium does not contain any known genes. The nearest genes are *TNP1* (181 kb), *IGFBP5* (345 kb), and *IGFBP2* (376 kb) upstream and *TNS1* (761 kb) downstream. Although this signal was not detected in the other genome-wide scans, we have replicated the association in a UK series of familial breast cancer cases [per allele OR = 1.17 (95% CI 1.07–1.27), $P = 0.0004$] (C. Turnbull and N. Rahman, unpublished data) and it is being evaluated by the Breast Cancer Association Consortium.

5p12. Stacey and coworkers (111) noted that one of the ten top-ranked SNPs in the Icelandic genome-wide study was on chromosome 5p12 (rs7703618). The CGEMS study had reported an association of borderline significance on 5p12 at rs4866929, a SNP in tight linkage disequilibrium with rs7703618, while Easton and coworkers (46) had reported a tentative signal at 5p45 (rs981782), 371 kb away, which was also in loose linkage disequilibrium with rs7703618 ($r^2 = 0.10$). Struck by the coincidence of tentative signals, they examined 21 SNPs in this region for association with breast cancer and found rs10941679 to be the most significantly associated SNP [per allele OR 1.19 (1.13–1.26), P = 2.9×10^{-11}]. The only gene in the region of linkage disequilibrium is *MRPS30* (*PDCD9*, programmed cell death protein 9), which encodes a component of the small subunit of the mitochondrial ribosome and has been implicated in apoptosis.

5q11. Association has been found for rs889312 [per allele OR = 1.13 (1.10–1.16), $P = 7 \times 10^{-20}$]. This SNP tags a 280-kb block of linkage disequilibrium on 5q11 that contains the genes *MAP3K1* (*MEKK*), *MGC33648*, and *MIER3*. *MAP3K1* is the most plausible candidate gene therein and encodes mitogen-activated protein kinase kinase kinase 1, which is involved in cell signaling (46).

2q33 (CASP8 D302H). This association was ascertained through the biological candidacy of *CASP8*, which encodes caspase-8, a protein involved in apoptosis [per allele OR = 0.88 (0.84–0.92), $P = 1.1 \times 10^{-7}$] (34). However, the region of linkage disequilibrium tagged by this SNP is approximately 290 kb in length and includes a number of other genes: *TRAK2* (encoding trafficking protein kinesin binding 2), three genes identified as candidates for juvenile amyotrophic lateral sclerosis 2 (*ALS2CR12*, *ALS2CR2*, and *ALS2CR11*), and two hypothetical genes (*LOC389286* and *LOC729191*). It is currently unclear whether D302H is the causal variant or whether it tags a distinct causal variant, which may or may not mediate its effect through CASP8. The relatively low minor allele frequency and risk of this allele mean that the power to detect its signal at the genome-wide level is comparatively low. This is consistent with failure of the allele to reach the arbitrary significance thresholds required for further investigation in the genome-wide association scans performed to date. It is also consistent with the presumed existence of many further alleles of comparable risk-prevalence profiles.

8q24. Association was found for a SNP within an 110-kb block of linkage disequilibrium on 8q24, which contains no known genes [per allele OR = 1.08 (1.05–1.11), $P = 5 \times 10^{-12}$] (46). It is interesting that in the first wave of genome-wide association studies in common cancers, 8q24 has also yielded multiple independent prostate cancer loci, one of which is also a colon cancer risk allele (rs6983267) (59, 60). The tag SNP associated with breast cancer does not demonstrate association with colon cancer or prostate cancer and is 60 kb proximal to rs6983267. Clustering of these predisposition alleles may be a coincidence or may indicate a common or related mechanism of cancer predisposition. The nearest gene to the breast cancer predisposition locus is *MYC* and it is plausible that the susceptibility occurs through some unknown mechanism of activation of this oncogene.

11p15. Another associated tag SNP lies in intron 10 of *LSP1* (*WP43*), which encodes lymphocyte-specific protein 1 [per allele OR = 1.07 (1.04–1.11), $P = 3 \times 10^{-9}$], an F-actin bundling cytoskeleton protein that is expressed in hematopoetic and endothelial cells (46). Other genes within this region of linkage disequilibrium include *TNNT3*, troponin T type 3; *MRPL23*, mitochondrial ribosomal protein; *H19*, an imprinted, maternally expressed, untranslated mRNA; and *LOC728008*, a hypothetical gene.

Distinctive features of low-penetrance breast cancer predisposition alleles. A fascinating aspect of the recent genome-wide scans has been the opaqueness of the relationship of the identified variants with known protein-coding genes. The patterns of linkage disequilibrium suggest that the causal variants need not lie in the coding region of a gene, as evidenced by the predisposition variants at 2q and 8q that are tens of kilobases away from the nearest protein-encoding gene. Any significance to the roles of genes upstream or downstream of the associated blocks of linkage disequilibrium is currently speculative.

The interpretation of subgroup analyses of the immunohistopathologic phenotypic variation of risk alleles requires caution on account of the limited power of some analyses. Initial analyses by the Icelandic group (110, 111), on identification of the predisposition SNPs at 16q12, 2q35, and 5p12, showed the effects of these alleles to be confined to ER-positive tumors. Although subsequent analyses of the SNPs at 10q26, 16q12, 8q24, 5q11, and 11p15 from the United Kingdom genome-wide association data also revealed that for all five alleles, the estimates of effect were stronger in ER-positive than ER-negative disease, the difference was only significant for those at 10q26 and 8q24. These two SNPs were also independently associated with a lower grade of disease. Adjusted analyses of these five SNPs did not reveal significant association with lymph node status nor survival (53).

Because these alleles occur at a high frequency in the general population, their population attributable risks (etiologic fractions) are relatively high (13%–16% for the alleles of stronger effects) (67, 110). However, this figure represents only the proportion of breast cancer cases in which the variant has played some causal role in development of disease. The associated risks are low and it is estimated that the five loci characterized by Easton and coworkers (46) account for a modest 3.6% of the excess familial risk of breast cancer in European populations.

INTERACTIONS BETWEEN BREAST CANCER PREDISPOSITION FACTORS

Interaction is an important aspect of risk calculation, particularly in familial disease clusters in which multiple predisposition factors are likely active. The combined risk of two factors is strongly dependent upon the nature of the interaction between them. Typically the default model assumes that they act independently and multiplicatively (as per terms in a multiple regression analysis). The true situation may be more complex and comprise a mixture of multiplicative, additive, antagonistic, synergistic, and/or complex intermediate interactions.

The study of co-occurrence of genetic predisposition factors and exposition of gene-gene interactions has become more viable since the identification of convincing common predisposition alleles. Preliminary analyses from genome-wide association studies suggest that the low-penetrance risk alleles act multiplicatively with each other (46, 110). However, association studies of the SNPs at 10q26 and 5q11 performed within *BRCA1* and *BRCA2* mutation-positive families suggest that these SNPs confer additional risk in the presence of *BRCA2* but not *BRCA1* mutations (12). This interesting disparity may in part reflect the association of the SNPs with ER-positive tumors because *BRCA1* is typically associated with ER-negative tumors. In a recent candidate-based experiment, it was reported that homozygosity for a variant in the 5′ untranslated region of *RAD51* confers an increased risk of breast cancer to *BRCA2* mutation carriers [hazard ratio = 3.18 (95% CI = 1.39–7.27)]. The modifying effect was not significant in *BRCA1* mutation carriers or when only a single copy of the risk allele was present (11). By contrast, *CHEK2*_1100delC has not been shown to confer an elevated risk on the background of mutations in either *BRCA1* or *BRCA2* (85). It has been proposed that this reflects the common pathway of the encoded proteins; abrogation of CHK2 function might have little additional impact on a pathway already radically subverted by a mutation in *BRCA1* or *BRCA2*. However, there is currently no biological proof of this hypothesis. Further analyses are required to establish whether similar interactions are observed for *ATM*, *BRIP1*, and *PALB2*, which also interact with BRCA1 and/or BRCA2 in DNA repair pathways.

Clarity regarding gene-environment interactions is even more limited. Recognized environmental risk factors for breast cancer in the general population are well established and predominantly relate to estrogen exposure. Endogenous estrogen-related risk factors include timing of menarche and menopause, parity, age of first live birth, and breast-feeding, whereas exogenous factors include administration of contraceptives and hormone replacement therapy (30, 31, 33). Studies of hormonal/reproductive factors in *BRCA1* and *BRCA2* mutation carriers are challenging, not just in the assembly of sufficient families, but because of the biases inherent in observation of the behaviors of a group of individuals who know themselves to be at elevated risk. However, well-powered studies of gene-environment interactions are becoming possible through collaboration and are allowing the effects of *BRCA1* and *BRCA2* to be studied independently. Significant alteration of breast cancer risk in *BRCA1* and *BRCA2* mutation carriers has been demonstrated for oral contraceptive usage, age at first pregnancy, and degree of parity, whereas significant effects were not demonstrated for age of menarche, age of natural menopause, nulliparity, and breast feeding (3, 23, 25). These observations warrant further investigation and confirmation in prospective studies. Other large studies will be required to explore gene-environment interactions for genetic predisposition factors of lower penetrance.

IDENTIFICATION OF FURTHER GENETIC BREAST CANCER PREDISPOSITION FACTORS

There is a clear discontinuity in the risks associated with the three categories of breast cancer predisposition factors identified to date: The high-penetrance genes confer a risk that is elevated >10-fold, the known intermediate-penetrance genes 2–4-fold, and the low-penetrance alleles <1.5-fold. There may be some biological significance to these strata or they may be an artifact of the limited methods of ascertainment. More than 70% of genetic predisposition to breast cancer remains unaccounted for and as further genetic predisposition factors are identified, new categories may emerge and/or the apparent distinctions between extant classes may blur or disappear. Nevertheless, risk and prevalence provide a useful framework for considering how technological and intellectual advances may allow extension of the repertoire of breast cancer predisposition factors.

High-Penetrance Genes

Any further high-penetrance dominant predisposition genes are likely to be very rare causes of familial breast cancer. Segregation analyses generated no evidence for further dominant genes of a risk-penetrance profile comparable to *BRCA1* or *BRCA2* and this has been corroborated by linkage studies. In a recently published example, genome-wide linkage analysis was undertaken in 149 nonsyndromic breast cancer families negative for mutations in *BRCA1* and *BRCA2* and the number of linkage peaks detected under parametric (dominant and recessive) and nonparametric (allele-sharing) models did not differ significantly from that expected by chance (109). However, linkage analyses such as this have been undertaken in families largely ascertained on the basis of multiple-generation breast cancer pedigrees. A highly penetrant recessive predisposition gene would not produce this pattern of disease in families. Two independent segregation analyses found evidence of a recessive pattern of inheritance (8, 35). Thus, there may be utility in genome-wide recessive linkage analyses in more suitable series, in particular families from understudied population isolates, and/or in families with higher levels of consanguinity. Further, unidentified breast cancer-associated syndromes may also exist. However, if these syndromes have not yet come to medical/scientific attention, they are likely to be very rare; linkage is likely to represent the optimal method for mapping any such underlying predisposition genes.

Intermediate-Penetrance Genes

Further intermediate-penetrance breast cancer predisposition genes likely exist, although it is difficult to predict how many there may be and what proportion of the total excess familial risk is attributable to them. Resequencing is currently the optimal approach by which to identify genes of this nature. The known intermediate-penetrance genes are all involved in DNA repair and function in pathways with BRCA1 and/or BRCA2; this commonality may reflect the underlying biology of this class or may just be an artifact of the groups of genes that have been most intensely investigated. Further genes involved in DNA repair and other relevant pathways represent plausible candidates and are being investigated by us and other groups. However, technological advances are facilitating increasingly high-throughput mutational screening such that genome-wide resequencing is becoming viable. This will allow interrogation of regions of the genome not previously investigated and may reveal further intermediate-penetrance genes that function in pathways not predictable from current paradigms.

Intermediate-penetrance predisposition to breast cancer may also be an unrecognized component of known pleomorphic cancer predisposition syndromes. Large, collaborative epidemiological studies of rare syndromes can optimize power and minimize bias and may represent the optimal strategy by which to establish accurate estimates of these breast cancer risks. Epidemiological studies of recessive syndromes, particularly those associated with childhood cancer, may reveal elevated frequency of breast cancers in relatives of affected individuals and may lead to the identification of further intermediate-penetrance breast cancer genes, as in the case of *ATM*.

Low-Penetrance Variants

Common low-penetrance variants. Extant genome-wide association data offer clear evidence that many further common low-penetrance breast cancer predisposition variants exist. Further risk alleles may be identified from these data by mining the variants of borderline significance for further signals and through the use of imputation techniques. However, comprehensive study of this (likely) extensive repertoire of low-penetrance alleles will require further, larger, genome-wide association experiments. The effect sizes of subsequent rounds of risk alleles are likely to be of progressively diminishing magnitude and will require commensurate increases in power for detection. It is currently unclear what proportion of common low-penetrance alleles will be detectable by the feasible studies in the immediate future.

Strongly associated tag SNPs from genome-wide scans may be utilized for genetic epidemiologic analyses and clinical risk estimation in their own right. However, the way in which these common variants contribute to cancer is largely unknown and exploration of the biological mechanisms that underlie these signals offers exciting new avenues of study. Resequencing and fine-association mapping of the region tagged by a reporter SNP can refine the association to a minimum set of SNPs in a fixed block of linkage disequilibrium. The cancer-causing components within these blocks are cryptic. Attempts to identify them may challenge current paradigms of the relationship between genes and disease and require innovative methods.

To date, genome-wide studies have been undertaken in simple series of breast cancer cases of European descent. Novel risk alleles may be identifiable through genome-wide studies of different ethnic groups in which the minor allele frequencies and phenotypic spectrum of disease differ from Europeans. For example, a risk locus at 6q22 was identified in a recent genome-wide study performed exclusively in Ashkenazi Jewish breast cancer cases and controls; verification of this association in other populations is awaited (54). Broader insight into genetic etiology may be gained through the expansion of genome-wide studies into subgroups of breast cancer cases. For example, analysis of *BRCA*-positive individuals may allow identification of modifier alleles. Studies focusing on well-characterized histopathological subgroups of breast cancers may advance our understanding of the genetic basis of disease heterogeneity. The use of quantitative intermediate phenotypes, such as mammographic density, as the outcome for genome-wide studies may also represent an alternative approach.

Rare low-penetrance variants. The risk-prevalence profiles already identified suggest that rare variants of low penetrance represent another plausible category of breast cancer predisposition factor. However, there are currently no reliable methods by which to quantify the risk of breast cancer associated with individual, rare variants. Therefore, although it is possible that some of the rare variants detected during mutational screening experiments are low-penetrance predisposition factors, our ability to demonstrate this is limited. Genome-wide resequencing will inevitably detect many further rare variants, some of which will likely be associated with small increases in risk of breast cancer. Developing robust methods for the identification of the cancer-associated rare variants among the numerous innocuous variants will be a major challenge for the future.

Optimizing the Power of Gene Identification Studies

A priority for all future studies will likely be the harnessing of sufficient statistical power at acceptable cost. An important tool for gaining power without increasing the experiment size/cost is to assemble a case series enriched for genetic predisposition. The best-recognized of these enrichment parameters, successfully used in many experiments to date, include family history, bilaterality, and early age of onset of disease. Their relative efficacies have been compared through modeling the sample size required to demonstrate association with disease of a variant of specified frequency and effect. The use of cases with a single affected first-degree relative affords a greater than twofold reduction in the required sample size; for cases with two affected first-degree relatives the reduction is more than fourfold. The sample size reduction when using bilateral cases is also fourfold (and thus equivalent to using cases with two first-degree relatives). However, the sample size required for an association study that uses cases diagnosed at age 35 is only 40% less than one that uses cases diagnosed at 65 years. Relative efficiencies are all magnified for experiments in which rarer alleles are studied (6).

An alternative strategy is geographical enrichment, namely to first screen genes in population isolates in which a higher prevalence of founder mutations might be anticipated, such as the Ashkenazim, the Finnish, or the Icelanders. This has proved very successful for the identification and characterization of known genes. However, as is currently the case for *RAD50*, the generalizability of findings from specific populations may prove challenging.

There has been longstanding expectation that evolution in molecular profiling may result in pathological classifications that more directly reflect the genetic etiology of tumors. Although the basal phenotype clearly enriches for cases that arise because of *BRCA1* mutations, it remains unclear whether novel immunohistopathological profiles or other phenotypic surrogates may emerge that can distinguish further subsets of breast cancers that occur because of genetic predisposition.

CLINICAL TRANSLATION

Risk Estimation and Management

Risk estimation is currently the primary clinical application for genetic factors that predispose to breast cancer. Ongoing improvement of clinical risk assessment tools is necessary to ensure that individuals at truly elevated risk are identified so that they might benefit from advances in surveillance techniques and prophylactic interventions. Advances in the understanding of the polygenic basis of breast cancer may have different effects on three groups: families positive for mutations in *BRCA1* or *BRCA2*, other breast cancer families, and the general population.

Risk estimation in BRCA-positive breast cancer families. Detection in a family of a mutation in *BRCA1* or *BRCA2* has afforded relative clarity in risk estimation. Unaffected females, at high prior risk of cancer on account of their family history, can undertake predictive testing to determine whether they carry the high-risk mutation. Risk-reducing prophylactic surgery and/or intensive surveillance may be offered to mutation carriers whereas noncarriers can be reassured that their risk is not as high. However, studies have found wide variation in the penetrance of mutations in *BRCA1* and *BRCA2* (see above) and this has raised questions regarding the appropriate estimates of risk for clinical usage. Some of this apparent variation in penetrance between families may be the result of differing doses of additional lower-risk modifying variants; recently discovered examples include SNPs in *RAD51*, *FGFR2*, and at 5q (11, 12). Thus, one of the early clinical applications of low-penetrance alleles may be to offer individualized refinement of risk to *BRCA1* and *BRCA2* mutation carriers.

Risk estimation in BRCA-negative breast cancer families. The majority of breast cancer families do not harbor mutations in *BRCA1* or *BRCA2*. Clinical breast cancer risk estimation in these families is currently based empirically upon family history of cancer. For unaffected individuals in breast cancer families negative for *BRCA1* and *BRCA2* mutations, it would be clinically useful to have better discrimination of risk to distinguish high-risk individuals in whom radical surgical prophylaxis may be justified and individuals at lower/population risk who could be spared some anxiety and intervention. It remains unclear whether the underlying genetic architecture is such that this discrimination could be provided by genotyping multiple genetic predisposition factors (presuming sufficient numbers were identified). If the clustering of breast cancer in a family has occurred because of multiple low-penetrance genetic factors, typing of these factors would result in many possible risk categories. Genotyping may place the majority of unaffected members of the family into intermediate risk categories that differ little from their risk based on family history. If family clustering is the result of fewer factors of greater penetrance, genotyping these may adjust risk estimation sufficiently to alter clinical management for unaffected individuals.

Risk estimation in the general population. By contrast, in the absence of a family history of breast cancer, genotyping is the only means of identifying individuals at increased genetic risk. The discrimination of risk that could be afforded by population-level genotyping of common low-penetrance variants is influenced by two factors: firstly, the true extent to which breast cancer risk varies across the population (which is currently unclear) and secondly, the proportion of the risk variants that is available for genotyping (8, 9, 94). Although the logistical, ethical, social, and economic considerations are numerous, in principle it seems plausible that population-based genotyping could eventually be used to aid stratification and resource allocation. Surveillance, primary antiestrogen chemoprophylaxis, prophylactic surgery, and other emerging interventions may be appropriate for the high-risk upper tail of the population. The low-risk tail may require little or no additional intervention.

Targeted Therapies

Understanding of the biological mechanisms underlying breast cancer predisposition genes is beginning to offer exciting opportunities for new therapies. The roles of BRCA1 and BRCA2 in DNA repair and the resultant sensitivity of BRCA1- and BRCA2-deficient cells to agents that cross-link DNA are being exploited by studies in which mutation carriers are treated with platinum-based drugs. The inherent DNA repair defect in BRCA-deficient cells has also provided a rationale for a further therapeutic approach. Poly (ADP-ribose) polymerase (PARP) is an enzyme involved in base excision repair. Inhibition of PARP results in an increase in DNA lesions that are normally repaired through homologous recombination, which requires BRCA1 and BRCA2. In a background deficient for either BRCA1 or BRCA2 protein, cells are profoundly sensitive to inhibition by PARP, which results in cell cycle arrest, chromosome instability, and cell death. Thus in *BRCA* mutation carriers, PARP inhibitors are synthetically lethal to tumor cells but confer no demonstrable toxicity to normal heterozygous cells. Studies are also underway to investigate whether PARP inhibitors have similar effects in individuals with basal tumors similar to those occurring in *BRCA1* mutation carriers and/or cancers associated with defects in other DNA repair proteins (50, 84). Understanding the molecular basis of the increased cancer risk associated with low-penetrance alleles is in its infancy. However, this understanding may offer opportunities for novel therapies. For example, if some risk alleles drive tumorigenesis through the upregulation of oncogenes, such as *FGFR2* and/or *MYC*, these genes may represent potential new targets for therapeutic interventions.

CONCLUSION

Although recent breakthroughs have resulted in the identification of distinct new groups of genetic breast cancer predisposition factors, more than 70% of the genetic predisposition to breast cancer remains unexplained. Technologies are emerging rapidly that may allow us to identify many further predisposition factors within these classes and of other risk-prevalence profiles. The identification of novel predisposition factors offers exciting challenges, but ongoing clarification and characterization of known genetic risk factors is also important. A formidable and ongoing challenge is to marshal disparate experimental results to capture, validate, qualify, and organize into an integrated schema all the emerging components of risk. This is particularly important if we are to advance clinical risk estimation to optimally apportion surveillance and prophylactic interventions.

SUMMARY POINTS

1. Multiple strategies have been used to identify breast cancer–predisposing genetic factors.
2. Linkage analysis has been successful in mapping high-penetrance genes such as *BRCA1* and *BRCA2*.
3. Mutational screening of candidate genes has been used to identify genes such as *CHEK2*, *ATM*, *BRIP1*, and *PALB2*, mutations in which are rare and confer intermediate penetrance of breast cancer.
4. Genome-wide association studies have revealed several low-penetrance alleles.
5. More than 70% of breast cancer predisposition remains unexplained.
6. The outstanding predisposition is likely to be polygenic and may include multiple further common low-penetrance alleles, rare intermediate-penetrance genes, and rare low-penetrance alleles.
7. *BRCA1* and *BRCA2* mutation testing allows identification of individuals at elevated risk of breast cancer who can be offered risk-reducing interventions.
8. Targeted therapies are being developed that exploit the biological functions of BRCA1 and BRCA2.

FUTURE ISSUES

1. Further genome-wide association studies are required to identify further common variants.

2. Genome-wide resequencing is likely to detect numerous novel rare variants, some of which may predispose to breast cancer.
3. Understanding how novel variants result in breast cancer predisposition may require innovative strategies.
4. Quantification of gene-gene interactions and gene-environment interactions, which may be heterogeneous in nature, may be used to improve risk estimation.
5. Judicious clinical translation of genetic factors in addition to *BRCA1* and *BRCA2* may assist in risk estimation, optimization of management, and development of therapies.

DISCLOSURE STATEMENT

The authors are not aware of any biases that might be perceived as affecting the objectivity of this review.

ACKNOWLEDGMENTS

We are grateful to Mike Stratton, Anthony Renwick, and Richard Scott for their critical reading of the manuscript and to Peter Donnelly, Julian Maller, and Paul Pharoah for their assistance in analyses of the low penetrance alleles.

LITERATURE CITED

1. Alessi DR, Sakamoto K, Bayascas JR. 2006. LKB1-dependent signaling pathways. *Annu. Rev. Biochem.* 75:137–63
2. Allinen M, Huusko P, Mantyniemi S, Launonen V, Winqvist R. 2001. Mutation analysis of the *CHK2* gene in families with hereditary breast cancer. *Br. J. Cancer* 85:209–12
3. Andrieu N, Goldgar DE, Easton DF, Rookus M, Brohet R, et al. 2006. Pregnancies, breast-feeding, and breast cancer risk in the International BRCA1/2 Carrier Cohort Study (IBCCS). *J. Natl. Cancer Inst.* 98:535–44
4. Anglian Breast Cancer Study Group. 2000. Prevalence and penetrance of *BRCA1* and *BRCA2* mutations in a population-based series of breast cancer cases. *Br. J. Cancer* 83:1301–8
5. Antoniou A, Pharoah PD, Narod S, Risch HA, Eyfjord JE, et al. 2003. Average risks of breast and ovarian cancer associated with *BRCA1* or *BRCA2* mutations detected in case series unselected for family history: a combined analysis of 22 studies. *Am. J. Hum. Genet.* 72:1117–30
6. Antoniou AC, Easton DF. 2003. Polygenic inheritance of breast cancer: Implications for design of association studies. *Genet. Epidemiol.* 25:190–202
7. Antoniou AC, Gayther SA, Stratton JF, Ponder BA, Easton DF. 2000. Risk models for familial ovarian and breast cancer. *Genet. Epidemiol.* 18:173–90
8. Antoniou AC, Pharoah PD, McMullan G, Day NE, Ponder BA, Easton D. 2001. Evidence for further breast cancer susceptibility genes in addition to *BRCA1* and *BRCA2* in a population-based study. *Genet. Epidemiol.* 21:1–18
9. Antoniou AC, Pharoah PD, McMullan G, Day NE, Stratton MR, et al. 2002. A comprehensive model for familial breast cancer incorporating BRCA1, BRCA2 and other genes. *Br. J. Cancer* 86:76–83
10. Antoniou AC, Pharoah PP, Smith P, Easton DF. 2004. The BOADICEA model of genetic susceptibility to breast and ovarian cancer. *Br. J. Cancer* 91:1580–90

11. Antoniou AC, Sinilnikova OM, Simard J, Leone M, Dumont M, et al. 2007. *RAD51* 135G–>C modifies breast cancer risk among *BRCA2* mutation carriers: results from a combined analysis of 19 studies. *Am. J. Hum. Genet.* 81:1186–200
12. Antoniou AC, Spurdle AB, Sinilnikova OM, Healey S, Pooley KA, et al. 2008. Common breast cancer-predisposition alleles are associated with breast cancer risk in BRCA1 and BRCA2 mutation carriers. *Am. J. Hum. Genet.* 82:937–48
13. Bell DW, Varley JM, Szydlo TE, Kang DH, Wahrer DC, et al. 1999. Heterozygous germ line *hCHK2* mutations in Li-Fraumeni syndrome. *Science* 286:2528–31
14. Bignell GR, Barfoot R, Seal S, Collins N, Warren W, Stratton MR. 1998. Low frequency of somatic mutations in the *LKB1*/Peutz-Jeghers syndrome gene in sporadic breast cancer. *Cancer Res.* 58:1384–86
15. Birch JM, Alston RD, McNally RJ, Evans DG, Kelsey AM, et al. 2001. Relative frequency and morphology of cancers in carriers of germline TP53 mutations. *Oncogene* 20:4621–28
16. Bishop DT, Cannon-Albright L, McLellan T, Gardner EJ, Skolnick MH. 1988. Segregation and linkage analysis of nine Utah breast cancer pedigrees. *Genet. Epidemiol.* 5:151–69
17. Bogdanova N, Enssen-Dubrowinskaja N, Feshchenko S, Lazjuk GI, Rogov YI, et al. 2005. Association of two mutations in the *CHEK2* gene with breast cancer. *Int. J. Cancer* 116:263–66
18. Borresen AL, Andersen TI, Garber J, Barbier-Piraux N, Thorlacius S, et al. 1992. Screening for germ line *TP53* mutations in breast cancer patients. *Cancer Res.* 52:3234–36
19. Breast Cancer Assoc. Consort. 2006. Commonly studied single-nucleotide polymorphisms and breast cancer: results from the Breast Cancer Association Consortium. *J. Natl. Cancer Inst.* 98:1382–96
20. Breast Cancer Link. Consort. 1997. Pathology of familial breast cancer: differences between breast cancers in carriers of BRCA1 or BRCA2 mutations and sporadic cases. *Lancet* 349:1505–10
21. Breast Cancer Link. Consort. 1999. Cancer risks in BRCA2 mutation carriers. *J. Natl. Cancer Inst.* 91:1310–16
22. Broca P. 1866. *Traité des Tumeurs*. Paris: Asselin
23. Brohet RM, Goldgar DE, Easton DF, Antoniou AC, Andrieu N, et al. 2007. Oral contraceptives and breast cancer risk in the international BRCA1/2 carrier cohort study: a report from EMBRACE, GENEPSO, GEO-HEBON, and the IBCCS Collaborating Group. *J. Clin. Oncol.* 25:3831–36
24. Cantor SB, Bell DW, Ganesan S, Kass EM, Drapkin R, et al. 2001. BACH1, a novel helicase-like protein, interacts directly with BRCA1 and contributes to its DNA repair function. *Cell* 105:149–60
25. Chang-Claude J, Andrieu N, Rookus M, Brohet R, Antoniou AC, et al. 2007. Age at menarche and menopause and breast cancer risk in the International *BRCA1/2* Carrier Cohort Study. *Cancer Epidemiol. Biomark. Prev.* 16:740–46
26. CHEK2 Breast Cancer Case-Control Consort. 2004. CHEK2*1100delC and susceptibility to breast cancer: a collaborative analysis involving 10,860 breast cancer cases and 9,065 controls from 10 studies. 2004. *Am. J. Hum. Genet.* 74:1175–82 (WAS 8)
27. Chen J, Lindblom P, Lindblom A. 1998. A study of the PTEN/MMAC1 gene in 136 breast cancer families. *Hum. Genet.* 102:124–25
28. Chompret A, Brugieres L, Ronsin M, Gardes M, Dessarps-Freichey F, et al. 2000. p53 germline mutations in childhood cancers and cancer risk for carrier individuals. *Br. J. Cancer* 82:1932–37
29. Claus EB, Risch N, Thompson WD. 1991. Genetic analysis of breast cancer in the cancer and steroid hormone study. *Am. J. Hum. Genet.* 48:232–42
30. Collab. Group Horm. Factors Breast Cancer. 1996. Breast cancer and hormonal contraceptives: collaborative reanalysis of individual data on 53,297 women with breast cancer and 100,239 women without breast cancer from 54 epidemiological studies. *Lancet* 347:1713–27
31. Collab. Group Horm. Factors Breast Cancer. 1997. Breast cancer and hormone replacement therapy: collaborative reanalysis of data from 51 epidemiological studies of 52,705 women with breast cancer and 108,411 women without breast cancer. *Lancet* 350:1047–59
32. Collab. Group Horm. Factors Breast Cancer. 2001. Familial breast cancer: collaborative reanalysis of individual data from 52 epidemiological studies including 58,209 women with breast cancer and 101,986 women without the disease. *Lancet* 358:1389–99

33. Collab. Group Horm. Factors Breast Cancer. 2002. Breast cancer and breastfeeding: collaborative reanalysis of individual data from 47 epidemiological studies in 30 countries, including 50,302 women with breast cancer and 96,973 women without the disease. *Lancet* 360:187–95
34. Cox A, Dunning AM, Garcia-Closas M, Balasubramanian S, Reed MW, et al. 2007. A common coding variant in *CASP8* is associated with breast cancer risk. *Nat. Genet.* 39:352–58
35. Cui J, Antoniou AC, Dite GS, Southey MC, Venter DJ, et al. 2001. After BRCA1 and BRCA2-what next? Multifactorial segregation analyses of three-generation, population-based Australian families affected by female breast cancer. *Am. J. Hum. Genet.* 68:420–31
36. Cully M, You H, Levine AJ, Mak TW. 2006. Beyond PTEN mutations: the PI3K pathway as an integrator of multiple inputs during tumorigenesis. *Nat. Rev. Cancer* 6:184–92
37. Cybulski C, Górski B, Huzarski T, Masojć B, Mierzejewski M, et al. 2004. *CHEK2* is a multiorgan cancer susceptibility gene. *Am. J. Hum. Genet.* 75:1131–35
38. Dong X, Wang L, Taniguchi K, Wang X, Cunningham JM, et al. 2003. Mutations in *CHEK2* associated with prostate cancer risk. *Am. J. Hum. Genet.* 72:270–80
39. Dufault MR, Betz B, Wappenschmidt B, Hofmann W, Bandick K, et al. 2004. Limited relevance of the *CHEK2* gene in hereditary breast cancer. *Int. J. Cancer* 110:320–25
40. Dunning AM, Healey CS, Pharoah PD, Teare MD, Ponder BA, Easton DF. 1999. A systematic review of genetic polymorphisms and breast cancer risk. *Cancer Epidemiol. Biomark. Prev.* 8:843–54
41. Easton DF. 1994. Cancer risks in A-T heterozygotes. *Int. J. Radiat. Biol.* 66:S177–82
42. Easton DF. 1999. How many more breast cancer predisposition genes are there? *Breast Cancer Res.* 1:14–47
43. Easton DF, Deffenbaugh AM, Pruss D, Frye C, Wenstrup RJ, et al. 2007. A systematic genetic assessment of 1433 sequence variants of unknown clinical significance in the BRCA1 and BRCA2 breast cancer-predisposition genes. *Am. J. Hum. Genet.* 81:873–83
44. Easton DF, Ford D, Bishop DT. 1995. Breast and ovarian cancer incidence in BRCA1-mutation carriers. Breast Cancer Linkage Consortium. *Am. J. Hum. Genet.* 56:265–71
45. Easton DF, Matthews FE, Ford D, Swerdlow AJ, Peto J. 1996. Cancer mortality in relatives of women with ovarian cancer: the OPCS Study. Office of Population Censuses and Surveys. *Int. J. Cancer* 65:284–94
46. Easton DF, Pooley KA, Dunning AM, Pharoah PD, Thompson D, et al. 2007. Genome-wide association study identifies novel breast cancer susceptibility loci. *Nature* 447:1087–93
47. Erkko H, Xia B, Nikkila J, Schleutker J, Syrjakoski K, et al. 2007. A recurrent mutation in *PALB2* in Finnish cancer families. *Nature* 446:316–19
48. Evans DG, Birch JM, Thorneycroft M, McGown G, Lalloo F, Varley JM. 2002. Low rate of *TP53* germline mutations in breast cancer/sarcoma families not fulfilling classical criteria for Li-Fraumeni syndrome. *J. Med. Genet.* 39:941–44
49. Evers B, Jonkers J. 2006. Mouse models of BRCA1 and BRCA2 deficiency: past lessons, current understanding and future prospects. *Oncogene* 25:5885–97
50. Farmer H, McCabe N, Lord CJ, Tutt AN, Johnson DA, et al. 2005. Targeting the DNA repair defect in *BRCA* mutant cells as a therapeutic strategy. *Nature* 434:917–21
51. Ford D, Easton DF, Stratton M, Narod S, Goldgar D, et al. 1998. Genetic heterogeneity and penetrance analysis of the BRCA1 and BRCA2 genes in breast cancer families. The Breast Cancer Linkage Consortium. *Am. J. Hum. Genet.* 62:676–89
52. Foulkes WD, Ghadirian P, Akbari MR, Hamel N, Giroux S, et al. 2007. Identification of a novel truncating *PALB2* mutation and analysis of its contribution to early-onset breast cancer in French Canadian women. *Breast Cancer Res.* 9:R83
53. Garcia-Closas M, Hall P, Nevanlinna H, Pooley K, Morrison J, et al. 2008. Heterogeneity of breast cancer associations with five susceptibility loci by clinical and pathological characteristics. *PLoS Genet.* 4:e1000054
54. Gold B, Kirchhoff T, Stefanov S, Lautenberger J, Viale A, et al. 2008. Genome-wide association study provides evidence for a breast cancer risk locus at 6q22.33. *Proc. Natl. Acad Sci. USA* 105:4340–5
55. Górski B, Cybulski C, Huzarski T, Byrski T, Gronwald J, et al. 2005. Breast cancer predisposing alleles in Poland. *Breast Cancer Res. Treat.* 92:19–24

56. Gudmundsdottir K, Ashworth A. 2006. The roles of BRCA1 and BRCA2 and associated proteins in the maintenance of genomic stability. *Oncogene* 25:5864–74
57. Guénard F, Labrie Y, Ouellette G, Beauparlant CJ, Bessette P, et al. 2007. Germline mutations in the breast cancer susceptibility gene *PTEN* are rare in high-risk non-*BRCA1/2* French Canadian breast cancer families. *Fam. Cancer* 6:483–90
58. Guilford P, Hopkins J, Harraway J, McLeod M, McLeod N, et al. 1998. E-cadherin germline mutations in familial gastric cancer. *Nature* 392:402–5
59. Haiman CA, Le ML, Yamamato J, Stram DO, Sheng X, et al. 2007. A common genetic risk factor for colorectal and prostate cancer. *Nat. Genet.* 39:954–56
60. Haiman CA, Patterson N, Freedman ML, Myers SR, Pike MC, et al. 2007. Multiple regions within 8q24 independently affect risk for prostate cancer. *Nat. Genet.* 39:638–44
61. Hall JM, Lee MK, Newman B, Morrow JE, Anderson LA, et al. 1990. Linkage of early-onset familial breast cancer to chromosome 17q21. *Science* 250:1684–89
62. Heikkinen K, Rapakko K, Karppinen SM, Erkko H, Knuutila S, et al. 2006. *RAD50* and *NBS1* are breast cancer susceptibility genes associated with genomic instability. *Carcinogenesis* 27:1593–99
63. Hemminki A, Markie D, Tomlinson I, Avizienyte E, Roth S, et al. 1998. A serine/threonine kinase gene defective in Peutz-Jeghers syndrome. *Nature* 391:184–87
64. Hemminki A, Tomlinson I, Markie D, Jarvinen H, Sistonen P, et al. 1997. Localization of a susceptibility locus for Peutz-Jeghers syndrome to 19p using comparative genomic hybridization and targeted linkage analysis. *Nat. Genet.* 15:87–90
65. Hopper JL, Carlin JB. 1992. Familial aggregation of a disease consequent upon correlation between relatives in a risk factor measured on a continuous scale. *Am. J. Epidemiol.* 136:1138–47
66. Howlett NG, Taniguchi T, Olson S, Cox B, Waisfisz Q, et al. 2002. Biallelic inactivation of *BRCA2* in Fanconi anemia. *Science* 297:606–9
67. Hunter DJ, Kraft P, Jacobs KB, Cox DG, Yeager M, et al. 2007. A genome-wide association study identifies alleles in *FGFR2* associated with risk of sporadic postmenopausal breast cancer. *Nat. Genet.* 39:870–74
68. Iselius L, Slack J, Littler M, Morton NE. 1991. Genetic epidemiology of breast cancer in Britain. *Ann. Hum. Genet.* 55:151–59
69. Jacquemier J, Padovani L, Rabayrol L, Lakhani SR, Penault-Llorca F, et al. 2005. Typical medullary breast carcinomas have a basal/myoepithelial phenotype. *J. Pathol.* 207:260–68
70. Jenne DE, Reimann H, Nezu J, Friedel W, Loff S, et al. 1998. Peutz-Jeghers syndrome is caused by mutations in a novel serine threonine kinase. *Nat. Genet.* 18:38–43
71. Keller G, Vogelsang H, Becker I, Hutter J, Ott K, et al. 1999. Diffuse type gastric and lobular breast carcinoma in a familial gastric cancer patient with an E-cadherin germline mutation. *Am. J. Pathol.* 155:337–42
72. Kilpivaara O, Vahteristo P, Falck J, Syrjakoski K, Eerola H, et al. 2004. CHEK2 variant I157T may be associated with increased breast cancer risk. *Int. J. Cancer* 111:543–47
73. Lakhani SR. 1999. The pathology of familial breast cancer: Morphological aspects. *Breast Cancer Res.* 1:31–35
74. Lakhani SR, Reis-Filho JS, Fulford L, Penault-Llorca F, van der Vijver M, et al. 2005. Prediction of *BRCA1* status in patients with breast cancer using estrogen receptor and basal phenotype. *Clin. Cancer Res.* 11:5175–80
75. Lakhani SR, van der Vijver M, Jacquemier J, Anderson TJ, Osin PP, et al. 2002. The pathology of familial breast cancer: predictive value of immunohistochemical markers estrogen receptor, progesterone receptor, HER-2, and p53 in patients with mutations in *BRCA1* and *BRCA2*. *J. Clin. Oncol.* 20:2310–18
76. Lalloo F, Varley J, Moran A, Ellis D, O'dair L, et al. 2006. *BRCA1*, *BRCA2* and *TP53* mutations in very early-onset breast cancer with associated risks to relatives. *Eur. J. Cancer* 42:1143–50
77. Le Dran H. 1757. Memoire avec un precis de plusieurs observations sur le cancer. *Mem. Acad. R. Chir.* 3:1–54
78. Levitus M, Waisfisz Q, Godthelp BC, de Vries Y, Hussain S, et al. 2005. The DNA helicase BRIP1 is defective in Fanconi anemia complementation group. *J. Nat. Genet.* 37:934–35

79. Levran O, Attwooll C, Henry RT, Milton KL, Neveling K, et al. 2005. The BRCA1-interacting helicase BRIP1 is deficient in Fanconi anemia. *Nat. Genet.* 37:931–33
80. Lichtenstein P, Holm NV, Verkasalo PK, Iliadou A, Kaprio J, et al. 2000. Environmental and heritable factors in the causation of cancer–analyses of cohorts of twins from Sweden, Denmark, and Finland. *N. Engl. J. Med.* 343:78–85
81. Litman R, Peng M, Jin Z, Zhang F, Zhang J, et al. 2005. BACH1 is critical for homologous recombination and appears to be the Fanconi anemia gene product FANCJ. *Cancer Cell* 8:255–65
82. Malkin D, Li FP, Strong LC, Fraumeni JF Jr, Nelson CE, et al. 1990. Germ line p53 mutations in a familial syndrome of breast cancer, sarcomas, and other neoplasms. *Science* 250:1233–38
83. Masciari S, Larsson N, Senz J, Boyd N, Kaurah P, et al. 2007. Germline E-cadherin mutations in familial lobular breast cancer. *J. Med. Genet.* 44:726–31
84. McCabe N, Turner NC, Lord CJ, Kluzek K, Bialkowska A, et al. 2006. Deficiency in the repair of DNA damage by homologous recombination and sensitivity to poly(ADP-ribose) polymerase inhibition. *Cancer Res.* 66:8109–15
85. Meijers-Heijboer H, van den Ouweland A, Klijn J, Wasielewski M, de Snoo A, et al. 2002. Low-penetrance susceptibility to breast cancer due to *CHEK2**1100delC in noncarriers of *BRCA1* or *BRCA2* mutations. *Nat. Genet.* 31:55–59
86. Miki Y, Swensen J, Shattuck-Eidens D, Futreal PA, Harshman K, et al. 1994. A strong candidate for the breast and ovarian cancer susceptibility gene *BRCA1*. *Science* 266:66–71
87. Nelen MR, Padberg GW, Peeters EA, Lin AY, van den Helm B, et al. 1996. Localization of the gene for Cowden disease to chromosome 10q22–23. *Nat. Genet.* 13:114–16
88. Nelen MR, van Staveren WC, Peeters EA, Hassel MB, Gorlin RJ, et al. 1997. Germline mutations in the *PTEN/MMAC1* gene in patients with Cowden disease. *Hum. Mol. Genet.* 6:1383–87
89. Parmigiani G, Berry D, Aguilar O. 1998. Determining carrier probabilities for breast cancer-susceptibility genes BRCA1 and BRCA2. *Am. J. Hum. Genet.* 62:145–58
90. Peng M, Litman R, Jin Z, Fong G, Cantor SB. 2006. BACH1 is a DNA repair protein supporting BRCA1 damage response. *Oncogene* 25:2245–53
91. Peto J, Collins N, Barfoot R, Seal S, Warren W, et al. 1999. Prevalence of BRCA1 and BRCA2 gene mutations in patients with early-onset breast cancer. *J. Natl. Cancer Inst.* 91:943–49
92. Peto J, Easton DF, Matthews FE, Ford D, Swerdlow AJ. 1996. Cancer mortality in relatives of women with breast cancer: the OPCS Study. Office of Population Censuses and Surveys. *Int. J. Cancer* 65:275–83
93. Peto J, Mack TM. 2000. High constant incidence in twins and other relatives of women with breast cancer. *Nat. Genet.* 26:411–14
94. Pharoah PD, Antoniou A, Bobrow M, Zimmern RL, Easton DF, Ponder BA. 2002. Polygenic susceptibility to breast cancer and implications for prevention. *Nat. Genet.* 31:33–36
95. Pharoah PD, Dunning AM, Ponder BA, Easton DF. 2004. Association studies for finding cancer-susceptibility genetic variants. *Nat. Rev. Cancer* 4:850–60
96. Pharoah PD, Guilford P, Caldas C. 2001. Incidence of gastric cancer and breast cancer in CDH1 (E-cadherin) mutation carriers from hereditary diffuse gastric cancer families. *Gastroenterology* 121:1348–53
97. Rahman N, Scott RH. 2007. Cancer genes associated with phenotypes in monoallelic and biallelic mutation carriers: new lessons from old players. *Hum. Mol. Genet.* 16(Spec. No. 1):R60–66
98. Rahman N, Seal S, Thompson D, Kelly P, Renwick A, et al. 2007. *PALB2*, which encodes a BRCA2-interacting protein, is a breast cancer susceptibility gene. *Nat. Genet.* 39:165–67
99. Ramus SJ, Harrington PA, Pye C, Dicioccio RA, Cox MJ, et al. 2007. Contribution of *BRCA1* and *BRCA2* mutations to inherited ovarian cancer. *Hum. Mutat.* 28:1207–15
100. Reid S, Schindler D, Hanenberg H, Barker K, Hanks S, et al. 2007. Biallelic mutations in *PALB2* cause Fanconi anemia subtype FA-N and predispose to childhood cancer. *Nat. Genet.* 39:162–64
101. Renwick A, Thompson D, Seal S, Kelly P, Chagtai T, et al. 2006. *ATM* mutations that cause ataxia-telangiectasia are breast cancer susceptibility alleles. *Nat. Genet.* 38:873–75
102. Savitsky K, Bar-Shira A, Gilad S, Rotman G, Ziv Y, et al. 1995. A single ataxia telangiectasia gene with a product similar to PI-3 kinase. *Science* 268:1749–53

103. Schutte M, Seal S, Barfoot R, Meijers-Heijboer H, Wasielewski M, et al. 2003. Variants in *CHEK2* other than 1100delC do not make a major contribution to breast cancer susceptibility. *Am. J. Hum. Genet.* 72:1023–28
104. Seal S, Thompson D, Renwick A, Elliott A, Kelly P, et al. 2006. Truncating mutations in the Fanconi anemia J gene *BRIP1* are low-penetrance breast cancer susceptibility alleles. *Nat. Genet.* 38:1239–41
105. Seppala EH, Ikonen T, Mononen N, Autio V, Rokman A, et al. 2003. *CHEK2* variants associate with hereditary prostate cancer. *Br. J. Cancer* 89:1966–70
106. Sharan SK, Wims M, Bradley A. 1995. Murine *Brca1*: sequence and significance for human missense mutations. *Hum. Mol. Genet.* 4:2275–78
107. Shugart YY, Cour C, Renard H, Lenoir G, Goldgar D, et al. 1999. Linkage analysis of 56 multiplex families excludes the Cowden disease gene *PTEN* as a major contributor to familial breast cancer. *J. Med. Genet.* 36:720–21
108. Smid M, Wang Y, Klijn JG, Sieuwerts AM, Zhang Y, et al. 2006. Genes associated with breast cancer metastatic to bone. *J. Clin. Oncol.* 24:2261–67
109. Smith P, McGuffog L, Easton DF, Mann GJ, Pupo GM, et al. 2006. A genome wide linkage search for breast cancer susceptibility genes. *Genes Chromosomes Cancer* 45:646–55
110. Stacey SN, Manolescu A, Sulem P, Rafnar T, Gudmundsson J, et al. 2007. Common variants on chromosomes 2q35 and 16q12 confer susceptibility to estrogen receptor-positive breast cancer. *Nat. Genet.* 39:865–69
111. Stacey SN, Manolescu A, Sulem P, Thorlacius S, Gudjonsson SA, et al. 2008. Common variants on chromosome 5p12 confer susceptibility to estrogen receptor-positive breast cancer. *Nat. Genet.* doi:10.1038/ng.131
112. Swift M, Sholman L, Perry M, Chase C. 1976. Malignant neoplasms in the families of patients with ataxia-telangiectasia. *Cancer Res.* 36:209–15
113. Thompson D, Duedal S, Kirner J, McGuffog L, Last J, et al. 2005. Cancer risks and mortality in heterozygous ATM mutation carriers. *J. Natl. Cancer Inst.* 97:813–22
114. Thompson D, Easton D. 2001. Variation in cancer risks, by mutation position, in BRCA2 mutation carriers. *Am. J. Hum. Genet.* 68:410–19
115. Thompson D, Easton D. 2002. Variation in *BRCA1* cancer risks by mutation position. *Cancer Epidemiol. Biomark. Prev.* 11:329–36
116. Thompson D, Easton DF. 2002. Cancer incidence in BRCA1 mutation carriers. *J. Natl. Cancer Inst.* 94:1358–65
117. Thompson D, Seal S, Schutte M, McGuffog L, Barfoot R, et al. 2006. A multicenter study of cancer incidence in CHEK2 1100delC mutation carriers. *Cancer Epidemiol. Biomark. Prev.* 15:2542–45
118. Tommiska J, Seal S, Renwick A, Barfoot R, Baskcomb L, et al. 2006. Evaluation of RAD50 in familial breast cancer predisposition. *Int. J. Cancer* 118:2911–16
119. Walsh T, Casadei S, Coats KH, Swisher E, Stray SM, et al. 2006. Spectrum of mutations in *BRCA1*, *BRCA2*, *CHEK2*, and *TP53* in families at high risk of breast cancer. *JAMA* 295:1379–88
120. Ware MD, DeSilva D, Sinilnikova OM, Stoppa-Lyonnet D, Tavtigian SV, Mazoyer S. 2006. Does nonsense-mediated mRNA decay explain the ovarian cancer cluster region of the *BRCA2* gene? *Oncogene* 25:323–28
121. Williams WR, Anderson DE. 1984. Genetic epidemiology of breast cancer: segregation analysis of 200 Danish pedigrees. *Genet. Epidemiol.* 1:7–20
122. Wooster R, Bignell G, Lancaster J, Swift S, Seal S, et al. 1995. Identification of the breast cancer susceptibility gene *BRCA2*. *Nature* 378:789–92
123. Wooster R, Neuhausen SL, Mangion J, Quirk Y, Ford D, et al. 1994. Localization of a breast cancer susceptibility gene, BRCA2, to chromosome 13q12–13. *Science* 265:2088–90
124. Xia B, Dorsman JC, Ameziane N, de Vries Y, Rooimans MA, et al. 2007. Fanconi anemia is associated with a defect in the BRCA2 partner PALB2. *Nat. Genet.* 39:159–61
125. Xia B, Sheng Q, Nakanishi K, Ohashi A, Wu J, et al. 2006. Control of BRCA2 cellular and clinical functions by a nuclear partner, PALB2. *Mol. Cell* 22:719–29

From Linkage Maps to Quantitative Trait Loci: The History and Science of the Utah Genetic Reference Project

Stephen M. Prescott,[1] Jean Marc Lalouel,[2] and Mark Leppert[2]

[1]Oklahoma Medical Research Foundation, Oklahoma City, Oklahoma 73104; email: Steve-Prescott@omrf.ouhsc.edu

[2]Department of Human Genetics, University of Utah, Salt Lake City, Utah 84112; email: jml@genetics.utah.edu; mleppert@genetics.utah.edu

Annu. Rev. Genomics Hum. Genet. 2008. 9:347–58

First published online as a Review in Advance on June 10, 2008

The *Annual Review of Genomics and Human Genetics* is online at genom.annualreviews.org

This article's doi:
10.1146/annurev.genom.9.081307.164441

1527-8204/08/0922-0347$20.00

Key Words

genetics, polymorphism, mapping, CEPH

Abstract

One of the early decisions in what became the Human Genome Project was to recruit families that would serve as a reference set, thereby focusing efforts to create human genetic maps on the same sets of DNA samples. The families recruited from Utah provided the most widely used samples in the Centre d'Etudes du Polymorphisme Humain (CEPH) set, were instrumental in generating human linkage maps, and often serve as the benchmark for establishing allele frequency when a new variant is identified. In addition, the immortalized cell lines created from the peripheral blood cells of these subjects are a broadly used resource and have yielded insights in many areas, from the genetics of gene expression to the regulation of telomeres. More recently, these families were recontacted and underwent extensive, protocol-based evaluation to create a phenotypic database, which will aid in the study of the genetic basis of quantitative traits. As with the earlier efforts, this project involved collaborations among many investigators and has yielded insights into multiple traits.

ORIGINS OF THE HUMAN LINKAGE MAP

Until the late 1970s, a limited set of protein polymorphisms (~30) was available to perform linkage studies in families in which a single mendelian trait, generally an inherited disorder of metabolism, was segregating (25). Typically, linkage studies relied on the tests of cosegregation between a trait behaving as a single mendelian and a genetic marker; these tests had been developed and first applied by Newton Morton (23, 24). The likelihood of success was low, given the limited coverage of the genome afforded by the markers then available.

New technologies emerging from molecular biology that afford the definition and use of an unlimited number of genetic markers, together with the development of algorithms to model segregation of multiple loci, have revolutionized the field of genetic mapping. A University of Utah–sponsored genetics retreat at Alta Summit, Utah, in April 1978 changed the view of how to approach the creation of a human linkage map and laid the foundation to map systematically the human genome. Discussions at this meeting culminated in the paper by David Botstein, Raymond White, Mark Skolnick, and Ronald Davis (3) in which they proposed the use of randomly derived restriction fragment polymorphisms (RFLPs) to generate a linkage map of the human genome. Because RFLPs had already been identified in human mitochondrial DNA and DNA from several human cell lines (14, 27), the authors predicted that prior knowledge of specific gene sequences was not necessary to generate this type of linkage map. Botstein and colleagues also suggested that only 150 RFLPs were needed to generate the linkage map; however, these RFLPs should be of high quality and be highly informative. To establish linkage at any given locus, Botstein and colleagues calculated that DNA from several hundred individuals was needed and concurred with Thompson and coworkers (28) that DNA collected from multigenerational families was more useful than DNA from nuclear families. Thus, they proposed a blind search for RFLPs in DNA from healthy individuals in extended Utah pedigrees (3).

The discovery of RFLPs and their use in linkage mapping efforts proceeded at a rapid pace; early successes with well-defined mendelian disorders amply demonstrated the power of the linkage strategy using these markers. Indeed, James Gusella from Harvard Medical School, working with a team from Venezuela (12), found an RFLP on chromosome 4 that linked to the Huntington's disease gene. Additional studies yielded DNA markers in genetic linkage with Duchenne (26) and Becker (16) muscular dystrophies, X-linked retinitis pigmentosa (2), and fragile X–linked mental retardation (5).

IMPLEMENTATION OF THE CEPH COLLABORATION

As multiple genetic research groups became quite active in this area, it soon became apparent that the efficiency of genetic mapping using this wealth of new RFLP markers could be markedly enhanced if these markers were characterized on a common source of DNA. Indeed, genetic linkage relationships between RFLPs developed and genotyped in distinct pedigrees at various laboratories could not readily be assembled into linear genetic maps of the human chromosomes using the pooled genotypic information.

Jean Dausset, who had won the 1980 Nobel Prize in Physiology or Medicine for his work on human leukocyte antigen (HLA) and received a large private donation, proposed in 1983 to develop the Center d'Etude du Polymorphisme Humain (CEPH) (11) as a nonprofit research institute to fill this specific gap. He had conducted his work on histocompatibility antigens and their roles in graft rejection through the participation of dedicated families of volunteers with large sibship size. He understood the value that a common typing panel would have in genetic mapping and wished to contribute his resources to the international scientific community. In concert with Daniel Cohen and Jean-Marc Lalouel, a program was developed that began with formal invitations from CEPH to genetic researchers from around the world to participate in the first major collaborative research project in human genetics, the first international genome project. CEPH proposed to act as the centralized clearing house to produce and supply DNA from a common panel of reference families, collate genotyping data from participating laboratories into a central database, and redistribute these data to collaborators.

The key premise that a "human genetic map will be efficiently achieved by collaborative research on DNA from the same sample of families" (8) was enthusiastically endorsed by leading geneticists engaged in gene mapping at the time. Of particular significance to CEPH was the participation of Ray White, one of the authors of the key concept paper that introduced RFLPs to gene mapping (3), and his research team in Salt Lake City. Ray White had earnestly joined the project with extended pedigrees and kindreds with large sibship size collected in Salt Lake City and was leading a major effort to identify new RFLPs. His participation was crucial for the success of the CEPH initiative. Daniel Cohen and Jean-Marc Lalouel met with Ray White at a Miami Symposium in early 1984. Ray White generously agreed to participate and contribute his family resources as necessary; leading concepts of the collaboration were delineated in this productive encounter. They agreed that specifics of the CEPH operational features would be defined in the course of a meeting of collaborators from the international community sponsored by CEPH in Paris later in the year.

The meeting occurred in Paris on October 20, 1984. A sample of planning notes from Lalouel's notebook is shown in **Figure 1**. Howard Cann had left Stanford University to join CEPH permanently. Scientific, strategic, and logistic issues were discussed and settled in the course of this meeting that set the stage for the future of a most successful international endeavor to generate a linkage map of the human chromosomes. A set of 40 families was selected as a reference panel on the basis of the consensus that emerged among collaborators. Emphasis was put on nuclear families with large sibships (mean sibship size was 8.3) and their parents, and wherever feasible grandparents would be included because they would provide phase information. The large sibships afforded replication of segregating events within the family and the genetic knowledge of maternal and paternal grandparents provided the ability to determine phase of loci (19, 30). The families consisted of 10 French families contributed by Jean Dausset, 27 families contributed by Ray White, one Amish family, and the two core families from a large Huntington pedigree from Venezuela. The panel, evidently a compromise to achieve overall consensus, was subsequently extended by inclusion of an additional set of 20 Utah families with grandparents, so the latter were available in 44 of these 60 families. This extension was largely justified by the need to refine mapping with enhanced power once linkage with a particular mendelian disorder in select pedigrees was established. The higher the map density, the smaller the interval into which a candidate locus could be narrowed, thereby markedly enhancing the power of positional cloning.

Other elements of the CEPH collaboration were specified at that time, including production and distribution issues for DNA, typing strategies, procedures for data collection and redistribution among collaborators, and protocols for publication. Much of this was briefly described years later in a review article (8).

Following this initial meeting, all reference lymphoblastoid cell lines from members of the panel were assembled and placed in culture at CEPH, DNA was produced in large quantities, and aliquots were forwarded to participating laboratories. In parallel, software was provided to share data through the CEPH database and the LINKAGE computer program (17, 18) was provided to all members to analyze genotypic data upon assembly into a defined format. This major undertaking, which required the establishment of large production facilities at CEPH to store cell lines, expand them in culture, and perform large-scale extraction of DNA,

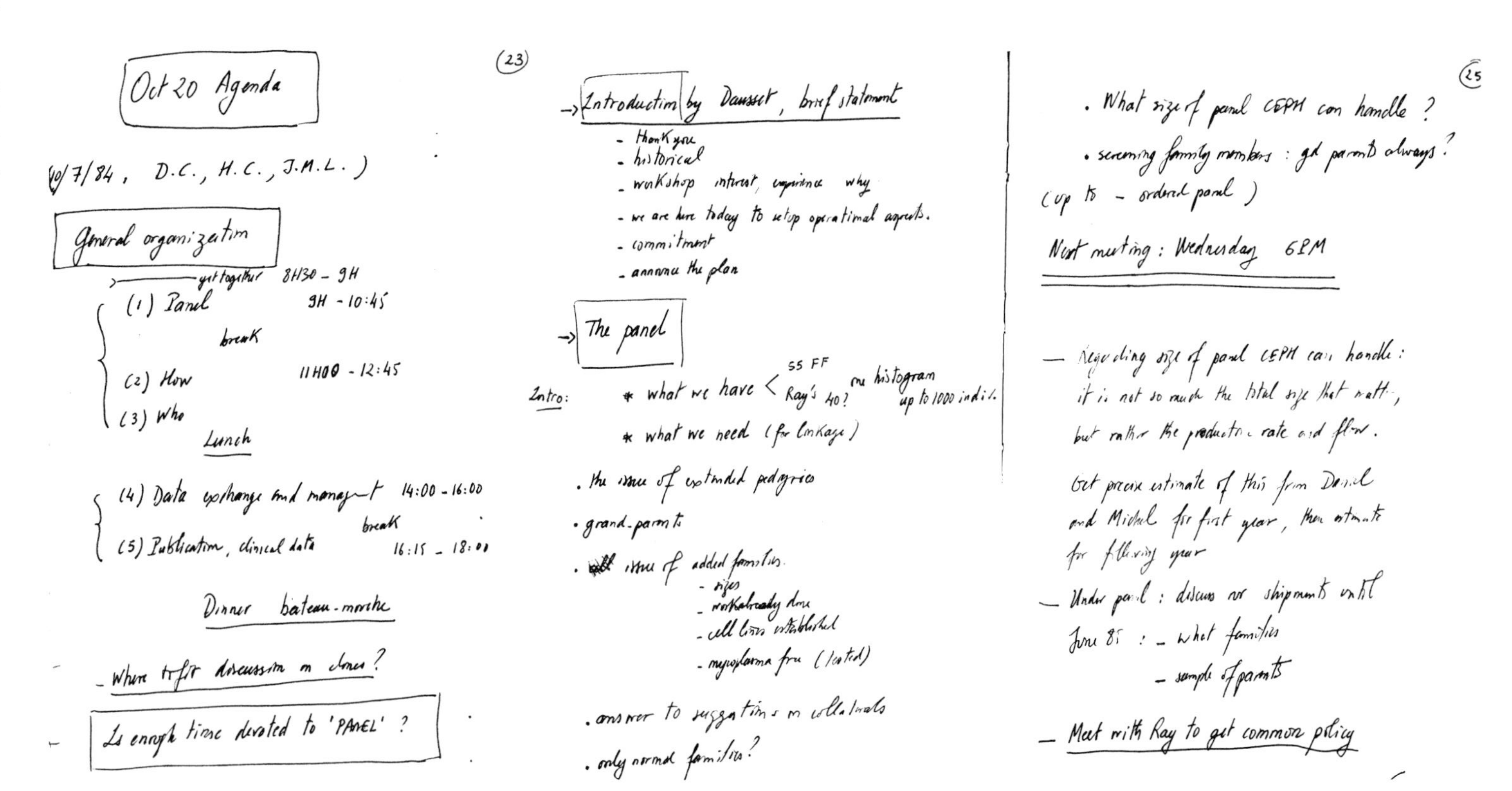

Figure 1
Original notes from Jean Marc Lalouel's notebook on the planning of the first Center d'Etude du Polymorphisme Humain (CEPH) Meeting.

proceeded smoothly, and the CEPH community expanded at a rapid pace to include 63 scientific laboratories by 1989.

LINKAGE MAPS OF THE HUMAN CHROMOSOMES

By 1987, with the support of the Howard Hughes Medical Institute, a complete set of genetic maps of the human chromosomes was assembled by Ray White and Jean-Marc Lalouel and their associates in Salt Lake City; these maps were made available to the scientific community in an internally produced mimeograph document handed out prior to publication to scientists participating to the Human Gene Mapping Conference held in Paris in 1987 (31). These maps were subsequently released in formal publications. Another set of genetic maps generated internally at Collaborative Research, Inc. was published independently the same year (9). Many investigators developed and published the results of their own efforts, which were often focused on particular chromosomes where linkage with specific mendelian disorders was established. One of the first comprehensive human genetic maps was published in 1998 (4). This map utilized 8000 microsatellite markers developed by three centers and constructed a map from eight CEPH families. This map further documented the difference in female versus male recombination across chromosomes, and the increased female recombination yielded an overall length (cM) ratio of female to male of 1.6 for autosomes. In 2003 a 3.9-cM resolution map was published using 56 of the CEPH families and nearly 3000 single nucleotide polymorphism (SNP) markers (20). This map compared favorably with other previously published linkage maps and the human physical map. The 3.9-cM resolution map showed broad trends in sex-specific recombination rates across all chromosomes and demonstrated that every autosome showed an increased recombination in females compared with males, whereas the average female-to-male length ratio (1.7:1 cM) varied by chromosome and specific chromosome region.

ACHIEVEMENTS OF CEPH

In retrospect it is clear that the CEPH project, which arose from the independent initiatives of investigators willing to advance a common goal, was the first successful international genome collaboration. This project markedly accelerated the production of genetic maps, which enabled investigators to map and thereafter proceed toward positional cloning of the unknown genes underlying a host of mendelian disorders. The resource was further enhanced by providing reference lymphoblastoid cell lines from a subset of the CEPH panel to the global scientific community through the Coriell Cell Repository in Camden, New Jersey. Currently, cell lines from 61 CEPH families are available to researchers from the repository. Forty-eight of these are from Utah CEPH families and include parents and for the most part all four members of the grandparental generation. A total of 665 cell lines from the Utah CEPH families are deposited in Coriell, with an average of 8.3 sibs per family. Details for all CEPH families with pedigrees can be found on the Coriell Institute for Medical Research Web site at **http://www.coriell.org.**

The CEPH genotype database V10.0 (November 2004) now contains genotypes for 32,356 genetic markers, 21,480 bi-allelic markers, and 9900 microsatellite markers. This resource has more than 6 million genotypes and is available at **http://www.cephb.fr.**

The CEPH collaboration can claim several long-lasting cultural achievements in science, including a data-sharing strategy among the CEPH collaborators that used a single format and is still in use today. In addition, the ethical standards developed for the treatment of the confidential information for the CEPH families remain intact after 25 years; these standards were such a success that when asked to participate in the next phase of data collection most families readily agreed.

The CEPH panel also facilitated at least two subsequent collaborative endeavors. One consisted of using this genotypic resource to investigate extensive phenotypes of members of the panel as part of the Utah Genetic

Reference Project, described below. The other was to provide the HapMap project with reference sets of offspring and parents trios of Caucasian ancestry (7). The reconsent process of individuals for use of these cell lines in the HapMap project was facilitated by the continual contact of the Utah investigators with the original Utah CEPH families through the Utah Genetic Reference Project.

THE UTAH GENETIC REFERENCE PROJECT

By the early 1990s, genotypic data were known for more than 300 individuals from the 47 original CEPH families from Utah, and linkage maps had been generated for most of the human genome (11). But more value could be extracted from these data through the development of phenotypic information from the panel members that originated from Utah and southern Idaho.

Examination of the original institutional review board (IRB)-approved informed consent documents for these 47 families revealed that family members had provided informed consent to be recontacted for further studies. We decided to capitalize on the plethora of genetic information already obtained from these families and began a process to collect phenotypic data; thus began the Keck/Utah Genetic Reference Project (UGRP). At the time, we reasoned that this type of study at worst would result in a clinical analysis with no molecular basis and at best would provide a powerful tool to begin to understand the genetics of quantitative traits.

Because the original goals of the UGRP project were to identify the genes involved in normal variation in humans and to ascertain the genes involved in the predisposition to common human diseases, this was a high-risk proposal that could not readily be supported through conventional public agencies. In 1995, we sought funding from the Keck Foundation to begin to recontact 16 of the original 47 Utah families.

Initial funding by the Keck Foundation for the first phase of the phenotypic characterization assisted in the development of the infrastructure, the identification of the phenotypic characteristics to be measured, and the determination of the feasibility for such an undertaking. More than 50 local, national, and international scientists and clinicians were recruited during the planning stages for the development of testing and information planning protocols. These collaborators were recruited because they had expertise in ethics and consent issues or in clinically and scientifically relevant fields, including autoimmune, eye, pulmonary, or cardiovascular diseases; metabolic disorders; hematology; speech pathology; neuropsychiatric disorders; or cancer. These collaborators helped to develop an 85-page questionnaire that each family answered, administered and monitored the physical testing of the family members, conducted testing on biological samples, interpreted test data, and analyzed genetic data for linkages with quantitative traits and intermediate phenotypes.

Of the multitude of biochemical, clinical, and physical characteristics that could be measured, a final 180 characteristics were chosen on the basis of ease of data collection, cost of performing the test, clinical significance, and overall scientific interest (**Table 1**). These CEPH families were originally considered 'normal' because they were selected on the basis of the lack of phenotypic symptoms for genetic diseases. However, we did expect that rates of common diseases and traits would be normal in this population.

We initiated the recontact of the original families in 1997. By the time the original families were recontacted and asked to participate, many of the original grandparents had passed away, but a fourth generation had begun. Most families were both eager to participate and to include the fourth generation in our studies.

UTAH GENETIC REFERENCE PROJECT FAMILY REUNIONS

The most efficient method to rigorously collect the data in which we were interested in was to host a reunion of sorts for each participating

Table 1 Phenotypes collected from Utah Genetic Reference Project (UGRP)/Utah Center d'Etude du Polymorphisme Humain (CEPH) families

Phenotypes
Vital signs at rest
Oxygen saturation by pulse oximetry
Blood pressure/urine and blood hormones and metabolites
Catecholamine metabolism
Pulmonary function tests [forced expiratory volume (FEV) in one second and forced vital capacity (FVC)]
Attention span, attention deficit disorder [test of variables of attention (TOVA) on the computer]
Bone density, indirect muscle mass measurements
Taste and smell
Handedness
Voice pitch, vocal senescence amplitude, and frequency measurements
Audiology testing
Facial photography and facial/cranial/cranial 3-D laser scan
Dental exam with teeth impressions
Electrocardiogram
Echocardiography
Ophthalmological testing, including accommodation and visual acuity
Reflectometric assay of skin melatonin content and composition
Lipids, lipoproteins by subtype
Telomere length and aging
Susceptibility to various common human viruses
Blood coagulation proteins
White cell subtypes, innate immunity parameters
Personality traits
Anthropometric measurements

family. Each family reunion was planned for a weekend. On Friday evening, the family arrived at a Salt Lake City hotel, received dinner, and met with the study coordinator and Drs. Mark Leppert and Andy Peiffer to ask any questions and provide informed consent. On Saturday, the family was transported to the University of Utah General Clinical Research Center and submitted to a day-long collection of quantitative and qualitative data (**Table 1**). We arranged the testing schedule such that fasting blood samples were collected from each family member before 9:30 AM, after which the participants were fed breakfast and escorted to the various locations within the General Clinical Research Center to finish their testing. Study collaborators or representatives also attended these clinic days to oversee the collection of the data specific to their respective projects that were outside the scope of our funding. All data collection concluded by 5:00 PM, and family members who were not local residents were provided with another night's stay at the hotel. Collection of all measurements and samples was carried out at the same time and followed an identical protocol for all family visits to ensure consistency in phenotypic measurement between families.

By 1999, we secured a second round of funding from the Keck Foundation to support recontact and data collection for the remaining families that originally participated in the CEPH database. Amazingly, 42 out of the original 47 CEPH families agreed to participate in the UGRP; one family refused to participate and four families were lost to follow-up. Data from the last family were collected in September of 2005. At this time new biosamples collected from fasting family members included plasma and DNA from peripheral blood; no new lymphoblastoid cell lines were established. A set of 537 microsatellite markers was genotyped on these 42 families with high heterozygosity, with an average of 75%, by laboratories at Utah and Marshfield, WI. Very recently, all

Table 2 Heritabilities of selected anthropomorphic measures in 42 Utah Genetic Reference Project families

Variable name	Heritability
Body mass index	0.30
Height	0.97
Elbow breadth	0.72
Biacromal diameter	0.76
Biygomatic diameter	0.70
Face height	0.71
Height anterior-superior iliac spine	0.90
Sitting height/crown-rump	0.84
Stature/supine length	0.96
Total arm length	0.79

members of the three-generation UGRP were genotyped on the Affymetrix SNP array 6.0 chip.

HERITABILITY ESTIMATES FROM 42 UTAH GENETIC REFERENCE PROJECT FAMILIES

Prior to any genetic linkage analysis, we estimated heritabilities of quantitative traits to assess the role of genes for a given phenotype. As an example, **Table 2** provides a list of select anthropomophric measurements that document high levels of heritabilities in the 42 UGRP families. These heritability estimates were computed using the polygenic function available in the SOLAR program (1). All measures were adjusted for gender, age, age^2, and age^3. In the following sections, a few practical applications are provided to document the significance and practical usefulness of the UGRP program.

MAPPING QUANTITATIVE TRAIT LOCI FOR LYMPHOCYTE SUBPOPULATIONS

Levels of defined lymphocyte subpopulations are commonly used in the prognosis and monitoring of a variety of human diseases. These subpopulations can be separated readily by fluorescence-activated cell sorting (FACS) using epitopes that tag each subset of cells, affording counts of total lymphocytes, CD4 T cells, CD8 T cells, the CD4/CD8 cell ratio, CD19 B cells, and natural killer cells (21). This investigation was performed in each member of the first 15 CEPH families included in the UGRP project to ascertain the genetic variation among these phenotypes. The calculated heritabilities ranged from 0.46 to 0.61, consistent with a substantial genetic component. To test for major effects, a whole-genome scan was performed that identified significant quantitative trait loci (QTL) on chromosomes 1, 2, 3, 4, 8, 9, 11, 12, and 18. Each QTL accounted for a significant proportion of the phenotypic variance of lymphocyte subpopulations (13). To test whether candidate genes underlying such QTL could be identified, the chromosome 18 QTL for CD4 T cells was selected because it encompasses genes implicated in T cell function, particularly *Bcl-2*. A multiallelic short tandem repeat polymorphism (STR) identified at this locus was used in further tests of association, and high significance was obtained for one allele in this series. This work provided clear proof of principle that QTL mapping could be applied in these families as a first step toward the identification of genes underlying common genetic variation.

PHENYLTHIOCARBAMIDE TASTING

The inheritance of the ability to taste phenylthiocarbamide (PTC) was long considered to be a classical problem in human genetics because it did not quite follow Mendelian rules for a single locus. In the late 1980s some groups suggested that a multifactorial effect could explain the trait. Previous studies analyzed linkage data based on taster versus nontaster status using a demonstrated threshold. These data indicated strong support for linkage to the Kell blood group antigen and other chromosomal regions, working under the assumption of recessive inheritance. UGRP families were asked to taste a range of 14 PTC concentrations starting at 1 μM and increasing twofold to 8.54 mM, the results of which were ultimately presented as a range of PTC scores from 0 (least sensitive to taste) to 14 (most sensitive). Analysis of the UGRP family members' ability to taste a range of concentrations of PTC indicated a surprising bimodal distribution of PTC scores (means 3.16 and 9.26; $\chi^2 = 93.27$, df = 3, $P < 0.001$). There was considerable variation in this phenotype, and the use of a threshold approach to analyze the data, regardless of where the threshold was placed, would inevitably misclassify some individuals. Linkage analyses using taste sensitivity as a quantitative trait revealed a major locus on chromosome 7q and a secondary locus on chromosome 16p (10). Subsequent to the localization of the major locus and additional fine mapping on chromosome 7q, a gene for PTC tasting was discovered in the families contributing to the linkage signal on chromosome 7q (15). This study identified a small region of strong linkage disequilbrium on chromosome 7q that contained a single gene, the TAS2R taste receptor gene. Further analysis led to the discovery of three missense SNPs that gave rise to five major haplotypes specifically associated with PTC taster or nontaster phenotypes. This study demonstrated the power of these UGRP families in the detection of quantitative trait loci with highly significant logarithm of the odds (LOD) scores even when locus heterogeneity was present. Moreover, the large number of UGRP family members included in the phenotypic test proved to be sufficient for purposes of gene discovery.

GENETICS OF GENE EXPRESSION

Warren and colleagues (29) used vaccine virus vectors for transient induction of HA-8, a minor histocompatibility antigen, in CEPH lymphoblastoid cell lines, followed by detection of the antigen with a cytotoxic T cell clone by standard in vitro cytotoxicity assays. By performing linkage analysis of surface markers on the Epstein-Barr virus (EBV)-transformed cell lines induced by this approach, these authors were able to map genes that control the induction of this histocompatibility antigen.

In a seminal experiment that would lay the stage for subsequent studies, a group led by Vogelstein and Kinzler (32) showed that allelic variation in human gene expression could be unraveled using CEPH reference cell lines as a source. After validation of their approach in a preliminary experiment, they proceeded to screen SNPs for 13 genes using RNA from 96 unrelated CEPH founders. They applied a quantitative assay based on fluorescent dideoxy terminators to quantitate the levels of each of two alleles at each locus. They found that 17 of 37 individuals were heterozygous for any given gene. Significant differences in allelic expression were observed for 6 of the 13 genes studied. They next examined the families of nine individuals who exhibited allelic variation. Three families were informative and displayed expression patterns consistent with Mendelian inheritance. The researchers concluded that *cis*-acting inherited variation in gene expression is relatively common among normal individuals.

Taking advantage of the high throughput afforded by microarrays to investigate gene expression profiles, two groups have independently used part of the CEPH panel to further establish that common genetic variation underlying gene expression could be mapped by linkage analysis. Cheung and Spielman and their associates (22) investigated mRNA from lymphoblastoid cell lines of 14 Utah CEPH families obtained from Coriell. Using Affymetrix Genome Focus Arrays, they characterized gene expression for 8500 genes in 94 unrelated CEPH grandparents and restricted subsequent analysis to 3554 genes that exhibited greater variation between individuals than between replicates. Expression profiling and genotypes at 2756 SNP loci previously mapped in 56 CEPH reference families (20) afforded linkage mapping of 142 to 984 expression profiles, depending on the stringency of the linkage detection test. Linkage occurred with markers in the genomic region both encompassing or not encompassing the gene coding for the transcript examined, suggesting *cis* or *trans* effects. In further tests of 17 genes that exhibited linkage in *cis* under the most stringent conditions,

SNP typing within or near the target gene revealed significant association by quantitative transmission disequilibrium test (QTDT) in 14 (82%) of these genes, strongly supporting the hypothesis that a common variant in the region accounts for the differential gene expression.

Similar work was pursued by Eric Schadt and colleagues (21) in 15 Utah CEPH families also obtained from Coriell. Expression profiling was performed for 23,499 genes, of which 2430 passed a criterion for differential expression in at least half of the offspring tested. Linkage analysis was performed using 346 autosomal markers obtained from the CEPH database (version 9.0) and yielded significant linkage for 333 to 132 phenotypes, depending on the stringency of the pointwise significance selected. In this study, 13 of the 333 and 25 of the 132 mapped expression profiles could be classified as occurring in *cis*. The differences between the two studies were essentially methodological in nature. Taken together, these data provided strong evidence that common genetic variation underlying gene expression can be detected by linkage analysis in CEPH reference families.

AGING AND TELOMERE LENGTH

A recent study (6) demonstrated a significant association ($p = 0.004$) between telomere length in white blood cells and mortality in people aged 60 years or older. This study utilized biosamples from the original Utah CEPH families and cause of death information from the Utah Population Database and the Social Security death index. Individuals with shorter telomeres had a mortality rate of nearly twice that of those with longer telomeres, and individuals from the bottom half of the telomere length distribution had a heart disease mortality rate that was more than three times that of those in the top half of the distribution. The benefit of continued connection to the families over many decades for aging studies is obvious. Another potential advantage of conducting a long-term longitudinal study with a large cohort of families is the ability to collect DNA samples from individuals over the course of several decades and then compare telomere length decline with mortality.

SUMMARY AND FUTURE DIRECTIONS

The efforts of the Utah Genetics Reference Project have affected multiple areas of genetics. The initial goal, which was to create a resource for the generation of linkage maps early in the human genome project, was successful from the outset. The availability of DNA samples from three-generation families with large sibships and grandparents fulfilled needs throughout the different stages of map making. It was a pleasant surprise that the effort to collect these samples also resulted in additional immediate benefits. Indeed, these individuals, via the DNA samples and the immortalized cell lines they generously contributed, became a worldwide reference resource as investigators seeking to find mutant alleles associated with disease used this set of DNA to determine the frequencies of alleles in a normal population. Subsequently, it was obvious that this was a key resource to be used in the HapMap project.

The initial decision to create immortalized cell lines proved to be prescient as these lines have had multiple uses beyond the initial goal of serving as a permanent source of DNA. The cell lines have been used for studies of the genetic basis for gene expression, tissue transplantation, and genomic stability over time in cell culture, among other efforts.

In the most recent phase of this project, the rerecruitment of the subjects for phenotypic analysis to study the genetic basis of quantitative traits has likewise yielded results that highlight the value of the resource. The approach, which was to make precise measurements according to a rigorously standardized protocol, offered advantages over studies that have used clinical records because the data in those circumstances may be flawed as a result of variation in measurement methods. Additionally, the decision to treat all traits as continuous variables added an important dimension to the power of the project.

One crucial element for the success of this project that has extended for more than 25 years has been the dedicated involvement of the families. It is interesting to note that they are tremendously proud of their contributions. They are informed about the confidential use of their data through regular communications (including a newsletter), and although few if any of them have scientific training, they have followed the studies with interest and pride. Likewise it is a source of pride to the investigators that the confidentiality and privacy of the subjects have been maintained over this long period of time and with many collaborators. This provides reassurance that large scale, longitudinal projects with genetic information can be conducted in a manner that does not harm the participants.

In conclusion, the Utah Reference Genetics Project offers a distinct example of the utility of a long-range, multifaceted study of large multigenerational families to unravel genetic variation underlying health and disease.

DISCLOSURE STATEMENT

The authors are not aware of any biases that might be perceived as affecting the objectivity of this review.

ACKNOWLEDGMENTS

The investigation described herein was supported by a Public Health Services research grant to the Huntsman General Clinical Research Center at the University of Utah, grant number M01-RR00064, from the National Center for Research Resources. Research was also supported by generous gifts from the W.M. Keck Foundation and from the George S. and Delores Doré Eccles Foundation. We would like to extend our sincere thanks to all family members who participated in the Utah Genetic Reference Project. Thanks also to Andreas P. Peiffer, M.D., Ph.D., UGRP Medical Director, and Melissa M. Dixon, UGRP Study Coordinator.

LITERATURE CITED

1. Almasy L, Blangero J. 1998. Multipoint quantitative-trait linkage analysis in general pedigrees. *Am. J. Hum. Genet.* 62:1198–211
2. Bhattacharya SS, Wright AF, Clayton JF, Price WH, Phillips CI, et al. 1984. Close genetic linkage between X-linked retinitis pigmentosa and a restriction fragment length polymorphism identified by recombinant DNA probe L1.28. *Nature* 309:253–55
3. Botstein D, White RL, Skolnick M, Davis RW. 1980. Construction of a genetic linkage map in man using restriction fragment length polymorphisms. *Am. J. Hum. Genet.* 32:314–31
4. Broman KW, Murray JC, Sheffield VC, White RL, Weber JL. 1998. Comprehensive human genetic maps: Individual and sex-specific variation in recombination. *Am. J. Hum. Genet.* 63:861–69
5. Camerino G, Mattei MG, Mattei JF, Jaye M, Mandel JL. 1983. Close linkage of fragile X-mental retardation syndrome to haemophilia B and transmission through a normal male. *Nature* 306:701–4
6. Cawthon RM, Smith KR, O'Brien E, Sivatchenko A, Kerber RA. 2003. Association between telomere length in blood and mortality in people aged 60 years or older. *Lancet* 361:393–95
7. Consort. The Int. HapMap. 2005. A haplotype map of the human genome. *Nature* 437:1299–320
8. Dausset J, Cann H, Cohen D, Lathrop M, Lalouel JM, White R. 1990. Centre d'etude du polymorphisme humain (CEPH): collaborative genetic mapping of the human genome. *Genomics* 6:575–77
9. Donis-Keller H, Green P, Helms C, Cartinhour S, Weiffenbach B, et al. 1987. A genetic linkage map of the human genome. *Cell* 51:319–37
10. Drayna D, Coon H, Kim UK, Elsner T, Cromer K, et al. 2003. Genetic analysis of a complex trait in the Utah Genetic Reference Project: a major locus for PTC taste ability on chromosome 7q and a secondary locus on chromosome 16p. *Hum. Genet.* 112:567–72

11. Group, NIH/CEPH Collab. Mapp. 1992. A comprehensive genetic linkage map of the human genome. *Science* 258:67–86
12. Gusella JF, Wexler NS, Conneally PM, Naylor SL, Anderson MA, et al. 1983. A polymorphic DNA marker genetically linked to Huntington's disease. *Nature* 306:234–38
13. Hall MA, Norman PJ, Thiel B, Tiwari H, Peiffer A, et al. 2002. Quantitative trait loci on chromosomes 1, 2, 3, 4, 8, 9, 11, 12, and 18 control variation in levels of T and B lymphocyte subpopulations. *Am. J. Hum. Genet.* 70:1172–82
14. Hutchison CA 3rd, Newbold JE, Potter SS, Edgell MH. 1974. Maternal inheritance of mammalian mitochondrial DNA. *Nature* 251:536–38
15. Kim UK, Jorgenson E, Coon H, Leppert M, Risch N, Drayna D. 2003. Positional cloning of the human quantitative trait locus underlying taste sensitivity to phenylthiocarbamide. *Science* 299:1221–25
16. Kingston HM, Thomas NS, Pearson PL, Sarfarazi M, Harper PS. 1983. Genetic linkage between Becker muscular dystrophy and a polymorphic DNA sequence on the short arm of the X chromosome. *J. Med. Genet.* 20:255–58
17. Lathrop GM, Lalouel JM. 1984. Easy calculations of LOD scores and genetic risks on small computers. *Am. J. Hum. Genet.* 36:460–65
18. Lathrop GM, Lalouel JM, Julier C, Ott J. 1984. Strategies for multilocus linkage analysis in humans. *Proc. Natl. Acad. Sci. USA* 81:3443–46
19. Malhotra A, Cromer K, Leppert MF, Hasstedt SJ. 2005. The power to detect genetic linkage for quantitative traits in the Utah CEPH pedigrees. *J. Hum. Genet.* 50:69–75
20. Matise TC, Sachidanandam R, Clark AG, Kruglyak L, Wijsman E, et al. 2003. A 3.9-centimorgan-resolution human single-nucleotide polymorphism linkage map and screening set. *Am. J. Hum. Genet.* 73:271–84
21. Monks SA, Leonardson A, Zhu H, Cundiff P, Pietrusiak P, et al. 2004. Genetic inheritance of gene expression in human cell lines. *Am. J. Hum. Genet.* 75:1094–105
22. Morley M, Molony CM, Weber TM, Devlin JL, Ewens KG, et al. 2004. Genetic analysis of genome-wide variation in human gene expression. *Nature* 430:743–47
23. Morton NE. 1955. Sequential tests for the detection of linkage. *Am. J. Hum. Genet.* 7:277–318
24. Morton NE. 1956. The detection and estimation of linkage between the genes for elliptocytosis and the Rh blood type. *Am. J. Hum. Genet.* 8:80–96
25. Mourant AE, Kopeć AC, Domaniewska-Sobczak K. 1976. *The Distribution of the Human Blood Groups and Other Polymorphisms*. London: Oxford Univ. Press
26. Murray JM, Davies KE, Harper PS, Meredith L, Mueller CR, Williamson R. 1982. Linkage relationship of a cloned DNA sequence on the short arm of the X chromosome to Duchenne muscular dystrophy. *Nature* 300:69–71
27. Potter SS, Newbold JE, Hutchison CA 3rd, Edgell MH. 1975. Specific cleavage analysis of mammalian mitochondrial DNA. *Proc. Natl. Acad. Sci. USA* 72:4496–500
28. Thompson EA. 1978. Linkage and the power of a pedigree structure. In *Genetic Epidemiology*, ed. NE Morton, CS Chung, pp. 247–53. New York: Academic
29. Warren EH, Otterud BE, Linterman RW, Brickner AG, Engelhard VH, et al. 2002. Feasibility of using genetic linkage analysis to identify the genes encoding T cell-defined minor histocompatibility antigens. *Tissue Antigens* 59:293–303
30. White R, Leppert M, O'Connell P, Nakamura Y, Julier C, et al. 1986. Construction of human genetic linkage maps: I. Progress and perspectives. *Cold Spring Harbor Symp. Quant. Biol.* 51(Pt. 1):29–38
31. White R, Lalouel JM, Leppert M, O'Connell P, Nakamura Y, Lathrop GM. 1989. Linkage maps of human chromosomes. *Genome* 31:1066–72
32. Yan H, Yuan W, Velculescu VE, Vogelstein B, Kinzler KW. 2002. Allelic variation in human gene expression. *Science* 297:1143

Disorders of Lysosome-Related Organelle Biogenesis: Clinical and Molecular Genetics*

Marjan Huizing,[1] Amanda Helip-Wooley,[2] Wendy Westbroek,[2] Meral Gunay-Aygun,[2] and William A. Gahl[2]

[1]Cell Biology of Metabolic Disorders Unit and [2]Section on Human Biochemical Genetics, Medical Genetics Branch, National Human Genome Research Institute, National Institutes of Health, Bethesda, Maryland 20892; email: mhuizing@mail.nih.gov; ahwooley@mail.nih.gov; wwestbro@mail.nih.gov; mgaygun@mail.nih.gov; bgahl@helix.nih.gov

Annu. Rev. Genomics Hum. Genet. 2008. 9:359–86

First published online as a Review in Advance on June 10, 2008

The *Annual Review of Genomics and Human Genetics* is online at genom.annualreviews.org

This article's doi:
10.1146/annurev.genom.9.081307.164303

1527-8204/08/0922-0359$20.00

Key Words

Chediak-Higashi syndrome, Griscelli syndrome, Hermansky-Pudlak syndrome, melanosome, platelet

Abstract

Lysosome-related organelles (LROs) are a heterogeneous group of vesicles that share various features with lysosomes, but are distinct in function, morphology, and composition. The biogenesis of LROs employs a common machinery, and genetic defects in this machinery can affect all LROs or only an individual LRO, resulting in a variety of clinical features. In this review, we discuss the main components of LRO biogenesis. We also summarize the function, composition, and resident cell types of the major LROs. Finally, we describe the clinical characteristics of the major human LRO disorders.

INTRODUCTION

Lysosome-related organelle (LRO): an organelle that shares features with the lysosome but has a distinct morphology, composition, and/or function

Lysosomes are membrane-bound cytoplasmic organelles that serve as major degradative compartments in eukaryotic cells (65). Genetic, biochemical, and structural data have demonstrated that certain specialized cell types contain lysosome-related organelles (LROs) that share features with lysosomes but have distinct morphology, composition, and/or functions. Such organelles include melanosomes in melanocytes, lytic granules in lymphocytes, delta granules in platelets, lamellar bodies in lung type 2 epithelial cells, and other variants of acidic granules (**Table 1**). Features that LROs share with lysosomes include a low intraorganellar pH, specific membrane proteins or other components, and a common pathway of formation (20). Recently, cell biologists have studied LROs because of the lessons they impart concerning vesicle formation. Abnormalities in both lysosomes and LROs occur in certain human genetic diseases (**Table 2**), including the Hermansky-Pudlak (HPS), Griscelli (GS), and Chediak-Higashi syndromes (CHS), further demonstrating the close relationship between such organelles. Cells of patients with these rare disorders (or their mouse, fly, zebrafish, or yeast counterparts) are important tools for the investigation of membrane trafficking. Here we discuss the machinery of LRO biogenesis, the major recognized LROs, and the known human disorders of LRO biogenesis.

Table 1 **Function, resident cell types, and clinical relevance of the major identified lysosome-related organelles (LROs)**

LRO	Function	Resident cell type	Major clinical features when defective
Melanosome	Intracellular melanin biosynthesis and storage; melanin transfer to keratinocytes	Melanocytes, iris, and retinal pigment epithelial cells	Ocular and cutaneous hypopigmentation
Delta granule	Storage of small molecules, released for blood coagulation	Platelets, megakaryocytes	Bleeding diathesis
Lytic granule	Intracellular degradation of macromolecules; upon release, extracellular destruction of virally infected or cancerous target cells	Cytotoxic T lymphocytes, natural killer cells	Immune deficiency, viral infections
Azurophil granule	Storage of hydrolytic enzymes, destruction of phagocytosed bacteria; upon secretion, support of various pathological processes, including inflammation	Neutrophils, eosinophils	Neutropenia, immune deficiency, bacterial infections
Basophil granule	Storage of histamine, serotonin, heparin, IL-4, and lysosomal proteases; released for regulation of inflammation	Basophils, mast cells	Immune deficiency, allergies
Lamellar body	Storage and secretion of surfactants for lung function	Lung type II epithelial cells	Interstitial lung disease, lung inflammation and fibrosis
MHC class II compartment	Intracellular processing and incorporation of antigens into cell membranes	B lymphocytes, macrophages, dendritic cells, other antigen presenting cells	Immune deficiency
Neuromelanin granule	Storage of the insoluble pigment neuromelanin, which binds iron	Catecholaminergic neurons of the brainstem	Unknown
Ruffled border	Bone resorption and remodeling; storage, activation and/or release of acid hydrolases	Osteoclasts	Osteopetrosis
Weibel-Palade body	Storage and regulated release of hemostatic and proinflammatory factors (von Willebrand Factor, P-selectin)	Endothelial cells	Bleeding diathesis

Table 2 Protein complexes, genes, and their encoded proteins involved in human or animal lysosome-related organelle (LRO) disorders

Protein complex	Mutant gene	Human locus	Mutant protein	Human disease[a]	Animal model[b]
BLOC-1	*DTNBP1*	6p22.3	Dysbindin	HPS-7	*sandy (m)*
	PLDN	15q21.1	Pallidin	-	*pallid (m)*
	CNO	4p16.1	Cappuccino	-	*cappuccino (m)*
	MUTED	6p25.1-p24.3	Muted	-	*muted (m)*
	SNAPAP	1q21.3	Snapin	-	-
	BLOC1S1	12q13-q14	BLOS1	-	-
	BLOC1S2	10q24.31	BLOS2	-	-
	BLOC1S3/HPS8	19q13.32	BLOS3/HPS8	HPS-8	*reduced pigmentation (m)*
BLOC-2	*HPS3*	3q24	HPS3	HPS-3	*cocoa (m)*
	HPS5	11p14	HPS5	HPS-5	*ruby-eye 2 (m)* *pink (f)*
	HPS6	10q24.32	HPS6	HPS-6	*ruby-eye (m)*
BLOC-3	*HPS1*	10q23.1-23.3	HPS1	HPS-1	*pale ear (m)*
	HPS4	22q11.2-q12.2	HPS4	HPS-4	*light ear (m)*
AP3	*AP3B1*	5q14.1	AP3 beta 3A	HPS-2	*pearl (m)* *orange (f)*
	AP3D1	19p13.3	AP3 delta	-	*mocha (m)* *garnet (f)*
	AP3M1	10q22.2	AP3 mu 3A	-	*carmine (f)*
	AP3S1/AP3S2	5q22/15q26.1	AP3 sigma 3A/3B	-	*ruby (f)*
-	*CHS1/LYST*	1q42.1-q42.2	CHS1/LYST	CHS	*beige (m,r)*[c]
Unnamed complex[d]	*MYO5A*	15q21	Myosin Va	GS1	*dilute (m)*
	RAB27A	15q15-q21.1	Rab27A	GS2	*ashen (m)*
	MLPH	2q37.3	Melanophilin	GS3	*leaden (m)*
HOPS complex[e]	*VPS11*	11q23	VPS11	-	*pale gray eyes (jm)*
	VPS16	20p13-p12	VPS16A	-	-
	VPS18	15q14-q15	VPS18	-	*deep orange (f)* *vps18(hi2499A) (z)*
	VPS33A	12q24.31	VPS33A	-	*buff (m)* *carnation (f)*
	VPS33B	15q26.1	VPS33B	ARC	-
	VPS39/Vam6	15q15.1	VPS39/Vam6	-	*leberknödel (z)*
	VPS41	7p14-p13	VPS41	-	*light (f)*
-	*RAB38*	11 q14	Rab38	-	*chocolate (m)* *lightoid (f)*[f,g]
-	*RABGGTA*	14q11.2	RABGGTA	-	*gunmetal (m)*

(*Continued*)

Table 2 (*Continued*)

Protein complex	Mutant gene	Human locus	Mutant protein	Human disease[a]	Animal model[b]
-	*SLC7A11*	4q28-q32	SLC7A11	-	*subtle gray (m)*
-	*p14/MAPBPIP*	1q22	MAPBPIP/p14	MIM 610798[h]	-

[a]HPS, Hermansky-Pudlak syndrome; CHS, Chediak-Higashi syndrome; GS, Griscelli syndrome; ARC, arthrogryposis-renal dysfunction-cholestasis.
[b]Naturally occurring LRO models; *m*, mouse; *r*, rat; *f*, fruit fly; *z*, zebrafish; *jm*, Japanese Medaka ricefish.
[c]Other species with CHS mutants are Aleutian mink, cats, killer whale, and Japanese cattle (58).
[d]These three proteins form an unnamed tripartite complex (**Figure 3**).
[e]Homotypic vacuolar sorting (HOPS) complex; a conserved protein complex that functions as a tethering factor in membrane fusion events involving lysosomes and/or endosomes (26).
[f]The fly Rab-RP1 protein is defective in *lightoid*. Fly Rab-RP1 has high homology to the human Rab38 and Rab32 proteins (26, 121).
[g]The rat models *Ruby (R)*, *Fawn-hooded (FH)*, and *Tester-Moriyama (TM)* also carry *RAB38* mutations (90).
[h]Unnamed syndrome comprising partial albinism, immunodeficiency, congenital neutropenia, B-cell and cytotoxic T-cell deficiency, and short stature (6).

BIOGENESIS OF LYSOSOME-RELATED ORGANELLES

Lysosomes and LROs are produced through the interaction of ubiquitous trafficking mechanisms with cell-specific machinery that targets cargo to a particular compartment. Whereas classic secretory granules form directly from the trans-Golgi network (TGN), some or all LRO contents derive from the endosomal system. This group of individual compartments relies on sorting and trafficking of membrane structures, assisted by a network of filaments, tubules, motor proteins, Rabs and other small GTPases, and numerous other components such as biogenesis of lysosome-related organelles complexes (BLOCs), soluble N-ethylmaleimide-sensitive-factor attachment protein (SNAP) receptors (SNAREs), syntaxins, and vacuolar protein for sorting (VPS) proteins.

Trans-Golgi network (TGN): a membranous network of vesicles and tubules that assists in the sorting and targeting of secreted proteins to the correct cellular destinations

BLOC: biogenesis of lysosome-related organelles complex

Soluble N-ethylmaleimide-sensitive-factor attachment protein (SNAP) receptors (SNAREs): form bridging complexes that are required for intracellular membrane fusion in eukaryotes

Vacuolar protein for sorting (VPS): defects in yeast affect vacuolar biogenesis; homologs are candidates for causing LRO disorders in higher organisms

The Endosomal System

Lysosome-related organelles maintain their structure, composition, and function by means of a continuous flow of proteins and membranes/vesicles among endosomal compartments (**Figure 1**). The endocytic pathway takes up molecules from outside of the cell and internalizes membrane receptors, whereas the exocytic pathway sorts newly synthesized proteins from the endoplasmic reticulum toward the endosomal system (9). The two pathways connect at the early endosome, a tubulovesicular network with a pH of 5.9–6.0 that contains distinct resident proteins, including early endosome–associated protein (EEA)-1, Rab5, and Rabaptin-5. Material is sorted from the early endosome to the cell surface, recycling endosomes, the biosynthetic pathway (of LROs), or late endosomes/lysosomes (9, 36, 98). Components destined for lysosomes are sorted to late endosomes, which have an acidic pH (5.0–6.0), contain whorls of membranes and vesicles, include multivesicular bodies, and recycle mannose-6-phosphate receptors (MPRs) back to the trans-Golgi membranes. Other late endosome-specific markers are Rab7 and lysobisphosphatidic acid (LBPA). Late endosomes morph into lysosomes, which lack multivesicular structures and MPRs and have a pH of 5.0–5.5. Components destined for specific LROs possess unique cell-type and LRO-specific sorting and trafficking pathways (98). To support the endosomal system, cell-type specific chaperones exist for vesicle targeting, transport, and fusion events. These include a cytoskeletal system of tubules and filaments, coat-associated proteins, SNAREs, syntaxins, Rabs, motor proteins, and specific membrane lipids (9, 36, 94, 104).

Membrane Budding and Fusion

Tight control of membrane flux through the endosomal system occurs through continued vesicular budding and fusion. Budding is largely mediated by coat components, whereas

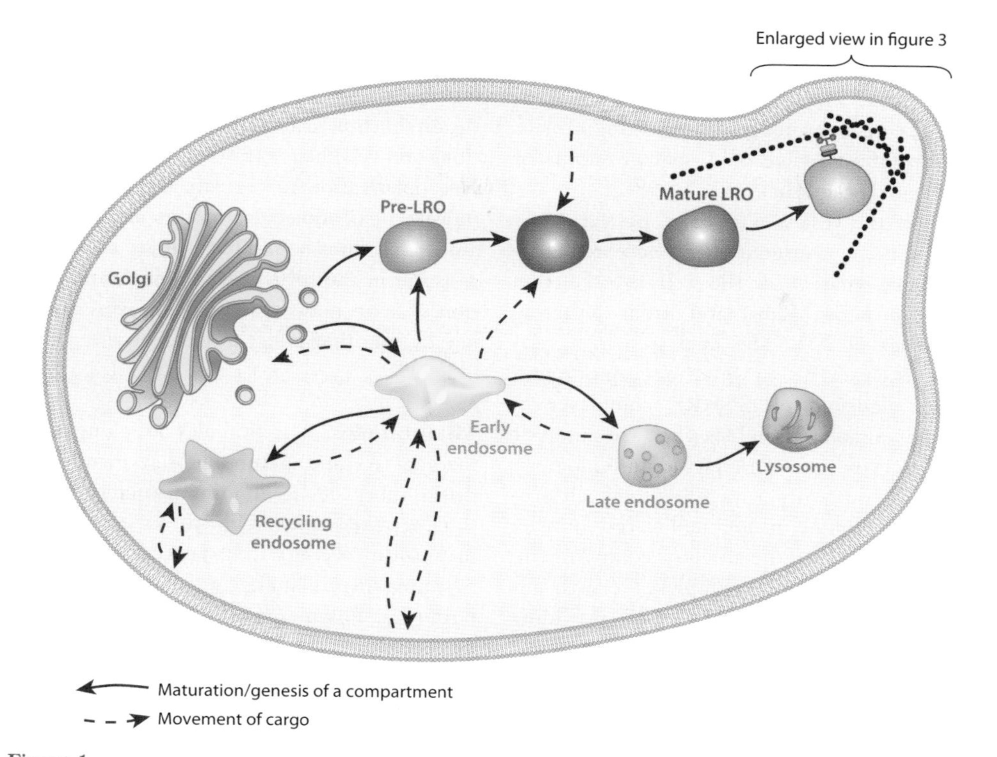

Figure 1

The endosomal system and lysosome-related organelle (LRO) biogenesis. The endosomal system is a collection of highly dynamic compartments defined by their morphology, contingent of marker proteins, function, and accessibility to endocytic tracers. Rather than focus on a specific cell type, this model shows the basic endosomal elements involved in generic LRO biogenesis. Solid arrows depict maturation/genesis of a compartment whereas dashed arrows represent the movement of cargo. The early (sorting) endosome is the major sorting center of the cell, where Golgi-derived biosynthetic cargo destined for LROs or late endosomes/lysosomes as well as receptors and other molecules internalized from the cell membrane are sorted. Some receptors return to the cell membrane by way of recycling endosomes, whereas others continue on to late endosomes and lysosomes for degradation. Specialized LROs coexist in the same cell as nonspecialized lysosomes and contents destined for each compartment must be sorted appropriately. In some LRO-containing cells, LRO formation involves sequential delivery of LRO-specific proteins. The mature LRO acquires specific accessory proteins (Rabs, motor proteins, SNAREs) that assist in its function and/or localization.

targeting and fusion are generally mediated by SNARE complexes (9, 11, 94). Vesicular structures required for LRO formation arise at the plasma membrane, the TGN, or other intracellular membranous compartments and involve the recruitment of coat components. At least three major vesicle coat proteins exist: COPI (coatomer) and COPII, which function in the early endocytic pathway (ER and Golgi), and clathrin, which functions in post-Golgi locations including the TGN, endosomes, and plasma membrane (11). The coat proteins trigger the generation of highly curved membrane areas where vesicle-specific cargo recruitment takes place, including recruitment of adaptor proteins and SNAREs. A variety of clathrin-specific adaptors exist, including the heterotetrameric adaptor protein (AP) complexes AP1, AP2, AP3, and AP4 and the monomeric Golgi-localized gamma-ear containing ADP ribosylation factor–binding (GGA) proteins, the hepatocyte growth factor–regulated tyrosine

Mannose 6-phosphate receptor (MPR): sorts and delivers newly synthesized lysosomal enzymes at the TGN, assists in delivery to endosomes, and recycles back to the TGN

kinase substrate (Hrs), Epsin-1, and autosomal recessive hypercholesterolemia (ARH) proteins. These adaptors bind to transmembrane cargo proteins by recognizing sorting signals, e.g., specific tyrosine or di-leucine motifs or conjugated ubiquitin (9, 11, 73, 98).

Once the vesicle is released by the donor membrane, it is targeted to its acceptor compartment, guided by the cell cytoskeleton, motor proteins, Rabs, and other tethering molecules (9, 11). The final step in a vesicle's existence is fusion with the acceptor membrane, mediated by SNARE complexes. A SNARE complex assembles when a monomeric v-SNARE on a vesicle binds to an oligomeric t-SNARE on a target membrane, which promotes membrane fusion and cargo transfer from the vesicle to the acceptor compartment. Different v-/t-SNARE complexes form at different steps of intracellular transport, providing specificity of time and place to the membrane fusion process (11, 94). The role of the SNARE and adaptor complexes in LRO biogenesis became apparent from mutant animal and human models, which provided an invaluable resource for elucidating these complex pathways. For example, defects in subunits of adaptor protein complex-3 (AP3) result in human HPS-2 (21, 54) and the HPS mouse mutants *pearl* and *mocha* (67). These genetic models demonstrate that AP3 regulates the sorting of proteins specific for the lysosome, melanosome, and platelet granule.

Tubules and Filaments

The cytoskeleton, unique to eukaryotic cells, is a dynamic, organized, three-dimensional network of fibers that maintains cell shape and motility. It consists of actin networks, microtubules, and intermediate filaments.

Actin networks. Actin networks, also called microfilaments, consist of fine, thread-like protein fibers that are 3–7 nm in diameter and composed of actin monomers. Actin networks serve short-range movements in the cell periphery and in the cell body surrounding the Golgi complex, including muscle contraction, cell motility, cell division and cytokinesis, vesicle and organelle movement, cell signaling, and the establishment and maintenance of cell junctions and cell shape (127). Actin subunits provide polarization to regulate the direction of movement of molecular motors along the filaments. Actin is bound by a variety of proteins assisting in the formation, stability, and function of actin networks (127). Myosins are the key motors that effect movement along actin (22, 127).

Microtubules. Microtubules are long polymers of α- and β-tubulin that form cylindrical tubes 20–25 nm in diameter. Microtubules provide a backbone for long-range (μm to mm) intracellular transport processes such as mitosis, cytokinesis, and vesicular transport (12). Polymerization of microtubules is nucleated in a microtubule organizing center (MTOC), away from which the microtubule grows in the plus direction. The motor protein kinesin moves toward the plus end of the microtubule (79), whereas dynein moves toward the minus end (95). A variety of microtubule-associated proteins (MAPs) assure proper function, formation, and stability of microtubules (12). Actin filaments, microtubules, and their molecular motors play critical roles in LRO biogenesis (12). For example, kinesin-2 is a plus end–directed motor for late endosomes and lysosomes (79) and Rab7 (defective in Charcot-Marie-Tooth disease type 2B) regulates recruitment of dynein/dynactin and movement of late endosomal compartments on microtubules (48, 57). The motor myosin Va and its interactors Rab27A and melanophilin are essential for the tethering and actin-based movement of melanosomes in the periphery of the melanocyte (112). Finally, the biogenesis of lysosome-related organelles complex-1 (BLOC-1) (**Figure 2**; see below) associates with actin filaments (27).

Intermediate filaments. Intermediate filaments, approximately 10 nm in diameter, are tissue-specific structures often interconnected

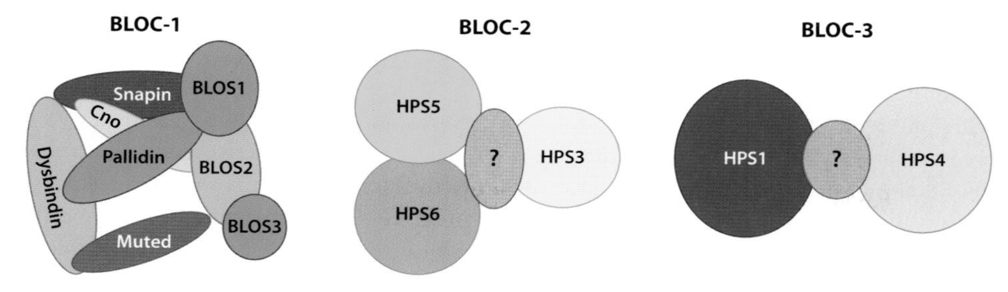

Figure 2

Biogenesis of lysosome-related organelles complexes (BLOCs). The currently recognized proteins identified as components of each BLOC are indicated. Most of the BLOC subunits are associated with subtypes of Hermansky-Pudlak syndrome. Subunits indicated by question marks are predicted by molecular weight analyses of the entire complex, and may represent a yet-to-be-identified subunit.

with other filamentous systems for mechanical stability. Intermediate filament proteins include keratin, vimentin, desmin, and lamin (46). Human disorders of intermediate filaments, which function as mechanical scaffolds (18), number at least 70 (**http://www.interfil.org**), including diseases of the skin, heart, muscle, liver, brain, adipose tissue, and even premature aging. Intermediate filaments assist in endosomal trafficking and LRO biogenesis pathways; defects in specific intermediate filaments affect the endocytic pathway of low-density lipoprotein-cholesterol, the formation of autophagocytic vacuoles, endosome-Golgi recycling of glycosphingolipids, and AP3-dependent endosomal sorting (17, 18, 33, 113).

Motor Proteins

Kinesins, dyneins, and myosins are the motors that drive most active transport of proteins and endosomal vesicles (4, 12).

Myosins. Myosins are actin-associated motor proteins that use ATP hydrolysis to generate mechanical force. Myosins consist of an actin- and ATP-binding head domain, a calmodulin-binding neck domain, and a cargo-interaction C-terminal tail domain. Myosins regulate organelle and vesicle movement, cytokinesis, muscle contraction, adhesion, and migration (4, 128). The dimeric myosin Va (encoded by *MYO5A*) achieves actin-dependent transport of LROs, specifically melanosomes, by binding of its cargo receptor, melanophilin (4, 22, 93, 126, 129) (**Figure 3**).

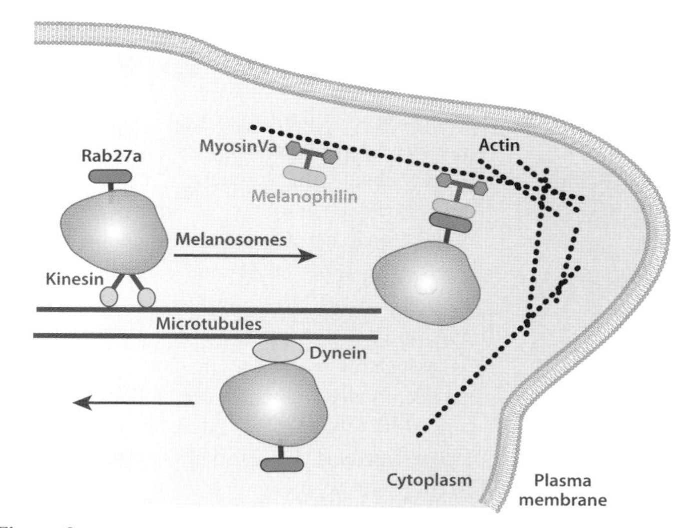

Figure 3

Functional mechanism of the Rab27A-melanophilin-myosinVa tripartite complex. In melanocytes, Rab27A resides on mature melanosomes that travel to the cell periphery on microtubules via a kinesin motor protein. Once in the periphery, Rab27A recruits the motor protein myosinVa via a direct interaction with melanophilin. The Rab27A-melanophilin-myosinVa tripartite complex is then responsible for the accumulation of mature melanosomes in the actin-rich dendritic tips. This localization is necessary for efficient transfer of melanosomes to keratinocytes for normal pigmentation. If any member of the tripartite complex is defective, melanosomes are not captured in the periphery and return to the cell center on microtubules in a dynein-mediated process. Defects in any member of the tripartite complex result in Griscelli syndrome.

Tyrosinase (TYR) and tyrosinase-related protein-1 (TYRP1): melanocyte-specific enzymes that catalyze distinct reactions in melanin synthesis in melanosomes

Kinesins. Kinesins are a superfamily of microtubule-associated motor proteins; their catalytic head motor domains bind to microtubules and hydrolyze ATP to produce force. Three subfamilies of kinesins differ in the position of their motor domain (79). Conventional kinesin is an anterograde motor formed by two light chains and two heavy chains that possess the ATPase and microtubule-binding activity. The kinesin coiled-coil stalk domain is involved in dimer formation and the C-terminal tail participates in cargo binding (4, 79). Kinesins bind LROs, in particular melanosomes, through their binding partner kinectin, which associates with melanosomal membranes (92). Studies in *Xenopus* melanophores show that kinesin II interacts with p150^{Glued}, a subunit of dynactin, to disperse melanosomes (19). Inhibition of kinesin caused inhibition of the bidirectional movement of melanosomes along microtubules, leading to perinuclear accumulation of melanosomes (40).

Dyneins. Dyneins regulate retrograde movement toward microtubule minus ends. Dyneins are large complexes composed of two to three heavy chains that contain a motor domain and variable numbers of intermediate and light chains (4). Cytoplasmic dyneins are expressed in most eukaryotic cells and are implicated in mRNA transport, vesicle trafficking, and cell division (95). Cytoplasmic dynein is associated with the protein complex dynactin, which contains ten subunits including p150^{Glued}, p135^{Glued}, dynamitin, and actin-related protein 1 (47). The role of dynein in retrograde melanosome transport is well established (4, 47) and its role in melanosome maturation is emerging (57).

Rabs and Other Small GTPases

The Rab family of small GTPases consists of more than 60 members in mammals (104). The unique subcellular distribution of each Rab protein is thought to determine membrane identity (81, 104). Rabs are synthesized as soluble proteins and recognized by a Rab escort protein (REP) that presents them to Rab geranylgeranyl transferase (RabGGTase) for prenylation (1, 81, 104), which anchors them to a membrane. Rab proteins cycle between GTP-bound (active) and GDP-bound (inactive) forms, assisted by GTPase-activating protein (GAP), guanine nucleotide dissociation inhibitor (GDI), and guanine nucleotide exchange factor (GEF) proteins (1, 104). This on/off switch regulates Rab protein binding to downstream effectors and execution of cellular functions. At least three Rab proteins (Rab27A, Rab38, and Rab32) are involved in LRO biogenesis.

Rab27A assists in the transfer of melanosomes from microtubules to actin filaments (**Figure 3**). Rab38 is mutated in the mouse *chocolate* mutant, which displays mild oculocutaneous albinism with small, round, hypopigmented melanosomes but no platelet granule deficiency (70). Rab38 is also mutated in three rat models (**Table 2**) of pigment dilution and platelet granule deficiency (90). Mouse Rab38 and the closely related Rab32 (67% amino acid identity) localize to mature melanosomes and to perinuclear vesicles that contain melanosome-specific proteins [tyrosinase (TYR), tyrosinase-related protein-1 (TYRP1)] (121). In *chocolate* melanocytes treated with Rab32 small interfering RNA (siRNA), TYR and TYRP1 fail to exit the TGN, resulting in a complete loss of pigment, suggesting that these Rabs are functionally redundant; indeed, only one homolog exists in other species (121). Interestingly, both Rab38 and Rab32 are expressed in melanocytes and bone marrow mast cells but only Rab38 is present in rat basophil leukemia cells and alveolar type II-derived mouse lung epithelial (MLE)-12 cells (121).

Defects in universal regulators of Rab function such as RabGGTase and REP can also lead to defects in LRO biogenesis. The *gunmetal* mouse has reduced RabGGTase activity owing to a mutated alpha subunit (*RABGGTA*) (23). *Gunmetal* mice have hypopigmentation, macrothrombocytopenia, decreased numbers of platelet alpha and delta granules, and reduced geranylgeranylation and membrane association of Rab27A (104). Deficient REP1 (one of two REP isoforms) in humans leads to choroideremia (CHM) (MIM 303100), a form of X-linked retinal degeneration. Rab27A prenylation is reduced in CHM but it is unclear if Rab27A alone or a combination of Rabs is responsible for the disease (104).

Biogenesis of Lysosome-Related Organelle Complexes

BLOCs are unique groups of proteins that together function in the formation and/or trafficking of lysosome-related endosomal compartments (24, 27, 68, 73, 83, 109) (**Figure 2**). BLOC-1 consists of the pallidin, muted, cappuccino, HPS7/dysbindin, BLOS1, BLOS2, BLOS3, and Snapin proteins. Hermansky-Pudlak syndrome subtype 7 (HPS-7, due to defective dysbindin) and HPS-8 (defective BLOC1S3) are BLOC-1 defects (68, 80); some other BLOC-1 subunits are associated with a mutant mouse (**Table 2**) (67).

The identification of potential binding partners of BLOC-1 suggested that this complex regulates SNARE-mediated membrane fusion. Pallidin, by interacting with syntaxin 13, may function as a premelanosomal t-SNARE to assist TYRP1- and TYRP2-containing vesicles to fuse with the premelanosome (27, 71). Snapin interacts with SNAP-25/23, SNAREs implicated in membrane fusion events (109). Dysbindin binds alpha and beta dystrobrevins, i.e., dystrophin-related protein components of large dystrophin-associated membrane-spanning complexes (85). However, recent studies indicate that when assembled into the BLOC-1 complex, endogenous dysbindin cannot bind dystrobrevins because of an occupied dystrobrevin binding site (85). BLOC-1 also associates with F-actin (27), supporting a role for the complex in vesicle movement along actin filaments. Moreover, BLOC-1 apparently mediates the egress of cargo from early endosomes toward LROs (105) and interacts with other LRO biogenesis complexes, including AP3 and BLOC-2 (25).

BLOC-2 contains the HPS3, HPS5, and HPS6 proteins and BLOC-3 consists of the HPS1 and HPS4 proteins. The cell biology of these BLOCs is discussed below in the section on Hermansky-Pudlak syndrome.

Other Endosomal Chaperones

Endosomal trafficking and LRO formation require several additional components and regulatory mechanisms. Human disorders and animal models have helped elucidate the pathways involved. For example, the granule group of *Drosophila melanogaster* eye color mutants (69) and murine models of hypopigmentation and bleeding (67, 116) have helped researchers uncover novel players in LRO biogenesis (**Table 2**). These include RABGGTA, defective in the *gunmetal* mouse (23); Slc7a11, a cystine/glutamate exchanger defective in the *subtle gray* mouse (15); lysosomal trafficking regulator (LYST)/Chediak-Higashi Syndrome 1 (CHS1) protein, defective in the *beige* mouse, rat, and human Chediak-Higashi syndrome (58); the endosomal adaptor protein p14, defective in a syndrome of albinism and immunodeficiency (6); and VPS proteins (homologs of vacuolar sorting proteins in yeast) such as VPS18, 33, and 41, defective in the fly models *deep orange*, *carnation*, and *light*, respectively (**Table 2**). Apart from protein defects, alterations in membrane lipids, rafts, cholesterol, and other cellular components can affect LRO biogenesis (36, 104).

TYPES OF LYSOSOME-RELATED ORGANELLES

LROs share features with lysosomes but they also exhibit cell type–specific components responsible for specialized functions. **Table 1** provides a list of the major LROs.

Lysosomes

Lysosomes are membrane-bound, acidic, cytoplasmic organelles responsible for intracellular protein degradation (65). Lysosomes contain mature acid-dependent hydrolases and

Lysosome-associated membrane protein (LAMP): heavily glycosylated protein that is localized in and targeted to lysosomal membranes

Multiple vesicular body (MVB): endosomal structure that contains multiple internal vesicles; intermediate structures in the endosomal sorting pathway to lysosomes and/or lysosome-related organelles

highly glycosylated integral membrane proteins (lysosome-associated membrane proteins or LAMPs) but lack MPRs (65). LROs share most or all of these characteristics with lysosomes. In some cell types, LROs such as secretory lysosomes perform both lysosomal functions and specialized LRO functions (98, 111). In other cell types, such as melanocytes and platelets, highly specialized LROs (melanosomes and delta granules, respectively) coexist with conventional lysosomes, indicating the presence of distinct biogenesis and sorting pathways (98).

Melanosomes

Melanosomes are specialized LROs responsible for the synthesis and storage of melanin in melanocytes (4, 71, 98). The insoluble black to dark-brown eumelanins and the red, brown, and yellow pheomelanins provide color for human skin, eye, and hair (71, 97). Morphologically, pheomelanosomes are rich in vesicles but lack an organized structure and will not be further discussed (71). Melanosomes mature through four stages (**Figure 4**). Stage I premelanosomes are clathrin-coated endosomal compartments with intralumenal vesicles (ILV) that resemble multivesicular bodies (MVBs). Stage II melanosomes are elongated and have intralumenal striations spanning the length of the organelle. Newly synthesized melanin particles are deposited on the striated fibrils, resulting in the black streaks of stage III melanosomes that eventually become fully melanized stage IV melanosomes (71, 97) (**Figure 4**). Melanin synthesis requires distinct, nonlysosomal sorting and enrichment of melanosome-specific proteins to different melanosomal stages (97, 99).

The melanosome-specific protein PMEL17 is the major component of intralumenal striations (5). PMEL17 is activated when Golgi-glycosylated PMEL17 undergoes proteolytic cleavage by a furin-like proprotein convertase

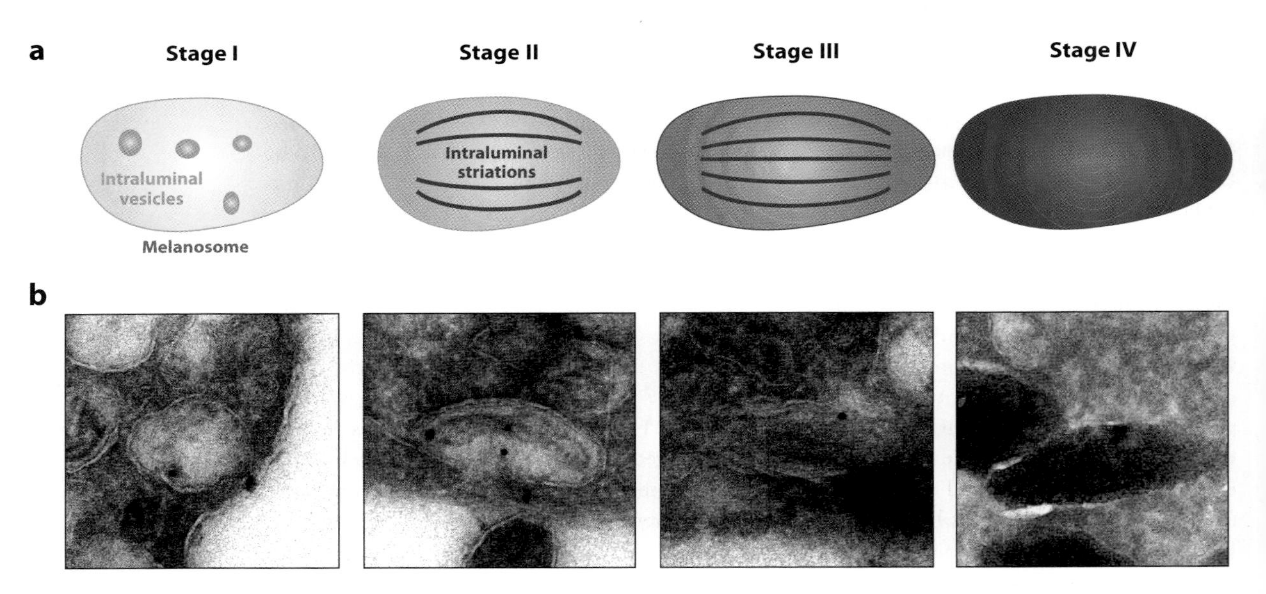

Figure 4

Morphological characterization of the four stages of melanosome biogenesis. (*a*) A drawing of each of the four stages of melanosome biogenesis. (*b*) An immune-electron microscopy study of human primary epidermal melanocytes. Ultrathin cryosections were labeled with anti-NKI beteb, a PMEL17-specific marker, conjugated to 20-nm gold particles. Stage I melanosomes, electron-lucent multivesicular bodies (MVBs) without melanin, contain multiple intralumenal vesicles. Stage II melanosomes are more elongated and have PMEL17-containing intralumenal striations that run the length of the organelle. Stage III melanosomes exhibit melanin deposits on the striations. These striations are masked in stage IV melanosomes owing to complete melanization.

in a post-Golgi compartment. Active PMEL17 is sorted into clathrin-coated MVBs where it becomes enriched in ILVs and initiates striation formation (5). Sorting of melanogenic enzymes (TYR, TYRP1, TYRP2) to melanosomes occurs through a pathway different from that for PMEL17; TYR and TYRP1 are absent from PMEL17-rich clathrin-coated endosomal compartments (98, 99). Melanosomal sorting of TYR and TYRP1 occurs via a distinct population of early endosomes (25, 44, 105). The adaptor protein complex AP3 functions in TYR trafficking from the early endosomes to melanosomes (53). BLOC-1 helps regulate the sorting and budding of TYRP1-containing vesicles from early endosomal compartments, whereas BLOC-2 is involved in targeting of TYRP1-positive vesicles to melanosomes (25, 44, 105). Human syndromes and animal models continue to shed light on the process of melanosome biogenesis (50, 67, 104).

Platelet Delta Granules

Platelets deliver prohemostatic proteins and other mediators to sites of vessel injury. In addition to lysosomes, alpha and delta (also called dense) granules are the major platelet organelles and they release mature secretory proteins and small molecules such as ADP and serotonin upon stimulation (100). Alpha granules, the most abundant vesicles in platelets, store proteins that promote platelet adhesiveness and wound healing (100). Delta granules, which are platelet LROs, number only three to eight per platelet and store nonprotein small molecules; their membranes contain the lysosomal membrane proteins LAMP2 and CD63 (granulophysin or LAMP3), but not LAMP1 (38, 74). Intralumenal delta granule constituents include calcium, serotonin, ADP, ATP, and polyphosphates (38, 74). The high calcium content of delta granules gives them an intrinsic electron density (**Figure 6*a*,*b***) and their highly osmophilic environment causes them to appear dark on transmission electron microscopy images (74).

Unlike other secretory cells, circulating platelets do not form their own vesicles (100). Instead, bone marrow megakaryocytes create the granules and transport them into developing proplatelets in the periphery via microtubules and actin networks. During megakaryocytopoiesis, alpha and delta granules arise from the Golgi complex (130), with MVBs as intermediate structures (42, 130). Young megakaryocytes contain numerous MVBs but few alpha and delta granules (42, 130). As megakaryocytes mature, the number of MVBs decreases, the number of alpha and delta granules increases, a demarcating membrane system develops, and the granules enter nascent proplatelets (42, 62, 130). Proplatelets released from the megakaryocyte fragment into individual platelets (62).

PMEL17: also called gp100; a melanocyte-specific glycoprotein enriched in the lumen of premelanosomes, where it forms characteristic melanosomal striations

Lytic Granules

Cytotoxic T lymphocytes (CTLs) kill virus-infected cells and tumor cells by releasing their lytic granule contents (28). Lytic granules are specialized secretory LROs that degrade proteins at acidic pH. Lytic granules also contain proteins such as perforin, which becomes functional upon secretion into the neutral pH of the extracellular milieu (111). Recognition of major histocompatibility complex (MHC) peptide complexes on target cells by CTL receptors results in polarization of the secretory machinery and formation of the MTOC at the contact point between the CTL and target cell. There, lytic granules accumulate and attack the target cell, sparing other cells including the CTL itself (28, 111). Released perforin makes pores in the target cell membrane, allowing granzymes (and other lytic granule constituents) to enter the target cells, where they cleave many different substrates and kill the cell. Fas ligand, released from the CTL on small exosome-like vesicles of the lytic granule (102), binds to the Fas receptor on the target cell, beginning a cascade of caspase cleavage events that initiates apoptosis. Lytic granules also contain the lysosomal membrane proteins CD63, LAMP1, and LAMP2 (28, 111).

Other Lysosome-Related Organelles

The exact definition and the precise number of LROs have not been established. However, LROs must possess an acidic lumen, a contingent of characteristic lysosomal proteins, and a specialized function, usually related to storage or secretion. Accessibility to endocytic markers may be another criterion. Over the past decade, several specialized vesicles, previously considered orphan organelles, have been declared to be LROs. The list may expand as more proteomic studies are performed and additional disease phenotypes are recognized.

Lamellar bodies. Lamellar bodies are secretory granules found in pulmonary type II cells that store and secrete surfactant phospholipids and proteins (122). The phospholipids form a film at the air-liquid interface, reducing surface tension in the alveoli and preventing alveolar collapse during expiration. Lamellar bodies maintain a pH of ~5.5 and contain characteristic lysosomal enzymes (acid phosphatases and cathepsins) and membrane proteins (CD63 and LAMP1). Surfactant protein B (SP-B) is necessary for the formation of lamellar bodies; SP-B (–/–) mice have morphologically and functionally abnormal lamellar bodies (108). Giant lamellar bodies have been reported in disorders of LRO biogenesis both in patients with HPS (82) and in the *beige* mouse, a model of CHS (13). The relationship between enlarged lamellar bodies and the pulmonary fibrosis of HPS remains speculative.

Azurophilic granules. Azurophilic granules are one of three types of granules found in neutrophils, the most abundant phagocytic cell acting in the innate immune response (37). Like lysosomes, azurophil granules contain proteins (e.g., serine proteases, antibiotic proteins, and myeloperoxidase) that undergo proteolytic processing into mature, active proteins. Azurophil granules are affected in at least two disorders of LRO biogenesis. Chediak-Higashi syndrome neutrophils contain giant, CD63-positive azurophil granules that do not properly exocytose myeloperoxidase upon stimulation (63). AP3-deficient individuals with HPS-2 have low intracellular neutrophil elastase, an azurophil granule constituent, resulting in a granulocyte-colony stimulating factor (G-CSF)-responsive neutropenia and a diathesis toward bacterial infection (29).

Basophil granules. Basophil granules are uniform acidic granules of basophils. Basophils represent 0.5%–3% of circulating leukocytes and, together with mast cells, mediate allergic inflammation (72). Upon activation, basophil granules release their contents, which include histamine, serotonin, heparin, and lysosomal proteases (tryptase and chymase). Basophil granules contain lysosomal membrane proteins such as LAMP1, LAMP2, CD63/LAMP3, and LIMPIV/5G10 antigen (20, 114). Basophils are an important source of the cytokine interleukin-4, which is critical in the development of allergies and the production of IgE antibodies.

Other candidate lysosome-related organelles. Other candidate LROs include MHC class II compartments (MIICs), neuromelanin (NM) granules, ruffled borders, and Weibel-Palade bodies (WPBs). MIICs are important for antigen processing and are found in B cells, macrophages, and dendritic cells. MIICs have multiple internal vesicles and/or membranous whorls that contain lysosomal enzymes and lysosomal membrane markers (41). NM granules, pigmented organelles found in certain catecholaminergic neurons, were recently described as LROs on the basis of proteomics studies (119). Ruffled borders are the site of bone resorption in osteoclasts and may be LRO variants on the basis of their lysosomal proteins and secretory function (110). WPBs are lysosome-related only in that they carry the lysosomal marker CD63; this marker reaches the WPB via an endocytic rather than a direct biosynthetic route (39).

HUMAN DISORDERS OF LYSOSOME-RELATED ORGANELLE BIOGENESIS AND TRAFFICKING

Disorders of LRO biogenesis typically affect melanosomes and platelet dense granules, and may or may not affect other LROs. Some LRO disorders have murine (67), fruit fly (26, 66, 69), zebrafish (103), as well as yeast (50) counterparts. The major LRO disorders are listed in **Table 2**.

Hermansky-Pudlak Syndrome

Eight human HPS genes are known, each associated with a murine model (**Table 2**). The eight HPS gene products operate in distinct complexes, e.g., BLOC-1 (containing HPS7 and HPS8); BLOC-2 (HPS3, HPS5, and HPS6), mutations in which cause relatively mild clinical involvement (3, 49, 52, 123); BLOC-3 (HPS1 and HPS4), mutations in which cause more severe complications (2, 45, 123); and AP3 (HPS-2 gene product), mutations in which involve immune deficiency (54, 123). *pale ear* (Hps1-deficient), *light ear* (Hps4-deficient), and double-homozygous *pale ear/light ear* mice have identical coat color and a characteristic hypopigmentation of ears, tail, and feet, and *light-ear* tissues (Hps4 deficient) express no Hps4 protein, indicating a physical and/or functional interaction between Hps1 and Hps4 (67, 115). Human HPS subtypes that involve the same BLOC generally resemble each other.

Clinical manifestations of Hermansky-Pudlak syndrome. All HPS (MIM 203300) patients exhibit some degree of hypopigmentation and a diathesis toward bleeding. The hypopigmentation derives from impaired melanosome formation, trafficking, or transfer to keratinocytes. In contrast, classical albinism is caused by deficiency of melanosomal proteins such as tyrosinase, the *P* protein, or TYRP1 (61). Patients with albinism have abnormal neuronal crest cell migration during development, with ~90% of their optic nerve fibers decussated (crossed), compared with 55% normally (61). Patients display horizontal nystagmus, an involuntary, painless lateral movement of the eyes in both directions. The retinal fundus appears pale and visual acuity is generally stable at 20/200 (legally blind in the United States) or worse. The eyes appear light or blue, and iris transillumination is readily apparent. In this phenomenon, light shone into the pupil is transmitted through the iris owing to a decrease in iris pigment. In albinism, the hair is white and the skin is light- and sun-sensitive. Solar keratoses and melanocytic nevi are common and patients are at increased risk of developing squamous cell carcinoma, basal cell carcinoma, and possibly melanoma (118). Sun avoidance is critical.

HPS patients with BLOC-3 defects, i.e., HPS-1 or HPS-4, generally have features of classical albinism, such as light hair (**Figure 5*a*,*b***), solar keratoses (**Figure 5*c***), significant iris transillumination (**Figure 5*d***), and a hypopigmented retina (**Figure 5*e***). HPS-2 patients have less hair hypopigmentation than HPS patients with BLOC-3 defects (**Figure 5*f***). Compared with BLOC-3 patients, individuals with BLOC-2 defects, i.e., HPS-3, HPS-5, or HPS-6, have mild findings, including minimal iris transillumination (**Figure 5*g***) and hypopigmentation of the retina (**Figure 5*h***), skin, and hair (**Figure 5*i***). Visual acuity can be as good as 20/50. However, there is enormous variability in each HPS subtype. HPS-7 and HPS-8 patients have not been well characterized with respect to pigmentation (68, 80).

The bleeding diathesis of HPS involves defective platelet aggregation due to absence of dense granules, which is apparent on whole mount electron microscopy images (**Figure 6*a*,*b***). Specifically, the secondary aggregation response of platelets to exogenous stimuli is absent. Bleeding manifestations include spontaneous bruising (**Figure 6*c***), epistaxis, menorrhagia, and prolonged oozing after trauma or minor surgery such as a tooth extraction (30). A platelet transfusion essentially cures this bleeding diathesis, and other interventions

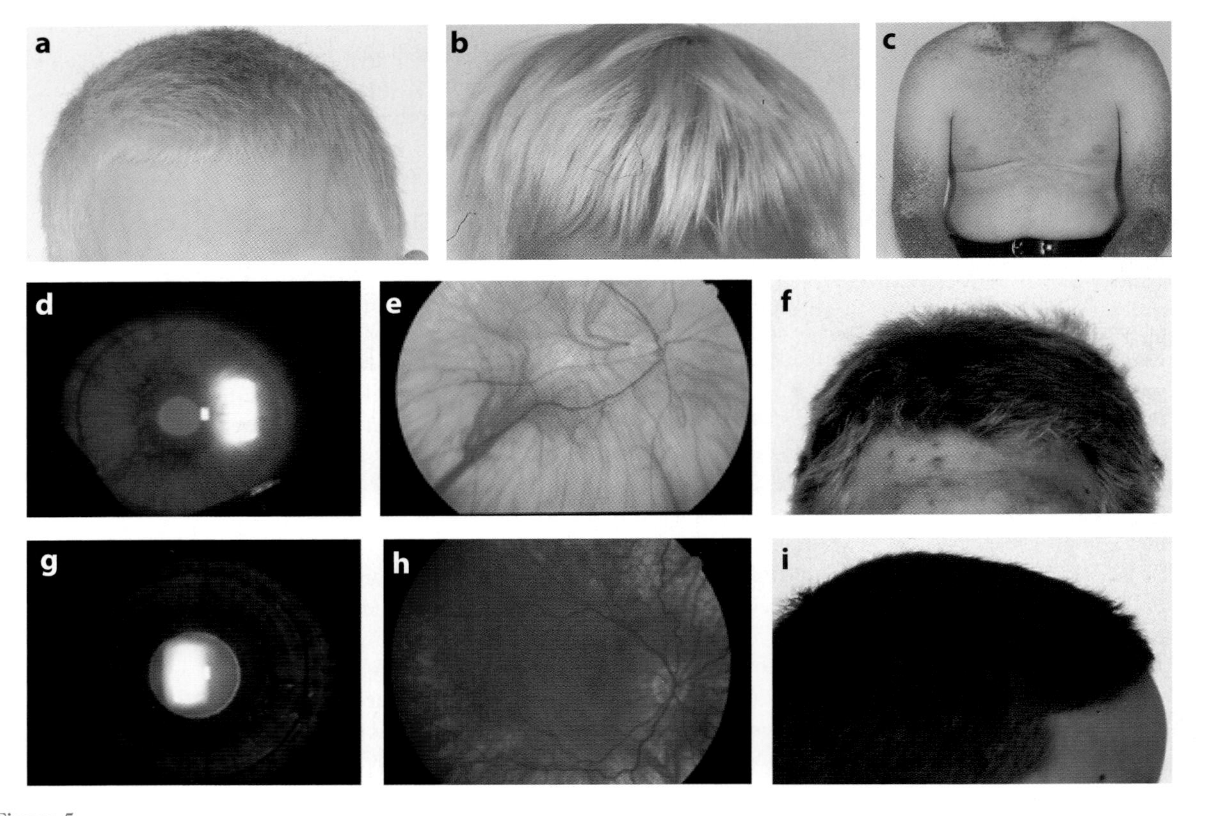

Figure 5

Hair, skin and eye pigmentation in Hermansky-Pudlak syndrome (HPS). (*a*) Light colored hair of a boy with HPS-1. (*b*) White hair of a boy with HPS-4. (*c*) Keratoses in sun-exposed areas of a Puerto Rican patient with HPS-1. (*d*) Significant iris transillumination in an HPS-1 patient. Orange light, abnormally present, appears because the iris contains insufficient melanin to block it. (*e*) Pale retinal fundus in HPS-1, with vessels clearly visible owing to lack of retinal pigment epithelium. (*f*) Tan-blond hair of a patient with HPS-2. (*g*) Mild iris transillumination in a patient with HPS-3. (*h*) Mild hypopigmentation of the retina in HPS-3. (*i*) Dark hair of a Puerto Rican boy with HPS-3. (Parts *d*, *e*, *g*, and *h* courtesy of Dr. E. Tsilou, National Eye Institute.)

such as intravenous 1-desamino-8D-arginine vasopressin and topical thrombin ameliorate the situation. HPS patients vary enormously in their bleeding tendencies.

Besides hypopigmentation and bleeding, the complications of HPS appear somewhat subtype-specific. A colitis resembling Crohn's disease occurs in up to one-third of patients with BLOC-3 defects, i.e., HPS-1 and HPS-4 (55). The colitis involves granulomas, erosions (**Figure 6*d***), and/or inflammatory cells and generally responds to the anti-TNF-α agent, Remicade, as well as to steroids. In HPS-2 lymphocytes, lytic granules are reduced in number and function, leading to infections in childhood. A G-CSF-responsive neutropenia is also specific for HPS-2 (29, 52, 54). Ceroid lipofuscin, an amorphous, autofluorescent lipid-protein material (**Figure 6*e***), has been found in the lysosomes of HPS-1 cells, including renal tubular cells, alveolar macrophages, and cells of the gastrointestinal tract, bone marrow, liver, spleen, lymph nodes, and heart (51). Ceroid lipofuscin may accumulate because cells cannot rapidly degrade mistargeted vesicle membranes. Pulmonary fibrosis occurs only in HPS patients with subtypes 1 or 4; in HPS-1, more than of 80% of patients develop this fatal

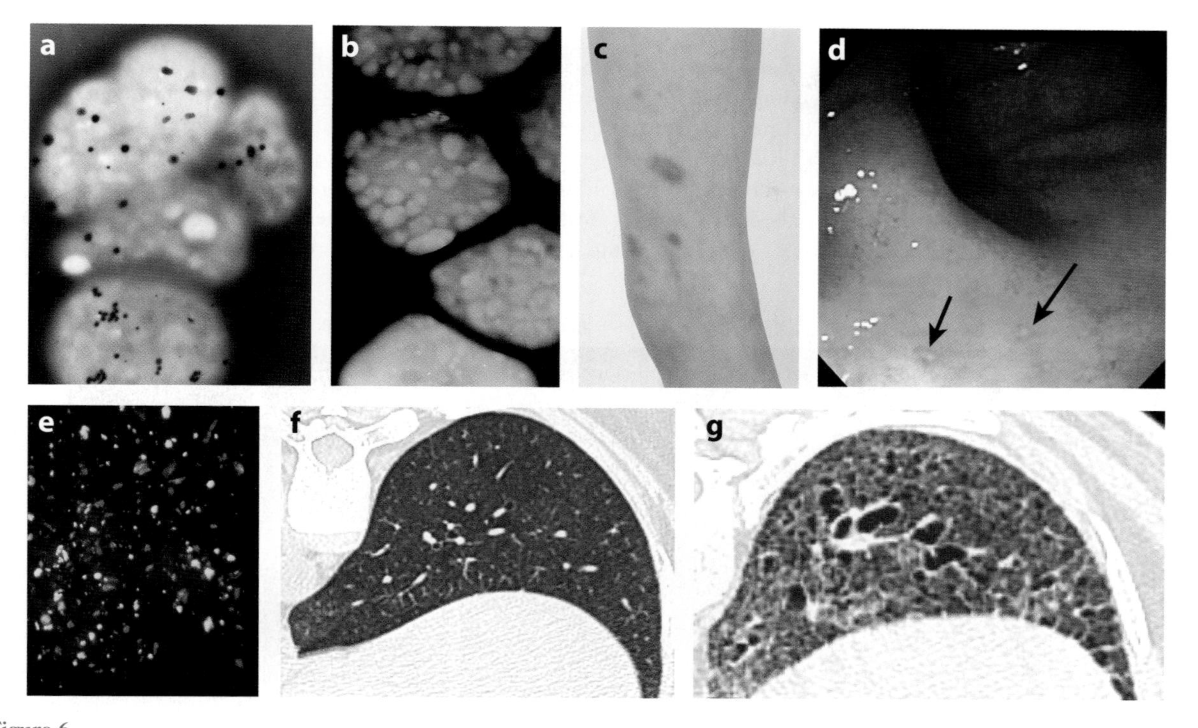

Figure 6

Clinical features of Hermansky-Pudlak syndrome (HPS). (*a*) Whole mount electron micrographs of normal platelets, each containing several delta granules (*small black dots*). (*b*) HPS platelets, devoid of delta granules. (*c*) Spontaneous bruising in a boy with HPS. (*d*) HPS intestine showing mucosal ulcerations (*arrows*). (*e*) Autofluorescent ceroid lipofucsin (*yellow*) in renal tubular cells sloughed into the urine of an HPS-1 patient. (*f*) High-resolution CT scan of a normal lung. (*g*) High-resolution CT scan of an HPS-1 lung, showing bullae, fibrosis, and loss of alveoli. [(*a*) and (*b*) courtesy of Dr. James G. White, University of Minnesota; (*f*) and (*g*) courtesy of Dr. Thomas Markello, NHGRI.]

complication in their thirties, forties, or fifties (**Figure 6*f***). The cause of restrictive lung disease in HPS remains unknown; there is no effective therapy, although pirfenidone may have a salutary effect (31). BLOC-2 patients do not develop pulmonary fibrosis (52), and there is insufficient information to know if other HPS subtypes are prone to this complication (123).

Cell biology of Hermansky-Pudlak syndrome. BLOC-2 consists of the HPS3, HPS5, and HPS6 proteins and its molecular mass (340 ± 64 kDa) is consistent with it being a heterotrimer (324 kDa), though additional subunits cannot be excluded (24). The HPS5 and HPS6 proteins interact directly, but no direct interactions with HPS3 have been reported (24). HPS3, 5, and 6 have no homology to known proteins and the only recognizable motif is a clathrin-binding domain (LLDFE) in HPS3 at residues 172–176. HPS3 coimmunoprecipitates with clathrin in melanocytes and GFP-HPS3 colocalizes with clathrin on small vesicles in the perinuclear area only when this motif is intact (43). BLOC-2 interacts with BLOC-1; immunoelectron microscopy was used to localize components of both BLOC-2 (HPS3 and HPS6) and BLOC-1 (pallidin and dysbindin) to tubulovesicular elements of EEA1-positive early endosomes (25). In human HPS-3 and HPS-5 melanocytes, melanosome-targeted cargo (TYRP1, TYR) does not efficiently traffic to melanosomes, but instead resides in small vesicles throughout the cell (8, 44). BLOC-2-deficient melanocytes demonstrate normal early melanosome formation (measured by

PMEL17 staining) but later delivery of TYR and TYRP1 to melanosomes is perturbed, causing reduced staining of these markers in the dendritic tips (**Figure 7**). Further, increased trafficking of TYRP1 via the cell membrane, demonstrated by increased internalization of anti-TYRP1 antibodies, occurs in BLOC-2-, BLOC-1-, and AP3-deficient melanocytes (25, 44). These findings support a role for BLOC-2 in trafficking LRO components (e.g., TYRP1) from an early endosomal compartment to the developing LRO. In the absence of BLOC-2, these components are misdirected to the cell membrane where they are internalized and subsequently degraded. BLOC-2 may also assist in the secretion of lysosomes and

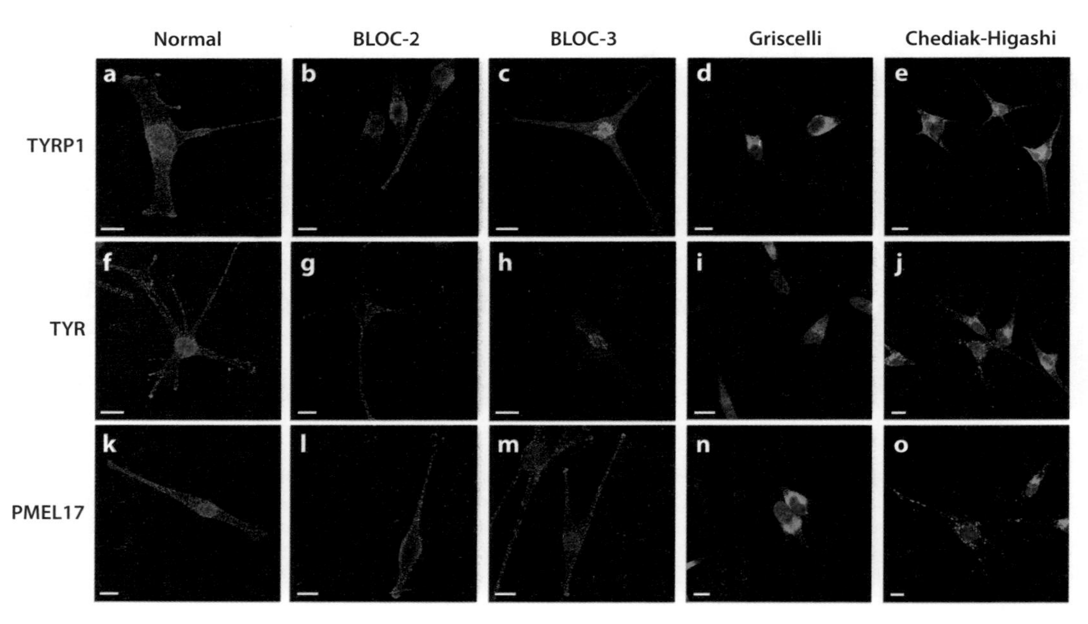

Figure 7

Distribution of melanogenic proteins tyrosinase-related protein-1 (TYRP1), tyrosinase (TYR1), and PMEL17 in melanocytes of patients with lysosome-related organelle (LRO) disorders. Confocal immunofluorescence microscopy of primary epidermal melanocytes derived from a normally pigmented individual and patients with LRO biogenesis defects stained for TYRP1 (anti-mouse MEL5) (*a–e*), TYR (anti-mouse tyrosinase) (*f–j*), and PMEL17 (anti-mouse HMB45) (*k–o*). All melanocytes were costained with TO-PRO-3 to visualize the nucleus. (*a,f,k*) Normal melanocytes demonstrate punctate TYRP1 (*a*), TYR (*f*), and PMEL17 (*k*) staining in the perinuclear region and throughout the dendrites with TYRP1 and TYR accumulation in the dendritic tips. (*b,g,l*) Biogenesis of lysosome-related organelles complex (BLOC)-2-deficient melanocytes were derived from an Hermansky-Pudlak syndrome (HPS)-5 patient compound heterozygous for a nonsense mutation (c.2624C>T; p.R865X) and a one-base-pair deletion (c.2264delT: P.L875fsX19) in exon 18 of *HPS5*. BLOC-2-deficient melanocytes display punctate staining of TYRP1 (*b*) and TYR (*g*) in the perinuclear region, extending into the dendrites, but lack pronounced accumulation of TYRP1 and TYR in the tips. (*l*) PMEL17 staining in BLOC-2 deficient melanocytes is distributed throughout the cell as in normal melanocytes. (*c,h,m*) BLOC-3 deficient melanocytes were derived from an HPS-1 patient homozygous for a 16-base-pair duplication (c.1472_1487dup16) in exon 15 of *HPS1*. In the absence of BLOC-3 TYRP1 (*c*) and TYR (*h*) are concentrated in the perinuclear/trans-Golgi network (TGN) region, whereas PMEL17 (*m*) extends into the dendrites and appears normally distributed. (*d,i,n*) Griscelli syndrome (GS2) melanocytes were derived from a patient compound heterozygous for two nonsense mutations (c.550C>T; p.R184X and c.598C>T; p.R200X) in *RAB27A*. In GS2 melanocytes, TYRP1 (*d*), TYR (*i*), and PMEL17 (*n*) localize to the perinuclear area and do not occupy the dendritic tips. (*e,j,o*) Chediak-Higashi syndrome (CHS) melanocytes were derived from a patient carrying a nonsense mutation (c.1540C>T; p.R514X) in exon 5 and a one-base-pair deletion (c.9893delT; p.F3298fsX3304) in exon 43 of *CHS1*. In CHS melanocytes, TYRP1 (*e*) localizes to vesicular structures in the perinuclear area with some dendritic localization, TYR accumulates in large granules that localize to the dendrites but not their tips (*j*), and PMEL17 localizes to enlarged vesicular structures that are present in the perinuclear and dendritic area but not in the dendritic tips (*o*). All images are 1D projections of confocal z-sections. Scale bars = 10 μm.

related organelles on the basis of reduced lysosomal enzyme secretion by kidney and platelets from both *ruby eye* (HPS-6) and *ruby eye-2* (HPS-5) mice (116). However, fibroblasts from *cocoa* (HPS-3) and *ruby eye* (HPS-6) mice showed normal secretory levels of the enzyme β-hexosaminidase, indicating that BLOC-2 is not critical for secretion of this lysosomal enzyme (24). In addition, an increase in the frequency and duration of transient fusion events in *ruby eye* mast cells during stimulation suggests a function for the *ruby eye* gene product in regulating closure of cell fusion pores (88).

HPS1 and HPS4 are components of BLOC-3, but no direct interaction between them has been demonstrated. Although the existence of another small subunit cannot be excluded (73, 83, 84), the molecular mass of BLOC-3 (140 ± 30 kDa) is most consistent with BLOC-3 being a heterodimer (156 kDa). HPS1 and HPS4 contain no recognizable homology to each other or to other known proteins nor do they contain structural motifs that would help predict the function of BLOC-3. BLOC-3 is largely cytosolic, with a small fraction associated with membranes (73). In BLOC-3-deficient human melanocytes, TYRP1 and TYR are not properly trafficked to developing, early stage melanosomes and largely remain in the perinuclear/TGN region (**Figure 7*c*,*h***). The mislocalization of TYR and TYRP1 results in reduced melanin synthesis (101). The formation of early stage melanosomes, marked by PMEL17 staining, appears unaffected in BLOC-3-deficient cells (**Figure 7*m***). Ultrastuctural studies of the choroids of BLOC-3 deficient *pale ear* (HPS-1) and *light ear* (HPS-4) mice demonstrated the presence of macromelanosomes, suggested to result from abnormal melanosome fusion events (67). BLOC-3 may also be involved in general events effecting lysosome/late endosome biogenesis or movement. *Light ear* and *pale ear* mice have enlarged kidney lysosomes, containing increased amounts of hydrolases, with decreased secretion into the urine (75, 116). *Pale ear* fibroblasts showed reduced perinuclear localization of lysosomal and late endosomal markers compared with control fibroblasts (83). The possible existence of BLOC-4 and BLOC-5 complexes that share some BLOC-3 subunits has been suggested (14).

AP3 recognizes and sorts proteins with tyrosine-based sorting signals (such as LAMPs and TYR) for LRO targeting (89). Consequently, AP3-deficient (i.e., HPS-2) fibroblasts exhibit enhanced abnormal routing of LAMPs through the plasma membrane (21, 54) and AP3 deficient human and mouse melanocytes have an abnormal distribution of TYR (117). HPS-2 is the only HPS subtype that manifests immunodeficiency (21, 53, 54), apparently owing to the severely impaired killing ability of HPS-2 CTLs. The secretory lysosomes of HPS-2 CTLs remain in the periphery and the CTLs fail to polarize upon activation; presumably AP3 fails to sort a protein required for granule polarization (111).

Chediak-Higashi Syndrome

CHS (MIM 214500) results from mutations in the *CHS1* gene, whose murine analog is the *beige* gene (56, 58). The clinical phenotype of CHS correlates somewhat with the molecular genotype and cell biological phenotype (59, 125).

Clinical manifestations of Chediak-Higashi syndrome. Individuals with CHS manifest decreased pigmentation (**Figure 8*a***), giant intracellular granules that are pathognomonic of the disease (**Figure 8*b***), pigment clumping in hair shafts (**Figure 8*c***), and a bleeding diathesis related to platelet dense bodies that are absent or reduced in number (56, 58). The granules, which are azurophilic and contain acid hydrolases and myeloperoxidase (56), are also present in CHS eosinophils, basophils, and monocytes. Children with CHS have life-threatening infections, primarily of the skin and respiratory systems. The pathogens are commonly *Staphylococcus aureus*, *Streptococcus*, gram-negative organisms, *Candida*, and *Aspergillus*. Immunoglobulins, antibody production, and phagocytosis are normal, but neutropenia is

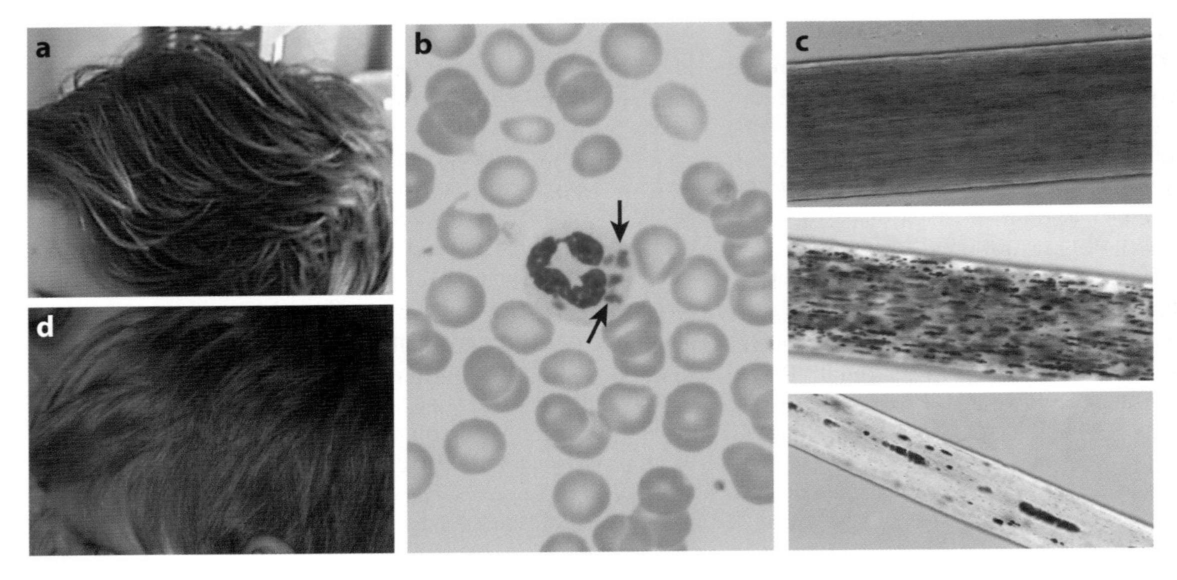

Figure 8

Clinical characteristics of Chediak-Higashi syndrome (CHS) and Griscelli syndrome (GS). (*a*) Light brown hair in a girl with CHS, after bone marrow transplantation. (*b*) Wright stain of a peripheral blood smear from a classic CHS patient, showing abnormal giant granules in a polymorphonuclear leukocyte (*arrows*). (*c*) Light microscopy of hair shafts from a normal individual (*top*), a patient with CHS (*middle*), and a patient with GS2 (*bottom*). Uneven granularity characterizes the pigmentation pattern of the bottom two hair shafts. (*d*) Silvery gray hair of a boy with GS2.

common and leucocytes display impaired migration (56, 58). Natural killer (NK) cells are decreased in function, causing impaired cellular immunity. Platelet counts in CHS are normal (56), but the secondary wave of platelet aggregation is impaired and shows an increased ATP-to-ADP ratio and decreased platelet serotonin and calcium. Humans affected by CHS, as well as CHS mice (*beige*) and CHS cattle, have absent or markedly reduced platelet dense bodies (38, 56, 76).

The curious phenomenon of lymphohistiocytosis occurs in approximately 85% of CHS patients. In this fatal complication also called the accelerated phase, the unfettered proliferation of lymphocytes creates a lymphoma-like situation with fever, anemia, neutropenia, and occasionally thrombocytopenia, hepatosplenomegaly, and lymphadenopathy. Liver function tests, serum ferritin, and fibronectin may be elevated and cellular immunity is decreased (56). Other disorders involving lymphohistiocytosis, such as autoimmune lymphoproliferative syndrome (MIM 601859) and familial hemophagocytic lymphohistiocytosis (FHL2) (MIM 603553) (60), have as their cause genetic defects in perforin or in the perforin receptor.

If CHS patients survive until adulthood, they develop signs and symptoms of neurological involvement, including neuropathies, autonomic dysfunction, atrophy, sensory deficits, areflexia, cerebellar signs, seizures, decreased ambulation, and cognitive defects. These findings, which appear independent of lymphohistocytic infiltration of the nervous system, are not prevented by bone marrow transplantation, which essentially cures the diathesis toward infections, lymphohistiocytosis, and immunological disorders of CHS.

Cell biology of Chediak-Higashi syndrome. The 429-kDa CHS1 protein is a member of the Beige and Chediak (BEACH) protein family and contains several tandemly repeated N-terminal Armadillo (ARM) and HEAT motifs, a C-terminal WD40 domain, and a BEACH motif (56, 58). The exact cell biological function of CHS1 remains unknown. However, studies on other members of the BEACH protein family, and the identification of CHS1-interacting proteins such as v-SNAREs, t-SNAREs, signaling protein 14-3-3, casein kinase II, and Hrs, suggest a role in membrane trafficking and/or organelle size regulation (58). Many cell types in CHS manifest giant lysosomes and LROs. The melanosomes of CHS melanocytes are enlarged (56) and melanosome-specific markers such as PMEL17, TYRP1, and TYR mislocalize to large vesicular structures present in the cell body and occasionally within the dendrites, but not in the dendritic tips (125) (**Figure 7*e*,*j*,*o***).

Griscelli Syndrome

Griscelli syndrome (GS) is a rare autosomal recessive disorder characterized by mild skin hypopigmentation and by immunological impairment, lymphohistiocytosis, or defects in the central nervous system (77, 78, 93). GS patients do not have an obvious bleeding tendency; whether they have a defect in platelet delta granules is not known. Three types of human GS are recognized, all displaying pigment dilution of the hair and skin (77, 78, 93).

Griscelli syndrome type 1. Griscelli syndrome type 1 (GS1) (MIM 214450) is caused by mutations in the *MYO5A* gene on chromosome 15q21, which encodes the actin-associated myosin Va motor protein (93). Most GS1 patients display primary neurological impairment but no immunological deficiencies (22), which resembles the phenotype of the *dilute* mouse. *MYO5A* undergoes alternative splicing that involves six exons; in melanocytes, the presence of the alternative exon F is required for actin-dependent melanosome transport (126, 129). Because *MYO5A* exon F transcripts are abundantly expressed in melanocytes but not in brain, the phenotype of patients with mutations in *MYO5A* exon F is restricted to hypopigmentation (77). Elejalde syndrome (ES) is an autosomal recessive disorder associated with silver-colored hair, pigment abnormalities of the skin, and severe neurological dysfunction, i.e., clinical features comparable to those of GS1 (10). ES patients do not manifest immunological impairment or the hemophagocytic syndrome. ES is thought to correspond to GS1 (10).

Griscelli syndrome type 2. Griscelli syndrome type 2 (GS2) (MIM 607624) results from mutations in the *RAB27A* gene on chromosome 15 (78). Clinical features include silver hair (**Figure 6*d***) with pigment clumping in hair shafts (**Figure 6*c***), infections, and lymphohistiocytosis. Infiltration of leukocytes in the brain can cause secondary neurological impairment. Primary central nervous system defects have not been observed in GS2 patients (10, 78). The *ashen* mouse is the murine counterpart of GS2.

Griscelli syndrome type 3. Griscelli syndrome type 3 (GS3) (MIM 609227) is caused by mutations in *Melanophilin* (*MLPH*), a member of the Rab effector family and the human homolog of the murine *leaden* gene. The only described GS3 patient has a homozygous R35W missense mutation in *MLPH* (77).

The clinical manifestations of GS result from defects in melanosome transport in melanocytes and lytic granule targeting in CTLs (7, 10, 111). In melanocytes, Rab27A in its GTP-bound form interacts with the membrane of mature melanosomes through its C20 geranylgeranyl lipid tail, which is added posttranslationally. Within the melanosomal membrane, Rab27A acts as a receptor for myosin Va through an indirect interaction with the Rab27A effector melanophilin (126, 129). The formation of the Rab27A-melanophilin-myosinVa complex results in accumulation of mature melanosomes in the actin-rich dendritic tips, a process necessary for transfer of melanosomes to keratinocytes for normal pigmentation. When myosin Va, Rab27A, or melanophilin is mutated, the tripartite complex (**Figure 3**) fails to form and melanosome transport is impaired, resulting in perinuclear accumulation of melanosomes (**Figure 7*d*,*i*,*n***). Skin hypopigmentation and silvery-gray hair in GS are not caused by defects in the melanosome biogenesis pathway but by ineffective capturing of melanosomes by the peripheral actin network (126, 129).

Platelet-Granule Specific Disorders

Storage pool deficiency (SPD) (MIM 185050) is the term used to describe the heterogeneous group of disorders characterized by deficiency of granule-bound substances in platelets (38, 62, 87, 124). The newer term hypogranular platelet disorders encompasses partial forms of SPD. Combined alpha/delta-SPD defines deficiency of both alpha and delta granules, whereas delta-SPD, which can be isolated or syndromic (as in HPS), refers to patients with delta granule defects.

Nonsyndromic delta-SPD. Nonsyndromic delta-SPD, or isolated delta-SPD, is characterized by mild to moderate bleeding due to deficiency of platelet dense granules without other systemic findings (38, 62). Isolated delta-SPD appears to be underdiagnosed. Nieuwenhuis and coworkers (87) reviewed 390 patients with a bleeding tendency; among the 145 with a prolonged bleeding time and normal platelet counts, 27 (18%) had congenital delta-SPD. Platelet aggregation studies were completely normal in 23% of the delta-SPD patients (87). Definitive diagnosis of delta-SPD requires electron microscopy and/or measurement of adenine nucleotides and serotonin to demonstrate the absence of dense granule limiting membranes and contents (124). Autosomal dominant inheritance appears to be the primary mode of transmission (124). Mutations in RUNX1 (AML1, CBFA2) are associated with familial predisposition to acute myeloid leukemia, thrombocytopenia, and dense granule deficiency (MIM 601399) (32).

Combined alpha/delta-SPD. Combined alpha/delta-SPD is characterized by decreased numbers of delta granules associated with a variable deficiency of alpha granules (38, 62). The bleeding tendency resembles that of delta-SPD. The frequency of alpha/delta-SPD is not known, but it is probably less common than isolated delta-SPD. The mode of inheritance appears to be autosomal dominant. The basic defect in most patients remains unknown. Mutations in the X-linked transcription factor GATA1 result in a form of hypogranular thrombocytopenia (MIM 305371) characterized by a generalized decrease in platelet organelles including alpha and delta granules (86).

Wiskott-Aldrich syndrome. Wiskott-Aldrich syndrome (WAS) (MIM 301000) is an X-linked microthrombocytopenia associated with immunodeficiency and eczema and is caused by mutations in the *WASP* gene (120), which codes for a protein that regulates signal-mediated actin cytoskeleton rearrangement. WAS platelets have markedly reduced delta granules, alpha granules, and mitochondria.

Thrombocytopenia absent radius syndrome. Thrombocytopenia absent radius syndrome (TAR) (MIM 274000) is characterized by thrombocytopenia and bilateral absence of the radii in the presence of thumbs (35). TAR is associated with a microdeletion on chromosome 1q21.1, but the phenotype develops only in the presence of an additional, as-yet-unknown modifier (64). Platelet counts of TAR patients are extremely low (15 to 30×10^{12}/L) in infancy; consequently, 90% of cases are symptomatic within the first four months of life (35). However, platelet counts increase with age and may improve to almost normal by adulthood. Defects in platelet granules are reported in a subset of TAR patients.

Other Lysosome-Related Organelle Disorders

Newly reported LRO-related disorders involve mutations in SNAREs or associated proteins, Rabs or Rab effectors, VPS proteins or their interactors, microtubules or actin-binding proteins and molecular motors, membrane lipid regulators, and in genes of unknown function (91). For example, mutations in *Munc13-4* and *syntaxin 11*, encoding SNARE proteins, cause familial hemophagocytic lymphohistocytosis (FHL), i.e., FHL3 (MIM 608898) and FHL4 (MIM 603552), respectively (60, 131). Arthrogryposis-renal dysfunction-cholestasis (ARC) (MIM 208085) is caused by mutations in *VPS33B*, which codes for a Sec1/Munc18-related SNARE-interacting protein involved in late endosomal membrane dynamics (34). Mutated *SNAP-29* underlies the cerebral dysgenesis, neuropathy, ichthyosis, and keratoderma (CEDNIK) syndrome (MIM 609528), in which developmental nervous system and skin lamellar granule defects occur due to impaired SNARE machinery (107). Defective kinesin molecular motors *KIF1B* and *KIF5A* cause Charcot-Marie-Tooth disease type 2A1 (MIM 118210) and spastic paraplegia 10 (MIM 604187), respectively (47). Mutated small GTPases or their regulating factors cause Charcot-Marie-Tooth disease type 2B (MIM 600882), defective in the late endosomal GTPase *Rab7* (48); choroideremia (MIM 303100), defective in Rab escort protein-1 (*REP-1*), which functions in prenylation of Rabs GTPases (96); and periventricular heterotopia with microcephaly (MIM 608097), defective in *ARFGEF2*, a guanine nucleotide exchange factor for ADP-ribosylation factor (ARF) (106). Finally, Charcot-Marie-Tooth disease type 4J (MIM 611228) and the *pale tremor* mutant mouse have late endosome-lysosome defects due to mutations in *FIG4*. *FIG4* codes for a homolog of a yeast SAC (suppressor of actin) domain phosphatidylinositol-3,5-bisphosphate 5-phosphatase located in the vacuolar membrane (16).

Various case reports and genetically unclassified disorders of hypopigmentation and/or bleeding diathesis and/or immunodeficiency have been described, including Cross or Kramer syndrome (MIM 257800), Vici syndrome (MIM 242840), Preus syndrome (MIM 257790), and osteoporosis and oculocutaneous hypopigmentation (OOCH) syndrome (MIM 601220).

CONCLUDING REMARKS

Defining the cellular and molecular events that regulate LRO biogenesis and trafficking remains a work in progress, aided enormously by mutant animal and human pigmentation models. The catalog of components involved in LRO biogenesis will expand with evolving experimental tools such as whole-genome sequencing, gene silencing, fluorescent imaging, and in vitro expression and transport assays. Insights into mechanisms should translate into rational therapeutic interventions for these intriguing disorders of lysosome-related organelles.

SUMMARY POINTS

1. Lysosome-related organelles (LROs) share features with lysosomes but have distinct morphology, composition, and/or functions.
2. LROs are maintained by the regulated flow of proteins and membranes/vesicles among a complex endosomal machinery, which provides specificity of time and place. Chaperones for this machinery include a cytoskeletal system of tubules and filaments, coat associated proteins, biogenesis of lysosome-related organelles complexes (BLOCs), soluble N-ethylmaleimide-sensitive-factor attachment protein (SNAP) receptors (SNAREs), syntaxins, Rabs, motor proteins, and membrane lipids.
3. The biogenesis of LROs employs a common machinery whose genetic defects can affect all LROs or only an individual LRO. Consequently, a variety of clinical features occur in affected individuals, including hypopigmentation, bleeding, and immune deficiency.

4. Cell biologists study LROs because of the lessons they can impart with respect to vesicle formation in general.
5. Abnormalities in both lysosomes and LROs have been observed in certain human genetic diseases, including Hermansky-Pudlak syndrome, Griscelli syndrome, and Chediak-Higashi syndrome.
6. Novel players in LRO biogenesis are revealed by human and animal mutations. These models continue to provide invaluable resources for elucidating the complex pathways involved.

FUTURE ISSUES

1. An exact definition of LROs has not been established, so the precise number of LROs remains unknown. To be considered an LRO, an organelle must possess an acidic lumen, characteristic lysosomal proteins, and a specialized function, usually related to storage or secretion. Accesibility to endocytic markers may be included as another criterion.
2. In the future, newly recognized organelles may be considered LROs.
3. More human and animal models of albinism/bleeding/immune deficiency/other clinical features will be recognized as LRO disorders. Cells of these patients (or their rodent, fly, zebrafish, or yeast counterparts) will continue to be important tools for analyzing membrane trafficking.
4. The functions of several proteins in LRO biogenesis, including the BLOC-1, BLOC-2, and BLOC3 subunits and the lysosomal trafficking regulator (LYST) protein, await elucidation. Understanding these functions will move us toward rational therapeutics for Hermansky-Pudlak syndrome, Chediak-Higashi syndrome, and related disorders.

DISCLOSURE STATEMENT

Dr. Gahl is a member of the Scientific Review Board of the Hermansky-Pudlak Syndrome Network.

ACKNOWLEDGMENTS

This work was supported by the Intramural Research program of the National Human Genome Research Institute (NHGRI), National Institutes of Health, Bethesda, Maryland, USA.

LITERATURE CITED

1. Ali B, Seabra M. 2005. Targeting of Rab GTPases to cellular membranes. *Biochem. Soc. Trans.* 33:652–56
2. Anderson PD, Huizing M, Claassen DA, White J, Gahl WA. 2003. Hermansky-Pudlak syndrome type 4 (HPS-4): clinical and molecular characteristics. *Hum. Genet.* 113:10–17
3. Anikster Y, Huizing M, White J, Shevchenko YO, Fitzpatrick DL, et al. 2001. Mutation of a new gene causes a unique form of Hermansky-Pudlak syndrome in a genetic isolate of central Puerto Rico. *Nat. Genet.* 28:376–80

4. Barral DC, Seabra MC. 2004. The melanosome as a model to study organelle motility in mammals. *Pigment Cell Res.* 17:111–18

5. Berson JF, Theos AC, Harper DC, Tenza D, Raposo G, Marks MS. 2003. Proprotein convertase cleavage liberates a fibrillogenic fragment of a resident glycoprotein to initiate melanosome biogenesis. *J. Cell Biol.* 161:521–33

6. Bohn G, Allroth A, Brandes G, Thiel J, Glocker E, et al. 2007. A novel human primary immunodeficiency syndrome caused by deficiency of the endosomal adaptor protein p14. *Nat. Med.* 13:38–45

7. Bohn G, Welte K, Klein C. 2007. Severe congenital neutropenia: new genes explain an old disease. *Curr. Opin. Rheumatol.* 19:644–50

8. Boissy RE, Richmond B, Huizing M, Helip-Wooley A, Zhao Y, et al. 2005. Melanocyte-specific proteins are aberrantly trafficked in melanocytes of Hermansky-Pudlak syndrome-type 3. *Am. J. Pathol.* 166:231–40

9. Bonifacino JS, Glick BS. 2004. The mechanisms of vesicle budding and fusion. *Cell* 116:153–66

10. Cahali JB, Fernandez SA, Oliveira ZN, Machado MC, Valente NS, Sotto MN. 2004. Elejalde syndrome: report of a case and review of the literature. *Pediatr. Dermatol.* 21:479–82

11. Cai H, Reinisch K, Ferro-Novick S. 2007. Coats, tethers, Rabs, and SNAREs work together to mediate the intracellular destination of a transport vesicle. *Dev. Cell* 12:671–82

12. Caviston JP, Holzbaur EL. 2006. Microtubule motors at the intersection of trafficking and transport. *Trends Cell Biol.* 16:530–37

13. Chi E, Pruiett J, Lagunoff D. 1975. Abnormal lamellar bodies in type II pneumocytes and increased lung surface active material in the beige mouse. *J. Histochem. Cytochem.* 23:863–66

14. Chiang PW, Oiso N, Gautam R, Suzuki T, Swank RT, Spritz RA. 2003. The Hermansky-Pudlak syndrome 1 (HPS1) and HPS4 proteins are components of two complexes, BLOC-3 and BLOC-4, involved in the biogenesis of lysosome-related organelles. *J. Biol. Chem.* 278:20332–37

15. Chintala S, Li W, Lamoreux ML, Ito S, Wakamatsu K, et al. 2005. *Slc7a11* gene controls production of pheomelanin pigment and proliferation of cultured cells. *Proc. Natl. Acad. Sci. USA* 102:10964–69

16. Chow CY, Zhang Y, Dowling JJ, Jin N, Adamska M, et al. 2007. Mutation of *FIG4* causes neurodegeneration in the pale tremor mouse and patients with CMT4J. *Nature* 448:68–72

17. Cordonnier MN, Dauzonne D, Louvard D, Coudrier E. 2001. Actin filaments and myosin I alpha cooperate with microtubules for the movement of lysosomes. *Mol. Biol. Cell* 12:4013–29

18. Coulombe PA, Bousquet O, Ma L, Yamada S, Wirtz D. 2000. The 'ins' and 'outs' of intermediate filament organization. *Trends Cell Biol.* 10:420–28

19. Deacon SW, Serpinskaya AS, Vaughan PS, Lopez Fanarraga M, Vernos I, et al. 2003. Dynactin is required for bidirectional organelle transport. *J. Cell Biol.* 160:297–301

20. Dell'Angelica EC, Mullins C, Caplan S, Bonifacino JS. 2000. Lysosome-related organelles. *FASEB J.* 14:1265–78

21. Dell'Angelica EC, Shotelersuk V, Aguilar RC, Gahl WA, Bonifacino JS. 1999. Altered trafficking of lysosomal proteins in Hermansky-Pudlak syndrome due to mutations in the β3A subunit of the AP-3 adaptor. *Mol. Cell* 3:11–21

22. Desnos C, Huet S, Darchen F. 2007. 'Should I stay or should I go?': myosin V function in organelle trafficking. *Biol. Cell* 99:411–23

23. Detter JC, Zhang Q, Mules EH, Novak EK, Mishra VS, et al. 2000. Rab geranylgeranyl transferase αmutation in the *gunmetal* mouse reduces Rab prenylation and platelet synthesis. *Proc. Natl. Acad. Sci. USA* 97:4144–49

24. Di Pietro SM, Falcon-Perez JM, Dell'Angelica EC. 2004. Characterization of BLOC-2, a complex containing the Hermansky-Pudlak syndrome proteins HPS3, HPS5 and HPS6. *Traffic* 5:276–83

25. Di Pietro SM, Falcon-Perez JM, Tenza D, Setty SR, Marks MS, et al. 2006. BLOC-1 interacts with BLOC-2 and the AP-3 complex to facilitate protein trafficking on endosomes. *Mol. Biol. Cell* 17:4027–38

26. Falcon-Perez JM, Romero-Calderon R, Brooks ES, Krantz DE, Dell'Angelica EC. 2007. The *Drosophila* pigmentation gene *pink* (*p*) encodes a homologue of human Hermansky-Pudlak syndrome 5 (HPS5). *Traffic* 8:154–68

5. Describes MVBs as intermediates in the generation of stage II melanosomes; identifies PMEL17 as the main component of the intralumenal fibrils.

21. Demonstrates that mutations of an AP3 subunit cause Hermansky-Pudlak syndrome type 2, which involves defects in intracellular vesicle transport pathways.

25. Demonstrates that BLOC-1 interacts with AP3 and BLOC-2, and that BLOC-1 and BLOC-2 localize mainly to early endosome–associated tubules.

27. Falcon-Perez JM, Starcevic M, Gautam R, Dell'Angelica EC. 2002. BLOC-1, a novel complex containing the pallidin and muted proteins involved in the biogenesis of melanosomes and platelet-dense granules. *J. Biol. Chem.* 277:28191–99
28. Fischer A, Latour S, de Saint Basile G. 2007. Genetic defects affecting lymphocyte cytotoxicity. *Curr. Opin. Immunol.* 19:348–53
29. Fontana S, Parolini S, Vermi W, Booth S, Gallo F, et al. 2006. Innate immunity defects in Hermansky-Pudlak type 2 syndrome. *Blood* 107:4857–64
30. Gahl WA, Brantly M, Kaiser-Kupfer MI, Iwata F, Hazelwood S, et al. 1998. Genetic defects and clinical characteristics of patients with a form of oculocutaneous albinism (Hermansky-Pudlak syndrome). *N. Engl. J. Med.* 338:1258–64
31. **Gahl WA, Brantly M, Troendle J, Avila NA, Padua A, et al. 2002. Effect of pirfenidone on the pulmonary fibrosis of Hermansky-Pudlak syndrome. *Mol. Genet. Metab.* 76:234–42**

31. A clinical treatment study of pirfenidone, which slows the progression of pulmonary fibrosis in a subset of Hermansky-Pudlak syndrome patients.

32. Ganly P, Walker LC, Morris CM. 2004. Familial mutations of the transcription factor RUNX1 (AML1, CBFA2) predispose to acute myeloid leukemia. *Leuk. Lymphoma* 45:1–10
33. Gillard BK, Clement R, Colucci-Guyon E, Babinet C, Schwarzmann G, et al. 1998. Decreased synthesis of glycosphingolipids in cells lacking vimentin intermediate filaments. *Exp. Cell Res.* 242:561–72
34. Gissen P, Johnson CA, Morgan NV, Stapelbroek JM, Forshew T, et al. 2004. Mutations in *VPS33B*, encoding a regulator of SNARE-dependent membrane fusion, cause arthrogryposis-renal dysfunction-cholestasis (ARC) syndrome. *Nat. Genet.* 36:400–4
35. Greenhalgh KL, Howell RT, Bottani A, Ancliff PJ, Brunner HG, et al. 2002. Thrombocytopenia-absent radius syndrome: a clinical genetic study. *J. Med. Genet.* 39:876–81
36. Gruenberg J. 2001. The endocytic pathway: a mosaic of domains. *Nat. Rev. Mol. Cell Biol.* 2:721–30
37. Gullberg U, Bengtsson N, Bülow E, Garwicz D, Lindmark A, Olsson I. 1999. Processing and targeting of granule proteins in human neutrophils. *J. Immunol. Methods* 232:201–10
38. Gunay-Aygun M, Huizing M, Gahl WA. 2004. Molecular defects that affect platelet dense granules. *Semin. Thromb. Hemost.* 30:537–47
39. Hannah MJ, Williams R, Kaur J, Hewlett LJ, Cutler DF. 2002. Biogenesis of Weibel-Palade bodies. *Semin. Cell Dev. Biol.* 13:313–24
40. Hara M, Yaar M, Byers HR, Goukassian D, Fine RE, et al. 2000. Kinesin participates in melanosomal movement along melanocyte dendrites. *J. Invest. Dermatol.* 114:438–43
41. Harding C. 1995. Intracellular organelles involved in antigen processing and the binding of peptides to class II MHC molecules. *Semin. Immun.* 7:355–60
42. Heijnen HF, Debili N, Vainchencker W, Breton-Gorius J, Geuze HJ, Sixma JJ. 1998. Multivesicular bodies are an intermediate stage in the formation of platelet α-granules. *Blood* 91:2313–25
43. Helip-Wooley A, Westbroek W, Dorward H, Mommaas M, Boissy RE, et al. 2005. Association of the Hermansky-Pudlak syndrome type-3 protein with clathrin. *BMC Cell Biol.* 6:33
44. **Helip-Wooley A, Westbroek W, Dorward HM, Koshoffer A, Huizing M, et al. 2007. Improper trafficking of melanocyte-specific proteins in Hermansky-Pudlak syndrome type-5. *J. Invest. Dermatol.* 127:1471–78**

44. Employs BLOC-2-deficient human melanocytes to demonstrate that early stage melanosome formation and PMEL17 trafficking are preserved in these cells.

45. Hermos CR, Huizing M, Kaiser-Kupfer MI, Gahl WA. 2002. Hermansky-Pudlak syndrome type 1: gene organization, novel mutations, and clinical-molecular review of non-Puerto Rican cases. *Hum. Mutat.* 20:482
46. Herrmann H, Bar H, Kreplak L, Strelkov SV, Aebi U. 2007. Intermediate filaments: from cell architecture to nanomechanics. *Nat. Rev. Mol. Cell Biol.* 8:562–73
47. Hirokawa N, Takemura R. 2003. Biochemical and molecular characterization of diseases linked to motor proteins. *Trends Biochem. Sci.* 28:558–65
48. Houlden H, King RH, Muddle JR, Warner TT, Reilly MM, et al. 2004. A novel RAB7 mutation associated with ulcero-mutilating neuropathy. *Ann. Neurol.* 56:586–90
49. Huizing M, Anikster Y, Fitzpatrick DL, Jeong AB, D'Souza M, et al. 2001. Hermansky-Pudlak syndrome type 3 in Ashkenazi Jews and other non-Puerto Rican patients with hypopigmentation and platelet storage-pool deficiency. *Am. J. Hum. Genet.* 69:1022–32
50. Huizing M, Boissy RE, Gahl WA. 2002. Hermansky-Pudlak syndrome: vesicle formation from yeast to man. *Pigment Cell Res.* 15:405–19

51. Huizing M, Gahl WA. 2002. Disorders of vesicles of lysosomal lineage: the Hermansky-Pudlak syndromes. *Curr. Mol. Med.* 2:451–67
52. Huizing M, Hess R, Dorward H, Claassen DA, Helip-Wooley A, et al. 2004. Cellular, molecular and clinical characterization of patients with Hermansky-Pudlak syndrome type 5. *Traffic* 5:711–22
53. Huizing M, Sarangarajan R, Strovel E, Zhao Y, Gahl WA, Boissy RE. 2001. AP-3 mediates tyrosinase but not TRP-1 trafficking in human melanocytes. *Mol. Biol. Cell* 12:2075–85
54. Huizing M, Scher C, Strovel E, Fitzpatrick D, Hartnell L, et al. 2002. Nonsense mutations in *ADTB3A* cause complete deficiency of the β3A subunit of adaptor complex-3 and severe Hermansky-Pudlak syndrome type 2. *Pediatr. Res.* 51:150–58
55. Hussain N, Quezado M, Huizing M, Geho D, White JG, et al. 2006. Intestinal disease in Hermansky-Pudlak syndrome: occurrence of colitis and relation to genotype. *Clin. Gastroenterol. Hepatol.* 4:73–80
56. Introne W, Boissy RE, Gahl WA. 1999. Clinical, molecular, and cell biological aspects of Chediak-Higashi syndrome. *Mol. Genet. Metab.* 68:283–303
57. Jordens I, Fernandez-Borja M, Marsman M, Dusseljee S, Janssen L, et al. 2001. The Rab7 effector protein RILP controls lysosomal transport by inducing the recruitment of dynein-dynactin motors. *Curr. Biol.* 11:1680–85
58. Kaplan J, De Domenico I, Ward DM. 2008. Chediak-Higashi syndrome. *Curr. Opin. Hematol.* 15:22–29
59. Karim MA, Suzuki K, Fukai K, Oh J, Nagle DL, et al. 2002. Apparent genotype-phenotype correlation in childhood, adolescent, and adult Chediak-Higashi syndrome. *Am. J. Med. Genet.* 108:16–22
60. Katano H, Cohen JI. 2005. Perforin and lymphohistiocytic proliferative disorders. *Br. J. Haematol.* 128:739–50
61. King RA, Hearing VJ, Creel DJ, Oetting WS. 2001. Albinism. In *The Metabolic and Molecular Bases of Inherited Disease*, ed. CR Scriver, AL Beaudet, WS Sly, DL Valle. pp. 5587–627. New York: McGraw-Hill
62. King SM, Reed GL. 2002. Development of platelet secretory granules. *Semin. Cell Dev. Biol.* 13:293–302
63. Kjeldsen L, Calafat J, Borregaard N. 1998. Giant granules of neutrophils in Chediak-Higashi syndrome are derived from azurophil granules but not from specific and gelatinase granules. *J. Leukoc. Biol.* 64:72–77
64. Klopocki E, Schulze H, Strauss G, Ott CE, Hall J, et al. 2007. Complex inheritance pattern resembling autosomal recessive inheritance involving a microdeletion in thrombocytopenia-absent radius syndrome. *Am. J. Hum. Genet.* 80:232–40
65. Kornfeld S, Mellman I. 1989. The biogenesis of lysosomes. *Annu. Rev. Cell Biol.* 5:483–525
66. Kramer H. 2002. Sorting out signals in fly endosomes. *Traffic* 3:87–91
67. Li W, Rusiniak ME, Chintala S, Gautam R, Novak EK, Swank RT. 2004. Murine Hermansky-Pudlak syndrome genes: regulators of lysosome-related organelles. *BioEssays* 26:616–28
68. Li W, Zhang Q, Oiso N, Novak EK, Gautam R, et al. 2003. Hermansky-Pudlak syndrome type 7 (HPS-7) results from mutant dysbindin, a member of the biogenesis of lysosome-related organelles complex 1 (BLOC-1). *Nat. Genet.* 35:84–89
69. Lloyd V, Ramaswami M, Kramer H. 1998. Not just pretty eyes: *Drosophila* eye-colour mutations and lysosomal delivery. *Trends Cell Biol.* 8:257–59
70. Loftus S, Larson DM, Baxter LL, Antonellis A, Chen Y, et al. 2002. Mutation of melanosome protein RAB38 in chocolate mice. *Proc. Natl. Acad. Sci. USA* 99:4471–76
71. Marks MS, Seabra MC. 2001. The melanosome: membrane dynamics in black and white. *Nat. Rev. Mol. Cell Biol.* 2:738–48
72. Marone G, Casolaro V, Patella V, Florio G, Triggiani M. 1997. Molecular and cellular biology of mast cells and basophils. *Int. Arch. Allergy Immunol.* 114:207–17
73. Martina J, Moriyama K, Bonifacino J. 2003. BLOC-3, a protein complex containing the Hermansky-Pudlak syndrome gene products HPS1 and HPS4. *J. Biol. Chem.* 278:29376–84
74. McNicol A, Israels SJ. 1999. Platelet dense granules: structure, function and implications for haemostasis. *Thromb. Res.* 95:1–18
75. Meisler M, Levy J, Sansone F, Gordon M. 1980. Morphologic and biochemical abnormalities of kidney lysosomes in mice with an inherited albinism. *Am. J. Pathol.* 101:581–93
76. Menard M, Meyers KM. 1988. Storage pool deficiency in cattle with the Chediak-Higashi syndrome results from an absence of dense granule precursors in their megakaryocytes. *Blood* 72:1726–34

77. Menasche G, Ho CH, Sanal O, Feldmann J, Tezcan I, et al. 2003. Griscelli syndrome restricted to hypopigmentation results from a melanophilin defect (GS3) or a MYO5A F-exon deletion (GS1). *J. Clin. Invest.* 112:450–56

78. Menasche G, Pastural E, Feldmann J, Certain S, Ersoy F, et al. 2000. Mutations in *RAB27A* cause Griscelli syndrome associated with haemophagocytic syndrome. *Nat. Genet.* 25:173–76

78. The first demonstration that Griscelli syndrome is heterogeneous and most patients with the classic syndrome have mutations in *RAB27A*.

79. Miki H, Okada Y, Hirokawa N. 2005. Analysis of the kinesin superfamily: insights into structure and function. *Trends Cell Biol.* 15:467–76

80. Morgan NV, Pasha S, Johnson CA, Ainsworth JR, Eady RAJ, et al. 2006. A germline mutation in *BLOC1S3/reduced pigmentation* causes a novel variant of Hermansky-Pudlak syndrome (HPS8). *Am. J. Hum. Genet.* 78:160–66

81. Munro S. 2002. Organelle identity and the targeting of peripheral membrane proteins. *Curr. Opin. Cell Biol.* 14:506–14

82. Nakatani Y, Nakamura N, Sano J, Inayama Y, Kawano N, et al. 2000. Interstitial pneumonia in Hermansky-Pudlak syndrome: significance of florid foamy swelling/degeneration (giant lamellar body degeneration) of type-2 pneumocytes. *Virchows. Arch.* 437:304–13

83. Nazarian R, Falcon-Perez JM, Dell'Angelica EC. 2003. Biogenesis of lysosome-related organelles complex 3 (BLOC-3): a complex containing the Hermansky-Pudlak syndrome (HPS) proteins HPS1 and HPS4. *Proc. Natl. Acad. Sci. USA* 100:8770–75

84. Nazarian R, Huizing M, Helip-Wooley A, Starcevic M, Gahl W, Dell'Angelica E. 2007. An immunoblotting assay to facilitate the molecular diagnosis of Hermansky-Pudlak syndrome. *Mol. Genet. Metab.* 93:134–44

85. Nazarian R, Starcevic M, Spencer MJ, Dell'Angelica EC. 2006. Reinvestigation of the dysbindin subunit of BLOC-1 (biogenesis of lysosome-related organelles complex-1) as a dystrobrevin-binding protein. *Biochem. J.* 395:587–98

86. Nichols KE, Crispino JD, Poncz M, White JG, Orkin SH, et al. 2000. Familial dyserythropoietic anaemia and thrombocytopenia due to an inherited mutation in *GATA1*. *Nat. Genet.* 24:266–70

87. Nieuwenhuis HK, Akkerman JW, Sixma JJ. 1987. Patients with a prolonged bleeding time and normal aggregation tests may have storage pool deficiency: studies on one hundred six patients. *Blood* 70:620–23

88. Oberhauser A, Fernandez J. 1996. A fusion pore phenotype in mast cells of the ruby-eye mouse. *Proc. Natl. Acad. Sci. USA* 93:14349–54

89. Ohno H, Aguilar RC, Yeh D, Taura D, Saito T, Bonifacino JS. 1998. The medium subunits of adaptor complexes recognize distinct but overlapping sets of tyrosine-based sorting signals. *J. Biol. Chem.* 273:25915–21

90. Oiso N, Riddle SR, Serikawa T, Kuramoto T, Spritz RA. 2004. The rat Ruby (*R*) locus is *Rab38*: identical mutations in Fawn-hooded and Tester-Moriyama rats derived from an ancestral Long Evans rat substrain. *Mamm. Genome* 15:307–14

91. Olkkonen VM, Ikonen E. 2006. When intracellular logistics fails–genetic defects in membrane trafficking. *J. Cell Sci.* 119:5031–45

92. Ong LL, Lim AP, Er CP, Kuznetsov SA, Yu H. 2000. Kinectin-kinesin binding domains and their effects on organelle motility. *J. Biol. Chem.* 275:32854–60

93. Pastural E, Barrat FJ, Dufourcq-Lagelouse R, Certain S, Sanal O, et al. 1997. Griscelli disease maps to chromosome 15q21 and is associated with mutations in the myosin-Va gene. *Nat. Genet.* 16:289–92

94. Pelham HR. 2001. SNAREs and the specificity of membrane fusion. *Trends Cell Biol.* 11:99–101

95. Pfister KK, Shah PR, Hummerich H, Russ A, Cotton J, et al. 2006. Genetic analysis of the cytoplasmic dynein subunit families. *PLoS Genet.* 2:e1

96. Rak A, Pylypenko O, Niculae A, Pyatkov K, Goody RS, Alexandrov K. 2004. Structure of the Rab7:REP-1 complex: insights into the mechanism of Rab prenylation and choroideremia disease. *Cell* 117:749–60

97. Raposo G, Marks MS. 2007. Melanosomes–dark organelles enlighten endosomal membrane transport. *Nat. Rev. Mol. Cell Biol.* 8:786–97

98. Raposo G, Marks MS, Cutler DF. 2007. Lysosome-related organelles: driving post-Golgi compartments into specialisation. *Curr. Opin. Cell Biol.* 19:394–401

99. Raposo G, Tenza D, Murphy DM, Berson JF, Marks MS. 2001. Distinct protein sorting and localization to premelanosomes, melanosomes, and lysosomes in pigmented melanocytic cells. *J. Cell Biol.* 152:809–24

100. Rendu F, Brohard-Bohn B. 2001. The platelet release reaction: granules' constituents, secretion and functions. *Platelets* 12:261–73
101. Sarangarajan R, Budev A, Zhao Y, Gahl WA, Boissy RE. 2001. Abnormal translocation of tyrosinase and tyrosinase-related protein 1 in cutaneous melanocytes of Hermansky-Pudlak syndrome and in melanoma cells transfected with antisense *HPS1* cDNA. *J. Invest. Dermatol.* 117:641–46
102. Schneider P, Holler N, Bodmer JL, Hahne M, Frei K, et al. 1998. Conversion of membrane-bound Fas(CD95) ligand to its soluble form is associated with downregulation of its proapoptotic activity and loss of liver toxicity. *J. Exp. Med.* 187:1205–13
103. Schonthaler HB, Fleisch VC, Biehlmaier O, Makhankov Y, Rinner O, et al. 2008. The zebrafish mutant *lbk/vam6* resembles human multisystemic disorders caused by aberrant trafficking of endosomal vesicles. *Development* 135:387–99
104. Seabra MC, Mules EH, Hume AN. 2002. Rab GTPases, intracellular traffic and disease. *Trends Mol. Med.* 8:23–30
105. Setty SR, Tenza D, Truschel ST, Chou E, Sviderskaya EV, et al. 2007. BLOC-1 is required for cargo-specific sorting from vacuolar early endosomes toward lysosome-related organelles. *Mol. Biol. Cell* 18:768–80
106. Sheen VL, Ganesh VS, Topcu M, Sebire G, Bodell A, et al. 2004. Mutations in *ARFGEF2* implicate vesicle trafficking in neural progenitor proliferation and migration in the human cerebral cortex. *Nat. Genet.* 36:69–76
107. Sprecher E, Ishida-Yamamoto A, Mizrahi-Koren M, Rapaport D, Goldsher D, et al. 2005. A mutation in *SNAP29*, coding for a SNARE protein involved in intracellular trafficking, causes a novel neurocutaneous syndrome characterized by cerebral dysgenesis, neuropathy, ichthyosis, and palmoplantar keratoderma. *Am. J. Hum. Genet.* 77:242–51
108. Stahlman M, Gray M, Falconieri M, Whitsett J, Weaver T. 2000. Lamellar body formation in normal and surfactant B-deficient mice. *Lab. Invest.* 80:395–402
109. Starcevic M, Dell'Angelica EC. 2004. Identification of snapin and three novel proteins (BLOS1, BLOS2, and BLOS3/reduced pigmentation) as subunits of biogenesis of lysosome-related organelles complex-1 (BLOC-1). *J. Biol. Chem.* 279:28393–401
110. Stenbeck G. 2002. Formation and function of the ruffled border in osteoclasts. *Semin. Cell Dev. Biol.* 13:285–92
111. Stinchcombe J, Bossi G, Griffiths GM. 2004. Linking albinism and immunity: the secrets of secretory lysosomes. *Science* 305:55–59
112. Strom M, Hume AN, Tarafder AK, Barkagianni E, Seabra MC. 2002. A family of Rab27-binding proteins. Melanophilin links Rab27a and myosin Va function in melanosome transport. *J. Biol. Chem.* 277:25423–30
113. Styers ML, Salazar G, Love R, Peden AA, Kowalczyk AP, Faundez V. 2004. The endo-lysosomal sorting machinery interacts with the intermediate filament cytoskeleton. *Mol. Biol. Cell* 15:5369–82
114. Suarez-Quian CA. 1987. The distribution of four lysosomal integral membrane proteins (LIMPs) in rat basophilic leukemia cells. *Tissue Cell* 19:495–504
115. Suzuki T, Li W, Zhang Q, Karim A, Novak EK, et al. 2002. Hermansky-Pudlak syndrome is caused by mutations in *HPS4*, the human homolog of the mouse light-ear gene. *Nat. Genet.* 30:321–24

116. Swank RT, Novak EK, McGarry MP, Rusiniak ME, Feng L. 1998. Mouse models of Hermansky-Pudlak syndrome: a review. *Pigment Cell Res.* 11:60–80

117. Theos AC, Tenza D, Martina JA, Hurbain I, Peden AA, et al. 2005. Functions of adaptor protein (AP)-3 and AP-1 in tyrosinase sorting from endosomes to melanosomes. *Mol. Biol. Cell* 16:5356–72
118. Toro J, Turner M, Gahl WA. 1999. Dermatologic manifestations of Hermansky-Pudlak syndrome in patients with and without a 16-base pair duplication in the *HPS1* gene. *Arch. Dermatol.* 135:774–80
119. Tribl F, Marcus K, Meyer HE, Bringmann G, Gerlach M, Riederer P. 2006. Subcellular proteomics reveals neuromelanin granules to be a lysosome-related organelle. *J. Neural. Transm.* 113:741–49

99. Demonstrates that melanosomes are distinct from lysosomes, with a common precursor at the stage I melanosome/vacuolar early endosome.

116. With Reference 67, thoroughly describes and classifies the mouse models of hypopigmentation and bleeding.

120. Villa A, Notarangelo L, Macchi P, Mantuano E, Cavagni G, et al. 1995. X-linked thrombocytopenia and Wiskott-Aldrich syndrome are allelic diseases with mutations in the WASP gene. *Nat. Genet.* 9:414–17

121. Wasmeier C, Romao M, Plowright L, Bennett DC, Raposo G, Seabra MC. 2006. Rab38 and Rab32 control post-Golgi trafficking of melanogenic enzymes. *J. Cell Biol.* 175:271–81

122. Weaver T, Na C, Stahlman M. 2002. Biogenesis of lamellar bodies, lysosome-related organelles involved in storage and secretion of pulmonary surfactant. *Semin. Cell Dev. Biol.* 13:263–70

123. Wei ML. 2006. Hermansky-Pudlak syndrome: a disease of protein trafficking and organelle function. *Pigment Cell Res.* 19:19–42

123. Review of Hermansky-Pudlak syndrome subtypes, including a valuable listing of all the HPS gene mutations reported to date.

124. Weiss HJ, Witte LD, Kaplan KL, Lages BA, Chernoff A, et al. 1979. Heterogeneity in storage pool deficiency: studies on granule-bound substances in 18 patients including variants deficient in α-granules, platelet factor 4, β-thromboglobulin, and platelet-derived growth factor. *Blood* 54:1296–319

125. Westbroek W, Adams D, Huizing M, Koshoffer A, Dorward H, et al. 2007. Cellular defects in Chediak-Higashi syndrome correlate with the molecular genotype and clinical phenotype. *J. Invest. Dermatol.* 127:2674–77

126. Westbroek W, Lambert J, Bahadoran P, Busca R, Herteleer MC, et al. 2003. Interactions of human Myosin Va isoforms, endogenously expressed in human melanocytes, are tightly regulated by the tail domain. *J. Invest. Dermatol.* 120:465–75

126. Shows alternative splicing of MYO5A, involving six exons; the alternative exon F is required for actin-dependent melanosome transport in melanocytes.

127. Winder SJ. 2003. Structural insights into actin-binding, branching and bundling proteins. *Curr. Opin. Cell Biol.* 15:14–22

128. Wu X, Jung G, Hammer JA 3rd. 2000. Functions of unconventional myosins. *Curr. Opin. Cell Biol.* 12:42–51

129. Wu X, Rao K, Zhang H, Wang F, Sellers J, et al. 2002. Identification of an organelle receptor for myosin-Va. *Nat. Cell Biol.* 4:271–78

130. Youssefian T, Cramer EM. 2000. Megakaryocyte dense granule components are sorted in multivesicular bodies. *Blood* 95:4004–7

131. zur Stadt U, Schmidt S, Kasper B, Beutel K, Diler AS, et al. 2005. Linkage of familial hemophagocytic lymphohistiocytosis (FHL) type-4 to chromosome 6q24 and identification of mutations in syntaxin 11. *Hum. Mol. Genet.* 14:827–34

Next-Generation DNA Sequencing Methods

Elaine R. Mardis

Departments of Genetics and Molecular Microbiology and Genome Sequencing Center, Washington University School of Medicine, St. Louis MO 63108; email: emardis@wustl.edu

Annu. Rev. Genomics Hum. Genet. 2008. 9:387–402

First published online as a Review in Advance on June 24, 2008

The *Annual Review of Genomics and Human Genetics* is online at genom.annualreviews.org

This article's doi: 10.1146/annurev.genom.9.081307.164359

1527-8204/08/0922-0387$20.00

Key Words

massively parallel sequencing, sequencing-by-synthesis, resequencing

Abstract

Recent scientific discoveries that resulted from the application of next-generation DNA sequencing technologies highlight the striking impact of these massively parallel platforms on genetics. These new methods have expanded previously focused readouts from a variety of DNA preparation protocols to a genome-wide scale and have fine-tuned their resolution to single base precision. The sequencing of RNA also has transitioned and now includes full-length cDNA analyses, serial analysis of gene expression (SAGE)-based methods, and noncoding RNA discovery. Next-generation sequencing has also enabled novel applications such as the sequencing of ancient DNA samples, and has substantially widened the scope of metagenomic analysis of environmentally derived samples. Taken together, an astounding potential exists for these technologies to bring enormous change in genetic and biological research and to enhance our fundamental biological knowledge.

INTRODUCTION

The sequencing of the reference human genome was the capstone for many years of hard work spent developing high-throughput, high-capacity production DNA sequencing and associated sequence finishing pipelines. The approach used >20,000 large bacterial artificial chromosome (BAC) clones that each contained an approximately 100-kb fragment of the human genome, which together provided an overlapping set or tiling path through each human chromosome as determined by physical mapping (31). In BAC-based sequencing, each BAC clone is amplified in bacterial culture, isolated in large quantities, and sheared to produce size-selected pieces of approximately 2–3 kb. These pieces are subcloned into plasmid vectors, amplified in bacterial culture, and the DNA is selectively isolated prior to sequencing. By generating approximately eightfold oversampling (coverage) of each BAC clone in plasmid subclone equivalents, computer-aided assembly can largely recreate the BAC insert sequence in contigs (contiguous stretches of assembled sequence reads). Subsequent refinement, including gap closure and sequence quality improvement (finishing), produces a single contiguous stretch of high-quality sequence (typically with less than 1 error per 40,000 bases). Since the completion of the human genome project (HGP) (26, 51), substantive changes have occurred in the approach to genome sequencing that have moved away from BAC-based approaches and toward whole-genome sequencing (WGS), with changes in the accompanying assembly algorithms. In the WGS approach, the genomic DNA is sheared directly into several distinct size classes and placed into plasmid and fosmid subclones. Oversampling the ends of these subclones to generate paired-end sequencing reads provides the necessary linking information to fuel whole-genome assembly algorithms. The net result is that genomes can be sequenced more rapidly and more readily, but highly polymorphic or highly repetitive genomes remain quite fragmented after assembly.

Despite these dramatic changes in sequencing and assembly approaches, the primary data production for most genome sequencing since the HGP has relied on the same type of capillary sequencing instruments as for the HGP. However, that scenario is rapidly changing owing to the invention and commercial introduction of several revolutionary approaches to DNA sequencing, the so-called next-generation sequencing technologies. Although these instruments only began to become commercially available in 2004, they already are having a major impact on our ability to explore and answer genome-wide biological questions; more than 100 next-generation sequencing–related manuscripts have appeared to date in the peer-reviewed literature. These technologies are not only changing our genome sequencing approaches and the associated timelines and costs, but also accelerating and altering a wide variety of types of biological inquiry that have historically used a sequencing-based readout, or effecting a transition to this type of readout, as detailed in this review. Furthermore, next-generation platforms are helping to open entirely new areas of biological inquiry, including the investigation of ancient genomes, the characterization of ecological diversity, and the identification of unknown etiologic agents.

NEXT-GENERATION DNA SEQUENCING

Three platforms for massively parallel DNA sequencing read production are in reasonably widespread use at present: the Roche/454 FLX (30) (**http://www.454.com/enabling-technology/the-system.asp**), the Illumina/Solexa Genome Analyzer (7) (**http://www.illumina.com/pages.ilmn?ID=203**), and the Applied Biosystems SOLiD^{TM} System (**http://marketing.appliedbiosystems.com/images/Product/Solid_Knowledge/flash/102207/solid.html**). Recently, another two massively parallel systems were announced: the Helicos Heliscope^{TM} (**www.helicosbio.com**) and Pacific Biosciences SMRT (**www.pacificbiosciences.com**) instruments. The

Helicos system only recently became commercially available, and the Pacific Biosciences instrument will likely launch commercially in early 2010. Each platform embodies a complex interplay of enzymology, chemistry, high-resolution optics, hardware, and software engineering. These instruments allow highly streamlined sample preparation steps prior to DNA sequencing, which provides a significant time savings and a minimal requirement for associated equipment in comparison to the highly automated, multistep pipelines necessary for clone-based high-throughput sequencing. By different approaches outlined below, each technology seeks to amplify single strands of a fragment library and perform sequencing reactions on the amplified strands. The fragment libraries are obtained by annealing platform-specific linkers to blunt-ended fragments generated directly from a genome or DNA source of interest. Because the presence of adapter sequences means that the molecules then can be selectively amplified by PCR, no bacterial cloning step is required to amplify the genomic fragment in a bacterial intermediate as is done in traditional sequencing approaches. Importantly, both the Helicos and Pacific Biosystems instruments mentioned above are so-called "single molecule" sequencers and do not require any amplification of DNA fragments prior to sequencing.

Another contrast between these instruments and capillary platforms is the run time required to generate data. Next-generation sequencers require longer run times of between 8 h and 10 days, depending upon the platform and read type (single end or paired ends). The longer run times result mainly from the need to image sequencing reactions that are occurring in a massively parallel fashion, rather than a periodic charge-coupled device (CCD) snapshot of 96 fixed capillaries. The yield of sequence reads and total bases per instrument run is significantly higher than the 96 reads of up to 750 bp each produced by a capillary sequencer run, and can vary from several hundred thousand reads (Roche/454) to tens of millions of reads (Illumina and Applied Biosystems SOLiD). The combination of streamlined sample preparation and long run times means that a single operator can readily keep several next-generation sequencing instruments at full capacity. The following sections aim to introduce the reader to the primary features of each of the three most widely used next-generation platforms and to discuss strengths and weaknesses.

Charge-coupled device (CCD): a capacitor array used in optical scanners to capture images

Emulsion PCR (ePCR): method for DNA amplification that uses a water in oil emulsion to isolate single DNA molecules in aqueous microreactors

Roche/454 FLX Pyrosequencer

This next-generation sequencer was the first to achieve commercial introduction (in 2004) and uses an alternative sequencing technology known as pyrosequencing. In pyrosequencing, each incorporation of a nucleotide by DNA polymerase results in the release of pyrophosphate, which initiates a series of downstream reactions that ultimately produce light by the firefly enzyme luciferase. The amount of light produced is proportional to the number of nucleotides incorporated (up to the point of detector saturation). In the Roche/454 approach (**Figure 1**), the library fragments are mixed with a population of agarose beads whose surfaces carry oligonucleotides complementary to the 454-specific adapter sequences on the fragment library, so each bead is associated with a single fragment. Each of these fragment:bead complexes is isolated into individual oil:water micelles that also contain PCR reactants, and thermal cycling (emulsion PCR) of the micelles produces approximately one million copies of each DNA fragment on the surface of each bead. These amplified single molecules are then sequenced en masse. First the beads are arrayed into a picotiter plate (PTP; a fused silica capillary structure) that holds a single bead in each of several hundred thousand single wells, which provides a fixed location at which each sequencing reaction can be monitored. Enzyme-containing beads that catalyze the downstream pyrosequencing reaction steps are then added to the PTP and the mixture is centrifuged to surround the agarose beads. On instrument, the PTP acts as a flow cell into which each pure nucleotide solution is introduced in a stepwise fashion, with an imaging step after each

a

DNA library preparation

4.5 hours

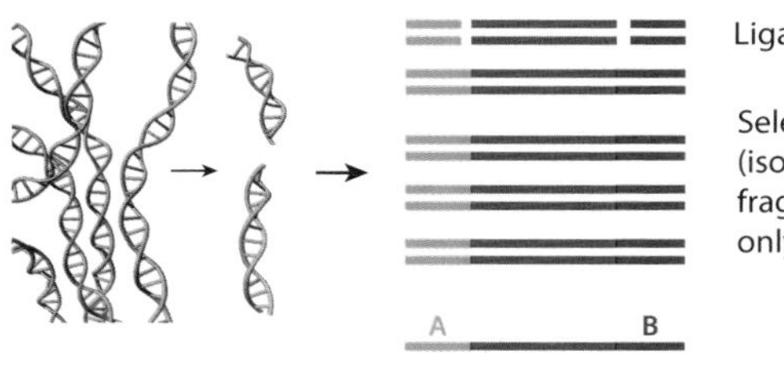

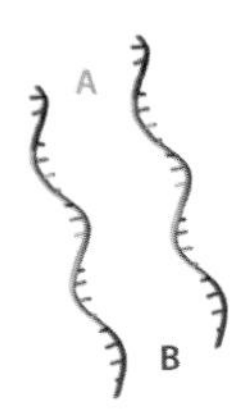

•Genome fragmented by nebulization
•No cloning; no colony picking
•sstDNA library created with adaptors
•A/B fragments selected using avidin-biotin purification

gDNA → sstDNA library

b

Emulsion PCR

8 hours

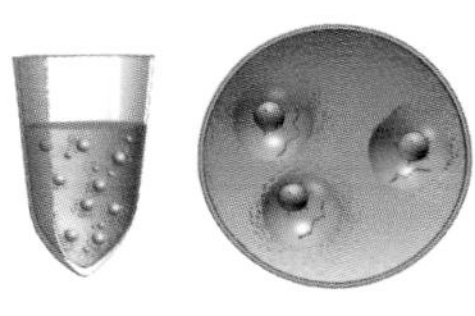

Anneal sstDNA to an excess of DNA capture beads

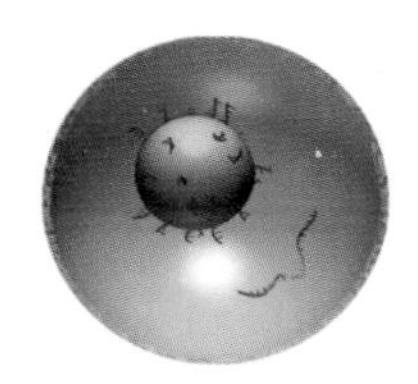

Emulsify beads and PCR reagents in water-in-oil microreactors

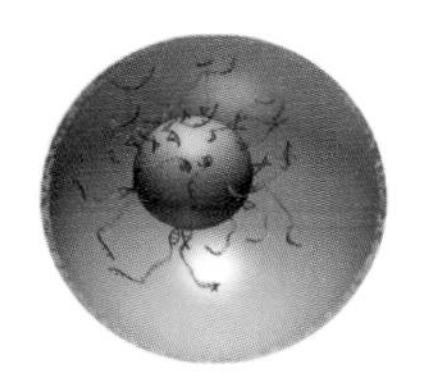

Clonal amplification occurs inside microreactors

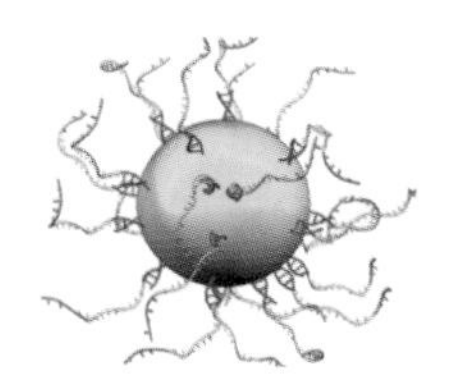

Break microreactors and enrich for DNA-positive beads

sstDNA library → Bead-amplified sstDNA library

c

Sequencing

7.5 hours

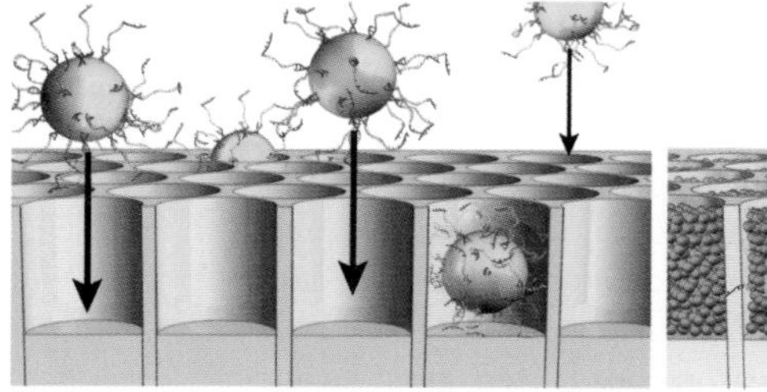

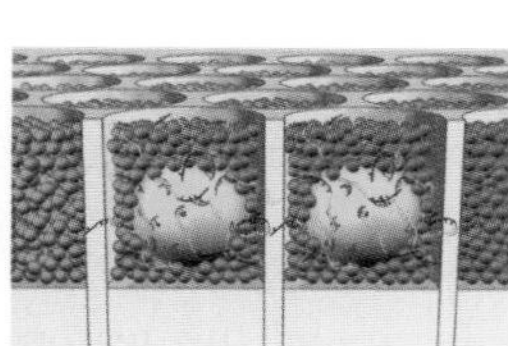

•Well diameter: average of 44 µm
•400,000 reads obtained in parallel
•A single cloned amplified sstDNA bead is deposited per well

Amplified sstDNA library beads → Quality filtered bases

nucleotide incorporation step. The PTP is seated opposite a CCD camera that records the light emitted at each bead. The first four nucleotides (TCGA) on the adapter fragment adjacent to the sequencing primer added in library construction correspond to the sequential flow of nucleotides into the flow cell. This strategy allows the 454 base-calling software to calibrate the light emitted by a single nucleotide incorporation. However, the calibrated base calling cannot properly interpret long stretches (>6) of the same nucleotide (homopolymer run), so these areas are prone to base insertion and deletion errors during base calling. By contrast, because each incorporation step is nucleotide specific, substitution errors are rarely encountered in Roche/454 sequence reads.

The FLX instrument currently provides 100 flows of each nucleotide during an 8-h run, which produces an average read length of 250 nucleotides (an average of 2.5 bases per flow are incorporated). These raw reads are processed by the 454 analysis software and then screened by various quality filters to remove poor-quality sequences, mixed sequences (more than one initial DNA fragment per bead), and sequences without the initiating TCGA sequence. The resulting reads yield 100 Mb of quality data on average. Downstream of read processing, an assembly algorithm (Newbler) can assemble FLX reads. Although shorter than reads derived from capillary sequencers, FLX reads are of sufficient length to assemble small genomes such as bacterial and viral genomes to high quality and contiguity. As mentioned, the lack of a bacterial cloning step in the Roche/454 process means that sequences not typically sampled in a WGS approach owing to cloning bias will be more likely represented in a FLX data set, which contributes to more comprehensive genome coverage.

Illumina Genome Analyzer

The single molecule amplification step for the Illumina Genome Analyzer starts with an Illumina-specific adapter library, takes place on the oligo-derivatized surface of a flow cell, and is performed by an automated device called a Cluster Station. The flow cell is an 8-channel sealed glass microfabricated device that allows bridge amplification of fragments on its surface, and uses DNA polymerase to produce multiple DNA copies, or clusters, that each represent the single molecule that initiated the cluster amplification. A separate library can be added to each of the eight channels, or the same library can be used in all eight, or combinations thereof. Each cluster contains approximately one million copies of the original fragment, which is sufficient for reporting incorporated bases at the required signal intensity for detection during sequencing.

Bridge amplification: allows the generation of in situ copies of a specific DNA molecule on an oligo-decorated solid support

The Illumina system utilizes a sequencing-by-synthesis approach in which all four nucleotides are added simultaneously to the flow cell channels, along with DNA polymerase, for incorporation into the oligo-primed cluster fragments (see **Figure 2** for details). Specifically, the nucleotides carry a base-unique fluorescent label and the 3′-OH group is chemically blocked such that each incorporation is a unique event. An imaging step follows each base incorporation step, during which each flow cell lane is imaged in three 100-tile segments by the instrument optics at a cluster density per tile of 30,000. After each imaging step, the 3′ blocking group is chemically removed

Figure 1

The method used by the Roche/454 sequencer to amplify single-stranded DNA copies from a fragment library on agarose beads. A mixture of DNA fragments with agarose beads containing complementary oligonucleotides to the adapters at the fragment ends are mixed in an approximately 1:1 ratio. The mixture is encapsulated by vigorous vortexing into aqueous micelles that contain PCR reactants surrounded by oil, and pipetted into a 96-well microtiter plate for PCR amplification. The resulting beads are decorated with approximately 1 million copies of the original single-stranded fragment, which provides sufficient signal strength during the pyrosequencing reaction that follows to detect and record nucleotide incorporation events. sstDNA, single-stranded template DNA.

to prepare each strand for the next incorporation by DNA polymerase. This series of steps continues for a specific number of cycles, as determined by user-defined instrument settings, which permits discrete read lengths of 25–35 bases. A base-calling algorithm assigns sequences and associated quality values to each read and a quality checking pipeline evaluates the Illumina data from each run, removing poor-quality sequences.

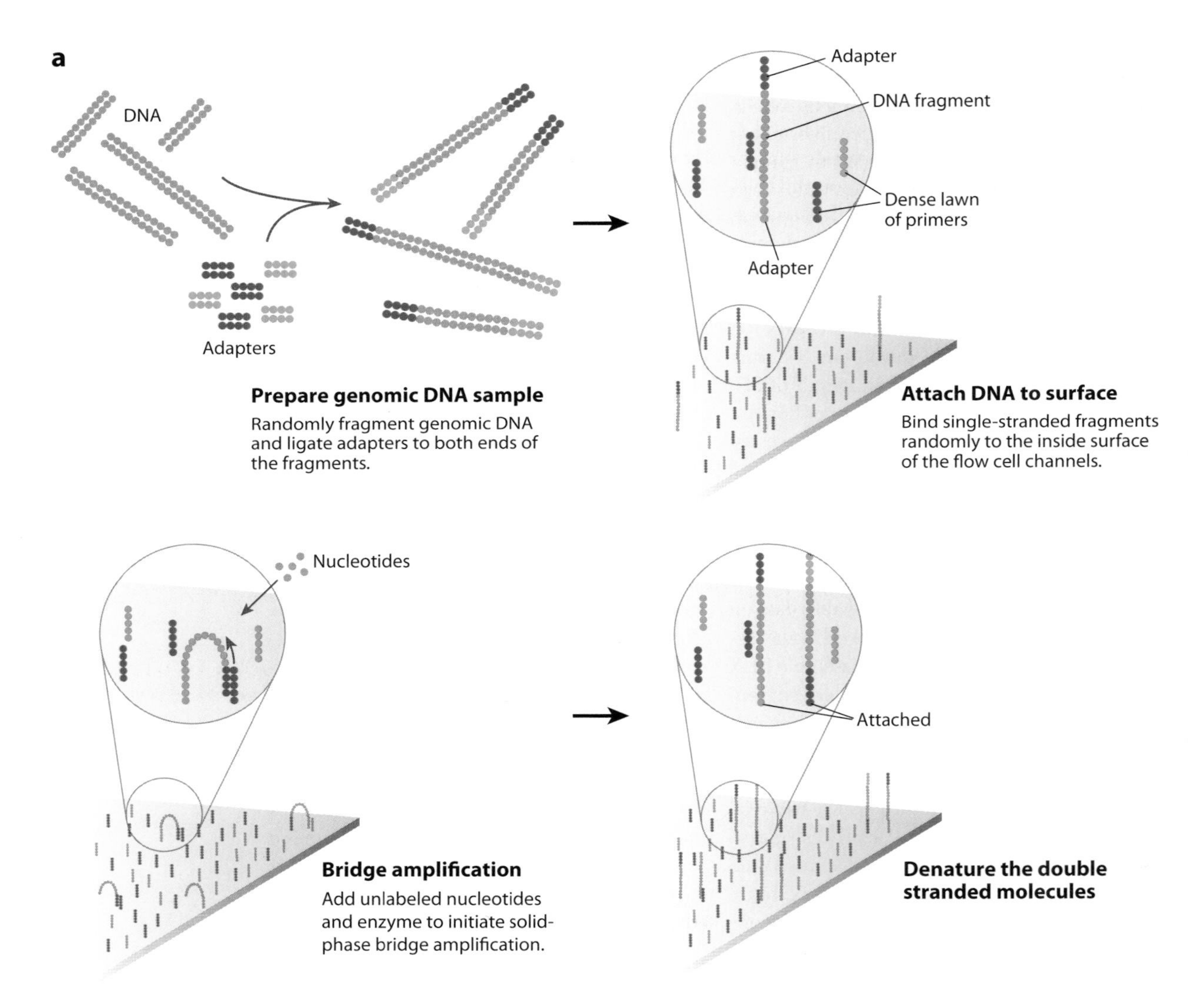
a
DNA
Adapters
Prepare genomic DNA sample
Randomly fragment genomic DNA and ligate adapters to both ends of the fragments.
Adapter
DNA fragment
Dense lawn of primers
Adapter
Attach DNA to surface
Bind single-stranded fragments randomly to the inside surface of the flow cell channels.
Nucleotides
Bridge amplification
Add unlabeled nucleotides and enzyme to initiate solid-phase bridge amplification.
Attached
Denature the double stranded molecules

Figure 2

The Illumina sequencing-by-synthesis approach. Cluster strands created by bridge amplification are primed and all four fluorescently labeled, 3′-OH blocked nucleotides are added to the flow cell with DNA polymerase. The cluster strands are extended by one nucleotide. Following the incorporation step, the unused nucleotides and DNA polymerase molecules are washed away, a scan buffer is added to the flow cell, and the optics system scans each lane of the flow cell by imaging units called tiles. Once imaging is completed, chemicals that effect cleavage of the fluorescent labels and the 3′-OH blocking groups are added to the flow cell, which prepares the cluster strands for another round of fluorescent nucleotide incorporation.

Applied Biosystems SOLiD™ Sequencer

The SOLiD platform uses an adapter-ligated fragment library similar to those of the other next-generation platforms, and uses an emulsion PCR approach with small magnetic beads to amplify the fragments for sequencing. Unlike the other platforms, SOLiD uses DNA ligase and a unique approach to sequence the amplified fragments, as illustrated in **Figure 3*a***. Two flow cells are processed per instrument run, each of which can be divided to contain different libraries in up to four quadrants. Read lengths for SOLiD are user defined between 25–35 bp, and each sequencing run yields between 2–4 Gb of DNA sequence data. Once the reads are base called, have quality values, and low-quality sequences have been removed, the reads are aligned to a reference genome to enable a second tier of quality evaluation called two-base encoding. The principle of two-base encoding is shown in **Figure 3*b***, which illustrates how this approach works to differentiate true single base variants from base-calling errors.

Two key differences that speak to the utility of next-generation sequence reads are (*a*) the length of a sequence read from all current next-generation platforms is much shorter than that from a capillary sequencer and (*b*) each next-generation read type has a unique error model different from that already established for

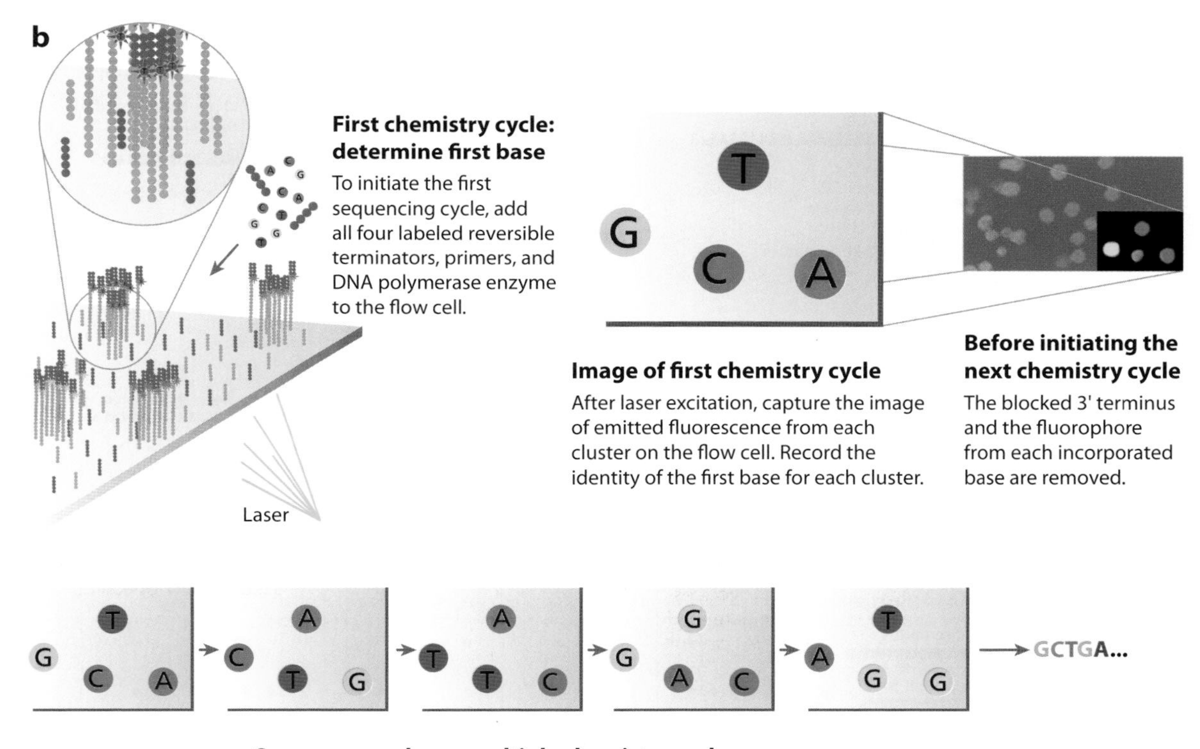

b
First chemistry cycle: determine first base
To initiate the first sequencing cycle, add all four labeled reversible terminators, primers, and DNA polymerase enzyme to the flow cell.
Laser
Image of first chemistry cycle
After laser excitation, capture the image of emitted fluorescence from each cluster on the flow cell. Record the identity of the first base for each cluster.
Before initiating the next chemistry cycle
The blocked 3' terminus and the fluorophore from each incorporated base are removed.
GCTGA...
Sequence read over multiple chemistry cycles
Repeat cycles of sequencing to determine the sequence of bases in a given fragment a single base at a time.

Figure 2

(*Continued*)

a

SOLiD™ substrate

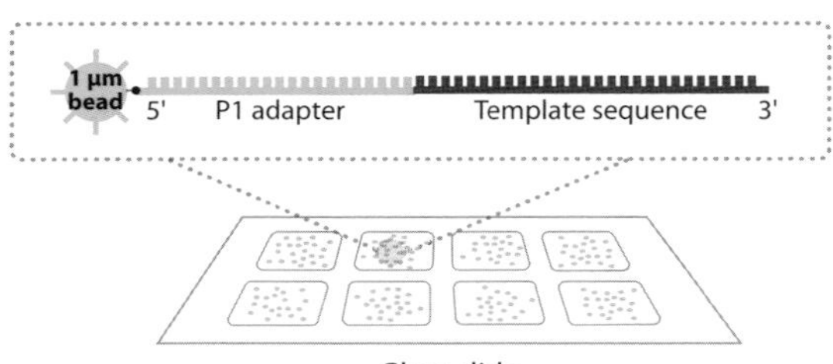

1. Prime and ligate

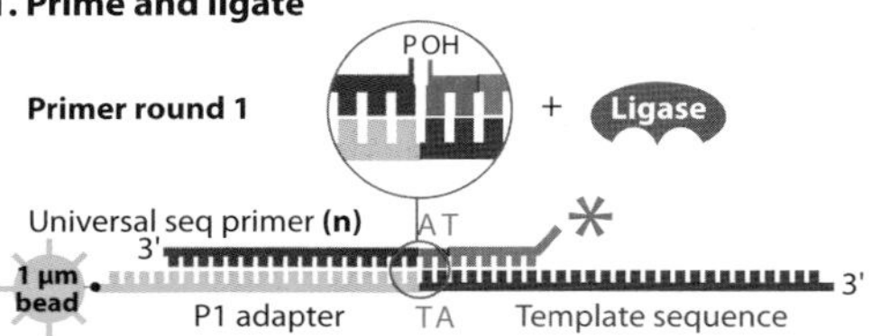

2. Image

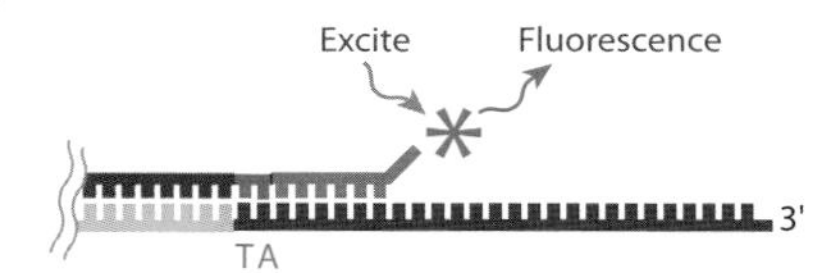

3. Cap unextended strands

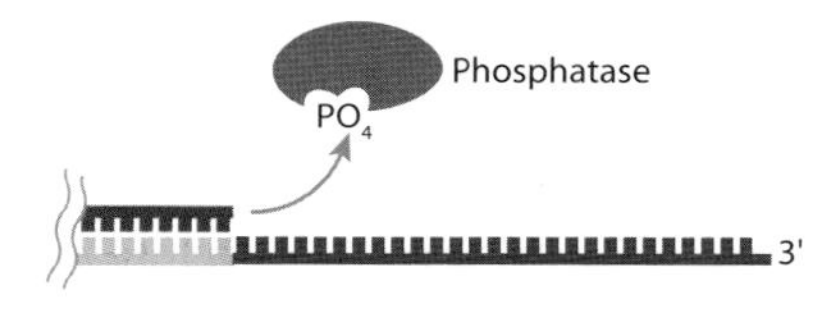

4. Cleave off fluor

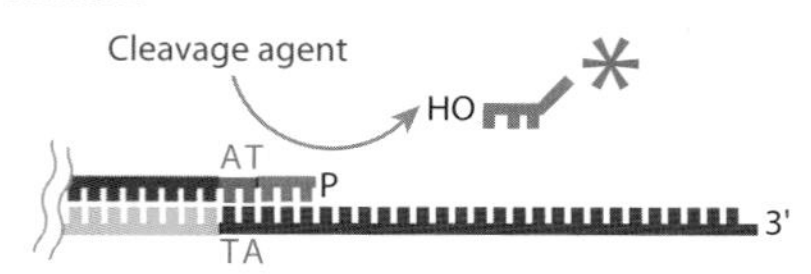

Di base probes

3' TTnnnzzz5'

3' TGnnnzzz5'

3' TCnnnzzz5'

3' TAnnnzzz5'

Cleavage site

Template

2nd base

A C G T

A C G T

1st base

5. Repeat steps 1–4 to extend sequence

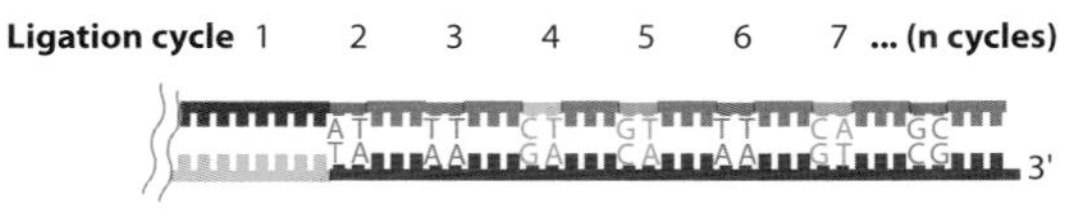

6. Primer reset

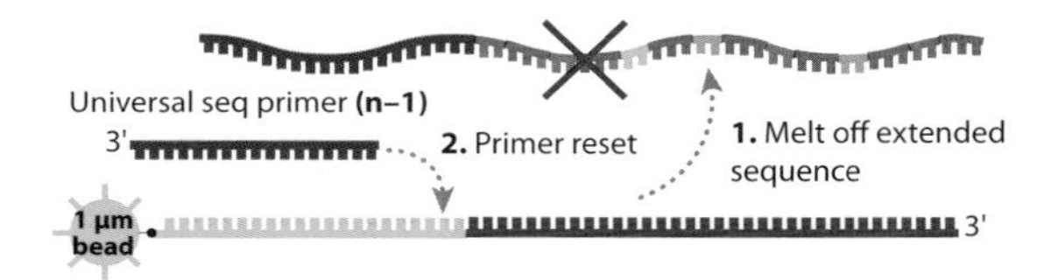

7. Repeat steps 1–5 with new primer

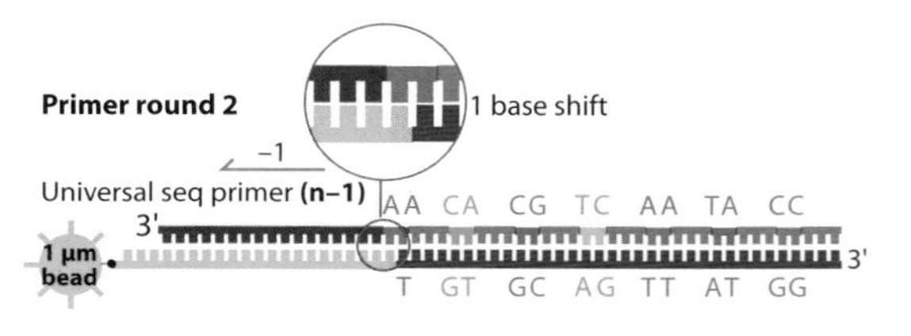

8. Repeat Reset with , n–2, n–3, n–4 primers

Primer round	Read position	0	1	2	3	4	5	6	7	8	9	10	11	12	13	14	15	16	17	18	19	20	21	22	23	24	25	26	27	28	29	30	31	32	33	34	35
1	Universal seq primer (n) 3'		•	•				•	•				•	•				•	•				•	•				•	•				•	•			
2	Universal seq primer (n–1) 3'	•	•				•	•				•	•				•	•				•	•				•	•				•	•				
3	Universal seq primer (n–2) 3'	Bridge probe				•	•				•	•				•	•				•	•				•	•				•	•				•	•
4	Universal seq primer (n–3) 3'	Bridge probe			•	•				•	•				•	•				•	•				•	•				•	•				•	•	
5	Universal seq primer (n–4) 3'	Bridge probe		•	•				•	•				•	•				•	•				•	•				•	•				•	•		

● Indicates positions of interrogation Ligation cycle 1 2 3 4 5 6 7

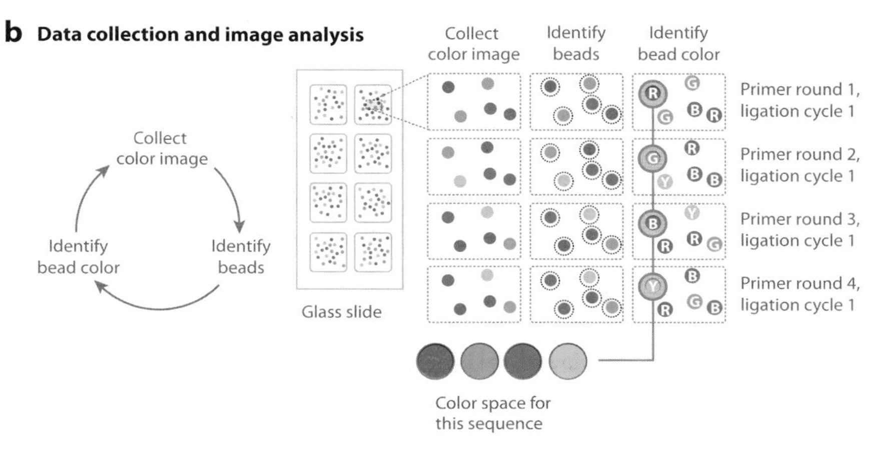

Possible dinucleotides encoded by each color

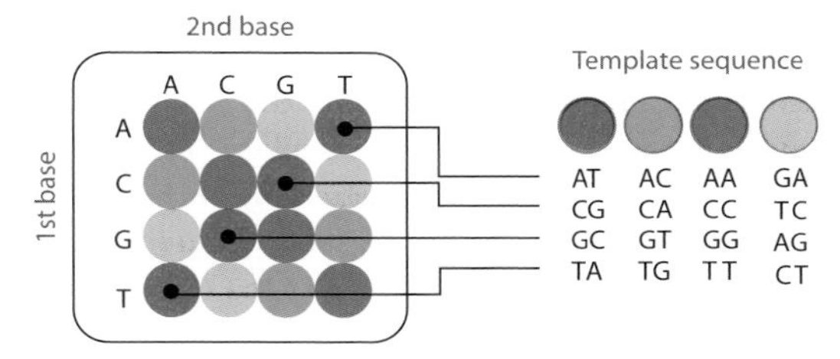

Double interrogation

With 2 base encoding each base is defined twice

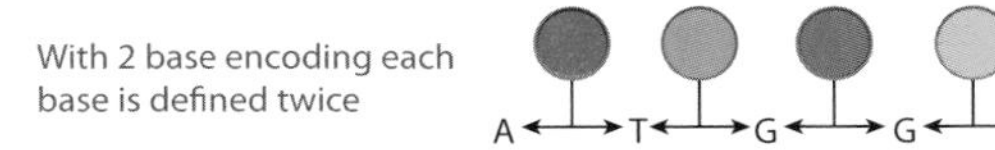

Decoding

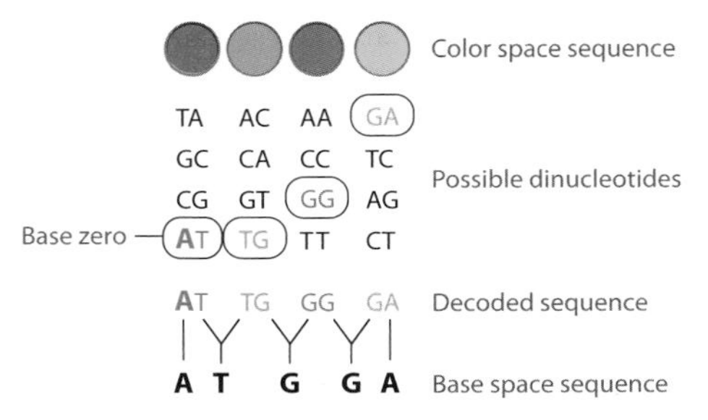

Figure 3

(*a*) The ligase-mediated sequencing approach of the Applied Biosystems SOLiD sequencer. In a manner similar to Roche/454 emulsion PCR amplification, DNA fragments for SOLiD sequencing are amplified on the surfaces of 1-μm magnetic beads to provide sufficient signal during the sequencing reactions, and are then deposited onto a flow cell slide. Ligase-mediated sequencing begins by annealing a primer to the shared adapter sequences on each amplified fragment, and then DNA ligase is provided along with specific fluorescent-labeled 8mers, whose 4th and 5th bases are encoded by the attached fluorescent group. Each ligation step is followed by fluorescence detection, after which a regeneration step removes bases from the ligated 8mer (including the fluorescent group) and concomitantly prepares the extended primer for another round of ligation. (*b*) Principles of two-base encoding. Because each fluorescent group on a ligated 8mer identifies a two-base combination, the resulting sequence reads can be screened for base-calling errors versus true polymorphisms versus single base deletions by aligning the individual reads to a known high-quality reference sequence.

Chromatin immunoprecipitation (ChIP): chemical crosslinking of DNA and proteins, and immunoprecipitation using a specific antibody to determine DNA:protein associations in vivo

Quantitative polymerase chain reaction (qPCR): rapidly measures the quantity of DNA, cDNA, or RNA present in a sample through cycle-by-cycle measurement of incorporated fluorescent dyes

capillary sequence reads. Both differences affect how the reads are utilized in bioinformatic analyses, depending upon the application. For example, in strain-to-reference comparisons (resequencing), the typical definition of repeat content must be revised in the context of the shorter read length. In addition, a much higher read coverage or sampling depth is required for comprehensive resequencing with short reads to adequately cover the reference sequence at the depth and low gap size needed.

Some applications are more suitable for certain platforms than others, as detailed below. Furthermore, read length and error profile issues entail platform- and application-specific bioinformatics-based considerations. Moreover, it is important to recognize the significant impacts that implementation of these platforms in a production sequencing environment has on informatics and bioinformatics infrastructures. The massively parallel scale of sequencing implies a similarly massive scale of computational analyses that include image analysis, signal processing, background subtraction, base calling, and quality assessment to produce the final sequence reads for each run. In every case, these analyses place significant demands on the information technology (IT), computational, data storage, and laboratory information management system (LIMS) infrastructures extant in a sequencing center, thereby adding to the overhead required for high-throughput data production. This aspect of next-generation sequencing is at present complicated by the dearth of current sequence analysis tools suited to shorter sequence read data; existing data analysis pipelines and algorithms must be modified to accommodate these shorter reads. In many cases, and certainly for new applications of next-generation sequencing, entirely new algorithms and data visualization interfaces are being devised and tested to meet this new demand. Therefore, the next-generation platforms are effecting a complete paradigm shift, not only in the organization of large-scale data production, but also in the downstream bioinformatics, IT, and LIMS support required for high data utility and correct interpretation. This paradigm shift promises to radically alter the path of biological inquiry, as the following review of recent endeavors to implement next-generation sequencing platforms and accompanying bioinformatics-based analyses serves to substantiate.

ELUCIDATING DNA-PROTEIN INTERACTIONS THROUGH CHROMATIN IMMUNOPRECIPITATION SEQUENCING

The association between DNA and proteins is a fundamental biological interaction that plays a key part in regulating gene expression and controlling the availability of DNA for transcription, replication, and other processes. These interactions can be studied in a focused manner using a technique called chromatin immunoprecipitation (ChIP) (43). ChIP entails a series of steps: (*a*) DNA and associated proteins are chemically cross-linked; (*b*) nuclei are isolated, lysed, and the DNA is fragmented; (*c*) an antibody specific for the DNA binding protein (transcription factor, histone, etc.) of interest is used to selectively immunoprecipitate the associated protein:DNA complexes; and (*d*) the chemical crosslinks between DNA and protein are reversed and the DNA is claimed for downstream analysis. In early applications, typical analyses examined the specific gene of interest by qPCR (quantitative PCR) or Southern blotting to determine if corresponding sequences were contained in the captured fragment population. Recently, genome-wide ChIP-based studies of DNA-protein interactions became possible in sequenced genomes by using genomic DNA microarrays to assay the released fragments. This so-called ChIP-chip approach was first reported by Ren and coworkers (38). Although utilized for a number of important studies, it has several drawbacks, including a low signal-to-noise ratio and a need for replicates to build statistical power to support putative binding sites.

Many of these drawbacks were addressed by shifting the readout of ChIP-derived DNA sequences onto next-generation sequencing platforms. The precedent-setting paper for this paradigm was published by Johnson and colleagues (21), who used the model organism *Caenorhabditis elegans* and the Roche platform to elucidate nucleosome positioning on genomic DNA. This study established that sequencing the micrococcal nuclease–derived digestion products of genomic DNA carefully isolated from mixed stage hermaphrodite populations of *C. elegans* was sufficient to generate a genome-wide, highly precise positional profile of chromatin. This capability enables studies of specific physiological conditions and their genome-wide impact on nucleosome positioning, among other applications. Subsequent studies have utilized a ChIP-based approach and the Illumina platform to provide insights into transcription factor binding sites in the human genome such as neuron-restrictive silencer factor (NRSF) (20) and signal transducer and activator of transcription 1 (STAT1) (39). In a landmark study, Mikkelsen and coworkers (32) explored the connection between chromatin packaging of DNA and differential gene expression using mouse embryonic stem cells and lineage-committed mouse cells (neural progenitor cells and embryonic fibroblasts), providing a next-generation sequencing-based framework for using genome-wide chromatin profiling to characterize cell populations. This group demonstrated that trimethylation of lysine 4 and lysine 27 determines genes that are either expressed, poised for expression, or stably repressed, effectively reflecting cell state and lineage potential. Also, lysine 4 and lysine 9 trimethylation mark imprinting control regions, whereas lysine 9 and lysine 20 trimethylation identifies satellite, telomeric, and active long-terminal repeats. These early studies demonstrated that the ability to map genome-wide changes in transcription factor binding or chromatin packaging under different environmental conditions offers a profound opportunity to couple evidence of altered DNA:protein interactions to expression level changes of specific genes in the context of specific environmental stimuli, thereby enhancing our understanding of gene expression–based cellular responses.

GENE EXPRESSION: SEQUENCING THE TRANSCRIPTOME

Historically, mRNA expression has been gauged by microarray or qPCR-based approaches; the latter is most efficient and cost-effective for a genome-wide survey of gene expression levels. Even the exquisite sensitivity of qPCR, however, is not absolute, nor is it straightforward or reliable to evaluate novel alternative splicing isoforms using either technology. In the past, serial analysis of gene expression (SAGE) (50) and variants have provided a digital readout of gene expression levels using DNA sequencing. These approaches are powerful in their ability to report the expression of genes at levels below the sensitivity of microarrays, but have been limited in their application by the cost of DNA sequencing.

By contrast, the rapid and inexpensive sequencing capacity offered by next-generation sequencing instruments meshes perfectly with SAGE tagging or conventional cDNA sequencing approaches, as evidenced by several studies that used Roche/454 technology (6, 11, 46, 53). Undoubtedly, the shorter read lengths offered by the Illumina and Applied Biosystems instruments will be utilized with these approaches in the future, offering the advantage of sequencing individual SAGE tags rather than requiring concatenation of the tags prior to sequencing. Indeed, one might imagine combining the data obtained from isolating and sequencing ChIP-derived DNA bound by a transcription factor of interest to the corresponding coisolated and sequenced mRNA population from the same cells. Such experiments will be entirely feasible with next-generation technologies, especially given the low input amount of each type of biomolecule required for a suitable library and the high sensitivity afforded by the sequencing method.

Serial analysis of gene expression (SAGE): measures the quantitative expression of genes in an mRNA population sample by generating SAGE tags that are then sequenced

DISCOVERING NONCODING RNAs

Noncoding RNA (ncRNA): any native RNA that is transcribed but not translated into a protein

microRNA (miRNA): 21–23-base RNA molecules that participate in RNA-induced silencing of gene expression

One of the most exciting areas of biological research in recent years has been the discovery and functional analysis of noncoding RNA (ncRNA) systems in different organisms. First described in plants, ncRNAs are providing new insights into gene regulation in animal systems as well, as recognized by the awarding of the Nobel Prize in Medicine and Physiology to Andrew Fire and Craig Mello in 2006. Perhaps the most profound impact of next-generation sequencing technology has been on the discovery of novel ncRNAs belonging to distinct classes in an extraordinarily diverse set of species (3, 8, 9, 18, 22, 29, 41, 55). In fact, this approach has been responsible for the discovery of ncRNA classes in organisms not previously known to possess them (41). These discoveries are being coupled with an ever-expanding comprehension of the functions embodied by these unique RNA species, including gene regulation by a variety of mechanisms. In this regard, studying the roles of specific microRNAs (miRNAs) in cancer is helping to uncover certain aspects of the disease (10, 44).

Noncoding RNA discovery is best accomplished by sequencing because the evolutionary diversity of ncRNA gene sequences makes it difficult to predict their presence in a genome with high certainty by computational methods alone. The unique structures of the processed ncRNAs pose difficulties for converting them into next-generation sequencing libraries (29), but remarkable progress has already been made in characterizing these molecules. With these barriers dissolving, the high capacity and low cost of next-generation platforms ensure that discovery of ncRNAs will continue at a rapid pace and that sequence variants with important functional impacts will also be determined. Because the readout from next-generation sequencers is quantitative, ncRNA characterization will include detecting expression level changes that correlate with changes in environmental factors, with disease onset and progression, and perhaps with complex disease onset or severity, for example. Importantly, the discovery and characterization of ncRNAs will enhance the annotation of sequenced genomes such that, especially in model organisms and humans, the impact of mutations will become more broadly interpretable across the genome.

ANCIENT GENOMES RESURRECTED

Attempts to characterize fossil-derived DNAs have been limited by the degraded state of the sample, which in the past permitted only mitochondrial DNA sequencing and typically involved PCR amplification of specific mitochondrial genome regions (1, 15, 23, 24, 35, 40). The advent of next-generation sequencing has for the first time made it possible to directly sample the nuclear genomes of ancient remains from the cave bear (34), mammoth (37), and the Neanderthal (17, 33). Several non-trivial technical complications arise in these inquiries, most notably the need to identify contaminating DNA from modern humans in the case of Neanderthal remains. Although even next-generation sequencing of these sample remains is quite inefficient, owing largely to bacterial DNA that is coisolated with the genomic preparation and the degraded nature of the ancient genome, important characterizations are being made. So far, one million bases of the Neanderthal genome have been sequenced, starting from DNA obtained from a single fossil bone (17).

METAGENOMICS EMERGES

Characterizing the biodiversity found on Earth is of particular interest as climate changes reshape our planet. DNA- or RNA-based approaches for this purpose are becoming increasingly powerful as the growing number of sequenced genomes enables us to interpret partial sequences obtained by direct sampling of specific environmental niches. Such investigations are referred to as metagenomics, and are typically aimed at answering the question: who's there? Conventionally, this question is addressed by isolating DNA from an environmental sample, amplifying the collective of 16S ribosomal RNA (rRNA) genes with degenerate PCR primer sets, subcloning the PCR products that result, and classifying the taxa present according to a database of assigned 16S rRNA sequences. As an alternative, DNA (or RNA) is isolated, subcloned, and then sequenced to produce a fragment pool representative of the existing population. These sequences can then be translated in silico into protein fragments and compared with the existing database of annotated genome sequences to identify community members. In both approaches, deep sequencing of the population of subclones is necessary to obtain the full spectrum of taxa present, and is limited by potential cloning bias that can result from the use of bacterial cloning. By sampling RNA sequences from a metagenomic isolate, one can attempt to reconstruct metabolic pathways that are active in a given environment (45, 47). Several early metagenomic studies utilized DNA sequence sampling by capillary sequencing to investigate an acidophilic biofilm (49), an acid mine site (12) and the Sargasso Sea (52). Although these studies defined metagenomics as a scientific pursuit, they were limited in the breadth of diversity that could be sampled owing to the expense of the conventional sequencing process. By contrast, the rapid, inexpensive, and massive data production enabled by next-generation platforms has caused a recent explosion in metagenomic studies. These studies include previously sampled environments such as the ocean (2, 19, 42) and an acid mine site (13), but soil (14, 27) and coral reefs (54) also were studied by Roche/454 pyrosequencing.

Another metagenomic environment that is being characterized by next-generation sequencing is the human microbiome; the human body contains several highly specific environments that are inhabited by various microbial, fungal, viral, and eukaryotic symbiont communities, the inhabitants of which may vary according to the health status of the individual. These environments include the skin, the oral and nasal cavities, the gastrointestinal tract, and the vagina, among others. Particularly well-studied is the lower intestine of humans, first characterized by 16S rRNA classification (4, 5, 28) and more recently by 454 pyrosequencing in adult humans (16, 36, 48) and in infants (25, 36). Supporting these characterization efforts will be a large-scale project to sequence hundreds of isolated microbial genomes that are known symbionts of humans as references (**http://www.genome.gov/25521743**). The early successes in human microbiome characterization and the apparent interplay between the human host and its microbial census have resulted in the inclusion of a Human Microbiome Initative in the NIH Roadmap (**http://nihroadmap.nih.gov/hmp/**). The associated funding opportunities, with the advantages offered by next-generation sequencing instrumentation, should initiate a revolution in our understanding of how the human microbiome influences our health status.

Metagenomics: the genomics-based study of genetic material recovered directly from environmentally derived samples without laboratory culture

Epigenomics: seeks to define the influence of changes to gene expression that are independent of gene sequence

FUTURE POSSIBILITIES

As this review has described, the advent and widespread availability of next-generation sequencing instruments has ushered in an era in which DNA sequencing will become a more universal readout for an increasingly wide variety of front-end assays. However, more applications of next-generation sequencing, beyond those covered here, are yet to come. For example, genome resequencing will likely be used to characterize strains or isolates relative to high-quality reference genomes such as *C. elegans*, *Drosophila*, and human. Studies of this type will identify and catalog genomic variation on a wide scale, from single nucleotide polymorphisms (SNPs) to copy number variations in large sequence blocks (>1000 bases). Ultimately, resequencing studies will help to better characterize, for example, the range of normal variation in complex genomes such as the human genome, and aid in our ability to comprehensively view the range of genome variation in clinical isolates of pathogenic microbes, viruses, etc.

Epigenomic variation, as an extension of genome resequencing applications, also will be investigated using next-generation sequencing approaches that enable the ascertainment of genome-wide patterns of methylation and how these patterns change through the course of an organism's development, in the context of disease, and under various other influences.

Perhaps the most exciting possibility engendered by the ability to use DNA sequencing to rapidly read out experimental results is the enhanced potential to combine the results of different experiments—correlative analyses of genome-wide methylation, histone binding patterns, and gene expression, for example—owing to the similar data type produced. The power in these correlative analyses is the power to begin unlocking the secrets of the cell.

DISCLOSURE STATEMENT

The author serves as a Director of the Applera Corporation.

LITERATURE CITED

1. Adcock GJ, Dennis ES, Easteal S, Huttley GA, Jermiin LS, et al. 2001. Mitochondrial DNA sequences in ancient Australians: implications for modern human origins. *Proc. Natl. Acad. Sci. USA* 98:537–42
2. Angly FE, Felts B, Breitbart M, Salamon P, Edwards RA, et al. 2006. The marine viromes of four oceanic regions. *PLoS Biol.* 4:e368
3. Axtell MJ, Snyder JA, Bartel DP. 2007. Common functions for diverse small RNAs of land plants. *Plant Cell* 19:1750–69
4. Bäckhed F, Ding H, Wang T, Hooper LV, Koh GY, et al. 2004. The gut microbiota as an environmental factor that regulates fat storage. *Proc. Natl. Acad. Sci. USA* 101:15718–23
5. Bäckhed F, Ley RE, Sonnenburg JL, Peterson DA, Gordon JI. 2005. Host-bacterial mutualism in the human intestine. *Science* 307:1915–20
6. Bainbridge MN, Warren RL, Hirst M, Romanuik T, Zeng T, et al. 2006. Analysis of the prostate cancer cell line LNCaP transcriptome using a sequencing-by-synthesis approach. *BMC Genomics* 7:246
7. Bentley DR. 2006. Whole-genome resequencing. *Curr. Opin. Genet. Dev.* 16:545–52
8. Berezikov E, Thuemmler F, van Laake LW, Kondova I, Bontrop R, et al. 2006. Diversity of microRNAs in human and chimpanzee brain. *Nat. Genet.* 38:1375–77
9. Brennecke J, Aravin AA, Stark A, Dus M, Kellis M, et al. 2007. Discrete small RNA-generating loci as master regulators of transposon activity in *Drosophila*. *Cell* 128:1089–103
10. Calin GA, Liu CG, Ferracin M, Hyslop T, Spizzo R, et al. 2007. Ultraconserved regions encoding ncRNAs are altered in human leukemias and carcinomas. *Cancer Cell* 12:215–29
11. Cheung F, Haas BJ, Goldberg SM, May GD, Xiao Y, Town CD. 2006. Sequencing *Medicago truncatula* expressed sequenced tags using 454 Life Sciences technology. *BMC Genomics* 7:272
12. Edwards KJ, Bond PL, Gihring TM, Banfield JF. 2000. An archaeal iron-oxidizing extreme acidophile important in acid mine drainage. *Science* 287:1796–99
13. Edwards RA, Rodriguez-Brito B, Wegley L, Haynes M, Breitbart M, et al. 2006. Using pyrosequencing to shed light on deep mine microbial ecology. *BMC Genomics* 7:57
14. Fierer N, Breitbart M, Nulton J, Salamon P, Lozupone C, et al. 2007. Metagenomic and small-subunit rRNA analyses of the genetic diversity of bacteria, archaea, fungi, and viruses in soil. *Appl. Environ. Microbiol.* 73:7059–66
15. Gilbert MT, Tomsho LP, Rendulic S, Packard M, Drautz DI, et al. 2007. Whole-genome shotgun sequencing of mitochondria from ancient hair shafts. *Science* 317:1927–30
16. Gill SR, Pop M, Deboy RT, Eckburg PB, Turnbaugh PJ, et al. 2006. Metagenomic analysis of the human distal gut microbiome. *Science* 312:1355–59
17. Green RE, Krause J, Ptak SE, Briggs AW, Ronan MT, et al. 2006. Analysis of one million base pairs of Neanderthal DNA. *Nature* 444:330–36
18. Houwing S, Kamminga LM, Berezikov E, Cronembold D, Girard A, et al. 2007. A role for Piwi and piRNAs in germ cell maintenance and transposon silencing in zebrafish. *Cell* 129:69–82

19. Huber JA, Mark Welch DB, Morrison HG, Huse SM, Neal PR, et al. 2007. Microbial population structures in the deep marine biosphere. *Science* 318:97–100
20. Johnson DS, Mortazavi A, Myers RM, Wold B. 2007. Genome-wide mapping of in vivo protein-DNA interactions. *Science* 316:1497–502
21. Johnson SM, Tan FJ, McCullough HL, Riordan DP, Fire AZ. 2006. Flexibility and constraint in the nucleosome core landscape of *Caenorhabditis elegans* chromatin. *Genome Res.* 16:1505–16
22. Kasschau KD, Fahlgren N, Chapman EJ, Sullivan CM, Cumbie JS, et al. 2007. Genome-wide profiling and analysis of *Arabidopsis* siRNAs. *PLoS Biol.* 5:e57
23. Krause J, Dear PH, Pollack JL, Slatkin M, Spriggs H, et al. 2006. Multiplex amplification of the mammoth mitochondrial genome and the evolution of Elephantidae. *Nature* 439:724–27
24. Krings M, Geisert H, Schmitz RW, Krainitzki H, Paabo S. 1999. DNA sequence of the mitochondrial hypervariable region II from the Neandertal type specimen. *Proc. Natl. Acad. Sci. USA* 96:5581–85
25. Kurokawa K, Itoh T, Kuwahara T, Oshima K, Toh H, et al. 2007. Comparative metagenomics revealed commonly enriched gene sets in human gut microbiomes. *DNA Res.* 14:169–81
26. Lander ES, Linton LM, Birren B, Nusbaum C, Zody MC, et al. 2001. Initial sequencing and analysis of the human genome. *Nature* 409:860–921
27. Leininger S, Urich T, Schloter M, Schwark L, Qi J, et al. 2006. Archaea predominate among ammonia-oxidizing prokaryotes in soils. *Nature* 442:806–9
28. Ley RE, Bäckhed F, Turnbaugh P, Lozupone CA, Knight RD, Gordon JI. 2005. Obesity alters gut microbial ecology. *Proc. Natl. Acad. Sci. USA* 102:11070–75
29. Lu C, Meyers BC, Green PJ. 2007. Construction of small RNA cDNA libraries for deep sequencing. *Methods* 43:110–17
30. Margulies M, Egholm M, Altman WE, Attiya S, Bader JS, et al. 2005. Genome sequencing in microfabricated high-density picolitre reactors. *Nature* 437:376–80
31. McPherson JD, Marra M, Hillier L, Waterston RH, Chinwalla A, et al. 2001. A physical map of the human genome. *Nature* 409:934–41
32. Mikkelsen TS, Ku M, Jaffe DB, Issac B, Lieberman E, et al. 2007. Genome-wide maps of chromatin state in pluripotent and lineage-committed cells. *Nature* 448:553–60
33. Noonan JP, Coop G, Kudaravalli S, Smith D, Krause J, et al. 2006. Sequencing and analysis of Neanderthal genomic DNA. *Science* 314:1113–18
34. Noonan JP, Hofreiter M, Smith D, Priest JR, Rohland N, et al. 2005. Genomic sequencing of Pleistocene cave bears. *Science* 309:597–99
35. Ovchinnikov IV, Gotherstrom A, Romanova GP, Kharitonov VM, Liden K, Goodwin W. 2000. Molecular analysis of Neanderthal DNA from the northern Caucasus. *Nature* 404:490–93
36. Palmer C, Bik EM, Digiulio DB, Relman DA, Brown PO. 2007. Development of the human infant intestinal microbiota. *PLoS Biol.* 5:e177
37. Poinar HN, Schwarz C, Qi J, Shapiro B, Macphee RD, et al. 2006. Metagenomics to paleogenomics: large-scale sequencing of mammoth DNA. *Science* 311:392–94
38. Ren B, Robert F, Wyrick JJ, Aparicio O, Jennings EG, et al. 2000. Genome-wide location and function of DNA binding proteins. *Science* 290:2306–9
39. Robertson G, Hirst M, Bainbridge M, Bilenky M, Zhao Y, et al. 2007. Genome-wide profiles of STAT1 DNA association using chromatin immunoprecipitation and massively parallel sequencing. *Nat. Methods* 4:651–57
40. Rogaev EI, Moliaka YK, Malyarchuk BA, Kondrashov FA, Derenko MV, et al. 2006. Complete mitochondrial genome and phylogeny of Pleistocene mammoth *Mammuthus primigenius*. *PLoS Biol.* 4:e73
41. Ruby JG, Jan C, Player C, Axtell MJ, Lee W, et al. 2006. Large-scale sequencing reveals 21U-RNAs and additional microRNAs and endogenous siRNAs in *C. elegans*. *Cell* 127:1193–207
42. Sogin ML, Morrison HG, Huber JA, Welch DM, Huse SM, et al. 2006. Microbial diversity in the deep sea and the underexplored "rare biosphere". *Proc. Natl. Acad. Sci. USA* 103:12115–20
43. Solomon MJ, Larsen PL, Varshavsky A. 1988. Mapping protein-DNA interactions in vivo with formaldehyde: evidence that histone H4 is retained on a highly transcribed gene. *Cell* 53:937–47
44. Stahlhut Espinosa CE, Slack FJ. 2006. The role of microRNAs in cancer. *Yale J. Biol. Med.* 79:131–40

45. Strous M, Pelletier E, Mangenot S, Rattei T, Lehner A, et al. 2006. Deciphering the evolution and metabolism of an anammox bacterium from a community genome. *Nature* 440:790–94
46. Torres TT, Metta M, Ottenwälder B, Schlötterer C. 2007. Gene expression profiling by massively parallel sequencing. *Genome Res.* 18:172–77
47. Tringe SG, von Mering C, Kobayashi A, Salamov AA, Chen K, et al. 2005. Comparative metagenomics of microbial communities. *Science* 308:554–57
48. Turnbaugh PJ, Ley RE, Mahowald MA, Magrini V, Mardis ER, Gordon JI. 2006. An obesity-associated gut microbiome with increased capacity for energy harvest. *Nature* 444:1027–31
49. Tyson GW, Chapman J, Hugenholtz P, Allen EE, Ram RJ, et al. 2004. Community structure and metabolism through reconstruction of microbial genomes from the environment. *Nature* 428:37–43
50. Velculescu VE, Zhang L, Vogelstein B, Kinzler KW. 1995. Serial analysis of gene expression. *Science* 270:484–87
51. Venter JC, Adams MD, Myers EW, Li PW, Mural RJ, et al. 2001. The sequence of the human genome. *Science* 291:1304–51
52. Venter JC, Remington K, Heidelberg JF, Halpern AL, Rusch D, et al. 2004. Environmental genome shotgun sequencing of the Sargasso Sea. *Science* 304:66–74
53. Weber AP, Weber KL, Carr K, Wilkerson C, Ohlrogge JB. 2007. Sampling the *Arabidopsis* transcriptome with massively parallel pyrosequencing. *Plant Physiol.* 144:32–42
54. Wegley L, Edwards R, Rodriguez-Brito B, Liu H, Rohwer F. 2007. Metagenomic analysis of the microbial community associated with the coral *Porites astreoides*. *Environ. Microbiol.* 9:2707–19
55. Zhao T, Li G, Mi S, Li S, Hannon GJ, et al. 2007. A complex system of small RNAs in the unicellular green alga *Chlamydomonas reinhardtii*. *Genes Dev.* 21:1190–203

African Genetic Diversity: Implications for Human Demographic History, Modern Human Origins, and Complex Disease Mapping

Michael C. Campbell[1] and Sarah A. Tishkoff[1,2]

[1]Department of Genetics, University of Pennsylvania School of Medicine, Philadelphia, Pennsylvania 19107; email: mcam@mail.med.upenn.edu

[2]Department of Biology, University of Pennsylvania, School of Arts and Sciences, Philadelphia, Pennsylvania 19104; email: tishkoff@mail.med.upenn.edu

Annu. Rev. Genomics Hum. Genet. 2008. 9:403–33

First published online as a Review in Advance on July 1, 2008

The *Annual Review of Genomics and Human Genetics* is online at genom.annualreviews.org

This article's doi:
10.1146/annurev.genom.9.081307.164258

1527-8204/08/0922-0403$20.00

Key Words

disease susceptibility, African populations, genetic variation, human evolution, linkage disequilibrium

Abstract

Comparative studies of ethnically diverse human populations, particularly in Africa, are important for reconstructing human evolutionary history and for understanding the genetic basis of phenotypic adaptation and complex disease. African populations are characterized by greater levels of genetic diversity, extensive population substructure, and less linkage disequilibrium (LD) among loci compared to non-African populations. Africans also possess a number of genetic adaptations that have evolved in response to diverse climates and diets, as well as exposure to infectious disease. This review summarizes patterns and the evolutionary origins of genetic diversity present in African populations, as well as their implications for the mapping of complex traits, including disease susceptibility.

INTRODUCTION

One of the "grand challenges" of the post-genome era is to "develop a detailed understanding of the heritable variation in the human genome" (34). By characterizing genetic variation among individuals and populations, we may gain a better understanding of differential susceptibility to disease, differential response to pharmacological agents, human evolutionary history, and the complex interaction of genetic and environmental factors in producing phenotypes. Africa is an important region to study human genetic diversity because of its complex population history and the dramatic variation in climate, diet, and exposure to infectious disease, which result in high levels of genetic and phenotypic variation in African populations. A better understanding of levels and patterns of variation in African genomes, together with phenotype data on variable traits, including susceptibility to disease and drug response, will be critical for reconstructing modern human origins, the genetic basis of adaptation to diverse environments, and the development of more effective vaccines and other therapeutic treatments for disease. This information will also be important for identifying variants that play a role in susceptibility to a number of complex diseases in people of recent African ancestry (170, 185, 204).

HUMAN EVOLUTIONARY HISTORY IN AFRICA

Africa is a region of considerable genetic, linguistic, cultural, and phenotypic diversity. There are more than 2000 distinct ethnolinguistic groups in Africa, speaking languages that constitute nearly a third of the world's languages (**http://www.ethnologue.com/**) (**Figure 1**). These populations practice a wide range of subsistence patterns including various modes of agriculture, pastoralism, and hunting-gathering. Africans also live in climates that range from the world's largest desert and second largest tropical rainforest to savanna, swamps, and mountain highlands, and these climates have, in some cases, undergone dramatic changes in the recent past (103, 170).

According to the Out of Africa (OOA) model of modern human origins, anatomically modern humans originated in Africa and then spread across the rest of the globe within the past ~100,000 years (202). The transition to modern humans within Africa was not sudden; rather, the paleobiological record indicates an irregular mosaic of modern, archaic, and regional morphological and behavioral traits that occurred over a substantial period of time and across a broad geographic range within Africa (124). The earliest known derived suite of morphological traits associated with modern humans appears in fossil remains from Ethiopia, dated to ~150–190 kya (125, 224). However, this finding does not rule out the existence of modern morphological traits in other regions of Africa before 100 kya, particularly where specimens may be less well preserved and/or where extensive archaeological and paleobiological investigations have not been conducted (170). Indeed, a multiregional origin model for modern humans within Africa is not as unlikely as it would be for global populations, considering the greater potential for migration and admixture within a single continental region (170, 237). A more fully modern suite of traits appears in East Africa and Southwest Asia around 90 kya, followed by a rapid spread of modern humans throughout the rest of Africa and Eurasia within the past 40,000–80,000 years (117, 170) (**Figure 2**).

Two migration routes of modern humans out of Africa have been proposed. The presence of modern humans in Oceania as early as ~50 kya (63, 64), which predates their presence in Europe ~40 kya, has suggested a southern coastal route around the Indian Ocean in which modern humans first left Africa (possibly via Ethiopia) by crossing the Bab-el-Mandeb strait at the mouth of the Red Sea and then rapidly migrated to Southeast Asia and Oceania (61, 170) (**Figure 1**). This migration model is supported by the presence of very old mtDNA haplotypes in South Asia and their absence in the Levant (117, 166, 194). Other models

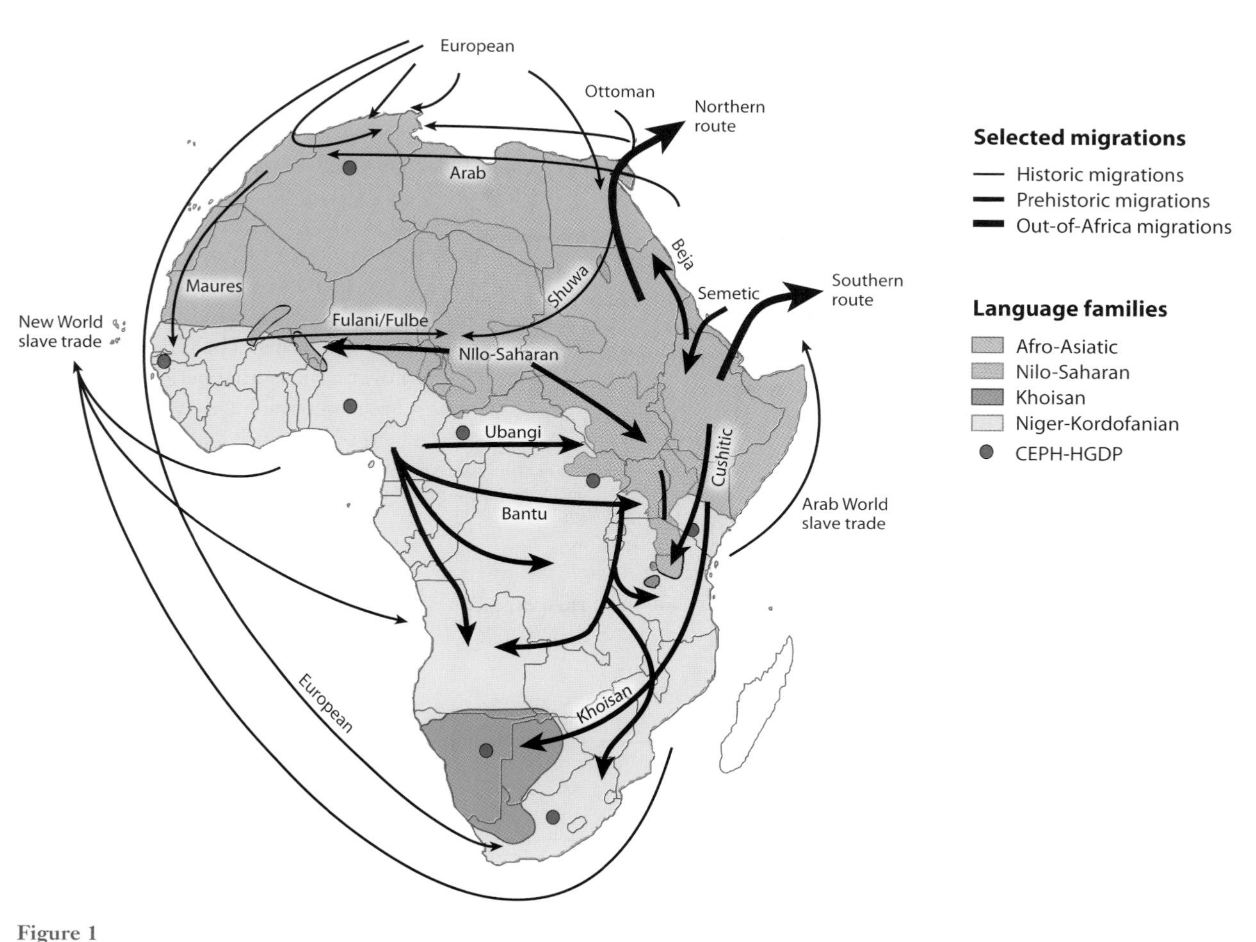

Figure 1

A map of African language family distributions and hypothesized migration events within and out of Africa. African languages have been classified into four major language families: Niger-Kordofanian (spoken predominantly by agriculturalist populations across a broad geographic distribution in Africa), Afro-Asiatic (spoken predominantly by northern and eastern Africa pastoralists and agro-pastoralists), Nilo-Saharan (spoken predominantly by eastern and central African pastoralists), and Khoisan (a language containing click-consonants, spoken by southern and eastern African hunter-gatherer populations). Also plotted are the geographic origins of African samples included in the Center d'Etude du Polymorphisme Humain (CEPH) Human Genome Diversity Panel (CEPH-HGDP). Diagram adapted from Reference 170.

have traditionally favored a second (or single) northern route via the Sinai Peninsula into the Levant (61, 170) (**Figure 1**). Regardless of the route of migration of modern humans out of Africa, the shared patterns of genetic diversity among non-African populations [e.g., at the CD4 locus (197)] and the divergent patterns of genetic variation among African populations argue against repeated sampling of African diversity from multiple source populations (170, 197, 202). However, analyses of more independent loci and a larger number of African populations, particularly from East Africa, will be necessary to better estimate the number and source of migration events out of Africa (170). After modern humans migrated from Africa, there could have been some admixture of modern humans with archaic populations in Eurasia, such as Neanderthals. This hypothesis remains a topic of considerable interest and debate and is the

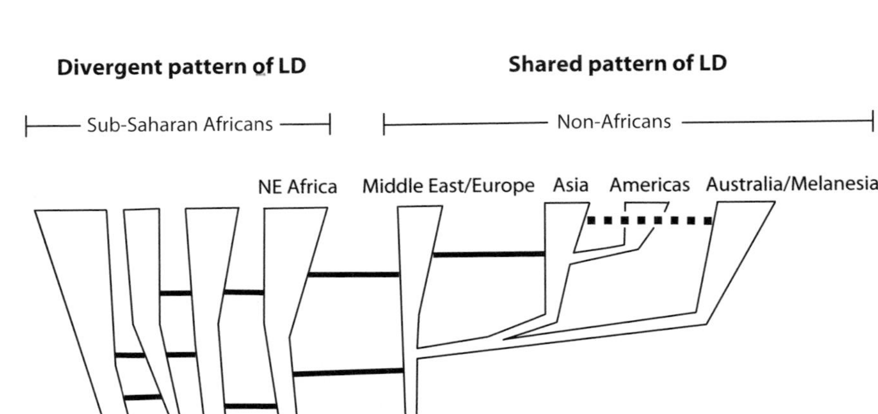

Figure 2

Ancestral Africans have maintained a large and subdivided population structure and have experienced complex patterns of population expansions, contractions, migration, and admixture during their evolutionary history. The bottleneck associated with the founding of non-African populations (~50–100 kya) resulted in lower levels of genetic diversity, an increase in linkage disequilibrium (LD), and more similar patterns of LD. In addition, several recent studies have suggested that a serial founder model of migration occurred in the history of non-Africans in which the geographic expansion of these populations occurred in many small steps, and each migration involved a sampling of variation from the previous population (36, 90, 109, 167). Solid horizontal lines indicate gene flow between populations and the dashed horizontal line indicates recent gene flow from Asia to Australia/Melanesia.

subject of a number of recent studies and reviews (44, 58, 69, 71, 74, 142, 157, 170, 182, 183, 219).

The migration of modern humans out of Africa is thought to be accompanied by a population bottleneck. The size of the population(s) migrating out of Africa is estimated to be ~600 effective founding females (i.e., census size of ~1800 females) on the basis of mtDNA evidence (61, 117), to be ~1000 effective founding males and females (i.e., census size of ~3000 individuals) based on the analysis of 783 autosomal microsatellites genotyped in the Center d'Etude du Polymorphisme Humain (CEPH) human genome diversity panel (HGDP) (110), and to be ~1500 (i.e., a census size of ~4500 individuals) based on a combined analysis of mtDNA, Y chromosome, and X chromosome nucleotide diversity data (70). These estimates imply that Eurasians must have rapidly expanded to a larger size to account for estimates of a long-term effective population size (N_e) of ~10,000 individuals (census size of ~30,000 individuals) for global populations (170, 239).

Indeed, several recent studies indicate a rapid expansion of Eurasian populations within the past ~50,000 years, whereas Africans have maintained a large effective population size (70, 122, 239).

PATTERNS OF GENETIC VARIATION IN AFRICA

The pattern of genetic variation in modern African populations is influenced by demographic history (e.g., changes in population size, short- and long-range migration events, and admixture) as well as locus-specific forces such as natural selection, recombination, and mutation. For example, the migration of agricultural Bantu speakers from West Africa throughout sub-Saharan Africa within the past ~4000 years and subsequent admixture with indigenous populations have had a major impact on patterns of variation in modern African populations (156, 165, 170, 198, 231) (**Figure 1**). Although Africa is critical for understanding modern human origins and genetic risk factors for disease, it has been under-represented in human genetic studies. Much of what we currently know about genetic diversity is from a limited number of the ~2000 ethno-linguistic groups in Africa, and the majority of these data are from mtDNA and Y chromosome studies. Large-scale autosomal studies of African genetic diversity are only now beginning to become available.

Mitochondrial DNA and Y Chromosome Variation

Phylogenetic analyses of both mtDNA and Y chromosome DNA indicate that the oldest lineages are specific to Africa and have a Time to Recent Common Ancestry (TMRCA) of ~200 kya (72, 202). Interestingly, the most ancient mtDNA lineage (L0d) [dated to ~106 kya (72)], which is common in click-speaking southern African Khoisan (SAK) populations, has recently been identified at low frequency (5%) in the click-speaking Sandawe population from Tanzania (72, 198). A maximum likelihood estimate for the time of divergence of these populations based on all mtDNA lineages is ~44 kya, indicating that any common ancestry is quite old. This finding supports studies of classical polymorphisms as well as archeological data that suggest that Khoisan-speaking populations may have originated in eastern Africa and subsequently migrated into southern Africa (24), although a southern African origin of Khoisan speakers cannot be ruled out.

Phylogenetic analysis indicates that the most recent African specific mtDNA haplogroup lineage, L3, is the likely precursor of modern European and Asian mtDNA haplotypes (221). Indeed a subset of this lineage (M1) is observed at high frequency in Ethiopian populations (98, 166) and may have expanded out of Africa ~60 kya (166). This observation adds strength to the proposal that the dispersal of modern humans out of Africa may have occurred via Ethiopia (114, 197). However, more recent analysis of whole mtDNA genomes suggests that the M1 lineage may have originated in southwestern Asia and then was introduced into East Africa from Asia ~40–45 kya (148), whereas others have argued for a much more recent introduction of the M1 lineage into Africa from the Middle East (62).

Nucleotide and Haplotype Variation

The migration of modern humans out of Africa resulted in a population bottleneck and a concomitant loss of genetic diversity (110, 167). Numerous studies have shown higher levels of nucleotide and haplotype diversity in Africans compared to non-Africans in both nuclear and mitochondrial genomes (38, 70, 90, 109, 197, 199, 202, 204). Non-African populations appear to have a subset of the genetic diversity present in sub-Saharan Africa and more private alleles and haplotypes are observed in Africa relative to other regions (36, 90, 109, 167, 197, 199, 202, 204, 239), as expected under an OOA model. For example, a resequencing study of 3873 genes in 154 chromosomes from European, Latino/Hispanic, Asian, and

African American populations observed that African Americans had the highest percentage of rare single nucleotide polymorphisms (SNPs) (64%) and the lowest percentage of common SNPs (36%). Additionally, 44% of all SNPs in this population were private (75). The high level of genetic diversity in African populations is also consistent with a larger long-term effective population size (N_e) compared to non-Africans (70, 192, 193, 199, 202); N_e is estimated to be ~15,000 for Africans and ~7500 for non-Africans based on a resequencing analysis of several 10-kb regions (239) (see **Supplemental Material**; follow the **Supplemental Material link** from the Annual Reviews home page at **http://www.annualreviews.org**).

Structural Variation

Although most studies of genetic variation in humans have focused on nucleotide and microsatellite diversity, a number of recent studies have demonstrated considerable amounts of structural variation (SV) in the human genome, including both copy number variation (which can include insertions and deletions as well as gene duplications) and inversions (15, 35, 188, 206) (**http://projects.tcag.ca/variation/**). Some of these structural variants are also associated with phenotypic variability (35, 169, 190). For example, variation in copy number of the amylase gene, which plays a role in digestion of starch, is correlated with enzyme activity level and with diet in ethnically diverse human populations (155). Additionally, SVs may play an important role in susceptibility to common disease (107, 121). A recent study that used high-resolution paired-end mapping to identify SVs in the genomes of a single African (Yoruba from Nigeria) individual and an individual of European descent led to the identification of 1175 insertions/deletions (INDELs) and 122 inversions (101). By extrapolation, these researchers predicted 761 and 887 SVs in the full genomes of these European and African individuals, respectively. Additionally, 45% of the SVs were shared between these samples, suggesting that a large proportion of SV events occurred prior to the divergence of African and non-African populations. The majority of these SVs were smaller than 10 kb in size, but at least 15% were larger than 100 kb and some SVs were predicted to be several megabases in size in both the European and African samples, indicating that the genomes of healthy individuals may differ by megabases of nucleotide sequence (101). To date, few population genetic studies of SVs across ethnically diverse populations have been performed (35). Instead, most studies have focused on the European American, Japanese, Chinese, and African (Yoruba) HapMap populations (35). A study of 67 common copy number variants (CNVs) in these populations indicated that 11% of the variation was due to differences among populations and that many of the variants were shared among populations from different regions, further supporting the argument that these variants existed prior to migration of modern humans out of Africa (169). There are currently no studies of SV variability within and between ethnically diverse African populations. Such knowledge will be informative for reconstructing human evolutionary history and for understanding the role of SVs in normal phenotypic diversity and in susceptibility to disease.

POPULATION STRUCTURE IN AFRICA

Measures of population structure on a global level indicate that only ~10–16% (Wright's fixation index, F_{ST} = 0.10–0.16) of observed genetic variation is due to differences among populations from Africa, Europe, and Asia (24, 38, 202, 223). Analysis of population structure using the program STRUCTURE (161), based on 1048 individuals from the CEPH human diversity panel genotyped for 993 genome-wide microsatellite and insertion/deletion markers, indicates that individuals cluster into five major geographic regions: Africa, Europe/Middle East, East Asia, Oceania, and the New World (172, 173). Two recent studies of >500,000 SNPs genotyped in the CEPH diversity panel support these initial findings (90, 109).

Analyses of African populations indicate that additional substructure exists, particularly between hunter-gatherer and agriculturalist populations (90, 109). However, the CEPH diversity panel includes just eight African populations, four of which are agricultural Bantu-speakers likely to share recent common ancestry (**Figure 1**). Thus, results from these studies may not reflect the full extent of population structure within Africa.

Several studies of nucleotide and haplotype variation have indicated that ancestral African populations were geographically structured prior to the migration of modern humans out of Africa (70, 71, 79, 157, 197, 237). Additionally, a recent study of 800 short tandem repeat polymorphisms (STRPs) and 400 INDELs genotyped in more than 3000 geographically and ethnically diverse Africans indicates the presence of at least 13 genetically distinct ancestral populations in Africa and high levels of population admixture in many regions (F.A. Reed & S.A Tishkoff, unpublished data). Population clusters are correlated with self-described ethnicity and shared cultural and/or linguistic properties (e.g., Pygmies, Khoisan-speaking hunter-gatherers, Bantu speakers, Cushitic speakers). This study reveals extensive admixture between inferred ancestral populations in most African populations. One exception is among West African Niger-Kordofanian (i.e., Bantu) speakers who are more genetically homogeneous compared with other African populations, likely reflecting the recent and rapid spread of Bantu speakers from a common origin in Cameroon/Nigeria (although fine-scale genetic structure can be detected amongst these populations). Thus, the pattern of genetic diversity in Africa indicates that African populations have maintained a large and subdivided population structure throughout much of their evolutionary history (**Figure 2**). Historic subdivision among African populations is likely due to ethnic and linguistic barriers, as well as a number of geographic, ecological, and climatic factors (including periods of glaciation and warming) that could have contributed to population expansions, contractions, fragmentations, and extinctions during recent human evolution in Africa (170, 202).

PATTERNS OF LINKAGE DISEQUILIBRIUM

Linkage disequilibrium (LD), the nonrandom association between alleles at different loci, is typically measured using two different estimators: D' and r^2 (160). Levels and patterns of LD depend on a number of demographic factors including population size and structure, as well as locus-specific factors such as selection, mutation, recombination (1, 160, 202, 203), and gene conversion (see **Supplemental Material**). LD is particularly useful for inferring evolutionary and demographic processes, as well as for mapping disease susceptibility loci. Therefore, an understanding of levels and patterns of LD has broader implications for studies of human evolutionary history and disease.

Empirical Studies of Linkage Disequilibrium

Several haplotype studies have indicated lower levels of LD in African populations compared to non-Africans (197, 199, 202, 203). Studies of long-range LD between SNP markers at multiple nuclear loci confirmed these initial results and demonstrated that haplotype blocks (where SNPs are in strong LD) extend over greater genomic distances and are more uniform in non-Africans compared to African populations (67, 113, 171, 181) (**Figure 3**).

Given that recombination is an important determinant of the extent of LD, an alternative way to assess LD is to estimate the population recombination rate ($\rho = 4N_e r$, where N_e is effective population size and r is the meiotic recombination rate/kb) (160). Empirical studies have shown that African Americans have higher ρ estimates compared to Europeans and Asians (31, 45, 59), consistent with the results of previous studies that described less LD in Africans relative to non-African populations. The divergent patterns of LD can be explained by the distinct demographic histories of African and

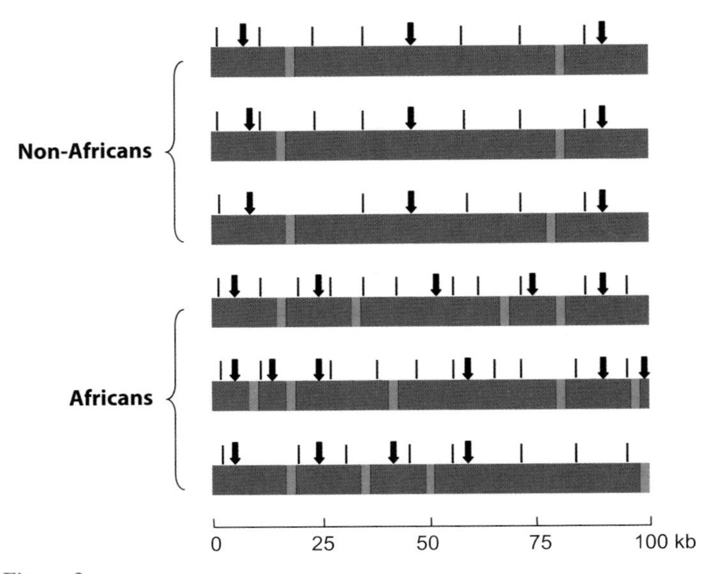

Figure 3

Several analyses have indicated that haplotype blocks [where single nucleotide polymorphisms (SNPs) are in strong linkage disequilibrium (LD)] extend over greater genomic distances and are more uniform in non-Africans compared to African populations. Additionally, the size and location of haplotype blocks can vary among African samples owing to the distinct demographic histories of populations from different geographic regions in Africa. The blue bars represent haplotype blocks and the thin orange bars denote regions of recombination. Vertical lines indicate SNPs and vertical arrows indicate haplotype tag SNPs (htSNPs). Because haplotype blocks are more variable in Africans compared to non-Africans, identification of htSNPs in diverse African ethnic groups and more dense tag htSNP coverage are needed to detect an association between marker(s) and disease loci in association studies.

non-African populations (202, 204) (**Figure 2**). Specifically, African populations have shorter blocks of LD because ancestral Africans maintained a larger effective population size (N_e) and because there has been more time for recombination to decrease levels of LD. Greater LD in non-African populations is likely the result of a founding event during expansion of modern humans out of Africa within the past 100,000 years (197, 202, 204).

However, an ongoing challenge has been to characterize patterns of LD among populations within continental regions, especially in Africa. Some evidence has suggested variance in levels and patterns of LD among subpopulations in Africa. Tishkoff and colleagues (197) noted that African populations have divergent patterns of LD; specifically, alleles that were in positive association in one population were in negative association in another. Additionally, a resequencing analysis of the IL-13 gene in 126 geographically diverse Africans identified divergent patterns of LD across West and East African populations (192). These observations suggest that not all African populations are characterized by a single discrete pattern of LD and each may have distinct haplotype block structures (**Figure 3**). Theoretically, under a model of population subdivision, allelic associations can differ between populations due to the stochastic effects of genetic drift (1).

Recombination Hotspots

Recombination hotspots, where historical crossing-over events are clustered and separate relatively large haplotype blocks, have been a topic of considerable interest in the scientific literature (10, 40, 41). The occurrence of recombination hotspots in human DNA has been demonstrated empirically from studies of single sperm DNA and from pedigree analyses (10, 41, 91). However, the extent to which this pattern is a general feature of the genome remains unknown, particularly because genetic drift can result in a similar pattern of LD (10, 45).

Several recent studies have observed that the locations of most hotspots tend to be shared between diverse populations (36, 45, 73). However, several datasets have suggested that some hotspots may be population specific (32, 36, 45, 73) and that African and African American populations have more recombination hotspots relative to non-Africans (32, 45). Given that recombination rates vary between species (163, 218) and even individuals (41, 99, 100), it is possible that hotspots could also differ among ethnically distinct populations, including Africans. Indeed the identification of haplotypes at the *RNF212* gene associated with recombination rate, which occur at different frequencies in the HapMap populations (100), raises the possibility that population-specific genetic variants may influence recombination rates.

Implications of Linkage Disequilibrium for Association Studies

The mapping of complex disease genes relies on the identification of an association of polymorphic markers, either individually or as haplotypes, with disease susceptibility loci (203). The International HapMap Project (**http://www.hapmap.org/**) has characterized patterns of haplotype structure and LD across the human genome to facilitate mapping of complex disease genes (38, 39, 126). Another goal of this project has been to identify haplotype tag SNPs (htSNPs) that distinguish major haplotypes, thereby reducing the number of SNPs needed for association studies (203).

To date, 3.4 million SNPs have been characterized in 270 individuals from four populations: Yoruba from Nigeria, European Americans, Japanese, and Chinese. Knowledge of the frequency and distribution of these SNPs across ethnically diverse populations is important to assess their usefulness as markers for gene mapping studies in diverse ethnic groups (202, 203). A survey of 3024 SNPs spaced across 36 genomic regions genotyped in 927 unrelated individuals from the CEPH human genetic diversity panel indicates that although haplotype block sharing with the HapMap populations is high in European and East Asian populations, sharing for most other populations is low, particularly for haplotypes in African hunter-gatherer populations (36). These results suggest that development of distinct panels of htSNPs and more dense coverage of SNPs will be needed for African populations (36, 203). Seven additional populations have been added to the HapMap inititative: Luhya (Bantu) and Maasai (Nilo-Saharan) from Kenya, Tuscans from Italy, Gujarati Indians from Texas, metropolitan Chinese in Denver, people of Mexican ancestry in Los Angeles, and African Americans from the southwest United States. The characterization of SNP and haplotype diversity in these additional populations will be important for the identification of htSNPs that are more informative across ethnically diverse populations.

Although the SNPs used in the HapMap study have been highly informative for use in association mapping studies, the initial identification of SNPs in one or a few populations can result in an ascertainment bias (AB) toward high-frequency, presumably older, SNPs. Several studies have shown that AB can distort estimates of migration rate (217), mutation rates (138), recombination rates (31, 138), and LD (5, 31). Although the effects of AB can sometimes be corrected (31, 140, 141), these correction methods make a number of assumptions that are not applicable to African populations, including the assumption of no population substructure (140). To more accurately infer human genetic variation it will be necessary to characterize the entire frequency distribution of nucleotide variants in diverse populations. Additionally, because variants associated with disease could be geographically restricted due to new mutation, genetic drift, or region-specific selection pressure, de novo identification of genetic variation in diverse African populations will be important. The HapMap ENCODE (**http://www.hapmap.org**) and the proposed "1000 genomes" (**http://www.genome.gov**) resequencing projects aim to discover novel variation, including rare SNPs and structural variants, in targeted regions of the genome (as well as in whole genomes for a subset of samples) from the extended HapMap populations and other ethnically diverse populations. However, the extensive levels of substructure identified in Africa will likely require analysis of additional ethnically and geographically diverse African populations.

NATURAL SELECTION

Natural selection, the process by which favorable heritable traits become more common in successive generations, operates to either increase or decrease the frequency of mutations that have an effect on an individual's fitness. When a mutation is advantageous it can rapidly increase in frequency, together with linked variants (i.e., genetic hitchhiking), due to positive selection and replace pre-existing variation in a given population (i.e., a selective sweep) (80, 139, 178, 202). The strength of selection and local rates of recombination dictate how large of a genomic region is affected by a selective sweep. If selection is recent, there may not be enough time for the selected variant to become fixed in the population, resulting in an incomplete selective sweep. The genetic signatures of a selective sweep include a region of extensive LD [extended haplotype homozygosity (EHH)] and low variation on high-frequency chromosomes with the derived beneficial mutation relative to chromosomes with the ancestral allele (177, 200, 215). After this selective sweep, given enough time, new mutations and recombination will occur, leading to an excess of rare variants and a decrease in the extent of LD. Weak purifying selection is also expected to result in an increase of low-frequency variants. Under this scheme of selection, deleterious mutations entering the population generally remain at low frequencies because their adverse effect on fitness makes it unlikely that they will reach high frequencies. In contrast, long-term balancing selection (resulting from greater fitness of heterozygotes or when maintenance of multiple alleles in a population is adaptively advantageous) is expected to result in an excess of alleles at intermediate frequency.

Demographic processes can also cause similar skews in the frequency of polymorphisms. For example, when population size rapidly increases, genetic drift has less effect in a rapidly expanding population, leading to an excess of rare polymorphisms (mimicking the pattern seen under positive or purifying selection). In contrast, a population bottleneck is expected to cause the loss of low-frequency variants, and thus produce an excess of intermediate-frequency variants (mimicking the pattern observed under balancing selection) (139). Although natural selection and demographic history can cause similar departures from a neutral equilibrium model, it is possible to distinguish these forces either by simulating the expected pattern of variation under different demographic scenarios or by using an outlier approach in which targets of selection are identified because they show an unusual pattern of variation or population differentiation compared with empirical data collected from other loci across the genome (95, 139). Given the vast number of studies published on natural selection, the next sections focus on a few case studies of genetic and phenotypic adaptation in sub-Saharan Africa.

Malaria Resistance in Africa

Malaria (caused by infection with the *Plasmodium falciparum* parasite) is a major cause of mortality in sub-Saharan Africa, resulting in more than 1 million deaths (primarily children) each year (104). Given the enormous impact of malaria in Africa, it is not surprising that this disease has exerted strong selective pressure on African populations during recent human evolutionary history.

A number of genetic variants in African populations have been shown to confer resistance to malaria. One of the best known genetic adaptations is the HbS mutation in the β-globin gene, which causes sickle cell anemia in homozygous individuals. Individuals who are heterozygous for the sickle cell trait are protected against malarial infection and have higher reproductive fitness (104), which results in the maintenance of the HbS allele at high frequency in many malaria endemic regions. A recent genetic study observed long-range LD extending over 400 kb at the β-globin locus on haplotypes with the HbS mutation in West African and Afro-Caribbean populations, consistent with the pattern expected under positive selection (78). Other well-known hemoglobin variants associated with malaria resistance in African populations include hemoglobins C (HbC) and E (HbE). Studies have also identified pattens of LD on chromosomes that contain either the HbC variant (232) or the HbE variant (145) that are consistent with recent positive selection.

Glucose-6-phosphate dehydrogenase (*G6PD*) mutations that result in reduced enzyme activity are also associated with malaria resistance (201). The most common *G6PD* mutation in Africa, *G6PD* A-, occurs at a frequency of ~25% in malaria endemic regions (201). Several empirical studies have found evidence for recent selection at the *G6PD* locus in African populations. For example, a study of SNP and microsatellite haplotype variation revealed low microsatellite variability and high LD among loci on chromosomes with the A- variant, indicative of positive selection (201). On the basis of the breakdown of LD between the microsatellite markers and the *G6PD* A- variant, the age of this variant was estimated to be between 3840 and 11,760 years (201). Similarly, nucleotide sequence analyses of the *G6PD* locus in Africa also showed patterns of variation consistent with recent positive selection (179, 213). A signature of a recent partial selective sweep was also supported by two studies that showed extensive LD extending >400 kb on chromosomes with the *G6PD* A- mutation (177, 180). A comparative analysis of human and nonhuman primates suggested that signatures of selection at the A- allele are unique to humans (214). Overall, these data are consistent with other evidence suggesting that the malaria parasite has had a significant impact on humans only within the past 10,000 years (93, 216), possibly corresponding with the development of agriculture and/or pastoralism in Africa (201).

The Duffy gene on chromosome 1 confers resistance to malaria caused by *Plasmodium vivax*, which is not prevalent in Africa today but may have been in the past. The Duffy gene encodes a receptor on the surface of erythrocytes and is characterized by three alleles (*FY*A*, *FY*B*, and *FY*O*). The frequency of the *FY*O* allele is at or close to fixation in most sub-Saharan African populations, but is very rare outside of Africa (76). A resequencing study of nucleotide variation at the FY locus in five sub-Saharan African populations and in a comparative Italian population (76) reported that variation at this locus is two- to threefold lower in Africans than in the Italian sample, which is the opposite pattern observed at most loci. A more extensive resequencing analysis of this locus (77) also revealed reduced sequence variation around the *FY*O* mutation and an excess of high-frequency derived alleles at linked sites, consistent with a selective sweep of this region. Additionally, researchers observed unusually large F_{ST} values for the *FY*A* and *FY*O* variants at this locus across African, European, and Asian populations, consistent with local adaptation in different geographic regions. These results have led to the conclusion that positive selection has been a dominant force in shaping the distribution of Duffy alleles among human populations.

Dietary Adaptations in African Populations

Lactase persistence (LP), the ability to digest milk and other dairy products into adulthood, is a classic example of a genetic adaptation in humans. LP varies in frequency in different human populations; it is most common in northern Europeans and certain African and Arabian nomadic tribes that practice pastoralism and is at low frequency in East Asians and most West African populations (85). A number of studies have demonstrated a strong association between LP and the presence of the T allele at a C/T SNP located −13910 kb upstream from the lactase gene (*LCT*) in European populations (56, 158). In vitro studies also showed that the T-13910 variant, in combination with the *LAC* promoter, enhances gene transcription (108, 146). In a study of long-range LD in Europeans, Bersaglieri and colleagues (20) found that haplotypes containing the LP-associated T-13910 variant were largely identical over nearly a 1-Mb region, consistent with a strong selective sweep. In genome-wide scans of selection in the HapMap samples, the *LCT* gene showed the strongest signal of positive selection in European Americans (178, 215).

Although the T-13910 variant is likely the causal mutation of the lactase persistence trait in Europeans, analyses of this SNP in ethnically and geographically diverse African populations indicated that it is present in only a few West African pastoralist populations, such as the

Fulani (or Fulbe) and Hausa from Cameroon (33, 132, 133). These results suggested that the T-13910 allele may not be a strong predictor of lactase persistence in most sub-Saharan Africans. A more recent genotype and phenotype association study in a sample of 43 populations from Tanzania, Kenya, and the Sudan identified three novel SNPs located ~14 kb upstream of *LCT* that are significantly associated with the LP trait in African populations (200). These SNPs are located within 100 bp of the European LP-associated variant (C/T-13910). One LP-associated SNP (G/C-14010) is common in Tanzanian and Kenyan pastoralist populations, whereas the other two (T/G-13915 and C/G-13907) are common in northern Sudanese and Kenyans. The derived alleles at these loci (C-14010, G-13915, and G-13907) were shown to enhance transcription in combination with the *LCT* promoter in vitro (200). Genotyping of 123 SNPs across a 3-Mb region in these populations demonstrated that these African LP-associated variants exist on haplotype backgrounds that are distinct from the European LP-associated variant and from each other. In addition, haplotype homozygosity extends >2 Mb on chromosomes with the LP-associated C-14010 variant, consistent with an ongoing selective sweep over the past 3000–7000 years. An independent study of Sudanese populations also identified a significant association of the G-13915 allele with LP in that region (87) and a recent study confirmed the enhancer effect of the G-13915 variant, together with a C-3712 variant, on the same haplotype background (55). These data indicate a striking example of convergent evolution and local adaptation due to strong selective pressure resulting from shared cultural traits (e.g. cattle domestication and adult milk consumption) in Europeans and Africans. These studies also demonstrate the effect of local adaptation on patterns of genetic variation and the importance of resequencing across geographically and ethnically diverse African populations to identify population-specific variants associated with variable traits, including disease susceptibility.

Another important dietary adaptation in human populations is the ability to taste bitter compounds. A hypothesized selective advantage of bitter taste is that it helps individuals avoid ingesting toxic substances in plants. Variation at the *TAS2R* genes is associated with sensitivity to bitter taste substances (52). An analysis of nucleotide and haplotype variation at 24 *TASR2* genes in 55 globally diverse individuals also identified substantial amino acid diversity, an excess of nonsynonymous substitutions, and high levels of population differentiation at variable sites, suggesting that amino acid variability at these loci may be maintained due to natural selection (96).

There have also been a number of in-depth studies at individual *TAS2R* loci for which there are known associations with bitter taste perception. For example, the ability to taste phenylthiocarbamide (PTC), a synthetic bitter substance, is a highly variable trait in humans (80). Although several TAS2R loci contribute to variability in PTC taste perception (52, 53), 50–85% of the phenotypic variance in PTC sensitivity is attributed to variation at the *TAS2R38* gene (233). Studies have identified three amino acid substitutions at *TAS2R38* that are in nearly complete LD in non-African populations and that form two common amino acid haplotypes (a taster haplotype PAV and a nontaster haplotype AVI; PAV is dominant). Furthermore, considerably more haplotype variability has been observed in Africa (234, M.C. Campbell & S.A. Tishkoff, unpublished data) and these haplotypes are associated with a broad range of taste perception phenotypes (M.C. Campbell & S.A. Tishkoff, unpublished data).

Genetic analyses of both African and non-African populations have detected signatures of balancing selection at the *TAS2R38* locus, including an excess of intermediate-frequency variants, a low amount of genetic differentiation between continental populations (F_{ST} = 0.056), and an ancient divergence between the major taster and nontaster haplotypes (234). Furthermore, Wooding and coworkers (233) showed that PTC taste sensitivity in chimpanzees is associated with different amino acid haplotypes at the *TAS2R38* gene compared to humans, implying a unique origin of the taster/nontaster variants in humans and chimpanzees.

It has also been suggested that low sensitivity to bitter taste may provide a selective advantage against malarial infection in some African populations. Recent data have shown that the K172 allele at the *TAS2R16* gene (which is associated with low sensitivity to bitter taste substances, including salicin) occurs at moderately high frequencies in malaria endemic regions in central Africa (187). Furthermore, sequence analysis of the entire *TAS2R16* coding region, and part of the 5′ and 3′ untranslated regions (UTRs), in 997 individuals from 60 human populations detected a signature of positive selection on chromosomes with the K172N variant (187). Although the variant driving the signal of positive selection was not conclusively determined, the authors speculated that differential selection in malarial (favoring the K variant) and non-malarial (favoring the N variant) environments has maintained both alleles at relatively high frequencies in Africa. An earlier study suggested that the higher dietary intake of naturally occurring bitter substances, such as organic cyanogens, may be protective against malarial infection in populations from Central and Southeast Africa (88). Additionally, an inhibitory effect of cyanide on the normal development of the *P. falciparum* parasite has been observed in vitro (135). Thus, it has been hypothesized that individuals with a low sensitivity to the bitter taste of cyanogenic compounds may have a survival advantage against malarial infection through a higher intake of these bitter compounds, leading to the maintenance of the K allele in malaria-endemic regions (187). However, this hypothesis remains to be tested.

Selection in African Versus Non-African Populations

Several studies have reported more evidence for positive selection in populations outside of Africa relative to those in Africa. For example, Akey and colleagues (4) examined 132 genes in 24 African Americans and 23 European Americans and found evidence for selection at eight genes only in the European-derived population. A number of genome scans for selection (which aim to identify de novo targets of selection) have identified differential patterns of selection in Africans and non-Africans (23, 81, 94, 95, 134, 178, 189, 215, 220, 229). Several of these studies have observed more loci under recent selection in non-African relative to African populations (23, 94, 134, 189, 229). Furthermore, it has been hypothesized that non-African populations have experienced more recent strong local adaptation as modern humans migrated out of Africa into novel and diverse environments (4, 81, 189, 229).

Although an increase in positive selection might conceivably occur in populations that have migrated into new environments, it is premature to conclude that the amount of recent positive selection is greater in non-African versus African populations. For example, Voight and colleagues (215) identified widespread evidence of recent selection in each of the HapMap samples in a genome-wide scan. Furthermore, they observed the strongest signals of selection in the HapMap Nigerian population compared to HapMap European American and Asian populations (although the power to detect selection may be greater in larger African populations). Additionally, a recent study demonstrated that non-Africans have an excess proportion of nonsynonymous variation, including many variants that are likely to be deleterious (111), which is attributed to population demographic history rather than increased adaptive evolution. Therefore, demographic events, such as a population bottle neck during the migration of modern humans out of Africa, may have influenced the differential pattern of selection observed in African and non-African populations.

However, the primary limitation in comparing the frequency of selective events in African and non-African populations is that African populations have been severely understudied. For instance, many studies have used African

Americans as the sole representative of African populations. However, the statistical power to detect selective sweeps is likely to be lower in studies using African American samples because of their recent admixture with Europeans (23, 229). Indeed, one might predict that Africans would have relatively high amounts of local adaptation, considering that Africa has the highest levels of genetic diversity and contains populations living in a wide range of environments and with high exposure to infectious disease. However, signatures of selection in African populations may be missed because studies have relied mainly on one or two African populations. To gain a clearer understanding of genetic and phenotypic adaptations in Africa, it is important to scan for genetic signatures of selection in a broad range of ethnically diverse African populations living in distinct environments. Additionally, it will be important to identify functional variants that are likely to be targets of selection and to verify their impact on phenotypic variation (30) before the relative number of selection events in African and non-African populations can be clearly determined.

INFECTIOUS DISEASE

Host genetic variation plays a key role in influencing susceptibility to many infectious diseases in humans. Through recurrent exposure to different pathogens, a number of genetic adaptations have evolved that provide resistance to infection. Although the number of known candidate genes related to infectious disease has expanded, progress in the identification of genes that influence infectious disease susceptibility and/or resistance in diverse African populations has been slow. Understanding the genetic basis of infectious disease in Africans may provide useful insight into devising effective strategies to combat these diseases that have a large impact on African populations. Here we focus on three infectious diseases that cause the highest number of deaths in Africa: acquired immune deficiency syndrome (AIDS), tuberculosis (TB), and malaria.

Acquired Immune Deficiency Syndrome

It is estimated that 42 million people are infected worldwide with HIV, the virus that causes AIDS (144). Moreover, greater than 75% of HIV-1 infections and 84% of all AIDS-related deaths occur in sub-Saharan Africa (230). Although most individuals exposed to HIV rapidly progress to more advanced stages of this disease, researchers have identified a number of Africans who do not progress to AIDS despite exposure to HIV (144). This observation suggests that polymorphisms associated with disease susceptibility and resistance may be present in African populations (**Table 1**).

Chemokine receptors, which aid in the entry of HIV into the host cell, play a role in AIDS susceptibility (144). The protective effect of the $\Delta 32$ mutation at the chemokine receptor 5 (*CCR5*) gene has been well established in populations of Northern European ancestry (105). Because this mutation occurs at a relatively low frequency in sub-Saharan Africa, it is unlikely to play a major role in disease resistance in that region. However, several studies have demonstrated that mutations at the *CCR2* gene play a role in resistance to HIV in Africans; for example, the *CCR2-64I* allele is associated with delayed HIV disease progression in African and African American populations (115, 228), although this effect was not observed in a large number of individuals from Uganda (168), indicating that underlying genetic and/or environmental factors that affect resistance may vary across African populations. See **Table 1** for more details about other genes associated with HIV susceptibility in African populations.

Tuberculosis

Mycobacterium tuberculosis infection leading to TB causes significant mortality throughout the world, particularly in resource-poor countries. Furthermore, HIV infection is strongly associated with an increased risk for TB in sub-Saharan Africa (43, 116). In Africa, the rates of tuberculosis range from 50 to greater than 300 cases per 100,000 individuals (210). Genetic variation has been shown to influence susceptibility to TB in African populations. For example, polymorphisms in the *NRAMP1* gene

Table 1 Infectious disease loci

Infectious disease	Known genes associated with disease susceptibility in Africans	Effect of genetic loci on infection or disease progression	References
HIV/AIDS	Chemokine receptor 2 (CCR2)	CCR-2 64I allele associated with ↓ disease progression	(144, 230)
	Human leukocyte antigen (HLA) locus	HLA A2/6802 supertype and HLA DRB1*01 associated with ↓ HIV infection in female sex workers from East Africa	(118)
	Killer immunoglobin-like receptors (KIR)	KIR3DS1 in combination with HLA-B Bw4–80Ile associated with ↓ disease progression in African Americans and West African sex workers	(92, 123)
	Interferon regulatory factor (IRF-1)	Three polymorphisms (located at 619, microsatellite region, and 6516 of genes) associated with ↓ susceptibility to infection in East African sex workers	(14)
	TRIM5α	136Q and 43Y alleles protect against HIV infection in African Americans	(105)
	APOBEC3G	186R allele associated with ↑ disease progression in African Americans	(6)
	CUL5	Haplotype10 associated with ↑ disease progression of HIV-1 in African Americans	(7)
Tuberculosis	Human homolog of natural-resistance-associated macrophage protein 1 (NRAMP1)	Several polymorphisms (5′ microsatellite repeats, a SNP in intron 4, and a deletion in 3′ UTR) associated with ↑ risk for tuberculosis in East and West Africans	(17, 19, 186)
	UBE3A	7-bp deletion at 5′ end of UBE3A gene associated with ↑ susceptibility in West African families	(25)
	Chromosome 15 and X chromosome	Unknown loci on these chromosomes associated with ↑ tuberculosis susceptibility in West and South African populations	(18)
	Human leukocyte antigen (HLA) locus	DRB1*1302 allele and DQB1*0301–0304 alleles associated with ↑ tuberculosis susceptibility in Venda population from South Africa	(112)
	Vitamin D receptor (VDR) locus	F-b-A-T haplotype associated with ↓ susceptibility in a South African population	(112)
		Polymorphism at codon 352 of *VDR* gene associated with ↓ susceptibility in a Gambian population	(18)
		Polymorphisms (FokI and ApaI) associated with ↑ susceptibility in West Africans	(147)
	CD209	Promoter -336G allele associated with ↓ susceptibility in West Africa and Malawi	(210)
		Promoter -336A and -871G alleles associated with ↓ susceptibility in an admixed colored population from South Africa	(16)
		Intron 6 variant associated with ↑ susceptibility in West Africans	(147)
	Pentraxin 3 (PTX3)	G-A-G haplotype associated with ↓ susceptibility in West Africans	(147)

(*Continued*)

Table 1 (*Continued*)

Infectious disease	Known genes associated with disease susceptibility in Africans	Effect of genetic loci on infection or disease progression	References
Malaria	β-globin gene	HbS, HbC, and HbE alleles associated with ↓ malaria susceptibility in East and West Africans	(12, 212)
	α-globin gene	α^+-thalassemias associated with ↓ malaria susceptibility in East Africans	(57)
	β-globin/α-globin genes	HbS and α^+-thalassemia variants inherited together associated with ↑ susceptibility in East Africans	(227)
	Duffy gene	FY*O allele associated with ↓ malaria susceptibility in Africans	(80, 83)
	Glucose 6-phosphate dehydrogenase (G6PD)	G6PD A- allele associated with ↓ malaria susceptibility in West Africans	(65)
	Human leukocyte antigen (HLA) locus	HLA-B53 and HLA-DR.B1_1302 alleles associated with ↓ malaria susceptibility in West Africans	(65)
	Chromosome 10	Chromosome 10p15.3–10p14 associated with ↑ malaria susceptibility in West Africans	(196)
	Interleukin-4 (IL-4)	IL-4 -590 C allele associated with ↓ malaria susceptibility in the Fulani of Mali and Burkina Faso	(208)

have been associated with increased TB susceptibility in ethnically diverse populations from the Gambia (19) and a single population in northern Tanzania (186). Other genes that are thought to play a role in TB susceptibility are shown in **Table 1**.

Malaria

As previously mentioned, malaria infection has been a strong selective force in recent human evolution. Approximately 40% of the world is at risk for malaria infection, and approximately 90% of all malaria deaths occur in sub-Saharan Africa (65). It is estimated that 500 million new cases of malarial illness caused by the *Plasmodium falciparum* parasite occur every year in Africa (**http://www.rbm.who.int/amd2003/amr2003/ch1.htm**).

Most of the common variants associated with resistance to malarial infection in Africans are expressed in red blood cells or play a role in immune response. These variants include hemoglobin HbS, HbC, HbE and α^+-thalassemia, the G6PD A- allele, the FY*O Duffy allele (which prevents *P. vivax* from invading erythrocytes), and a number of HLA alleles (65, 80, 83).

Interestingly, several studies have shown that ethnically diverse African populations may differ in regard to genetic susceptibility to malarial infection. A variant in the promoter region of the *IL4* gene, for example, is associated with a decrease in *P. falciparum* infection in the pastoralist Fulani from Mali, as evidenced by lower parasite load, but no such genetic association is observed in the neighboring agriculturalist Dogon population (50, 208). Other studies also reported lower prevalence of malaria parasites and fewer clinical attacks of malaria among the Fulani compared to other ethnically distinct populations in neighboring villages (104, 130, 151). Differences in the expression profile of genes involved in immune response in the Fulani have been suggested to explain the distinct resistance to malaria in this population, but not in neighboring African populations (205). Therefore, novel genetic adaptations to malaria have evolved in genetically distinct populations in response to differential exposure to pathogens. These studies demonstrate that ethnically diverse African populations may have different resistance or susceptibility alleles (**Table 1**), motivating the need to include a large number of genetically distinct populations in studies of infectious disease susceptibility in Africa.

OTHER COMPLEX DISEASES IN AFRICAN POPULATIONS

There are a number of useful approaches for mapping complex disease traits that involve analyses of the association of markers and disease traits in pedigrees or parent/offspring trios [i.e., linkage and transmission disequilibrium test (TDT) analyses], and/or in populations (i.e., case/control association studies) (84). Several genome-wide SNP panels have been developed for use in genome-wide association studies (GWAS) in populations. Association mapping for complex disease relies on having some a priori knowledge of genetic diversity, population structure, and LD in both case and control populations to identify polymorphisms associated with disease genes and to avoid erroneous associations (119). Given the high levels of substructure and admixture between genetically distinct populations in Africa, even within small geographic regions, it is particularly important to control for population heterogeneity and substructure in GWAS of African populations [using, for example, programs such as PLINK (164)]. Another useful method for mapping complex traits in highly admixed populations (e.g., African Americans who have African and European ancestry) is mapping by admixture linkage disequilibrium (MALD) (see **Supplemental Material**). The MALD approach assumes that the genomic region that contains the susceptibility allele for a given disease will have enriched ancestry from populations in which the disease phenotype is more prevalent. Thus, detailed characterization of allelic variation across ancestral West African populations, which will enable accurate inference of African ancestry, will be important for the success of this approach.

Overall, the genetic factors underlying complex diseases are still poorly understood. To date, two models of complex disease have been proposed. The common disease/common variant (CD/CV) hypothesis posits that alleles influencing complex diseases are relatively common and are therefore found in multiple populations (49, 202). In contrast, some data have suggested that complex disease is caused by rare susceptibility alleles at many loci with small effect (48, 159). Additionally, gene × environment interactions as well as epistatic interactions among loci likely influence complex disease susceptibility (including infectious disease susceptibility) (42, 225, 227). Due to local adaptation, there may be population- or region-specific susceptibility alleles underlying some complex diseases in Africa. Here we discuss three complex diseases that are common in populations of recent African descent (see **Supplemental Material** for a review of genetic susceptibility to prostate cancer in African descent populations and Reference 185 for a detailed review of additional genetic disease susceptibility studies in Africa).

Obesity

Obesity is a multifactorial disease that disproportionately affects African Americans and Afro-Caribbeans living in the United States (149). Moreover, this disease is increasing in prevalence in sub-Saharan Africa (11), particularly among urban residents. A recent study reported that the incidence of obesity in urban West Africa has more than doubled (114%) over the past 15 years (2). Obesity is a serious health concern because it is closely associated with other common disorders, such as type 2 diabetes and hypertension (204).

Although environmental factors are important determinants of obesity (89, 129), studies have also identified candidate loci that contribute to the onset of this disease. For example, the human uncoupling gene *UCP3* has been correlated with obesity and lower resting energy expenditure in African Americans and the Mende tribe of Sierra Leone (8, 97); in contrast, risk for obesity associated with this gene varied among non-African populations (46, 131,

150). Another closely linked human uncoupling gene, *UCP2*, was also correlated with obesity in African American children (236). However, whether or not these tightly linked genes exert an independent effect on obesity in populations of African ancestry has not been firmly established.

Other candidate gene studies have reported associations between polymorphisms in promoter regions and weight-related phenotypes in populations of African descent. For example, polymorphisms in the promoter of the angiotensin-converting enzyme (*ACE*) gene are correlated with obesity in Nigerians and African Americans (102). Also, allelic variation in the promoter of the agouti-related protein (*AGRP*) gene was found to affect gene expression and was implicated in the regulation of body weight in people of African origin (9, 21).

Genome-wide linkage analyses of obesity-related phentoypes have indicated strong linkage to chromosomes 1, 2, 5, 7, 8, and 11 in West Africans (3, 27). These results demonstrate that multiple loci, together with environmental factors, likely contribute to the phenotypic variance of obesity-related traits.

Type 2 Diabetes Mellitus

Type 2 diabetes mellitus (T2DM) is a late-onset metabolic disorder characterized by reduced insulin secretion and insulin action (86). African Americans have a twofold increase in risk for T2DM compared to other populations in the United States. Furthermore, the prevalence of T2D is lower in Africa (~1–2%) than among people of African descent in industrialized nations (~11–13%) (175).

One of the earliest T2DM susceptibility loci implicated in this disease is the calpain 10 (*CAPN10*) gene on chromosome 2 in Mexican Americans (86). Subsequent association studies, however, have yielded inconsistent results among geographically distinct populations (68). In addition, the risk associated with this gene also differs between ethnically diverse populations from West Africa. Specifically, a CAPN10 haplotype defined by known risk polymorphisms was associated with T2DM in populations from Nigeria, but not in distinct populations from Ghana (29). Further study in a larger set of Africans will be needed to help resolve this discrepancy.

Other recently identified candidate loci associated with diabetes or diabetes-related traits in Africans and African Americans include the *AGRP* gene (21) the transcription factor 7-like 2 (*TCF7L2*) gene (82), and the proprotein convertase subtilisin/kexin-type 2 (*PCSK2*) gene (106). Genome-wide linkage analyses in West African families have also indicated suggestive linkage to diabetes on chromosomes 12 and 19 as well as stronger evidence of linkage on chromosome 20 (175). Other studies have reported strong linkage for several quantitative traits that contribute to diabetes on chromosomes 4, 6, 8, 10, 15, 16, 17, and 18 (26, 28, 176).

Increased risk for type 2 diabetes in Africans and other indigenous populations is also suggested to be due to changes in selective pressure (i.e., the thrifty gene hypothesis) (137). Prior to 10,000 years ago, all modern humans subsisted as hunter-gatherers who likely experienced frequent cycles of feast and famine. According to the thrifty gene hypothesis, ancestral genetic variants that once promoted the efficient absorption, storage, or utilization of nutrients in this ancestral environment are now maladaptive in more modern environments, increasing risk for disease (49, 152). Although genetic evidence for this hypothesis has been inconclusive (32, 66, 209), recent data from the Macaque Genome Project indicated that a number of polymorphisms in the macaque correspond to known disease-predisposing alleles in humans (37). These results suggest that ancestral variants may influence disease susceptibility in humans.

Hypertension

Hypertension (HT) disproportionately affects people of African descent living in Western environments. For example, African Americans have a 1.6-fold higher prevalence of HT than European Americans (47). However, HT was not considered to be a major disease in sub-Saharan Africa until recently. Studies have now reported a growing incidence of HT, especially in urban settings in Nigeria, Cameroon, and Tanzania (54, 207).

To date, candidate gene studies have shown inconsistent associations with HT. For example, increased risk for disease was correlated with the G protein β3-subunit (*GNB3*) gene, angiotensin (AGT), and angiotensin II receptor (AGTR1) in some populations of African ancestry (51, 120, 184, 241), but these results were not replicated in other studies (60, 211, 226). A study of the heritability of AGT and ACE levels in Nigerians and African Americans indicated that heritability is high in Nigerians (77% for AGT and 67% for ACE) but low in African Americans (18% for AGT and ACE), suggesting a strong environmental component to variability in African American populations (42).

Other evidence has suggested that interactions between susceptibility loci may play an important role in HT susceptibility. For example, polymorphisms in the ACE, ACE4, and ACE8 genes were jointly associated with blood pressure through epistatic interaction in Nigerian families (240). Moreover, interaction between genes [*ACE* and G protein–coupled receptor kinase 4 (*GRK4*)] on different chromosomes was also found to influence blood pressure in a cohort of West African families (226). These studies suggest a complex model of disease susceptibility that involves epistatic interactions and may account for inconsistent associations in previous studies.

The results of genome-wide linkage studies have also suggested that chromosome 6q24 and 21q21 likely contain genes that influence risk for hypertension in African Americans. In another study, fine mapping of HT susceptibility loci on chromosome 6 uncovered polymorphisms in the vanin 1 (*VNN1*) gene that were correlated with HT in African Americans and Mexican Americans (242).

Salt retention (a characteristic of HT) has been proposed to represent a phenotypic adaptation to heat. Specifically, ancient human populations living in hot, humid areas, who consumed low levels of dietary salt, were theorized to have adapted to their environment by retaining salt (13). Under this scenario, polymorphisms that promote salt retention will increase in frequency in hot and humid environments due to selective pressure. This hypothesis was recently supported by several studies that observed the highest frequency of variants associated with salt retention and/or high blood pressure in Africans and decreasing allele frequency outside of Africa (136, 195, 238). Therefore, genetic adaptations to a low-salt environment in ancestral African populations may contribute to the risk for HT in African-derived populations living in a high-salt environment (13).

Pharmacogenetics

Individuals with distinct ancestry are known to vary in drug response. For example, drugs commonly used to treat heart disease are known to be less effective in individuals of African descent relative to individuals of European descent (see **Supplemental Material**). To better understand variation in drug response, it is important to characterize levels and patterns of genetic diversity at genes that may influence drug response, drug metabolism, and/or transport in ethnically diverse populations (22). To date, some of the most intensively studied drug metabolism enzyme (DME) loci include cytochrome P450s (CYPs), N-acetyl transferases (NATs), and multidrug transporters (MDR). These genes are highly polymorphic and their variation results in enzymes with increased, normal, or decreased activity (185).

Studies have shown differences in the distribution of variation at these DME loci in African and non-African populations. For example, the *CYP2B6* gene is involved in the metabolism of several clinically important drugs, including artemisinin and efavirenz, used to treat malaria and HIV infection, respectively. Common *CYP2B6* polymorphisms vary in frequency between ethnically diverse populations from West Africa, as well as between African and non-African populations. The functional significance of most of these variants is not yet known (128). However, recent studies have shown that two polymorphisms (983T>C and 516G>T), found at a higher frequency in African populations relative to non-Africans, are associated with a reduction in CYP2B6 protein expression as well as a large increase in levels of the antiretroviral drug efavirenz in the plasma of African HIV patients (127, 143, 235).

Additionally, the frequency of functional variants at the *NAT2* gene, known to play a role in the metabolism of the drug isoniazid (used to treat TB), was found to vary among ethnically diverse African populations. In particular, haplotypes associated with the fast-acetylation and the slow-acetylation phenotypes at the *NAT2* gene differ in frequency among African populations, particularly between hunter-gatherer and agriculturalist populations (H.M. Mortensen & S.A.Tishkoff, unpublished data; 153, 154). However, Africans have high levels of haplotype diversity and the effect of many of these haplotypes remains unknown. These studies imply that ethnically diverse Africans may differ in response to drugs used to treat infectious disease due to variation at genes involved in drug metabolism or transport. Given the high numbers of deaths due to infectious disease in Africa, the study of variation at genes that play a role in response to these drugs across ethnically diverse Africans is of critical importance.

FUTURE DIRECTIONS

To date, only a fraction of the ~2000 linguistically distinct ethnic groups in Africa has been extensively studied for genome-wide variation. Extensive sampling of East African populations will be informative for testing models of the origin and dispersal of modern humans out of Africa, whereas in-depth sampling of West African populations will be informative for determining African American ancestry and for the identification of markers and populations useful in MALD mapping. It is important to use an ethical approach when collecting samples to be used for genetic diversity studies, particularly if those samples and genetic data will be made publicly available. In addition to obtaining research permits from the local African governments and informed consent from individual participants (including benefits and risks involved in use of samples in current and future studies), results should ideally be made available to participants after translation into the local language. Additionally, efforts should be made to train local African scientists and to build resources across Africa for independent human genetics research. The African Society of Human Genetics was recently formed in 2004 to help achieve many of these goals (174) (**http://www.afshg.org**).

As we begin to build diverse sets of samples, a shift toward genome-wide studies of genetic diversity will be particularly informative for making inferences about African demographic history and the genetic basis of complex traits. A better understanding of the distribution and frequency of structural variation and its role in phenotypic diversity will be of particular interest. The development of high-throughput SNP genotyping methodology (i.e., the Affymetrix 6.0 and Illumina 1M SNP chips), which is rapidly becoming less expensive, will facilitate the possibility of GWAS in a large number of Africans. Indeed, several recent GWAS in Europeans that have identified genetic variants associated with traits such as height, skin pigmentation, and eye color indicate that very large sample sizes (>3000–5000) are required to identify variants associated with complex traits influenced by multiple loci and the environment (191, 222). However, to date no GWAS have been perfomed on a comparable scale in Africa. Such studies will be highly informative for identifying genes that play a role in a number of common traits (e.g., height), as well as for identifying genes that play a role in susceptibility to infectious and other complex diseases. The high levels of genetic substructure in Africa, even within small geographic regions (F.A. Reed & S.A. Tishkoff, unpublished data), require determination of individual ancestry and proper correction for substructure in association studies.

Genotyping ethnically diverse African populations living in distinct climates and with distinct subsistence patterns for these high density SNP panels will also be useful for conducting whole-genome scans of selection to identify genes that have played a role in local adaptation and disease. The continued development of statistical and computational methods for inferring demographic history and natural selection will shed light on human evolutionary history in Africa. Approaches that incorporate detailed geographic information such as natural boundaries (i.e., mountain ranges, rivers, deserts, etc.) will be particularly useful for inferring African demographic history (162, 170).

Given that African populations possess a large fraction of population-specific alleles and may have experienced local adaptation, resequencing across diverse African populations will be important for identifying population-specific functional variants. Targeted resequencing of genes that play a key role in disease susceptibility and drug response will be particularly important for the design of more effective treatments in individuals of recent African descent. Additionally, whole-genome resequencing (a goal of the "1000 genomes" project) will be informative for identifying large-scale structural variants and rare variants that may play an important role in disease and for reconstructing human evolutionary history.

DISCLOSURE STATEMENT

The authors are not aware of any biases that might be perceived as affecting the objectivity of this review.

ACKNOWLEDGMENTS

We thank S.M. Williams, J.M. Akey, J.S. Friedlaender, and N.A. Rosenberg for critical review of sections of the manuscript and/or figures and for helpful suggestions. The authors are funded by the U.S. National Science Foundation (NSF) grant BSC-0552486, U.S. National Institutes of Health (NIH) grant R01GM076637, and a David and Lucile Packard Career Award to S.A.T.

LITERATURE CITED

1. Abecasis GR, Ghosh D, Nichols TE. 2005. Linkage disequilibrium: ancient history drives the new genetics. *Hum. Hered.* 59:118–24
2. Abubakari AR, Lauder W, Agyemang C, Jones M, Kirk A, Bhopal RS. 2008. Prevalence and time trends in obesity among adult West African populations: a meta-analysis. *Obes. Rev.* doi:10.1111/j.1467-789X.2007.00462.x
3. Adeyemo A, Luke A, Cooper R, Wu X, Tayo B, et al. 2003. A genome-wide scan for body mass index among Nigerian families. *Obes. Res.* 11:266–73
4. Akey JM, Eberle MA, Rieder MJ, Carlson CS, Shriver MD, et al. 2004. Population history and natural selection shape patterns of genetic variation in 132 genes. *PLoS Biol.* 2:e286
5. Akey JM, Zhang K, Xiong M, Jin L. 2003. The effect of single nucleotide polymorphism identification strategies on estimates of linkage disequilibrium. *Mol. Biol. Evol.* 20:232–42
6. An P, Bleiber G, Duggal P, Nelson G, May M, et al. 2004. *APOBEC3G* genetic variants and their influence on the progression to AIDS. *J. Virol.* 78:11070–76
7. An P, Duggal P, Wang LH, O'Brien SJ, Donfield S, et al. 2007. Polymorphisms of *CUL5* are associated with $CD4^+$ T cell loss in HIV-1 infected individuals. *PLoS Genet.* 3:e19
8. Argyropoulos G, Brown AM, Willi SM, Zhu J, He Y, et al. 1998. Effects of mutations in the human uncoupling protein 3 gene on the respiratory quotient and fat oxidation in severe obesity and type 2 diabetes. *J. Clin. Invest.* 102:1345–51

9. Argyropoulos G, Rankinen T, Bai F, Rice T, Province MA, et al. 2003. The agouti-related protein and body fatness in humans. *Int. J. Obes. Relat. Metab. Disord.* 27:276–80
10. Arnheim N, Calabrese P, Nordborg M. 2003. Hot and cold spots of recombination in the human genome: the reason we should find them and how this can be achieved. *Am. J. Hum. Genet.* 73:5–16
11. Aspray TJ, Mugusi F, Rashid S, Whiting D, Edwards R, et al. 2000. Rural and urban differences in diabetes prevalence in Tanzania: the role of obesity, physical inactivity and urban living. *Trans. R. Soc. Trop. Med. Hyg.* 94:637–44
12. Ayodo G, Price AL, Keinan A, Ajwang A, Otieno MF, et al. 2007. Combining evidence of natural selection with association analysis increases power to detect malaria-resistance variants. *Am. J. Hum. Genet.* 81:234–42
13. Balaresque PL, Ballereau SJ, Jobling MA. 2007. Challenges in human genetic diversity: demographic history and adaptation. *Hum. Mol. Genet.* 16(Spec. No. 2):R134–39
14. Ball TB, Ji H, Kimani J, McLaren P, Marlin C, et al. 2007. Polymorphisms in *IRF-1* associated with resistance to HIV-1 infection in highly exposed uninfected Kenyan sex workers. *AIDS* 21:1091–101
15. Bansal V, Bashir A, Bafna V. 2007. Evidence for large inversion polymorphisms in the human genome from HapMap data. *Genome Res.* 17:219–30
16. Barreiro LB, Neyrolles O, Babb CL, Tailleux L, Quach H, et al. 2006. Promoter variation in the DC-SIGN-encoding gene *CD209* is associated with tuberculosis. *PLoS Med.* 3:e20
17. Bellamy R. 2003. Susceptibility to mycobacterial infections: the importance of host genetics. *Genes Immun.* 4:4–11
18. Bellamy R, Beyers N, McAdam KP, Ruwende C, Gie R, et al. 2000. Genetic susceptibility to tuberculosis in Africans: a genome-wide scan. *Proc. Natl. Acad. Sci. USA* 97:8005–9
19. Bellamy R, Ruwende C, Corrah T, McAdam KP, Whittle HC, Hill AV. 1998. Variations in the *NRAMP1* gene and susceptibility to tuberculosis in West Africans. *N. Engl. J. Med.* 338:640–44
20. Bersaglieri T, Sabeti PC, Patterson N, Vanderploeg T, Schaffner SF, et al. 2004. Genetic signatures of strong recent positive selection at the lactase gene. *Am. J. Hum. Genet.* 74:1111–20
21. Bonilla C, Panguluri RK, Taliaferro-Smith L, Argyropoulos G, Chen G, et al. 2006. Agouti-related protein promoter variant associated with leanness and decreased risk for diabetes in West Africans. *Int. J. Obes.* 30:715–21
22. Brockmoller J, Tzvetkov MV. 2008. Pharmacogenetics: data, concepts and tools to improve drug discovery and drug treatment. *Eur. J. Clin. Pharmacol.* 64:133–57
23. Carlson CS, Thomas DJ, Eberle MA, Swanson JE, Livingston RJ, et al. 2005. Genomic regions exhibiting positive selection identified from dense genotype data. *Genome Res.* 15:1553–65
24. Cavalli-Sforza LL, Piazza A, Menozzi P. 1994. *History and Geography of Human Genes*. Princeton: Princeton Univ. Press
25. Cervino AC, Lakiss S, Sow O, Bellamy R, Beyers N, et al. 2002. Fine mapping of a putative tuberculosis-susceptibility locus on chromosome 15q11-13 in African families. *Hum. Mol. Genet.* 11:1599–603
26. Chen G, Adeyemo A, Zhou J, Chen Y, Huang H, et al. 2007. Genome-wide search for susceptibility genes to type 2 diabetes in West Africans: potential role of C-peptide. *Diabetes Res. Clin. Pract.* 78:e1-6
27. Chen G, Adeyemo AA, Johnson T, Zhou J, Amoah A, et al. 2005. A genome-wide scan for quantitative trait loci linked to obesity phenotypes among West Africans. *Int. J. Obes.* 29:255–59
28. Chen G, Adeyemo AA, Zhou J, Chen Y, Doumatey A, et al. 2007. A genome-wide search for linkage to renal function phenotypes in West Africans with type 2 diabetes. *Am. J. Kidney Dis.* 49:394–400
29. Chen Y, Kittles R, Zhou J, Chen G, Adeyemo A, et al. 2005. Calpain-10 gene polymorphisms and type 2 diabetes in West Africans: the Africa America Diabetes Mellitus (AADM) Study. *Ann. Epidemiol.* 15:153–59
30. Cho MK, Sabeti PC, Tishkoff SA. 2008. Natural selection in humans: ethical issues and methodological challenges. *Trends Genet.* In review
31. Clark AG, Nielsen R, Signorovitch J, Matise TC, Glanowski S, et al. 2003. Linkage disequilibrium and inference of ancestral recombination in 538 single-nucleotide polymorphism clusters across the human genome. *Am. J. Hum. Genet.* 73:285–300

32. Clark VJ, Ptak SE, Tiemann I, Qian Y, Coop G, et al. 2007. Combining sperm typing and linkage disequilibrium analyses reveals differences in selective pressures or recombination rates across human populations. *Genetics* 175:795–804
33. Coelho M, Luiselli D, Bertorelle G, Lopes AI, Seixas S, et al. 2005. Microsatellite variation and evolution of human lactase persistence. *Hum. Genet.* 117:329–39
34. Collins FS, Green ED, Guttmacher AE, Guyer MS. 2003. A vision for the future of genomics research. *Nature* 422:835–47
35. Conrad DF, Hurles ME. 2007. The population genetics of structural variation. *Nat. Genet.* 39:S30-36
36. Conrad DF, Jakobsson M, Coop G, Wen X, Wall JD, et al. 2006. A worldwide survey of haplotype variation and linkage disequilibrium in the human genome. *Nat. Genet.* 38:1251–60
37. Consort. RMGSAA. 2007. Evolutionary and biomedical insights from the rhesus macaque genome. *Science* 316:222–34
38. Consort. Int. HapMap. 2005. A haplotype map of the human genome. *Nature* 437:1299–320
39. Consort. Int. HapMap. 2007. A second generation human haplotype map of over 3.1 million SNPs. *Nature* 449:851–62
40. Coop G, Przeworski M. 2007. An evolutionary view of human recombination. *Nat. Rev. Genet.* 8:23–34
41. Coop G, Wen X, Ober C, Pritchard JK, Przeworski M. 2008. High-resolution mapping of crossovers reveals extensive variation in fine-scale recombination patterns among humans. *Science* 319:1395–98
42. Cooper RS, Guo X, Rotimi CN, Luke A, Ward R, et al. 2000. Heritability of angiotensin-converting enzyme and angiotensinogen: A comparison of US blacks and Nigerians. *Hypertension* 35:1141–47
43. Corbett EL, Marston B, Churchyard GJ, De Cock KM. 2006. Tuberculosis in sub-Saharan Africa: opportunities, challenges, and change in the era of antiretroviral treatment. *Lancet* 367:926–37
44. Cox MP, Mendez FL, Karafet TM, Pilkington MM, Kingan SB, et al. 2008. Testing for archaic hominin admixture on the X chromosome: model likelihoods for the modern human *RRM2P4* region from summaries of genealogical topology under the structured coalescent. *Genetics* 178:427–37
45. Crawford DC, Bhangale T, Li N, Hellenthal G, Rieder MJ, et al. 2004. Evidence for substantial fine-scale variation in recombination rates across the human genome. *Nat. Genet.* 36:700–6
46. de Luis DA, Aller R, Izaola O, González Sagrado M, Conde R, Pérez Castrillón JL. 2006. Lack of association of -55CT polymorphism of *UCP3* gene with fat distribution in obese patients. *Ann. Nutr. Metab.* 51:374–78
47. Deo RC, Patterson N, Tandon A, McDonald GJ, Haiman CA, et al. 2007. A high-density admixture scan in 1,670 African Americans with hypertension. *PLoS Genet.* 3:e196
48. Di Rienzo A. 2006. Population genetics models of common diseases. *Curr. Opin. Genet. Dev.* 16:630–36
49. Di Rienzo A, Hudson RR. 2005. An evolutionary framework for common diseases: the ancestral-susceptibility model. *Trends Genet.* 21:596–601
50. Dolo A, Modiano D, Maiga B, Daou M, Dolo G, et al. 2005. Difference in susceptibility to malaria between two sympatric ethnic groups in Mali. *Am. J. Trop. Med. Hyg.* 72:243–48
51. Dong Y, Zhu H, Sagnella GA, Carter ND, Cook DG, Cappuccio FP. 1999. Association between the C825T polymorphism of the G protein β3-subunit gene and hypertension in blacks. *Hypertension* 34:1193–6
52. Drayna D. 2005. Human taste genetics. *Annu. Rev. Genomics Hum. Genet.* 6:217–35
53. Drayna D, Coon H, Kim UK, Elsner T, Cromer K, et al. 2003. Genetic analysis of a complex trait in the Utah Genetic Reference Project: a major locus for PTC taste ability on chromosome 7q and a secondary locus on chromosome 16p. *Hum. Genet.* 112:567–72
54. Edwards R, Unwin N, Mugusi F, Whiting D, Rashid S, et al. 2000. Hypertension prevalence and care in an urban and rural area of Tanzania. *J. Hypertens.* 18:145–52
55. Enattah NS, Jensen TG, Nielsen M, Lewinski R, Kuokkanen M, et al. 2008. Independent introduction of two lactase-persistence alleles into human populations reflects different history of adaptation to milk culture. *Am. J. Hum. Genet.* 82:57–72
56. Enattah NS, Sahi T, Savilahti E, Terwilliger JD, Peltonen L, Jarvela I. 2002. Identification of a variant associated with adult-type hypolactasia. *Nat. Genet.* 30:233–37

57. Enevold A, Alifrangis M, Sanchez JJ, Carneiro I, Roper C, et al. 2007. Associations between α+-thalassemia and *Plasmodium falciparum* malarial infection in northeastern Tanzania. *J. Infect. Dis.* 196:451–59
58. Eswaran V, Harpending H, Rogers AR. 2005. Genomics refutes an exclusively African origin of humans. *J. Hum. Evol.* 49:1–18
59. Evans DM, Cardon LR. 2005. A comparison of linkage disequilibrium patterns and estimated population recombination rates across multiple populations. *Am. J. Hum. Genet.* 76:681–87
60. Fejerman L, Wu X, Adeyemo A, Luke A, Zhu X, et al. 2006. The effect of genetic variation in angiotensinogen on serum levels and blood pressure: a comparison of Nigerians and US blacks. *J. Hum. Hypertens.* 20:882–87
61. Forster P, Matsumura S. 2005. Evolution. Did early humans go north or south? *Science* 308:965–66
62. Forster P, Romano V. 2007. Timing of a back-migration into Africa. *Science* 316:50–53
63. Friedlaender JS, Friedlaender FR, Hodgson JA, Stoltz M, Koki G, et al. 2007. Melanesian mtDNA complexity. *PLoS ONE* 2:e248
64. Friedlaender JS, Friedlaender FR, Reed FA, Kidd KK, Kidd JR, et al. 2008. The genetic structure of Pacific islanders. *PLoS Genetics* 4:e19
65. Frodsham AJ, Hill AV. 2004. Genetics of infectious diseases. *Hum. Mol. Genet.* 13 Spec. No. 2:R187–94
66. Fullerton SM, Bartoszewicz A, Ybazeta G, Horikawa Y, Bell GI, et al. 2002. Geographic and haplotype structure of candidate type 2 diabetes susceptibility variants at the calpain-10 locus. *Am. J. Hum. Genet.* 70:1096–106
67. Gabriel SB, Schaffner SF, Nguyen H, Moore JM, Roy J, et al. 2002. The structure of haplotype blocks in the human genome. *Science* 296:2225–29
68. Garant MJ, Kao WH, Brancati F, Coresh J, Rami TM, et al. 2002. SNP43 of *CAPN10* and the risk of type 2 diabetes in African-Americans: the Atherosclerosis Risk in Communities Study. *Diabetes* 51:231–37
69. Garrigan D, Hammer MF. 2006. Reconstructing human origins in the genomic era. *Nat. Rev. Genet.* 7:669–80
70. Garrigan D, Kingan SB, Pilkington MM, Wilder JA, Cox MP, et al. 2007. Inferring human population sizes, divergence times and rates of gene flow from mitochondrial, X and Y chromosome resequencing data. *Genetics* 177:2195–207
71. Garrigan D, Mobasher Z, Severson T, Wilder JA, Hammer MF. 2004. Evidence for archaic Asian ancestry on the human X chromosome. *Mol. Biol. Evol.* 22:189–92
72. Gonder MK, Mortensen HM, Reed FA, de Sousa A, Tishkoff SA. 2007. Whole-mtDNA genome sequence analysis of ancient African lineages. *Mol. Biol. Evol.* 24:757–68
73. Graffelman J, Balding DJ, Gonzalez-Neira A, Bertranpetit J. 2007. Variation in estimated recombination rates across human populations. *Hum. Genet.* 122:301–10
74. Green RE, Krause J, Ptak SE, Briggs AW, Ronan MT, et al. 2006. Analysis of one million base pairs of Neanderthal DNA. *Nature* 444:330–36
75. Guthery SL, Salisbury BA, Pungliya MS, Stephens JC, Bamshad M. 2007. The structure of common genetic variation in United States populations. *Am. J. Hum. Genet.* 81:1221–31
76. Hamblin MT, Di Rienzo A. 2000. Detection of the signature of natural selection in humans: evidence from the Duffy blood group locus. *Am. J. Hum. Genet.* 66:1669–79
77. Hamblin MT, Thompson EE, Di Rienzo A. 2002. Complex signatures of natural selection at the Duffy blood group locus. *Am. J. Hum. Genet.* 70:369–83
78. Hanchard N, Elzein A, Trafford C, Rockett K, Pinder M, et al. 2007. Classical sickle beta-globin haplotypes exhibit a high degree of long-range haplotype similarity in African and Afro-Caribbean populations. *BMC Genet.* 8:52
79. Harding RM, McVean G. 2004. A structured ancestral population for the evolution of modern humans. *Curr. Opin. Genet. Dev.* 14:667–74
80. Harris EE, Meyer D. 2006. The molecular signature of selection underlying human adaptations. *Am. J. Phys. Anthropol. Suppl.* 43:89–130
81. Hawks J, Wang ET, Cochran GM, Harpending HC, Moyzis RK. 2007. Recent acceleration of human adaptive evolution. *Proc. Natl. Acad. Sci. USA* 104:20753–58

82. Helgason A, Palsson S, Thorleifsson G, Grant SF, Emilsson V, et al. 2007. Refining the impact of *TCF7L2* gene variants on type 2 diabetes and adaptive evolution. *Nat. Genet.* 39:218–25
83. Hill AV. 2006. Aspects of genetic susceptibility to human infectious diseases. *Annu. Rev. Genet.* 40:469–86
84. Hirschhorn JN, Daly MJ. 2005. Genome-wide association studies for common diseases and complex traits. *Nat. Rev. Genet.* 6:95–108
85. Hollox EJ, Poulter M, Zvarik M, Ferak V, Krause A, et al. 2001. Lactase haplotype diversity in the Old World. *Am. J. Hum. Genet.* 68:160–72
86. Horikawa Y, Oda N, Cox NJ, Li X, Orho-Melander M, et al. 2000. Genetic variation in the gene encoding calpain-10 is associated with type 2 diabetes mellitus. *Nat. Genet.* 26:163–75
87. Ingram CJ, Elamin MF, Mulcare CA, Weale ME, Tarekegn A, et al. 2007. A novel polymorphism associated with lactose tolerance in Africa: multiple causes for lactase persistence? *Hum. Genet.* 120:779–88
88. Jackson F. 1990. Two evolutionary models for the interactions of dietary organic cyanogens, hemoglobins, and falciparum malaria. *Am. J. Hum. Biol.* 2:521–32
89. Jackson M, Walker S, Cruickshank JK, Sharma S, Cade J, et al. 2007. Diet and overweight and obesity in populations of African origin: Cameroon, Jamaica and the UK. *Public Health Nutr.* 10:122–30
90. Jakobsson M, Scholz SW, Scheet P, Gibbs JR, VanLiere JM, et al. 2008. Genotype, haplotype and copy-number variation in worldwide human populations. *Nature* 451:998–1003
91. Jeffreys AJ, Kauppi L, Neumann R. 2001. Intensely punctate meiotic recombination in the class II region of the major histocompatibility complex. *Nat. Genet.* 29:217–22
92. Jennes W, Verheyden S, Demanet C, Adje-Toure CA, Vuylsteke B, et al. 2006. Cutting edge: resistance to HIV-1 infection among African female sex workers is associated with inhibitory KIR in the absence of their HLA ligands. *J. Immunol.* 177:6588–92
93. Joy DA, Feng X, Mu J, Furuya T, Chotivanich K, et al. 2003. Early origin and recent expansion of *Plasmodium falciparum*. *Science* 300:318–21
94. Kayser M, Brauer S, Stoneking M. 2003. A genome scan to detect candidate regions influenced by local natural selection in human populations. *Mol. Biol. Evol.* 20:893–900
95. Kelley JL, Madeoy J, Calhoun JC, Swanson W, Akey JM. 2006. Genomic signatures of positive selection in humans and the limits of outlier approaches. *Genome Res.* 16:980–89
96. Kim U, Wooding S, Ricci D, Jorde LB, Drayna D. 2005. Worldwide haplotype diversity and coding sequence variation at human bitter taste receptor loci. *Hum. Mutat.* 26:199–204
97. Kimm SY, Glynn NW, Aston CE, Damcott CM, Poehlman ET, et al. 2002. Racial differences in the relation between uncoupling protein genes and resting energy expenditure. *Am. J. Clin. Nutr.* 75:714–19
98. Kivisild T, Reidla M, Metspalu E, Rosa A, Brehm A, et al. 2004. Ethiopian mitochondrial DNA heritage: tracking gene flow across and around the gate of tears. *Am. J. Hum. Genet.* 75:752–70
99. Kong A, Gudbjartsson DF, Sainz J, Jonsdottir GM, Gudjonsson SA, et al. 2002. A high-resolution recombination map of the human genome. *Nat. Genet.* 31:241–47
100. Kong A, Thorleifsson G, Stefansson H, Masson G, Helgason A, et al. 2008. Sequence variants in the *RNF212* gene associate with genomewide recombination rate. *Science* 319:1398–1401
101. Korbel JO, Urban AE, Affourtit JP, Godwin B, Grubert F, et al. 2007. Paired-end mapping reveals extensive structural variation in the human genome. *Science* 318:420–26
102. Kramer H, Wu X, Kan D, Luke A, Zhu X, et al. 2005. Angiotensin-converting enzyme gene polymorphisms and obesity: an examination of three black populations. *Obes. Res.* 13:823–28
103. Kuper R, Kropelin S. 2006. Climate-controlled Holocene occupation in the Sahara: motor of Africa's evolution. *Science* 313:803–7
104. Kwiatkowski DP. 2005. How malaria has affected the human genome and what human genetics can teach us about malaria. *Am. J. Hum. Genet.* 77:171–92
105. Lama J, Planelles V. 2007. Host factors influencing susceptibility to HIV infection and AIDS progression. *Retrovirology* 4:52
106. Leak TS, Keene KL, Langefeld CD, Gallagher CJ, Mychaleckyj JC, et al. 2007. Association of the proprotein convertase subtilisin/kexin-type 2 (*PCSK2*) gene with type 2 diabetes in an African American population. *Mol. Genet. Metab.* 92:145–50

107. Lee JA, Lupski JR. 2006. Genomic rearrangements and gene copy-number alterations as a cause of nervous system disorders. *Neuron* 52:103–21
108. Lewinsky RH, Jensen TG, Moller J, Stensballe A, Olsen J, Troelsen JT. 2005. T-13910 DNA variant associated with lactase persistence interacts with Oct-1 and stimulates lactase promoter activity in vitro. *Hum. Mol. Genet.* 14:3945–53
109. Li JZ, Absher DM, Tang H, Southwick AM, Casto AM, et al. 2008. Worldwide human relationships inferred from genome-wide patterns of variation. *Science* 319:1100–4
110. Liu H, Prugnolle F, Manica A, Balloux F. 2006. A geographically explicit genetic model of worldwide human-settlement history. *Am. J. Hum. Genet.* 79:230–37
111. Lohmueller KE, Indap AR, Schmidt S, Boyko AR, Hernandez RD, et al. 2008. Proportionally more deleterious genetic variation in European than in African populations. *Nature* 451:994–97
112. Lombard Z, Brune AE, Hoal EG, Babb C, Van Helden PD, et al. 2006. HLA class II disease associations in southern Africa. *Tissue Antigens* 67:97–110
113. Lonjou C, Zhang W, Collins A, Tapper WJ, Elahi E, et al. 2003. Linkage disequilibrium in human populations. *Proc. Natl. Acad. Sci. USA* 100:6069–74
114. Lovell A, Moreau C, Yotova V, Xiao F, Bourgeois S, et al. 2005. Ethiopia: between sub-Saharan Africa and western Eurasia. *Ann. Hum. Genet.* 69:275–87
115. Ma L, Marmor M, Zhong P, Ewane L, Su B, Nyambi P. 2005. Distribution of *CCR2-64I* and *SDF1-3′A* alleles and HIV status in 7 ethnic populations of Cameroon. *J. Acquired Immune Defic. Syndr.* 40:89–95
116. Maartens G, Wilkinson RJ. 2007. Tuberculosis. *Lancet* 370:2030–43
117. Macaulay V, Hill C, Achilli A, Rengo C, Clarke D, et al. 2005. Single, rapid coastal settlement of Asia revealed by analysis of complete mitochondrial genomes. *Science* 308:1034–36
118. MacDonald KS, Fowke KR, Kimani J, Dunand VA, Nagelkerke NJ, et al. 2000. Influence of HLA supertypes on susceptibility and resistance to human immunodeficiency virus type 1 infection. *J. Infect. Dis.* 181:1581–89
119. Marchini J, Cardon LR, Phillips MS, Donnelly P. 2004. The effects of human population structure on large genetic association studies. *Nat. Genet.* 36:512–17
120. Markovic D, Tang X, Guruju M, Levenstien MA, Hoh J, et al. 2005. Association of angiotensinogen gene polymorphisms with essential hypertension in African-Americans and Caucasians. *Hum. Hered.* 60:89–96
121. Marshall CR, Noor A, Vincent JB, Lionel AC, Feuk L, et al. 2008. Structural variation of chromosomes in autism spectrum disorder. *Am. J. Hum. Genet.* 82:477–88
122. Marth GT, Czabarka E, Murvai J, Sherry ST. 2004. The allele frequency spectrum in genome-wide human variation data reveals signals of differential demographic history in three large world populations. *Genetics* 166:351–72
123. Martin MP, Gao X, Lee JH, Nelson GW, Detels R, et al. 2002. Epistatic interaction between *KIR3DS1* and *HLA-B* delays the progression to AIDS. *Nat. Genet.* 31:429–34
124. McBrearty S, Brooks A. 2000. The revolution that wasn't: a new interpretation of the origin of modern human behavior. *J. Hum. Evol.* 39:453–563
125. McDougall IBF, Fleagle JG. 2005. Stratigraphic placement and age of modern humans from Kibish, Ethiopia. *Nature* 433:733–36
126. McVean G, Spencer CC, Chaix R. 2005. Perspectives on human genetic variation from the HapMap Project. *PLoS Genet.* 1:e54
127. Mehlotra RK, Bockarie MJ, Zimmerman PA. 2007. *CYP2B6* 983T>C polymorphism is prevalent in West Africa but absent in Papua New Guinea: implications for HIV/AIDS treatment. *Br. J. Clin. Pharmacol.* 64:391–95
128. Mehlotra RK, Ziats MN, Bockarie MJ, Zimmerman PA. 2006. Prevalence of *CYP2B6* alleles in malaria-endemic populations of West Africa and Papua New Guinea. *Eur. J. Clin. Pharmacol.* 62:267–75
129. Mendez MA, Wynter S, Wilks R, Forrester T. 2003. Under- and overreporting of energy is related to obesity, lifestyle factors and food group intakes in Jamaican adults. *Public Health Nutr.* 7:9–19
130. Modiano D, Luoni G, Sirima BS, Lanfrancotti A, Petrarca V, et al. 2001. The lower susceptibility to *Plasmodium falciparum* malaria of Fulani of Burkina Faso (west Africa) is associated with low frequencies of classic malaria-resistance genes. *Trans. R. Soc. Trop. Med. Hyg.* 95:149–52

131. Mottagui-Tabar S, Hoffstedt J, Brookes AJ, Jiao H, Arner P, Dahlman I. 2008. Association of *ADRB1* and *UCP3* gene polymorphisms with insulin sensitivity but not obesity. *Horm. Res.* 69:31–36
132. Mulcare CA, Weale ME, Jones AL, Connell B, Zeitlyn D, et al. 2004. The T allele of a single-nucleotide polymorphism 13.9 kb upstream of the lactase gene (*LCT*) (*C-13.9kbT*) does not predict or cause the lactase-persistence phenotype in Africans. *Am. J. Hum. Genet.* 74:1102–10
133. Myles S, Bouzekri N, Haverfield E, Cherkaoui M, Dugoujon JM, Ward R. 2005. Genetic evidence in support of a shared Eurasian-North African dairying origin. *Hum. Genet.* 117:34–42
134. Myles S, Tang K, Somel M, Green RE, Kelso J, Stoneking M. 2008. Identification and analysis of genomic regions with large between-population differentiation in humans. *Ann. Hum. Genet.* 72:99–110
135. Nagel RL, Raventos C, Tanowitz HB, Wittner M. 1980. Effect of sodium cyanate on *Plasmodium falciparum* in vitro. *J. Parasitol.* 66:483–87
136. Nakajima T, Wooding S, Sakagami T, Emi M, Tokunaga K, et al. 2004. Natural selection and population history in the human angiotensinogen gene (*AGT*): 736 complete *AGT* sequences in chromosomes from around the world. *Am. J. Hum. Genet.* 74:898–916
137. Neel JV. 1962. Diabetes mellitus: a "thrifty" genotype rendered detrimental by "progress"? *Am. J. Hum. Genet.* 14:353–62
138. Nielsen R. 2000. Estimation of population parameters and recombination rates from single nucleotide polymorphisms. *Genetics* 154:931–42
139. Nielsen R, Hellmann I, Hubisz M, Bustamante C, Clark AG. 2007. Recent and ongoing selection in the human genome. *Nat. Rev. Genet.* 8:857–68
140. Nielsen R, Hubisz MJ, Clark AG. 2004. Reconstituting the frequency spectrum of ascertained single-nucleotide polymorphism data. *Genetics* 168:2373–82
141. Nielsen R, Signorovitch J. 2003. Correcting for ascertainment biases when analyzing SNP data: applications to the estimation of linkage disequilibrium. *Theor. Popul. Biol.* 63:245–55
142. Noonan JP, Coop G, Kudaravalli S, Smith D, Krause J, et al. 2006. Sequencing and analysis of Neanderthal genomic DNA. *Science* 314:1113–18
143. Nyakutira C, Röshammar D, Chigutsa E, Chonzi P, Ashton M, et al. 2007. High prevalence of the *CYP2B6* 516G–>T(*6) variant and effect on the population pharmacokinetics of efavirenz in HIV/AIDS outpatients in Zimbabwe. *Eur. J. Clin. Pharmacol.* 64:357–65
144. O'Brien SJ, Nelson GW. 2004. Human genes that limit AIDS. *Nat. Genet.* 36:565–74
145. Ohashi J, Naka I, Patarapotikul J, Hananantachai H, Brittenham G, et al. 2004. Extended linkage disequilibrium surrounding the hemoglobin E variant due to malarial selection. *Am. J. Hum. Genet.* 74:1198–208
146. Olds LC, Sibley E. 2003. Lactase persistence DNA variant enhances lactase promoter activity in vitro: functional role as a cis regulatory element. *Hum. Mol. Genet.* 12:2333–40
147. Olesen R, Wejse C, Velez DR, Bisseye C, Sodemann M, et al. 2007. DC-SIGN (*CD209*), pentraxin 3 and vitamin D receptor gene variants associate with pulmonary tuberculosis risk in West Africans. *Genes Immun.* 8:456–67
148. Olivieri A, Achilli A, Pala M, Battaglia V, Fornarino S, et al. 2006. The mtDNA legacy of the Levantine early Upper Palaeolithic in Africa. *Science* 314:1767–70
149. Osei K. 1999. Metabolic consequences of the West African diaspora: lessons from the thrifty gene. *J. Lab. Clin. Med.* 133:98–111
150. Otabe S, Clement K, Dina C, Pelloux V, Guy-Grand B, et al. 2000. A genetic variation in the 5′ flanking region of the *UCP3* gene is associated with body mass index in humans in interaction with physical activity. *Diabetologia* 43:245–49
151. Paganotti GM, Babiker HA, Modiano D, Sirima BS, Verra F, et al. 2004. Genetic complexity of *Plasmodium falciparum* in two ethnic groups of Burkina Faso with marked differences in susceptibility to malaria. *Am. J. Trop. Med. Hyg.* 71:173–78
152. Paradies YC, Montoya MJ, Fullerton SM. 2007. Racialized genetics and the study of complex diseases: the thrifty genotype revisited. *Perspect. Biol. Med.* 50:203–27
153. Patin E, Barreiro LB, Sabeti PC, Austerlitz F, Luca F, et al. 2006. Deciphering the ancient and complex evolutionary history of human arylamine N-acetyltransferase genes. *Am. J. Hum. Genet.* 78:423–36
154. Patin E, Harmant C, Kidd KK, Kidd J, Froment A, et al. 2006. Sub-Saharan African coding sequence variation and haplotype diversity at the *NAT2* gene. *Hum. Mutat.* 27:720

155. Perry GH, Dominy NJ, Claw KG, Lee AS, Fiegler H, et al. 2007. Diet and the evolution of human amylase gene copy number variation. *Nat. Genet.* 39:1256–60
156. Pilkington MM, Wilder JA, Mendez FL, Cox MP, Woerner A, et al. 2008. Contrasting signatures of population growth for mitochondrial DNA and Y chromosomes among human populations in Africa. *Mol. Biol. Evol.* 25:517–25
157. Plagnol V, Wall JD. 2006. Possible ancestral structure in human populations. *PLoS Genet.* 2:e105
158. Poulter M, Hollox E, Harvey CB, Mulcare C, Peuhkuri K, et al. 2003. The causal element for the lactase persistence/non-persistence polymorphism is located in a 1 Mb region of linkage disequilibrium in Europeans. *Ann. Hum. Genet.* 67:298–311
159. Pritchard JK. 2001. Are rare variants responsible for susceptibility to complex diseases? *Am. J. Hum. Genet.* 69:124–37
160. Pritchard JK, Przeworski M. 2001. Linkage disequilibrium in humans: models and data. *Am. J. Hum. Genet.* 69:1–14
161. Pritchard JK, Stephens M, Donnelly P. 2000. Inference of population structure using multilocus genotype data. *Genetics* 155:945–59
162. Prugnolle F, Manica A, Balloux F. 2005. Geography predicts neutral genetic diversity of human populations. *Curr. Biol.* 15:R159–60
163. Ptak SE, Hinds DA, Koehler K, Nickel B, Patil N, et al. 2005. Fine-scale recombination patterns differ between chimpanzees and humans. *Nat. Genet.* 37:429–34
164. Purcell S, Neale B, Todd-Brown K, Thomas L, Ferreira MA, et al. 2007. PLINK: a tool set for whole-genome association and population-based linkage analyses. *Am. J. Hum. Genet.* 81:559–75
165. Quintana-Murci L, Quach H, Harmant C, Luca F, Massonnet B, et al. 2008. Maternal traces of deep common ancestry and asymmetric gene flow between Pygmy hunter-gatherers and Bantu-speaking farmers. *Proc. Natl. Acad. Sci. USA* 105:1596–601
166. Quintana-Murci L, Semino O, Bandelt HJ, Passarino G, McElreavey K, Santachiara-Benerecetti AS. 1999. Genetic evidence of an early exit of *Homo sapiens sapiens* from Africa through eastern Africa. *Nat. Genet.* 23:437–41
167. Ramachandran S, Deshpande O, Roseman CC, Rosenberg NA, Feldman MW, Cavalli-Sforza LL. 2005. Support from the relationship of genetic and geographic distance in human populations for a serial founder effect originating in Africa. *Proc. Natl. Acad. Sci. USA* 102:15942–47
168. Ramaley PA, French N, Kaleebu P, Gilks C, Whitworth J, Hill AV. 2002. HIV in Africa (Communication arising): chemokine-receptor genes and AIDS risk. *Nature* 417:140
169. Redon R, Ishikawa S, Fitch KR, Feuk L, Perry GH, et al. 2006. Global variation in copy number in the human genome. *Nature* 444:444–54
170. Reed FA, Tishkoff SA. 2006. African human diversity, origins and migrations. *Curr. Opin. Genet. Dev.* 16:597–605
171. Reich DE, Cargill M, Bolk S, Ireland J, Sabeti PC, et al. 2001. Linkage disequilibrium in the human genome. *Nature* 411:199–204
172. Rosenberg NA, Mahajan S, Ramachandran S, Zhao C, Pritchard JK, Feldman MW. 2005. Clines, clusters, and the effect of study design on the inference of human population structure. *PLoS Genet.* 1:e70
173. Rosenberg NA, Pritchard JK, Weber JL, Cann HM, Kidd KK, et al. 2002. Genetic structure of human populations. *Science* 298:2381–85
174. Rotimi CN. 2004. Inauguration of the African Society of Human Genetics. *Nat. Genet.* 36:544
175. Rotimi CN, Chen G, Adeyemo AA, Furbert-Harris P, Parish-Gause D, et al. 2004. A genome-wide search for type 2 diabetes susceptibility genes in West Africans: the Africa America Diabetes Mellitus (AADM) Study. *Diabetes* 53:838–41
176. Rotimi CN, Chen G, Adeyemo AA, Jones LS, Agyenim-Boateng K, et al. 2006. Genomewide scan and fine mapping of quantitative trait loci for intraocular pressure on 5q and 14q in West Africans. *Invest. Ophthalmol. Vis. Sci.* 47:3262–67
177. Sabeti PC, Reich DE, Higgins JM, Levine HZ, Richter DJ, et al. 2002. Detecting recent positive selection in the human genome from haplotype structure. *Nature* 419:832–37

178. Sabeti PC, Varilly P, Fry B, Lohmueller J, Hostetter E, et al. 2007. Genome-wide detection and characterization of positive selection in human populations. *Nature* 449:913–18
179. Saunders MA, Hammer MF, Nachman MW. 2002. Nucleotide variability at *G6pd* and the signature of malarial selection in humans. *Genetics* 162:1849–61
180. Saunders MA, Slatkin M, Garner C, Hammer MF, Nachman MW. 2005. The extent of linkage disequilibrium caused by selection on *G6PD* in humans. *Genetics* 171:1219–29
181. Sawyer SL, Mukherjee N, Pakstis AJ, Feuk L, Kidd JR, et al. 2005. Linkage disequilibrium patterns vary substantially among populations. *Eur. J. Hum. Genet.* 13:677–86
182. Serre D, Langaney A, Chech M, Teschler-Nicola M, Paunovic M, et al. 2004. No evidence of Neandertal mtDNA contribution to early modern humans. *PLoS Biol.* 2:E57
183. Shimada MK, Panchapakesan K, Tishkoff SA, Nato AQ Jr, Hey J. 2007. Divergent haplotypes and human history as revealed in a worldwide survey of X-linked DNA sequence variation. *Mol. Biol. Evol.* 24:687–98
184. Siffert W, Forster P, Jockel KH, Mvere DA, Brinkmann B, et al. 1999. Worldwide ethnic distribution of the G protein β3 subunit 825T allele and its association with obesity in Caucasian, Chinese, and black African individuals. *J. Am. Soc. Nephrol.* 10:1921–30
185. Sirugo G, Hennig BJ, Adeyemo A, Matimba A, Newport MJ, et al. 2008. Genetic studies of African populations: an overview on disease susceptibility and response to vaccines and therapeutics. *Hum. Genet* 123:557–98
186. Soborg C, Andersen AB, Range N, Malenganisho W, Friis H, et al. 2007. Influence of candidate susceptibility genes on tuberculosis in a high endemic region. *Mol. Immunol.* 44:2213–20
187. Soranzo N, Bufe B, Sabeti PC, Wilson JF, Weale ME, et al. 2005. Positive selection on a high-sensitivity allele of the human bitter-taste receptor *TAS2R16*. *Curr. Biol.* 15:1257–65
188. Stefansson H, Helgason A, Thorleifsson G, Steinthorsdottir V, Masson G, et al. 2005. A common inversion under selection in Europeans. *Nat. Genet.* 37:129–37
189. Storz JF, Payseur BA, Nachman MW. 2004. Genome scans of DNA variability in humans reveal evidence for selective sweeps outside of Africa. *Mol. Biol. Evol.* 21:1800–11
190. Stranger BE, Forrest MS, Dunning M, Ingle CE, Beazley C, et al. 2007. Relative impact of nucleotide and copy number variation on gene expression phenotypes. *Science* 315:848–53
191. Sulem P, Gudbjartsson DF, Stacey SN, Helgason A, Rafnar T, et al. 2007. Genetic determinants of hair, eye and skin pigmentation in Europeans. *Nat. Genet.* 39:1443–52
192. Tarazona-Santos E, Tishkoff SA. 2005. Divergent patterns of linkage disequilibrium and haplotype structure across global populations at the interleukin-13 (*IL13*) locus. *Genes Immun.* 6:53–65
193. Tenesa A, Navarro P, Hayes BJ, Duffy DL, Clarke GM, et al. 2007. Recent human effective population size estimated from linkage disequilibrium. *Genome Res.* 17:520–26
194. Thangaraj K, Chaubey G, Kivisild T, Reddy AG, Singh VK, et al. 2005. Reconstructing the origin of Andaman Islanders. *Science* 308:996
195. Thompson EE, Kuttab-Boulos H, Witonsky D, Yang L, Roe BA, Di Rienzo A. 2004. *CYP3A* variation and the evolution of salt-sensitivity variants. *Am. J. Hum. Genet.* 75:1059–69
196. Timmann C, Evans JA, Konig IR, Kleensang A, Ruschendorf F, et al. 2007. Genome-wide linkage analysis of malaria infection intensity and mild disease. *PLoS Genet.* 3:e48
197. Tishkoff SA, Dietzsch E, Speed W, Pakstis AJ, Kidd JR, et al. 1996. Global patterns of linkage disequilibrium at the *CD4* locus and modern human origins. *Science* 271:1380–87
198. Tishkoff SA, Gonder MK, Henn BM, Mortensen H, Knight A, et al. 2007. History of click-speaking populations of Africa inferred from mtDNA and Y chromosome genetic variation. *Mol. Biol. Evol.* 24:2180–95
199. Tishkoff SA, Kidd KK. 2004. Implications of biogeography of human populations for 'race' and medicine. *Nat. Genet.* 36:S21–27
200. Tishkoff SA, Reed FA, Ranciaro A, Voight BF, Babbitt CC, et al. 2007. Convergent adaptation of human lactase persistence in Africa and Europe. *Nat. Genet.* 39:31–40
201. Tishkoff SA, Varkonyi R, Cahinhinan N, Abbes S, Argyropoulos G, et al. 2001. Haplotype diversity and linkage disequilibrium at human *G6PD*: recent origin of alleles that confer malarial resistance. *Science* 293:455–62

202. Tishkoff SA, Verrelli BC. 2003. Patterns of human genetic diversity: implications for human evolutionary history and disease. *Annu. Rev. Genomics Hum. Genet.* 4:293–340
203. Tishkoff SA, Verrelli BC. 2003. Role of evolutionary history on haplotype block structure in the human genome: implications for disease mapping. *Curr. Opin. Genet. Dev.* 13:569–75
204. Tishkoff SA, Williams SM. 2002. Genetic analysis of African populations: human evolution and complex disease. *Nat. Rev. Genet.* 3:611–21
205. Torcia MG, Santarlasci V, Cosmi L, Clemente A, Maggi L, et al. 2008. Functional deficit of T regulatory cells in Fulani, an ethnic group with low susceptibility to *Plasmodium falciparum* malaria. *Proc. Natl. Acad. Sci. USA* 105:646–51
206. Tuzun E, Sharp AJ, Bailey JA, Kaul R, Morrison VA, et al. 2005. Fine-scale structural variation of the human genome. *Nat. Genet.* 37:727–32
207. Unwin N, Setel P, Rashid S, Mugusi F, Mbanya JC, et al. 2001. Noncommunicable diseases in sub-Saharan Africa: where do they feature in the health research agenda? *Bull. WHO* 79:947–53
208. Vafa M, Maiga B, Berzins K, Hayano M, Bereczky S, et al. 2007. Associations between the *IL-4* -590 T allele and *Plasmodium falciparum* infection prevalence in asymptomatic Fulani of Mali. *Microbes Infect.* 9:1043–48
209. Vander Molen J, Frisse LM, Fullerton SM, Qian Y, Del Bosque-Plata L, et al. 2005. Population genetics of *CAPN10* and *GPR35*: implications for the evolution of type 2 diabetes variants. *Am. J. Hum. Genet.* 76:548–60
210. Vannberg FO, Chapman SJ, Khor CC, Tosh K, Floyd S, et al. 2008. *CD209* genetic polymorphism and tuberculosis disease. *PLoS ONE* 3:e1388
211. Velez DR, Guruju M, Vinukonda G, Prater A, Kumar A, Williams SM. 2006. Angiotensinogen promoter sequence variants in essential hypertension. *Am. J. Hypertens.* 19:1278–85
212. Verra F, Simpore J, Warimwe GM, Tetteh KK, Howard T, et al. 2007. Haemoglobin C and S role in acquired immunity against *Plasmodium falciparum* malaria. *PLoS ONE* 2:e978
213. Verrelli BC, McDonald JH, Argyropoulos G, Destro-Bisol G, Froment A, et al. 2002. Evidence for balancing selection from nucleotide sequence analyses of human *G6PD*. *Am. J. Hum. Genet.* 71:1112–28
214. Verrelli BC, Tishkoff SA, Stone AC, Touchman JW. 2006. Contrasting histories of G6PD molecular evolution and malarial resistance in humans and chimpanzees. *Mol. Biol. Evol.* 23:1592–601
215. Voight BF, Kudaravalli S, Wen X, Pritchard JK. 2006. A map of recent positive selection in the human genome. *PLoS Biol.* 4:e72
216. Volkman SK, Barry AE, Lyons EJ, Nielsen KM, Thomas SM, et al. 2001. Recent origin of *Plasmodium falciparum* from a single progenitor. *Science* 293:482–84
217. Wakeley J, Nielsen R, Liu-Cordero SN, Ardlie K. 2001. The discovery of single-nucleotide polymorphisms–and inferences about human demographic history. *Am. J. Hum. Genet.* 69:1332–47
218. Wall JD, Frisse LA, Hudson RR, Di Rienzo A. 2003. Comparative linkage-disequilibrium analysis of the β-globin hotspot in primates. *Am. J. Hum. Genet.* 73:1330–40
219. Wall JD, Hammer MF. 2006. Archaic admixture in the human genome. *Curr. Opin. Genet. Dev.* 16:606–10
220. Wang ET, Kodama G, Baldi P, Moyzis RK. 2006. Global landscape of recent inferred Darwinian selection for *Homo sapiens*. *Proc. Natl. Acad. Sci. USA* 103:135–40
221. Watson E, Forster P, Richards M, Bandelt HJ. 1997. Mitochondrial footprints of human expansions in Africa. *Am. J. Hum. Genet.* 61:691–704
222. Weedon MN, Lettre G, Freathy RM, Lindgren CM, Voight BF, et al. 2007. A common variant of *HMGA2* is associated with adult and childhood height in the general population. *Nat. Genet.* 39:1245–50
223. Weir BS, Cardon LR, Anderson AD, Nielsen DM, Hill WG. 2005. Measures of human population structure show heterogeneity among genomic regions. *Genome Res.* 15:1468–76
224. White T, Asfaw B, DeGusta D, Gilbert H, Richards G, et al. 2003. Pleistocene *Homo sapiens* from Middle Awash, Ethiopia. *Nature* 423:742–47
225. Williams SM, Haines JL, Moore JH. 2004. The use of animal models in the study of complex disease: all else is never equal or why do so many human studies fail to replicate animal findings? *BioEssays* 26:170–79
226. Williams SM, Ritchie MD, Phillips JA 3rd, Dawson E, Prince M, et al. 2004. Multilocus analysis of hypertension: a hierarchical approach. *Hum. Hered.* 57:28–38

227. Williams TN, Mwangi TW, Wambua S, Peto TE, Weatherall DJ, et al. 2005. Negative epistasis between the malaria-protective effects of α^{+}-thalassemia and the sickle cell trait. *Nat. Genet.* 37:1253–57
228. Williamson C, Loubser SA, Brice B, Joubert G, Smit T, et al. 2000. Allelic frequencies of host genetic variants influencing susceptibility to HIV-1 infection and disease in South African populations. *AIDS* 14:449–51
229. Williamson SH, Hubisz MJ, Clark AG, Payseur BA, Bustamante CD, Nielsen R. 2007. Localizing recent adaptive evolution in the human genome. *PLoS Genet.* 3:e90
230. Winkler C, An P, O'Brien SJ. 2004. Patterns of ethnic diversity among the genes that influence AIDS. *Hum. Mol. Genet.* 13:R9-19
231. Wood ET, Stover DA, Ehret C, Destro-Bisol G, Spedini G, et al. 2005. Contrasting patterns of Y chromosome and mtDNA variation in Africa: evidence for sex-biased demographic processes. *Eur. J. Hum. Genet.* 13:867–76
232. Wood ET, Stover DA, Slatkin M, Nachman MW, Hammer MF. 2005. The β-globin recombinational hotspot reduces the effects of strong selection around HbC, a recently arisen mutation providing resistance to malaria. *Am. J. Hum. Genet.* 77:637–42
233. Wooding S, Bufe B, Grassi C, Howard MT, Stone AC, et al. 2006. Independent evolution of bitter-taste sensitivity in humans and chimpanzees. *Nature* 440:930–34
234. Wooding S, Kim UK, Bamshad MJ, Larsen J, Jorde LB, Drayna D. 2004. Natural selection and molecular evolution in PTC, a bitter-taste receptor gene. *Am. J. Hum. Genet.* 74:637–46
235. Wyen C, Hendra H, Vogel M, Hoffmann C, Knechten H, et al. 2008. Impact of *CYP2B6* 983T>C polymorphism on non-nucleoside reverse transcriptase inhibitor plasma concentrations in HIV-infected patients. *J. Antimicrob. Chemother.* 61:914–18
236. Yanovski JA, Diament AL, Sovik KN, Nguyen TT, Li H, et al. 2000. Associations between uncoupling protein 2, body composition, and resting energy expenditure in lean and obese African American, white, and Asian children. *Am. J. Clin. Nutr.* 71:1405–20
237. Yotova V, Lefebvre JF, Kohany O, Jurka J, Michalski R, et al. 2007. Tracing genetic history of modern humans using X-chromosome lineages. *Hum. Genet.* 122:431–43
238. Young JH, Chang YP, Kim JD, Chretien JP, Klag MJ, et al. 2005. Differential susceptibility to hypertension is due to selection during the out-of-Africa expansion. *PLoS Genet.* 1:e82
239. Zhao Z, Yu N, Fu YX, Li WH. 2006. Nucleotide variation and haplotype diversity in a 10-kb noncoding region in three continental human populations. *Genetics* 174:399–409
240. Zhu X, Bouzekri N, Southam L, Cooper RS, Adeyemo A, et al. 2001. Linkage and association analysis of angiotensin I-converting enzyme (*ACE*)-gene polymorphisms with ACE concentration and blood pressure. *Am. J. Hum. Genet.* 68:1139–48
241. Zhu X, Chang YP, Yan D, Weder A, Cooper R, et al. 2003. Associations between hypertension and genes in the renin-angiotensin system. *Hypertension* 41:1027–34
242. Zhu X, Cooper RS. 2007. Admixture mapping provides evidence of association of the *VNN1* gene with hypertension. *PLoS ONE* 2:e1244

Cumulative Indexes

Contributing Authors, Volumes 1–9

F

G

H

I

J

K

L

M

Chapter Titles, Volumes 1–9

Volume 1 (2000)

Volume 3 (2002)

Volume 4 (2003)

Volume 5 (2004)

Volume 6 (2005)

Volume 7 (2006)

Volume 8 (2007)

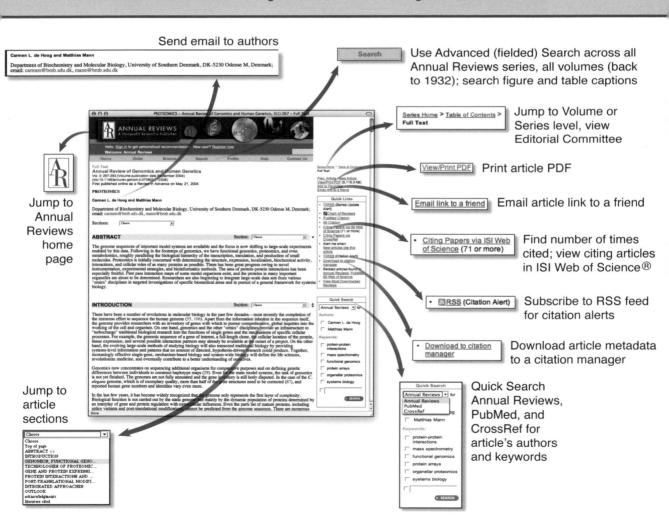